Lacustrine Petroleum
Source Rocks

Geological Society Special Publications
Series Editor K. COE

GEOLOGICAL SOCIETY SPECIAL PUBLICATION NO 40

Lacustrine Petroleum Source Rocks

EDITED BY

A. J. FLEET
Exploration and Production Division
B.P. Research International

K. KELTS
Geology Section
EAWAG/ETH
Switzerland

M. R. TALBOT
Geology Institute
University of Bergen

1988

Published for
The Geological Society by
Blackwell Scientific Publications

OXFORD LONDON EDINBURGH
BOSTON PALO ALTO MELBOURNE

Published for
The Geological Society by
Blackwell Scientific Publications
Osney Mead, Oxford OX2 0EL
 (*Orders:* Tel. 0865 240201)
8 John Street, London WC1 2ES
23 Ainslie Place, Edinburgh EH3 6AJ
3 Cambridge Center, Suite 208, Cambridge,
 Massachusetts 02142, USA
667 Lytton Avenue, Palo Alto,
 California 94301, USA
107 Barry Street, Carlton, Victoria 3053,
 Australia

First published 1988

© 1988 The Geological Society. Authorization to photocopy items for internal or personal use, or the internal or personal use of specific clients, is granted by The Geological Society for libraries and other users registered with the Copyright Clearance Center (CCC) Transactional Reporting Service, providing that a base fee of $03.00 per copy is paid directly to CCC, 27 Congress Street, Salem, MA 01970, USA. 0305-8719/88 $03.00

Typeset, printed and bound in Great Britain by
William Clowes Limited, Beccles and London

DISTRIBUTORS

USA
 Blackwell Scientific Publications Inc.
 PO Box 50009, Palo Alto
 California 94303
 (*Orders:* Tel. (415) 965-4081)

Canada
 Oxford University Press
 70 Wynford Drive
 Don Mills
 Ontario M3C 1J9
 (*Orders:* Tel. (416) 441-2941)

Australia
 Blackwell Scientific Publications
 (Australia) Pty Ltd.
 107 Barry Street, Carlton,
 Victoria 3053
 (*Orders:* Tel. (03) 347 0300)

British Library
Cataloguing in Publication Data

Lacustrine petroleum source rocks.—
 (Geological Society special publication
 ISSN 0305-8719).
 1. Lacustrine sediments. Petroleum deposits
 I. Fleet, A. J. (Andy J) II. Kelts, K.
 III. Talbot, M. R. IV. Series
 553.2'82

 ISBN 0-632-01803-8

Library of Congress Cataloguing-in-Publication Data

Lacustrine petroleum source rocks.
 (Geological Society special publication,
 ISSN 0305-8719; no. 40).
 Includes index.
 1. Petroleum—Geology. 2. Paleolimnology.
 I. Fleet, A. J. II. Kelts, K. R. III. Talbot,
 M. R. IV. Geological Society of London. V.
 Series.
 TN870.5.L22 1988 553.2'82 88-2870

 ISBN 0-632-01803-8

Contents

Introduction	vii
Acknowledgements	xi

Part I: Tectonic, Geological, Geochemical and Biological Framework

KELTS, K. Environments of deposition of lacustrine petroleum source rocks: an introduction	3
TALLING, J. F. Modern phytoplankton production in African lakes*	27
TALBOT, M. R. The origins of lacustrine oil source rocks: evidence from the lakes of tropical Africa	29
DE DECKKER, P. Large Australian lakes during the last 20 million years: sites for petroleum source rock or metal ore deposition, or both?	45
OREMLAND, R. S., CLOERN, J. E., SOFER, Z., SMITH, R. L., CULBERTSON, C. W., ZEHR, J., MILLER, L., COLE, B., HARVEY, R., IVERSEN, N., KLUG, M., DES MARAIS, D. J. & RAU, G. Microbial and biogeochemical processes in Big Soda Lake, Nevada	59
SUMMERHAYES, C. P. Predicting palaeoclimates*	77

Part II: Palaeoenvironmental Indicators

KATZ, B. J. Clastic and carbonate lacustrine systems: an organic geochemical comparison (Green River Formation and East African lake sediments)	81
VANDENBROUCKE, M. & BEHAR, F. Geochemical characterization of the organic matter from some recent sediments by a pyrolysis technique	91
VOLKMAN, J. K. Biological marker compounds as indicators of the depositional environments of petroleum source rocks	103
TEN HAVEN, H. L., DE LEEUW, J. W., SINNINGHE DAMSTÉ, J., SCHENCK, P. A., PALMER, S. E. & ZUMBERGE, J. Application of biological markers in the recognition of palaeo-hypersaline environments	123
DAVISON, W. Interactions of iron, carbon and sulphur in marine and lacustrine sediments	131
YURETICH, R. F. Possible relationships of stratigraphy and clay mineralogy to source rock potential in lacustrine sequences	139
BAHRIG, B. Palaeo-environment information from deep water siderite (Lake of Laach, West Germany)	153
JIANG DE-XIN. Spores and pollen in oils as indicators of lacustrine source rocks	159

Part III: Case Studies

DUNCAN, A. D. & HAMILTON, R. F. M. Palaeolimnology and organic geochemistry of the Middle Devonian in the Orcadian Basin	173

* Extended abstract.

HILLIER, S. J. & MARSHALL, J. E. A. Hydrocarbon source rocks, thermal maturity and burial history of the Orcadian basin, Scotland*   203

PARNELL, J. Significance of lacustrine cherts for the environment of source-rock deposition in the Orcadian Basin, Scotland   205

LOFTUS, G. W. F. & GREENSMITH, J. T. The lacustrine Burdiehouse Limestone Formation—a key to the deposition of the Dinantian Oil Shales of Scotland   219

PARNELL, J. Lacustrine petroleum source rocks in the Dinantian Oil Shale Group, Scotland: a review   235

GORE, P. J. W. Lacustrine sequences in an early Mesozoic rift basin: Culpeper Basin, Virginia, USA   247

FU JIAMO, SHENG GUOYING & LIU DEHAN. Organic geochemical characteristics of major types of terrestrial petroleum source rocks in China   279

LUO BINJIE, YANG XINGHUA, LIN HEJIE & ZHENG GUODONG. Characteristics of Mesozoic and Cenozoic non-marine source rocks in north-west China   291

BRASSELL, S. C., SHENG GUOYING, FU JIAMO & EGLINTON, G. Biological markers in lacustrine Chinese oil shales   299

WANG TIEGUAN, FAN PU & SWAIN, F. M. Geochemical characteristics of crude oils and source beds in different continental facies of four oil-bearing basins, China   309

MCKIRDY, D. M., COX, R. E. & MORTON, J. G. G. Biological marker, isotopic and geological studies of lacustrine crude oils in the western Otway Basin, South Australia*   327

HUTTON, A. C. The lacustrine Condor oil shale sequence   329

GIBLING, M. R. Cenozoic lacustrine basins of South-east Asia, their tectonic setting, depositional environment and hydrocarbon potential   341

ANADÓN, P., CABRERA, L. & JULIÀ, R. Anoxic–oxic cyclical lacustrine sedimentation in the Miocene Rubielos de Mora Basin, Spain   353

CROSSLEY, R. & OWEN, B. Sand turbidites and organic-rich diatomaceous muds from Lake Malawi, Central Africa   369

Index   375

* Extended abstract.

Introduction

Lacustrine Petroleum Source Rocks is a collection of papers arising from a meeting held at the Geology Society, London, in September 1985. The meeting was organized by the IGCP Project 219, 'Comparative lacustrine sedimentology in space and time', and the Petroleum Group of the Geological Society.

Organic-rich lacustrine sediments, potential sources of oil and/or gas, represent a group of lacustrine sediments whose interpretation is not only intellectually challenging but whose subsurface prediction, in terms of location, nature and lateral variation, is economically important. The papers in this volume represent an attempt to bring together synthesized concepts, techniques and real examples in order to provide ideas for both interpretation and prediction.

Petroleum source rocks deposited in lakes have come more into focus over recent years as petroleum exploration has shifted to new areas and as more detailed analysis of known petroleum provinces has become an exploration necessity. New areas include the multifarious basins of onshore China, for instance as described in this volume by Fu Jiamo *et al.*, Brassell *et al.*, Wang Tieguan *et al.* and Luo Binjie *et al.*, and the rift basins of Africa (e.g. Sudan: Schull 1984; Frostick *et al.* 1986). Lacustrine sources of petroleum must also be accounted for in some established petroleum provinces ranging from passive margin sequences, such as offshore Gabon (e.g. Brice *et al.* 1980), to the North Sea (e.g. Duncan & Hamilton, this volume). Lacustrine source rocks are often unsampled, being among the first deposits of a syn-rift sequence, in which case evidence for them is indirect, provided by oils themselves (e.g. McKirdy *et al.*, this volume).

Lacustrine petroleum source rocks represent one suite of the very varied lithologies which can accumulate in lakes. This variety reflects the broad range of dissolved and detrital inputs to lakes and the large spectrum of environmental conditions which can occur in lakes. Rapidly fluctuating conditions in individual lakes and, in some cases, the ephemeral nature of lakes often complicates particular sequences further. The variety of lacustrine sediments in general, and the complexity of individual deposits, has meant that lacustrine sediments have, with notable exceptions (e.g. the Green River basins, western USA), frequently received scant attention. IGCP Project 219 has set out to rectify this by gathering data and interpreting the distribution and variability of lacustrine sediments in space and time.

The volume is divided into three parts. The papers of the first part provide the overall framework and background against which to consider the more specific studies presented in Parts II and III. No attempt has been made here to give a precis of the nature of petroleum source rocks: the reader should refer to standard texts such as Tissot & Welte (1984) for such fundamentals. Part II of the volume details a range of techniques and approaches which can be used when interpreting the palaeoenvironments of lakes. Part III provides various case studies from around the world arranged in stratigraphic sequence.

Frameworks

The first six papers of the volume provide an overview of lacustrine sedimentation, in particular the factors which control organic matter deposition and preservation. These factors are founded on fundamental tectonic and hydrologic controls which cause a lake to exist and allow it to persist. Organic matter supply is both autochthonous and allochthonous. Autochthonous supply depends on nutrient availability and resulting biological productivity in the surface waters of the lake. Allochthonous input reflects bordering vegetation and transport mechanisms into the lake. Organic preservation is governed by various factors. For labile, potentially oil-prone aquatic detritus it depends most significantly on oxygen-deficient or -depleted bottom-water conditions and thus on lake hydrodynamics. However, rapid sedimentation may also lead to preservation, while sulphate reducing organisms may cause significant destruction even after deposition under anaerobic–anoxic conditions. The interrelated factors of geography, tectonics, hydrology, lake circulation, inorganic and organic sediment supply and preservational conditions, therefore, all need consideration when assessing the presence, nature and variability of lacustrine source rocks.

The scene is set by Kelts, who spans from the essential conditions for a lake (a topographic depression and a hydrological balance adequate to support surface water) to an outline of currently perceived problems relating to lacustrine kerogens, the sources of petroleum. Modern and ancient lakes and their tectonic settings are put in context and the variable chemistry of lakes is outlined. The stratigraphic distribution of lacustrine source rocks and published models of their formation are briefly reviewed. This leads to a survey of how carbon cycles and deposition vary between lakes with differing chemistries.

Recent African lakes provide natural laboratories. Talling's brief contribution focuses on a first step in the formation of lacustrine source rocks: biological productivity. Talbot follows the process further by considering the nature and accumulation of Pleistocene organic-rich sediments in six tropical African lakes. The preserved organic matter is a mixture of aquatic and terrigenous higher-plant detritus. The richest potential oil-prone source rocks accumulated in both stratified (meromictic) and annually mixed (monomictic) lakes when the climate was humid and winds slack. The humid climate ensured two things: firstly, dense land vegetation and thus minimal clastic input to the lakes which would have diluted accumulating organic matter; secondly, intense chemical weathering and thus a supply of nutrients. Stable stratification in the lakes allowed organic matter preservation under oxygen-deficient or -depleted bottom-water conditions; occasional mixing may have occurred in those lakes where organic accumulation rates were high. Overall Talbot finds no evidence from his studies to support the idea that shallow saline lakes are especially favourable as sites for the deposition of potential oil-prone source rocks.

Organic matter preservation in various types of lake is discussed by De Deckker. He considers three types of large Neogene lake from Australia: deep lakes, lakes with ephemeral (and shallow) water and 'dry' lakes. He thus extends discussion to more saline conditions. His work suggests that petroleum source rocks are principally limited to deep lakes in which organic matter is preserved under anaerobic bottom-water conditions. In ephemeral lakes significant algal mat bioherms can develop under hypersaline conditions but these are subject to attack during early diagenesis by sulphate-reducing organisms which destroy their organic content. Some ephemeral and 'dry' Australian lakes therefore tend to be sites of metalliferous deposits rather than petroleum source rocks.

Saline conditions are considered further in the next paper. This illustrates the kind of detailed study necessary to understand organic productivity and deposition in just one such lake. It discusses the ongoing work of Oremland et al. who are investigating microbial and biogeochemical processes in Big Soda Lake, Nevada, which is stratified, 65 m deep and alkaline (pH 9.7).

A final aspect of understanding the framework of lacustrine deposits is outlined by Summerhayes who discusses recent work on predicting palaeoclimates which can be used to assess the regional environments of lake basins.

Palaeoenvironmental indicators

Organic, inorganic, mineralogical and isotopic indicators of organic matter input and environmental conditions are all considered in Part II of the volume. (Information on the range and ecological significance of many limnic biota in the geologic record is generally lacking.) Organic geochemistry is dealt with first. The first two papers are concerned with bulk characterization of the organic matter of organic-rich lacustrine sediments and the second two with molecular parameters. Katz compares clastic lacustrine sediments from three modern African lakes with lacustrine carbonates represented by the ancient, but thermally immature, Green River Formation of Utah and Colorado. The former contain more inert material than the latter but both types of sediments are capable of generating oils, indicating that hydrogen-rich, oil-prone material can be preserved in both types of lacustrine environment. Katz's pyrolysis–gas chromatography results indicate that the oils derived from either type of potential source rock on maturation would have high wax contents. Vandenbroucke and Behar also discuss a pyrolysis technique. This involves fractionation of the pyrolysate into groups of organic compounds and analysis of the compounds by gas chromatography and gas chromatography–mass spectrometry. Their results tentatively suggest that lacustrine kerogens can be distinguished from other kerogens and are indeed characterized by Type I kerogen.

In contrast to bulk analyses, molecular or biological markers are very specific indicators of either organic matter inputs (types of organisms or even specific genera) or depositional environment. They can be derived from either sediments or oils. The first paper on this topic, by Volkman, is a comprehensive state-of-the-art review. It highlights the pitfalls associated with blindly using some published biological markers and suggests future directions for relevant research. In contrast, the second biological marker paper, by ten Haven et al., is specific in scope dealing with the molecular characterization of sediments or oils from hypersaline environments.

Inorganic chemical or mineralogical indicators of palaeoenvironment are the subject of the next three papers. Because of the lack of dissolved sulphate in freshwater lakes relative to seawater, carbon–sulphur ratios and the partitioning of sulphur between pyrite and organic matter have been used to distinguish lacustrine and marine environments. Davison reviews this approach and, while accepting that it is soundly based, urges cautious use because of our limited understanding, particularly of pyrite formation in freshwater environments. Yuretich also suggests that there are gaps in our knowledge of lacustrine systems. He argues that the inherently heterogeneous nature of lacustrine clays and the extremes of porewater chemistries likely to occur beneath lakes probably mean that lacustrine clays behave very differently from their marine counterparts during diagenesis. He suggests this may affect the catalytic influence of clays during kerogen breakdown to petroleum and thus explain the localized migration in lacustrine oilfields and their small size. Bahrig's contribution on siderite illustrates how isotopic evidence can be used to interpret the origin of mineral phases and so help build up the depositional and diagenetic histories of lacustrine sediments.

Finally in this Part of the volume, Jiang De-xin presents a novel technique for studying the origin and migration history of oils. This involves using spore and pollen recovered from oils as indicators of both the source rock of the oils and the carrier beds through which the oils have migrated. From an investigation of the spores and pollen of Chinese oils, Jiang De-xin concludes that wet, hot climatic conditions are conducive for the deposition of lacustrine oil-prone source rocks.

Case studies

The final part of the volume is devoted to fifteen examples of actual or potential lacustrine source rocks. The ordering of these case studies is stratigraphic. The first three deal with the Devonian Orcadian basin of Scotland. Duncan and Hamilton extend a depositional model for organic-rich lacustrine laminites, based on outcrop data, to the ill-defined subsurface of the offshore Moray Firth basin. They illustrate the strengths of molecular marker geochemistry by using the approach to argue for a component of Orcadian lacustrine-derived oil in the Beatrice field. Hillier and Marshall briefly summarize points concerning the distribution and maturity of source rocks in the Orcadian Basin. Parnell interprets cherts from the Orcadian Basin as evidence of saline, alkaline conditions of source-rock deposition.

The lacustrine Carboniferous Oil Shales of the Scottish Midland Valley, the foundations of the ancestral UK petroleum industry, are the subject of the next two papers. Loftus and Greensmith identify a series of freshwater stratified lakes in a wet tropical climate. These underwent cyclic changes and some connection to the open sea. Oil-shale deposition occurred at intervals. Parnell reviews the petroleum potential and composition of the oil-shales and the associated sediments.

The Newark Rift System of eastern North America provides analogues of lacustrine sediments for some passive margin and intra-continental rift sequences. Gore provides a detailed review and interpretation of one of the basins of the Rift System: the Culpeper Basin. Late Triassic to Early Jurassic lacustrine sediments occur in the Culpeper Basin, which is a half-graben, along with fluvial and alluvial fan deposits. Gore recognizes freshwater, saline and hypersaline lakes as both lateral equivalents and successive water bodies. Organic-rich laminated shales were deposited where the lakes were stratified and bottom waters 'anoxic'. Gore suggests that climatic variations, controlled by Milankovitch-type cyclicity, influenced chemical sedimentation in the basin and thus the hydrology of the area including lake formation.

The next four papers deal with petroleum-bearing Chinese continental basins. These basins constitute the most significant area of lacustrine-sourced petroleum production in the world. They vary in age from the Late Palaeozoic to Cenozoic and cover the full range of possible lacustrine environments. This wide spectrum is illustrated by Fu Jiamo *et al.* and Luo Binjie *et al.* The former authors emphasize the relationship between tectonic setting and lacustrine sedimentation, while the latter group suggests that humid or semi-humid climatic conditions and deep to intermediate lakes most favoured oil-prone source-rock deposition. Both discuss the natures of the source rocks and their oils. Brassell *et al.* and Wang Tieguan *et al.* expand on these themes with detailed geochemical studies.

Two Australian studies are considered next. McKirdy *et al.* use the indirect evidence provided by the chemical compositions of surface bitumens to argue for Early Cretaceous lacustrine source rocks in the rift-valley sequence of the Otway basin, offshore of southeastern Australia. The nature of Australian Tertiary oil shales as freshwater lacustrine deposits is then illustrated by Hutton who describes the Condor Oil Shale of Queensland as an example.

Finally, three sets of Cenozoic lacustrine basins are considered. Gibling describes and interprets oil shales and associated coals and sediments which accumulated in various sized strike-slip basins of Thailand during Oligocene and Miocene times. In general, the oil shales represent early episodes of aquatic productivity during the history of the lakes, but in one case deposition occurred in a deep, persistent, 'perennially stagnant' lake. The existence of lacustrine carbonates and siliciclastic sediments in adjacent basins points to the strong influence of localized sediment supply and water compositions. Local and temporally-variable controls on lacustrine deposition are further emphasized by Anadon *et al.*'s study of anoxic–oxic lacustrine cycles in the Miocene Rubielos de Mora basin of Spain. Lastly, Crossley and Owen return the volume to Africa with a detailed study of cores from Lake Malawi which illustrates the interplay of mass flow sedimentation and productivity events in lakes.

Our understanding of the formation of lacustrine petroleum source rocks is more diffuse than that of their marine counterparts (e.g. Brooks and Fleet 1987). This volume should provide stimulation on directions to follow in attempting to improve this understanding. We believe that future prospects are exciting. There is a new challenge for geologists to formulate sophisticated models for lake deposits, integrating results from diverse modern lake environments with those from ancient lake deposits.

References

BRICE, S. E., KELTS, K. R. & ARTHUR, M. A. 1980. Lower Cretaceous lacustrine source beds from the early rifting phase of the South Atlantic. *Bulletin of the American Association of Petroleum Geologists*, **64**, 680–681.

BROOKS, J. & FLEET, A. J. 1987. *Marine Petroleum Source Rocks*. Geological Society Special Publication, **26**, Blackwell Scientific Publications, Oxford.

FROSTICK, L. E., RENAUT, R. W., REID, I. & TIERCELIN, J.-J. 1986. *Sedimentation in the African Rifts*. Geological Society Special Publication, **25**, Blackwell Scientific Publications, Oxford.

SCHULL, T. J. 1984. Oil exploration in non-marine rift basins of Interior Sudan. *Bulletin of the American Association of Petroleum Geologists*, **68**, 526.

TISSOT, B. & WELTE, D. 1984. *Petroleum Formation and Occurrence*. 2nd edn. Springer-Verlag, Heidelberg.

A. J. FLEET, Exploration & Production Division, B.P. Research International, Chertsey Road, Sunbury-on-Thames, Middlesex, TW16 7IN, UK.

K. KELTS, Geology Section, EAWAG/ETH, C8600 Dübendorf, Switzerland.

M. R. TALBOT, Geologisk Institutt Avd A, Universitetet i Bergen, Allegt. 41, 5014, Bergen, Norway.

Acknowledgements

The meeting on 'Lacustrine Petroleum Source Rocks', and hence this volume, would not have been possible without the financial support of the International Geological Correlation Programme (IGCP) and the organization and help of the Geological Society. We are particularly grateful to Caroline Symonds and her colleagues in the Society Office and Mr and Mrs Ewens and their helpers for the smooth running of the meeting.

The volume itself reflects not only the efforts of the authors but also the contributions of the questioning and constructive reviewers and the patience of Mary-Clare Swatman of Blackwell Scientific Publications who pulled the book together.

Part I
Tectonic, Geological, Geochemical and Biological Framework

Environments of deposition of lacustrine petroleum source rocks: an introduction

K. Kelts

Introduction

F. A. Forel (1892–1902) established the scientific discipline of limnology with his integrated study of biology, chemistry, circulation and sedimentation in modern Lake Geneva. Subsequent literature on lakes is largely separate from geology in contrast to the historical development of marine geology. Strakhov (1970) is one exception and Pia (1933) another. Classic texts which summarize the limnological view of lakes include: F. Forel (1901), F. Ruttner (1963), G. E. Hutchinson (1957), D. Frey (1974), R. Wetzel (1983), and F. Taub (1984a). None of these however emphasize the burial and preservation of reactive organic carbon.

A lake may be defined as an inland body of standing water occupying a depression in the earth's crust. It is larger than a pond. As such, lakes can exhibit a wide range of possible settings, sizes, chemistries, concentrations and morphologies. The Greek 'Limne' which is at the root of the scientific disciplines Limnology and Limnogeology means marsh, lake or pool and by implication has been generally applied to freshwater environments. Thirteen of the world's 40 largest lakes, and innumerable smaller ones are however without outlets and commonly quite saline. Of the largest lakes, only 20 are deeper than 400 m but these hold most of the world's fresh lake water (23 000 km^3 in Baikal). Because traditionally the study of modern lakes is linked to securing water resources for man, only a minor emphasis has been given the brackish to hypersaline ecosystems. These, however, form important parts of the geological record of lake basins. Geology textbooks gloss over lakes as ephemeral geological features, but in recent years it has become increasingly obvious that vast lake deposits are preserved in the geological record and many of the ancient lake basins hold great promise for economic deposits including metals, salts and hydrocarbons. The situation is perhaps similar to the Dolomite Problem in the 1960s. Very skimpy data are available from modern systems to help formulate general models that can be useful in an exploration strategy. Wherever one begins; whether in palaeontology, petrology, geochemistry, seismic stratigraphy, or sedimentology, there is simply insufficient synthetic information on lake deposits.

The distribution of large lakes in our present world is dominated by settings related to Pleistocene glaciations (Fig. 1). During various geological time slices; climatic, tectonic and physiographic conditions supported the long-term existence of lacustrine giants covering thousands of square kilometres. Preferred sites were broad tectonic depressions and along the trends of continental rifting. Such targets are the current focus of most of the oil exploration in continental basins. In order to interpret the record from ancient lake basins it is essential to view them in their correct palaeogeographic and tectonic setting.

This chapter aims to provide a matrix (Fig. 2) with which to analyse environments of deposition of lacustrine petroleum source rocks. It attempts to simplify some of the complexities of lacustrine systems with models based on modern lake studies. Lakes are dynamic systems with rates of change greater than marine environments. A few centimetres of sediment in some outcrop or core sections may record the changes from a dilute, deep-water setting to a shallow, hypersaline brine. These changes may be archived by the geochemical signals in organic matter and lacustrine carbonates, as well as changes in biota.

We examine the lacustrine carbon cycle for various settings in terms of productivity, sedimentation and preservation potential. The source-rock potentials of lacustrine basins are promising. Per unit time, lakes probably bury more carbon than oceanic environments and the preservation potential is commonly high. The importance of understanding lacustrine source-rock environments is to discern which type, or types of lakes are likely to contain important petroleum reserves.

The essential conditions for a lake are simply a topographical depression, and a hydrological balance that is adequate to support surface water. The hydrological balance (*input–output*), in turn, is a function of the prevailing climate which is influenced by latitude-dependent zonal winds, altitude, continentality and orbital parameters (Street-Perrott & Harrison 1984; Kutzbach & Street-Perrott 1985). Arid zones are common in both cold and warm regions, particularly in the continental interiors. Monsoon patterns determine the amount of moisture excess reaching interior Africa and Asia. Global moisture and

From: FLEET, A. J., KELTS, K. & TALBOT, M. R. (eds), 1988, *Lacustrine Petroleum Source Rocks*, Geological Society Special Publication No. 40, pp. 3–26.

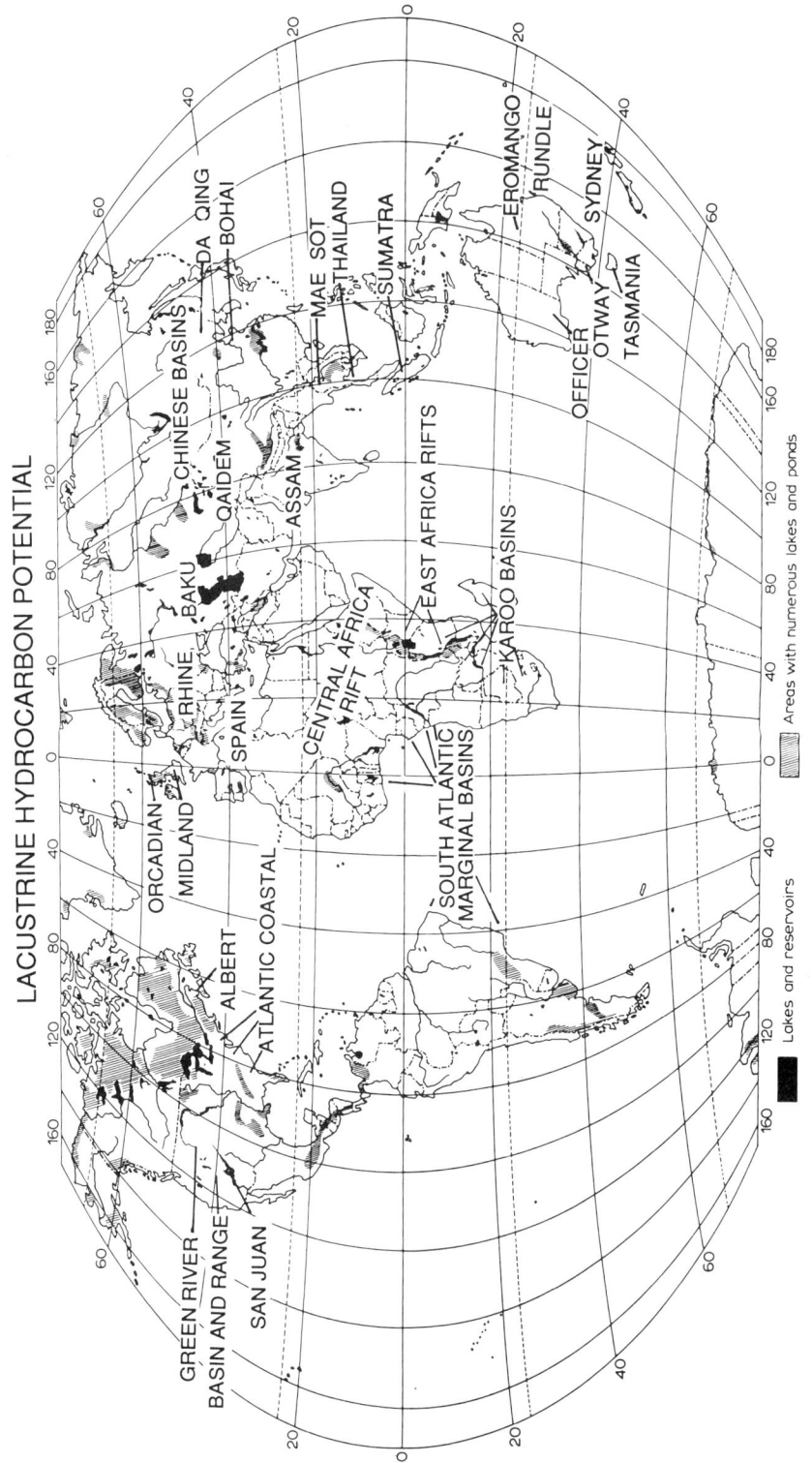

FIG. 1. Global map of modern lake distribution and locations of some potential lacustrine hydrocarbon provinces. Base map after Taub (1984a).

FIG. 2. Matrix for lacustrine systems.

lake distribution therefore would be radically altered by past plate positions (Parrish & Barron 1986).

Depending on the balance of *input* vs *evaporation*, a lake may be considered hydrologically *open* or *closed*. An open system is characterized by relatively stable shorelines, and a limited residence time for solutes (Fig. 3a). Open systems may also be linked such as in the Pleistocene pluvial lakes of Panamint Valley, California (Eugster & Hardie 1978).

Topographically subdivided basins may exhibit reverse sediment facies sequences. High lake levels may represent closed basin conditions whereas low levels re-establish open conditions (Fig. 3b). The Great Salt Lake provides an example (Spencer *et al.* 1984). The lake is fed from the southern end. As lake level drops to less than 5 m water depth, flow is restricted to a shallow channel and reflux stops. The result is that the northern basin becomes increasingly concentrated toward halite saturation. Synchronously, the southern basin becomes more dilute due to the smaller reservoir with respect to input, and the extraction of salts in the northern basin. The sediment histories of this basin would therefore show parallel development to a point, and then opposite signs caused by the same environmental change (lowering water level) (Fig. 3b).

Lake basin

Lakes are formed by numerous mechanisms that have been classified by Hutchinson (1957). For geological purposes we may simplify this scheme with *event lake basins* which are formed by short-term processes and thus less likely to preserve thick lake deposits in the geological record (see Table 1). These include meteorite and volcanic craters, landslide and other dams, glacial lakes, karst sink-holes, groundwater lakes and river meanders. *Paralic lake basins* include cut-off marine embayments and shoreline depressions which are controlled by sea level fluctuations. *Tectonic lake basins* have the highest preservation potential for the geological record. These may be broad crustal warps, foreland deeps, intermontane basins, or rifts and strike-slip. Pull-apart basins and continental rifts have the maximum preservation potential because of their rapid subsidence and long histories (Biddle & Christie-Blick 1985). Ancient lacustrine basin deposits often record the conditions during the first stages of continental rifting of a plate.

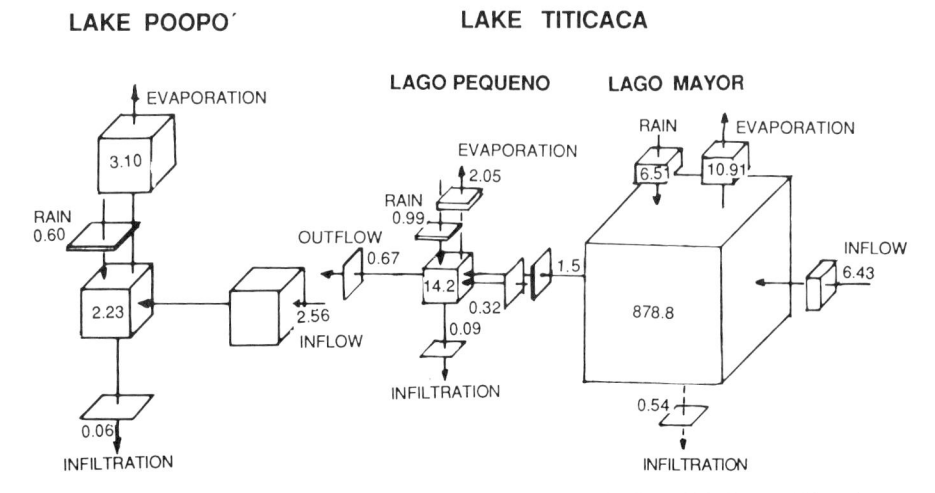

FIG. 3a. Open Lake Titicaca to closed Poopo Salt Lake as an example of hydrologic balance. Although open, only a small fraction of the inflow leaves the Titicaca basin. Long residence times lead to evolved water masses. Units of 10^9 m^3. After Serruya & Pollingher (1983).

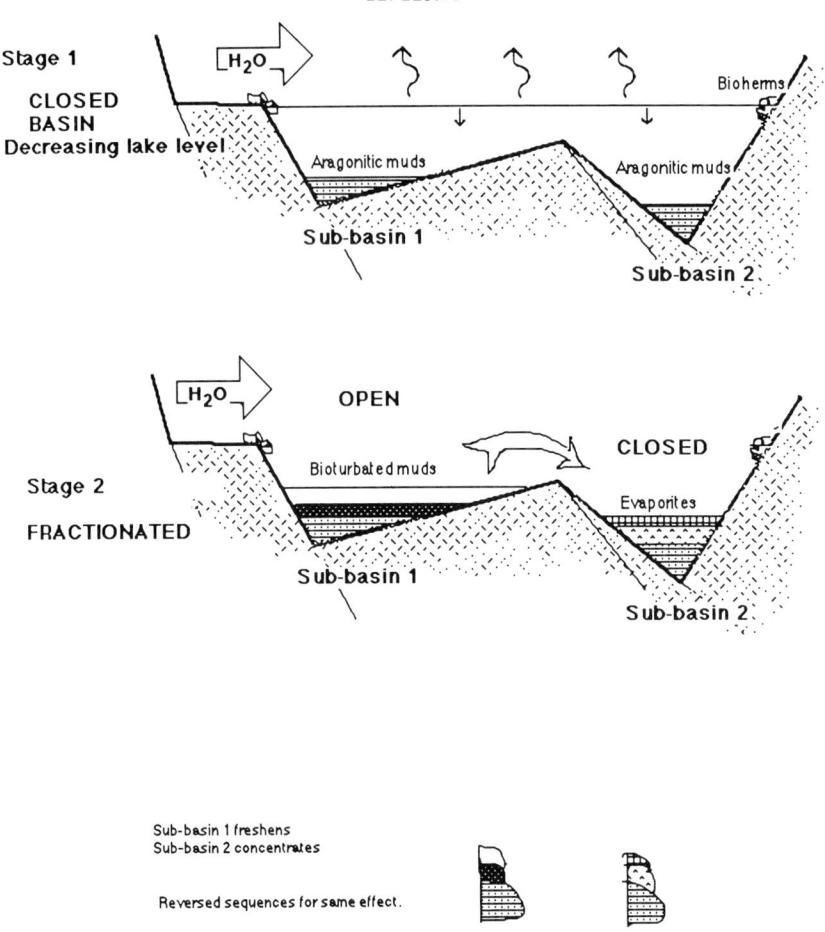

FIG. 3b. Diagrammatic cross section illustrating the reversal of concentration trends as water level falls in a larger closed basin with topographic complexity. Where lake levels intersect and sub-basin margins have asymetric water supply, one sub-basin may freshen while another desiccates. These lead to initially equivalent sediment sequence trends then reversal of salinity signals.

Geochemistry

Rainwater composition is a function of distance to oceans, wind strengths, and evolution of the vapour. Near oceans, rain may contain 60 ppm TDS with seawater ratios. Continental interiors may have less than 6 ppm TDS of a very different ionic ratio than seawater. Thus the starting point for filling a basin may differ. Isotopic compositions for example are directly affected. A lake with drinking water (e.g. 200 ppm TDS) may show $\delta^{18}O$ values of extreme evaporation that are the same as a saline brine in another region.

One of the main differences between marine and lacustrine systems is that the initial ionic balance of waters is not fixed; salinity and composition are not the same. Mobilization of elements from the drainage basin geology by weathering determines the initial ionic ratios of most lakes fed by surface water (Eugster & Hardie 1978). This fingerprint is already present in the early dilute to freshwater phases but becomes particularly important as concentration increases. In closed basins, the concentration of brines may fluctuate widely, but the chemical type will remain relatively uniform. As brines evolve, a main control is the calcium/carbonate ionic ratio and the early precipitation of calcium carbonates (calcite, aragonite) (Eugster & Hardie 1978). If the calcium ion concentration is less than that of the carbonate ion, it may be depleted before significant brine concentration is reached, and thus reduce the buffering capacity of a lake. Without buffering components, lakes may have

TABLE 1. *Origin of lake basins and examples*

Mechanism	Selected examples
Event basins	
Meteorite	L. Bosumtwi, Africa; Miocene Nordlinger Ries, Germany
Volcanic crater	Laachersee, Germany; L. Toba & Maninjau, Sumatra; Crater Lake, USA; Nyos, Cameroon; Cretaceous Kimberlite pipes, S. Africa
Glacial	Perialpine Lakes; Great Lakes; L. Permian Karoo, Africa
Dam; meander	L. Van, Turkey; Tai Hu, China
Paralic basins (diverse origins)	L. Maricaibo; The Coorong; Black Sea; Cretaceous Bohai, China; Baltic Sea; L. Carpentaria, Quat. Australia
Tectonic basins	
Crustal warps	L. Victoria, E. Africa; L. Chad; Eocene Green River Fm., USA; Permian Gondwana Karoo
Forland basin	Caspian Sea; Aral Sea; Miocene Molasse
Intermontane	Qing Hai, China; Neogene Iberia, Spain; Great Basin; Salt Lake Urmia, Iran; Great Salt Lake, USA
Strike-slip	Walker Lake, Nev.; Dead Sea; Plio. Ridge Basin, Calif.; Miocene S.E. Asia; Bohai, China
Rift	Baikal; Tanganyika; Malawi; Albert; Turkana; Manyara; E. Africa; Triassic E. Coast USA; L. Cret. S. Atlantic margin; Mesoz. Cent. Africa

pH ranges from less than 1 in some sulphuric-acid-rich Japanese crater lakes to more than 11.0 in alkaline brines of several East African Rift lakes. There are, of course, concomitant effects in the corresponding ecosystems.

Eugster & Hardie (1978) provide a chemical classification based on solute species from modern lakes (Fig. 4a). It is however difficult to subdivide in most ancient deposits without trace element data or index minerals (e.g. trona). For the discussion of petroleum source rocks it is however important to differentiate between alkaline affinities and saline (i.e. oceanic composition) lakes. The Green River Formation, Western African, Lower Cretaceous coastal basins and several other lacustrine oil basins suggest alkaline affinities.

Lakes may range from dilute, monsoonal rainwater compositions (10 ppm) up to viscous chloride brines of 500 000 mg/kg. We can thus best classify the range of solute concentrations for lakes by a logarithmic scheme (Fig. 4b).

Marine vs non-marine environments

Table 2 is a schematic attempt to summarize some aspects to consider when comparing marine and lacustrine environments. Major differences lie in the variable geochemical signatures, carbonate mechanisms and biota. Differences between lacustrine and marine environments are due mainly to the environmental sensitivity of lakes to their regional setting and the accompanying biological diversity (cf. Fig. 5).

In contrast to the marine environment, where foraminifera and calcareous nannoplankton are the most abundant basinal, calcareous components; there are few planktonic calcareous fossils in lake deposits. Most fine-grained lacustrine basinal micrite is an inorganic precipitate, induced by biological or physical–chemical processes. Along marginal areas, oncolite, bioherms and stromatolitic carbonates are more prevalent than in marine environments. Charophyte chalks are typically lacustrine. Benthic ostracods, molluscs, and gastropods may form significant beds. Fish, mollusca, ostracods, gastropods and other limnic fauna are particularly subject to endemism. There are very few systematic palaeoecological studies of past non-marine environments so that future work is promising. Picard & High (1972, 1981) concluded that there are few criteria which by themselves are sufficient to indicate lacustrine deposits. A combination of positive and negative evidence is needed. Hardie (1984) indicates sedimentological and geochemical signatures that separate marine and non-marine evaporites.

Source-rock models

Extraction of hydrocarbons from lake deposits was already common in biblical times, for example, from the Dead Sea seeps. Chinese produced oil from bamboo wells (Mason 1980); Scots distilled lacustrine Carboniferous oil shales. It was not until the classic monographs of the vast oil shale deposits of the lacustrine Eocene Green River Formation, USA (Bradley 1929,

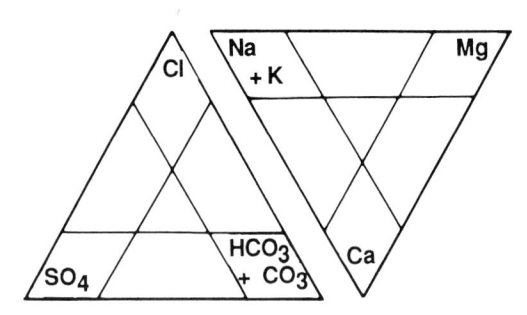

FIG. 4a. Ternary classification of lacustrine chemical types. Simplified after Eugster & Hardie (1978).

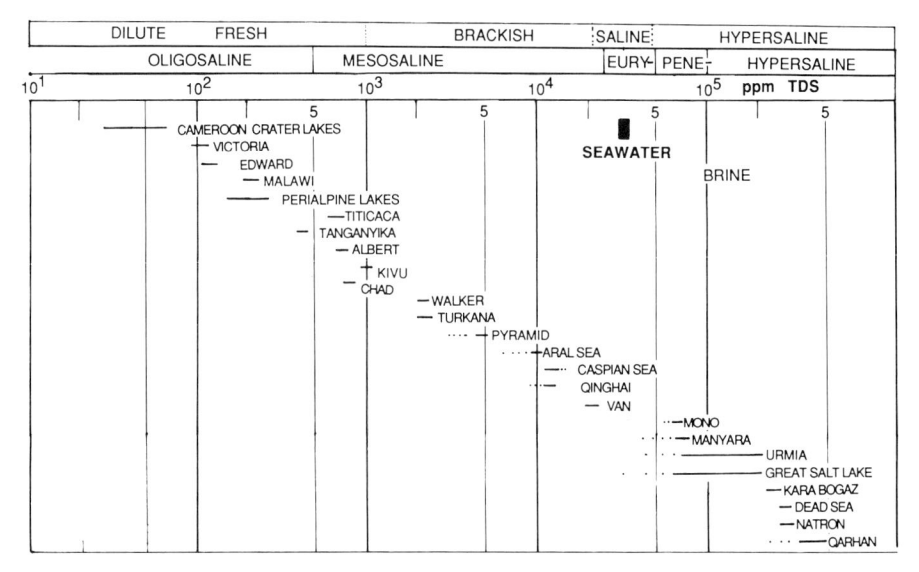

FIG. 4b. Log-range of solute concentrations for selected lakes. Includes a suggested standard division. TDS = total dissolved solids in ppm. Data after Hutchinson (1957), Talling & Talling (1965), Golterman (1975), Eugster & Hardie (1978), Beadle (1981), Serruya & Pollingher (1983), Taub (1984a) and Hammer (1986).

1964) that controversy among geologists arose concerning the kind of lakes conducive to accumulation of such hydrocarbon deposits.

The Green River Formation in Wyoming sits well-exposed and relatively undeformed in a shallow continental saucer. Bradley was struck by the long sequence of rhythmic laminations of carbonate and kerogen (up to 20% organic carbon) that are common in the basinal areas. He interpreted these as annual varves after visiting a young pharmacist, Fritz Nipkow, in Zurich, who had just completed a thesis on the recent sediments of freshwater, deep Lake Zurich (Bradley, pers. comm. 1971). Nipkow (1920, 1927) used a glass corer to show that the uppermost 20 cm of the lake were characterized by a regular succession of couplets of dark organic sludge (6% C-org.) plus diatom frustules and light-coloured calcite crystals. The couplets followed the seasonal productivity cycle with a pelagic rain of precipitated calcite in the summer months induced by biological depletion of CO_2 (cf. Kelts & Hsü 1978). Bradley's deep permanent lake model was later criticized partly because the Green River oil shales are associated with evaporites (trona: $NHCO_3 \cdot Na_2CO_3 \cdot 2H_2O$) and because the carbonates are predominantly dolomitic (Eugster & Surdam 1973). Numerous horizons with sedimentary structures characteristic of shallow to ephemeral conditions led to the formulation of a playa-lake complex model (Eugster & Hardie 1975; Surdam & Wolfbauer

TABLE 2. *Points of contrast for lacustrine vs marine environments of deposition*

Aspect	Lacustrine	Marine
Aqueous reservoir	Limited, variable	Immense, uniform
Chemistry	Highly variable, ion species function of drainage basin geology and climate	Uniform Na-Cl-
Salinity	Highly variable 10^1–10^5 mg TDS/l	35‰
pH	Variable 3.0–11.0	8.3 in surface waters 7.7 in bottom waters
Size	Highly variable, up to 80 000 km^3 today	Immense, but source-rock areas similar in size to lake basins
Deposits, thickness	Generally thin within continental clastics	Thin organic-rich zones
Sediment rates	0.1–2 m/ty rapid	0.001–0.35 m/ty modest
Tectonics	Event basins or sags or rifts + fault control	Sea floor spreading. Continental margin subsidence
Geodynamics	Includes altitude variations, drainage capture, sudden changes	Sea level, epirogeny; slower changes
Climate control	Zonal latitude dependent	
Climate change	Immediate, drastic response; level changes, composition tens of years	Long-term response; thousands of years
Residence time	1–500 yr	1000 yr +
Cycles	Annual, sun-spot, short-term climate	Long-term climate, palaeo-oceanography, Milankovitch
Tides	No tidal, seasonal level variation	Tidal dominated
Organic matter	Algae/bacteria; land plants. Type I common	Marine algae, or land plant. Type II & III
Productivity	Very high, high nutrient	Modest, upwelling zones
Preservation potential	High with high sed. rates. Anoxia, low sulphate, common	Requires high sed. rates, or anoxia
Bacteria/algae	Special adaptations. Photochemotroph, etc.	Marine
Palaeontology		
Silica microfossils	Diatoms dominant since Eocene, sponges	Diatoms since L. Cret., radiolaria
Calc. microfossils	No pelagic calcareous	Forams, Nannos dominant
Benthics	Ostracods. Local endemism	Forams. Worldwide index
Dinoflagellates	Abundant, but few cysts preserved	Cysts preserved
Vertebrates, etc.	Micromammals, reptiles, fish (kills), insects, chironomids	Vertebrates, insects rare
Littoral shelf	No corals, gastropods, stromatolites common, algal bioherms, charophyte chalks	Coral reefs, calc. algae, molluscs, strom. rare, subtidal, no marine charo.
Offshore	Macrofossils scarce	
Bioturbation	Worms, insects, vertebrates. Few burrowers	Numerous burrowers
Facies		
Evaporites	Derived, evaporitic concentration variable types, reworked, thin, fractionated basins	Marine, fixed sequence, giants, may be kms thick
Carbonates	No barrier reefs, no calcareous plankton oozes, mostly chemical, dolomite common	Mostly biogenic, calcite dominant
Oolites	Saline and brackish lakes	Turbulent upwelling
Silica	Abiotic chert common	Biogenic chert common
Basinal	Commonly anoxic sediments	Anoxia not common
Deltas	Short-term, rapid variance response to level changes	Long-term stability
Turbidites	Common in dilute waters	Rare events
Sands	Fan-delta complexes, flood plain, fluvial	Clastic shores, beach
Transgression/regression	Very short period	Long period phenomena
Stratigraphy	Often not by Walthers' law, rapid supplantation	Walthers' law, transitional
Life span	1 ma is old, up to 35 ma	1–100 ma
Kerogen	Variable Type I to III, common high wax, amorphous, low sulphur, high or low pour points	
Biomarkers	Botryococcane, bacterial common	

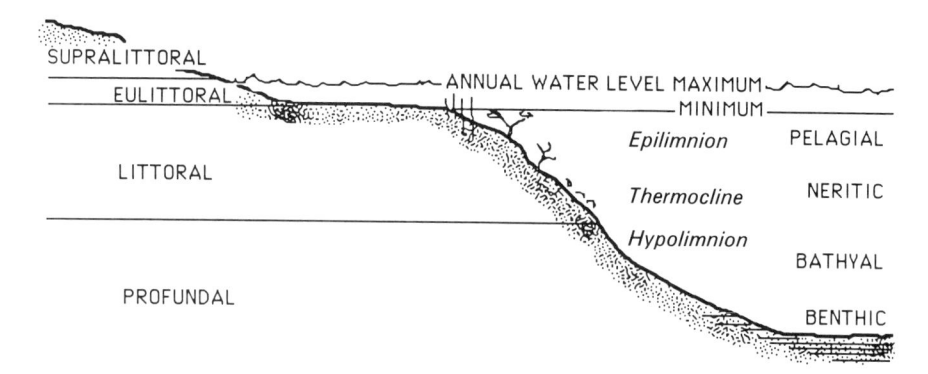

FIG. 5. Scheme of marginal and basinal depositional areas for lakes. Terminology modified after Hutchinson (1957). The eulittoral zone of annual flooding constitutes a major difference to marine coastal basins with their diurnal tides.

1975; Surdam & Stanley 1979) which commonly is viewed as an industry standard of lacustrine petroleum source-rock environments. Recent studies are more differentiated and emphasize the dynamic changes in such basins with rapid fluctuations of concentration and depth (e.g. Eugster 1985, Boyer 1982, Ryder et al. 1976, Desborough 1978). For example, dolomite in rhythmic laminites is interpreted as an early diagenetic product within the organic-rich sediments of a meromictic, brackish, alkaline lake rather than as a storm deposit from hypersaline mud flats (Kelts & McKenzie 1985).

For US geologists, the Green River Formation became the prototype environment for lacustrine oil shales. Organic geochemical analyses of Green River oil shales have also encouraged the standard view for a predominance of Type I amorphous kerogen in lacustrine settings (Tissot & Welte 1978) and the ensuing controversy as to its origin.

With the motto 'oil is where you find it' the industry paid little attention to the systematics of source-rock depositional environments although it was known that non-marine sources were important in Chinese mainland basins, West Africa marginal basins, Brazilian marginal basins, Sumatra and in Australia. Hedberg (1968) raised the question of the unusual waxiness of the Green River hydrocarbons and implied that non-marine oils are generally related to more saline environments and land-plant input.

Demaison & Moore (1980) developed the concept of large anoxic lakes as an ideal source environment. They used Lake Tanganyika as a model with aspects also applied to the Green River Formation oil shales. Kirkland & Evans (1981) stressed that hypersaline lakes have particularly high productivity of limited phytoplankton species and thus represent high-potential environments. This theme is further developed by Warren (1986) favouring organic matter, particularly from cyanobacterial mats, accumulating in shallow-water environments with anoxic, highly saline brines. Powell (1986) takes a more differentiated view based on Australian and Chinese conditions and recognizes multiple environments with source-rock potential including oligotrophic lakes, humid fluvial flood plains and saline to hypersaline lakes. His view is that most lacustrine source rock is formed in oligotrophic, meromictic lakes with a major component of land-plant organics that are biodegraded. Considerable confusion persists, partly because elements of the carbon cycle (particularly Type I kerogen environments) have not been adequately identified in modern lakes (see Talbot, this volume) and partly because detailed sedimentological models for many ancient and modern deposits remain unpublished. A starting look at the analyses of the East African Rift sedimentation (Frostick et al. 1986) shows the complexity in detail.

Time windows

The geological record of major lacustrine basins with hydrocarbon potential (Fig. 1), shows a preference for certain periods with a favourable combination of climate, palaeogeography and structural setting (Fig. 6).

Palaeogeographical reconstructions (Ziegler et al. 1985) are the starting point of a basin analysis. In the Palaeozoic, widespread Devonian dark lacustrine source beds are associated with continental, semi-arid basins near the Iapetus opening. Shallow Carboniferous lakes of Britain persisted in humid climate, with abundant land-derived

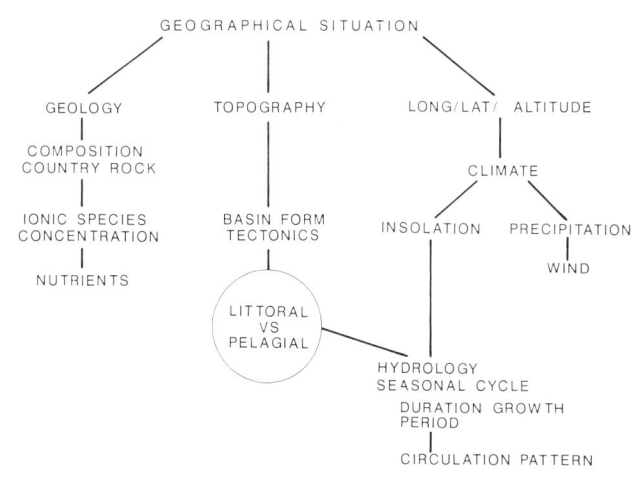

FIG. 6. Palaeogeographic controls on lacustrine environments of deposition and production/preservation of organic matter.

plant matter. The configuration of Gondwanaland in the Permian provides the setting for widespread, fluvial–lacustrine, giant lowland lake deposits in a taiga-like, cool–temperate climate. These include the Permian Gippsland and Cooper Basin area of Australia (Powell 1986) and Karoo of southern Africa as well as the Parana oil shale basin of South America. The Permian hypersaline Zechstein of Europe and the Delaware basin of Texas behaved as giant saline lakes in arid climates. Triassic lacustrine basins are common along the US East coast concomitant with N. Atlantic rifting. Many show playa characteristics (Demicco & Gierlowski-Kordesh 1986). Manspeizer (1981) attributes these partly to rift-associated orographic rain shadows. Africa has Upper Jurassic semi-arid lake basins, whereas Jurassic fluvial–lacustrine floodplain facies occur in Eromango Basin, Australia (Powell 1986) consistent with more temperate latitudes. Lower Cretaceous rift lake basins occur parallel with Gondwana break-up along South Atlantic margins (Brice et al. 1982) and the interior Central African Rift (Fairhead 1986) as well as in the Otway Basin Australia (McKirdy, this volume). The giant Songliao inland lake in Middle Cretaceous, China (Yang et al. 1985) was a dominantly humid zone feature as were the lake basins of the Bohai region. Eocene and Oligocene lacustrine source rocks of China are more associated with arid climates. The Eocene giant Uinta Complex, USA was in a subtropical, semi-arid environment with strong seasonality similar to the conditions expected in other Great Basin Tertiary lakes, or the Tertiary Qaidem Basin of interior China. More tropical wooded conditions surrounded lakes of Miocene Sumatra and Thailand (Gibling et al. 1985). Studies of Pleistocene evaporite lakes have provided important palaeoclimatic information as well as the models for such basins in the past (Hardie 1984; Warren 1985).

Questions that arise for these settings within a differentiated lacustrine matrix include:

—Origin, size, and duration of the structural basin?
—Lake depth, stability and duration of the lacustrine phase?
—Original lake chemistry and concentration at lake at the time of maximum organic burial? Open or closed system?
—Rates of production, sedimentation, and preservation of organic matter and sediments?
—Source of organic matter? Nature of flora and fauna?
—Sedimentary facies patterns?
—Criteria for depositional environment?

The carbon cycle in lakes

The amount of oil-prone organic carbon in a potential source rock is determined by the history of production, sedimentation and preservation in the precursor lacustrine environment.

Depending on the setting, size, facies, climate, and chemistries of the lake basins, organic matter with source-bed potential may be dominated by input of transported land-plant debris, or the endogenic production of algae and microbes. Organic carbon content is very much determined by input rates versus dilution by mineral input (chemical, biogenic or detrital sediments) as well as preservation and dissolution. Because most lakes bury organic carbon at high rates, their preservation potential is generally higher than most marine settings (see Fig. 7a, 7b).

Land-plants include woody plants and herbaceous plants with C-3 and C-4 photosynthetic pathways. For many lacustrine environments such as Lake Victoria one must evaluate the importance of fringing swamps, and littoral water plants (roots submerged and leaves emerged) as sources of buried organic matter.

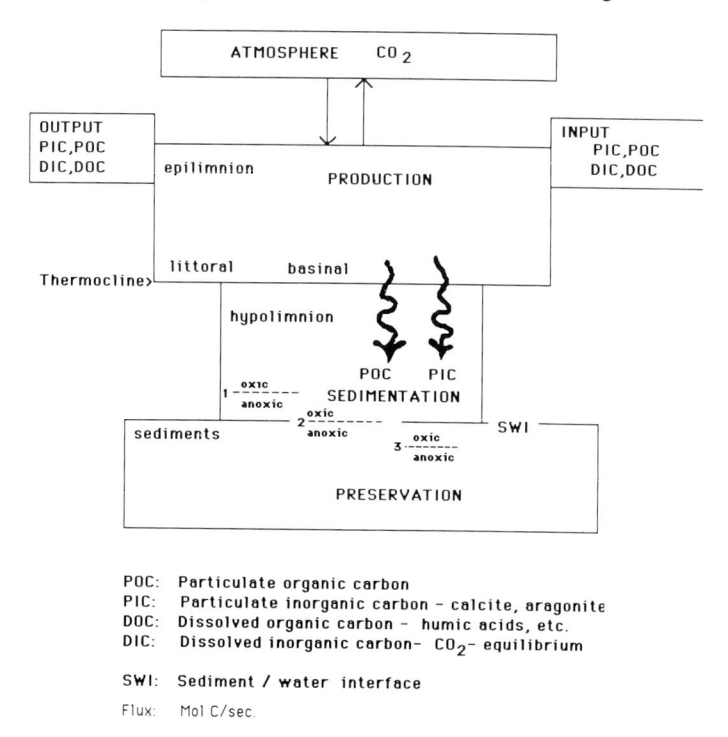

FIG. 7a. Elements of the carbon cycle in lakes.

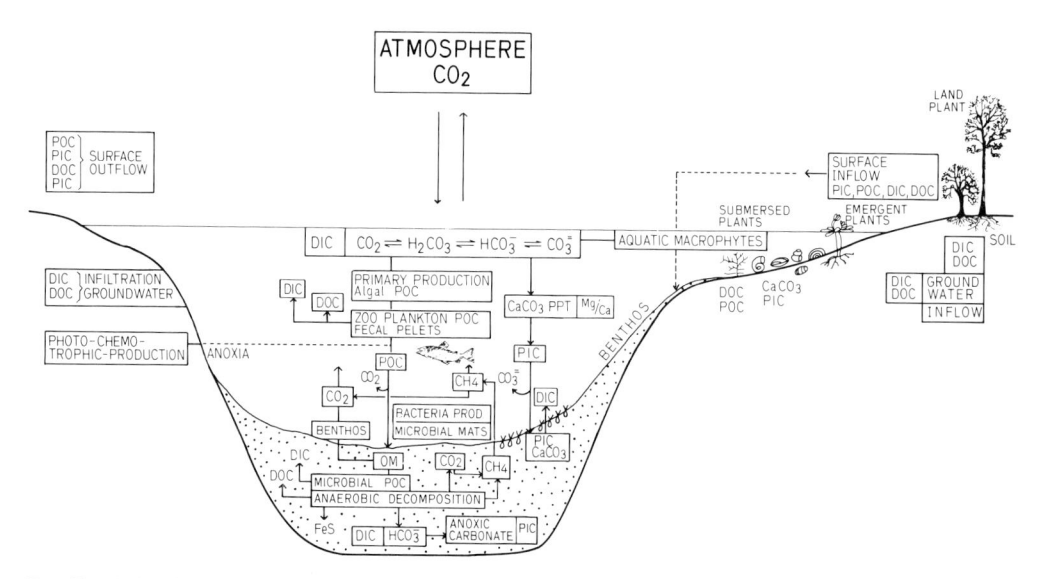

FIG. 7b. Model of lacustrine organic–inorganic carbon cycle.

The aquatic system may supply organic matter from emergent cyanobacteria mats, submerged macrophytes and microphytes, rhizopod algae, charophytes, phytoplankton, microbial plates, zooplankton, or even benthic organisms, fish and faeces from grazing or resting animals (e.g. hippos in L. George; cf. Beadle 1981).

In modern oligosaline and mesosaline lakes, diatoms are some of the most prolific plankton and incidentally the most important silica sink. We cannot trace their non-marine geological history further back than the Eocene. Cyanophytes and chlorophytes are thus considered the likely precursors to algal-sourced organic matter in ancient deposits such as the Green River Formation. Some algae are special to specific chemical environments. For example, *Spirulina* is a prolific blue-green algae in alkaline hypersaline lakes of East Africa. As one of the healthiest meals known with 60–65% protein, it is the preferred food for hoards of flamingos which consume tons per day. Flamingos may consume 90% of the primary production per day (Livingstone & Melack 1984). It is perhaps not a coincidence that fossil flamingo footprints are also common along the periphery of the Laney Shale basin of the Green River Formation. Cyanophyte blooms from Lake George in the western Rift and Lake Chad for example (Beadle 1981) provide a microbial sludge ooze which is locally harvested. Dinoflagellates (e.g. *Peridium*) are abundant in oligo- to mesosaline lakes, but their cysts are rare compared with marine environments. *Botryococcus* is a modern, lipid-rich green algae, which occurs in most brackish and freshwater lakes, but only rarely in significant amounts. Exceptions such as the shoreline deposits of the Aral Sea and the Coroong (Cane 1969) have been used to argue it as a precursor of Type I oils. *Botryococcus* floats when dead, and is washed up by winds onto shorelines. It has not, however, been reported as a major constituent of basinal deposits of modern lakes, although *Botryococcus* and *Pediastrum* traces are common in some oil shales (Cook *et al.* 1981).

Aquatic productivity

The carbon cycle in lakes (Fig. 7a, b) comprises external inputs and internal interactions among particulate inorganic carbon (PIC: e.g. calcite), particulate organic carbon (POC: wood fragments, chitin, algae, bacteria), dissolved organic carbon (DOC: humic acids, carbohydrates), and dissolved inorganic carbon (DIC: CO_2, bicarbonate, HCO_3^-, and carbonate, CO_3^{2-}). Environmental signals on lake conditions are stored in the various lacustrine carbonate facies given schematically in Fig. 8. In the following we examine aspects of *productivity, sedimentation and preservation* of organic matter in lakes that have some bearing on the interpretation of source-rock environments.

The main factors affecting the condition of a lake are climate, solar input, wind energy, precipitation, chemistry and temperature. Solar input provides light for primary production and heat for the vertical gradients that stratify lakes. Annual solar input is mainly controlled by latitude with values ranging from 0–1000 cal cm^{-2} d^{-1} (Taub 1984b). Arctic areas have only a short production season; temperate areas two, and tropical areas a long continuous season. Wind provides the energy for mixing. Annual rainfall patterns show global patterns modified by altitude and wind. The nutrient input is commonly a function of the drainage area geology.

It is generally accepted that inorganic orthophosphorous is one of the main nutrients limiting the primary production in lakes. For example, in freshwater Lake Zurich, an epilimnion productivity maximum of 500 mg C m^{-2} d^{-1} is

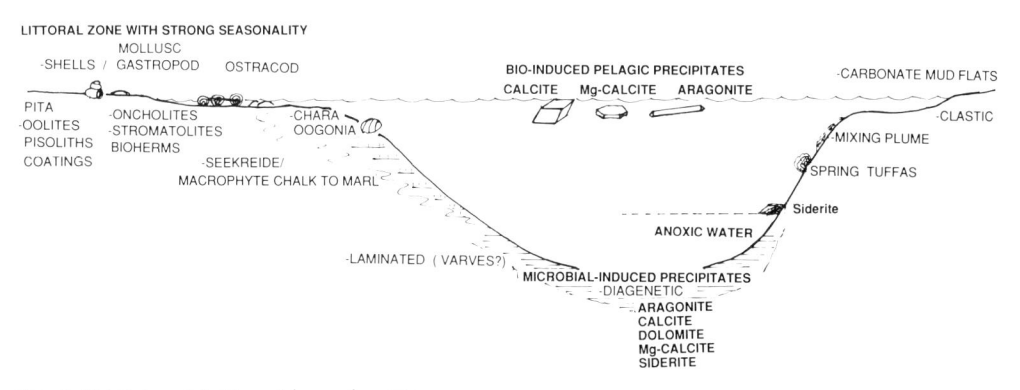

FIG. 8. Habitat model of lacustrine carbonates.

accompanied by 10 mg m^{-2} d^{-1} phosphorous uptake (Gächter 1968). The P:C ratio is 1:40, and at the above rates the entire PO$_4$ content of the lake must be recycled in a matter of days. A C:N:P ratio of 106:16:1 is considered a reasonable average for lake biomass (Stumm & Morgan 1970) although strong divergence occurs during the productive seasons (Gächter & Bloesch 1985).

Table 3 gives an overview of the relationship between phosphorous concentration and productivity in a number of modern lake environments. Most measurements in the temperate zone have been made on freshwater lakes that have had considerable anthropogenic phosphorous loading and thus represent maximums. The nutrient potential of an open system depends on the timing of major inflow, mixing and outflow. If the outflow maximum occurs when the lake is stratified, nutrient-poor waters exit. Lake waters are enriched in nutrients in contrast to lakes with outflow maximum during mixing. Oligotrophic lakes generally have 5–10 µg/l total phosphorous; mesotrophic lakes 10–30 µg/l; and eutrophic lakes over 30 µg/l total epilimnetic phosphorous (Wetzel 1983) (Table 3).

TABLE 3. *Some examples of aqueous productivity ranges in modern lakes*

Lake	Estimated g C m^{-2} y^{-1}
Dystrophic lakes, acid, bog, humic	20–70
Oligotrophic lakes 5–10 µg/l P	
L. Huron	100
L. Ontario	180
Mesotrophic lakes 10–30 µg/l P	
L. Esrom, Denmark	260
L. Erie	310
L. Zurich	200
Eutrophic Lakes >30 µg/l P	
L. Zug/Greifensee, Switzerland	400–440
L. Lugano	350–460
L. Victoria, E. Africa	680
L. Tanganyika	430
L. Kivu	
Hypereutrophic (closed basin, saline)[1]	
Borax L., CA. Alk	1050
Great Salt Lake (N. Arm)	1800
Great Salt Lake (S. Arm)	145
L. Turkana, Alk.	300–1500
L. Aranguadi, Alk. (3–7 mg/l P)	3000–6000
L. Natron, 2600 meq. Alk. (3 mg/l P)	900–1200
L. Nakuru, Alk. (11 mg/l P)	900–1200
Monolake, Alk.	1000

[1] Warren 1986, includes additional data for saline lakes. Data selected from Wetzel 1983; Talling & Talling 1965; Melack 1981; Gächter & Bloesch 1985; and references in Eugster 1985.

Arctic lakes

Arctic lakes show low species diversity but populations may be large. Most arctic lakes seem to be oligotrophic for lack of nutrients. Chemical weathering is limited by the cold. Commonly primary production ranges only from 0.6 to 21 g C m^{-2} y^{-1} but is higher than nearby Arctic Ocean waters during the season of maximum insolation. Shallow tundra lakes are part of a carbon ecosystem where emergent plants and shrubs play an important role along with bacteria. Insect larvae become a major component and benthic algae are more important than phytoplankton. Proglacial lakes have seasonally turbid waters that drastically limit production. Solar heated salt lakes, such as CaCl$_2$ L. Vanda, Antarctica, form exceptions that have higher productivity due to prolific microbial mats (Hobbie 1984).

It is not surprising that most of the fluvio-lacustrine lowlands of the Siberian taiga and Canadian shield are dominated by land-plant organic sources. These environments may be modern analogies to important parts of the Palaeozoic Gondwana deposits in Africa, Antarctica, and Australia which are likely to be dominated by Type III–II kerogen. The early post-glacial, paralic Baltic Sea had intermittent anoxia but the source of its buried organic matter is unclear.

Large temperate lakes

In large lakes, such as the Great Lakes, internal cycling of organic matter is usually of greater significance than land-plant influxes from the watershed. Production from the pelagic zone is commonly dominant over littoral/benthic zones. Biota are present in spatial and temporal patterns and vertical composition gradients.

Lake Lugano, Switzerland, serves as a useful temperate lake model recently studied by Niessen (1987). It is a complex, deep (270 m) perialpine, open, freshwater, temperate, eutrophic lake damned by glacial deposits. The Holocene sediments are clastic-dominated in contrast to the carbonate-dominated Holocene sediments of Lake Zurich. The sediment record shows that the lake circulation has been sluggish throughout the Holocene, with generally reducing sediments, and pigment evidence of intermittent anoxia in bottom waters. Bottom waters commonly received oxygen pulses, at least during part of an annual cycle, sufficient to support an impoverished benthic fauna. Surface productivity is estimated at up to 90 mg C m^{-2} y^{-1} throughout the Holocene. The sediments contain an average

2–4% organic carbon, at detrital sedimentation rates of 0.4–2.6 mm/yr. Turbidites with land-derived organics are common. Approximately 5–10% of the primary production appears to be preserved in the sediments (Niessen 1987) as a result of: (1) seasonal anoxia in deep waters, (2) bottom fauna inhibition by anoxia, (3) sedimentation pulses of organic matter leading to short residence time in the water column, (4) rapid burial by clastics leading to short exposure at the boundary layer, (5) low sulphate concentrations leading to low anaerobic degradation rates by sulphate reducers.

Acid lakes

Acid lakes are rare for the geological record, but modern volcanic lakes from Japan illustrate some aspects of the carbon cycle distinct from marine situations. These lakes have pH values as low as 1.5–2.0 as a result of sulphuric and hydrochloric emanations from volcanic sources (Mori et al. 1984). The productivity problem in acid lakes is the utilization of carbon sources. In other words, they may become carbon-limited because total carbon occurs only as dissolved CO_2 species which is less than 0.2 ppm C in atmospheric equilibrium. Fish cannot survive, but chironomids and diatoms may be abundant. Bacterial decomposers are limited. Dystrophic bog lakes are examples of such ecosystems with low productivity but high preservation potential that leads to thick accumulations of Type III lignitic kerogen.

Alkaline lakes

The high pH of many alkaline lakes has tremendous effects on the carbon cycle. The system is more open to atmospheric carbon dioxide and may store more DIC. Due to the shift in dominant carbonate species to more CO_3^{2-} ion, there is a rapid invasion of CO_2 if any carbon is fixed by photosynthesis. The OH^- ion accelerates the gas exchange of atmospheric CO_2 (Herczeg & Fairbanks 1987). The result is that the system can support a larger biomass than many corresponding but neutral pH lakes. The alkalinity effect on the availability of carbon can be illustrated by Lake Aranguadi, Ethiopia Rift. With 60 meq alkalinity at a pH 10.3, and an ionic strength of 0.08 at 20°C, a change of 0.1 pH unit means a change of 13 mg C/l (or 47 mg/l CO_2) (Talling & Talling 1965).

A second important factor is that solubility of phosphorus and other nutrients is highly pH dependent. The Gregory Rift lakes of East Africa derive solutions from terrane with reactive alkali volcanic ashes. Their higher pH values are thus concomitant with phosphorus concentrations that are higher than the most polluted temperate lakes. Even Lake Tanganyika, which has a low nutrient input, can maintain a rather large plankton biomass due to upwelling of deep waters having alkaline affinities (Burgis 1984). Shallow lake Manyara is an example near the Tanzanian/Kenyan border. The lake has a black organic ooze sediment, and TDS of about 70 g/l. The phosphorus concentration is one of the highest known at about 65 mg/l (Talling & Talling 1965) compared with about 0.2 mg/l for eutrophic Lake Zug, Switzerland. Lake Kivu also has alkaline affinities but mesosaline surface waters. It is also reported to have 70 mg/l phosphorous in the anoxic bottom waters (Degens et al. 1973). Even the surface waters contain 0.5 mg/l P which is ten times that of eutrophic lakes. These values can support tremendous algal blooms during overturn, or episodic mixing events which is reflected in the 15% TOC of the sediments. Similar high phosphorus contents are reported from other mesosaline, soda lakes; Lake Van (Kempe 1977), Walker Lake and Pyramid Lake (Koch et al. 1977).

Lake Van is 3574 km^2, over 450 m deep and contains 607 km^3 of alkaline water at pH 9.3, and with a brackish salinity 2/3 that of seawater. As a closed basin, the water residency can be expressed by a τ 1/2 of more than 100 years, or a renewal time of about 900 years (Kempe 1977). The evolution of carbonates, biota, and waters within this unique environment provide a basis for a geological model of an ancient deep, alkaline, brackish lake.

Lake Van is a soda lake, with practically no dissolved CO_2 gas species. Potassium is enriched relative to seawater. Sulphate is relatively low. No direct measurements have been made of primary or secondary productivity. The region is a continental steppe with only modest sources of land-derived (C-4?) organic matter. High pH values and volcanic source terrain result in extremely high phosphorus concentrations of 0.4 to 0.6 mg/l in Lake Van waters. High production, mainly of diatoms would thus be expected. Most diatom frustules dissolve in the high pH (9.55) water column. The sediments display stunning rhythmic laminations comprising aragonite, calcite, organic matter, and detrital clay.

Ash layers and evidence of slumping or turbidity currents are common. The laminae are cream, dark brown, green and red-brown. With an assumption of varve sedimentation, accumulation rates are estimated at 40–90 cm 10^{3}y. Calculations suggest 20% of this is authigenic carbonate. Several aspects are of interest in terms of source-rock potential. The organic carbon content ranges from 1.5 to 4% although the lake is not permanently stratified (Degens et al. 1984). Although C-org decreases with depth, spikes over 2% occur below the 10 000 bp level indicating a high preservation potential; also suggested by the retention of green chlorophyll pigment.

The Wadi Natrun, Egypt, is another example of a hypersaline alkaline lake. Extremely halophilic red-coloured bacteria and cyanobacteria, differing considerably from the Na–Cl type lakes, occur in abundance. Some are even alkaliophilic (pH optimum—8.5 to 10.5) (Larsen 1980). In some lakes, sulphate reduction is high, even contributing significantly to alkalinity (Abe-el-Malek & Rizk 1963). The hypersaline alkaline lakes contain no organisms above unicellular protozoa. Phototrophic bacteria therefore tend to form dense mats. Halobacteria remain suspended in the brine.

Hypersaline Na–Cl lakes

Warren (1986), Eugster (1985), and Kirkland & Evans (1981) review the high productivity of numerous saline lakes, although they do not differentiate alkaline types from Na–Cl type environments.

If a seawater sample is concentrated and fertilized in a laboratory experiment, the productivity of the media declines with salinity. This of course, is not surprising since salt is a common preservative. However, Zobell (1946) had already determined that samples of hypersaline waters (Great Salt Lake) were productive, but bacteria declined as the samples were diluted. The conclusion was that indigenous populations adapt, even prefer the conditions in a strong brine (Larsen 1980). Unfortunately not many data are available on the relative significance for productivity of specific elemental concentrations (Mg^{2+}, Ca^{2+}, $Na^+ \cdots$) in natural brine waters. Many organisms in hypersaline brines seem to require high magnesium concentrations.

The green alga *Dunaliella* attains a biomass of 150 g/m³ in 22% salt solutions of the Great Salt Lake (Larsen 1980). We must remember that these salt-tolerant algae have very high carbon/cell values because cells are filled with glycerol in order to counteract osmotic pressures. The glycerol returns to solution as the cell dies, and thus saline brines commonly have extremely high DOC values (up to 100 mg/l). Nearly salt-saturated waters of the North Basin showed very impressive concentrations of chemo-organo-trophic bacteria (biomass > 300 g/m³, Post 1977) which exceed the algae ten-fold. Growth was thought to be limited mainly by temperature. The food chain ends with masses of brine shrimp and fly larvae.

The Dead Sea waters support only *Dunaliella* algae and halobacterium. High Mg and Ca concentrations are considered toxic to zooplankton. In comparison with the Great Salt Lake, biomass is less than 1/10 and little seems to survive settling to the lake bottom due to sulphate reduction. Although primary productivity is high, the recycling of organic matter in hypersaline lakes is rapid. Low settling velocities in high density solutions also lead to longer retention of organic matter in the water mass.

Bacteria and lakes

The contribution of bacteria to the carbon budget in lakes with hydrocarbon potential has largely been ignored. In some lake environments, the production of organic matter by bacteria may exceed algal primary production. According to Golterman (1975), however, the growth rate of bacteria in natural waters is poorly known. Phototrophic bacteria (e.g. purple bacteria) living at pycnocline and sediment interfaces may photosynthesize via H_2 using reduced sulphur or organic compounds as an electron donor. They use CO_2 as a sole source of carbon. Chemolithotrophic bacteria derive their energy from the oxidation of inorganic compounds such as H_2, $NH_3NO_2^-$, S, and Fe^{2+}. There is little addition of organic matter to the system, because the energy released on oxidation is used to reduce compounds derived from precursor organic matter. This may either derive from allocthonous sources or primary algal production. Heterotrophic bacteria serve to change primary isotopic and chemical characteristics of the organic matter preserved in the sediments (Golterman 1975).

Desulphovibrio desulphuricans, a typical sulphate reducing bacterium, is not a strict chemolithotroph because cell carbon is not just derived from CO_2, it also derives its energy from inorganic sources:

$$H_2SO_4 + 4H_2 \rightarrow H_2S + 4H_2O.$$

Indigestible acetic acid is excreted. Rates of sulphate reduction are commonly high. In Lake Vechten with a sulphate concentration of 15 g/m³ for example, sulphate reduction utilizes 0.1 g C cm^{-2} d^{-1} which is sufficient to degrade the 25% of the annual algal production which reaches the sediment.

Many meromictic lakes commonly have a high concentration of phototrophic bacteria at a pycnocline in the photic zone. Examples include Lake Suigetsu, Japan, Big Soda Lake, Nevada, USA (Oremland et al., this volume), and Cadagno

Lake, Switzerland (Hanselmann 1986). A bacterial plate of 30 cm thickness commonly occurs at the interface of the chemocline and the monolimnion in small saline lakes of the Canadian Shield. In Waldsea Lake (Hammer 1984, 1986), the plate comprises green sulphur bacteria (*Chlorobacteriaeceae*). The annual production rate has been estimated at 32 g C m^{-2} which equals the total of the phytoplankton productivity zone which is 25 times thicker.

Lake Suigetsu in Japan, is a freshwater lake, meromictic because of the groundwater injection of saline water from neighbouring lakes. Mori *et al.* (1984) have modelled its carbon cycle. Bottom waters have a 7-year residence time. Sulphate reduction proceeds at 470–780 mg S m^{-2} d^{-1} near the sediment interface. Characteristically high rates of photo- and chemosynthesis occur at the boundary layer between oxic and anoxic water. Lake Suigetsu showed 133 mg C m^{-3} d^{-1} by photosynthesis, but over 485 mg C m^{-3} d^{-1} fixed by chemosynthesis.

Similarly, Cohen *et al.* (1977) describe productivity in Solar Lake, a hypersaline pond along the shores of the Gulf of Aquaba. The primary production of surface waters is only about 59 g C m^{-2} y^{-1} from the epilimnion. Almost all of the organic matter production of 923 g C m^{-2} y^{-1} derives from hypolimnial microbial mats and phototrophic bacterial plates.

Other bacterial effects on the carbon cycle, especially in higher concentration lakes, include iron and manganese reducers and oxidizers as well as chemotrophic methane-oxidizing bacteria (e.g. *Methanomonas*) which may produce characteristic lipids (Schleifer & Stackebrandt 1983).

Bacteria may also form deep-water microbial mats. Filamentous colonies of the chemotrophic sulphur bacteria; *Beggiatoa* colonize the sediment/water interface of the basin plain 70–110 m deep in Lake Zurich. *Beggiatoa* live at the boundary between oxic water and anoxic sediment where they utilize O$_2$ as an energy source to reduce H$_2$S to elemental sulphur which is stored in their cells. The loose, white, nerve-like, nodal network also serves to protect some areas from winnowing by weak bottom currents. This helps impart stability to gelatinous oozes and an irregular pillow like morphology to the sediment surface. *Thiobacillus*, a flagellate form, lives in the hypolimnion and uses either O$_2$ or NO$_3^-$ to reduce H$_2$O.

Sedimentation of organic matter

Organic matter settles in lake basins by much the same mechanisms as in marine basins. Reconstructions must however take into consideration the source, food chain, stratification, density gradient and depth differences to marine systems.

Most organic carbon in lakes is dissolved, with POC contributing only about 10%. Detrital input of particulate organic carbon can be carried by river and flood or be reworked from the littoral belt. River beds are flushed during highwater stages of their collected detrital organics and fine clastics. The density of flood-stage suspensions is commonly greater than the water in oligosaline lakes so that these flow as continuously-fed bottom turbidity currents that may efficiently transport large quantities of particulate organic matter to basin plains. Turbidite sedimentation is thus almost the rule in many deep oligosaline lakes. Commonly the lacustrine turbidites are silty to muddy with low porosities. In several perialpine lakes, these event-deposits contain 2–6% land-plant TOC which forms an ideal substrate for early methanogenesis. These form spongy layers charged with tiny exsolved gas bubbles which can be mapped by high-resolution seismic profiling.

At other times, particulate carbon input is distributed quickly in the epilimnion by overflows and interflows and settles out as part of the pelagic plankton rain. Turbulence above a viscous thermocline interface tends to homogenize the epilimnion content. The transfer is enhanced during overturn. Because most lakes are shallow in comparison with marine basins, the efficiency of transfer of organic matter from surface waters to the sediment surface is greater. The preservation potential is thus higher. The amount of primary production reaching sediments is estimated between 25% for small marl lakes (in Wetzel 1983) and about 10% in mesotrophic deep perialpine lakes (Niessen 1987).

Studies of modern oligosaline lakes show typical rates in large lakes for the sedimentation of POC ranging from 0.5 to 2 m/day (Bloesch & Sturm 1986). These are faster than Stoke's settling and suggest faecal pellet mechanisms. Although faecal pellets are found in sediment traps of modern freshwater lakes, they are rarely described in sediments (Sturm *et al.* 1982).

Few data are available for saline and alkaline systems. In systems with low predator populations, the percentage of primary organic matter reaching the sediments increases. In saline lakes bouyancy and fauna change the depositional rates for organic matter. Recycling in the epilimnion is increased for example by a large viscosity gradient at the thermocline or chemocline. Faecal pellets are important. *Artemia* brine shrimp occur in masses in many hypersaline environments and consume immense quantities of algae. They are prolific producers of millimetre size faecal pellets that characterize the sediments of hypersaline lakes Urmia and Great Salt Lake (cf. Kelts & Shahrabi 1986).

Preservation

Demaison & Moore (1980) stressed the significance of large anoxic lakes as a favourable environment for source-rock deposition. In water lacking oxygen, organic debris is more likely to be preserved, but this is more a result of limitations on bioturbation and irrigations than the efficiency of oxidation by aerobic vs anaerobic processes. Oxidation of 1 mole of organic matter consumes 138 moles of O_2; or about 1.25 g of oxygen per gram of organic matter (Redfield et al. 1963).

The TOC content of lake sediments reflects the balance of relative sedimentation rates of organic matter and mineral matter with the rates of decomposition or microbial additions. Lake Turkana, for example, has commonly been ignored as a productive system because only 1% organic carbon occurs in the sediment. With an overall sedimentation rate of centimetres per year, however, this represents an enormous burial rate of organic matter (Cerling 1986). Bioturbation is also reduced by high sedimentation rates. Clastic dilution is common in many mesotrophic, temperate lakes with modest organic production. In high-sulphate lakes with relatively slow sedimentation rates of organic and mineral matter such as the hypersaline Lake Urmia (0.1–0.3 mm/yr; Kelts & Shahrabi 1986), sulphate reduced in pore waters can be replenished by diffusion until most organic matter is destroyed.

Microbial degradation of organic matter utilizes oxidants in a definite order, proceeding finally to methanogenesis (Table 4). The microbiology of lacustrine environments, particularly alkaline, or brackish to saline is poorly known, but some differences to pathways in marine systems are expected. For example, oligosaline lakes show a tendency for biogenic methane production via the path of direct acetate fermentation rather than CO_2 reduction (Whiticar et al. 1986), which will affect the isotopic composition of the carbon reservoir.

Sediment surfaces may be aerobic, dysaerobic, or anaerobic depending on the relative position of the redox boundary (Fig. 7a). Anoxic processes may thus begin in the water column in meromictic lakes. The sequence of these processes is illustrated by their respective energy yield or free energy per mole of electrons exchanged (Fig. 9) and leads to a definite pattern of ion concentrations in pore waters of sediments (Fig. 10). Reactions governing the biologically mediated oxidation of organic matter are summarised in Table 4.

Most organic matter in oxic marine waters is degraded in the top 2–5 cm of sediment, not at the interface alone. The rate of bioturbation is undoubtedly the most important factor for efficient microbial degradation of organic matter. Typical marine oxidation rates are 0.084–0.025 μM O_2 cm^{-2} d^{-1}. Oxygen is quickly depleted in surface sediment layers but irrigation can replenish it sufficiently to reduce reactive organic matter within a season or two.

Hypersaline and penesaline alkaline systems do not support bottom fauna. In shallow lakes even with bottom anoxia, larger animals (hippopotamus, flamingos) and insects can be important bioturbators that need not be considered for marine systems. Submersible investigations of oxic environments at the bottom of deep lakes such as Lake Superior (Flood & Johnson 1984) show unimagined profusion of benthic life including thick carpets of Opossum Shrimp. Polychaete

TABLE 4. *Model reactions of microbial degradation*

		C:N:P
Organic matter (OM) = $(CH_2O)_{106}(NH_3)_{16}H_3PO_4$		106:16:1

Aerobic zone:	$OM + 106O_2 \rightarrow 106CO_2 + 16NH_3 + H_3PO_4$
	$[CH_4 + 2O_2 \rightarrow CO_2 + 2H_2O]$
Manganese reduction:	$OM + 212MnO_2 \rightarrow 438HCO_3^- + 16NH_4^+ + HPO_4^{2-} + 212Mn_4^{2+}$
Nitrate reduction:	$OM + 84.8NO_3^- \rightarrow 7.2CO_2 + 98.8HCO_3^- + 16NH_4^+ + 42.4N_2 + HPO_4^{2-} + 49H_2O$
Iron reduction:	$OM + 424 Fe(OH)_3 + 756CO_2 \rightarrow 862HCO_3^- + 16NH_4^+ + HPO_4^{2-} + 424Fe^{2+} + 304H_2O$
Sulphate reduction:	$OM + 53SO_4^{2-} \rightarrow 39CO_2 + 67HCO_3^- + 16NH_4^+ + HPO_4^{2-} + 53HS^- + 39H_2O$
Methane fermentation/carbonate reduction:	$OM + 14H_2O \rightarrow 39CO_2 + 14HCO_3^- + 53CH_4 + 16NH_4^+ + HPO_4^{2-}$
	$CO_2 + 4H_2 \rightarrow CH_4 + 2H_2O$

After Redfield 1983.

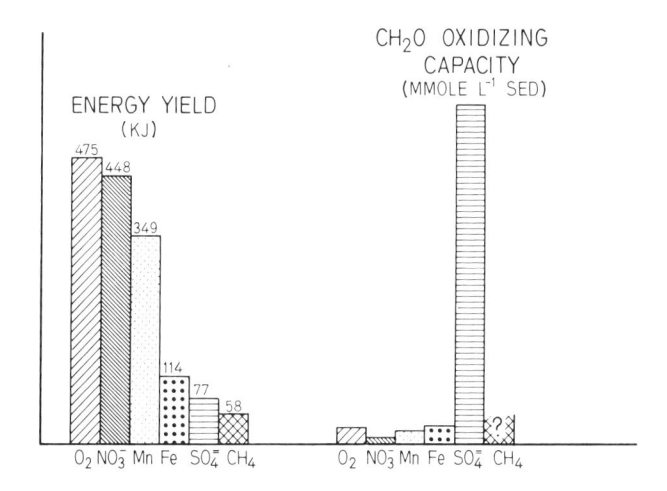

FIG. 9. Relative OM-oxidizing capacity of various electron donors. After data from Reeburgh (1980).

worms are common in oligo- to penesaline lakes and can survive at high population densities to very low oxygen concentrations ($\leqslant 0.1$ ml/l) as important benthic scavengers along with ostracods and molluscs. Seasonal stratification and short-term anoxia may however be sufficient to prevent their repopulation during months with oxic bottom water. This explains a common lack of trace fossils in lake deposits. Either muds are laminated, or else completely churned.

Algal mats along margins of oligo- to hypersaline lakes commonly exhibit a low preservation potential because of desiccation, wind-deflation and wave splash or storm events. If buried quickly, however, microbial mats have ideal precursor source-bed qualities. Microbial mats form dense multilayered communities that allow oxidation on the surface, and sulphate reduction in anoxic environments at depth. The ecosystem is only 2–5 mm thick, with up to five distinct sublayers.

Sulphate reduction is almost as efficient as oxygen utilization for microbial decomposition. Figure 9, for a marine example, illustrates clearly that because of the differences in initial concentration between oxygen and sulphate in systems, sulphate actually has the greatest potential OM-oxidizing capacity. This oxidizing capacity in lakes will thus greatly depend on the initial concentration of sulphate in lake waters (e.g. Table 5). Rates of sulphate reduction have also been shown to be dependent on the amount and type of organic matter and the concentration of sulphate. Some information suggests that sulphate reduction rates are significantly slowed by low temperatures (high altitude and arctic lakes). Recently, Lupton et al. (1984) indicate that sulphate reducing bacteria are severely limited above a critical hypersalinity or sulphate concentration.

Krom & Berner (1980) measured the Fick's Law diffusion transport rates of sulphate into FOAM marine sediments as $D_{s(SO_4)} = 5.0 \pm 1.2 \times 10^{-6}$ cm^2 s^{-1}, 20°C. Real values in lakes are dependent on salinity, concentration gradient and porosity, but order of magnitude estimates show that with sedimentation rates of 1 mm/yr which are common in many lake basins, sulphate replenishment is probably limited (Table 5a).

Howarth & Teal (1979) suggest that sulphate reduction alone is sufficient to consume the net production in highly productive marine salt marshes. Sulphate reduction of 75 mole/yr oxidizes 1800 g C m^{-2} yr^{-1} leaving almost no preservation potential. In these environments the sulphide is rapidly converted to pyrite and thus H$_2$S remains below toxic levels.

The Great Salt Lake and Lake Urmia, Iran, are shallow, perennial lakes with commonly hypersaline brines containing up to 20 000 ppm sulphate. It is thus not surprising to find that

TABLE 5a. *Rates of sulphate reduction*

Setting	g S m^{-1} y^{-1}	mmol m^{-2} d^{-1}
Black Sea	0.5–44	
Fresh waters	2.2–10	
Saline lakes		4–445
Tropical lowlands		77–2450
Chesapeake bay, estuary		6–335
Big Soda Lake	6.6	

After Trudinger 1979.

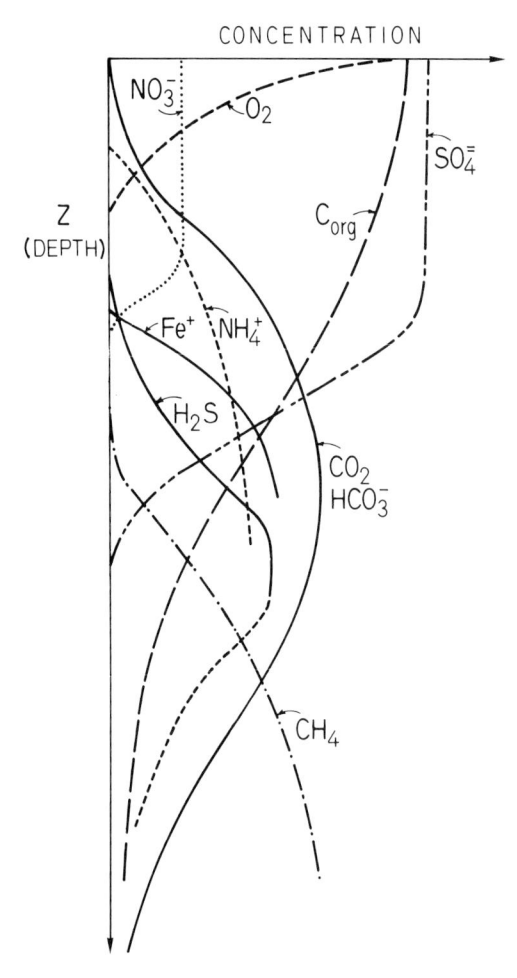

FIG. 10. Hypothetical gradients of pore water species in near-surface, organic-rich lake sediments. Sequence after Table 4. Z in millimetres to centimetres.

TABLE 5b. *Sulphate concentrations in some lakes and the ocean*

Lake	mg/l
L. Zurich	5
L. Greifensee	21
L. Urner	12
L. Baldegg	11–15
L. Constance	51
L. Cadagno	154
L. Kivu	25 surface water
	220 bottom water
Qing Hai Lake China	2402
Saline (Ocean)	2712
Big Soda Lake	5600
Great Salt Lake	16000
Urmia	22000
Dead Sea	450

of microbial decomposition reactions (Table 4), iron reduction, manganese reduction and sulphate reduction produce considerable alkalinity in pore waters, particularly if H_2S can escape. The result is that under such conditions, carbonate minerals may be well-preserved, or even precipitated. The phases and types of such 'anoxic carbonates' (Kelts & McKenzie 1982), should reflect the overlying conditions in a lake (Talbot & Kelts 1986).

Methanogenesis is one of the terminal reactions for the anaerobic carbon cycle. In lakes with low salinities, it is often however the dominant reaction. Many lakes with organic-rich deposits are known for the bubble ebullition of methane. This may even impart a spongy texture to sediments in tropical lakes. Methanogenesis may continue to considerable depth. Because of the specific substrate requirements for methanic bacteria, the efficiency of degradation of organic matter is limited. The process diminishes before total organic matter is consumed.

In deep meromictic oligo- to hypersaline lakes, anoxia in the water column commonly leads to complete reduction of sulphate, iron and manganese, as is observed in the Black and Dead Seas (Neev & Emery 1967). Gradients within the anoxic water mass are minimized by internal circulation. Organic matter arriving at the sediment thus has a very high preservation potential. This situation probably explains the very high TOC values of deep anoxic Lakes Tanganyika, Kivu, and Malawi (Talbot, this volume).

The ultimate content and quality of organic matter preserved in a sequence with potential for lacustrine source rocks is a balance between organic input, production and degradation rates, and the dilution by non-organic sedimentation (Table 6). This interaction is schematically given

beneath the sediment surface pyrite is quickly formed, organic content decreases rapidly, and the main bulk of Holocene age aragonitic, saline lake sediment has less than 1% organic carbon; probably refractile (Kelts & Shaharabi 1986).

Freshwater perialpine lakes have initial sulphate concentrations of only 10–51 mg/l with the exception of the bacterial, alpine, Lake Cadagno (Hanselmann 1986) which has 154 mg/l sulphate fed from springs from gypsum deposits (cf. Table 5b). These contrast markedly with the 2712 mg/l sulphate in the oceans.

Inorganic processes are coupled with the biogeochemical oxidative sequence in anoxic sediments. Authigenic phosphates, sulphides and carbonates may form (Fig. 8; Hanselmann 1986; Talbot & Kelts 1986). As is shown by the sequence in Fig. 11. Most smaller lake basins have relatively high clastic or chemical sedimentation rates with respect to marine basins. High contents of oil-prone organic matter thus suggests not only excellent preservation (anoxia), but also a large enough basin and situation to reduce clastic dilution.

Bralower & Thierstein (1984) proposed a model of preservation potential for marine environments based on various Holocene sites. Although only limited data is available, the theoretical considerations for lacustrine basins suggest that fresh, brackish, and hypersaline lakes would fall in the uppermost range of preservation potential (Fig. 12).

Lacustrine kerogen problems

One of the problems concerning lacustrine source rocks remains the origin and significance of Type I, alginite, low sulphur kerogen (Tissot & Welte 1978). A high hydrogen index suggests algae, but high wax suggests land-plant. All lake deposits certainly do not display Type I OM, but where it occurs, the sediments have been identified as lacustrine. The prevalent ideas on its origin may be summarized as follows:

1. Special kind of lacustrine lipid rich algae.
2. Bacterial reduction of land-derived organics to an amorphous state
3. Processes of selective preservation of lipid compounds.
4. Microbial tissue.
5. Late diagenetic feature.

The organic-rich sediments of modern prototype anoxic basins of E. Africa more generally display Type II characteristics (Talbot, this volume), although sediments closer to a Type I precursor have been identified from the Holocene in the basinal area of Lake Victoria (Talbot, private communication).

Lipid-rich green algae such as *Botryococcus* or *Pediastrum* have thick, robust cell walls, commonly preserved as characteristic organic fossil

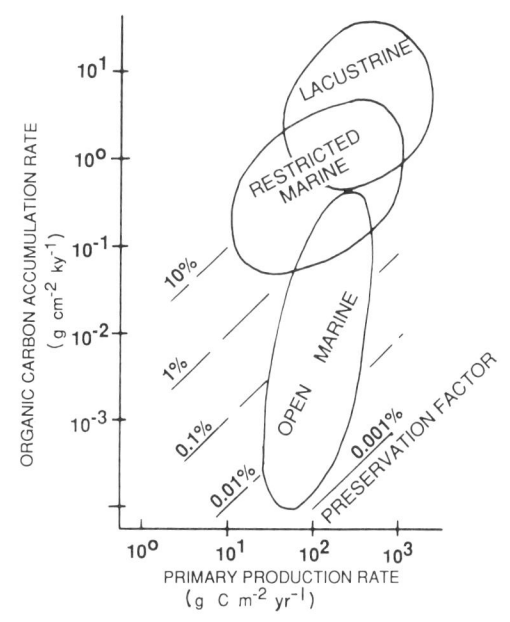

FIG. 12. General model of production, accumulation, and preservation of marine and lacustrine organic matter. Modified after Bralower and Thierstein (1984).

TABLE 6. *Total organic content of some lake sediments*

Dystrophic	2–30% Humic
Oligosaline	
L. Zurich (marl)	6–2% recent
	1–2% Holocene
L. Lugano (clastic)	2–4% Holocene
L. Walensee (clastic)	1%
L. Cadagno (bacteria)	1–10%
Alkaline affinity[1]	
L. Tanganyika	2–11%
L. Kivu	2–15%
L. Victoria	1–8%
L. Maninjau	4–8%
Mesosaline	
L. Van (carbonate)	2–4%
Hypersaline	
Dead Sea (CaCl$_2$)[2]	<0.4
L. Urmia	<1%
Great Salt Lake	<1% recent
Great Salt Lake	>6% late glacial

[1] Talbot (this volume).
[2] Warren (1986).

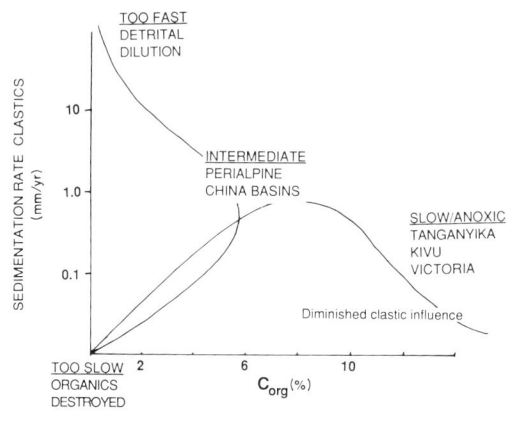

FIG. 11. Schematic model of sedimentation rate control on organic carbon burial in lacustrine basins.

shapes in thin sections. A few scattered specimens imbedded in amorphous OM stand out conspicuously. Pure laminae of these algae are, however, uncommon in most lacustrine oil shales (Cook et al. 1981).

Powell (1986) notes that the waxy Type I oils are paradoxically located farthest from the terrestrial plant sources, and proposes that microbial degradation of cellulose in mildly oxic basinal areas concentrates hydrogen rich components supplemented by bacterial biomass.

Proteins, carbohydrates and lipids are all highly susceptible to bacterial degradation under aerobic conditions. Lipids on the other hand, may be selectively concentrated by degradation under anaerobic conditions (Lewan 1986; Zobell 1946) and therefore Type I kerogens may also be most indicative of sedimentation in anoxic water masses. Because lipids tend to be isotopically lighter than proteins or carbohydrates (Degens et al. 1968), their selective preservation in sedimented algae may shift a bulk δC-13 signal more negative, causing confusion with land-derived sources.

Bacteria, particularly of the *Beggiatoaceae* family, have been proposed as a major source of organic matter in some laminated kerogenites (Williams 1984). Chemotrophic bacterial mats should inherit, to a large degree, the isotopic carbon ratios of precursor organic matter although few isotopic investigations have been published. Other microbial processes may complicate this interpretation. Phototrophic purple bacteria (*Chromatiaceae*) which fix carbon directly by a sulphur-based photosynthesis have also been shown to have carbon isotope signatures similar to C-3 land-plants (Schidlowski 1987).

Although few data are available, methanogens have been reported to produce a lipid-rich cell structure that may be preserved in sediments (Schleifer & Stakebrandt 1983). Several laminated Type I lacustrine oil shale occurrences have early diagenetic carbonate laminae with carbon isotope signatures indicating zones of extensive methanogenesis (Kelts & McKenzie 1982, 1985). Extensive methanogenesis near the sediment/water interface is one characteristic difference between marine and oligo- to mesosaline non-marine organic-rich systems.

Lewan (1986) suggests a model which links carbon isotopes in lake organic matter to morphology and reports that the following lacustrine units have amorphous kerogen that is characteristically light in carbon isotopes: Peace Valley, USA, Pliocene (Link & Osborne 1978), Mae Moh, Thailand, Oligocene (Sherwood et al. 1984), Elko Shale, USA, Eocene (Solomon et al. 1979), Green River Fm., Eocene (Bradley 1964), Scottish Lowlands Oil Shale, Carboniferous (Greensmith 1968). A parallel is drawn to the Precambrian Gunflint Iron Formation which also exhibits light-amorphous kerogen (-25 to $-30‰$) in shallow-water facies and heavy-amorphous kerogen (-15 to $-20‰$) in deeper water facies. Perhaps the light carbon values which are often considered land-plant derived in Phanerozoic sequences, may derive from archebacterial configurations that are common to the Precambrian and many modern lake environments.

The origin of Type I, amorphous waxy kerogen remains unclear. Progress will derive from further studies of floral, microbial, geochemical and isotopic signals from cores and ecosystems of modern lakes which display hydrogen-enriched organic-rich sediments.

Conclusions

Future studies on lacustrine source-rock environments will deal with the integration of production-preservation data in modern systems within a sedimentological framework of lacustrine basins such as given in Reading (1986). Perhaps the optimum, though not exclusive, setting for petroleum source-rock accumulation combines the features of a large, relatively deep, mesosaline, alkaline, closed basin lake, in subtropical, tectonic setting. This lake would have (1) adequate nutrients from chemical weathering for long productivity seasons, (2) a stressed environment with low predator numbers, (3) seasonal stratification and limited circulation, with probably permanent anoxia, (4) low sulphate concentrations, and concentration of OM by dissolution of silica and (5) a minimum dilution by clastic input. Evaporites are a closely related facies because of the delicate balance achieved in a closed basin environment and the rapid rate of response to any hydrological change.

ACKNOWLEDGEMENTS: Many colleagues have contributed to these ideas during excursions to various modern and ancient lake sites, and during the sessions of IGCP-219. In particular, I acknowledge M. Talbot, H. Eugster, M. Arthur, H. Faure, K. Yemane, F. Niessen, A. Losher, M. Sturm, K. Hsu, M. Gibling, G. Lister, M. Esteban, and J. McKenzie. A Swiss National Science Foundation fellowship in France was essential. Frau H. Bolliger and P. Schlup helped with illustrations. B. Schwertfeger and Macintosh are especially thanked for the text. A. Fleet, M. Talbot and F. Niessen kindly criticized versions of the manuscript.

References

ABE-EL-MALEK, Y. & RIZK, S. G. 1963. Bacterial sulfate reduction and the development of alkalinity. *Journal of Applied Bacteriology*, **26**, 7–26.

BEADLE, L. C. 1981. *The Inland Waters of Tropical Africa: An Introduction to Tropical Limnology*. 2nd edn. Longman, London.

BIDDLE, K. T. & CHRISTIE-BLICK, N. (eds), 1985. Strike-slip deformation basin formation, and sedimentation. *Special Publication. Society of Economic Paleontologists and Mineralogists*, **37**.

BLOESCH, J. & STURM, M. 1986. Settling flux and sinking velocities of particulate phosphorus and particulate organic carbon in Lake Zug, Switzerland. *In*: SLY, P. G. (ed.) *Sediments and Water Interactions*. Springer-Verlag, NY, 481–490.

BOYER, B. 1982. Green River laminites: does the playa-lake model really invalidate the stratified-lake model? *Geology*, **10**, 321–324.

BRADLEY, W. H. 1929. Varves and climate of the Green River epoch. *United States Geological Survey. Professional Paper*, **158-E**.

—— 1964. Geology of the Green River Formation and associated Eocene rocks in southwestern Wyoming and adjacent parts of Colorado and Utah. *United States Geological Survey. Professional Paper*, **496-A**.

BRALOWER, T. J. & THIERSTEIN, H. R. 1984. Low productivity and slow deep-water circulation in mid-Cretaceous oceans. *Geology*, **12**, 614–618.

BRICE, S., COCHRAN, M. D., PARDO, G. & EDWARDS, A. D. 1982. Tectonics and sedimentation of the South Atlantic rift sequence, Cabinda, Angola. *Memoir. American Association of Petroleum Geologists*, **34**, 5–18.

BURGIS, M. J. 1984. An estimate of zooplankton biomass for Lake Tanganyika. *Verhandlugen der Internationalen Vereinigung für Limnologie*, **22**, 1199–1203.

CANE, R. F. 1969. Coorongite and the genesis of oil shale. *Geochimica et Cosmochimica Acta*, **33**, 257–265.

CERLING, T. E. 1986. A Mass-balance approach to basin sedimentation: constraints on the recent history of the Turkana basin. *Palaeogeography, Palaeoclimatology, Palaeoecology*, **54**, 63–86.

CHIVAS, A. R., DE DECKKER, P. & SHELLEY, J. M. G. 1985. Strontium content of ostracods indicates lacustrine palaeosalinity. *Nature*, **316**, 251–253.

COHEN, A. S. & NIELSEN, C. 1986. Ostracodes as indicators of paleohydrochemistry in lakes: a Late Quarternary example from Lake Elmentia, Kenya. *Palaios*, **1**, 601–609.

COHEN, Y., KRUMBEIN, W. E. & SHILO, M. 1977. Solar Lake (Sinai) 3: Bacterial distribution and production. *Limnology and Oceanography*, **22**, 621–634.

COOK, A. C., HUTTON, A. C. & SHERWOOD, N. R. 1981. Classification of oil shales. *Bulletin des Centres de Recherches Exploration–Production Elf-Aquitaine*, **5**, 353–381.

—— & KURTMAN, F. (eds) 1978. The Geology of Lake Van. *Memoir. Mineral Research Exploration Institute Turkey, Ankara*, **169**.

——, BEHRENDT, M., GOTTHARDT, B. & REPPMANN, E. 1968. Metabolic fractionation of carbon isotopes in marine plankton. II. Data on samples collected off the coasts of Peru and Ecuador. *Deep Sea Research*, **15**, 11–20.

——, VON HERZEN, R. P., WONG HOW-KIN, DEUSER, W. G. & JANNASCH, H. W. 1973. Lake Kivu: structure, chemistry and biology of an East African rift lake. *Geologische Rundschau*, **62**, 245–277.

——, WONG, H. K., KEMPE, S. & KURTMAN, F. 1984. A geological study of Lake Van, Eastern Turkey. *Geologische Rundschau*, **73**, 701–734.

DEMAISON, G. J. & MOORE, G. T. 1980. Anoxic environments and oil source bed genesis. *Bulletin of the American Association of Petroleum Geologists*, **64**, 1179–1209.

DEMICCO, R. V. & GIERLOWSKI-KORDESCH, E. 1986. Facies sequences of a semi-arid closed basin: the Lower Jurassic East Berlin Formation of the Hartford Basin, New England, USA. *Sedimentology*, **33**, 107–118.

DESBOROUGH, G. A. 1978. A biogenic-chemical stratified lake model for the origin of oil shale of the Green River Formation: an alternative to the playa lake model. *Bulletin of the Geological Society of America*, **89**, 961–971.

EUGSTER, H. 1985. Oil shales, evaporites and ore deposits. *Geochimica et Cosmochimica Acta*, **49**, 619–635.

—— & HARDIE, L. A. 1975. Sedimentation in an ancient playa lake complex; the Wilkins Peak Member of the Green River Formation of Wyoming. *Bulletin of the Geological Society of America*, **86**, 319–334.

—— & HARDIE, L. A. 1978. Saline lakes. *In*: LERMAN, A. *et al.* (eds) *Lakes: Chemistry, Geology, Physics*. Springer-Verlag, NY, 237–293.

—— & SURDAM, R. 1973. Depositional environment of the Green River Formation: a preliminary report. *Bulletin of the Geological Society of America*, **84**, 1115–1120.

FAIRHEAD, J. D. 1986. Geophysical controls on sedimentation within the African Rift Systems. *In*: FROSTICK, L. E. *et al.* (eds) *Sedimentation in the African Rifts*. Special Publication of the Geological Society of London, **25**, 19–27. Blackwell Scientific Publications, Oxford.

FLOOD, R. D. & JOHNSON, T. C. 1984. Side-scan targets in Lake Superior—evidence for bedforms and sediment transport. *Sedimentology*, **31**, 311–323.

FOREL, F. A. 1901. *Handbuch der Seenkunde: Allgemeine Limnologie*, Stuttgart.

FREY, D. G. 1974. Paleolimnology. *Mitteilungen der Internationalen Vereinigung für Limnologie*, **20**, 95–123.

FROSTICK, L. E., RENAUT, R. W., REID, I. & TIERCELIN, J. J. (eds) 1986. Sedimentation in the African Rifts. Special Publication of the Geological Society of London, **25**. Blackwell Scientific Publications, Oxford.

GÄCHTER, R. 1968. Phosphathaushalt und planktische Primärproduktion im Vierwaldstättersee, Horwer

Bucht. *Schweizerische Zeitschrift für Hydrologie*, **30**, 1–66.

—— & BLOESCH, J. 1985. Seasonal and vertical variation in the C:P ratio of suspended and settling seston of lakes. *Hydrobiologia*, **128**, 193–200.

GIBLING, M. R., CHARN TANTISUKRIT, WUTTI UTTAMO, THEERAPONGS THANASUTHIPITAK & MUNGKORN HARALUCK, 1985. Oil shale sedimentology and geochemistry in Cenozoic Mae Sot Basin, Thailand. *Bulletin of the American Association of Petroleum Geologists*, **69**, 767–780.

GLIKSON, M. 1983. Microbiological precursors of coorongite and torbanite and the role of microbial degradation in the formation of kerogen. *Organic Geochemistry*, **4**, 161–172.

GOLTERMAN, H. L. 1975. Physiological limnology. An approach to the physiology of lake ecosystems. *Development in Water Science Vol. 2*, Elsevier, Amsterdam, 357–402.

GRAVESTOCK, D. I., MOORE, P. S. & PITT, G. M. 1986. Contributions to the Geology and Hydrocarbon Potential of the Eromanga Basin. *Special Publication. Geological Society of Australia*, **12**.

GREENSMITH, J. T. 1968. Palaeogeography and rhythmic deposition in the Scottish oil-shale group. *In: United Nations Symposium on the Development and Utilization of Oil Shale Resources*, Section B1, 1–16.

HAMMER, U. T. 1984. The saline lakes of Canada. *In*: TAUB, F. (ed.) *Lakes and Reservoirs. Ecosystems of the World*. Elsevier, Amsterdam, **23**, 521–540.

HAMMER, U. T. 1986. *Saline Lake Ecosystems of the World*. Junk, Amsterdam.

HANSELMANN, K. W. 1986. Microbially mediated processes in environmental chemistry: lake sediments as model systems. *Chimia*, **40**, 146–159.

HARDIE, L. A. 1984. Evaporites: marine or non-marine. *American Journal of Science*, **234**, 193–240.

HEDBERG, H. D. 1968. Significance of high-wax oils with respect to genesis of petroleum. *Bulletin of the American Association of Petroleum Geologists*, **52**, 736–750.

HERCZEG, A. L. & FAIRBANKS, R. G. 1987. Anomalous carbon isotopic fractionation between atmospheric CO_2 and dissolved inorganic carbon induced by intense photosynthesis. *Geochimica et Cosmochimica Acta*, **51**, 895–899.

HOBBIE, J. E. 1984. Polar Limnology. *In*: TAUB, F. (ed.) *Lakes and Reservoirs: Ecosystems of the World*. Elsevier, Amsterdam, **23**, 63–106.

HOWARTH, R. W. & TEAL, J. M. 1979. Sulfate reduction in a New England salt marsh. *Limnology and Oceanography*, **24**, 999–1013.

HUTCHINSON, G. E. 1957. *A Treatise on Limnology. I, Geography, Physics and Chemistry*. John Wiley and Sons, NY.

KELTS, K. & HSÜ, K. J. 1978. Freshwater carbonate sedimentation. *In*: LERMAN, A. *et al*. (eds) *Lakes: Chemistry, Geology, Physics*. Springer-Verlag, Berlin, 295–323.

—— & MCKENZIE, J. 1982. Diagenetic dolomite formation in Quaternary anoxic diatomaceous muds of Deep Sea Drilling Project Leg 64, Gulf of California. *In*: CURRAY, J. R., MOORE, D. G. *et al. Initial Reports of the Deep Sea Drilling Project*, **64** (**2**), Washington (US Government Printing Office), 527–534.

—— & MCKENZIE, 1985. Significance of anoxic dolomite in the Laney Shale Member of the Green River Formation. *Geological Society of America Annual Meeting, Orlando*, Abstract, **7**, 626.

—— & SHAHRABI, M. 1986. Holocene sedimentology of hypersaline Lake Urmia, Northwestern Iran. *Palaeogeography, Palaeoclimatology, Palaeoecology*, **54**, 105–130.

KEMPE, S. 1977. Hydrographie, Warven-Chronologie und Organische Geochemie des Van Sees, Ost-Türkei. *Mitteilunger Geologisch-paläontologischen Institut der Universität Hamburg*, **47**, 125–228.

KIRKLAND, D. W. & EVANS, R. 1981. Source-rock potential of evaporitic environment. *Bulletin of the American Association of Petroleum Geologists*, **65**, 181–190.

KOCH, D. L., MAHONEY, J. L., SPENCER, R. J., COOPER, J. J. & JACOBSON, R. L., 1977. Limnology of Walker Lake. *In*: GREER, D. C. (ed.) *Desertic Terminal Lakes*. Utah Water Research Laboratory, Logan, UT, 355–369.

KROM, M. D. & BERNER, R. A. 1980. The diffusion coefficients of sulfate, ammonium, and phosphate ions in anoxic marine sediments. *Limnology and Oceanography*, **25**, 327–337.

KUIVILA, K. M. & MURRAY, J. W. 1984. Organic matter diagenesis in freshwater sediments. The alkalinity and total CO_2 balance and methane production in the sediments of Lake Washington. *Limnology and Oceanography*, **29/6**, 1218–1230.

KUTZBACH, J. E. & STREET-PERROTT, F. A. 1985. Milankovitch forcing of fluctuations in the level of tropical lakes from 18 to 0 kyr BP. *Nature*, **317**, 130–134.

LARSEN, H. 1980. Ecology of hypersaline environments. *In*: NISSENBAUM, A. (ed.) Hypersaline brines and evaporitic environments. *Developments in Sedimentology*, **28**, 23–40.

LEWAN, M. D. 1986. Stable carbon isotopes of amorphous kerogens from Phanerozoic sedimentary rocks. *Geochica et Cosmochimica Acta*, **50**, 1583–1591.

LIKENS, G. E. 1975. Primary production of inland aquatic ecosystems. *In*: LIETH, H. & WHITTAKER, R. H. (eds) Primary productivity of the biosphere. *Ecological Studies*, **14**, Springer, Berlin, 179–202.

LINK, M. H. & OSBORNE, R. H. 1978. Lacustrine facies in the Pliocene Ridge Basin Group: Ridge Basin, California. *In*: MATTER, A. & TUCKER, M. E. (eds) Modern and Ancient Lake Sediments. *Special Publication, International Association of Sedimentologists*, **2**, 169–187.

LISTER, G. S. 1985. Late Pleistocene alpine deglaciation and post-glacial climatic developments in Switzerland: The record from sedimenta in a peri-alpine lake basin. *Dissertation ETH-Zurich*, No. 7753.

—— (in press). Stable isotopes from lacustrine ostracoda as tracers for continental palaeoenvironments. *In*: COLIN, J. P., DEDECCKER, P. and PEYPOUQUET, J. P. (eds) *Ostracods in Earth Sciences*, Elsevier.

LIVINGSTONE, D. A. & MELACK, J. M. 1984. Some lakes of subsaharan Africa. *In*: TAUB, F. (ed.) *Lakes and Reservoirs. Ecosystems of the World*. Elsevier, Amsterdam, **23**, 467–497.

LUPTON, F. S., PHELPS, T. J. & ZEIKUS, J. G. 1984. Methanogenesis, sulphate reduction, and hydrogen metabolism in hypersaline anoxic sediments of the Great Salt Lake, Utah. *Bass Becking Geobiological Laboratory Annual Report 1983*, 42–47.

MANSPEIZER, W. 1981. Early Mesozoic basins of the central Atlantic passive margins: *In*: BALLY, A. W. *et al.* (eds) Geology of passive continental margins. *American Association of Petroleum Geologists Education Course Note Series*, **19**, 4–1 to 4–60.

MASON, J. F. (ed.) 1980. *Petroleum Geology in China*. Penwell Books, Tulsa, OK.

MELACK, J. M. 1981. Photosynthetic activity of phytoplankton in tropical African soda lakes. *Hydrobiologia*, **81**, 71–85.

MORI, S., SAIJO, Y. & MIZUNO, T. 1984. Limnology of Japanese lakes and ponds. *In*: TAUB, F. (ed.) *Lakes and Reservoirs. Ecosystems of the World*. Elsevier, Amsterdam, **23**, 303–330.

NEEV, D. & EMERY, K. O. 1967. The Dead Sea—depositional processes and environments of evaporites. *Israel Geological Survey Bulletin*, **41**.

NIESSEN, F. 1987. Sedimentologische, geophysikalische und geochemische Untersuchungen zur entstehung und Ablagerungsgeschichte des Luganoseef. *Dissertation ETH*. Zürich No. 8354.

NIPKOW, F. 1920. Vorläufige Mitteilungen über Untersuchungen des Schlammabsatzes im Zürichsee. *Zeitschrift Hydrologie*, **1**, 1–27.

—— 1927. Ueber das Verhalten der Skelette planktischer Kieselalgen im geschichteten Tiefenschlamm des Zürich und Baldeggersees. *Dissertation ETH-Zürich*, No. 445.

PARRISH, J. T. & BARRON, E. J. 1986. Paleoclimates and economic geology. *Society of Economic Paleontologists and Mineralogists, Short Course Notes*, **18**.

PEYPOUQUET, J.-P. 1979. Ostracodes et paléoenvironments. Méthodologie et application aux domaines profonds du Cénozoïque. *Bulletin. Bureau de Recherches géologique*, **4**, 1, 3–79.

PIA, J. 1933. *Die rezenten Kalksteine*. Leipzig.

PICARD, M. D. & HIGH, L. R. 1972. Criteria for recognizing lacustrine rocks. *In*: RIGBY, J. K. & HAMBLIN, W. K. (eds) Recognition of Ancient Sedimentary Environments. *Special Publication, Society of Economic Paleontologists and Mineralogists*, **16**, 108–145.

—— & HIGH, L. R. 1981. Physical stratigraphy of ancient lacustrine deposits. *In*: ETHRIDGE, F. G. *et al.* (eds) Recent and ancient nonmarine depositional environments: models for exploration. *Special Publication, Society of Economic Paleontologists and Mineralogists*, **31**, 233–259.

POST, F. J. 1977. The microbial ecology of Great Salt Lake. *Microbiology and Ecology*, **3**, 143–365.

POWELL, T. G. 1986. Petroleum geochemistry and depositional setting of lacustrine source rocks. *Marine and Petroleum Geology*, **3**, 198–219.

READING, H. (ed.) 1986. *Sedimentary Environments and Facies*. 2nd edn, Blackwell Scientific Publications, Oxford.

REDFIELD, A. C., KETCHUM, B. H. & RICHARDS, F. A. 1963. The influence of organisms on the composition of sea water. *In*: HILL, M. N. (ed.) *The Sea Vol. 2*. Interscience, NY, 26–77.

REEBURGH, W. S. 1980. Anaerobic methane oxidation: rate depth distributions in Skan Bay sediments. *Earth and Planetary Science Letters*, **47**, 345–352.

RUTTNER, F. 1963. Fundamentals of Limnology (Transl. FREY, D. & FRY, F.), University of Toronto Press.

RYDER, R. T., FOUCH, T. D. & ELISON, J. H. 1976. Early Tertiary sedimentation in the western Uinta basin, Utah. *Bulletin of the Geological Society of America*, **87**, 496–512.

SCHIDLOWSKI, M. 1987. Application of stable carbon isotopes to early biochemical evolution on earth. *Annual Review of Earth and Planetary Science*, **15**, 47–72.

SCHLEIFER, K. H. & STACKEBRANDT, E. 1983. Molecular systematics of prokaryotes. *Annual Review of Microbiology*, **37**, 143–187.

SERRUYA, C. & POLLINGHER, U. 1983. *Lakes of the Warm Belt*. Cambridge University Press.

SHERWOOD, N. R., COOK, A. C., GIBLING, M. & TANTISUKRIT, C. 1984. Petrology of a suite of sedimentary rocks associated with some coal-bearing basins in northwestern Thailand. *International Journal of Coal Geology*, **4**, 45–71.

SOLOMON, B. J., MCKEE, E. H. & ANDERSEN, D. W. 1979. Eocene and Oligocene lacustrine and volcanic rocks near Elko, Nevada. *In*: NEWMAN, G. & GOODE, H. (eds) *Basin and Range Symposium*, Rocky Mountain Association of Geologists—Utah Geologist's Association, 325–337.

SPENCER, R. J., BAEDECKER, M. J., EUGSTER, H. P., FORESTER, R. M., GOLDHABER, M. B., JONES, B. F., KELTS, K., MCKENZIE, J., MADSEN, D. B., RETTIG, S. L., RUBIN, M. & BOWSER, C. J. 1984. Great Salt Lake, and precursors, Utah: the last 30 000 years. *Contributions to Mineralogy and Petrology*, **86**, 321–334.

STRAKHOV, N. M. 1970. *Principles of Lithogenesis*. Plenum, NY.

STREET-PERROTT, F. A. & HARRISON, S. P. 1984. Temporal variations in lake levels since 30 000 yrs BP—an index of the global hydrological cycle. *In*: HANSEN, J. E. & TAKAHASHI, T. (eds), Maurice Ewing Vol. 5, *American Geophysics Union Monograph*, **29**, 118–129.

STUMM, W. & MORGAN, J. 1970. *Aquatic Chemistry*. Wiley Interscience, NY.

STURM, M., ZEH, U., MÜLLER, J. & SIGG, L. 1982. Schwebstoffuntersuchungen im Bodensee. *Eclogae Geologicae Helvetiae*, **75**, 579–588.

SURDAM, R. & STANLEY, K. O. 1979. Lacustrine sedimentation during the culminating phase of Eocene Lake Gosiute, Wyoming, Utah and Colorado (Green River Formation). *Bulletin of the Geological Society of America*, **90**, 93–110.

—— & WOLFBAUER, C. A. 1975. Green River Formation, Wyoming: a playa lake complex. *Bulletin of the Geological Society of America*, **86**, 335–345.

TALBOT, M. R. & KELTS, K. 1986. Primary and diagenetic carbonates in the anoxic sediments of Lake Bosumtwi, Ghana. *Geology*, **14**, 912–916.

TALLING, J. F. & TALLING, I. B. 1965. The chemical composition of African lake waters. *Internationale Revue des gesamten Hydrobiologie*, **50**, 421–463.

TAUB, F. B. (ed.) 1984a. *Lakes and Reservoirs. Ecosystems of the World*, No. **23**, Elsevier, Amsterdam.

—— 1984b. Ecosystem processes. *In*: TAUB, F. (ed.) *Lakes and Reservoirs. Ecosystems of the World*. Elsevier, Amsterdam, **23**, 9–42.

TISSOT, B. & WELTE, D. H. 1978. *Petroleum Formation and Occurrence*. Springer, NY.

TRUDINGER, P. A. 1979. The biological sulphur cycle. *In*: TRUDINGER, P. A. & SWAINE, D. J. (eds) *Biochemical Cycling of Mineral-forming Elements*. Elsevier, Amsterdam, 293–313.

WARREN, J. K. 1982. The hydrological setting, occurrence, and significance of gypsum in Late Quaternary salt lakes in South Australia. *Sedimentology*, **29**, 609–637.

—— 1985. Coorong dolomite, South Australia: the depositional setting (Abstract). *Society of Economic Paleontologists and Mineralogists, Annual Midyear Meeting*. Boulder, Co.

—— 1986. Shallow-water evaporitic environments and their source rock potential. *Journal Sedimentary Petrology*, **56**, 442–454.

WETZEL, R. 1983. *Limnology*. 2nd edn, Saunders College Publication, Philadelphia.

WHITICAR, M. J., FABER, E. & SCHOELL, M., 1986. Biogenic methane formation in marine and freshwater environments: CO_2 reduction vs. acetate fermentation—isotope evidence. *Geochimica et Cosmochimica Acta*, **50**, 693–709.

WILLIAMS, W. D. 1984. Australian Lakes. *In*: TAUB, F. (ed.) *Lakes and Reservoirs. Ecosystems of the World*. Elsevier, Amsterdam, **23**, 499–519.

YANG, W., LI, Y. & GAO, R., 1985. Formation and evolution of non-marine petroleum in Songliao Basin, China. *Bulletin of the American Association of Petroleum Geologists*, **69**, 1112–1122.

ZIEGLER, A. M., TOWLEY, D. B., LOTTES, A. L., SAHAGIAN, D. L., HULVER, M. L. & GIERLOWSKI, T. C. 1985. Palaeogeographic interpretation: with an example from the Mid-Cretaceous. *Annual Review of Earth and Planetary Science*, **13**, 385–425.

ZOBELL, C. E. 1946. *Marine Microbiology*. Chronica Botanica Co., Waltham, MA.

KERRY KELTS, Geology Section, EAWAG/ETH, CH8600 Dübendorf, Switzerland.

Modern phytoplankton production in African lakes

J. F. Talling

Characteristics of modern phytoplankton production in African lakes reflect the great diversity of basin form, chemical composition (Talling & Talling 1965), climate, and flow regimes. The species assemblages involved are predominantly recruited from taxa of latitudinally widespread distribution (van Meel 1954; Kalff & Watson 1986), although there are distinctive tropical elements and some specifically African or apparently endemic forms (Compère 1981, Talling 1987).

Quantitatively, the population densities per unit water volume typically vary by up to three orders of magnitude (e.g. 1 to 1000 μg chlorophyll $a\, l^{-1}$) in different lakes, with implications for the associated amounts of organic C, N and P (Talling 1981). As the denser crops are usually associated with shallow basins or shallow upper strata, the corresponding variation of areal cover density is more limited. The seasonal *duration* of volumetric or areal abundance is responsive to factors both hydrographic (e.g. Viner 1985) and hydrologic (water-flow) (e.g. Prowse & Talling 1958), but an approach to year-long near constancy is uncommon even near the equator (Melack 1979b; Talling 1986).

Rates of photosynthetic carbon assimilation (g C m^{-2} day^{-1}) have been measured in many lakes and reservoirs (e.g. Talling 1957, 1965; Melack & Kilham 1974; Ganf 1975; Melack 1979a, 1981). High values are often encountered, for which a high specific activity per unit algal biomass is partly responsible (Talling *et al.* 1973). The depth-extent of activity varies widely, regulated by the varying quantities of background colour and seston (typically low in basins of long retention) and by the algal pigments themselves (e.g. Robarts 1979).

Transfer of phytoplankton biomass, and residues, from water-column to sediments is sometimes conspicuous and episodic. Returns from sediment surface to water-column may also be of ecological importance. Examples are provided by certain diatoms (e.g. in L. Victoria: Fish 1957; Talling 1966, 1969) and blue-green algae (e.g. in L. George: Ganf 1974).

References

COMPÈRE, P. 1981. Phytoplankton. Biogeography and systematics. *In:* BURGIS, M. & GAUDET, J. J. (eds) *The Ecology and Utilization of African Inland Waters.* United Nations Environment Programme, Nairobi, 37–39.

FISH, G. R. 1957. A seiche movement and its effect on the hydrology of Lake Victoria. *Colonial Office Fisheries Publications, London*, **10**, 1–68.

GANF, G. G. 1974. Phytoplankton biomass and distribution in a shallow eutrophic lake (Lake George, Uganda). *Oecologia*, **16**, 9–29.

—— 1975. Photosynthetic production and irradiance–photosynthesis relationships of the phytoplankton from a shallow equatorial lake (Lake George, Uganda). *Oecologia*, **18**, 165–183.

KALFF, J. & WATSON, S. 1986. Phytoplankton and its dynamics in two tropical lakes: a tropical and temperate zone comparison. *Hydrobiologia*, **138**, 161–176.

MELACK, J. M. 1979a. Photosynthetic rates in four tropical African fresh waters. *Freshwater Biology*, **9**, 555–571.

—— 1979b. Temporal variability of phytoplankton in tropical lakes. *Oecologica*, **44**, 1–7.

—— 1981. Photosynthetic activity of phytoplankton in tropical African soda lakes. *Hydrobiologia*, **81**, 71–85.

—— & KILHAM, P. 1974. Photosynthetic rates of phytoplankton in East African alkaline, saline lakes. *Limnology and Oceanography*, **19**, 743–755.

PROWSE, G. A. & TALLING, J. F. 1958. The seasonal growth and succession of plankton algae in the White Nile. *Limnology and Oceanography*, **3**, 223–238.

ROBARTS, R. D. 1979. Underwater light penetration, chlorophyll *a* and primary production in a tropical African lake (Lake McIlwaine, Rhodesia). *Archiv für Hydrobiologie*, **86**, 423–444.

TALLING, J. F. 1957. Diurnal changes of stratification and photosynthesis in some tropical African waters. *Proceedings of the Royal Society of London B*, **147**, 57–83.

—— 1965. The photosynthetic activity of phytoplankton in East African lakes. *Internationale Revue der gesamten Hydrobiologie*, **50**, 1–32.

—— 1966. The annual cycle of stratification and phytoplankton growth in Lake Victoria (East Africa). *Internationale Revue der gesamten Hydrobiologie*, **51**, 545–621.

—— 1969. The incidence of vertical mixing, and some biological and chemical consequences, in tropical African lakes. *Verhandlungen der internationale Vereinigung für theoretische und angewandte Limnologie*, **17**, 998–1012.

—— 1981. Phytoplankton. The conditions for high concentration of biomass. In: BURGIS, M. & GAUDET, J. J. (eds), *The Ecology and Utilization of African Inland Waters.* United Nations Environment Programme, Nairobi, 41–45.

—— 1986. The seasonality of phytoplankton in African lakes. *Hydrobiologia*, **138**, 139–160.

—— 1987. The phytoplankton of Lake Victoria (East Africa). In: MUNAWAR, M. (ed.), *Proceedings of the Symposium on the Phycology of Large Lakes of the World. Archiv fur Hydrobiologie Beihäfte Ergebnisse Limnologie,* **25**, 229–256.

—— & TALLING, I. B. 1965. The chemical composition of African lake waters. *Internationale Revue der gesamten Hydrobiologie,* **50**, 421–463.

——, WOOD, R. B., PROSSER, M. V. & BAXTER, R. M. 1973. The upper limit of photosynthetic productivity by phytoplankton: evidence from Ethiopian soda lakes. *Freshwater Biology*, **3**, 53–76.

VAN MEEL, L. 1954. Le phytoplankton. *Exploration Hydrobiologique du Lac Tanganika 1946–47,* **4** (1), 681 ff.

VINER, A. B. 1985. Thermal stability and phytoplankton distribution. *Hydrobiologia*, **125**, 47–69.

J. F. TALLING, Freshwater Biological Association, Ambleside, Cumbria LA22 0LP, UK.

The origins of lacustrine oil source rocks: evidence from the lakes of tropical Africa

M. R. Talbot

SUMMARY: New data from core samples are presented on organic matter accumulation during the last 10 000–15 000 years in six tropical African lakes. Pyrolysis and maceral studies indicate that in most of these lakes the preserved organic matter is a mixture of phytoplankton and higher plant remains showing varying degrees of bacterial degradation. Organic assemblages dominated by phytoplankton material have not been encountered. Sediments with the richest oil potential accumulate in meromictic or monomictic lakes during periods when the climate is humid and surface winds are relatively slack. The evidence from tropical Africa does not support current theories that shallow saline lakes may be especially favourable sites for the accumulation of oil prone sediment.

Introduction

Although it has long been recognized that oil can be generated from source rocks of non-marine origin, the global importance of hydrocarbons of this sort has only recently become apparent. Two factors in particular seem to be responsible for this development. The first is the great increase in information available from Chinese oil fields, bringing with it the realization that most production in this country comes from non-marine basins. The second has been the discovery of a succession of major oil fields in Mesozoic continental rift basins within the African continent and along the South Atlantic margins of Africa and South America. These new data, plus previously known occurrences in North America, Europe and Australia, and periodic bursts of interest in oil shales, demonstrate that non-marine oils come mainly from organic-rich lacustrine sediments. The main purpose of this paper is to attempt to identify the optimum environmental conditions for lacustrine petroleum source rock accumulation based on evidence from modern African lakes.

Models of lacustrine source rock accumulation

A variety of depositional models for lacustrine oil source rocks have been proposed, some based on studies of ancient non-marine basins, others on modern lakes. Three different models are prominent:

(1) The Green River model (Bradley & Eugster, 1969; Bradley, 1970; Eugster & Surdam, 1973; Eugster & Hardie, 1975). Arising out of a long series of studies on the sediments of the Green River Basins, this model suggests that source rock (oil shale) deposition occurs in the shallow but anoxic waters of stratified, brackish-saline, alkaline lakes developed within major playa systems. The oil shales are associated with carbonates, bedded evaporites and sediments showing abundant evidence of very shallow water or subaerial conditions. Others, impressed by the high productivity of some modern playa lakes, have suggested that this environment should generally provide an ideal setting for source rock deposition (Reyre 1984; Warren 1986).

A playa lake origin for the oil shales of the Green River Formation has been challenged, however; an alternative explanation being that these sediments represent periods in the basin's history when the lakes were deeper and fresher (Desborough 1978; Ryder 1980; Boyer 1982). Eugster (1985) has recently clarified his and his co-workers use of the playa lake model, indicating that it was specifically intended as an explanation for the Wilkins Peak member, where the facies associations mentioned above are well-established. Reference to the 'Green River Model' in the present account implies the original playa lake model. Tiercelin & Le Fournier (1980) suggest that Lake Bogoria (Kenya), a shallow, meromictic, hypersaline sodium carbonate lake, contains a suite of facies analogous to those regarded as typical for the Green River Model.

(2) Deep anoxic lake model (Demaison & Moore 1980; Fan Pu et al. 1980; Ryder 1980). Demaison & Moore (1980) chose Lake Tanganyika as the type basin for this model and suggest that organic-rich sediments will most readily accumulate in anoxic waters below the thermocline of deep, permanently stratified, fresh to mildly brackish water lakes. They suggest that the optimal conditions required to maintain high rates of surface productivity and a permanently anoxic epilimnion would occur in a warm, humid climate regime, with minimal season contrasts. Associated sediments are of deep water origin.

From: FLEET, A. J., KELTS, K. & TALBOT, M. R. (eds), 1988, *Lacustrine Petroleum Source Rocks*, Geological Society Special Publication No. 40, pp. 29–43.

Both these and the organic-rich muds themselves will generally show fine, primary lamination, due to lack of disturbance by waves, currents or benthic organisms.

(3) Ephemeral lake model (Bauld 1981; Burne & Ferguson 1983). Observations of extensive blue-green algal mat development around the margins of seasonally flooded coastal lakes and lagoons in various areas of Australia has led to the suggestion that such mats may be the precursors of some laminated oil shales. These mats contain up to 30% total organic carbon (TOC) and are associated with sediments of entirely shallow water to subaerial origin, often containing early diagenetic evaporites of an essentially marine character. Similar environments may also see periodic blooms of the planktonic green alga *Botryococcus* (Chlorophyceae), remains of which may aggregate in such numbers that they form a rubbery organic material known as coorongite (Hutton *et al.* 1980). Such occurrences have attracted attention because of the importance that is often attached to *Botryococcus* as a possible contributor to ancient lacustrine source rocks (see below).

Algal mats formed in ephemeral lakes seem unlikely analogues for the thick packages of organic-rich shales known to exist in many oil-producing continental basins, but both the other models seem plausible. At present no consensus seems to exist as to which of the latter is the more appropriate for major lacustrine source rock accumulations, nor indeed if there is a unique environment for these sediments. Common to both models, however, is the emphasis placed on stratified water bodies and there does seem to be general agreement that stratification is essential for source rock accumulation. Jones & Demaison (1982 p. 54), for example, state that 'lakes must be density stratified and stagnant on the bottom ...'. One objective of the present survey will therefore be to test the relative merits of the Green River and deep lake models by comparison with modern African lakes. The degree to which permanent stratification is essential for source rock development is also examined.

Composition of lacustrine oil source rocks

The principal features of some of the better known lacustrine source rocks have been reviewed by Hutton *et al.* (1980) and Powell (1986). TOC content varies from less than 1% to over 60%, although the bulk are in the range of 1–10%. Kerogen composition ranges from Type I to Type III, with bacterial, algal and higher plant remains all making important but variable contributions. Oils derived from such source rocks are typically rather paraffinic, waxy and sulphur-poor. The high wax content is generally thought to reflect the important contribution made by components from higher plants (Hedberg 1968; Hunt 1979; Barker 1982; Tissot & Welte 1984; Powell 1986). Because it is probably the best and certainly the most widely known, attention has been focussed in particular upon the Green River Formation, especially some of the richest oil shale units, which are commonly regarded as forming the 'type' Type I kerogens (Espitalié *et al.* 1977; Tissot & Welte 1984). This attention has in turn brought about a widespread belief that Type I kerogens are typical of lacustrine source rocks in general. However, a compilation of Rock-Eval analyses of source rocks from various Chinese basins shows a complete spectrum of compositions from Type I to Type III (Wanli *et al.* 1981; Fig. 1). Wanli *et al.* (1981) suggests that what they define as Type IIA kerogens (Hydrogen Index ca. 400–600) are probably the most characteristic Chinese lacustrine source materials.

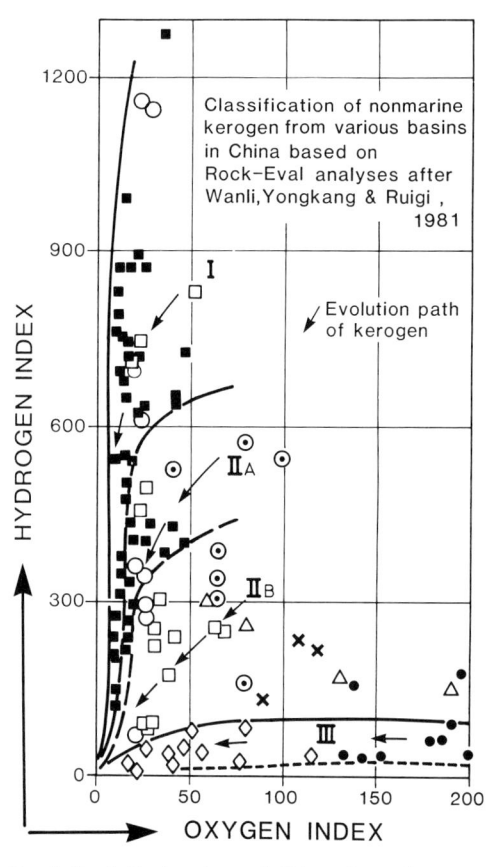

FIG. 1. Rock-Eval analyses of Chinese lacustrine organic shales. Symbols represent different basins (redrawn from Wanli *et al.* 1981).

These kerogens are apparently of mixed origin, containing significant qualities of planktonic and terrestial plant remains. Type I kerogens of lacustrine origin are composed of organic accumulations dominated by planktonic green algae, particularly *Botryococcus*, or of amorphous material, which may be of blue-green algal (cyanobacterial) or bacterial origin (Hutton *et al.* 1980; Cook *et al.* 1981; Jones & Demaison 1982; Powell 1986). In addition, as an attempt to explain the waxy nature of oils derived from some Type I kerogens, Powell (1986) has proposed that the latter can also be formed from those materials that remain after bacterial and fungal degradation of higher plant debris under mildly oxidizing conditions. He suggests that cellulose and lignins would be selectively removed, leaving a concentrated residue of waxy, oil prone cuticle, resin and spores supplemented by bacterial remains.

It is apparent from this brief review that lacustrine source rocks are highly varied in composition and origin and that some confusion probably exists as to what the typical composition of lacustrine source rocks might be.

As yet, comparatively little *detailed* use has been made of modern lakes as model basins where the accumulation of the likely precursor sediments to source rocks can be examined. The depositional models outlined above, refer to information from two African lakes. These, Lakes Tanganyika and Bogoria, are just two of a great diversity of lakes to be found in tropical Africa (see Livingstone & Melack 1984, for a recent review). New insights into the major controls on lacustrine oil source rock genesis can probably only be gained by examining organic matter accumulation in a variety of these lakes. Comprehensive data on modern productivity, organic matter types and preservation are lacking, but sufficient core material is available from enough lakes to give a clear indication of the sorts of source rocks that are likely to be formed in large tropical lakes. Cores used in this study span sufficiently long periods of time for major climatic changes to have occurred. The effects of environmental change on sedimentary and post-depositional processes must therefore be given careful attention when considering variations in organic matter accumulation through time.

Environmental change in tropical Africa and its influence on lacustrine systems

This survey examines organic matter accumulation under modern conditions and at various times in the more recent past. Extending the study back into the late Quaternary has several advantages. The tropics have been subject to major climatic changes throughout the Quaternary, changes which often had dramatic effects on lacustrine environments. Since hydrologic conditions rather different from the present existed in many lake basins at times in the past, understanding the late Quaternary deposits adds an extra dimension to the search for an optimal lacustrine source rock environment.

In recent years Quaternary climatic variations in tropical Africa have been reconstructed by a combination of palaeontological, geological and geomorphological evidence. (For comprehensive reviews see Livingstone 1975; Street 1981; Hamilton 1982.)

(a) Precipitation–evaporation balance

One of the most successful methods for tracing tropical climatic variations has been the study of lake level changes. The principles behind this approach are discussed in detail by Street-Perrott & Roberts (1983), who demonstrated that short-term (i.e., over periods of a few thousand years or less) lake level changes have in almost all cases been a direct result of variations in the precipitation–evaporation balance (P/E). Lakes have shown remarkably consistent behaviour across tropical Africa (Fig. 2). Levels were uniformly low during the late Pleistocene, such that several now open lakes seem to have been reduced to closed basins (e.g., Victoria: Kendall 1969; Stager *et al.* 1986; Tanganyika, Kivu: Hecky & Degens 1973). A major change occurred some time after 12 000–13 000 B.P., when lakes begin to rise in response to the advent of a major humid phase. By the early Holocene lake levels were high, many closed basins overflowed, at least episodically, and previously saline waterbodies became fresh.

Dramatic vegetational changes accompanied the transition to more humid conditions. Forests began to expand and by 10 000–8000 B.P. had achieved similar or greater extent than today. Humid conditions, interrupted by one or two drier intervals, continued until around 4500–4000 B.P., when a rather abrupt change to drier conditions occurred (Talbot 1982). Although there have been some notable oscillations, the P/E of tropical Africa has since that time generally approximated present-day conditions. Lake levels have therefore been somewhat lower than in early- to mid-Holocene times, and by assumption, more saline as well.

(b) Wind

One other factor in addition to P/E balance has a significant influence on tropical lakes. In the absence of major seasonal variations in surface

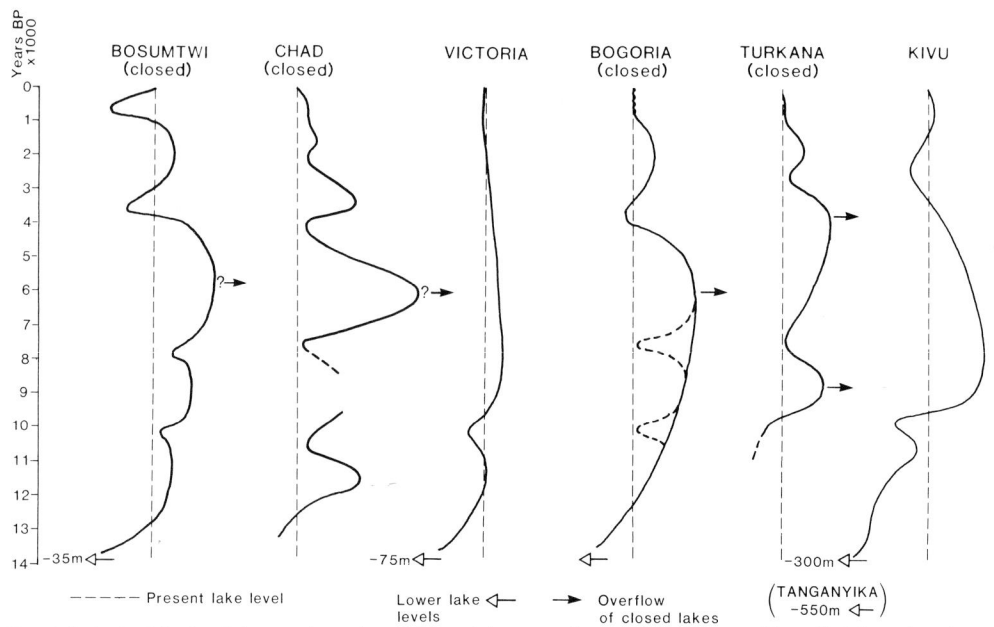

FIG. 2. Relative lake level changes in various African lakes over the past 14 000 years. (Compiled from data from Kendall 1969; Servant 1973; Hecky & Degens 1973; Tiercelin et al. 1981; Harvey & Grove 1982; Talbot et al. 1984).

temperature, the stability of any thermo- or halocline is often determined by wind shear. More or less permanent mixing of Lake Mobutu (Albert), for example, is due to the orientation of the basin parallel to the dominant wind direction, and in several otherwise stratified lakes (e.g., Victoria, Tanganyika), partial or complete mixing coincides with the windiest season (Beadle 1981). Livingstone & Melack (1984) have recently reviewed stratification and mixing regimes in tropical African lakes. They emphasize the important role played by the wind and note that basin shape and orientation with respect to the prevailing winds can have a decisive effect upon water column mixing. Wind shear clearly has a major influence on the maintenance or otherwise of stratification in African lakes; it is therefore important to give some consideration to possible long-term changes in windiness across the continent.

The most complete record of changes in wind regime comes from accumulations of aeolian dust in the Atlantic Ocean off north-west Africa (Bowles 1975; Sarnthein & Koopman 1980). Figure 3 summarizes the history of dust accumulation in that region for last ca. 100 000 years. Similar results have been obtained from other tropical oceans (Street 1981), so Fig. 3 can be assumed to present a record that is valid for the whole of tropical Africa. From this record it appears that the late Pleistocene was character-ized by generally windy conditions, while early- to mid-Holocene times were marked by a rather slack circulation system. The last 4000 years have seen a return to windier conditions.

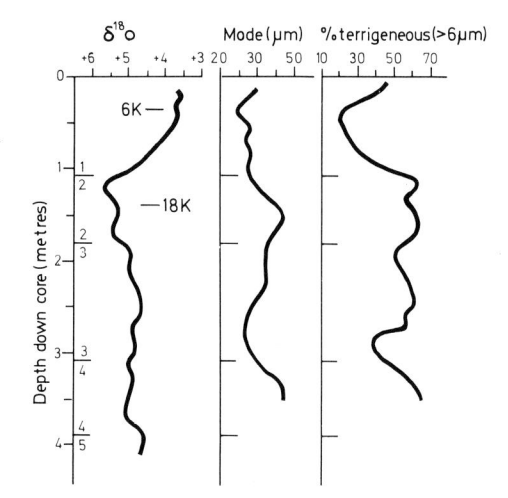

FIG. 3. Vertical distribution of terrigenous wind-blown sand and silt (>6 μm) and principal modal grain sizes (carbonate and opal-free) in sediments from Valdivia cores 13289-1 & 2 (18° 04′N, 18° 00′W). (Redrawn from Sarnthein & Koopman 1980.) Variations in modal grain size mainly indicate changes in wind speed and demonstrate that the late Pleistocene was generally a period of strong winds, while early- to mid-Holocene times were, in contrast, much calmer.

Origins of lacustrine oil source rocks

Organic matter accumulation and palaeoclimates

Three different time periods are selected for a comparison of organic matter accumulation and palaeoclimate:

(1) Late Pleistocene: Climate dry and windy; lakes generally shallow, often rather saline and alkaline.
(2) Early- to mid-Holocene: Warm, humid climate, slack winds; lakes deep, fresh to mildly alkaline.
(3) Late Holocene–Present: Sub-humid, moderately windy; lakes at intermediate levels and salinities.

Data base and methods

The lakes selected for study are shown on Fig. 4. Samples have come mainly from the core collections of Professor D. A. Livingstone at Duke University, and from the Woods Hole Oceanographic Institution.

Smear slides from all samples have been examined by standard petrographic techniques (transmitted and ultra-violet light). Total organic contents were determined using a LECO combustion furnace. Hydrocarbon potential was assessed by means of Rock-Eval pyrolysis (Espitalié et al. 1977), using a second generation Rock-Eval pyrolyser, supported in some cases by pyrolysis gas chromatography, performed on a Phillips-Pye Unicam PU 4500 gas chromatograph. Results from the latter are expressed as an oil–gas potential index (OGPI). Analyses were carried out at the Norsk Hydro Research Centre, Bergen.

Pyrolysis analyses on the sorts of samples used

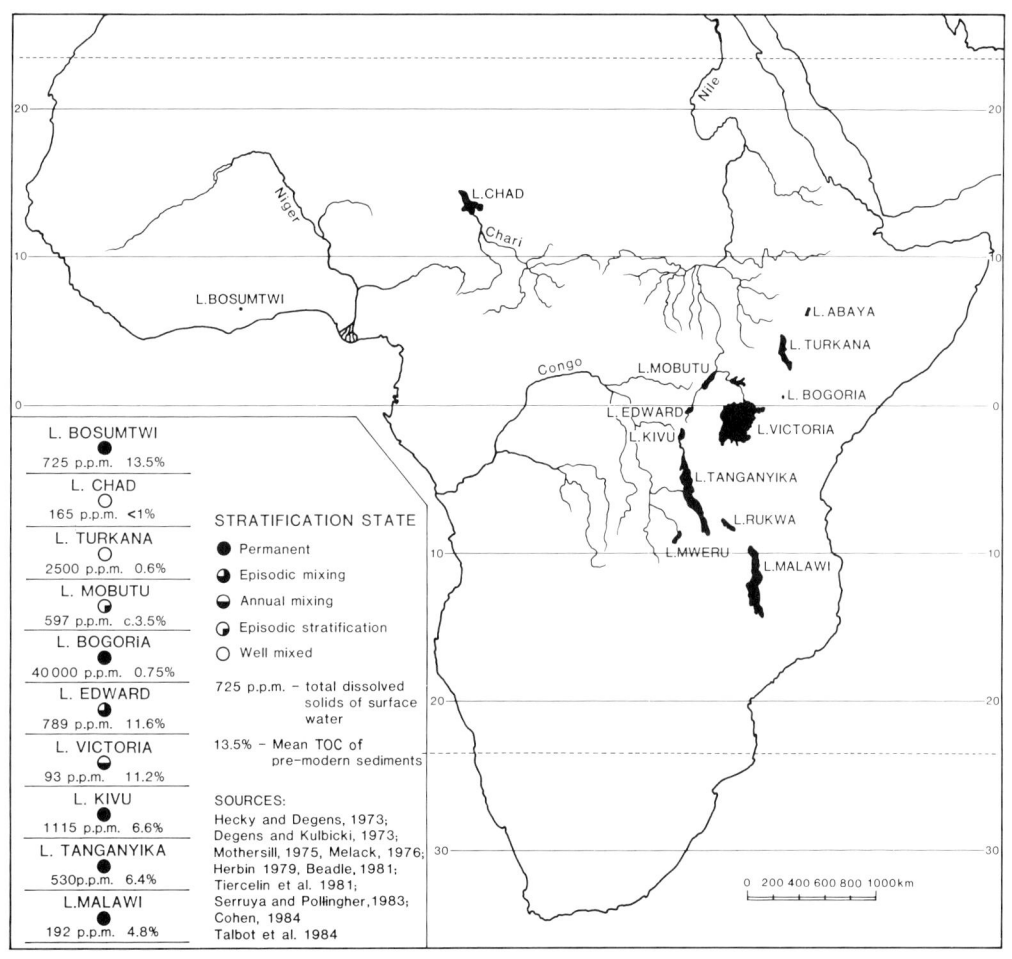

FIG. 4. Map of tropical Africa showing lakes covered in this survey, total dissolved solids of present surface waters, present stratification regime and mean organic carbon content of pre-modern sediments (see Table 1). Tanganyika, Malawi, Kivu, Edward, Mobutu, Bogoria and Turkana are rift lakes. Victoria and Chad occupy intracratonic sags and the Lake Bosumtwi basin is a meteorite impact crater.

in this study are not without drawbacks and may, in particular, be influenced by matrix effects (Herbin 1979; Espitalié et al. 1980; Katz 1983; Katz this volume). Clay minerals can be especially problematic. Given the high organic contents of the samples used in this study, however, it is thought unlikely that matrix effects will have seriously distorted the analytical results (Waples 1984; Katz this volume). In any case, relative differences between samples from any one basin are almost certainly real. Analytical results are summarized in Table 1.

TABLE 1. *Analytical data from samples used in this survey. Lake, core designation, core latitude and longitude, water depth at core site. TOC: total organic carbon content (weight percent dry sediment). Principal organic and biogenic components based upon microscopic examination: A, amorphous; B, Botryococcus sp.; C, plant cuticle; D, diatoms; S, spores, pollen; V, sponge spicules; W, wood; Y, blue-green algae. Rock-Eval analyses: S_1, hydrocarbon shows (kg t^{-1} sediment); S_2, hydrocarbons generated under heating (kg t^{-1} sediment); S_1+S_2, genetic potential (kg t^{-1} sediment); HI, hydrogen index; OI, oxygen index. OGPI, oil/gas potential index based upon pyrolysis gas chromatography*

	TOC %	Principal Components	Rock-Eval Analyses					OGPI
			S_1	S_2	S_1+S_2	HI	OI	
Edward Core 3PC 00° 26.9'S, 29° 27.5'E. 55 m								
100 cm	12.10	A,C,?Y,D	5.86	52.32	58.18	432	186	0.16
200 cm	14.46	A,C,S,?Y,D	7.43	69.85	77.28	483	155	0.18
300 cm	12.26	A,C,?Y,D	5.02	56.15	61.17	458	171	0.17
400 cm	11.24	A,?Y,D	7.30	58.43	65.73	520	161	0.17
504–515 cm	16.18	A,C,S,?Y,D	9.61	97.93	107.54	605	127	0.16
Edward Core 4 PC 00° 21.15'S, 29° 27.0'E. 100 m								
52 cm	3.46	A,C,S,?Y	0.50	4.84	5.34	139	262	0.31
Mobutu Core 2 PC 01° 29.2'N, 30° 42.9'E. 40 m								
Top	3.20	C,A,S,B,D,Y	0.48	4.56	5.04	142	220	0.36
100 cm	3.17	A,C,S,B,D	0.46	6.09	6.55	192	185	0.30
202 cm	3.97	A,C,P,S,B,D,S,	0.76	6.16	6.92	155	173	0.30
304 cm	2.91	A,C,B,S,D	0.43	2.72	3.15	93	178	0.35
405 cm	1.53	A,B,C,S,D	0.15	0.61	0.76	39	283	—
499 cm	1.70	A,P,B,C,D	0.15	0.74	0.89	43	239	—
525 cm	1.44	A,P,C,B,D	0.00	0.00	0.00	N.D.	8	—
Kivu Core 1 PC 02° 17'S, 29° 04.7'E. 225 m								
98–102 cm	5.88	A,S,C,B,Y,D	4.52	24.67	29.19	419	122	0.17
200 cm	6.00	A,C,S,B	1.79	11.57	13.36	192	106	0.30
296–298 cm	6.67	A,S,C,B,D	2.47	12.80	15.27	192	141	0.26
397–400 cm	3.12	A,S,C,D	0.63	1.13	1.76	36	146	—
Kivu Core 6 PC 02° 14'S, 29° 07'E. 310 m								
50 cm	8.40	A,C,D	8.91	35.57	44.48	423	157	0.13
150–152 cm	6.67	A,C,S,D	8.30	24.78	33.08	372	127	0.17
247–253 cm	7.94	A,S,C,P,D	6.61	34.51	41.12	434	123	0.17
301–303 cm	6.78	A,S,B,C,D	6.76	23.41	30.17	345	170	0.16
400 cm	7.95	A,C,S,D	4.85	23.81	28.81	301	105	0.19

TABLE 1 (cont.)

	TOC %	Principal Components	S_1	S_2	S_1+S_2	HI	OI	OGPI
Tanganyika Sta 15 05° 33′S, 29° 28′E. 663 m								
0 cm	7.53	A,S,D	3.71	34.29	38.0	455	99	0.24
100 cm	8.40	A,C,D,Y	5.21	36.81	42.0	438	85	—
200 cm	3.83	A,C,D,Y	0.78	7.40	8.2	193	147	—
Tanganyika Sta 18 05° 45′S, 29° 25′E. 64 m								
0 cm	3.70	A,C,S,D,Y	0.80	5.63	6.43	152	167	0.53
70 cm	3.37	A,C,D,Y	0.66	4.43	5.1	131	162	—
140 cm	3.73	A,C,S,D,Y	1.18	7.38	8.56	198	156	—
Tanganyika Sta 19 05° 56′S, 29° 32′E. 361 m								
0 cm	7.99	A,C,D,Y	3.84	38.6	42.24	483	93	—
75 cm	6.89	A,C,D,Y	2.98	12.98	15.96	188	101	—
150 cm	1.52	A,C,Y	0.35	0.27	0.62	18	130	—
200 cm	1.54	A,Y	0.27	0.46	0.73	30	122	1.86
Victoria Ibis C Damba Channel. 32 m								
0 cm	18.31	A,C,B,P,S,D	14.56	93.3	107.9	510	89	0.19
570 cm	19.05	A,P,B,C,S,D	10.56	108.5	119.1	570	69	0.18
860 cm	4.22	A,C,N,D	0.39	1.38	1.77	33	135	0.63
970 cm	3.34	A,B,C,V,D	0.37	4.41	4.78	132	102	0.47
Bosumtwi B-3 ~6° 30′N, 1° 25′W. 75 m								
~300 cm	17.08	Y,C,S	8.03	90.74	98.77	531	73	—
Botsumtwi B-6 ~6° 30′N, 1° 25′W. 78 m								
122 cm	11.91	A,C,Y,S	7.66	60.72	68.38	509	95	—
435 cm	18.96	C,A,S	2.31	55.55	57.86	292	95	—
1327 cm	5.35	A,C	2.00	15.98	17.98	299	98	—
Bosumtwi B-7 ~6° 30′N, 1° 25′W. 75 m								
43 cm	16.82	Y,A,C,S,V	8.09	82.76	90.85	492	108	—
53.5 cm	16.53	Y.A.C.S.	6.90	81.60	88.5	493	119	—
132 cm	12.49	A,Y,C	3.90	54.28	58.18	434	140	—
485 cm	8.51	C,A,Y	9.44	38.65	48.1	454	94	—
490 cm	9.32	C,A,S	6.41	43.51	49.92	466	99	—
788 cm	8.40	C.A.Y	9.61	40.3	49.91	479	93	—
1394 cm	8.43	A,C	2.17	49.42	51.59	586	73	0.22
1688 cm	18.60	A,C	9.27	101.79	111.06	547	60	0.25

Influence of stratification on sediment organic content

Data from this study and from published work have been used to assess the degree to which the stratification regime in African lakes controls the amount of organic matter preserved in the sediments. Figure 4 summarizes the relationship between these two parameters. With the exception of Bogoria, lakes with a permanent thermocline (Bosumtwi, Kivu, Tanganyika, Malawi) have high to very high TOC value, as does Lake Edward, which apparently shows complete mixing only episodically, during particularly severe wind storms (Beadle 1981). In addition, Lake Victoria, which mixes annually during the season of strongest trade winds (Beadle 1981), also contains highly organic-rich sediments. Even Lake Mobutu, with only rare periods of stratification, is accumulating sediment with TOCs of 1.4–4.0%. In sharp contrast to these waterbodies are lakes like Chad and Turkana, which despite high rates of primary productivity and moderately organic-rich surface sediments (Mothersill 1975; Beadle 1981; Serruya & Pollingher 1983; Cohen 1984) have less than 1% TOC preserved in the pre-modern sediments. Although high sedimentation rates may in part be responsible for the low TOC's, at least in the case of Lake Turkana (Yuretich, 1979), the permanently well-mixed nature of these waterbodies is likely to be the principal limitation on organic matter preservation.

The main conclusion to be drawn from these data is that although permanent stratification is favourable to the accumulation of organic-rich sediments, it seems to be by no means essential. Some systems (e.g., Lake Victoria) are capable of withstanding recurrent periods of mixing without any apparent adverse effects on the amount of organic matter that is ultimately preserved.

Organic matter composition

Bulk organic matter composition, and thus likely hydrocarbon source potential of the sediments, has been assessed by means of Rock-Eval pyrolysis, pyrolysis gas chromatography and smear slide microscopy (Table 1). Rock-Eval data are presented in Fig. 5, and show that the sediments would produce Type II and III kerogens. The prominence of Type II precursors in African lakes compares well with the relative abundance of source rock types encountered in Chinese basins. The richest samples analysed are

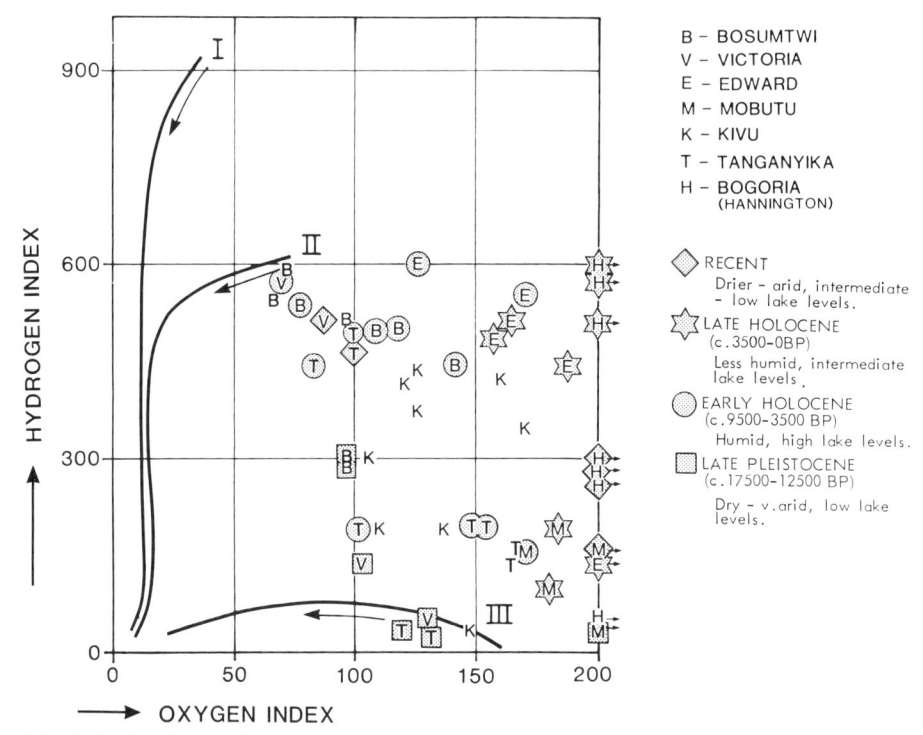

FIG. 5. Rock-Eval analyses of Quaternary–Recent sediments from African lakes (Bogoria data from Herbin 1979). Samples without age symbol are either of uncertain age or, in the case of Lake Bosumtwi, of pre-17 500 B.P. age.

comparable to what Wanli et al. (1981; 1985) have termed Type IIA kerogens. No Type I precursors have been identified in the present study.

Despite the absence of Type I assemblages, several of the analysed sediments would upon maturation nevertheless provide rich oil source rocks. Genetic potential as measured by Rock-Eval (S_1+S_2: Table 1) is in many samples greater than 20 and in some sediments from Lakes Edward, Victoria and Bosumtwi exceeds 100. Pyrolysis gas chromatography also reveals that some samples from Edward, Victoria and Kivu are highly oil prone (OGPI < 0.2: Table 1).

Precursor kerogen composition based upon Rock-Eval analyses are confirmed by visual inspection of sediment smears which show the identifiable organic components to be generally of mixed origin (Table 1). Virtually all samples contain at least some higher plant debris, together with varying amounts of autochthonous phytoplankton remains and amorphous material. Amorphous organic matter is dominant in most samples and has presumably been produced by bacterial degradation of the other components (Masran & Pocock 1981; Robert 1985).

Of the phytoplankton, diatoms are abundant in Tanganyika, Kivu and Edward sediments, and common in Mobotu and parts of the Bosumtwi sequence. Planktonic green algae do not generally appear to be an important component of the organic matter preserved in African lake sediments. *Botryococcus* and *Pediastrum* are relatively common in the Lake Victoria samples and are also said to be prominent in sediments that accumulated during some of the dilute phases in Lake Bogoria (Tiercelin et al. 1982), but are otherwise present in only small amounts. Recognizable blue-green algal remains are also uncommon; they have confidently been identified only from Lake Bogoria (Tiercelin et al. 1982) and in an early- to mid-Holocene sapropel unit from Lake Bosumtwi (Talbot et al. 1984).

Studies of modern phytoplankton in tropical African lakes indicate that the flora is typically dominated by diatoms and blue-green algae (Melack 1976; Hecky & Kling 1981; Beadle 1981). Sediment composition suggests that this was probably also the case in the late Quaternary. Infrequent observations of algal remains reflect their low preservation potential rather than any inherent scarcity. Some of the amorphous material may well represent bacterially reworked blue-green algae.

In sharp contrast to the generally rich source potential of many samples are the sediments with low to very low Hydrogen Indices (HI) and genetic potential (Fig. 5, Table 1). The reasons for the poor oil potential of some Victoria, Tanganyika and Kivu samples will be discussed in a later section. In the case of Lake Mobutu, none of the analysed samples would be of interest as potential oil source rocks. Although Mobutu seems capable of accumulating organic matter-rich sediments in spite of an active mixing regime, it is clear that the preserved organic matter must be dominated by humic material. Diatom frustules are common in the Lake Mobutu sediments but evidently little of the hydrogen-rich organic matter from these organisms has been preserved. It appears, therefore, that at least regular periods of stratification (monomixis) are a prerequisite for source rock accumulation.

Lake Mobutu possibly exhibits the sort of '... mildly oxidizing conditions...' that Powell (1986 p. 215) envisages as being favourable for the formation of Type I kerogens from higher plant remains. Although the lake can generally be regarded as well-mixed, periods of low oxygen concentrations occur regularly at the bottom in deeper waters (Beadle 1981). However, despite the abundant supply of plant debris from the heavily vegetated Albert rift to the deep waters of the lake, there is neither visual nor geochemical evidence for the selective oxidation of cellulose and lignin leading to preferential preservation of exinitic material.

Changes in organic matter accumulation through time

Figure 6 shows the range of TOC values for sediments of late Pleistocene to Recent age from five lakes. Although there is considerable variation, and some lakes consistently have very high TOC contents (e.g., Edward, Bosumtwi), highest values seem to occur in sediments of early- to mid-Holocene age. When the Rock-Eval analyses are subdivided into groups of different geologic age, it is clear that Hydrogen Indices as well as TOC content tends to be highest in the lower Holocene samples (Fig. 5). Some upper Holocene and Recent samples also have relatively high Hydrocarbon Indices. In contrast, samples with low TOC, HI and genetic potential are almost all of late Pleistocene age, even from lakes which otherwise contain sediments of relatively good source rock potential (Table 1, Fig. 5).

The reasons for these contrasts are environmental. The late Pleistocene was a period of low lake levels and strong winds. Frequent turnover in the relatively shallow waterbodies would have favoured higher bottom water oxygen contents and more persistent oxidizing conditions, thereby greatly reducing the preservation potential of the more labile organic components. In some basins,

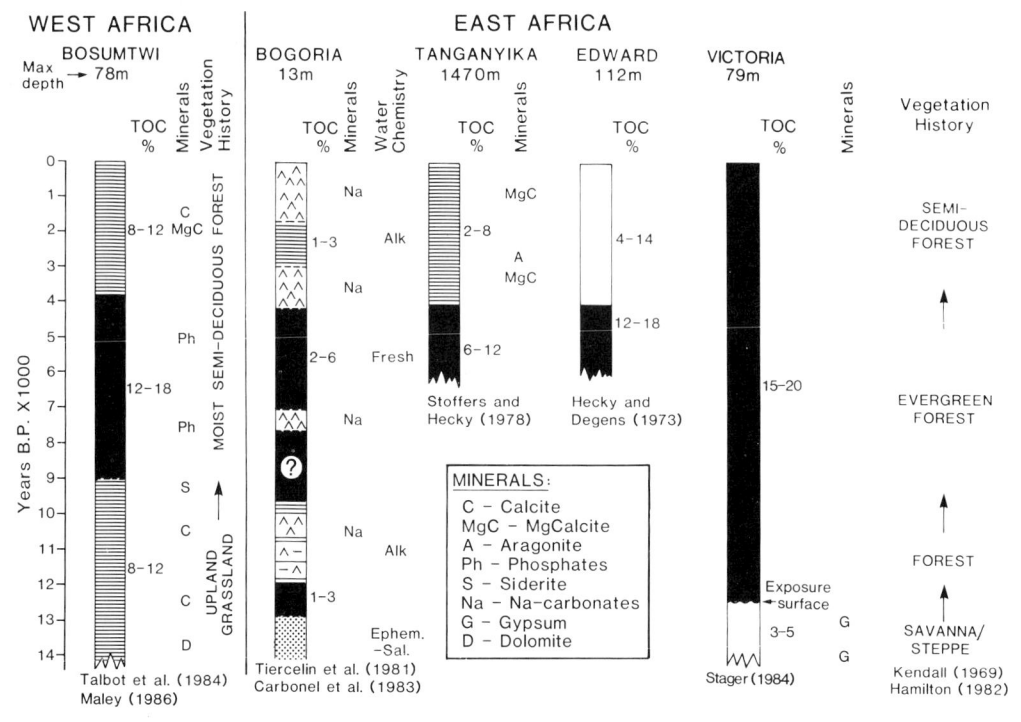

FIG. 6. Generalized stratigraphy, mineralogy and organic carbon content of cores from African lakes, compared with vegetational history of surrounding land areas.

low lake levels during the late Pleistocene exposed extensive areas of lake floor. Samples with very low HI from Victoria, Tanganyika, and probably also Kivu, came from areas which must have been shallow or subaerially exposed at that time. Oxidation, plus in some cases reworking by soil flora and fauna are almost certainly the reasons for the poor hydrocarbon potential of these sediments. The Tanganyika samples come from the base of a core taken in 361 m of water (Station 19, Table 1). Evidence for a dominance of degraded organic matter in these sediments thus provides independent support for Hecky's & Degens' (1973) contention that the lake was very much lower than present during late Pleistocene times.

During the early- to mid-Holocene, on the other hand, the climate seems to have been generally humid with rather slack winds. Under these conditions lakes were at high levels and meromixis probably existed in many waterbodies, even those which today mix regularly. The quantity and quality of organic matter preserved in lower- to mid-Holocene sediments indicates that in the tropics, at least, deep, dilute waterbodies are more favourable than shallow, more saline lakes for the accumulation oil prone sediments.

Two additional factors contributed to the widespread accumulation of especially rich sediments in early- to mid-Holocene times. As Fig. 2 shows, the terminal Pleistocene–early Holocene was marked by a dramatic deepening of the lakes. Transgressions on this scale would have led to starved basin conditions in many lakes, the bulk of the clastic sediment being trapped in newly formed estuaries. At the same time, major vegetational changes were occurring; forest replaced grassland or savanna over large areas of tropical Africa (Fig. 6). This transition led to reduced sediment yields. In Lakes Victoria, Mobutu and Bosumtwi, a marked reduction in sediment accumulation rate clearly coincided with the establishment of denser vegetation cover (Kendall 1969; Harvey 1976; Talbot et al. 1984). It is thus probable that most tropical African lakes received reduced supplies of clastic sediment during the first half of the Holocene.

Humid climatic conditions together with a stable vegetation cover would have favoured deep chemical weathering, so although clastic sediment yeilds were reduced, the abundant runoff must have carried a plentiful supply of dissolved nutrients. Lakes are therefore probably particularly productive during humid intervals like the early Holocene. A combination of high surface productivity, a deep, stratified water column and starved basin conditions must have enhanced both the relative organic content of the sediment and the accumulation and preservation of oil prone material.

Discussion

This survey of organic matter accumulation in tropical African lakes strongly suggests that the most favourable conditions for oil source rock accumulation occur during periods of humid climate with reduced surface wind activity. At such times lakes are deep, fresh to mildly alkaline with stable stratification regimes. This generalization includes the saline Lake Bogoria. During early to mid-Holocene times the lake was much deeper than it is at present (Tiercelin *et al.* 1981) and there are clear faunal indications of freshwater conditions at that time (Carbonel *et al.* 1983). Evidence presently available from Africa does not therefore favour the saline lake model for source rock accumulation. Previous discussions of this model have commonly placed much emphasis upon the high organic productivity of surface waters in some saline lakes (e.g., Warren 1986). This is certainly striking, but surface productivity is *not* a reliable guide to the amount of oil prone organic matter that may be preserved in the sediments. The nutrient content of a closed body of water is finite; without periodic renewal from external sources high levels of productivity in saline lakes can only be sustained by efficient recycling of nutrients. Much information now exists to indicate that nutrient recycling is of major importance in many productive playa lakes (see Melack 1976; Livingstone & Melack 1984 for review of data relevant to Africa). If high levels of productivity in closed lakes are sustained over long periods, it is clear that little organic matter can become immobilized in the bottom sediments.

Humid climatic conditions will promote dense terrestrial vegetation cover, so higher plant remains must be moved in relatively large amounts by the abundant runoff. A significant proportion of this material, together with debris from the extensive swamps which line the shores of many tropical African lakes (Livingstone & Melack 1984), is readily transported far into the lake basins. Inspection under low magnification of all the cores utilized in this study shows abundant plant remains, either concentrated in discrete laminae or intimately mixed with clastic material. These fragments have probably been transported by density inter- or underflows and resedimentation processes, principally as turbidites generated at delta fronts or other shallow water sites of accumulation. Such processes are apparently common in African lakes (Beadle 1981; Talbot, 1976 and unpubl., Tiercelin *et al.* 1980; Johnson 1984; Crossley *et al.* this volume) and it thus seems inevitable that the sediments will commonly contain a mixed assemblage of organic components. Upon maturation these would normally yield Type II kerogens.

Although Type II kerogens seem to be the typical lacustrine oil source material, it is worth reflecting on the likely conditions under which Type I kerogens may be formed. The absence of the latter from this survey may merely be a result of inadequate sampling, suitable sites for their accumulation simply not having been encountered during the coring programmes. However, the steep, often tectonically active margins to many of the basins covered here provide ideal settings for the initiation of mass flows, so, as Barker (1982) has previously noted, mixed organic assemblages are probably the norm for rift or similar basins. Lake Victoria, however, is distinctly different from such basins in being very large but relatively shallow and thus having only gentle lake floor gradients. Transport of material from the margins into the central part of this basin is therefore likely to be relatively inefficient. Samples from Victoria included in this study come from a core taken in water 32 m deep at a site only a few kilometres from the northern shore (Stager 1984), where sedimentation is probably influenced by the swamps that are common along this part of the lake margin (Livingstone & Melack 1984). It is probable that further out in the basin, beyond, for example, the 60 m depth contour (Fig. 7), higher plant remains make up a much smaller proportion of the TOC than at the core site. Sediments dominated by phytoplankton remains have probably been accumulating in the central part of the basin since the terminal Pleistocene. During the earlier Holocene, when winds were light, there is a strong possibility that Lake Victoria was meromictic, so by now a substantial deposit of phytoplankton-rich sediment should have accumulated in the central part of the basin.

Lake Victoria possibly provides one of the best modern analogues to palaeolake Gosiute, in which the Green River Formation accumulated. Although in contrast to Gosiute, carbonates seem to be relatively unimportant in Victoria, the lake waters are of the Na^+, HCO_3^- type (Talling & Talling 1965) and would, if concentrated sufficiently, produce carbonates, plus a suite of sodium carbonate-dominated evaporites similar to those found in the Green River Basin. In addition, both basins have been subject to periodic vulcan-

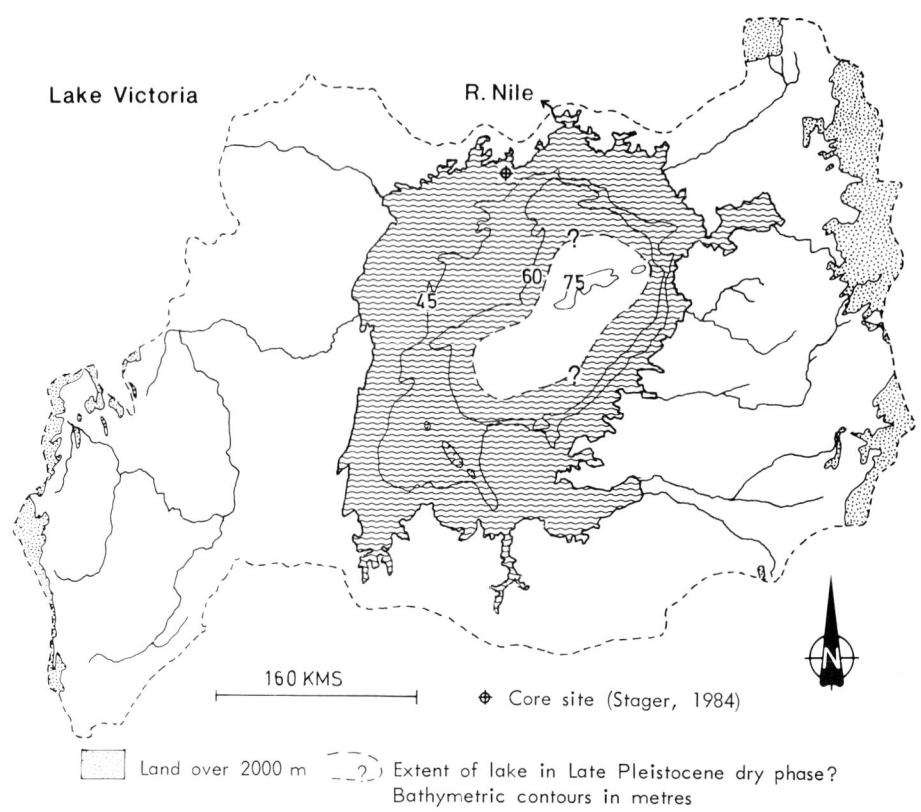

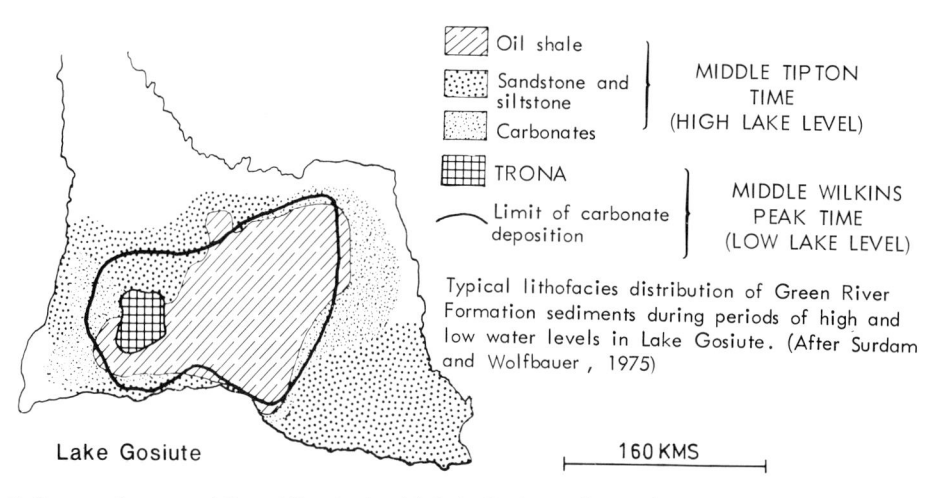

FIG. 7. Comparative maps of Green River basin with Lake Gosiute sediments (Paleocene–Eocene) and present basin of Lake Victoria.

icity, have surface areas of the same magnitude (Fig. 7) and carry evidence of major fluctuation in water depth. The occurrence in Victoria of Holocene organic-rich muds immediately overlying a desiccated and weathered surface of evaporite-bearing shallow water sediments (Kendall 1969; Stager 1984; Stager et al. 1986; Fig. 6) finds exact parallels in the Green River Formation (Bradley & Eugster 1969; Hardie et al. 1978; Eugster & Hardie 1978). The excellent chronological and palaeo-environmental control available in upper Quaternary sequences shows unequivocally that there is no genetic relationship between the sapropelic and shallow water facies. This fact, plus the general association of organic-rich muds with humid climatic intervals, provides strong support for those who have argued that accumulation of the rich oil shales of the Green River and similar basins must have occurred when the lakes were deep and relatively fresh (e.g., Desborough 1978; Ryder 1980; Boyer 1982).

Conclusions

(1) Based on examples from modern tropical African lakes, conditions favourable to the preservation of organic matter from which lacustrine oil source rocks may be formed are most likely to occur in deep, fresh to mildly alkaline lakes. Meromictic conditions enhance the preservation potential, but are not essential. Where rates of organic detritus supply are high enough, the system may be able to withstand periods of mixing without any significant deterioration in the quality of the preserved organic matter.

(2) Climate has a major influence upon organic matter preservation. Ideal environmental conditions are provided by a warm and humid climate with light surface winds. At such times lakes are deep with very stable stratification due to low surface wind shear. Surrounding land areas are densely vegetated, thus reducing clastic sediment supply and thereby enhancing the organic matter relative to clastic content of the lake sediments. At the same time, intense chemical weathering ensures a steady supply of dissolved nutrients to promote high rates of primary productivity within the lakes. Dry climatic intervals, when lakes are low and rather saline, seem much less favourable for the accumulation of oil prone organic matter.

(3) With a heavy vegetation cover in the basin, the lake sediments will normally contain a mixture of phytoplankton (alginite), swamp and terrestrial plant (exinite) material, reworked to varying degree by bacteria. Such sediments will typically yield Type II kerogens. The morphology of most rift basins will generally ensure that higher plant remains are distributed thoughout rift valley lakes. Type I kerogens should therefore be characteristic of low gradient basins where distribution of terrestrial components is relatively inefficient.

(4) Lake Victoria provides a possible modern analogue for those basins within which the Green River and related formations accumulated.

ACKNOWLEDGEMENTS: This study would not have been possible without the generous help of several individuals and institutions. Thanks are due in particular to: Professor Daniel Livingstone and Dr Curtis Stager, Duke University, for core material from Lakes Bosumtwi and Victoria; to Jim Broda and Paul Andrew of the McLean Laboratory, Woods Hole Oceanographic Institution, for assistance in sampling from the Lake Tanganyika, Kivu, Edward and Mobutu (Albert) cores; coring operations were supported by the U.S. National Science Foundation; to Gordon Speers and his staff at the Organic Geochemistry Unit, Norsk Hydro Research Centre, Bergen; to Aase Scheie for technical assistance; to Ellen Irgens and Jane Ellingsen for their drafting skills; and to Aslaug Pedersen for word processing. Dan Livingstone and Kerry Kelts are thanked for critical reading of an earlier version of this paper and many stimulating discussions on organic matter in lake sediments. This work was in part supported by a grant from the Royal Norwegian Research Council (NAVF).

References

BARKER, C. 1982. Oil and gas on passive continental margins. In: WATKINS, J. S. & DRAKE, C. L. (eds), *Studies in Continental Margin Geology,* Memoir, American Association of Petroleum Geologists, **34**, 549–565.

BAULD, J. 1981. Geobiological role of cyanobacterial mats in sedimentary environments: production and preservation of organic matter. *BMR Journal of Australian Geology and Geophysics,* **6**, 307–317.

BEADLE, L. C. 1981. *The Inland Waters of Tropical Africa.* 2nd edn. Longman, London, 475 pp.

BOWLES, F. A. 1975. Palaeoclimatic significance of quartz/illite variations in cores from the eastern equatorial North Atlantic. *Quaternary Research,* **5**, 225–235.

BOYER, B. W. 1982. Green River laminites: does the playa-lake model really invalidate the stratified lake model? *Geology,* **10**, 321–324.

BRADLEY, W. H. 1970. Green River oil shale—concept of origin extended. *Bulletin of the Geological Society of America,* **81**, 985–1000.

—— & EUGSTER, H. P. 1969. Geochemistry and

paleolimnology of the trona deposits and associated authigenic minerals of the Green River Formation of Wyoming. *United States Geological Survey Professional Paper,* **496-B**, 71 pp.

BURNE, R. V. & FERGUSON, J. 1983. Contrasting marginal sediments of a seasonally flooded saline lake—Lake Eliza, South Australia: Significance for oil shale genesis. *BMR Journal of Australian Geology and Geophysics,* **8**, 99–108.

CARBONEL, P., GROSDIDIER, E., PEYPOUQUET, J.-P. & TIERCELIN, J.-J. 1983. Les ostracodes, témoins de l'évolution hydrologique d'un lac de rift—exemple du lac Bogoria, Rift Gregory, Kenya. *Bulletin des Centres de Recherches Exploration–Production Elf-Aquitaine,* **7**, 301–313.

COHEN, A. S. 1984. Effect of zoobenthic standing crop on laminae preservation in tropical lake sediment, Lake Turkana, East Africa. *Journal of Paleontology,* **58**, 499–510.

COOK, A. C., HUTTON, A. C. & SHERWOOD, N. R. 1981. Classification of oil shales. *Bulletin des Centres de Recherches Exploration—Production Elf-Aquitaine,* **5**, 353–381.

DEGENS, E. T. & KULBICKI, G. 1973. Data file on metal distributions in East African rift sediments. *Woods Hole Oceanographic Institution, Technical Report WHOI,* **73-15**, 258pp.

DEMAISON, G. J. & MOORE, G. T. 1980. Anoxic environments and oil source bed genesis. *Organic Geochemistry,* **2**, 9–31.

DESBOROUGH, G. A. 1978. A biogenic-chemical stratified lake model for the origin of oil shale of the Green River Formation: an alternative to the playa-lake model. *Bulletin of the Geological Society of America,* **89**, 961–971.

ESPITALIÉ, J., LAPORTE, L. J., MADEC, M. *et al.* 1977. Méthode rapide de caractérization des roches méres de leur potential pétrolier et de leur degré d'evolution. *Revue de l'Institut Français du Pétrole,* **32**, 32–42.

——, MADEC, M. & TISSOT, B. 1980. Role of mineral matrix in kerogen pyrolysis: influence on petroleum generation and migration. *Bulletin of the American Association of Petroleum Geologists,* **64**, 59–66.

EUGSTER, H. P. 1985. Oil shales, evaporites and ore deposits. *Geochimica et Cosmochimica Acta,* **49**, 619–635.

—— & HARDIE, L. A. 1975. Sedimentation in an ancient playa-lake complex: the Wilkins Peak Member of the Green River Formation of Wyoming. *Bulletin of the Geological Society of America,* **86**, 319–334.

—— & HARDIE, R. 1978. Saline lakes. *In*: LERMAN, A. *et al.* (eds) *Lakes: Chemistry, Geology, Physics,* Springer-Verlag, New York, 237–293.

—— & SURDAM, R. C. 1973. Depositional environment of the Green River Formation: a preliminary report. *Bulletin of the Geological Society of America,* **86**, 319–334.

FAN P., LUO B., HUANG R., *et al.* 1980. Formation and migration of continental oil and gas in China. *Scientia Sinica,* **23**, 1286–1295.

HAMILTON, A. C. 1982. *Environmental History of East Africa,* Academic Press, London, 328 pp.

HARDIE, L. A., SMOOT, J. P. & EUGSTER, H. P. 1978. Saline lakes and their deposits: a sedimentological approach. *In*: MATTER, A. & TUCKER, M. E. (eds) *Modern and Ancient Lake Sediments,* International Association of Sedimentologists Special Publication, **2**, 7–41.

HARVEY, C. P. D. & GROVE, A. T. 1982. A prehistoric source of the Nile. *The Geographical Journal,* **148**, 327–336.

HARVEY, T. J. 1976. The paleolimnology of Lake Mobutu Sese Seko, Uganda-Zaire: the last 28000 years. Unpublished Ph. D. thesis, Duke University, 104 pp.

HECKY, R. E. & DEGENS, E. T. 1973. Late Pleistocene—Holocene chemical stratigraphy and paleolimnology of the rift valley lakes of central Africa. *Woods Hole Oceanographic Institution, Technical Report WHOI,* 73-28.

—— & KLING, H. J. 1981. The phytoplankton and protozooplankton of the euphotic zone of Lake Tanganyika: species composition, biomass, chlorophyll content, and spatio-temporal distribution. *Limnology and Oceanography,* **26**, 548–564.

HEDBERG, H. D. 1968. Significance of high-wax oils with respect to genesis of petroleum. *Bulletin of the American Association of Petroleum Geologists,* **52**, 736–750.

HERBIN, J. P. 1979. Sédimentation de rift: géochimie organique des sédiments recents du Lac Bogoria. *Institut Francais du Pétrole, Rapport,* 23250, 20 pp.

HUNT, J. M. 1979. *Petroleum Geochemistry and Geology.* W. H. Freeman, San Francisco, 617 pp.

HUTTON, A. C., KANTSLER, A. J., COOK, A. C. & MCKIRDY, D. M. 1980. Organic matter in oil shales. *Journal of the Australian Petroleum Exploration Association,* **20**, 44–67.

JOHNSON, T. C. 1984. Sedimentation in large lakes. *Annual Review of Earth and Planetary Sciences,* **12**, 179–204.

JONES, R. W. & DEMAISON, G. J. 1982. Organic facies—stratigraphic concept and exploration tool. *Proceedings ASCOPE Meeting, Phillipines,* 51–78.

KATZ, B. J. 1983. Limitations of Rock-Eval pyrolysis for typing organic matter. *Organic Geochemistry,* **4**, 195–199.

KENDALL, R. L. 1969. The ecological history of the Lake Victoria basin. *Ecological Monographs,* **39**, 121–176.

LIVINGSTONE, D. A. 1975. Late Quaternary climatic change in Africa. *Annual Review of Ecology and Systematics,* **6**, 249–280.

—— & MELACK, J. M. 1984. Some lakes of subsaharan Africa. *In*: TAUB, F. B. (ed.), *Lakes and Reservoirs. Ecosystems of the World.* Elsevier, Amsterdam, **23**, 467–497.

MALEY, J. 1986. Fragmentation et reconstitution de la forêt dense humide ouest-africaine au cours du Quaternaire recent: hypothèse sur la role des upwellings. *Abs. INQUA Symposium, Changements globaux en Afrique durant le Quaternaire,* Dakar, Sénégal.

MASRAN, T. C. & POCOCK, S. A. J. 1981. The

classification of plant-derived particulate organic matter in sedimentary rocks. *In*: BROOKS, J. (ed.), *Organic Maturation Studies and Fossil Fuel Exploration*, Academic Press, London, 45–175.

MELACK, J. M. 1976. *Limnology and Dynamics of Phytoplankton in Equatorial African Lakes*. Unpublished Ph.D. thesis, Duke University.

MOTHERSILL, J. S. 1975, Lake Chad: geochemistry and sedimentary aspects of a shallow polymictic lake. *Journal of Sedimentary Petrology*, **45**, 295–309.

POWELL, T. G. 1986. Petroleum geochemistry and depositional setting of lacustrine source rocks. *Marine and Petroleum Geology*, **3**, 200–219.

REYRE, D. 1984. Remarques sur l'origine et l'évolution des bassins sédimentaires africains de la côte atlantique. *Bullétin de la Societé géologique de France*, **26**, 1041–1059.

ROBERT, P. 1985. Histoire Geothermique et Diagenése Organique. *Bulletin des Centres de Recherches Exploration–Production Elf-Aquitaine Mémoire*, **8**, 345 pp.

RYDER, R. T. 1980. Lacustrine sedimentation and hydrocarbon occurrences with emphasis on Uinta Basin models. *American Association of Petroleum Geologists, Fall Education Conference*, Houston, Texas, 103 pp.

SARNTHEIN, M. & KOOPMAN, B. 1980. Late Quaternary deep-sea record of northwest African dust supply and wind circulation. *Palaeoecology of Africa*, **12**, 238–253.

SERRUYA, C. & POLLINGHER, U. 1983. *Lakes of the Warm Belt*. Cambridge University Press, Cambridge, 569 pp.

SERVANT, M. 1973. Séquences continentals et variations climatiques, évolution du bassin du Tchad au Cénozoïque supérieur. Unpublished thesis, University of Paris.

STAGER, J. C. 1984. The diatom record of Lake Victoria (East Africa): the last 17 000 years. *Proceedings of the 7th International Diatom Symposium, Philadelphia, 1982*, 455–476.

—— REINTHAL, P. R. & LIVINGSTONE, D. A. 1986. A 25 000 years history for Lake Victoria, East Africa, and its significance for the evolution of cichlid fishes. *Freshwater Biology*, **16**, 15–19.

STOFFERS, P. & HECKY, R. E. 1978. Late Pleistocene–Holocene evolution of the Kivu-Tanganyika Basin. *In*: MATTER, A. & TUCKER, M. E. (eds), *Modern and Ancient Lake Sediments*, International Association of Sedimentologists Special Publication, **2**, 43–55.

STREET, F. A. 1981. Tropical palaeoenvironments. *Progress in Physical Geography*, **5**, 157–185.

STREET-PERROTT, F. A. & ROBERTS, N. 1983. Fluctuations in closed-basin lakes as an indicator of past atmospheric circulation patterns. *In*: STREET-PERROTT, A., BERAN, M. & RATCLIFFE, R. (eds), *Variations in the Global Water Budget*, Reidel, Dordrecht, 331–345.

SURDAM, R. & WOLFBAUER, C. A. 1975. Green River Formation, Wyoming: a playa-lake complex. *Bulletin of the Geological Society of America*, **86**, 335–345.

TALBOT, M. R. 1976. Late Quaternary sedimentation in Lake Bosumtwi, Ghana. *20th Annual Report of the Research Institute of African Geology, University of Leeds*, 69–73.

—— 1982. Holocene chronostratigraphy of tropical Africa. *Striae*, **16**, 17–20.

—— LIVINGSTONE, D. A., PALMER, P. G., MALEY, J., MELACK, J. M., DELEBRIAS, G., GULLIKSEN, S. 1984. Preliminary results from sediment cores from Lake Bosumtwi, Ghana. *Palaeoecology of Africa*, **16**, 173–192.

TALLING, J. F. & TALLING, I. B. 1965. The chemical composition of African lake waters. *Internationale Revue des gesamten Hydrobiologie*, **50**, 421–463.

TIERCELIN, J. J. & LE FOURNIER, J. 1980. Un exemple de sédimentation récente dans un rift continental: le semi-graben de Baringo-Bogoria, Rift Gregory, Kenya. *Recherches géologiques en Afrique*, **5**, 133–140.

—— LE FOURNIER, J., HERBIN, J. P. & RICHERT, J. P. 1980. Continental rifts: modern sedimentation, tectonic and volcanic controls. Example from the Bogoria-Baringo graben, Gregory Rift, Kenya. In: *Geodynamic Evolution of the Afro-Arabian Rift System*, Accademia Nazionale dei Lincei, Roma, **47**, 143–163.

—— RENAUT, R. W., DELIBRIAS, G., LE FOURNIER, J. & BIEDA, S. 1981. Late Pleistocene and Holocene lake level fluctuations in the Lake Bogoria basin, northern Kenya Rift Valley. *Palaeoecology of Africa*, **13**, 105–120.

—— PERINET, G., LE FOURNIER, J., BIEDA, S., & ROBERT, P. 1982. Lacs du rift est-africain, exemples de transition eaux douces—eaux salées: le lac Bogoria, rift Gregory, Kenya. *Mémoire Societé géologue de France*, **144**, 217–230.

TISSOT, B. P. & WELTE, D. H. 1984. *Petroleum Formation and Occurrence*. 2nd edn. Springer-Verlag, Berlin.

WANLI, Y. 1985. Daging oil field, People's Republic of China: a giant field with oil of nonmarine origin. *Bulletin of the American Association of Petroleum Geologists*, **69**, 1101–1111.

——, YONGKANG, L. & RUIGI, G. 1981. Formation and evolution of Nonmarine Petroleum in the Songliao Basin, China. *Scientific Research and Design Institute of Daging Oil Field, China, Report*, 22 pp.

WAPLES, D. W. 1984. Modern approaches to source-rock evaluation. *In*: WOODWARD, J., MEISSNER, F. F. & CLAYTON, J. L. (eds), *Hydrocarbon Source Rocks of the Greater Rocky Mountain Region*, Rocky Mountain Association of Geologists, Denver, 35–49.

WARREN, J. K. 1986. Shallow-water evaporitic environments and their source rock potential. *Journal of Sedimentary Petrology*, **56**, 442–454.

YURETICH, R. F. 1979. Modern sediments and sedimentary processes in Lake Rudolf (Lake Turkana) eastern Rift Valley, Kenya. *Sedimentology*, **26**, 313–331.

M. R. TALBOT, Geologisk Instituttt Avd A, Universitetet i Bergen, Allegt. 41, 5007 Bergen, Norway.

Large Australian lakes during the last 20 million years: sites for petroleum source rock or metal ore deposition, or both?

P. De Deckker

SUMMARY: Three basic 'regimes' of large lakes are recognized: (1) deep lake, (2) lake with ephemeral (and shallow) water, and (3) dry lake. The characteristics of these three lake types are discussed with regard to the composition of their water and sediment chemistry, but also to their potential for the production and preservation of organic matter, as well as the adsorption and accumulation of metals. Large Australian lakes, some with a sedimentation record spanning the last 20 million years, are examined in line with the above concept. Because of the low sedimentation rates in those lakes and the present-day water chemistry predominantly leading to the precipitation of gypsum and halite (especially in the groundwater below the lakes), it appears that any organic matter that would have been produced in the lakes during their carbonate phase in Miocene time would have been consumed by bacterial activity mainly in the groundwater. It is therefore suggested that the search for sites of metalliferous accumulation may prove more rewarding than will exploration for oil within sediments deposited since the early Miocene. As a comparison with large lakes on other continents, suggestions are presented as to what would have happened if sedimentation rates were much greater and if the lakes were in proximity to volcanism.

Introduction

In recent years there has been increased interest in the economic aspects of large lakes. Not only have the evaporites received much attention, but there is now ample recognition that lake basins can host significant quantities of petroleum source rocks as well as being sites for metalliferous deposits of economic importance. Already, a number of models have been proposed on lacustrine sedimentation associated with petroleum source rocks (Demaison & Moore 1980; Friedman 1980; Bauld 1981; Dean 1981; Kirkland & Evans 1981; Eugster 1985) and ore deposits (Renfro 1974; Eugster 1985).

The data presented here looks at the situation of Australian lakes and their deposits spanning the last 20 million years to ascertain their economic importance with regard to petroleum and metalliferous deposits. Because of the prescribed short length of this article, emphasis is placed on a series of schematic diagrams describing the conditions under which organic matter and metal ions occur in lakes and also those circumstances under which they can be preserved and become economically important. Although all the examples presented here are based on Australian sites, conditions encountered on other continents (viz. high sedimentation rates and proximity to volcanism, neither of which occur in Australia), are discussed with reference to petroleum source rocks and metalliferous deposit potentials.

Three basic lake types

De Deckker (1987) distinguished three lake types (which, accordingly, relate to hydrological regimes) in his description of the biological and sedimentary features characteristic of Australian salt lake facies. The first type is a *permanent* lake because it retains perennial water; the second occasionally fills up and therefore retains *ephemeral* water; and thirdly, there is the lake which continuously remains *dry*. For brevity, those three lake types are mentioned throughout the text and are referred to graphically in a basic triangle (Fig. 1). Respective characteristics of these lake types are also presented on triangular diagrams (Figs 1, 4, 5). For example, an increase in aridity (viz. an increase in the evaporation/precipitation (E/P) over the lake) will change the status of a lake, from permanent to ephemeral, and with further increase in E/P the lake will finally become dry (see Fig. 1). In addition, permanent lakes are usually characterized by minimal physico-chemical changes in comparison to those witnessed in ephemeral lakes. Nevertheless, in the case of permanent lakes, some parameters do change over time like oxygen levels, temperature etc. Accordingly, this can have significant implications for the aquatic biota. Some organisms will have to withstand the rapid changes affecting their host environment; some may barely survive through these changes, whereas others will require these changes during their life cycle (e.g., to hatch in low salinity water and later on thrive and reproduce in high salinity water). On the

From: FLEET, A. J., KELTS, K. & TALBOT, M. R. (eds), 1988, *Lacustrine Petroleum Source Rocks*, Geological Society Special Publication No. 40, pp. 45–58.

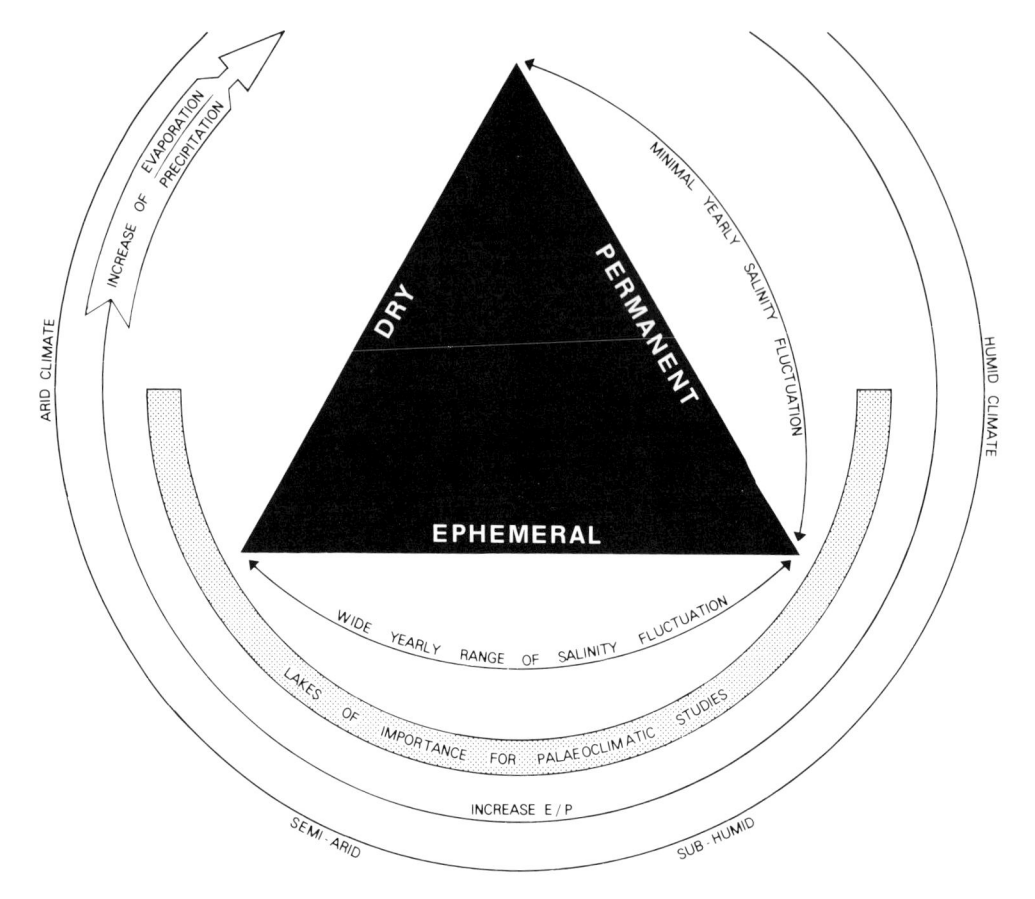

FIG. 1. Schematic triangular diagram showing the relationship between the three basic lake types and climatic conditions and salinity fluctuations (the latter summarizing other physico-chemical conditions like dissolved oxygen levels, temperature etc.).

other hand, organisms living in ephemeral water would tend to require particular and constant physico-chemical conditions to live and reproduce, and consequently would be unable to tolerate dramatic changes of their environment. However, it is more than likely that only a few species will be found inhabiting both permanent and ephemeral systems. Since a variety of organisms can be linked to the production and accumulation of organic matter and their association to metals in lakes (see below), it is necessary to identify the lake types in which those organisms live, and also where they occur in the lakes.

Facies and biota which help recognize the three different lake types are documented at length in De Deckker (1987). They will not be discussed here except for the biota of deep lakes which were not sufficiently documented in De Deckker (1987).

Figure 2 shows the distribution of organisms which readily fossilize in deep lakes [either fresh ($<3‰$ salinity) or saline ($>3‰$)] and where they can be transported by phenomena like turbidity flows/undercurrents. For more details on the latter see Sturm & Matter (1978). Ooids also deserve a mention here because they help recognize lake shore facies. Note also that calcareous organisms can become dissolved or partly etched in the hypolimnion if waters are undersaturated in calcium carbonate. Evidence of deep water may therefore disappear if calcareous fossils have dissolved. Calcareous nannoplankton (see Kelts & Hsü 1978, p. 297 and Dean 1981, p. 219) are not represented in this diagram since they are not restricted to any particular facies.

There are few differences between either a fresh or a saline deep-water lake since the major biological groups which fossilize are usually equally represented. However, with regard to diversity, differences do occur: in saline lakes,

FIG. 2. Diagram showing the distribution of organisms (with remains which can potentially preserve as fossils) in a deep lake situation, with either fresh or saline permanent water. The reworking of organism remains is also indicated. The location and relative abundance of organic matter in the lake is presented together with information on the likelihood of its preservation and decomposition.

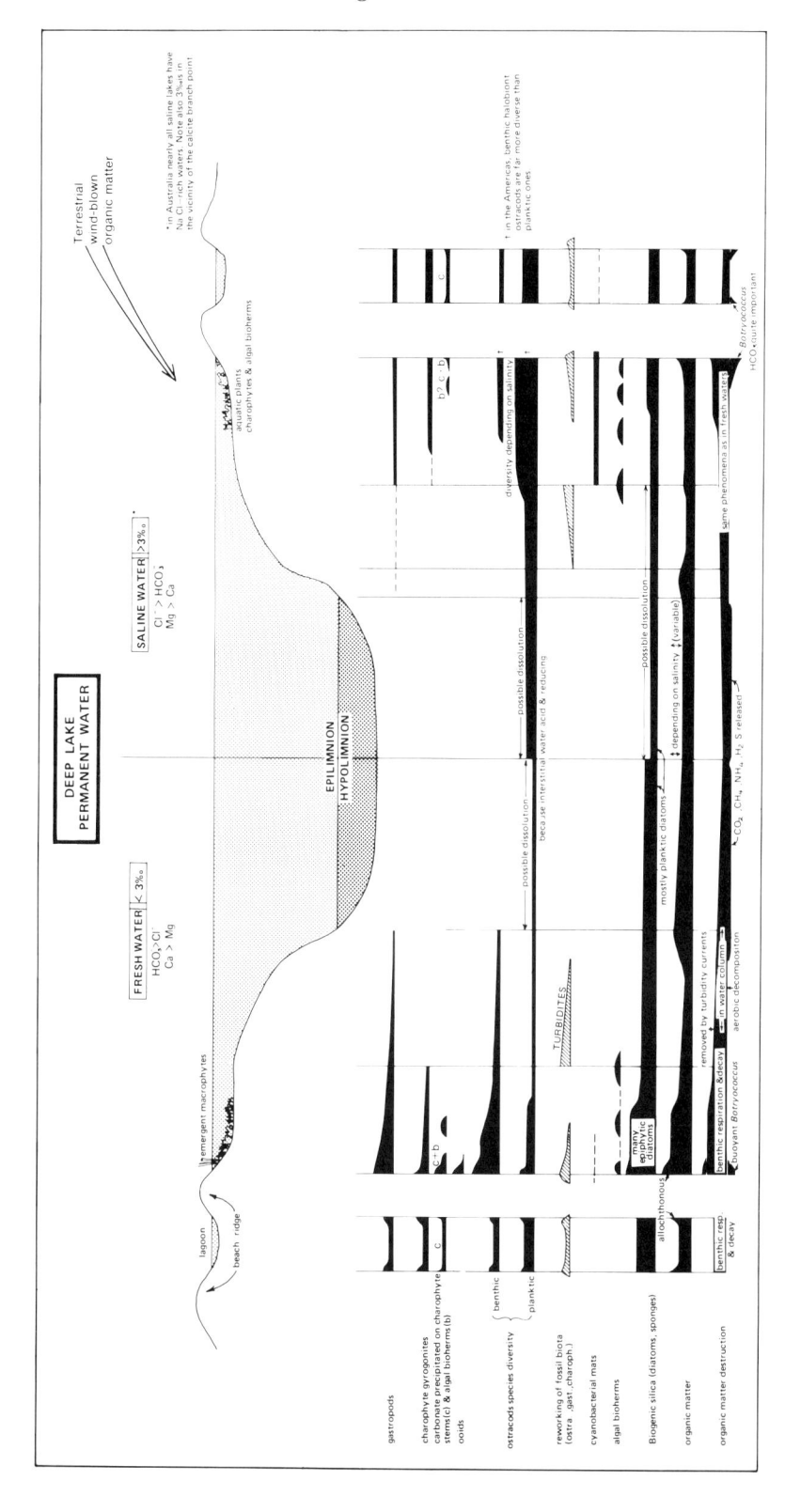

species diversity is generally much lower, but population density of some species may be extremely high. This applies to the flora, especially the algae, as well as the fauna. The former is of interest here because of its potential to become a source for petroleum. Note also the representation on Fig. 2 of where organic matter is produced, sometimes partly transported (in the case of turbidity flows etc.), and finally preserved or decayed. For more details on these phenomena refer to Dean (1981). Of note also is the influx of wind-blown terrestrial organic matter into lake systems, especially into saline lakes which occur in desertic regions. Galat *et al.* (1981), for example, have estimated that 50% of a tumble weed standing crop adjacent to Pyramid Lake in Nevada would be transported into it, thus contributing an input of over 2000 kg C yr^{-1} to the lake budget. Such input cannot be ignored.

Chemical characteristics: water and sediments

Nearly all saline lakes in Australia share a particular characteristic; Na and Cl are the dominant ions in their waters. They all belong to pathway II of Eugster & Hardie (1978) and, therefore, the most common salts precipitated from their waters, with an increase in water concentration, are $CaCO_3$, $CaSO_4 \cdot 2H_2O$ and NaCl. This chemical 'uniformity' in Australian waters, as already discussed in De Deckker (1983), is related to the stability of the Australian continent, as well as to the absence of active volcanism and rift-type sedimentation (and associated mineral springs), which otherwise would supply a greater variety of ions and minerals.

Before discussing the chemical properties which characterize the three lake types, it is important to note that nearly all lakes are the above-ground component of a hydrological basin. When the water table is below the lake floor, the lake remains dry, unless its surface (or near surface) is sealed by a hard layer (like indurated clay- or carbonate-band) for example and water enters the lake from a river or directly from rainfall. When the lake is dry, aeolian processes are most active and deflation of sediments and organism remains may occur, and more importantly oxidation of organic matter is extensive.

When the regional groundwater is obviously of different density from the brine pool below the lake, a Ghyben-Herzberg interface will occur (Freeze & Cherry 1979) (Fig. 3). As a consequence of this, potential flow lines converge towards the edge of the lake. This results in a marginal seepage-spring zone. There, a number of processes operate: they include evaporative pumping in the upper capillary fringe (see Hsü & Siegenthaler 1969), thus causing the formation of efflorescent salts (some of which may turn later into gypsum clay pellets, see Bowler 1973), polygonal cracking and associated with these are teepee structures and brecciation (see Muir *et al.* 1980; Warren 1982; De Deckker 1987).

The near-surface occurrence of the groundwater will also stimulate microbial activity resulting in cyanobacterial (algal) mat growth (Bauld 1981; De Deckker in press). Below these mats, bacterial sulphate reduction usually occurs (Teller *et al.* 1982, Skyring *et al.* 1983) and organic carbon is consumed in the process.

Permanent lakes (Fig. 4) most characteristically have carbonate-rich waters; this type of water chemistry also applies to the ephemeral coastal lakes in Australia, among which the lakes associated with the Coorong are the best known (von der Borch 1965). Some permanent lakes are also saline (>3‰ salinity) and have a more varied water chemistry, but, under the present-day (climatic) conditions in Australia, most permanent saline lakes have NaCl-dominant waters. If the lakes are of volcanic origin, like the well-studied maar lakes in Western Victoria (see Williams 1981), they will also have bicarbonate-rich waters. In any case, most permanent lakes are characterized by carbonate precipitates, some of which may form interstitially (Muir *et al.* 1980; De Deckker 1987).

Ephemeral lakes (Fig. 4) are usually characterized by the predominance of NaCl-rich waters and with gypsum and halite in their sediments. Interstitial precipitation of gypsum and halite also occurs as a result of groundwater concentration in association with evaporation, especially within the capillary fringe. Erosional phenomena are also commonly associated with the drying phase of these ephemeral lakes. Mechanical destruction of sedimentary structures, including algal mats (for more details see Walter *et al.* 1973; De Deckker 1987), formed under subaqueous conditions, commonly follows during the dry phase. Aeolian activity and associated phenomena (e.g., deflation, formation of clay lunettes) play a significant role with increasing aridity. Note, however, that aeolian deposits which originated from lakes may later act as reservoir rocks, hence it is important to recognize and locate them.

In the case of continuously *dry* lakes (Fig. 4) efflorescence of some salts operates, especially around the seepage zone when the groundwater is close to the surface (and also in association with a Ghyben-Herzberg interface). If the groundwater reaches saturation with respect to NaCl or $CaSO_4 \cdot 2H_2O$, and if there is little loss through the aquifer (viz. the basin being hydro-

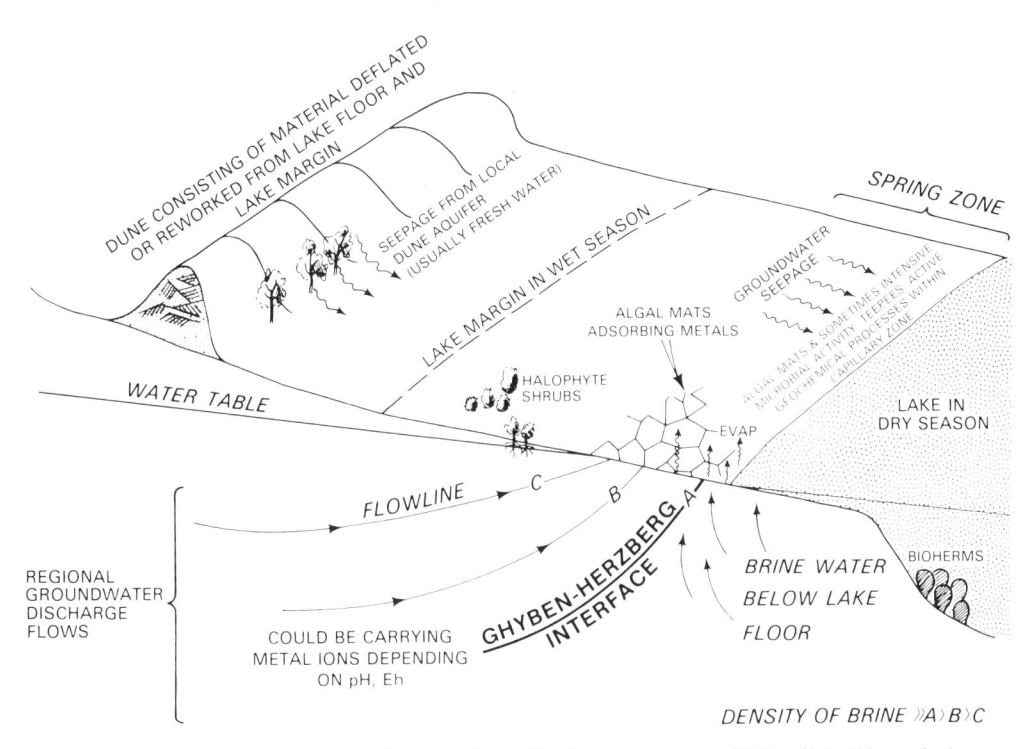

FIG. 3. Diagram representing a typical saline lake affected by the brine-water pool below it. In this particular case, a Ghyben-Herzberg interface operates between the regional groundwater and the brine pool, thus forcing both waters to converge towards the lake surface near the lake shore in an area characterized by seepage and springs. There, a number of processes operate, including the growth of algal mats which, in turn, could adsorb metal ions if they were transported there via the groundwater.

logically closed), a salt crust can form, thus preventing any further substantial sediment loss through deflation. Nevertheless, microbiological activity may continue below the salt crust, especially in regard to algal growth and bacterial activity.

Organic matter and metals in lakes

Phytoplankton productivity is usually very diversified in fresh and *permanent* lakes. An increase in salinity has an adverse effect on productivity except under certain circumstances when some algae grow under 'bloom' conditions. The alga *Botryococcus* is one of the best known examples, and is of interest here, because it contains high amounts of hydrocarbons (Hillen & Wake 1979). Accumulation of *Botryococcus*, along the shores of lakes or within lagoons connected to lakes, results from the positive buoyancy of this alga (Bauld 1983). In Australia, gelatinous deposits of *Botryococcus* are called Coorongite but elsewhere have received a variety of names (e.g., Balkashite, N'hangellite). Substantial accumulation of this alga, if adequately preserved, such as by rapid burial (e.g., during flash floods associated with adjacent alluvial fans), have the potential to form rich petroleum source rocks and oil shales.

Conditions favouring preservation of organic matter in deep lakes are highly variable. In general, with the lowering of water level in a deep lake, there is less likelihood of stratification within the lake. This prevents the occurrence of anaerobic conditions at the sediment/water interface. Thus the potential for organic matter preservation decreases. Similarly, as soon as a lake becomes *ephemeral* (or in the shallow parts of a deep lake) the potential for organic matter preservation is minimized because of mechanical erosion and transport during receding water levels, but also because benthic respiration and decay (caused by microbial organisms) are intensified during those circumstances (see Fig. 2 and Dean 1981).

Cyanobacterial (algal) mats, on the other hand, are best developed in *ephemeral* lake systems. They usually thrive under extreme physicochemical conditions (Bauld 1981) where they can

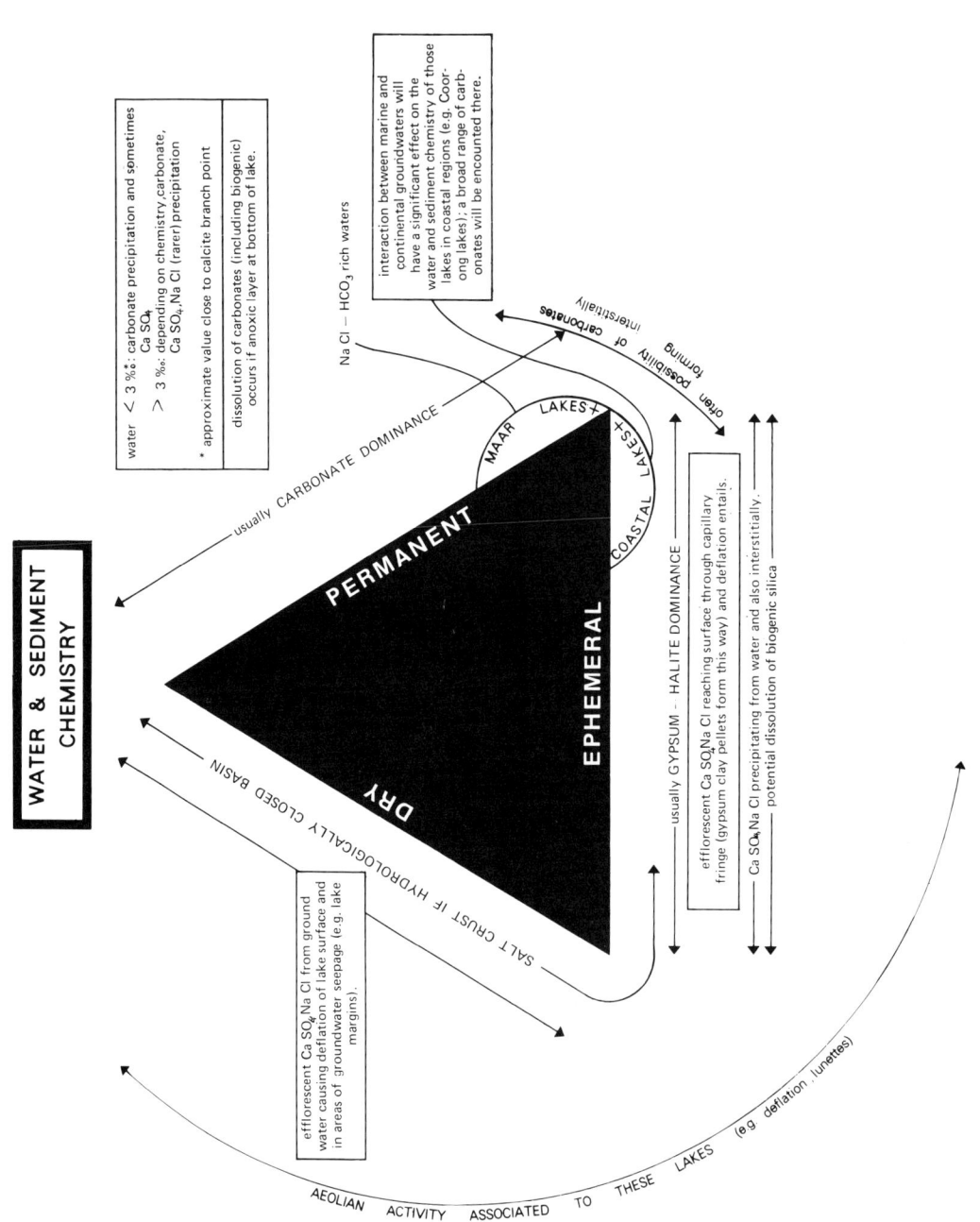

FIG. 4. Schematic triangular diagram showing the relationship between the three basic lake types and water and sediment chemistry for modern Australian lakes.

be less subject to grazing by organisms. They are also often aerially exposed and easily survive exposure. Note also that in hypersaline permanent water conditions, some cyanobacteria can grow to form larger bioherms (see De Deckker 1987) (up to several metres high and wide). Significantly, these bioherms may become potential hydrocarbon sources if they are adequately preserved. Specific conditions are indeed necessary for bioherm preservation since below algal mats sulphate reduction normally occurs with anoxic conditions and this causes the consumption of organic carbon (Bauld 1981, Skyring et al. 1983). An equilibrium between rate of sedimentation and conditions for (optimal) algal growth is therefore necessary to prevent the biological or mechanical (e.g., erosion by wave action) destruction of algal mats.

As mentioned earlier, algal mats are commonly found on the edge of lakes, especially in the areas associated with a Ghyben-Herzberg interface where a continuous, or near continuous, seepage of groundwater occurs (Fig. 3). They even commonly occur and grow below salt crusts or in association with teepee structures. The best documented example in Australia of large stromatolitic structures associated with polygonal cracking at Marion Lake in South Australia is published in von der Borch et al. (1977) and Warren (1982). There, the structures are well preserved but there is no organic matter left. A significant additional phenomenon is observed in association with cyanobacterial mats: commonly the latter can trap metal ions which become adsorbed on organic matter. This phenomenon was demonstrated experimentally by Disnar (1981) who showed the capability of cyanobacteria to select particular metal species under specific Eh and pH conditions. This is particularly relevant here because, as mentioned earlier, mats do grow along groundwater seepage zones where waters are often highly concentrated in many elements, and therefore may be relatively enriched in some metal ions. Degens & Ittekkot (1982) have also pointed out the common formation of metal–organic complexes in association with the growth of authigenic sulphide, phosphate, oxide and carbonate minerals. One of the processes which they described is the natural staining of organic matter by metals (best described by these authors as a kind of 'scavenging of metals from the environment by organics'). This could eventually lead to the possible formation of stratabound ore deposits. Already, Draper & Jensen (1976) have documented the close association between Mn and the presence of algal mats on the floor of the large playa Lake Frome. However, it must be noted that once some metals have been adsorbed by organic matter, they will not necessarily remain at the site of fixation. They can be remobilized depending on changing chemical conditions. This is the common case in Australia with uraniferous solutions from groundwater which eventually contribute to carnotite mineralization (see Arakel & McConchie 1982). Further, sulphides which also concentrate below algal mats as a result of bacterial activity (Skyring et al. 1983) may accumulate sufficient quantities of metals worthy of economic interest.

All the above mentioned phenomena are summarized in Fig. 5. It is important to know that petroleum source rocks are basically linked to *deep* lake systems and that *Botryococcus*-rich deposits probably originate in shallow water and perhaps *ephemeral* and slightly saline systems. Metalliferous deposits on the other hand are linked to *ephemeral* and *dry* lake systems, provided groundwater discharge areas occur and are enriched in metal ions. Association with algal mats is also necessary in the latter case. Nevertheless, it is important to realize that a close association exists between organic matter and metals in lacustrine systems.

Australian lakes, sites for hydrocarbon or metal deposits, or both?

Compared to the situation on other continents, some Australian lakes have been in existence for a long time. In fact, some lakes have been sites of sedimentation since the early Tertiary. For example, the elongated playa lakes in Western Australia which occupy ancient drainage systems have existed since the mid-Miocene (van de Graaf et al. 1978). In central Australia there are also mid-Miocene lacustrine carbonate sediments below Lakes Eyre and Frome (Callen 1977) and Eocene lacustrine sediments below Torrens (Johns 1968) (for more details see De Deckker 1983). Also, Lake George in eastern Australia has a well documented record of sedimentation that started in early Miocene (Singh et al. 1981).

Since the early Miocene, the Australian tectonic plate has moved northward, away from the Antarctic plate, across about 5–8° of latitude (see Fig. 6). Unfortunately, there are insufficient data to determine the climatic changes which occurred during the last 20 million years in Australia to postulate what would have happened to the large lakes. Nevertheless, there is evidence from lacustrine carbonate sediments and from the fossil biota recovered in the vicinity of Lake Frome that a warm, high rainfall climate prevailed during mid-Miocene time (Callen 1977). This is quite different from today, because now the lakes occur in the arid portion of Australia

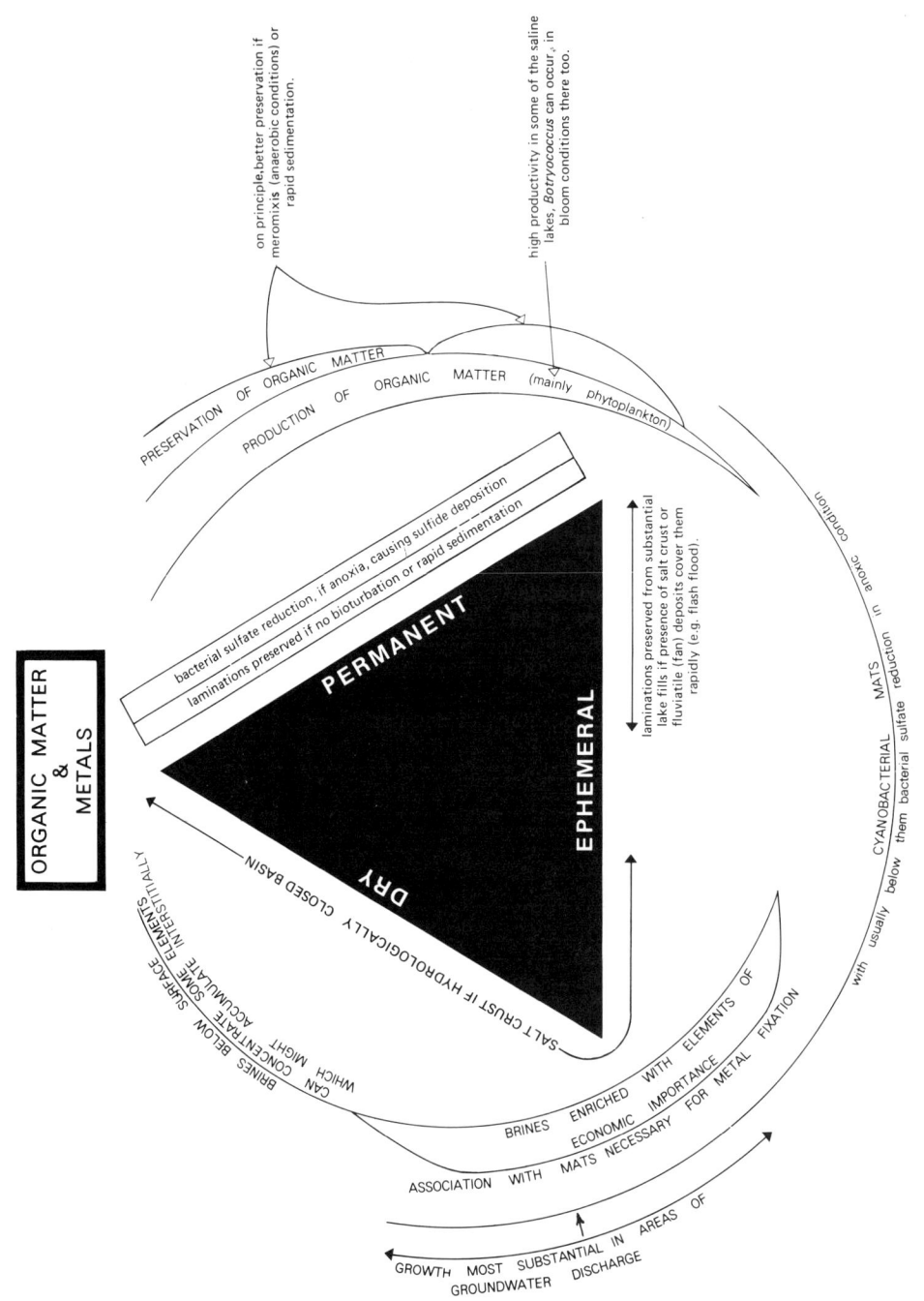

FIG. 5. Schematic triangular diagram showing the relationship between the three basic lake types and the presence of organic matter and metals, including their respective preservation.

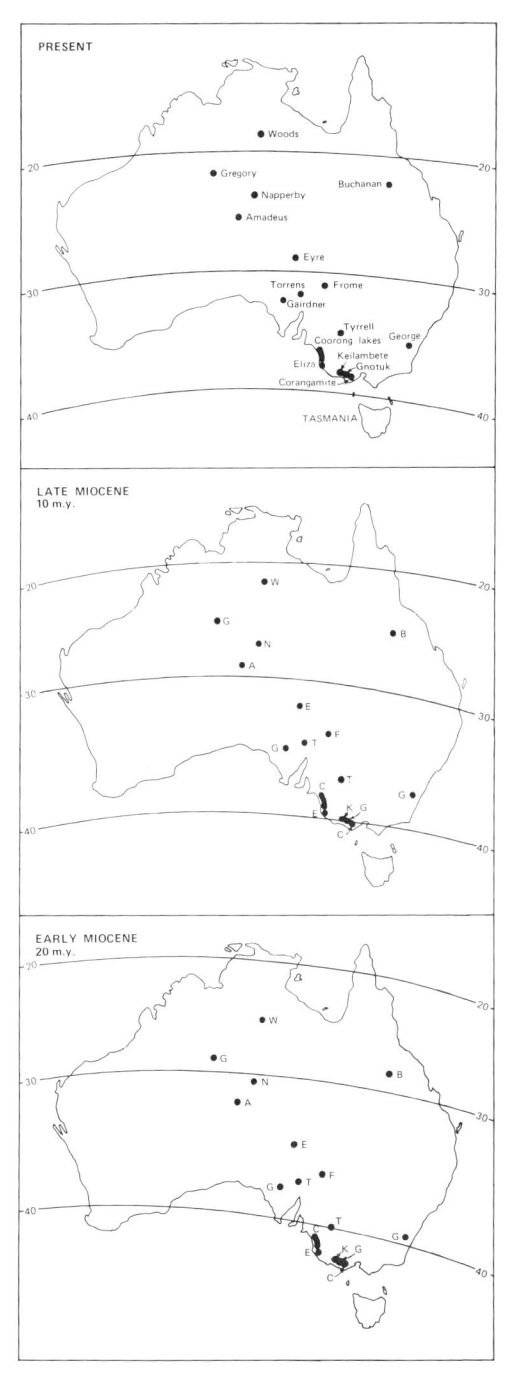

FIG. 6. Maps of Australia showing the location of the lakes mentioned in the text and in Fig. 7, and their respective latitudinal positions for three periods: early Miocene, late Miocene and Present.

and are the sites of gypsum and halite sedimentation.

An examination of the position occupied by most known Australian lakes on the basic triangle referring to water and sediment chemistry for three periods of the geological time scale (early Miocene, late Miocene and Present) (see Fig. 7) points to the following: nearly all lakes changed position (or would have if they had been in existence during the last 20 million years, e.g., the crater lakes and Lake Corangamite formed by damming of a lava flow are much too young) on the triangle and moved in an anticlockwise fashion. This, therefore, indicates that the lakes' water/sediment composition changed through time. Some lakes, however, remained groundwater-seepage areas (e.g., Lakes Amadeus and Napperby in central Australia) and would therefore only be sites of metal accumulation, but not hydrocarbons. The lakes which changed composition, and which would have been sites for hydrocarbon formation and storage (e.g., Lakes Eyre, Torrens and Frome: see Fig. 7), have undergone dramatic chemical changes through time, especially in the groundwater pool below them. Since the composition of the groundwater below most of these lakes is sulphate-rich today (and has been for some time), intense bacterial activity would have destroyed any organic carbon-rich sediments. However, the metals which originally were adsorbed by organic matter could have remained within the sediments, even if organic matter disappeared at a later stage. But they may also have been subsequently transported elsewhere by leaching processes via the groundwater.

As a consequence of the changes which occurred in Australia, especially as a result of an overall increase of aridity since the Miocene, the search for metals in large-lake sediments is likely to be more rewarding for the exploration geologist than the search for hydrocarbons.

Conclusions

Figure 8 provides a summary of the concepts presented in this article. The left hand portion of the diagram shows the evolution of a lake under Australian conditions with an increase in aridity (in an upward sequence on the diagram) with regard to water and sediment conditions. It shows the transition from sediment deposition to sediment deflation, and how this would favour either organic matter or metal preservation and accumulation. All the above phenomena relate to a system with low sedimentation rates. Water chemistry in Australian lakes which evolved

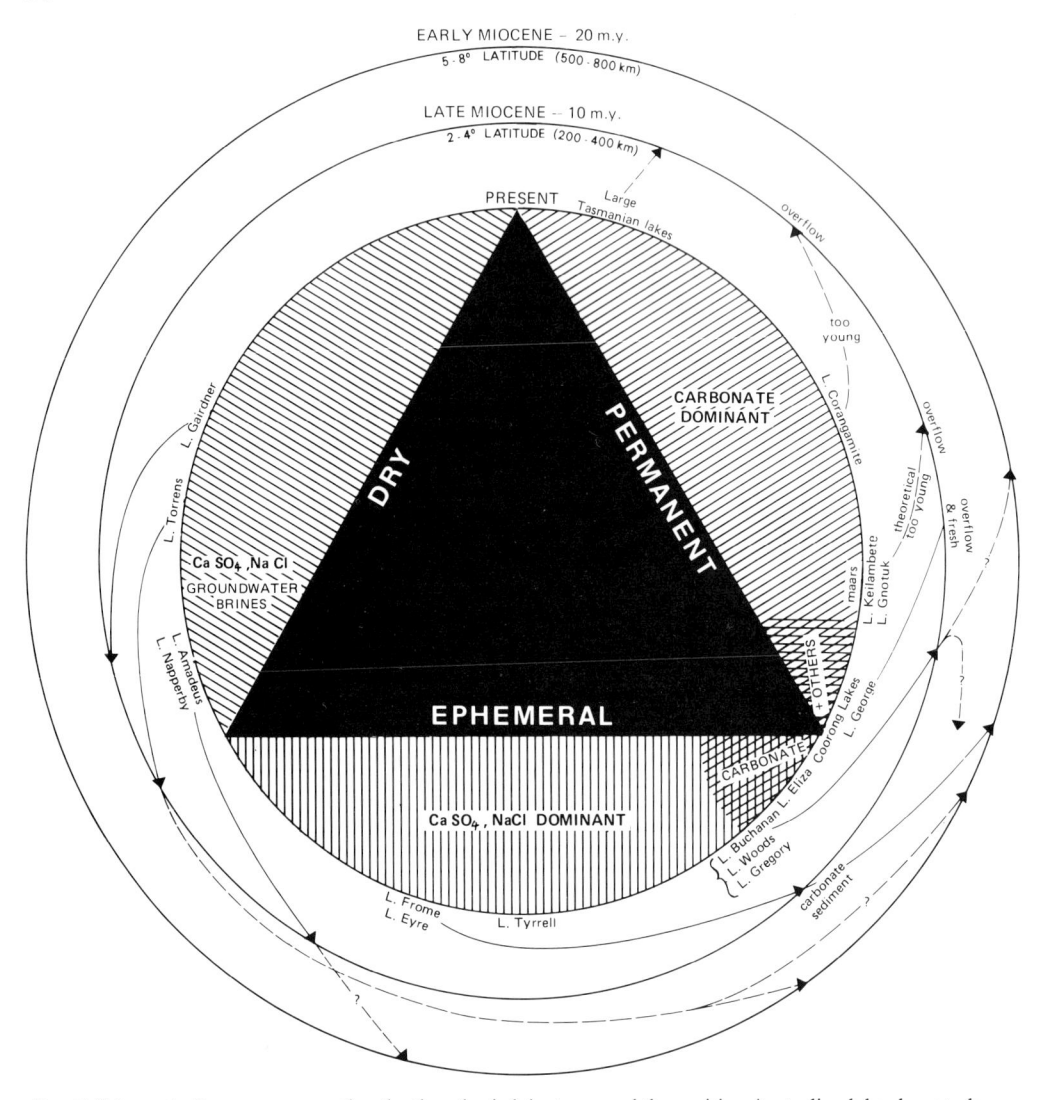

FIG. 7. Schematic diagram representing the three basic lake types and the position Australian lakes have today, but also had (or would theoretically have had if they were in existence then) during the early Miocene and late Miocene. Note also that since early Miocene, Australia has shifted 5–8° in latitude (= 500–800 km) and since late Miocene, it has moved 2–4° in latitude (or some 200–400 km).

through the $CaCO_3$–$CaSO_4$–$NaCl$ pathway, and which underwent an overall change during the last 20 million years from carbonate to gypsum–halite dominant sedimentation, has thus favoured the potential accumulation of metals and the destruction of organic matter.

Elsewhere on other continents, the proximity to volcanism, and high sedimentation rates due to high tectonic activity, would engender conditions favourable to the production and preservation of organic matter and accumulation of metalliferous deposits. By examining the right hand portion of Fig. 8, which fully documents those possibilities, it becomes obvious that high sedimentation rates are necessary for the preservation of organic matter, especially in a deep-lake situation with an anoxic hypolimnion. Thus, in general, a lake in a tropical region would become more favourable to organic matter preservation than in a temperate one because commonly deoxygenated waters occupy a much greater portion of the water column in tropical lakes (see Beadle (1974) for tropical African lakes and Kohzov (1963) for a comparison with Lake

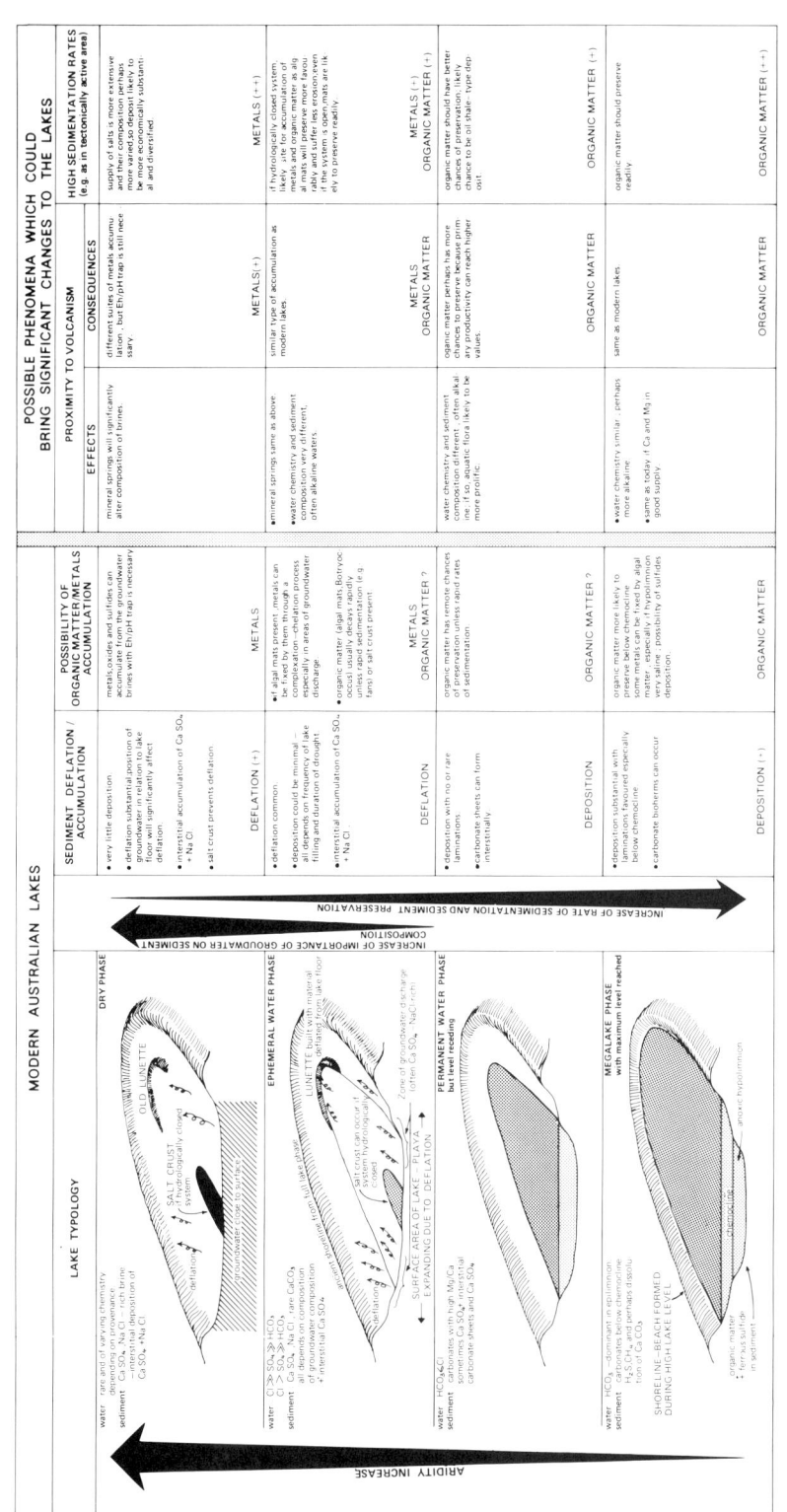

FIG. 8. Summary diagram showing on the left the possible evolution of a large Australian lake with an increase in aridity (from bottom of diagram to top). Information is given on sediment deflation and deposition in those systems, and also with regard to the possibility of organic matter and metals accumulation under the conditions operating today in Australia. Summary information is given in capital letters in the bottom portion of the boxes with the signs (+) or (++) indicating increasing degrees of amelioration of the conditions. On the right hand side, possible phenomena which may affect the lakes favourably with regard to metals and organic matter accumulation are presented. These apply to lakes outside Australia in volcanic provinces and tectonically active areas.

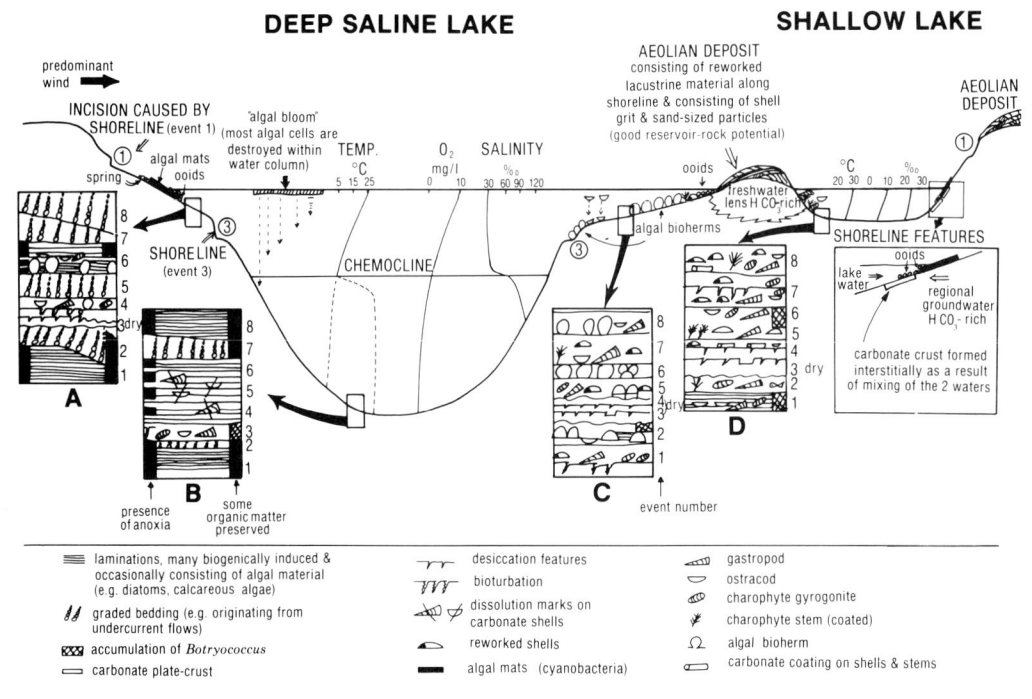

FIG. 9. Schematic diagram illustrating two adjacent lakes; one is a deep saline lake and the other is shallow and less saline. Distinctive physico-chemical characteristics of these two lakes (dissolved oxygen, temperature, salinity) are shown graphically along a depth profile. Sedimentological and palaeontological features which characterize facies within these two lakes are graphically presented and their association through a series of eight sequential, synchronous events, but operating in different parts of the lakes, are shown in four schematized stratigraphical columns (A–D). These events are briefly described below. *Event 1* (represented at the bottom of every column) relates to a phase during which both lakes were connected, thus with the larger lake margins forming an incision high up in the landscape. Extensive beach ridges also indicate (high) lake levels. The hypolimnion is enlarged during this phase at sites A and B and its presence favours the preservation or organic matter not decomposed or destroyed through the oxygenated part of the water column. *Botryococcus* blooms from the entire lake, due to the predominant winds, may accumulate in the shallower part of the large lake at site D. This hydrocarbon-rich alga should be preserved there if sedimentation rate is also rapid, thus preventing its decomposition by bacterial activity. During *event 2*, water level recedes and the lakes separate; stratification still occurs in the large lake and organic matter can preserve at sites A and B; in the former because of the high sedimentation rate related to undercurrent flows, and at site B because of anoxic conditions and the substantial supply of sediment from the turbidity flows. Note that the graded bedded sequences can act as hydrocarbon reservoirs especially along the margins of the lake due to their high porosity. *Botryococcus*-rich beds may accumulate along the right margin of the lake in among algal bioherms. The latter, if they are substantial (this is principally controlled by water depth) and lithified (most often by a carbonate matrix) they can act as both source rocks and reservoir rocks (the latter because of the large number of cavities they have). During *event 3*, lake level dropped to such an extent that stratification vanished in the large lake. Incision along the lake margin was substantial and this is visible below water level of the present lake (=represented by event 8). The shallow lake dried up and pedogenic phenomena destroyed evidence of past sedimentary structures down to a certain depth; fossil shells were also destroyed. If salinity is adequate for *Botryococcus* to bloom in the lake, it could accumulate at site B. Site C also suffered erosion, and material from this site as well as from site D is deflated and reworked, so as to become redeposited along the margins of the lake (on the right hand side of both lakes as controlled by the direction of the predominant winds). Some reworked material also accumulates there during the rise of water level during subsequent events. As lake level rises, anoxia reappears intermittently at site B and shells from oxygenated phases become partially dissolved during the anoxic ones. This intermittence of oxic–anoxic events causes the destruction of organic matter even in the deep parts of the lake. Elsewhere in the lake during that phase, and subsequent ones, one can see repetition of the events described previously. (Diagram modified from De Deckker (1987) and re-interpreted regarding the presence of organic matter in the two lakes.)

Baikal), but naturally there are exceptions to this rule. Demaison & Moore (1980) discuss this concept at length. Lakes in tectonically active areas, associated to mineral springs and high rates of erosion and weathering, would be more naturally propitious for the accumulation of metalliferous deposits, and in addition they may be sites of hydrocarbon source rocks.

To sum up, Fig. 9 provides a schematic view of the possible facies which can be recognized in association with a *deep* (and also *permanent*) lake, and a *shallow* (and occasionally *ephemeral*) lake, adjacent to one another. Four stratigraphic columns (A–D in Fig. 9), detailing schematically the evolution and changes of facies which might occur in both lakes during a series of eight events, relate to fluctuations in lake level and consequently to fluctuations of other physico-chemical parameters (e.g., presence or absence of anoxia, salinity changes). All these features are examined so as to distinguish the facies which are favourable for the production of organic matter and its preservation. Refer to the caption of Fig. 9 for a detailed description of the eight synchronous events relating to the evolution of both lakes. Figure 9 should also be examined while paying attention to the common association of metals to organic-rich sediments (as caused by the natural adsorption of metals to organic matter), and also to the presence of some conditions in lakes (like anoxia) which favour the formation of metalliferous sediments.

ACKNOWLEDGEMENTS: I am very grateful to Phil Macumber who introduced me to the concept of groundwater systems affecting lakes in Australia, and to Bill Last who critically went through a last version of the manuscript and suggested numerous changes to clarify the text.

References

ARAKEL, A. V. & MCCONCHIE, D 1982. Classification and genesis of calcrete and gypsite lithofacies in paleodrainage systems of inland Australia and their relationship to carnotite mineralization. *Journal of Sedimentary Petrology*, **52**, 1149–1170.

BAULD, J. 1981. Geobiological role of cyanobacterial mats in sedimentary environments: production and preservation of organic matter. *BMR Journal of Australian Geology and Geophysics*, **6**, 307–317.

—— 1983. Palaeoenvironment of *Botryococcus*-sourced oil shales. Abstracts, *Third International Symposium on Fossil Algae*, Golden, Colorado.

BEADLE, L. C. 1974. *The Inland Waters of Tropical Africa*, 1st edn. Longman, London, 365 pp.

BOWLER, J. M. 1973. Clay dunes: their occurrence, formation and environmental significance. *Earth Science Reviews*, **9**, 315–338.

CALLEN, R. A. 1977. Late Cainozoic environments of part of northeastern South Australia. *Journal of the Geological Society of Australia*, **24**, 151–169.

DEAN, W. E. 1981. Carbonate minerals and organic matter in sediments of modern north temperate hard-water lakes. *Special Publication. Society of Economic Paleontologists and Mineralogists*, **31**, 213–231.

DE DECKKER, P. 1983. Australian salt lakes: their history, chemistry, and biota—a review. *Hydrobiologia*, **105**, 231–244.

—— 1987. Biological and sedimentary facies of Australian salt lakes. *Palaeogeography, Palaeoclimatology, Palaeoecology*, **62**, 237–270.

DEGENS, E. T. & ITTEKKOT, V. 1982. *In situ* metal-staining of biological membranes in sediments. *Nature*, **298**, 262–264.

DEMAISON, G. J. & MOORE, G. T. 1980. Anoxic environments and oil source bed genesis. *Bulletin of the American Association of Petroleum Geologists'*, **64**, 1179–1209.

DISNAR, J.-R. 1981. Etude expérimentale de la fixation de métaux par un materiau sédimentaire actuel d'origine algaire.—II. Fixation 'in vitro' de UO^{2+}, Cu^{2+}, Ni^{2+}, Zn^{2+}, Pb^{2+}, Co^{2+}, Mn^{2+}, ainsi que de VO_3^-, MoO_4^{2-} et GeO_3^{2-}. *Geochimica et Cosmochimica Acta*, **45**, 363–379.

DRAPER, J. J. & JENSEN, A. R. 1976. The geochemistry of Lake Frome, a playa lake in South Australia. *BMR Journal of Australian Geology and Geophysics*, **1**, 83–104.

EUGSTER, H. P. 1985. Oil shales, evaporites and ore deposits. *Geochimica et Cosmochimica Acta*, **49**, 619–635.

—— & HARDIE, L. A. 1978. Saline lakes. *In*: Lerman, A. et al. (eds), *Lakes: Chemistry, Geology, Physics*, Springer-Verlag, New York, 237–293.

FREEZE, R. A. & CHERRY, J. A. 1979. *Groundwater*. Prentice-Hall Inc., Edgewood Cliffs, 604 pp.

FRIEDMAN, G. M. 1980. Review of depositional environments in evaporite deposits and the role of evaporites in hydrocarbon accumulation. *Bulletin des Centres de Recherche Exploration–Production Elf-Aquitaine*, **4**, 589–608.

GALAT, D. L., LIDER, E. L., VIGG, S. & ROBERTSON, S. R. 1981. Limnology of a large, deep, North American terminal lake, Pyramid Lake, Nevada, U.S.A. *Hydrobiologia*, **82**, 281–317.

HILLEN, L. W. & WAKE, L. V. 1979. 'Solar Oil'—liquid hydrocarbon fuels from solar energy via algae. *First National Conference, Energy Sources—Australia's Needs*, Australian Energy Institute, Newcastle, 18–25.

HSÜ, K. J. & SIEGENTHALER, C. 1969. Preliminary experiments on hydrodynamic movement induced by evaporation and their bearing on the dolomite problem. *Sedimentology*, **12**, 11–25.

JOHNS, R. K. 1968. Investigation of Lakes Torrens and

Gairdner. *Reports and Investigations of the Department of Mines of South Australia*, **31**, 1–90.

KELTS, K. & HSÜ, K. J. 1978. Freshwater carbonate sedimentation. *In*: Lerman, A. *et al.* (eds), *Lakes: Chemistry, Geology, Physics*, Springer-Verlag, New York, 295–323.

KIRKLAND, D. W. & EVANS, R. 1981. Source-rock potential of evaporitic environment. *Bulletin of the American Association of Petroleum Geologists*, **65**, 181–190.

KOHZOV, M. 1963. Lake Baikal and its life. *Monographiae Biologiae*, **11**, 1–352.

MUIR, M., LOCK, D., VON DER BORCH, C. C. 1980. The Coorong model for pene-contemporaneous dolomite formation in the Middle Proterozoic McArthur Group, Northern Territory, Australia. *Special Publication. Society of Economic Paleontologists and Mineralogists*, **28**, 51–67.

RENFRO, A. R. 1974. Genesis of evaporite-associated stratiform metalliferous deposits—a sabkha process. *Economic Geology*, **69**, 33–45.

SINGH, G., OPDYKE, N. D. & BOWLER, J. M. 1981. Late Cainozoic stratigraphy, palaeomagnetic chronology and vegetational history from Lake George, N.S.W. *Journal of the Geological Society of Australia*, **28**, 435–452.

SKYRING, G., CHAMBERS, L. A. & BAULD, J. 1983. Sulfate reduction in sediments colonized by cyanobacteria, Spencer Gulf, South Australia. *Australian Journal of Marine and Freshwater Research*, **34**, 359–374.

STURM, M. & MATTER, A. 1978. Turbidites and varves in Lake Brienz (Switzerland): deposition of clastic detritus by density currents. *In*: MATTER, A. & TUCKER, M. E. (eds), *Modern and Ancient Lake Sediments*. Special Publications. International Association of Sedimentologists, **2**, 147–168.

TELLER, J. T., BOWLER, J. M. & MACUMBER, P. G. 1982. Modern sedimentation and hydrology in Lake Tyrell, Victoria. *Journal of the Geological Society of Australia*, **29**, 159–175.

VAN DE GRAAF, W. J., CROWE, W. J. E., BUNTING, J. A. & JACKSON, M. J. 1978. Relict Early Cainozoic drainages in arid Western Australia. *Zeitschrift für Geomorphologie, Neue Folge*, **21**, 379–400.

VON DER BORCH, C. C. 1965. The distribution and preliminary geochemistry of modern carbonate sediments of the Coorong area, South Australia. *Geochimica et Cosmochimica Acta*, **29**, 781–799.

——, BOLTON, B. & WARREN J. K. 1977. Environmental setting and microstructure of subfossil lithified stromatolites associated with evaporites, Marion Lake, South Australia. *Sedimentology*, **24**, 693–708.

WALTER, M. R., GOLUBIC, S. & PREISS, W. V. 1973. Recent stromatolites from hydromagnesite and aragonite depositing lakes near the Coorong Lagoon, South Australia. *Journal of Sedimentary Petrology*, **43**, 1021–1030.

WARREN, J. K. 1982. The hydrological setting, occurrence and significance of gypsum in Late Quaternary salt lakes in South Australia. *Sedimentology*, **29**, 609–637.

WILLIAMS, W. D. 1981. The limnology of saline lakes in western Victoria: a review of some recent studies. *In*: Williams, W. D. (ed.), Salt Lakes, *Developments in Hydrobiology*, **5**, 233–259.

P. DE DECKKER, Department of Geography and Cenozoic Research Unit, Monash University, Clayton, Vic. 3168, Australia.

Microbial and biogeochemical processes in Big Soda Lake, Nevada

R. S. Oremland, J. E. Cloern, Z. Sofer, R. L. Smith, C. W. Culbertson,
J. Zehr, L. Miller, B. Cole, R. Harvey, N. Iversen, M. Klug,
D. J. Des Marais & G. Rau

SUMMARY: Meromictic, alkaline lakes represent modern-day analogues of lacustrine source rock depositional environments. In order to further our understanding of how these lakes function in terms of limnological and biogeochemical processes, we have conducted an interdisciplinary study of Big Soda Lake. Annual mixolimnion productivity (ca. 500 g m^{-2}) is dominated by a winter diatom bloom (60% of annual) caused by upward transport of ammonia to the epilimnion. The remainder of productivity is attributable to chemoautotrophs (30%) and photosynthetic bacteria (10%) present at the oxic–anoxic interface from May to November. Studies of bacterial heterotrophy and particulate fluxes in the water column indicate that about 90% of annual productivity is remineralized in the mixolimnion, primarily by fermentative bacteria. However, high rates of sulphate reduction (9–29 mmol m^{-2} yr^{-1}) occur in the monimolimnion waters, which could remineralize most (if not all) of the primary productivity. This discrepancy has not as yet been fully explained. Low rates of methanogenesis also occur in the monimolimnion waters and sediments. Most of the methane is consumed by anaerobic methane oxidation occurring in the monimolimnion water column. Other bacterial processes occurring in the lake are also discussed. Preliminary studies have been made on the organic geochemistry of the monimolimnion sediments. Carbon-14-dating indicates a lower depositional rate prior to meromixis and a downcore enrichment in ^{13}C of organic carbon and chlorophyll derivatives. Hydrous pyrolysis experiments indicate that the sediment organic matter is almost entirely derived from the water column with little or no contribution from terrestrial sources. The significance of the organics released by hydrous pyrolysis is discussed.

Introduction

Big Soda Lake is an ideal environment to study microbial reactions occurring in aquatic environments and to determine the impact of these reactions on geochemical processes. Because it is meromictic, the lake's monimolimnion provides a deep and unchanging anoxic water column where anaerobic bacterial processes may be quantified under conditions of elevated pH and salinity, as well as low redox potential. This greatly facilitates the study of microbial reactions occurring under anoxic conditions or at oxic–anoxic interfaces. In addition, because certain lacustrine petroleum deposits appear to have been derived from the sediments of alkaline meromictic lakes (Demaison & Moore 1980), the study of bacterial processes in this lake should be of importance to our understanding of the early diagenetic reactions related to oil and gas formation. Indeed, a detailed knowledge of the nature and function of water column microbial flora, as well as its downward flux into the sediments, should aid in deciphering the organic geochemistry of lacustrine source rocks (Didyk *et al.* 1978). This paper summarizes the findings of our interdisciplinary investigations of Big Soda Lake.

Study site hydrological properties

Big Soda Lake is located in the vicinity of the Carson Sink near Fallon, Nevada (Fig. 1). The lake became meromictic in this century as a consequence of irrigation practices (Kimmel *et al.* 1978). It occupies a closed-basin crater having a narrow (about 10–15 m) littoral zone and steep bottom slope. Surface area is 1.6 km^2, mean depth is 26 m, and maximum depth near the lake centre (the site of most studies described here) is 65 m. Salinity is 26 g l^{-1} in the surface layer (Kharaka *et al.* 1984), pH is 9.7, and the very sharp pycnocline–chemocline has persisted at a depth of 34.5 m throughout our studies (1981–1985). Salinity below the chemocline is 88 g l^{-1} and this vertical density gradient inhibits mixing between the lower monimolimnion and the upper mixolimnion.

The monimolimnion is permanently anoxic and has very high sulphide concentrations (ca. 7 mM), in addition to other reduced sulphur compounds (ca. 7 mM), ammonia (2.8 mM) and dissolved organic carbon (60 mg l^{-1}; Kharaka *et al.* 1984). Sediments of the pelagic zone are characterized by an overall green colour interspersed with numerous coloured laminations (see section on organic geochemistry of the pelagic

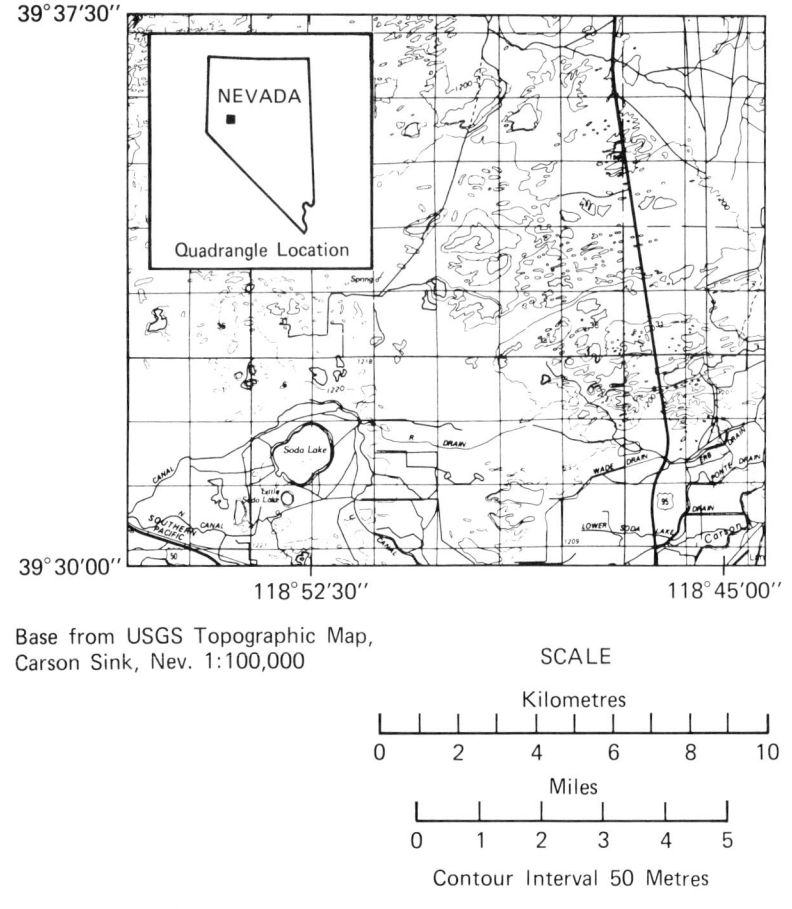

FIG. 1. A map of the location of Big Soda Lake.

sediments). Although the monimolimnion is relatively static (for example, temperature is constant at 12°C), the mixolimnion has large seasonal changes in microbial processes and the distribution of solutes (primarily nutrients) that result from seasonal mixing. From spring through autumn the mixolimnion is thermally stratified and partitioned into three distinct vertical zones (Fig. 2): the epilimnion, the aerobic hypolimnion, and the anaerobic hypolimnion. Thermocline depth is typically at 10–15 m during summer–autumn (Table 1), and the oxycline (separating the aerobic and anaerobic hypolimnion) is found at 20 m. However during winter, following surface cooling and wind mixing from winter storms, the thermocline falls almost to the depth of the chemocline (Table 1) and the mixolimnion becomes an aerobic well-mixed layer with a shallow (2–5 m) anaerobic zone. Hence the annual cycle is characterized by an alternation between thermal stratification in summer and rapid vertical mixing in winter (Fig. 2), with short transition periods between these two states.

Production and vertical fluxes of organic matter in the mixolimnion

The annual cycle of mixing controls the distribution of nutrients, dissolved gases, autotrophic bacteria, phytoplankton, and rates of production, all of which vary markedly between seasons of stratification and mixing (Cloern et al. 1983a, b; Cloern et al. 1987).

Summer

During the period of thermal stratification, dissolved inorganic nitrogen becomes depleted in the epilimnion and phytoplankton biomass is low (<30 mg m^{-2} chlorophyll a; Table 1). Bioassays confirm that phytoplankton are nitrogen limited during summer, although trace metal (Fe) limi-

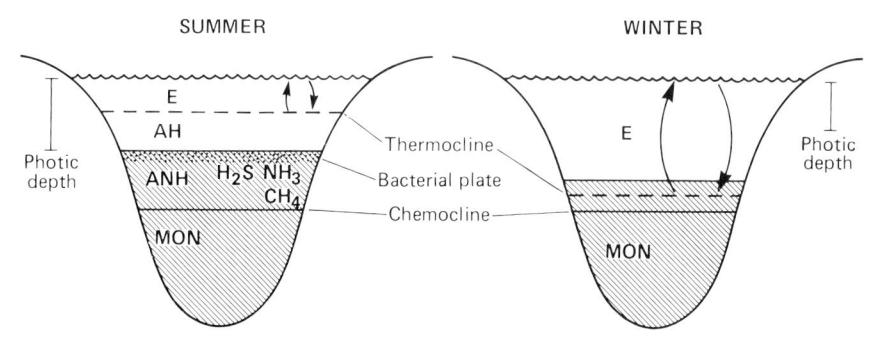

FIG. 2. Seasonal variation in the limnological properties of Big Soda Lake. E = epilimnion; AH = aerobic hypolimnion; ANH = anaerobic hypolimnion; Mon = monimolimnion.

tation is also important (Axler et al. 1978; Priscu et al. 1982). Oxygen disappears at the compensation depth of phytoplankton photosynthesis (≈ 20 m) and large gradients of other solutes coincide with the oxycline: the anaerobic hypolimnion has detectable concentrations of reduced sulphur compounds (ca. 0.2–1 mM), ammonia (2 μM), and methane (1–5 μM). Low light levels (<1% surface irradiance) penetrate to the depth of oxygen disappearance and sustain anoxygenic photosynthesis by purple sulphur bacteria (*Chromatium* sp. and *Ectothiorhodospira vacuolata*) that are confined to a dense plate (Fig. 2). Biomass of the photosynthetic bacteria is very high (500–1000 mg m^{-2} bacteriochlorophyll *a*; Table 1), and bacterial productivity (both anoxygenic photosynthesis and chemoautotrophy) exceeds phytoplankton productivity by about a factor of six. When production is dominated by bacteria, vertical fluxes of particulate organic matter (measured with sediment traps just above the chemocline; Cloern et al. 1987) are small, ranging from 45–110 mg C m^{-2} d^{-1} and 12–20 mg N m^{-2} d^{-1}, Table 1). These small vertical fluxes (about 10% of daily productivity) are presumably the result of slow sinking rates of autotrophic bacteria, and suggest that most new organic matter is mineralized in the water column before it sinks to the monimolimnion (see later section on microbial heterotrophy).

Winter

Vertical distributions of solutes and autotrophs are radically different in winter when the mixolimnion is isothermal to below 30 m. Erosion of

TABLE 1. *Properties of the Big Soda Lake mixolimnion contrasting the summer–autumn period of thermal stratification with the winter–spring period of mixing*

	Summer–Autumn	Winter–Spring
Thermocline depth (m)	10[c]–16[e]	26[g]–33[h]
DIN (= NH$_4^+$ + NO$_3^-$ + NO$_2^-$) in photic zone (μM)	< 1[c]	15[f]
Phytoplankton biomass (mg m^{-2} chl *a*)	11[a]–26[b]	100[h]–950[f]
Photosynthetic bacteria biomass (mg m^{-2} bacteriochlorophyll *a*)	500[d]–1040[b]	90[h]–130[g]
Productivity (mg C m^{-2} d^{-1}):		
Phytoplankton	90[b,c]	2800[f]
Bacteria	570[b]–620[c]	30[f]
Total	660[b]–710[c]	2830[f]
Vertical fluxes to the chemocline:		
mg C m^{-2} d^{-1}	45[e]–110[d]	400[h]
mg N m^{-2} d^{-1}	12[e]–20[d]	47[h]

[a] July 1981; [b] November 1981; [c] July 1982; [d] July 1984; [e] October 1984; [f] February 1982; [g] February 1985; [h] May 1985.
Data from Cloern et al. (1987).

the thermocline allows vertical mixing of NH_4^+ and other constituents from the anaerobic hypolimnion to the photic zone, and increased concentrations of DIN (Table 1) stimulate phytoplankton growth and lead to a bloom dominated by the pennate diatom *Nitzschia palea*. Phytoplankton biomass increases by a factor of 10–100 and phytoplankton productivity increases about thirty-fold (Table 1). Conversely the biomass and productivity of autotrophic bacteria decline. Vertical mixing disperses the plate of autotrophic bacteria, and it brings oxygen well below the photic zone (Fig. 2) so anoxygenic photosynthesis is not sustained in winter. Hence the winter period is characterized by increased community productivity (by a factor of four) that is dominated by planktonic diatoms, and greatly diminished significance of autotrophic bacteria. Sediment trap measurements following the decline of the winter bloom in 1985 show that vertical fluxes of particulate organic matter are also enhanced then. For example, vertical fluxes of particulate carbon and nitrogen increased about four-fold during May 1985, reflecting both the faster sinking rates of diatoms relative to bacteria and the higher rates of production during the bloom. As a consequence, there is seasonal variability in the sinking flux of biogenic materials (C, N, Si) to the monimolimnion, and the increased vertical flux during the winter bloom is a potential mechanism of layer formation in the pelagic sediments.

Water column microbial biomass

Profiles of microbial biomass (cell protein, adenosine triphosphate, cell counts, and turbidity) in the lake's water column during autumn and spring are shown in Fig. 3. The most apparent difference between these two seasons was the presence of a dense bacterial layer ('plate') at 21 m depth during October but not during May. The plate harboured a population of purple sulphur photosynthetic bacteria (as well as other

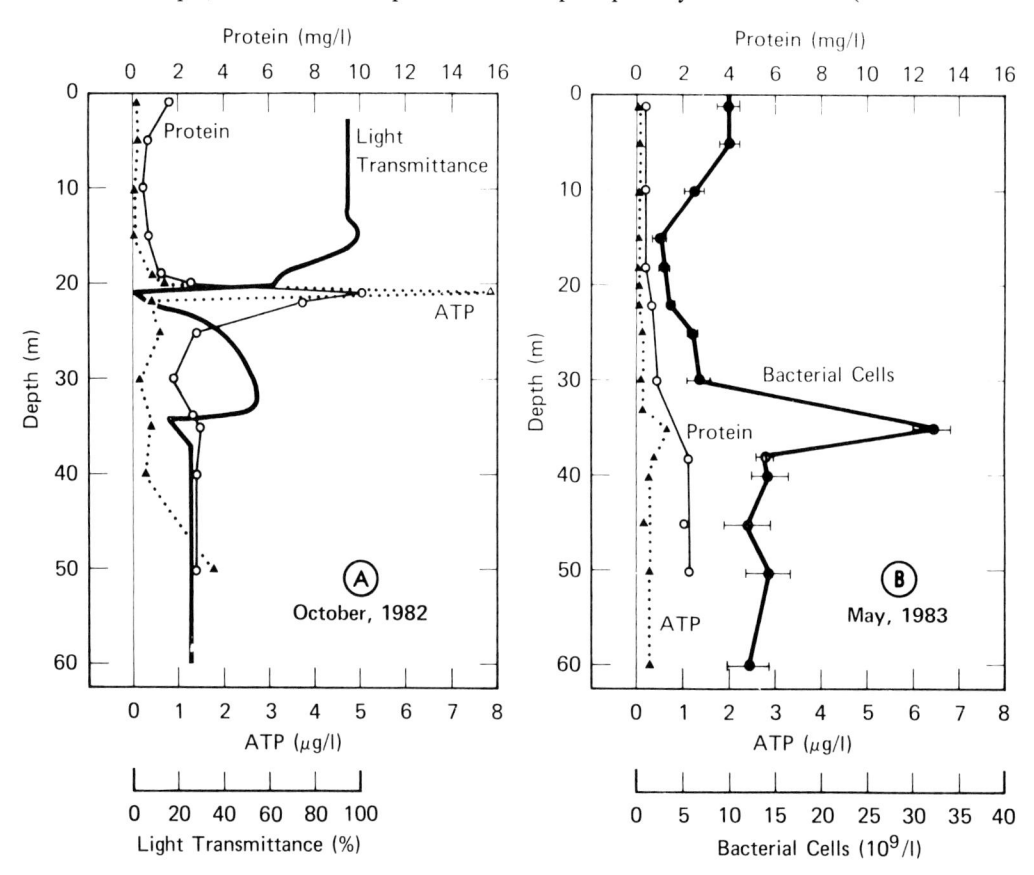

FIG. 3. Depth distributions of cell protein (○); adenosine triphosphate (△); bacterial cells (●) ± 1 standard deviation; and light transmittance determined during (A) October 1982, and (B) May 1983 in Big Soda Lake. The water column became anoxic at depths greater than 19 m in October and 16 m in May.

bacteria), which was reflected in high values of bacteriochlorophyll *a* and low values of light transmittance (Cloern *et al.* 1983a, b). Values of cell protein and adenosine triphosphate in the monimolimnion were similar for both seasons (Fig. 3). Direct counts of bacterial cells were made using acridine orange epifluorescence techniques (Harvey 1987). Bacterial density in the mixolimnion ranged from 2.5×10^9 to 10×10^9 cells l^{-1}. Cell density increased to 30×10^9 cells l^{-1} at the chemocline (35 m) and decreased to constant values (about 14×10^9 cells l^{-1}) in the water column of the monimolimnion (Fig. 3). The monimolimnion seasonally contained most of the total water column microbial biomass (ca. 60%), while lesser amounts were found in the anoxic mixolimnion (ca. 24–33%) and aerobic mixolimnion (ca. 15–17%) (Zehr *et al.* 1987).

Microbial heterotrophy

Uptake of ^{3}H-glucose by microbial assemblages in the water column of Big Soda Lake was linear at all depths studied during the course of a 5-day experiment. Incorporation rates were 20–60 times higher in the mixolimnion than in the monimolimnion (Fig. 4). Seasonal uptake profiles for either ^{14}C-glutamate or ^{3}H-thymidine exhibited maxima just beneath the photosynthetic plate (Zehr *et al.* 1987). Very low rates were usually observed in the monimolimnion. These results indicate that the lower uptake rates observed at or beneath the 35-m chemocline were due to a physiological response of the bacterial flora to the harsh chemical environment of the monimolimnion. Thus, although the monimolimnion harbours most (ca. 60%) of the water column microbial biomass, these cells exhibit little activity and are probably mainly derived from sinking out of the mixolimnion.

Chemical factors which may retard bacterial heterotrophic activity in the monimolimnion include high salinity and free sulphide levels. In contrast, the mixolimnion (especially the anoxic region) appears to be the zone in which most (ca. 90%) of the primary productivity is mineralized (Cloern *et al.* 1987; Zehr *et al.* 1987). The bulk of this activity appears to be linked to fermentative reactions, because of the low rates of sulphate reduction, methanogenesis and undetectable denitrification in this region.

Chemoautotrophy

From spring to autumn, high rates of chemoautotrophy (dark $^{14}CO_2$ fixation) were evident just beneath the oxycline (21 m). This activity coincided with the depth interval occupied by the bacterial plate (Cloern *et al.* 1983a). Addition of chemical inhibitors of nitrifying bacteria (nitrapyrin or acetylene) decreased dark CO_2 fixation by 40–80%. Addition of thiosulphate to washed cell suspensions taken from 21 m stimulated dark CO_2 fixation. These results indicated that oxidation of ammonia and reduced sulphur compounds at the oxic–anoxic interface (21 m) was responsible for the observed bacterial chemoautotrophic fixation of CO_2. Chemoautotrophy accounted for 30% of annual water column productivity (Cloern *et al.* 1983a). Estimates of rates of sulphide and ammonia oxidation, however, have not as yet been measured.

Hydrocarbons and $\delta^{13}CH_4$

Most of the methane formed in Big Soda Lake originates in the sediments of the monimolimnion at a depth of more than 1 m below the lakebed. However, geochemical evidence also suggests that small quantities of methane are produced by bacteria in the anoxic portion of the water column (Oremland & Des Marais 1983). Methane concentrations beneath 1 m in monimolimnion sediments were as high as 418 µmol kg^{-1} and dissolved methane concentrations in the monimolimnion waters were 50–60 µM. Methane concentrations decreased markedly (about ten-fold) above the chemocline and again above the oxycline. Surface waters were supersaturated

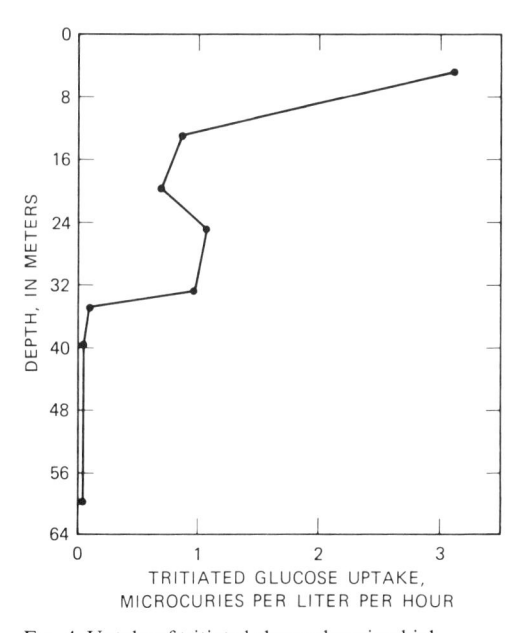

FIG. 4. Uptake of tritiated glucose by microbial assemblages in the water column of Big Soda Lake. Experiments were performed during May 1983.

with respect to the atmosphere and contained 0.2 μM methane. Methane efflux from the lake's surface was estimated to be about 36 μmol m^{-2} d^{-1} (Iversen et al. 1987). Ethane, propane and both normal and iso-butane were abundant in the monimolimnion and displayed maximum concentrations of 260, 80, 22 and 23 nM, respectively. A bacterial origin either in the monimolimnion water or surficial sediments was indicated for these gases (Oremland & Des Marais 1983), perhaps via mechanisms similar to those in estuarine sediments (Oremland 1981; Vogel et al. 1982).

Isotopically 'light' values of $\delta^{13}CH_4$ (−70 to −74‰) were encountered in the deeper (> 1 m) sediments of the monimolimnion. However, values became enriched in ^{13}C at the surface of these sediments (−55‰) and in the monimolimnion water column (−55 to −60‰). A further enrichment of water column $\delta^{13}CH_4$ values in ^{13}C was evident in the anoxic monimolimnion. In this zone, 92% of the samples had values between −48 and −20‰. Bacterial processes were cited as the cause of the ^{13}C enrichment (Oremland & Des Marais 1983). These processes include anaerobic methane oxidation and methanogenesis from isotopically 'heavy' substrates. A typical profile of $\delta^{13}CH_4$ is shown in Fig. 5.

Methanogenesis

Methanogenic activity was detected both in the monimolimnion sediments (Oremland et al. 1982a) and anoxic mixolimnion waters (Oremland & Des Marais 1983). Methanogenic substrates in these and other high sulphate

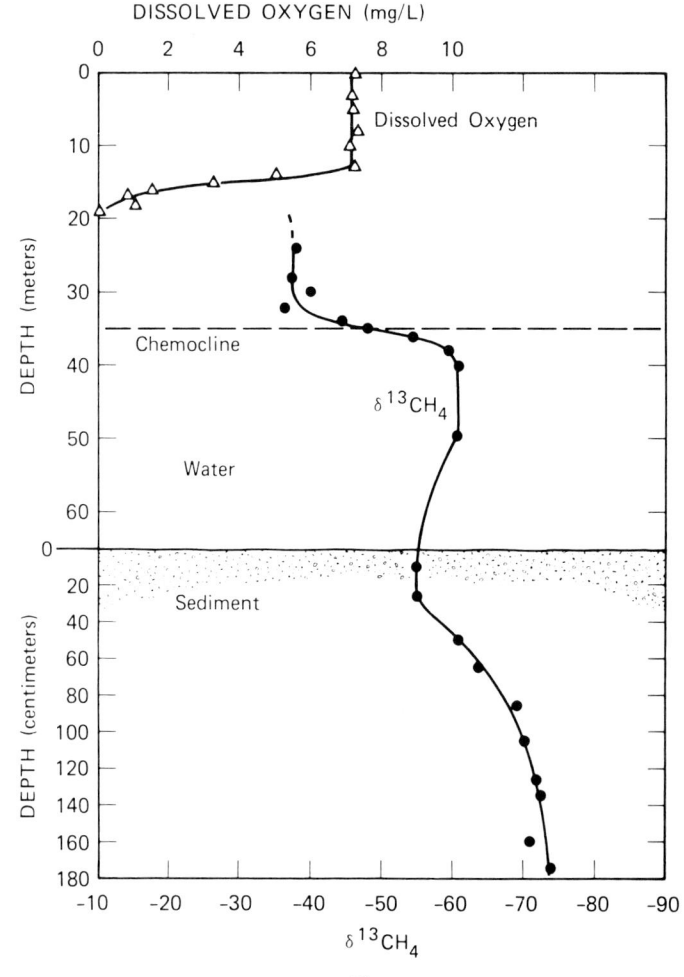

FIG. 5. Stable carbon isotopic composition of methane ($\delta^{13}CH_4$) determined in the anaerobic water column of Big Soda Lake during October 1982. Data is from Oremland & Des Marais (1983).

environments appear to be compounds such as methanol and methylamines rather than acetate or hydrogen (Oremland et al. 1982a, b; Oremland & Polcin 1982). This phenomenon is caused by the channelling of acetate and hydrogen to the metabolically more efficient sulphate-reducing bacteria. However, because sulphate reducers have relatively little affinity for methanol or methylated amines, methanogenic bacteria can metabolize these compounds, thereby allowing for both methanogenesis and sulphate reduction to occur simultaneously. Recently, dimethysulphide was identified as another possible methane precursor (Kiene et al. 1986).

In a preliminary study of methanogenesis in the water column of Big Soda Lake, lake water from 40 m depth was incubated in the laboratory. The collected water samples were stored for 3 months prior to the experiment at 12°C in completely filled, 4-l glass bottles fitted with ground glass stoppers. Water was degassed of methane and dispensed into serum bottles which contained a small headspace of N_2 (Iversen et al. 1987).

The results for some samples are shown in Fig. 6. Methanogenesis occurred in the unsupplemented bottles and was stimulated by methanol. Water containing BES, a specific inhibitor of methanogens (Gunsalus et al. 1978), formed much less of the gas, as did water supplemented with Ni^{2+}. Nickel addition caused an initial stimulation (< 10 days), after which time no further methane was produced. The results for all conditions after 83 days incubation are shown in Table 2. Only methanol and trimethylamine enhanced methanogenesis, while BES and Ni^{2+} inhibited the process.

The stimulation of methanogenesis by methanol and trimethylamine, as well as the absence of stimulation by acetate, was consistent with earlier observations with sediment slurries (Oremland et al. 1982a), as was the inhibition achieved by BES. Methanogenic bacteria have nutritional requirements for Ni^{2+}, Co^{2+}, and Fe^{2+} (Daniels et al. 1984). Because the chemistry of the lake water is such that these metals are probably present only at extremely low levels (Kharaka et al. 1984), these substances were added in an attempt to see if methanogenic bacteria in the monimolimnion were limited by the availability of trace metals. However, with the exception of slight initial stimulation by Co^{2+} and Ni^{2+}, no long-term enhancement was observed. Kaolinite was added as a control to determine if the presence of increased surface area would enhance activity, since addition of metals resulted in the formation of sulphide precipitates. Since no obvious enhancement occurred, additional surface area did not enhance methanogenesis.

Methanogenic activity in the water column of

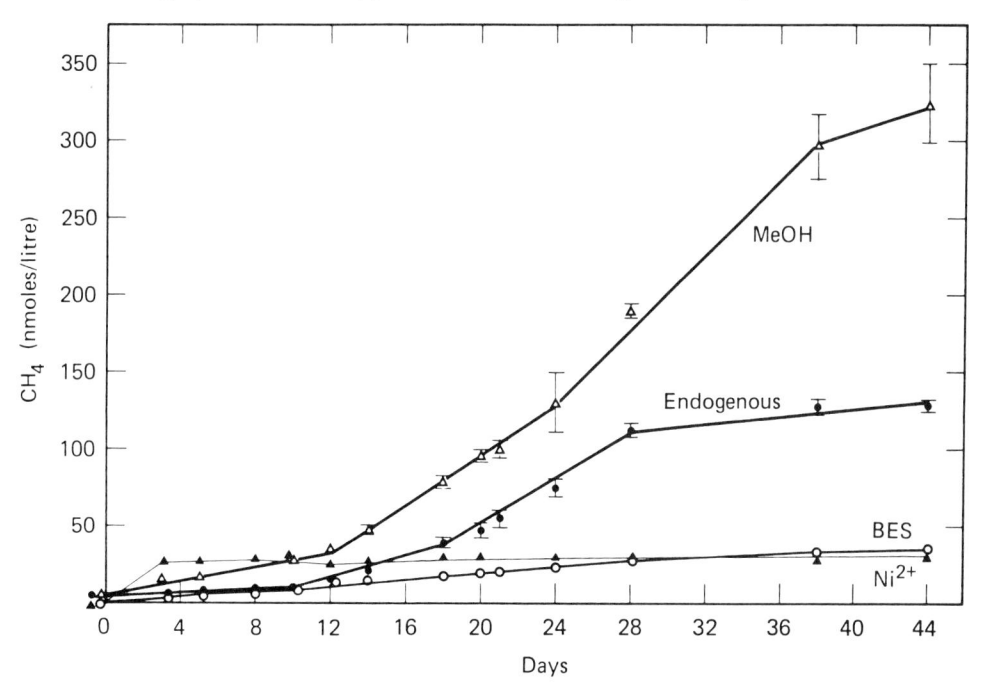

FIG. 6. Methane production by monimolimnion water. MeOH = methanol; BES = 2-bromoethanesulfonic acid. Results represent the mean of three bottles and bars = 1 standard deviation.

TABLE 2. *Production of methane by monimolimnion water (40 m) after 83 days of a laboratory incubation (12° C)*

Addition	Concentration	Methane (nmol l^{-1})
None	—	794 (84)
BES	5 mM	79 (11)
Methanol	50 μM	5129 (1331)
Trimethylamine	50 μM	6325 (2122)[a]
Acetate	50 μM	897 —[b]
FeCl$_2$ H$_2$O	100 mg l^{-1}	938 (116)
NiCl$_2$ 6H$_2$O	120 mg l^{-1}	30 (6.6)
CoCl$_2$ 6H$_2$O	100 mg l^{-1}	915 (127)
Kaolinite	1 g l^{-1}	1119 (287)

Values represent the mean of the three samples with a standard deviation indicated within brackets.
[a] represents average and range of two samples.
[b] represents one sample (due to breakage). After 20 days, triplicate acetate-amended samples were 49.5 ± 5.5 nmol l^{-1} and methanol-amended were 93.9 ± 4.0 nmol l^{-1}.

Big Soda Lake was measured with freshly recovered samples during October 1983 and July 1984 (Iversen *et al.* 1987). Methanogenic activity ranged between 0.1 and 1.0 nmol l^{-1} d^{-1} in the anaerobic mixolimnion and between 1.6 and 12 nmol l^{-1} d^{-1} in the monimolimnion. The experiments conducted under *in situ* conditions differed from the laboratory experiments in two respects. First, no enhancement of methane formation was observed when 40-m water samples were supplemented with either methanol or trimethylamine (each 50 μM). This suggests that these substances were not limiting methanogenic bacteria in the field experiments, although they were in the preliminary, long-term laboratory experiments. Second, although 5 mM BES inhibited methanogenesis in the preliminary, long-term laboratory experiments (Fig. 6), no inhibition was achieved using 10 mM BES at any of the eleven depths tested during October 1983. In the July 1984 experiments 37 mM BES caused a partial inhibition in mixolimnion samples, but did not inhibit monimolimnion samples (Fig. 7). These mixed results with inhibitor concentrations may have been caused by differential susceptibility by the methanogenic flora to BES. For example, acetoclastic methanogenesis in digestors was blocked by 1 mM BES, but that formed via H$_2$ reduction of CO$_2$ required 50 mM (Zinder *et al.* 1984). The fact that 37 mM BES worked in the mixolimnion, but not in the monimolimnion (Fig. 7) indicates that a situation of ineffective concentrations occurred in these experiments. Similar difficulties with BES (1.4–14 mM) were

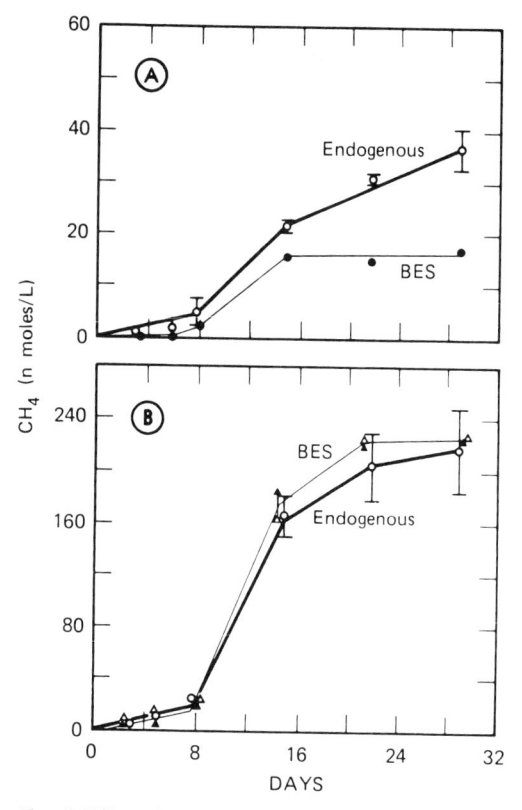

FIG. 7. Effect of BES (37 mM) on methanogenic activity in water samples taken from (A) the mixolimnion (28 m); and (B) the monimolimnion (40 m) of Big Soda Lake. Experiments were conducted during May 1984.

evident in some of the earlier reported Big Soda Lake sediment slurry-incubations (Oremland *et al.* 1982a).

Methane oxidation

Incubation of lake water under *in situ* conditions with ^{14}CH$_4$ demonstrated the production of ^{14}CO$_2$ with time (Iversen *et al.* 1987). Rates in the aerobic mixolimnion were very low (0.2–1.3 nmol l^{-1} d^{-1}) and accounted for only 0.04% of the bacterial methane oxidation occurring in the water column. Therefore, anaerobic methane oxidation was the process which accounted for over 99% of the water column methane consumption. Rates were first order with respect to methane and were higher in the monimolimnion (49–85 nmol l^{-1} d^{-1}) than in the anoxic mixolimnion (2–6 nmol l^{-1} d^{-1}). Rates of anaerobic oxidation always exceeded those of production, thus indicating a net consumption of the gas occurred in the anoxic water column. Anaerobic oxidation could be blocked by filter-sterilization, but was uninfluenced by Na_2WO_4 (an inhibitor of sulphate-reducing bacteria). Although a net consumption of methane occurred in the lake's anoxic zone, the daily rate of oxidation was about 1000-fold lower than the ambient levels of dissolved methane. Thus, a decrease in the dissolved methane content of incubated water samples could not be observed over a 97 h time period. However, anaerobic methane oxidation was probably responsible for the isotopically 'heavy' CH_4 detected in the anoxic mixolimnion (Fig. 5).

Sulphate reduction

Sulphate reduction is the apparent source of the high sulphide levels in the anoxic mixolimnion (ca. 0.7 mM) and monimolimnion (ca. 7 mM) as well as for the observed δ^{34}sulphate values in the monimolimnion ($-34‰$), mixolimnion ($-6‰$) and monimolimnion δ^{34}sulphide values ($-26‰$) (Kharaka et al. 1984). Estimates of the rates of water column sulphate reduction were achieved by performing in situ incubations with ^{35}S-sulphate (Smith & Oremland 1987). Rates of $^{35}S^{2-}$ production were linear over 5 days. Five depths sampled in the monimolimnion (35, 38, 40, 50 and 60 m) all had seasonal rates between 0.9 and 3 µmol SO_4^{2-} reduced l^{-1} d^{-1}. This agrees well with the overall estimate of 6.6 µmol SO_4^{2-} reduced l^{-1} d^{-1} deduced by analysis of past and current chemical data (Kharaka et al. 1984). Sulphate reduction rates decreased to ca. 1.4 and ca. 0.6 µmol l^{-1} d^{-1} at 33 and 30 m respectively. Rates in the anoxic mixolimnion were below 0.025 µmol l^{-1} d^{-1}. Annual monimolimnion sulphate reduction was estimated to be 9–29 mmol m^{-1} d^{-1} (Smith & Oremland 1987). This extrapolated to a monimolimnion sulphate turnover of 60–270 years, not accounting for oxidation of sulphide. Equivalent carbon mineralized was 210–700 g m^{-2} yr^{-1}, which could account for most, if not all, of the mixolimnion productivity. The apparent discrepancy between the carbon mineralized by monimolimnion sulphate reduction and that occurring via fermentative reactions in the mixolimnion has not been fully reconciled. Possibilities include downward flux of additional carbon from the littoral zone and the steeply-graded mixolimnion sediments. Sulphate reduction in monimolimnion waters was stimulated two-fold by addition of FeS (in contrast to methanogenesis) or H_2. Acetate or lactate caused only 28 and 16% enhancement, respectively. No stimulation occurred with MnS, methanol or kaolinite. Sulphate reduction was inhibited by about 67% when samples were incubated with Na_2WO_4 (20 mM). Rates of sulphate reduction were not influenced by either the removal or addition of methane to the samples. Results indicate that sulphate reduction is not directly coupled to anaerobic methane oxidation.

Nitrogen cycle

Attempts were made to measure nitrogen fixation and denitrification in the mixolimnion of the lake using the acetylene reduction and acetylene block assays, respectively. Seven depths (1, 5, 10, 15, 20, 25 and 30 m) were routinely assayed over the course of the four seasons during 1981–1983. Bottles containing 200 ml of lake water and ca. 50 ml N_2 (or air) gas phase were incubated in situ. Acetylene (10 ml) was added to the gas phase of all but the control bottles. Selected bottles were amended with $NaNO_3$ (1 mM) and/or glucose (1 g l^{-1}) to enhance denitrification, or with NH_4Cl (1 mM) to inhibit N_2 fixation. After 24 h incubation, headspace analyses revealed only background level of C_2H_4 (<0.1 nmol ml^{-1}) and N_2O (< 0.6 pmol ml^{-1}) in all the bottles, including the controls lacking C_2H_2. Therefore neither denitrification nor N_2 fixation were detected in the mixolimnion during the four seasonal sampling intervals. In addition, amendment of anoxic mixolimnion water with nitrate and/or glucose did not elicit any detectable denitrification (N_2O accumulation). It is preliminarily concluded that these two processes do not occur in the lake's water column. The aerobic mixolimnion of Big Soda Lake is severely nitrogen-limited for much of the year (Priscu et al. 1982; Cloern et al. 1983a).

Extremely high rates of nitrogen fixation (~ 100 µmol m^{-2} h^{-1}) were detected in the littoral zone. Fixation was caused by cyanobacterial epiphytes (Anabaena sp.) colonizing the abundant macrophyte (Ruppia sp.) present in the littoral zone. These cyanobacterial aggregates also evolved H_2 (6.8 µmol m^{-2} h^{-1}), however this activity was caused by fermentative bacteria living within the matrix and occurred only in the dark (Oremland 1983). Other littoral zone sites of nitrogen fixation include anaerobic, shallow 'cul-de-sacs' which are infested with purple sulphur photosynthetic bacteria. Areal rates of fixation (about 100 µmol m^{-2} h^{-1}) were equivalent to the high rates exhibited by the cyanobacteria (J. Duff et al. in prep.). The extent to which this littoral-zone-fixed nitrogen is transported to the pelagic region of the lake is not yet known. In addition, it is not entirely clear why the mixolimnion is devoid of cyanobacterial N_2-fixation, although possibly a trace element limitation could exist.

Active denitrification was found to occur in some littoral zone sediments. Sediments of the *Ruppia* beds had the greatest activity; however high concentrations of added nitrate ions (1 mM) were required to elicit such a response. Thus, the 'potential' for denitrification was about 100 µmol N_2 m^{-2} h^{-1}, which was equivalent to inputs via nitrogen fixation. However, the apparent K_m for nitrate was estimated to be 87 µM. Since nitrate + nitrite levels in the interstitial waters are 5 µM l^{-1} (or less), it is clear that *in situ* rates of denitrification are well below the 'potential' rate estimate (S. Paulsen *et al.* unpublished data). Estimates of littoral zone denitrification were also made using a novel N_2O reductase assay (Miller *et al.* 1986). A rate of 6 µmol N_2 m^{-2} h^{-1} was arrived at using this technique.

Nitrification occurs at the oxic–anoxic interface (21 m) of the mixolimnion (Cloern *et al.* 1983a). However, ammonia oxidation rates have as yet not been measured. In addition, a dissolved N_2O maximum that usually accompanies the zone of water column nitrification (McElroy *et al.* 1978) was not evident in Big Soda Lake (Fig. 8). It is possible that N_2O is not released by nitrifying bacteria under the highly alkaline conditions (pH = 9.7) characteristic of the lake.

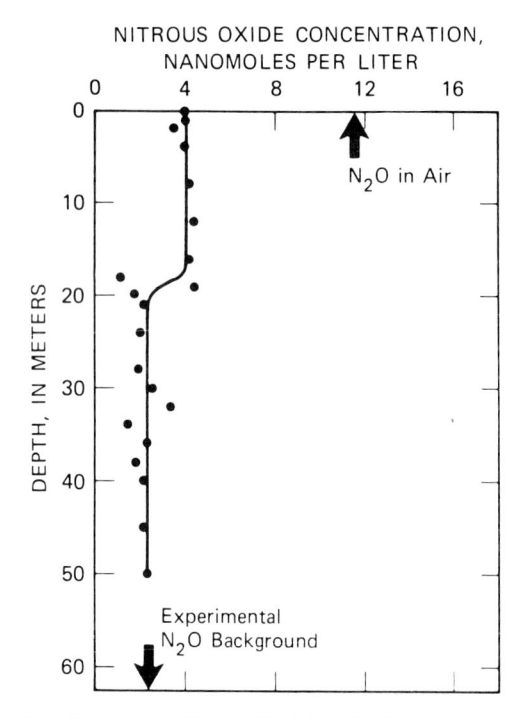

FIG. 8. Dissolved N_2O profile determined during November 1981. At this time of the year, the water column became anoxic at depths greater than 21 m. Arrows indicate the N_2O concentrations in the atmosphere (top of graph) as well as values of N_2O indistinguishable from background contamination (bottom).

Other processes

Anaerobic decomposition of oxalate was studied in the sediments of both the littoral and pelagic zones of Big Soda Lake (Smith & Oremland 1983). Although the total oxalate content of these sediments was similar (ca. 100 µM), degradation rates were about fifty-fold higher in the littoral zone. This lower activity in the monimolimnion sediments may also be a response to the harsh chemistry of the bottom waters.

Uprooted *Ruppia* sp. plants along with their associated decomposing cyanobacterial epiphytes eventually are forced into shallow, littoral-zone 'cul-de-sacs'. This rotting plant material can be as thick as 50 cm deep. Bubbles rich in methane (ca. 40%) are associated with this plant matter, and both ethane and propane are present as trace constituents. The bubbles have a distinct, obnoxious mercaptan odour. Anaerobic incubation of this material results in the formation of methane and traces of ethane. Methane formation was stimulated by addition of methanethiol, dimethylsulphide and dimethyldisulphide (Kiene *et al.* 1986), while stimulation of ethane production (traces) was achieved with diethylsulphide (Fig. 9) or ethanethiol. Because the formation of these gases was blocked by BES, methanogenic bacteria are responsible for the conversion of methylated and ethylated sulphur gases to methane and ethane (Kiene *et al.* 1986; R. Oremland *et al.* in prep.).

Organic geochemistry of the pelagic sediments

General characteristics

Only preliminary investigations have been conducted upon the monimolimnion sediments; nonetheless several interesting aspects are apparent. These sediments are unusual in that they have a distinctively green colour (from preserved chlorophylls and their degradation products). Usually, sediments in a highly sulphidic environment are black due to the presence of amorphous iron sulphides. However, because of the low iron content of the lake, iron sulphides are not present in sufficient abundance to mask the 'organic' colouration. Pennate diatoms, derived from the winter bloom (see pp. 61–62) are abundant in the upper 30 cm. Multicoloured layers are also evident within the sediments and occur from the surface to at least 2 m depth (Plate 1). A very low

PLATE 1. Core from monimolimnion (depth interval = 95–150 cm) used in the hydrous pyrolysis experiments.

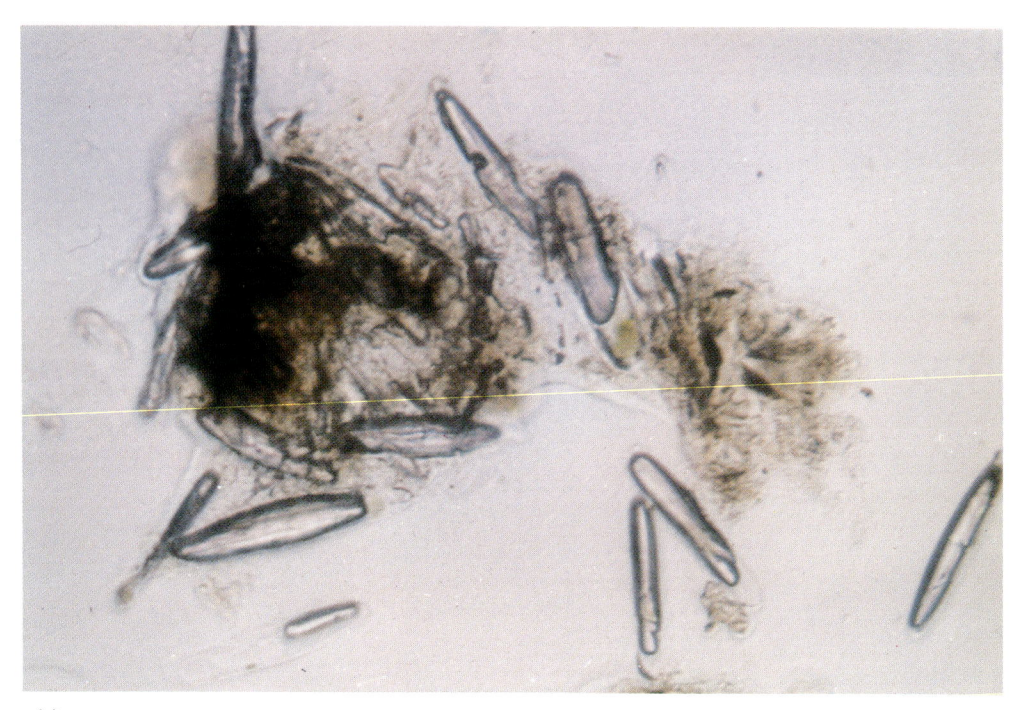

(a)

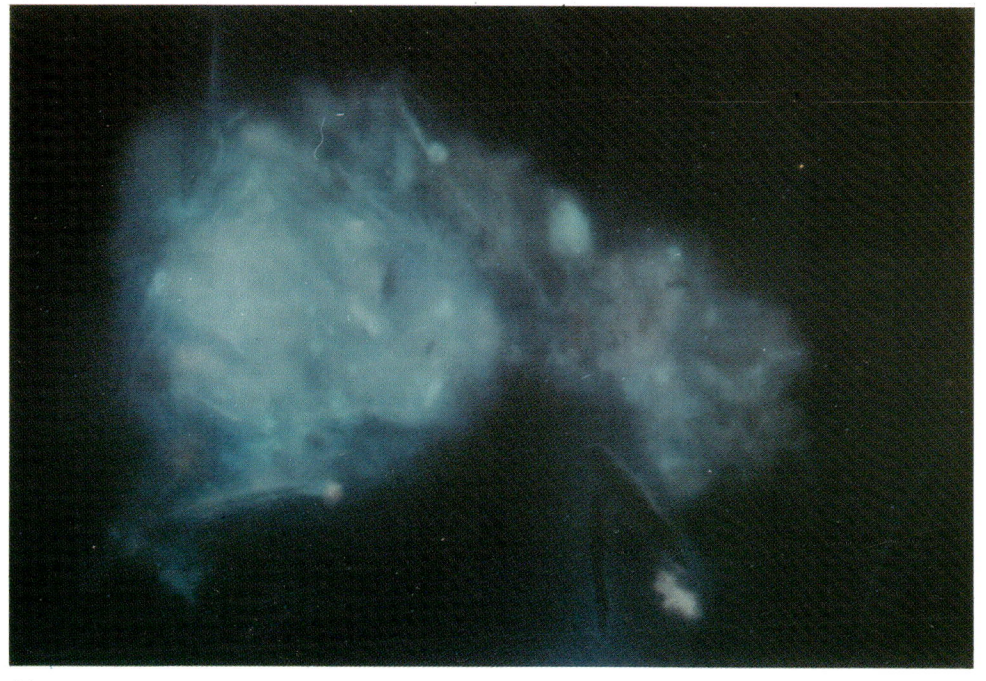

(b)

PLATE 2. Particulate organic matter in untreated lake sediments (scale: 1 cm = 25µ). Note pennate diatoms and amorphous organic matter under transmitted light (a) and UV light (b).

(a)

(b)

PLATE 3. Chitinous remains of zooplankton and sheet-like amorphous organic matter after removal of mineral matter by acid treatment (scale: 1 cm = 25µ). (a) transmitted light; (b) UV light.

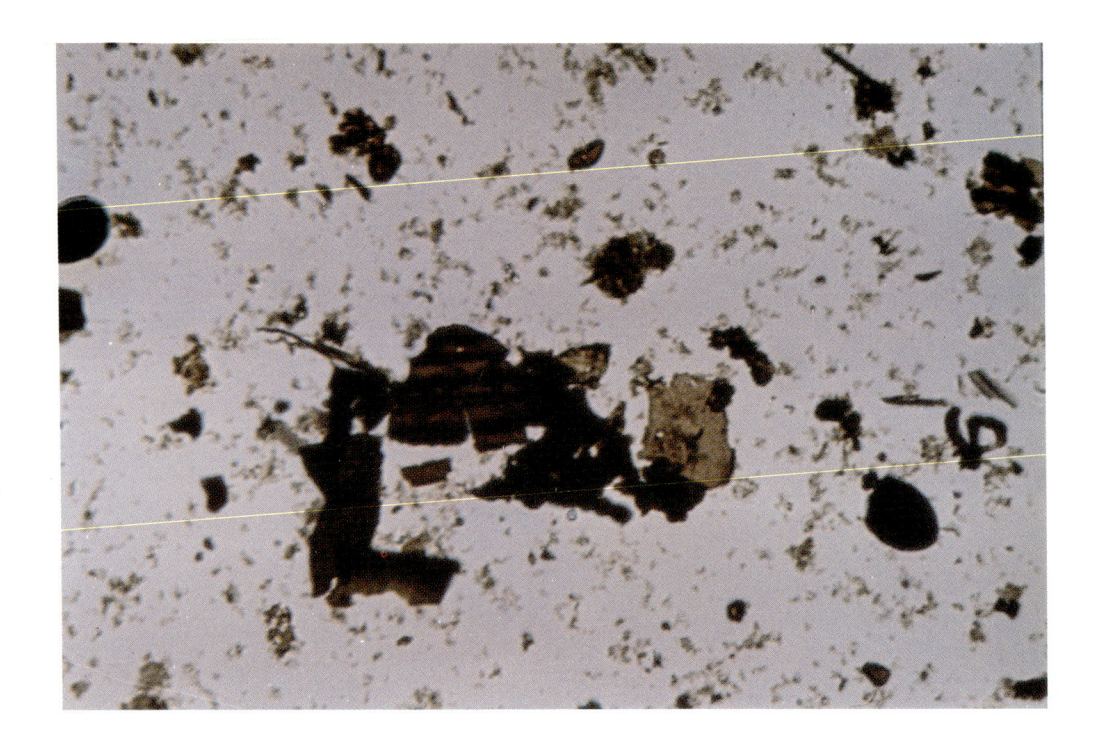

PLATE 4. Lake sediments after hydrous pyrolysis, solvent extraction and mineral matter removal (1 cm = 25μ). The relative amount of woody plant fragments and possibly altered chitinous material appear to increase in samples after hydrous pyrolysis. The fine grained nature of the amorphous material may be a result of thermal alteration. No fluorescence was observed after hydrous pyrolysis.

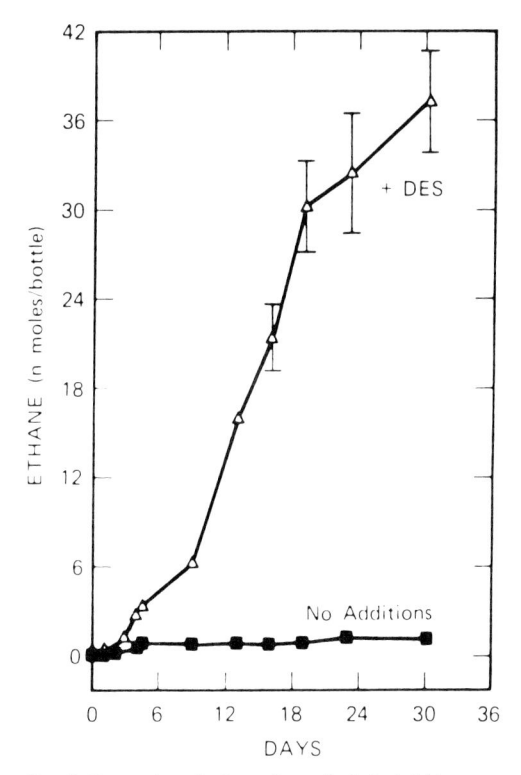

FIG. 9. Formation of ethane from diethylsulphide (10 mM) in Big Soda Lake littoral zone sediments. Results represent the mean of three bottles and bars indicate ±1 standard deviation.

pre-meromixis sedimentation rate (ca. 0.25 mm yr^{-1}) was indicated by performing ^{14}C-dating on core organic matter (Fig. 10). Stable carbon isotope values (δ^{13}C) of total organic carbon as well as chlorophyll degradation products exhibit a downcore shift from relatively heavy values at the surface to signficantly lighter values at 34-cm depth (Fig. 11). Because the surrounding watershed excluded significant input from terrestrial plant material (a fact reinforced by the organic chemistry data below), such isotopic variations must reflect changes in processes internal to the lake. Some possibilities include: (1) changes in phytoplankton communities and associated preferential isotopic abundance during carbon fixation (e.g., Wong & Sackett 1978); (2) alteration in phytoplankton isotope abundance induced by changes in epilimnetic temperature (e.g., Degens et al. 1968) or salinity; or (3) other possible effects (Deines 1980).

Nature of the organic matter (Z. Sofer)

The bottom sediments in the lake are highly reducing resulting in preservation of abundant organic matter. A portion of the sediment (55 cm representing the 95–150-cm interval below the sediment–water interface) was homogenized, freeze-dried, and studied microscopically and chemically. Based on the microscopical study (Plate 2) the organic matter is primarily amorphous; abundant pennate diatom frustules are also present and indicate that, along with bacteria, they are a major contributor to the organic matter. Table 3 shows that the dried sediment contains 2.65% organic carbon. After removal of the mineral matter from the sediment (by treatment with HCl and HF), the isolated organic matter consists of abundant amorphous material, spores, pollen, woody plant fragments and large, tan fragments, possibly the chitinous remains of crustaceans (Plate 3).

The freeze-dried sediment was soxhlet-extracted with chloroform: methanol (9:1) and the extractable matter was quantitatively separated into C_{15+} aliphatics, aromatics, heteroatomic (NSO) compounds, and asphaltenes (i.e., large heteroatomic compounds insoluble in cold pentane). As seen in Table 3, about 0.6% of the organic matter is soluble, and most of it is in the

TABLE 3. *Organic and isotopic composition of Big Soda Lake sediments before and after hydrous pyrolysis*

	% Organic carbon		C_{15+} Organic extract (ppm)				
	Total	Insoluble	Alkanes	Aromatics	NSO	Asphaltenes	Total C_{15+}
Organic composition							
Before HP*	2.65	2.04	32	278	1601	6937	8847
After HP	1.63	0.28	2253	2205	4458	34421	43337
Carbon isotopic composition (δ^{13}C–PDB ‰)							
Before HP	−23.80	−22.92	−26.48	−26.26	−26.26	−24.98	
After HP		−24.34	−25.76	−24.97	−24.42	−24.13	

*HP = hydrous pyrolysis.

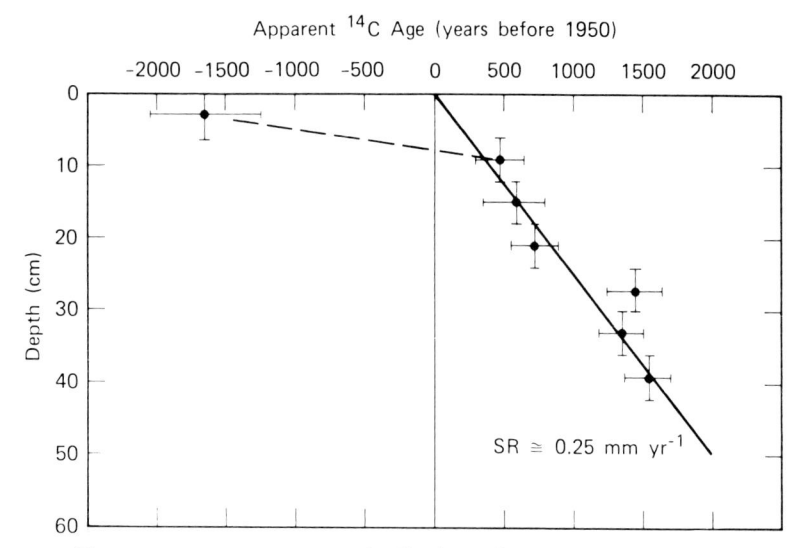

FIG. 10. Apparent ^{14}C age of organic carbon in monimolimnion sediment core.

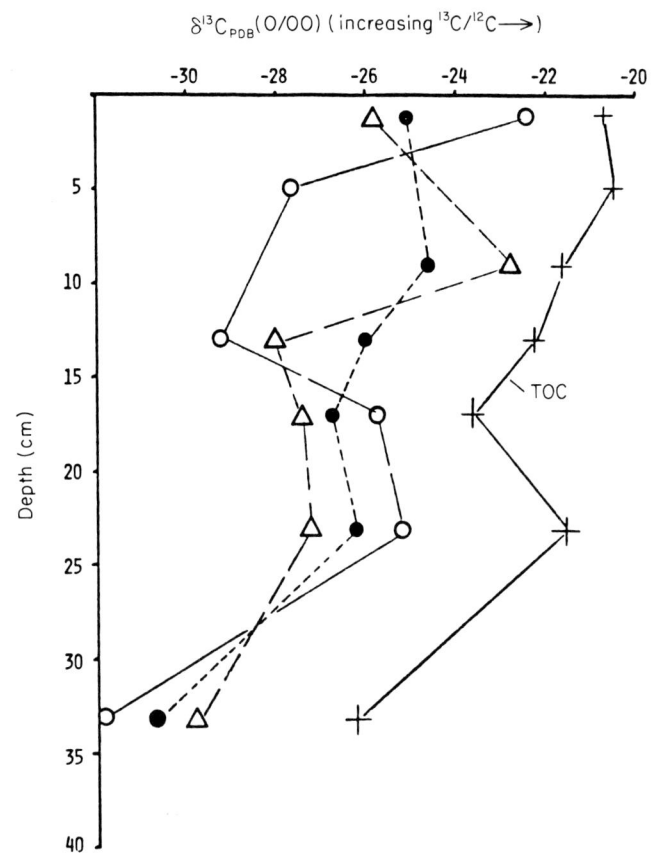

FIG. 11. Stable carbon isotope values (δ^{13}C) of total organic carbon (+), phaeophytins (△, ○) and phaeophorbides (○) in sediment core. Compounds were extracted with acetone/methanol solvent and separated by HPLC.

form of asphaltenes. In addition, the freeze-dried sediments were subjected to a mild thermal alteration by means of hydrous pyrolysis (300°C for 72 h, as described by Schiefelbein, 1982), in order to evaluate the 'petroleum-like' material that such sediments would generate.

After hydrous pyrolysis the total organic carbon in the lake sediments was reduced from 2.65 to 1.63% due to the formation of carbon dioxide, and hydrocarbon gases which are not accounted for in the total organic carbon. The resulting insoluble organic matter (kerogen) is reduced, as shown in Table 3, to approximately 1/7 of the original amount (from 2% to 0.3%). The thermally altered 'kerogen' (Plate 4) appears to be composed mainly of woody matter. The observed decrease in the amount of the amorphous and chitinous material suggests that these materials were transformed into organic soluble matter. Hydrous pyrolysis of the sediment resulted in a five-fold increase in the extractable C_{15+} compounds relative to the precursor material (see Table 3: aliphatics increased 70×, aromatics 8×, NSO 2.8×, and asphaltenes 5×). The amount of extractable matter relative to total organic carbon after hydrous pyrolysis is about 4.3-fold larger then what was observed in sediments from Mud Lake (Florida) where the dominant organic matter is peat derived from higher plant fragments (Schiefelbein 1982). This suggests that organic matter derived from vascular plants contributes only a small fraction to the total organic carbon in Big Soda Lake.

The gas chromatogram of the aliphatic fraction in the lake sediment (Fig. 12a) shows a very strong predominance of the odd numbered higher plant *n*-alkanes (Eglington & Hamilton 1963) indicating some contribution of land-derived organic matter. The gas chromatogram of the aliphatic fraction after thermal alteration (Fig. 12b), shows a waxy *n*-alkane distribution maximized at $n\text{-}C_{24}$ to $n\text{-}C_{25}$ and a slight even-over-odd *n*-alkane predominance has often been observed in carbonate environments (Dembicki *et al.* 1976); however, the waxy *n*-alkane distribution was unexpected in a depositional environment that is dominated by aquatic organisms such as bacteria and algae. A low Pr/Ph ratio is in agreement with a highly reducing depositional environment (Powell & McKirdy 1973). Another interesting feature in the aliphatic gas chromatogram of the hydrous pyrolysed sediment (Fig. 12b) is the presence of a homologous series of C_{21} to C_{35} branched alkanes which were positively identified by GC/MS mass spectra (Fig. 13) as 10-methylalkanes. The origin of these branched alkales (i.e., the organism in which the precursors are synthesized) is not yet clear.

GC/MS analysis of the di- and triterpanes (m/z=191) and steranes (m/z=217) in the ali-

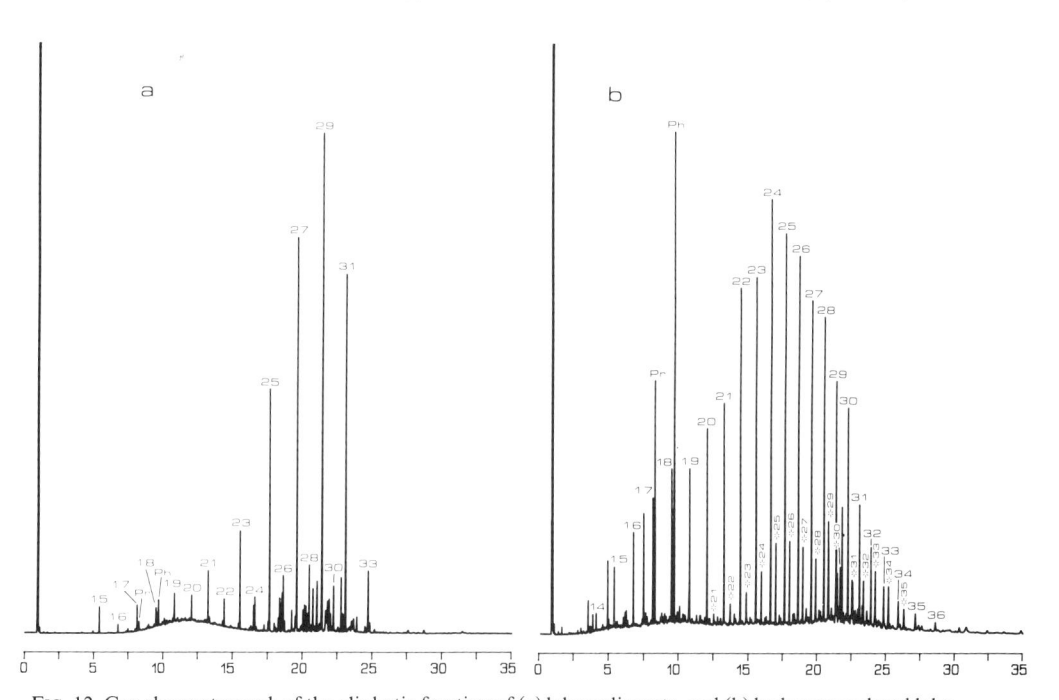

FIG. 12. Gas chromatograph of the aliphatic fraction of (a) lake sediments, and (b) hydrous pyrolysed lake sediments. Peaks with asterisk in (b) indicate 10-methylalkanes (see Fig. 5).

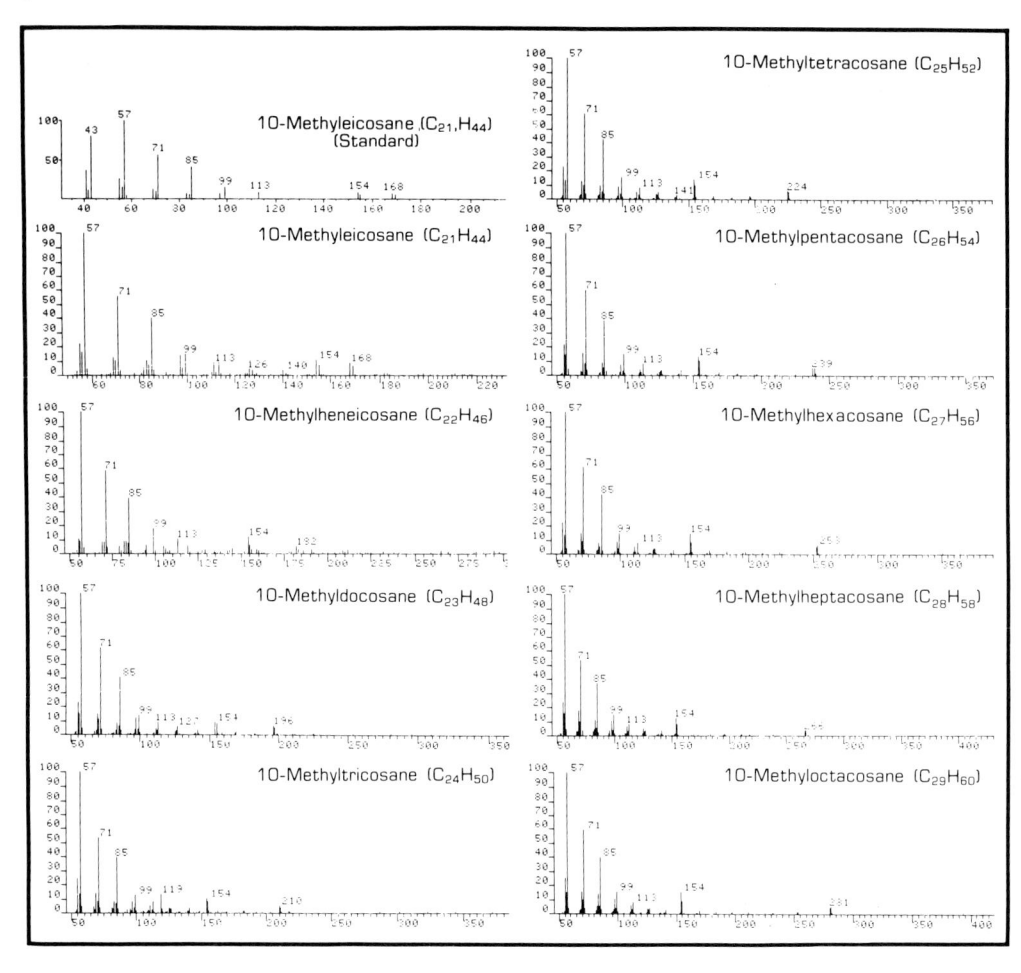

FIG. 13. Mass spectra of 10-methylalkanes (peaks with asterisks in Fig. 12b).

phatic fraction (Fig. 14) of the lake sediments shows only few of the compounds commonly observed in more mature sediments. The less thermally stable terpenoid compounds (such as the 17β(H)C$_{30}$ triterpane and other unidentified compounds) are diminished after hydrous pyrolysis; however, under the conditions of this experiment, the remaining terpanes (triterpanes and steranes) are still typical of low maturity. Because C$_{28}$ sterols are predominant in diatoms (Lee et al. 1980) it is very likely that the 5α(H) and 5β(H)C$_{28}$ steranes seen in the GC/MS patterns represent the abundant diatoms observed in the Soda Lake sediments. Small amounts of gammacereane were identified by mass spectra analysis of the pyrolizate. The diterpanes after hydrous pyrolysis show a predominance of the C$_{20}$ compounds. This predominance has also been observed in sediments containing non-marine organic matter as well as in non-marine oils (unpublished data).

The carbon isotopic composition of the various fractions (aliphatic, aromatic, etc.) before pyrolysis range from −26.5 to −24.9‰ and −25.8 to −24.1‰ after pyrolysis (Table 3). The isotopic composition of the total untreated organic matter is −23.8‰. This value is more positive (by about 4.5‰) than values reported for other recent non-marine organic matter (e.g., Schiefelbein 1982; Scalan & Morgan 1970) and again indicates that higher plant organic matter contributes only a small fraction to the total organic carbon. Mass balance calculations that include the TOC and the insoluble organic matter before hydrous pyrolysis indicate that the extractable organic matter should have an isotopic composition of −26.7‰. This is in good agreement with the measured isotopic composition of the individual

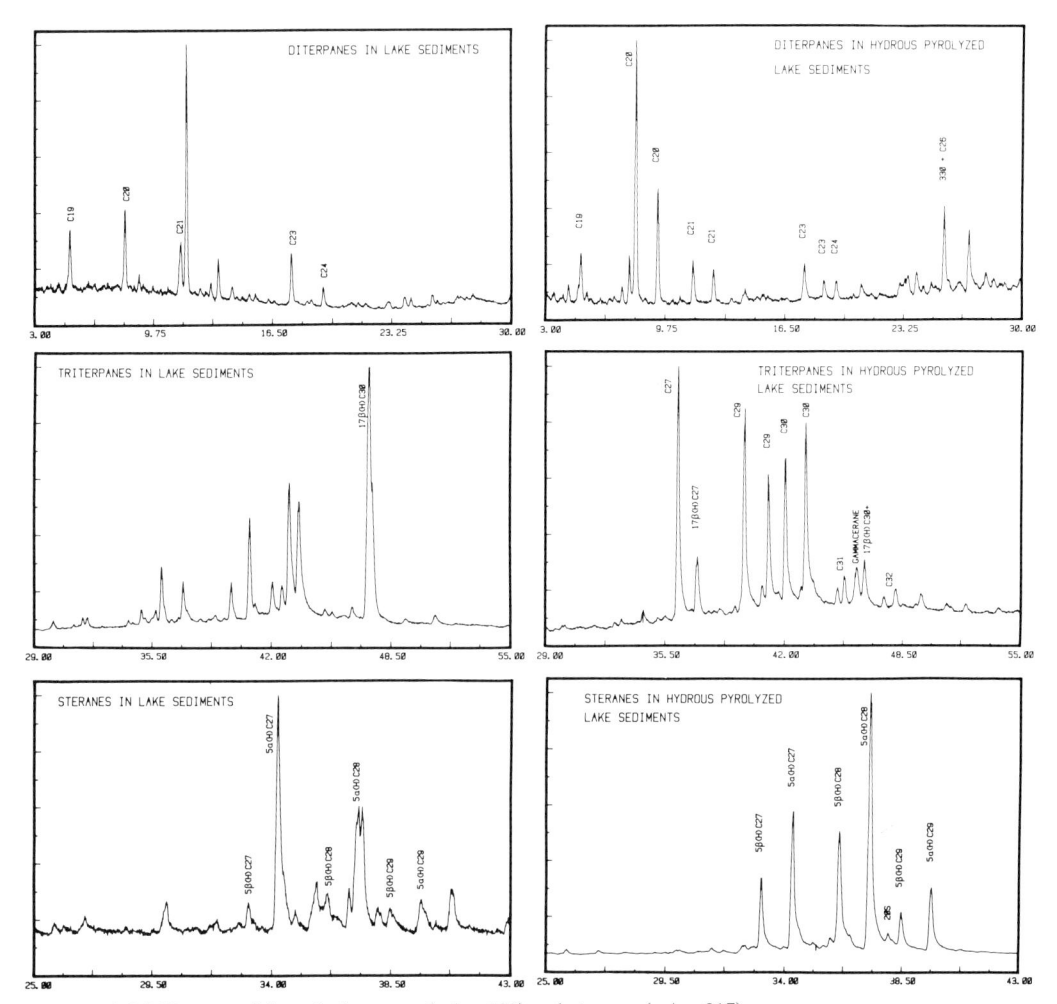

FIG. 14. GC/MS traces of di- and triterpanes (m/z = 191) and steranes (m/z = 217).

C_{15+} fractions (see Table 3). The isotopic composition of that extractable organic matter is more negative than the composition of the total organic carbon, suggesting that it contains larger proportions of higher-plant-derived organic matter, mainly in the aliphatic, aromatic, and NSO fractions. After hydrous pyrolysis the different C_{15+} fractions become isotopically more positive, indicating the addition of (soluble) compounds that were previously bonded to the (isotopically more positive) insoluble organic matter.

Summary and future work

Work on Big Soda Lake to date represents the first detailed data set conducted on modern day analogues of lacustrine oil source rocks as identified by Demaison & Moore (1980). Studies on the microbial geochemistry of Big Soda Lake have already made an important impact on our interpretation of $\delta^{13}CH_4$ values found in nature (Oremland & Des Marais 1983) and upon mechanisms of methanogenesis in high-sulphate environments (Oremland et al. 1982a, b; Oremland & Polcin, 1982). The lake has now been well-characterized in terms of its hydrogeochemistry, nutrient dynamics, mixing and productivity. Recent investigations have centered upon bacterial decomposition processes occurring seasonally in the water column of the lake, and how this relates to inputs of carbon derived from primary productivity. However, although methane oxidation in the water column has been investigated, estimates of sulphide and ammonia oxidation have not been made. Thus, the lake's carbon budget is the most comprehensive, while the sulphur and nitrogen budgets are as yet incomplete. All of these budgets require further studies. The contribution of the littoral zone to the C, N and S budgets must also be assessed.

In terms of the lake's sediments, only a preliminary data set exists with regard to its microbial carbon mineralization reactions, biogeochemistry and organic geochemistry. This aspect should be stressed in future investigations. It would be of particular interest to isolate alkalophilic anaerobes from the different zones of the lake because little is known about the physiology, ecology or bioenergetics of these types of bacteria (Horikoshi & Akiba 1982). This work is being pursued currently.

References

AXLER, R. P., GERSBER, R. M. & PAULSON, L. J. 1978. Primary productivity in meromictic Big Soda Lake, Nevada. *Great Basin Naturalist*, **38**, 187–192.

CLOERN, J. E., COLE, B. E. & OREMLAND, R. S. 1983a. Autotrophic processes in Big Soda Lake, Nevada. *Limnology and Oceanography*, **28**, 1049–1061.

——, —— & —— 1983b. Seasonal changes in the chemical and biological nature of a meromictic lake (Big Soda Lake, Nevada, USA). *Hydrobiologia*, **105**, 195–206.

——, —— & WIENKE, S. M. 1987. Big Soda Lake (Nevada). 4. Vertical fluxes of particulate matter in a meromictic lake: Seasonality and variations across the chemocline. *Limnology and Oceanography*, **32**, 815–824.

DANIELS, L., SPARLING, R. & SPROTT, G. D. 1984. The bioenergetics of methanogenesis. *Biochimica et Biophysica Acta*, **768**, 113–163.

DEGENS, E. T., GUILLARD, R. R. L., SACKETT, W. M. & HELLEBUST, J. A. 1968. Metabolic fractionation of carbon isotopes in marine plankton—I. Temperature and respiration experiments. *Deep Sea Research*, **15**, 1–9.

DEINES, P. 1980. The isotopic composition of reduced organic carbon. *In*: FRITZ, P. & FONTES, J. C. (eds), *Handbook of Environmental Isotope Geochemistry*, Elsevier, Amsterdam, 329–406.

DEMAISON, G. J. & MOORE, G. T. 1980. Anoxic environments and oil source bed genesis. *Bulletin of the American Association of Petroleum Geologists*, **64**, 1179–1209.

DEMBICKI, H. JR., MEINSCHEIN, W. G. & HATTIN, D. E. 1976. Possible ecological and environmental significance of the predominance of even carbon member C_{20}–C_{30} n-alkanes. *Geochimica et Cosmochimica Acta*, **40**, 203–208.

DIDYK, B. M., SIMONEIT, B. R. T., BRUSSEL, S. C. & EGLINGTON, G. 1978. Organic geochemical indicators of palaeoenvironmental conditions of sedimentation. *Nature*, **272**, 216–222.

EGLINGTON, G. & HAMILTON, R. J. 1963. The distribution of alkanes. *In*: SWAIN, T. (ed.), *Chemical Plant Taxonomy*, Academic Press, London, 187–217.

GUNSALUS, R. P., ROMESSER, J. A. & WOLFE, R. S. 1978. Preparation of coenzyme M analogues and their activity in the methyl coenzyme M reductase system of *Methanobacterium thermoautotrophicum*. *Biochemistry*, **17**, 2374–2377.

HARVEY, R. W. 1987. A fluorochrome-staining technique for enumeration of bacteria in saline, organically enriched, alkaline lakes. *Limnology and Oceanography*, **32**, 993–995.

IVERSEN, N., OREMLAND, R. S. & KLUG, M. J. 1987. Big Soda Lake (Nevada). 3. Pelagic methanogenesis and anaerobic methane oxidation. *Limnology and Oceanography*, **32**, 804–814.

KHARAKA, Y. K., ROBINSON, S. W., LAW, L. M. & CAROTHERS, W. W., 1984. Hydrogeochemistry of Big Soda Lake, Nevada: an alkaline, meromictic desert lake. *Geochimica et Cosmochimica Acta*, **48**, 823–835.

KIENE, R. P., OREMLAND, R. S., CATENA, A., MILLER, L. G. & CAPONE, D. G. 1986. Metabolism of reduced methylated sulfur compounds to methane and carbon dioxide by anaerobic sediments and a pure culture of an estuarine methanogen. *Applied and Environmental Microbiology*, **52**, 1037–1045.

KIMMEL, B. L., GERSBERG, R. M., PAULSON, L. J., AXLER, R. P. & GOLDMAN, C. R. 1978. Recent changes in the meromictic status of Big Soda Lake, Nevada. *Limnology and Oceanography*, **23**, 1021–1025.

LEE, C., GAGOSIAN, R. B. & FARRINGTON, J. W. 1980. Geochemistry of sterols in sediments from Black Sea and southwest African shelf and slope. *Organic Geochemistry*, **2**, 103–113.

MCELROY, M. B., ELKINS, J. W., WOFSY, S. C., KOLB, C. E., DURAN, A. P. & KAPLAN, W. A. 1978, Production and release of N_2O from the Potomac Estuary. *Limnology and Oceanography*, **23**, 1168–1182.

MILLER, L. G., OREMLAND, R. S. & PAULSEN, S. 1986. Measurement of nitrous oxide reductose activity in aquatic sediments. *Applied and Environmental Microbiology*, **51**, 18–24.

OREMLAND, R. S. 1981. Microbial formation of ethane in anoxic estuarine sediments. *Applied and Environmental Microbiology*, **42**, 122–129.

—— 1983. Hydrogen metabolism by decomposing cyanobacterial aggregates in Big Soda Lake, Nevada. *Applied and Environmental Microbiology*, **45**, 1519–1525.

—— & DES MARAIS, D. J. 1983. Distribution, abundance and carbon isotopic composition of gaseous hydrocarbons in Big Soda Lake, Nevada. An alkaline, meromictic lake. *Geochimica et Cosmochimica Acta*, **47**, 2107–2114.

—— & POLLIN, S. P. 1982. Methanogenesis and sulfate reduction: competitive and non-competitive substrates in estuarine sediments. *Applied and Environmental Microbiology*, **44**, 1270–1276.

——, MARSH, L. & DES MARAIS, D. J. 1982a. Methanogenesis in Big Soda Lake, Nevada: An alkaline, moderately hypersaline desert lake. *Applied and Environmental Microbiology*, **43**, 462–468.

——, —— & POLCIN, S. P. 1982b. Methane production and simultaneous sulfate reduction in anoxic, salt-marsh sediments. *Nature*, **296**, 143–145.

POWELL, T. G. & MCKIRDY, D. M. 1973. The effect of source material, rock type, and diagenesis on the n-alkane content of sediments. *Geochimica et Cosmochimica Acta*, **37**, 623–633.

PRISCU, J. C., AXLER, R. P., CARLTON, R. G., REUTER, J. E., ARNESON, P. A. & GOLDMAN, C. R. 1982. Vertical profiles of primary productivity, biomass and physiochemical properties in meromictic Big Soda Lake, Nevada, U.S.A. *Hydrobiologia*, **96**, 113–120.

SCALAN, R. S. & MORGAN, T. D. 1970. Isotope ratio mass spectrometer instrumentation and application to organic matter contained in recent sediments. *International Journal of Mass Spectrometry and Ion Physics*, **4**, 267–281.

SCHIEFELBEIN, C. F. 1982. *Hydrous Pyrolysis as a Thermal Alteration Simulation Technique.* M.Sc. thesis, University of Tulsa, Okla.

SMITH, R. L. & OREMLAND, R. S. 1983. Anaerobic oxalate degradation: Widespread natural occurrence in aquatic sediments. *Applied and Environmental Microbiology*, **46**, 106–113.

—— & OREMLAND, R. S. 1987. Big Soda Lake (Nevada). 2. Pelagic sulfate reduction. *Limnology and Oceanography*, **32**, 794–803.

VOGEL, T. M., OREMLAND, R. S. & KVENVOLDEN, K. K. 1982. Low-temperature formation of hydrocarbon gases in San Francisco Bay sediment. *Chemical Geology*, **37**, 289–298.

WONG, W. W. & SACKETT, W. M. 1978. Fractionation of stable carbon isotopes by marine phytoplankton. *Geochimica et Cosmochimica Acta*, **42**, 1809–1815.

ZEHR, J. P., HARVEY, R. W., OREMLAND, R. S., CLOERN, J. E., GEORGE, L., & LANE, J. L. 1987. Big Soda Lake (Nevada). 1. Pelagic bacterial heterotrophy and biomass. *Limnology and Oceanography*, **32**, 781–793.

ZINDER, S. H., ANGUISH, T. & CARDWELL, S. C. 1984. Selective inhibition by 2-bromoethanesulfonic acid of methanogenesis from acetate in thermophilic anaerobic digestor. *Applied and Environmental Microbiology*, **47**, 1343–1345.

R. S. OREMLAND, J. E. CLOERN, R. L. SMITH, C. W. CULBERTSON, J. ZEHR, L. MILLER, B. COLE & R. HARVEY, Water Resources Division, U.S. Geological Survey, Menlo Park, Ca. 94025, USA.

Z. SOFER, Cities Service Co., Tulsa, Okla. 74102, USA.

N. IVERSEN, Institute of Water, Soil and Environmental Technology, Aalborg, Denmark.

M. KLUG, Kellogg Biological Station, Hickory Corners, Mi. 49060, USA.

D. J. DES MARAIS & G. RAV, NASA Ames Research Center, Moffett Field, Ca. 94035, USA.

Predicting palaeoclimates

C. P. Summerhayes

The existence of lakes depends on several first order controls, one of the most important being climate*. Understanding climate to the point where it becomes predictable is an important facet in assessing how this first order control has operated in the geological past. Recently, geologists at the University of Chicago have shown how palaeoclimate models can be used to map the distribution of rainfall and of upwelling oceanic currents for selected time slices (Parrish 1982; Parrish & Curtis 1982; Parrish et al. 1982; Parrish et al. 1983). Use of the computer to perform this exercise is being perfected by other geologists (Barron 1984; Barron & Washington 1982, 1984). Furthermore, the approach of the Chicago school is now also computerized—allowing palaeopressure maps to be produced for any time slice for which palaeogeographic data are available (Scotese & Summerhayes 1986).

Predictions of upwelling made from palaeoclimate models can be tested against the distribution of climatically-sensitive geological variables (e.g. opaline silica, phosphorite, and black shales). Tests suggest that the models are reasonable (Parrish et al. 1982; Scotese & Summerhayes 1986). Accepting that palaeopressure maps produced by palaeoclimate modelling, and rainfall predictions made from such maps, are reasonable best guesses as to what the Earth's climate may have been like in the past, we can use this approach to make generalized predictive statements about the presence or absence of lakes of different kinds.

This extended abstract draws attention to the use of computers in enabling the rapid prediction of readily testable palaeoclimate maps for any chosen time slice. Such maps form crude working models, or thumbnail sketches of climate. While they lack subtle features, they are better than anything previously available. The user may choose a complex computer model (Barron & colleagues) or a simpler one (Scotese & colleagues).

Palaeoclimate maps have two main uses:

1. to help explain what we see in the rock record;
2. to help predict what we might expect to see.

The most recent tests that we have undertaken

* It is often forgotten that the key prerequisite for a lake is a hole in the ground (I. Price, personal communication 1985).

(Scotese & Summerhayes 1986) involve a comparison of predictions of coastal upwelling currents (made using computer-produced palaeopressure maps) against indicators of upwelling (opal-rich sediment detected in DSDP cores by Miskell et al. 1985).

For the Maastrichtian, upwelling was predicted off NW Africa, SW Africa, NW Australia, eastern USA, and Venezuela. Opal-rich sediment of this age was concentrated in all of these areas, as well as off the east coast of Africa. Although upwelling was also predicted for the west coast of the Americas, there were no DSDP sediments of that age from that area to test the model.

Why is there a discrepancy between upwelling predictions and opaline sediment distribution off eastern Africa? Does this mean the palaeoclimate model is wrong? To answer these questions we must recognize that not all upwelling is wind-induced. Geostrophic currents can induce upwelling, as in the Somali Current today. The Somali Current off northeastern Africa is the return flow for northern Indian Ocean water pushed south by the SW Monsoon. Seasonal accelerations in such currents tilt interior density surfaces up towards the coast to balance the current geostrophically, leading to upwelling at the coast. Such upwelling is geographically steady, seasonal, and independent of coastal winds (though it may be augmented by them). It seems likely that the east African discrepancy may reflect the influence of an ancestral Somali Current system, and does not invalidate the palaeoclimate model. For other examples of this kind see Scotese & Summerhayes (1986).

Are rainfall predictions made from palaeopressure maps reasonable? Parrish et al. (1982) predict from palaeoclimate models that in the Triassic, before Pangaea rifted apart, the continental interior was dry. The data show widespread desert sands, dunes, redbeds, and evaporites: thus the model is reasonable. Using Parrish's Triassic rainfall maps we would expect to find lakes typical of humid climates around the edges of Pangaea, and lakes typical of dry environments in the Pangaean interior.

After Pangaea broke up, seaways separated the different parts of Gondwanaland, bringing the former interiors of Pangaea closer to water. Parrish et al. (1982) predict that many of these formerly dry areas would have become wet (e.g. in the Cenomanian), depending on wind patterns and the location of pressure cells. We would expect more extensive lakes typical of humid climates, and greater restriction of lakes typical of dry climates, compared with the Triassic.

These global patterns derived from global palaeoclimate models obscure the effects of local variability that may be introduced by topography, for instance. Hay *et al.* (1982) show how local topography forces airstreams to rise and cool or to fall and warm. Air cooling adiabatically becomes saturated with moisture, leading to precipitation; sinking air warms adiabatically, becoming undersaturated, leading to evaporation. Thus some depressions become water-filled, while others become dry or accumulate playas. Hay uses this understanding to explain why lake deposits were typical of the western side of the Triassic–Liassic proto-North Atlantic, whereas evaporites accumulated in the centre.

Given (a) a global palaeoclimate model, showing wind directions, and a (b) detailed palaeogeographic map, showing main topographic highs and lows, local rainfall patterns might prove predictable within reasonable limits. Such models, tested or improved by feedback from the rock record, could then be used to explain and predict lake type with stated degrees of confidence.

References

BARRON, E. J. 1984. Numerical climate modeling, a frontier in petroleum source rock prediction: results based on Cretaceous simulations. *Bulletin of the American Association of Petroleum Geologists*, **69**, 448–459.

BARRON, E. J. & WASHINGTON, W. M. 1982. Cretaceous climate: a comparison of atmospheric simulations with the geological record. *Palaeogeography, Palaeoclimatology, Palaeoecology*, **40**, 103–133.

BARRON, E. J. & WASHINGTON, W. M. 1984. The role of geographic variables in explaining paleoclimates: results from Cretaceous climate model sensitivity studies. *Journal of Geophysical Research*, **89(D1)**, 1267–1279.

HAY, W. W., BEHENSKY, J. F., BARRON, E. J. & SLOAN, J. L. 1982. Late Triassic–Liassic paleoclimatology of the proto-central North Atlantic rift system. *Palaeogeography, Palaeoclimatology, Palaeoecology*, **40**, 13–30.

MISKELL, K. J., BRASS, G. W. & HARRISON, C. G. A. 1985. Global patterns in opal deposition from Late Cretaceous to Late Miocene. *Bulletin of the American Association of Petroleum Geologists*, **69**, 996–1012.

PARRISH, J. T. 1982. Upwelling and petroleum source beds with reference to Paleozoic. *Bulletin of the American Association of Petroleum Geologists*, **66**, 750–774.

PARRISH, J. T. & CURTIS, R. C. 1982. Atmospheric circulation, upwelling, and organic-rich rocks in the Mesozoic and Cenozoic Eras. *Palaeogeography, Palaeoclimatology, Palaeoecology*, **40**, 31–66.

PARRISH, J. T., ZIEGLER, A. M. & SCOTESE, C. R. 1982. Rainfall patterns and the distribution of coals and evaporites in the Mesozoic and Cenozoic. *Palaeogeography, Palaeoclimatology, Palaeoecology*, **40**, 67–101.

PARRISH, J. T., ZIEGLER, A. M. & HUMPHREVILLE, R. G. 1983. Upwelling in the Paleozoic Era. *In:* THIEDE, J. & SUESS, E. (eds) *Coastal Upwelling: Its Sediment Record*, B, NATO Conference Series IV, **10B**, Plenum Press, New York, 553–578.

SCOTESE, C. R. & SUMMERHAYES, C. P. 1986. A computer model of paleoclimate that predicts coastal upwelling in the Mesozoic and Cenozoic. *Geobyte*, **1**, 28–42.

C. P. SUMMERHAYES, Stratigraphy Branch, Exploration & Production Division, BP Research Centre, Sunbury-on-Thames, Middlesex.

Part II
Palaeoenvironmental Indicators

Clastic and carbonate lacustrine systems: an organic geochemical comparison (Green River Formation and East African lake sediments)

B. J. Katz

SUMMARY: Source rock properties were determined on samples obtained from three modern African lakes—Tanganyika, Kivu and Edward—and on thermally immature material from Eocene Lake Uinta of Utah and Colorado (Green River Formation). The modern rift lakes represent largely clastic lacustrine systems, whereas Lake Uinta is a carbonate-dominated system. The sediments from both systems exhibit a wide range of organic enrichment: up to ~ 13 wt.% TOC in the clastic lake samples and sometimes greater than 30 wt.% TOC in the carbonate-dominated system. Total hydrocarbon generation potentials, as determined by 'Rock-Eval' pyrolyses, also indicate higher yields in the carbonate system (up to ~ 200 mg HC g^{-1} rock) as compared to the clastic systems (up to ~ 90 mg HC g^{-1} rock). Differences in hydrocarbon yield are greater than can be explained by the organic content. They appear to be the combined result of mineral matrix effects and kerogen character. Elemental and visual analyses of isolated kerogen also support the idea of differences in organic character. The kerogen from the Green River Formation was used to define the Type I reference curve. The kerogen isolated from the clastic lakes appears intermediate between the Type I and Type II reference curves and contains varying and, in some cases, very significant proportions of Type III and/or residual (inert) organic material. In both cases, the principal product will be a high-wax crude oil. The primary differences in organic enrichment and character appear to be related to the relative quantity of allochthonous sedimentary material, both organic and inorganic.

Introduction

Conventional petroleum exploration strategies have typically relied upon the presence of marine source rocks. Exploration in parts of Africa, Australia, Brazil, China, Indonesia, Russia and the western United States have increasingly shown the presence of non-marine lacustrine sources. For example, lacustrine source facies account for 95% of China's hydrocarbon production (Halbouty 1980) and 54% of Brazil's known reserves (Ponte *et al.* 1980). Individual fields may be quite significant. The Taching field, located in the Sungliao basin of China, was sourced from Jurassic/Cretaceous lacustrine shales and has estimated reserves of up to 14 billion barrels (Chen, 1980). As a consequence, there has been increased interest in both the character and depositional settings of these oil generative sequences and their modern analogues.

Oil source rocks (potential, effective and/or exhausted) by definition either are presently, or have been capable of, generating and expelling commercial quantities of hydrocarbons. At low levels of thermal maturity they contain above average quantities of indigenous organic carbon and organically bound hydrogen. Upon pyrolysis they yield above average quantities of hydrocarbons (Bissada 1982).

The literature is ambiguous concerning the lower analytic thresholds for source rock definition. Numerous authors have attempted to define the threshold using organic carbon content (see Jones 1985, for a review). A more direct measure of generative capacity may be obtained through pyrolysis. The results from pyrolysis incorporate not only the level of organic enrichment but include the effects of organic matter type, diagenetic history, mineral matrix and level of thermal maturity. Based on a statistical study of pyrolysis data, Bissada (1982) suggested that a hydrocarbon yield of ~ 2.5 mg HC g^{-1} rock would be the minimum necessary for good source rocks. This yield threshold is independent of rock type or depositional environment.

The deposition of a source rock requires that the supply of organic matter is greater than oxidative demand. This organic matter is typically composed of a mixture of material derived from autochthonous (principally planktonic material) and allochthonous (higher plants and grasses) sources. The aim of this paper is to examine how the ratio of these two organic sources varies in carbonate and siliciclastic-dominated lacustrine systems and how this influences source rock potential and product type.

Study area

Samples were obtained from piston cores from three East African Rift valley lakes—Lakes

Tanganyika, Kivu and Edward (Fig. 1)—and outcrops and cores from oil shale horizons of the Green River Formation (Eocene Lake Uinta of Utah and Colorado; Fig. 2). The African lakes were selected to represent carbonate-poor systems. Calcium carbonate contents are typically less than 10%, but may occasionally dominate, as in the Bukavu basin of Lake Kivu, where $CaCO_3$ may be as high as 70% (Degens & Kulbicki 1973). No samples from the carbonate-rich areas of the African lakes were included in this study. The Green River Formation was selected to represent a carbonate-rich system. The carbonate mineralogy includes not only calcite and dolomite, but also ankerite, dawsonite and nahcolite (Desborough 1978).

The limnological characteristics of the three

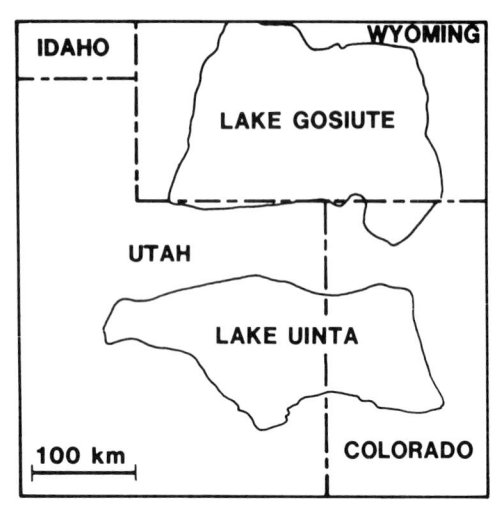

FIG. 2. Location of Lake Uinta.

African lakes have been well documented (see Serruya & Pollingher 1983, for a review). The lakes exhibit significant morphometric variation. Lakes Tanganyika and Kivu are permanently stratified (Degens et al. 1971; 1973). The water below 40 m in Lake Edward is typically devoid of oxygen except during the August (and possibly February) mixing period (Verbeke 1957). The high organic content of the sediment rapidly consumes the available oxygen during these turnover periods. All three lakes exhibit high levels of productivity, greater than 250 g C m^{-2} yr^{-1} (Degens & Kulbicki 1973).

The limnological characteristics of Lake Uinta during Green River deposition are poorly documented. Some authors have argued that the Green River Formation was deposited in a perennial, stratified lake with undefined water depths (Bradley & Eugster 1969; Desborough 1978), while others suggest a playa lake with a shallow central lake surrounded by carbonate mudflats acting as both a source and trap for sediments (Eugster & Surdam 1973). In either case, the lake basin was ~310 × ~130 km (Eugster 1985), larger than Lakes Edward and Kivu but smaller than Lake Tanganyika.

In the playa lake model the primary periods of organic deposition would have been associated with higher lake stands (i.e., more brackish water conditions). Temporary euxinic conditions would be established and maintained by elevated levels of productivity (Eugster 1985). In Lake Gosiute, another 'Green River basin' lake, Bradley (1963) estimated that primary productivity was ~270 g C m^{-2} yr^{-1}. Similar productivity values were probably achieved in Lake Uinta.

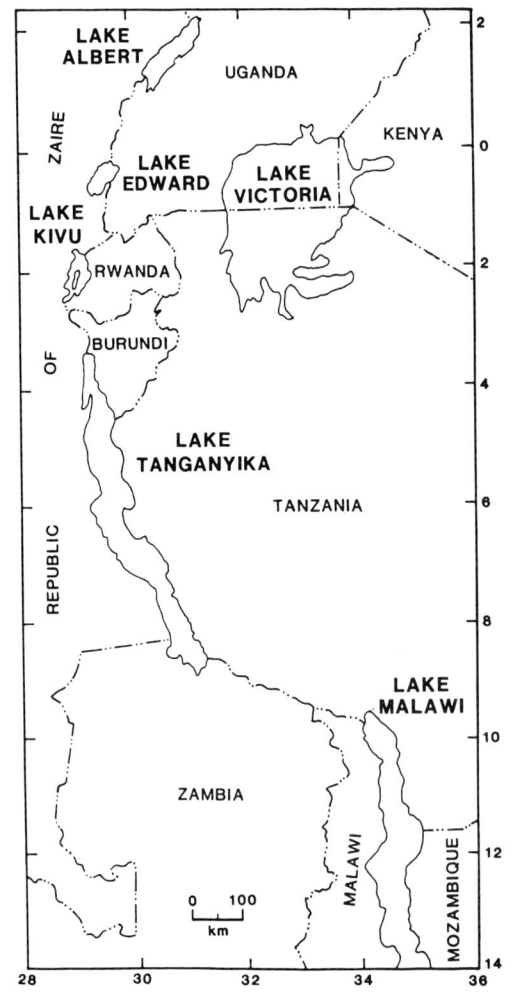

FIG. 1. Location of Lakes Tanganyika, Kivu and Edward.

Analytical methodology

Dried samples were ground to ~44 μm. The organic carbon content (TOC) was determined on each sample using a LECO combustion system after decarbonating. Samples with TOC > 0.5% were also subjected to pyrolytic assay using a 'Rock-Eval' system as described by Espitalié et al. (1977). The remainder of the analytical programme was determined principally upon the availability of sample material and the results of the two screening analyses. An attempt was made to examine both the soluble (bitumen) and insoluble (kerogen) organic fractions in all samples wherever possible. Samples were extracted using an azeotropic mixture of methanol, acetone and chloroform. The extracts were deasphaltened with pentane and fractionated for group-type analysis using high performance liquid chromatography. Kerogen was isolated following extraction using standard acid treatments and heavy liquid separation. Isolated kerogens underwent elemental analysis (C, H, O and N) and/or visual microscopic examination. Several samples underwent pyrolysis-gas chromatography to assist in the determination of the principal hydrocarbon products.

Analytical results

Organic carbon

Fifty-four samples were analysed from the African lake systems. Organic carbon contents ranged from 1.2 wt.% to 12.8 wt.% (Fig. 3). One-hundred-and-ninety-eight samples from several of the 'oil shale' intervals of the Green River Formation were also analysed. Organic carbon contents ranged from 0.2 wt.% to 29.6 wt.% (Fig. 4). All but five samples from the combined sample suites exhibit above-average levels of enrichment (>1.0 wt.%; Bissada 1982). While the maximum level of enrichment was significantly greater, more than double in the carbonate system, both sample suites exhibited modal values between 3 wt.% and 5 wt.%.

'Rock-Eval' pyrolysis

Pyrolysis measures the free (S_1) and generatable (S_2) hydrocarbons. The East African lake samples yielded values of $S_1 + S_2$ between 0.9 and 87.1 mg HC g^{-1} rock (Fig. 5). This compares with 4.2–204.7 mg HC g^{-1} rock for the Lake Uinta samples (Fig. 5). Nearly all of the pyrolysed samples have yields consistent with good or excellent source rocks (>6 mg HC g^{-1} rock; Tissot & Welte 1984).

Espitalié et al. (1977) suggested that pyrolysis may also be used to characterize the type of organic matter. Organic matter characterization is accomplished using a modified van Krevelen diagram, where the hydrogen index (S_2/TOC) is substituted for the H/C ratio and the oxygen index (S_3/TOC; S_3 is organically derived CO_2) is substituted for the O/C ratio (Fig. 6).

The African samples appear to be associated with the Type II and Type III reference curves. The high oxygen indices in many of the samples may be the result of large humic acid concentrations which are commonly encountered in modern sediments. In contrast, the data from the Green River Formation are closely aligned with the Type I reference curve.

Differences in the hydrogen indices of the two data sets are greater than can be explained as a result of rock matrix effects alone (Katz 1983). Carbonate rocks containing the same quantity and type of organic matter as argillaceous rocks have a higher hydrocarbon yield and consequently a higher hydrogen index value.

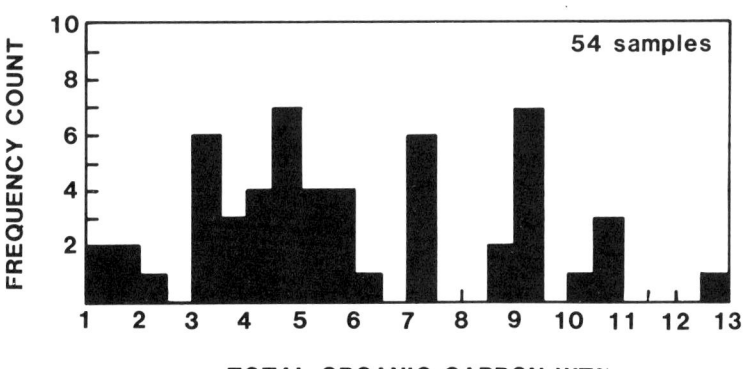

FIG. 3. Organic enrichment of East African sample suite.

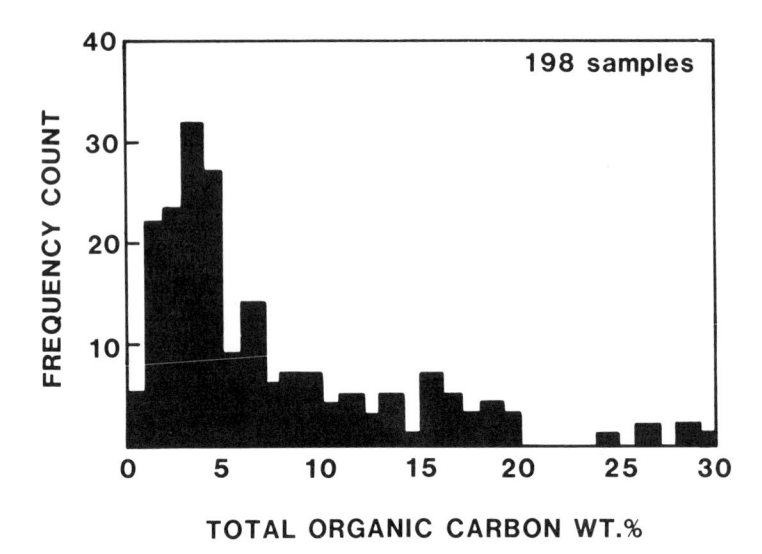

FIG. 4. Organic enrichment of Lake Uinta (Green River Formation) sample suite.

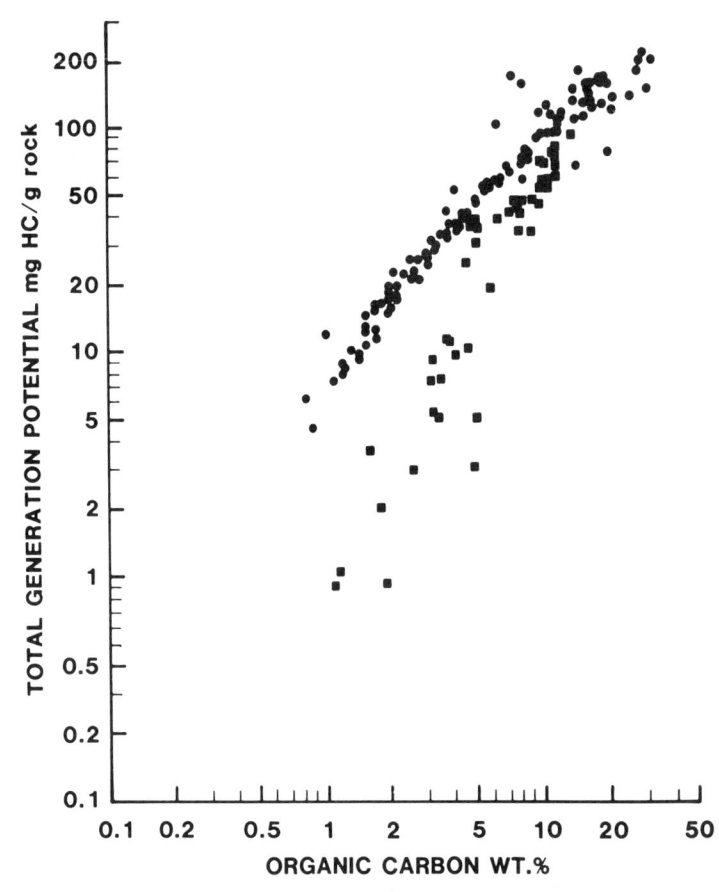

FIG. 5. Total generation potential vs organic carbon content, ■ East African lakes; ● Lake Uinta (Green River Formation).

Clastic and carbonate lacustrine systems   85

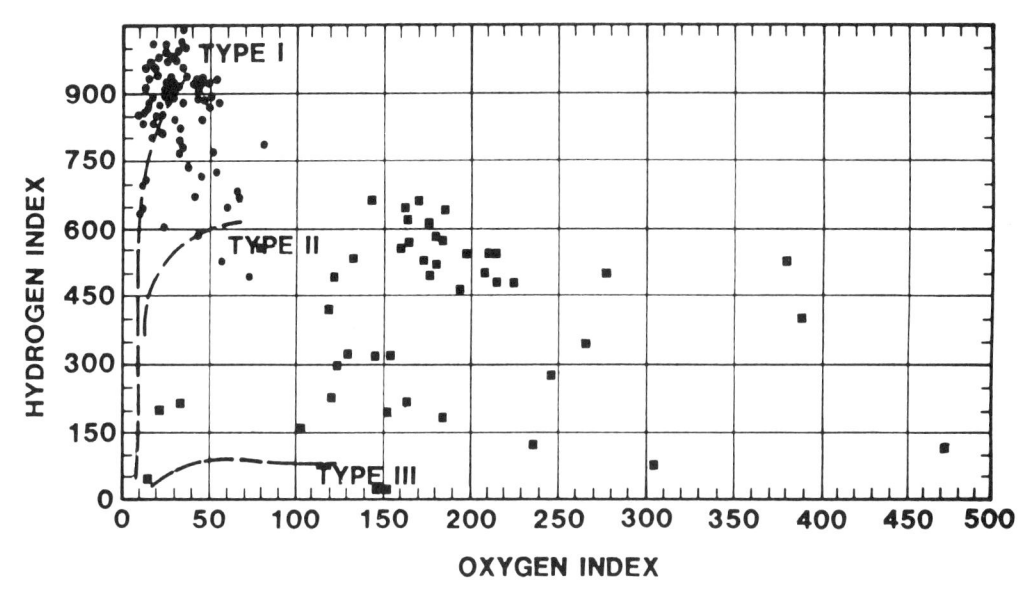

FIG. 6. Modified van Krevelen diagram, ■ East African lakes; ● Lake Uinta (Green River Formation).

Bitumen characterization

The abundance and character of bitumen are influenced by both the precursor material and the degree of thermal maturation. In shallowly buried rocks (present depth near burial maximum) and recent sediments, differences in both bitumen character and yield may be attributed to the character and quantity of the original organic matter. Total bitumen yields (TOE) in the East African field samples ranged from 486 ppm to 17 370 ppm (Fig. 7). All but one of these samples exhibited good to excellent potential source rock properties (TOE > 1000 ppm; GeoChem Laboratories, Inc. 1980). The TOE/TOC ratios are typically less than 0.2. Hydrocarbons (saturates and aromatics) commonly account for less than 15% of the extract (Fig. 8). The saturate to aromatic ratios are less than 1.0.

The Lake Uinta samples have total extract yields ranging from 11 430 to 42 022 ppm (Fig. 7). All of these samples would be considered excellent potential source rocks (TOE > 4000 ppm; GeoChem Laboratories, Inc. 1980). These higher yields relative to the African samples are, in part, attributable to higher levels of organic enrichment. However, the TOE/TOC ratios are elevated compared to the African samples, typically between 0.2 and 0.3 (Fig. 7). In addition, hydrocarbons account for ~30% of the extract (Fig. 8) and saturates dominate (saturate to aromatic ratios >1.0). These data imply that a greater proportion of saturated hydrocarbons were present in the Green River precursor material.

Elemental analysis

Elemental analysis provides a measure of the bulk chemical properties of an isolated kerogen. Commonly, the results are used to classify material according to type (Tissot *et al.* 1974).

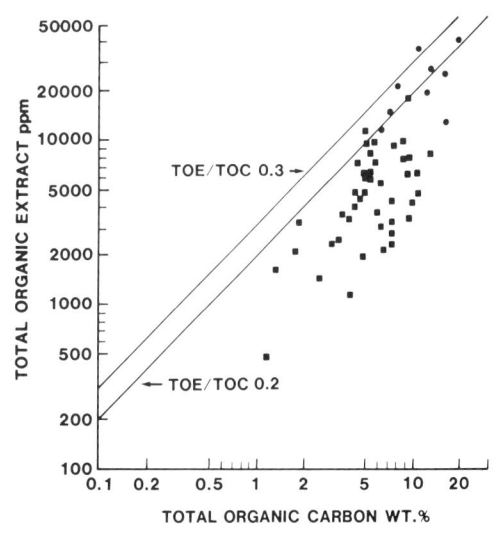

FIG. 7. Relationship between total organic extract and organic carbon content, ■ East African lakes; ● Lake Uinta (Green River Formation).

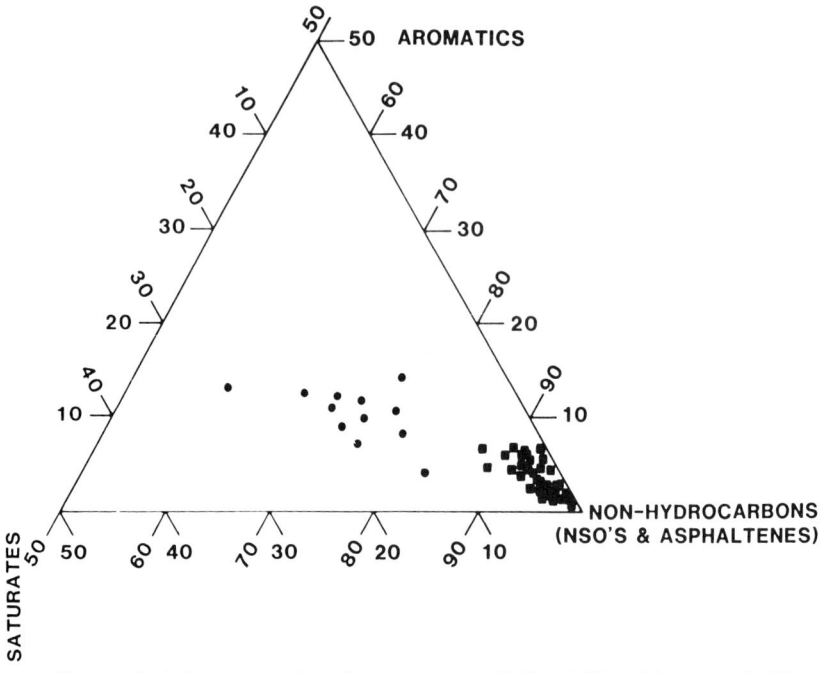

FIG. 8. Ternary diagram depicting composition of organic extract, ■ East African lakes; ● Lake Uinta (Green River Formation).

Although the literature commonly implies that specific depositional environments may be ascribed to each kerogen type (see for example Tissot & Welte 1984), kerogen composition should be viewed as a product of preservation state and source input.

The elemental analysis data are summarized in Fig. 9. The results obtained on the African lake samples show that they are less hydrogen-enriched and more oxygen-enriched than those obtained from the Green River material. The Green River kerogens are closely aligned with the Type I reference curve, with H/C ratios >1.3 and O/C ratios <0.1. Greater variability is observed in the African lake data; H/C ratios range from 1.05 to 1.56 and O/C ratios from 0.13 to 0.26. These data form a rather large scatter cloud about the Type II reference curve and suggest significant contributions of allochthonous organic matter. As with the pyrolysis data, the elevated oxygen levels in the African data suite may be due in part to the presence of humic acids.

Pyrolysis-gas chromatography

Pyrolysis-gas chromatography (PGC) provides a means of qualitatively and/or quantitatively characterizing the principal generation products of a sample utilizing the chromatographic 'fingerprint' of the pyrolysate (Dembicki *et al.* 1983). PGC is thought to be more sensitive to individual components than bulk kerogen analyses (e.g., elemental analysis; Larter & Douglas 1980).

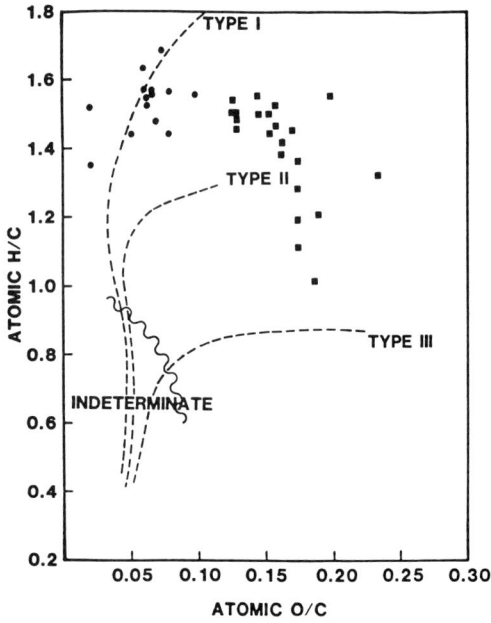

FIG. 9. van Krevelen diagram, ■ East African lakes; ● Lake Uinta (Green River Formation).

The chromatograms obtained from the two sample suites (Fig. 10) are similar. The pyrolysate is dominated by a series of doublets of normal alkanes and alkenes. Aromatics appear as minor components. The dominance by alkane–alkene doublets and the lack of aromatics is typical for alginite (Larter & Douglas 1980). Long-chain hydrocarbons are important contributors to the total pyrolysate. This pattern is clearly different from that obtained from Type II marine-derived kerogen (Fig. 10). This further supports the idea that although both the 'Rock-Eval' and elemental analysis data suggest that the African sample suite exhibits a strong Type II affinity, the PGC results indicate that the 'live' carbon has a greater affinity to the Type I kerogen of the Green River Formation.

Further examination of the two sample suites reveals that there are differences. Samples from Lake Uinta exhibit a slightly greater abundance of long-chain hydrocarbons compared to those obtained from the East African samples. This most probably can be attributed to the specific precursor material. There is also a predominance of even-numbered n-alkanes in the Lake Uinta material. This is typical of carbonate–evaporite environments (Tissot 1981). No such predominance was observed in the East African material.

Discussion

Bulk rock properties clearly differentiate between the two sample suites. The Lake Uinta samples exhibit both higher generative capacity and the potential for higher levels of organic enrichment when compared to the East African lake material (Fig. 5). This is, in part, a result of clastic input into the East African lakes which not only dilutes the organic matter present but also introduces allochthonous hydrogen-depleted organic matter.

The relative importance of this hydrogen-depleted organic matter may be estimated by examining the relative percent of the total carbon which remains after pyrolysis (i.e., residual carbon). These data are presented in Fig. 11. In the data from the East African samples the percentage of residual organic carbon was observed to be as high as 96%. The proportion is greater in the leaner rocks, indicating that there

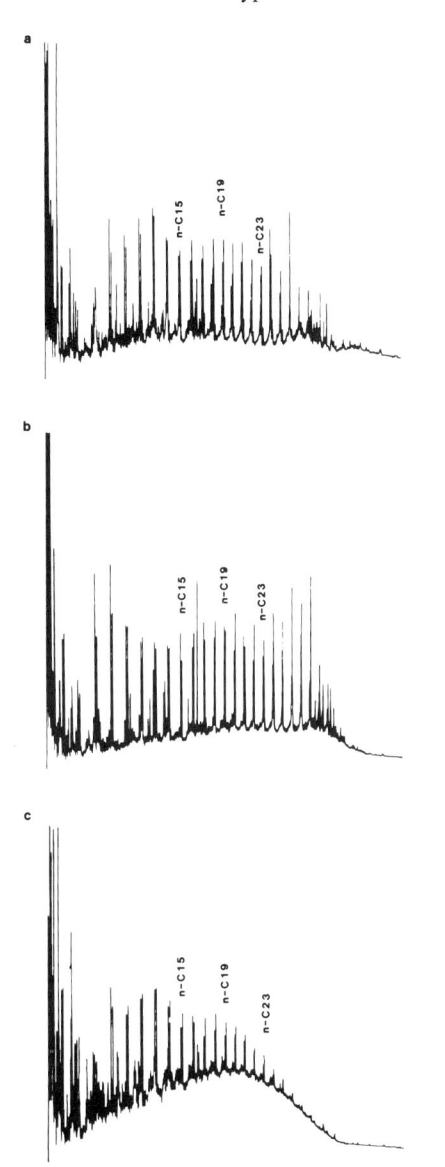

FIG. 10. Representative pyrolysis-gas chromatograms; (a) East African lakes; (b) Lake Uinta (Green River Formation); (c) an Albian 'black shale' (DSDP 93-603B-34-1-30/33 cm) with strong marine affinities.

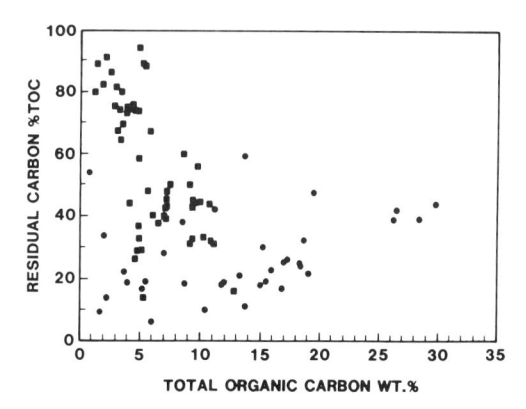

FIG. 11. Relationship between residual carbon content and level of organic enrichment, ■ East African lakes; ● Lake Uinta (Green River Formation).

is a background signal (~1–2 wt.%) of residual carbon.

Lake Uinta samples exhibit low residual carbon contents (typically <25% of the carbon content), indicating very little terrestrial input, as would be expected in a carbonate environment. The apparent increase in residual carbon with increasing carbon content as observed in these samples (Fig. 11) is an artifact of FID (flame ionization detector) saturation.

Differences in absolute organic enrichment between the two data suites appear to reflect differences in the sources for the autochthonous organic matter. In the East African lakes, autochthonous organic matter is principally derived from diatoms. The diatom tests dilute the organic matter and appear to place an upper limit on organic carbon content of ~15 wt.%. The organic matter for Lake Uinta was not associated with test-forming organisms. The dominant source of the organic matter, according to Bradley (1970), was blue-green algae, with contributions from yellow-green and green algae and minor contributions from bacteria, fungi, protozoa and wind-borne pollen and spores.

Differences in hydrocarbon yield relative to organic carbon content between the two sample suites (Fig. 5) decrease with increasing organic carbon content. This is the combined result of three related factors. The first is the ability of the clay matrix to retain hydrocarbons (Espitalié et al. 1980; Katz 1983). At low levels of enrichment, the quantity of generatable hydrocarbons is low and a significant proportion of these hydrocarbons may be retained by surface active clays. With increasing levels of enrichment, the quantity of generatable hydrocarbons increases, the surface sites become saturated and expulsion percentage increases. In the East African samples, with increasing carbon content, matrix clay content decreases and siliceous diatom tests become more abundant. The retentive capacity of silica is less than that of the clays and similar to that of carbonates. The third is a decrease in percent residual carbon with increasing carbon content in the East African lake samples. Additional carbon deposited in the system is hydrogen-enriched and adds substantially to the generative capacity.

The elemental and 'Rock-Eval' data are ambiguous with respect to expected products. These data suggest two distinctive populations (Figs 6 and 8) which would yield different hydrocarbon products upon thermal maturation. If classical interpretations are utilized, both systems would be considered capable of oil generation. The Lake Uinta material, being Type I, would produce an oil dominated by normal and iso-alkanes. The East African material, on the other hand, appears closely aligned with the Type II reference curve and the kerogen would be expected to produce an oil dominated by naphthenes and aromatics (Tissot & Welte 1984).

The bitumen data are consistent with the elemental and pyrolysis data. The Lake Uinta material yielded an extract richer in hydrocarbons than the African lake samples. The higher percentage of lipids in the Lake Uinta material may represent a greater environmental stress or variations due to differences in algal source. The hydrocarbons in the Lake Uinta material were dominated by saturates while aromatics played a greater role in the African extracts. However, these hydrocarbons represent indigenous material and not thermally generated products. These indigenous components are diluted by thermal products as generation occurs.

A more reliable indicator of product character can be obtained through PGC, which indicates that both lake systems will generate very similar products viz. high wax crudes. The primary difference between the two sample suites is their generative capacity. This is reflected in the bulk analyses. Minor compositional differences, as observed, would be expected as a result of the different contributors to the 'live' carbon.

The characteristics of organic matter appear to be associated with the availability of allochthonous input and autochthonous source. Consequently, while the data presented above are, in part, lake-specific, they should provide a general framework for the deposition of organic matter in carbonate or siliciclastic lacustrine systems. An examination of the literature tends to support this hypothesis. Published data from clastic lacustrine sequences appear very similar to the African lakes results, principally as a result of allochthonous input. Elemental data from the Tertiary of Queensland, Australia (Linder 1983) and from the Shahejie Formation, Bohai basin, China (Zhou 1981) suggest a Type II affinity, as do the 'Rock-Eval' data from the Fushun and Maoming regions of China (Nuttall et al. 1983). Similar values have been observed in the lacustrine units of the Mesozoic basins of the U.S. east coast, Cretaceous sequences in Brazil (Mello et al. 1984) and West Africa and the Tertiary of the Great Basin (Poole & Claypool 1984; Palmer 1984).

While other examples of argillaceous lacustrine rocks may be found in the literature, organic geochemical data associated with carbonate lacustrine rocks, other than the Green River Formation, are generally lacking. Limited data are available from the Elko Formation and Nuttall et al. (1983) published 'Rock-Eval' data on one sample from the Aleksinac Formation, Tertiary of Yugoslavia. These limited data do indicate a Type I affinity with a low residual carbon content ($\sim 30\%$). Other analyses performed on the Aleksinac Formation also indicated a strong Type I affinity with some minor amounts of what appear to be Type III material (Vitorovic 1980). Ross & Harwood (1979) also suggested that the organic matter contained in the lacustrine Todilto (Limestone) Formation of the San Juan basin is hydrogen-enriched, but no data are presented.

Conclusions

(1) Both depositional settings were capable of producing and preserving sufficient hydrogen-enriched organic matter for the sediments to be considered of good hydrocarbon source quality. While higher maximum levels of enrichment were observed in the carbonate environment, both sample suites had modal values between 3 and 5 wt.%.

(2) The organic matter contained in the clastic environments included a larger proportion of residual (inert) material compared with the carbonate environment, thus reducing the total generation potential of the system. The generation potential in the clastic system was also reduced through hydrocarbon retention by surface active clay minerals.

(3) Both environments will produce a principally paraffinic crude with a high wax content. The carbonate environment will produce a crude with an even carbon-number predominance in the n-alkane distribution. Such a preference will not be apparent in oil from clastic environments. Minor compositional differences result from differences in precursor material.

ACKNOWLEDGEMENTS: Analytical assistance was provided by the Organic Geochemistry Laboratory of the Texaco Houston Research Center. Earlier drafts of this manuscript were criticized by L. W. Elrod, J. E. Lacey and D. J. Schunk. Louise Jackson and Dianne DeLeon assisted with the preparation of the manuscript. Samples from Lakes Tanganyika, Kivu and Edward were supplied to the author by the Woods Hole Oceanographic Institution (NSF Grant OCE76-0262 and ONR Contract N00014-74-C-0262), DSDP sample 93-603B-34-1-30/33cm (Fig. 10) was supplied by the Deep Sea Drilling Project. This paper, Texaco Contribution Number 3030, has been published with the permission of Texaco Inc.

References

BISSADA, K. K. 1982. Geochemical constraints on petroleum generation and migration—a review. *Proceedings ASCOPE '81*, 69–87.

BRADLEY, W. H. 1963. Paleolimnology. *In*: FREY, D. G. (ed.), *Limnology of North America*, University of Wisconsin Press, Madison, 621–652.

—— 1970. Green River oil shale—concept of origin extended. *Bulletin of the Geological Society of America*, **81**, 985–1000.

—— & EUGSTER, H. P. 1969. Geochemistry and paleolimnology of the trona deposits and associated authigenic minerals of the Green River Formation of Wyoming. *U.S. Geological Survey Professional Paper*, **496-B**, 71 pp.

CHEN, C. 1980. Non-marine setting of petroleum in the Sungliao basin of northeastern China. *Journal of Petroleum Geology*, **2**, 233–264.

DEGENS, E. T. & KULBICKI, G. 1973. Data file on metal distribution in East African rift sediments. *Woods Hole Oceanographic Institution Technical Report*, **73–15**, 258 pp.

——, VON HERZEN, R. P. & WONG, H.-K. 1971. Lake Tanganyika; water chemistry, sediments, geological structure. *Naturwissenschaften*, **58**, 224–291.

——, ——, ——, DEUSER, W. G. & JANNASCH, H. W. 1973. Lake Kivu: structure, chemistry and biology of an East African rift lake. *Sonderdruck aus der Geologischen Rundschau*, **62**, 245–277.

DEMBICKI, H. JR., HORSFIELD, B. & HO, T. T. Y. 1983. Source rock evaluation by pyrolysis-gas chromatography. *Bulletin of the American Association of Petroleum Geologists*, **67**, 1094–1103.

DESBOROUGH, G. A. 1978. A biogenic-chemical stratified lake model for the origin of oil shale of the Green River Formation; an alternative to the playa-lake model. *Bulletin of the Geological Society of America*, **89**, 961–971.

ESPITALIÉ, J., LAPORTE, L. J., MADEC, M., MARQUIS, F., LEPLAT, P. J., PAULET, J. & BOUTEFEU, A. 1977. Méthode rapide de caractérization des roches mères de leur potentiel pétrolier et de leur degré d'évolution. *Revue de l'Institut Français du Pétrole*, **32**, 32–42.

——, MADEC, M. & TISSOT, B. 1980. Role of mineral matrix in kerogen pyrolysis: influence on petroleum generation and migration. *Bulletin of the American Association of Petroleum Geologists*, **64**, 59–66.

EUGSTER, H. P. 1985. Oil shales, evaporites and ore deposits. *Geochimica et Cosmochimica Acta*, **49**, 619–635.

—— & SURDAM, R. C. 1973. Depositional environment of the Green River Formation: a preliminary report. *Bulletin of the Geological Society of America*, **86**, 319–334.

GEOCHEM LABORATORIES, INC. 1980. *Source Rock Evaluation Reference Manual*, GeoChem Laboratories, Inc., Houston.

HALBOUTY, M. T. 1980. Methods used, and experience gained, in exploration for new oil and gas fields in highly explored (mature) areas. *Bulletin of the American Association of Petroleum Geologists*, **64**, 1210–1222.

JONES, R. W. 1985. Comparison of carbonate and shale source rocks. *American Association of Petroleum Geologists Studies in Geology*, **18**, 163–180.

KATZ, B. J. 1983. Limitations of 'Rock-Eval' pyrolysis for typing organic matter. *Organic Geochemistry*, **4**, 195–199.

LARTER, S. R. & DOUGLAS, A. G. 1980. A pyrolysis-gas chromatographic method for kerogen typing. *In*: DOUGLAS, A. G. & MAXWELL, J. R. (eds), *Advances in Organic Geochemistry*, Pergamon Press, New York, 579–583.

LINDER, A. W. 1983. Geology and geochemistry of some Queensland Tertiary oil shales. *In*: MIKNIS, F. P. & MCKAY, J. F. (eds), *Geochemistry and Chemistry of Oil Shales*, American Chemical Society, Washington, D.C., 97–118.

MELLO, M. R., ESTRELLA, G. D. O. & GAGLIANONE, P. C. 1984. Hydrocarbon source potential in Brazilian margin basins. *Bulletin of the American Association of Petroleum Geologists*, **68**, 506 (abstract).

NUTTALL, H. E., GUA, T.-M., SCHRADER, S. & THAKUR, D. S. 1983. Pyrolysis kinetics of several key world oil shales. *In*: MIKNIS, F. P. & MCKAY, J. F. (eds), *Geochemistry and Chemistry of Oil Shales*, American Chemical Society, Washington, D.C., 269–300.

PALMER, S. E. 1984. Hydrocarbon source potential of organic facies of the lacustrine Elko Formation (Eocene/Oligocene), northeast Nevada. *In*: WOODWARD, J., MEISSNER, F. & CLAYTON, J. L. (eds), *Hydrocarbon Source Rocks of the Greater Rocky Mountain Region*, Rocky Mountain Association of Geologists, Denver, 491–511.

PONTE, F. C., FONSECA, J. D. R. & CAROZZI, A. V. 1980. Petroleum habitats in the Mesozoic–Cenozoic of the continental margin of Brazil. *Memoir Canadian Society of Petroleum Geologists*, **6**, 857–886.

POOLE, F. G. & CLAYPOOL, G. E. 1984. Petroleum source-rock potential and crude-oil correlation in the Great Basin. *In*: WOODWARD, J., MEISSNER, F. F. & CLAYTON, J. L. (eds), *Hydrocarbon Source Rocks of the Greater Rocky Mountain Region*, Rocky Mountain Association of Geologists, Denver, 179–229.

ROSS, L. M. & HARWOOD, R. J. 1979. Organic geochemical correlation of San Juan basin oils and oil source sequences. *Geological Society of America Abstracts with Programs*, **11**, 506 (abstract).

SERRUYA, C. & POLLINGHER, U. 1983. *Lakes of the Warm Belt*. Cambridge University Press, Cambridge, 569 pp.

TISSOT, B. P. 1981. Generation of petroleum in carbonate rocks and shales of marine or lacustrine facies and its geochemical characteristics. *In*: MASON, J. F. (ed.), *Petroleum Geology in China*, PennWell Publishing Company, Tulsa, 71–82.

—— & WELTE, D. H. 1984. *Petroleum Formation and Occurrence*. 2nd edn. Springer-Verlag, New York, 699 pp.

——, DURAND, B., ESPITALIÉ, J. & COMBAZ, A. 1974. Influence of nature and diagenesis of organic matter in formation of petroleum. *Bulletin of the American Association of Petroleum Geologists*, **58**, 499–506.

VERBEKE, J. 1957. Recherches écologiques sur la faune des grand lacs de l'Est Congo Belge. *Exploration Hydrobiologique des lacs Kivu, Edouard et Albert (1952–54), Résultats scientifiques*, **3**, Institut Royal des Sciences Naturelles Belgique, Brussels, 177 pp.

VITOROVIC, D. 1980. Structure elucidation of kerogen by chemical methods. *In*: DURAND, B. (ed.), *Kerogen: Insoluble Organic Matter from Sedimentary Rocks*. Editions Technip, Paris, 301–338.

ZHOU, G. 1981. Character of organic matter in source rocks of continental origin and its maturation and evolution. *In*: MASON, J. F. (ed.), *Petroleum Geology in China*, PennWell Publishing Company, Tulsa, 26–47.

B. J. KATZ, Texaco Inc., E & P Technology Division, 3901 Briarpark, Houston, TX77042, USA.

Geochemical characterization of the organic matter from some recent sediments by a pyrolysis technique

M. Vandenbroucke & F. Behar

SUMMARY: A comparative study of humic acids and kerogens from recent sediments was performed using geochemical analyses of hydrocarbons released by pyrolysis of these fractions. Lacustrine samples came from the Eastern African Rift (Lake Bogoria, Lake Tanganyika) and Europe (Lake Leman), marine samples from upwelling zones (Mauritania, Arabia), and coastal land-derived sediments from Kerguelen and Svalbard islands and from a deltaic facies in Indonesia. The deposited aquatic organic matter consisted of clearly identifiable algae (diatomae, coccolithophoridae). C_{14+} saturated and unsaturated hydrocarbons generated by programmed temperature pyrolysis under an inert gas flow (argon) were recovered, hydrogenated and analysed by GC. The iso- + cyclo-alkane fraction was separated, then analysed by GC-MS. Pyrolysates of humic acids, kerogens and fulvic acids reveal diverse compositions important for comparative studies. Hydrogenated hydrocarbon distributions of kerogens and humic acids can be distinguished by diminished amounts of C_{14+} hydrocarbons in the latter, whereas fulvic acids generate unique hydrocarbon distributions atypical of both kerogens and humic acids. Major iso- + cyclo-alkanes in the kerogens, the main fraction of the organic matter, are isoprenoids (pristane and phytane), steranes (α cholestane) and triterpanes (C_{27} and C_{29} hopanes and moretanes), whatever the origin of the kerogen, whether lacustrine or marine.

Aims of the study

Geochemists classify sedimentary organic matter into three main types depending on its atomic ratios H/C and O/C (Tissot *et al.* 1974). Type I is typified by kerogen from lacustrine sediments from the Green River Shales in the USA; Type II is represented by the kerogen of early Jurassic marine sediments from West Europe and Type III by the terrigenous organic matter of deltaic sediments from West Africa and Indonesia.

This classification being based on bulk chemical analysis, similar atomic ratios may correspond to very different chemical structures. However, it is known that there is a relationship between the type of the deposited organic matter and its chemical structure (Tissot *et al.* 1974). Studies on recent sediments give a way to link chemical analysis and genetic typing, as the nature of the initial biomass and its alteration in the sedimentary environment can be studied.

However geochemical analyses are made difficult because of the polymeric nature and the variety of chemical functionality of organic matter in recent sediments. Detailed structural analyses are generally carried out on hydrocarbons separated from organic extracts. Organic fractions isolated from recent sediments, by alkaline extraction for instance, are analysed by bulk methods such as elemental analysis, IR, NMR etc.

Information on the structure of the carbon skeleton in the organic fractions from recent sediments, which is useful for estimating petroleum potential, can be obtained by breaking the complex macromolecules into smaller fragments. This can be achieved using programmed pyrolysis under an inert gas flow, as has already been done for kerogens and asphaltenes from ancient sediments (Bordenave *et al.* 1970; Behar & Pelet 1985; Pelet *et al.* 1986).

This method was applied to organic fractions separated from recent sediments (fulvic acids, humic acids, stable residues, i.e. the non-hydrolysable part of humin which can be physically recovered after acid digestion of minerals), so as to compare their composition. Various sediments containing known biomasses and coming from different sedimentary environments were analysed so as to search for chemical markers of the organic input and environmental influences. We present here analyses on three samples containing algae deposited in lacustrine environments, three samples containing algae deposited in marine environments, and three samples including land plant detritus deposited in nearshore marine environments.

Experimental

The analytical procedure used for fractionation of organic matter from recent sediments is shown in Fig. 1. Recovery of residual organic matter (humin and grey humic acids) is done by acid digestion of the mineral fraction. This procedure was initially developed for isolation of kerogen in ancient sediments (Durand & Nicaise 1980),

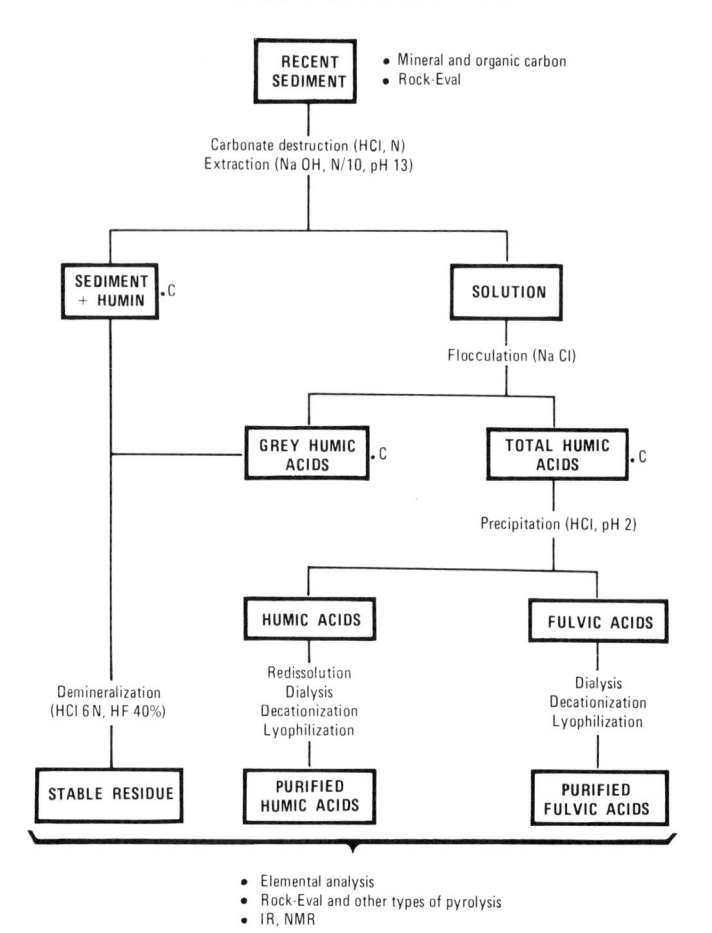

FIG. 1. Isolation procedure of organic fractions from recent sediments.

but it is less convenient with recent sediments, as humin is readily hydrolysable. The insoluble organic residue obtained after acid digestion is defined as the 'stable residue'. In some pyrolyses, stable residues were also extracted by chloroform in order to compare their pyrolysates with those of non-extracted stable residues. The complete procedure modified from Behar & Pelet, 1985, for pyrolysis and C_{14+} effluent analysis is summarized in Fig. 2.

Sample selection

Some geochemical data on the samples are given in Table 1. Lacustrine sediments come from the Eastern African Rift: Lake Tanganyika and Lake Bogoria, and from Europe: Lake Leman. Lake Bogoria is saline: up to 90 g/l in October 77 (Tiercelin *et al.* 1980), the two other lakes contain freshwater. The nature of the primary organic input has been assessed for all the samples except for Lake Leman. It consists of autochthonous algae without any significant detrital input. In Lake Leman, geochemical data show that there is probably a detrital contribution of inert carbon.

This lacustrine organic matter can be compared to that reported from marine sediments containing either marine or terrestrial organic matter. Marine sediments were considered from offshore Arabia (Pelet 1981) containing algae with carbonated (M1) or siliceous (M2) tests, and from the Black Sea (Huc *et al.* 1978) containing algae wih organic tests. Three nearshore marine sediments containing terrestrial organic matter were also studied. They contain debris from lower plants (terrestrial algae, mosses, lichens—sample T1 from Svalbard Islands) or from higher plants, (trees, etc.—sample T3 from the Mahakam River delta in Indonesia: Combaz & de Matharel 1978), or a mixture of higher and lower plants (sample T2 from the Kerguelen Islands).

Analytical Procedure on Pyrolysed Compounds

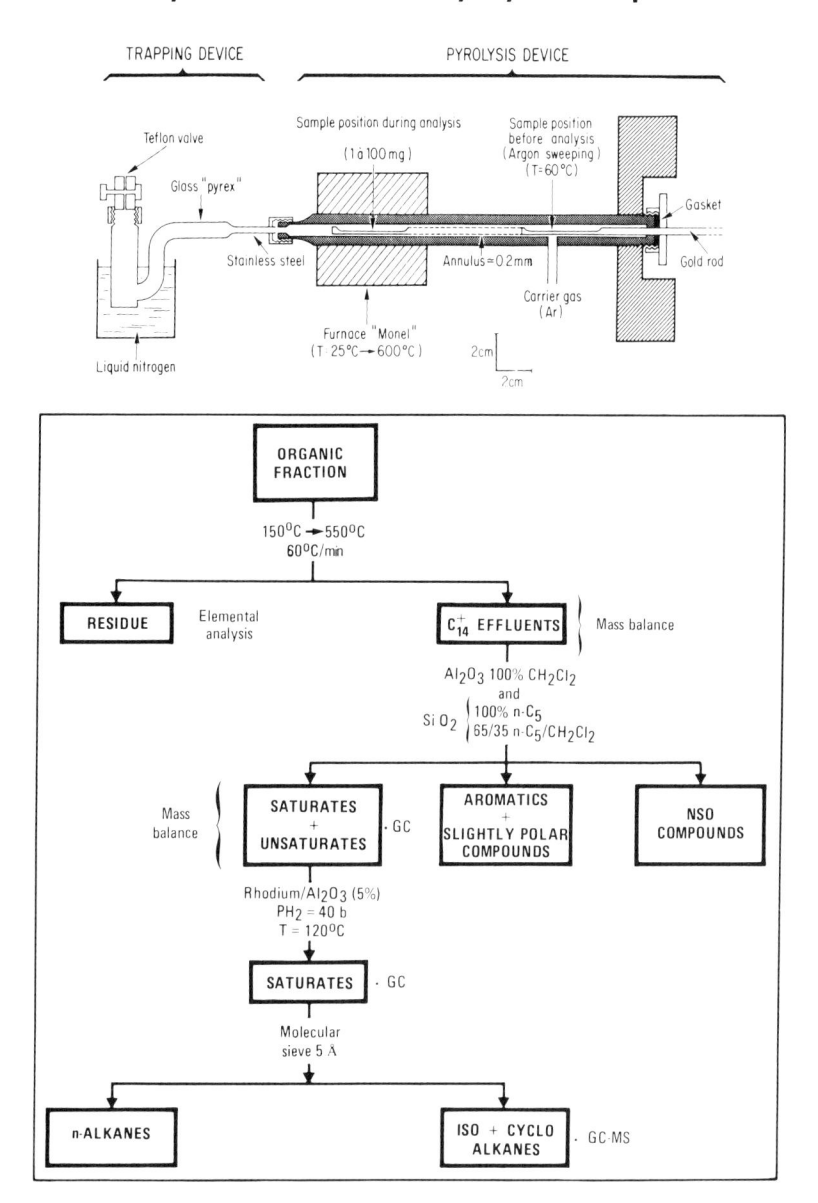

FIG. 2. Analytical procedure on pyrolysed compounds.

Results and discussion

Comparison of the different organic fractions

Quantitative data on pyrolyses performed on organic fractions (stable residues, humic acids, fulvic acids) isolated from samples L1, M1 and T1 are given in Table 2. Data are fairly similar for stable residues and humic acids, but those for fulvic acids seem to differ. This is supported by the chromatograms of saturates and unsaturates of the C_{14+} fraction released by pyrolysis (Fig. 3). In the fulvic acid fraction, which represents less than 10% of the sedimented carbon, saturates and unsaturates from pyrolysis are mainly normal alkanes with even carbon numbers in the C_{14}-

TABLE 1. *Geochemical data on initial stable residues*

Sample ident.	Source	Organic input	Climate	C* (%)	Atomic H/C	Atomic O/C	Rock-Eval HI (mg/g C)
L1	Lake Tanganyika	Diatomae	Hot	56.7	1.17	0.244	517
L2	Lake Bogoria	Spirulines	Hot	67.5	1.43	0.181	801
L3	Lake Leman	n.d.	Temperate	45.3	0.90	0.225	237
M1	Oman Sea	Coccolithophoridae	Hot	73.4	1.16	0.230	496
M2	Oman Sea	Diatomae	Hot	43.4	1.32	0.232	510
M3	Black Sea	Dinoflagellates	Hot	54.7	1.47	0.255	676
T1	Spitzbergen	Lower plants	Cold	49.6	1.12	0.455	460
T2	Kerguelen	Higher/lower plants	Cold	36.6	1.22	0.347	394
T3	Mahakam	Higher plants	Hot	32.6	1.12	0.338	313

*% weight of the stable residue, including some remaining ashes.

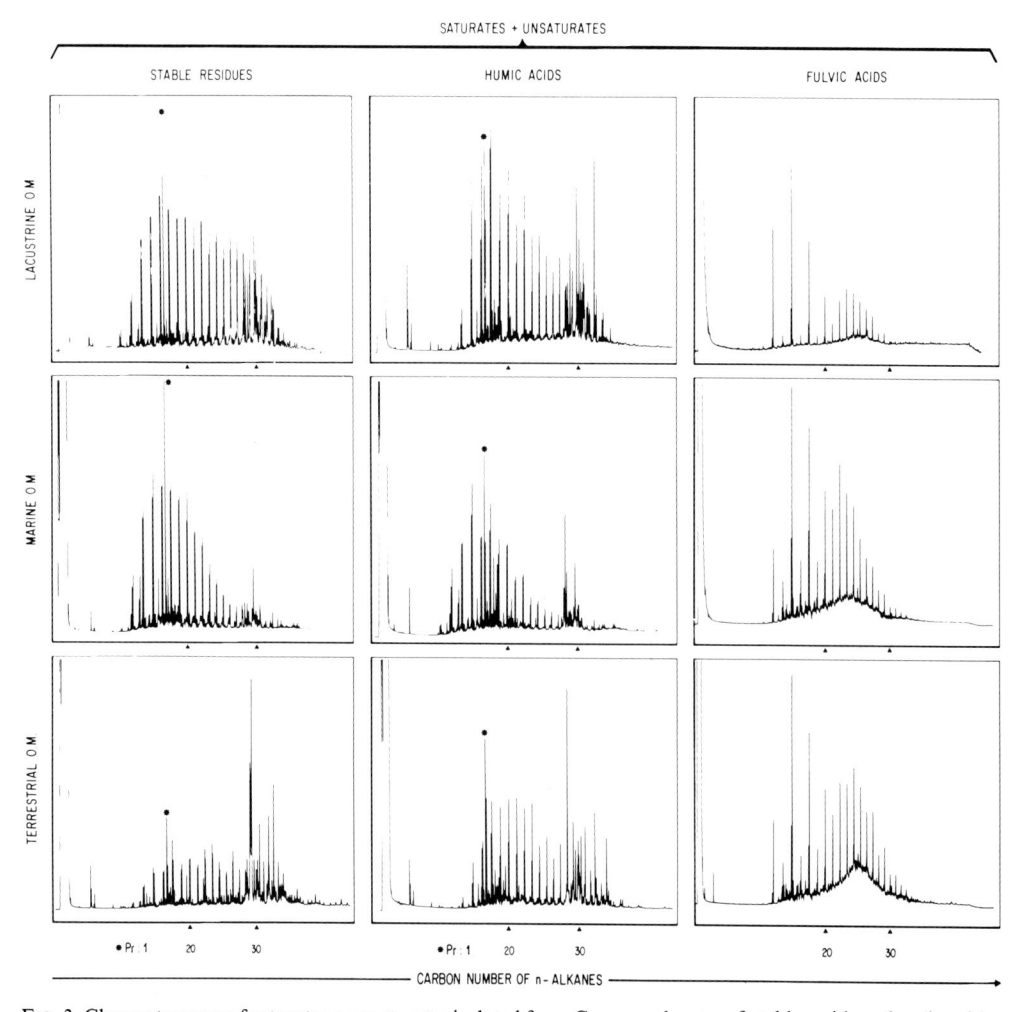

FIG. 3. Chromatograms of saturates + unsaturates isolated from C_{14+} pyrolysates of stable residues, humic acids, fulvic acids from lacustrine organic matter (Sample L1, Lake Tanganyika), marine organic matter (Sample M1, Oman Sea) and terrestrial organic matter (Sample T1, Svalbard Islands).

TABLE 2. *Quantitative data on pyrolyses performed on stable residues, humic acids, fulvic acids, from samples L1 (Lake Tanganyika), M1 (Oman Sea), T1 (Svalbard Island)*

Sample ident.	Organic fraction	Pyrolysate amount C_{14+} (1) (weight %)	(mg/g C)	Residue (2) (weight %)	Estimated C_{14-}* (weight %)	SAT† (%)	ARO† (%)	NSO† (%)	Residue analysis at. H/C	at. O/C
L1	SR	33.1	584	38.5	28.4	8.1	8.8	83.1	0.50	0.121
	HA	15.9	290	40.5	43.7	6.1	5.6	88.3	0.48	0.116
	FA	9.1	208	36.0	58.9	20.0	4.6	75.4	0.46	0.203
M1	SR	26.7	364	29.5	43.8	3.9	10.3	85.8	0.50	0.091
	HA	19.5	326	36.7	43.8	3.3	9.3	87.4	0.55	0.109
	FA	6.4	163	39.2	54.4	6.0	4.3	89.7	0.48	0.212
T1	SR	12.8	258	39.5	47.7	10.1	6.1	83.9	0.50	0.187
	HA	18.7	379	33.9	47.4	2.0	10.2	87.8	0.54	0.151
	FA	7.8	188	35.5	56.7	3.9	2.2	93.9	0.46	0.207

* Calculated as $100 - (1) - (2)$. Main constituents are CO, CO_2, H_2O and hydrocarbons $< C_{14}$.
† SAT = saturates + unsaturates, ARO = aromatics + slightly polar compounds, NSO = nitrogen-, sulphur- and oxygen-containing compounds.

C_{20} range, hence they could be derived from fatty acids or alcohols. Hydrocarbons released by pyrolysis of humic acids and stable residues are fairly similar, as shown by the mass balance and the chromatograms of Fig. 3. The carbon distributions in the chromatograms are typical of the nature of the organic matter which has been pyrolysed. This agrees with data reported by Verheyen *et al.* (1985) as their humic acid fractions were obtained after a prior extraction of all the lipidic fraction of the sediment. Thus the fulvic acid fraction being a minor part of the organic matter and showing atypical hydrocarbon pyrolysates seems not to be representative of the bulk structure of the organic matter and will not be further considered. Stable residues which make up most of the sedimented organic matter (often more than 50%) and generate hydrocarbon pyrolysates similar to those of humic acids will be considered as representative of the total organic matter.

Amounts and composition of pyrolysates from stable residues

Quantitative data on pyrolysates from the three lacustrine stable residues are compared to those of the six other stable residues (Table 3). It can be seen that data for sample L3 from Lake Leman are more similar to those of the terrestrial samples than to those of the other lacustrine samples. However it will be shown later than the GC of the hydrocarbons released during pyrolysis can indeed be related to algal organic matter. In this sample, organic matter is probably a mixture of algae and inert detrital carbon; all values normalized to total organic carbon are thus lowered. If for this reason, sample L3 is excluded, we can observe that the pyrolysate yield decreases when going from lacustrine organic matter to terrestrial organic matter. This follows the decrease of H/C ratio and hydrogen index in the initial samples. As confirmed by pyrolysis data on many other samples, the ratio of the fractions of the pyrolysate labelled SAT and ARO seems to be typical of the organic matter. This ratio SAT/ARO is around 1 in lacustrine organic matter, around 0.5 in marine organic matter and greater than 1 in terrestrial organic matter. The same relation is observed between saturates and aromatics in chloroform extracts from mature source rocks containing Type I, Type II or Type III organic matter. This suggests that the main structural features of kerogens are already present at an early stage in the recently deposited organic matter.

Distribution of alkanes in pyrolysates from stable residues

The distribution of saturates and unsaturates in the C_{14+} pyrolysates (Fig. 4) gives a picture of the carbon structures in the organic matter. Pristene-1 is often a major peak. Presence of the n-alk-l-ene just before each n-alkane peak is a by-product of the pyrolysis, but as each set of the two compounds have a related origin, the interpretation of carbon distributions based solely on n-alkane peaks does not alter the general picture. In lacustrine organic matter, the carbon distribution is fairly flat between C_{15} and C_{25}; in marine organic matter, the maximum is around C_{16} and the decrease beyond this carbon number is fairly rapid. In terrestrial organic matter, the carbon distribution encompasses a larger range and carbon numbers beyond C_{30} are abundant; moreover odd or even predominance can be

TABLE 3. *Quantitative data on pyrolysates from stable residues*

Sample ident.	Source	Pyrolysate amount C_{14+} (1)		Residue (2)	Estimated C_{14-}*	SAT†	ARO†	NSO
		(weight %)	(mg/g C)	(weight %)	(weight %)	(%)	(%)	(%)
L1	Lake Tanganyika	33.1	584	38.5	28.4	8.1	8.8	83.1
L2	Lake Bogoria	46.6	690	25.8	27.6	7.3	11.5	81.2
L3	Lake Leman	10.4	230	60.7	28.9	15.2	17.1	67.7
M1	Oman Sea	26.8	364	29.5	43.8	3.9	10.3	85.8
M2	Oman Sea	24.7	569	41.9	33.4	5.9	13.0	81.1
M3	Black Sea	37.1	678	28.5	34.4	5.4	10.4	84.2
T1	Spitzbergen	12.8	258	39.5	47.7	10.1	6.1	83.9
T2	Kerguelen	16.5	451	50.0	24.9	20.2	n.d.	n.d.
T3	Mahakam	9.3	285	57.0	33.7	21.3	14.7	64.0

*Calculated as $100-(1)-(2)$. Main constituents are CO, CO_2, H_2O and hydrocarbons $<C_{14}$.
†SAT = saturates + unsaturates, ARO = aromatics + slightly polar compounds, NSO = nitrogen-, sulphur- and oxygen-containing compounds.

observed. These C_{14+} carbon distributions of hydrocarbons evolved from pyrolysis of stable residues are similar to those of oils generated naturally.

The isolation of iso- + cyclo-alkanes obtained by hydrogenation and molecular sieving of saturates and unsaturates released by pyrolysis allows analysis of some biomarkers which can be discussed in terms of nature of the organic matter input or the depositional sedimentary environment. The similarity of GC traces (see Fig. 8, left) between algal material originating in lacustrine or marine environments is obvious. Pristane and phytane are generally major peaks, together with steroids and triterpenoids in the C_{30} range. In terrestrial organic matters, the picture is very different, and the major peaks are homologous series of iso-alkanes, every three carbons over a very large carbon range. The detailed structure of these iso-alkanes is not yet known. Other homologous series of iso-alkanes are also present in algal material of L3 and M3 samples, but their carbon range is shifted towards lower numbers, with a predominance of C_{16} and C_{18}.

GC-MS analysis of hopanes (m/z = 191, Fig. 5) and steranes (m/z = 217, Fig. 6) allows further identification of some peaks in the GC trace of iso- + cyclo-alkanes and hydrogenated alkenes from the pyrolysates of the stable residues. The presence of $\beta\beta$ isomers of hopanes is remarkable as these epimers are thermodynamically less stable than the $\alpha\beta$ epimers observed in the geological environment. It shows that, in our pyrolysis method, secondary reactions are not significant and thus, that the method is well adapted for structural studies of the organic matter in recent sediments.

We can obtain an idea of the relative amounts of steranes and hopanes in the GC traces, as the flame ionisation detector gives a linear response according to the carbon weight. A high amount of pristane (Pr) and phytane (Ph) compared to polycyclic biomarkers is observed in lacustrine and marine samples. The ratio of steranes to hopanes increases when passing from lacustrine to marine to terrestrial sediments. It also seems, although less clear, that the relative amounts of C_{27} and C_{29} molecules in steranes and triterpanes are a characteristic feature of lacustrine versus marine or terrestrial organic matter. To tell the truth we do not yet know exactly what parameters influence these various distributions in iso- + cyclo-alkanes from pyrolysates, but we suggest that the influence of the sedimentary environment, for instance the nature of the microorganisms typical of aerobic or anaerobic environments, is probably as important as the nature of the initial organic input (algae or land plants).

Comparison of iso- + cyclo-alkanes in pyrolysates of stable residues and related humic acids

These compounds have been considered as fairly similar previously on the basis of quantitative pyrolysis data and GC of the saturates + unsaturates fraction. When comparing mass fragmentograms of triterpanes (Fig. 5) and steranes (Fig. 6) in samples L1, M1 and T1, there is no clearcut difference between stable residues and related humic acids. Moreover the GC traces of the iso- + cyclo-alkanes (Fig. 7), which allow a quantitative estimation of these biomarkers, indicate that pyrolysates of humic acids are enriched in

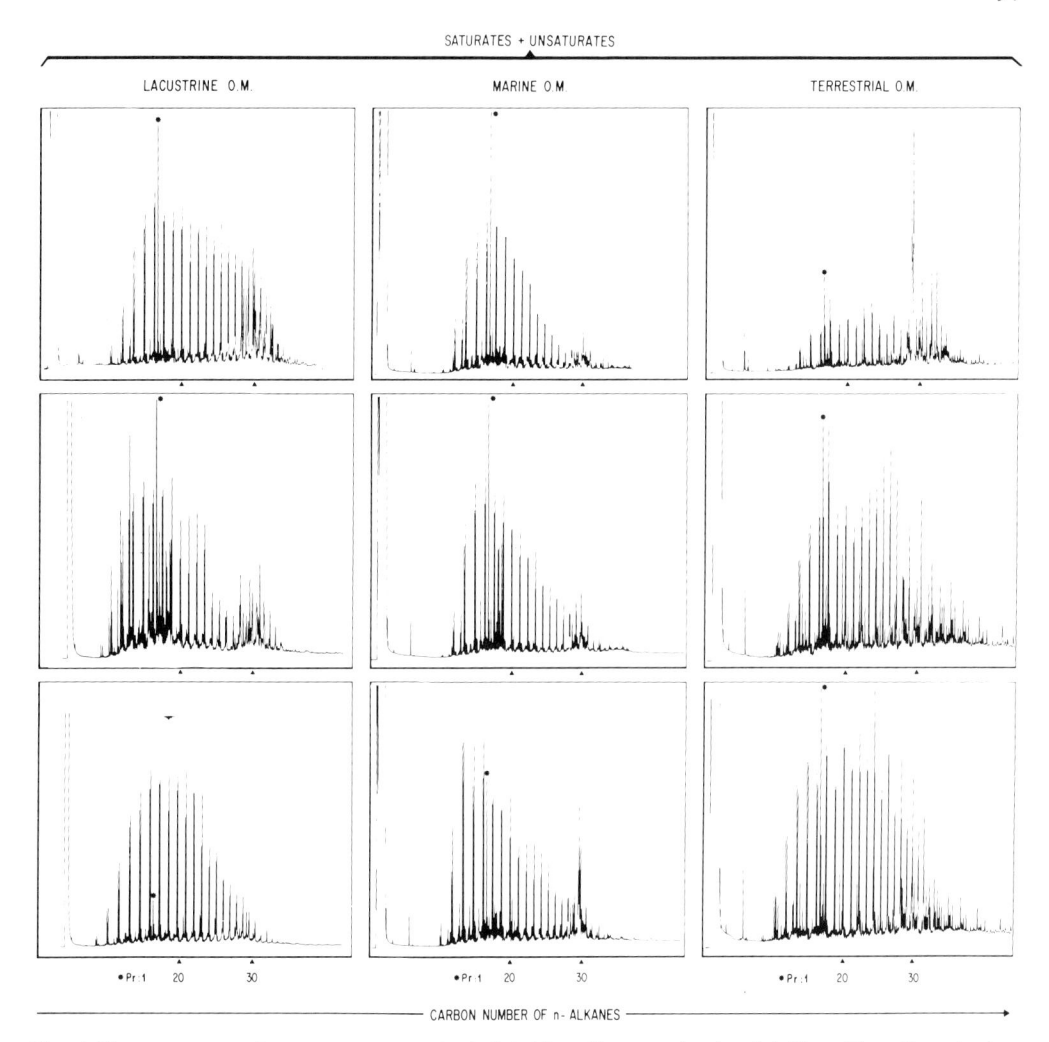

FIG. 4. Chromatograms of saturates + unsaturates isolated from C_{14+} pyrolysates of stable residues. From top to bottom: lacustrine organic matter from samples L1, L2, L3; marine organic matter from samples M1, M2, M3; terrestrial organic matter from samples T1, T2, T3.

biomarkers, particularly in steroids, compared to other iso- + cyclo-alkanes. An increase of Ph relatively to Pr is also noticeable. This could be due either to the procedure used for isolating the humic acids (differences due to molecular weight or functionality) or to different origins for these two organic fractions (condensation products in stable residues, bacterial degradation products in humid acids).

Conclusions

Our pyrolysis method allows us to characterize the carbon structures in humic acids and stable residues. They are similar to those which are generated naturally and found in extracts and oils. As in ancient sediments, three main types of organic matter can be differentiated on the basis of the structures from the point of deposition. This observation could not be deduced from bulk analysis of recent sediments, such as elemental analysis or Rock-Eval pyrolysis, because in very young sediments such data are very similar. The pyrolysis of fulvic acids shows that their carbon structures are very different from those of the other fractions, and that they do not allow us to identify the type of the organic matter from which they are formed.

This pyrolysis study seems to confirm that

m/z = 191

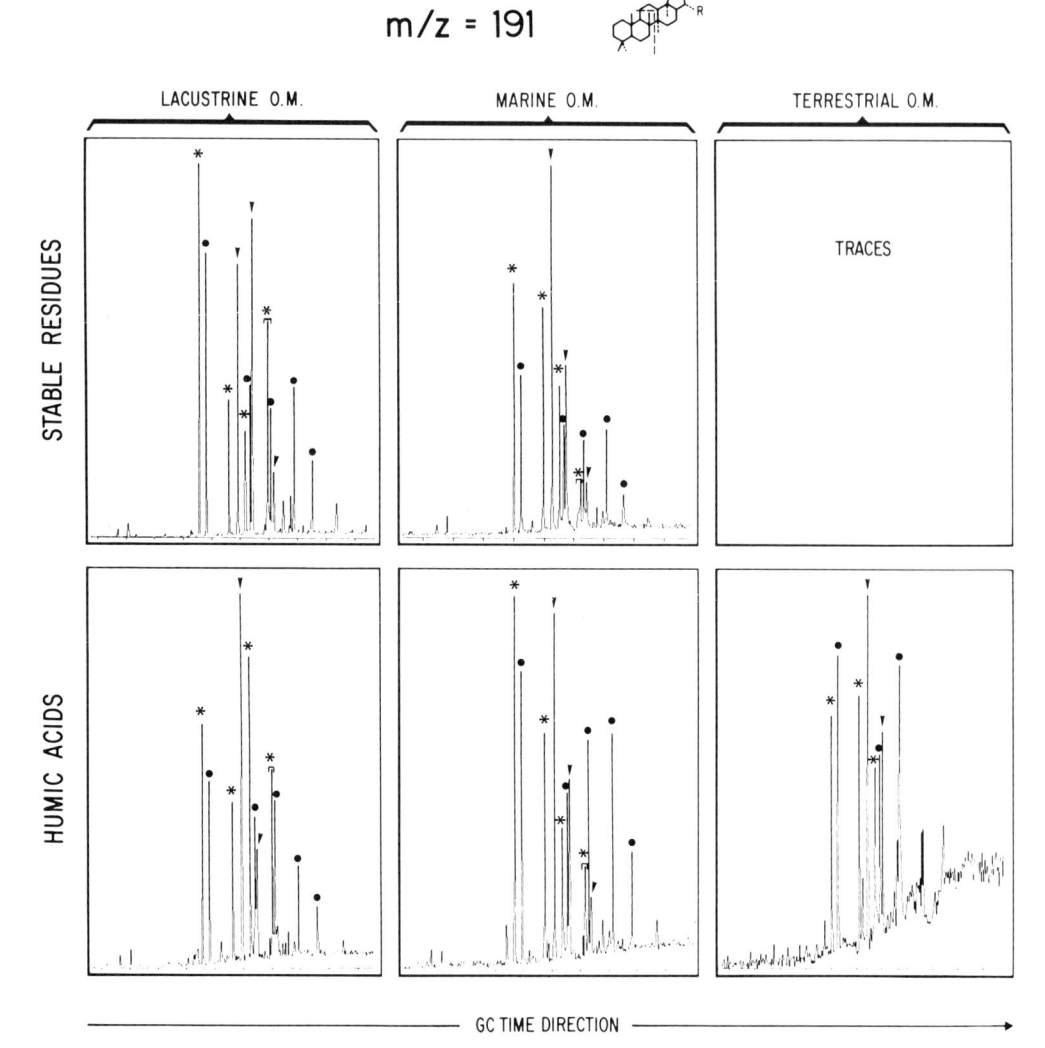

FIG. 5. Mass fragmentograms of triterpanes in pyrolysates of stable residues and humic acids from samples L1, M1, T1.
● $C_{27}\beta + (C_{29} + C_{30} + C_{31} + C_{32})\beta\beta$
▼ $(C_{29} + C_{30} + C_{31})\beta\alpha$
* $C_{27}\alpha + (C_{29} + C_{30} + C_{31})\alpha\beta$

Geochemical characterization 99

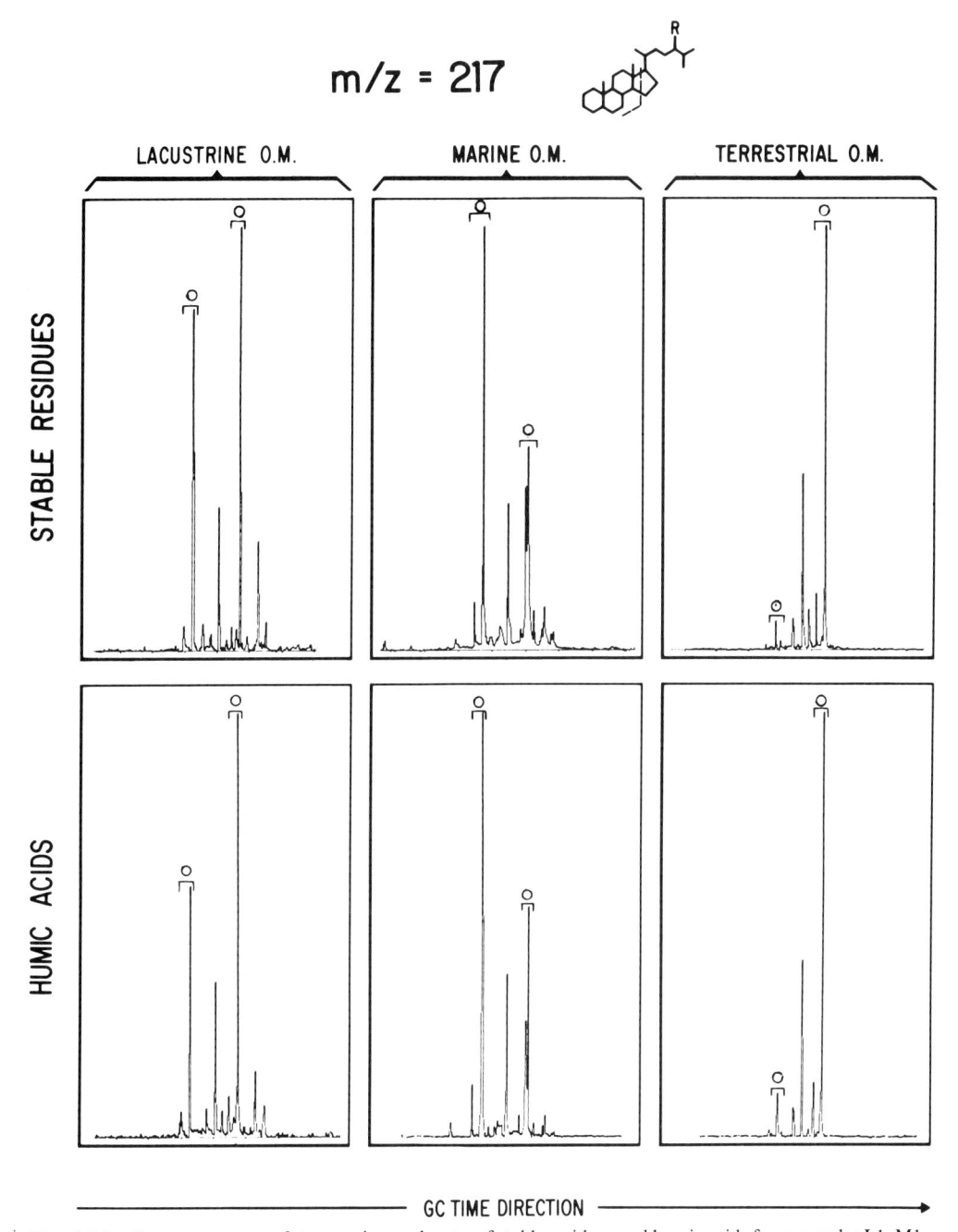

FIG. 6. Mass fragmentograms of steranes in pyrolysates of stable residues and humic acids from samples L1, M1, T1. The peaks under the circles are, from left to right, C_{27} and C_{29} steranes.

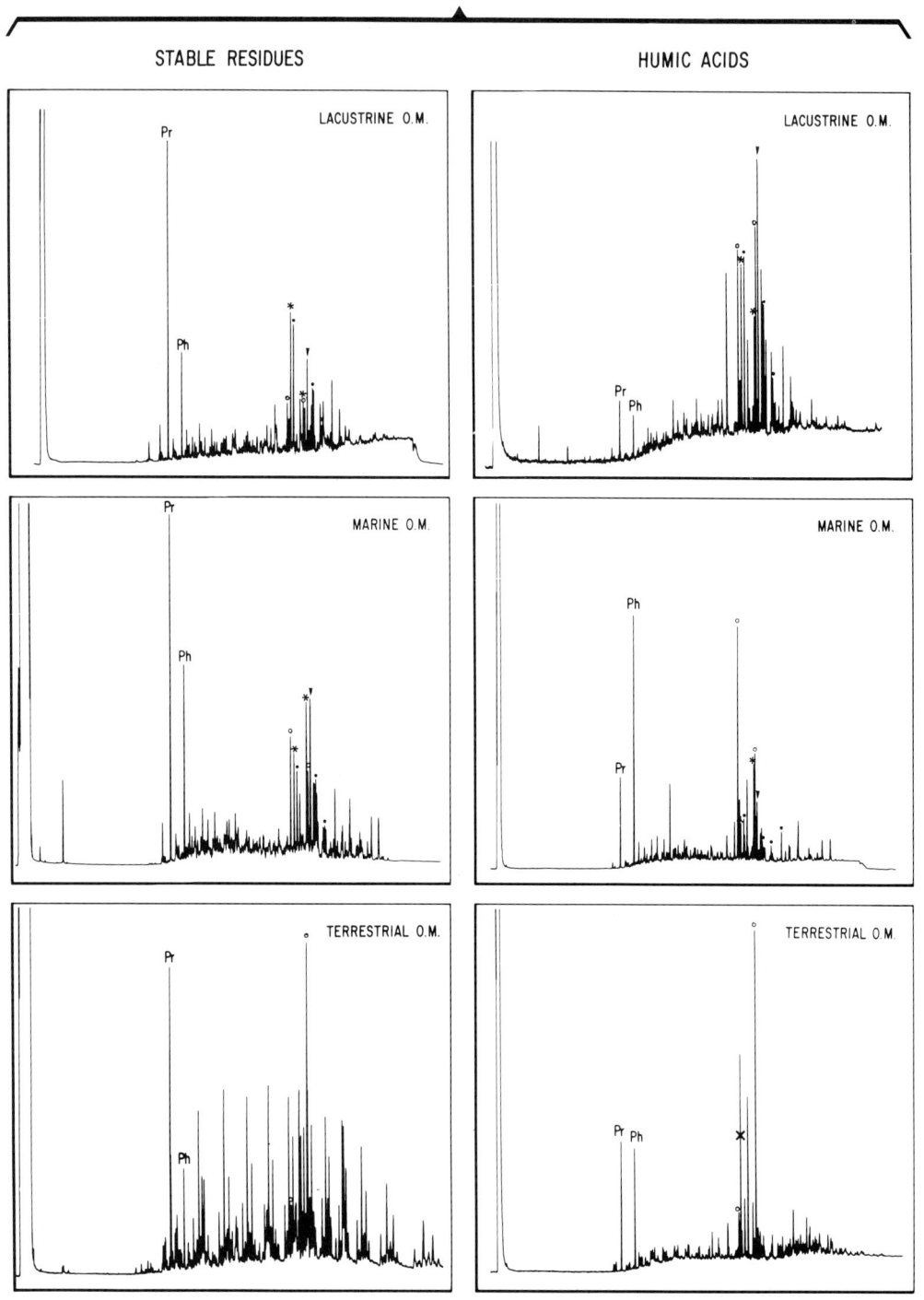

FIG. 7. Chromatograms of iso- + cyclo-alkanes isolated from pyrolysates of stable residues and humic acids from samples L1, M1, T1. Symbols are the same as on Figs 5 and 6. The cross is a solvent pollution.

lacustrine sediments are characterized by Type I kerogen. The hydrocarbon yield is high and the carbon distribution of the generated hydrocarbons is fairly flat in the range C_{15}–C_{25}, but a complementary study on other sediments from various lakes is needed to test the generality of this conclusion.

References

BEHAR, F. & PELET, R. 1985. Pyrolysis-gas chromatography applied to organic geochemistry. Structural similarities between kerogens and asphaltenes from related rock extracts and oils. *Journal of Applied and Analytical Pyrolysis*, **8**, 173–187.

BORDENAVE, M., COMBAZ, A. & GIRAUD, A. 1970. Influence de l'origine des matières organiques et de leur degré d'évolution sur les produits de pyrolyse du kérogène. In: HOBSON, G. D. & SPEERS, G. C. (eds), *Advances in Organic Geochemistry 1966*, Pergamon Press, Oxford, 389–405.

COMBAZ, A. & DE MATHAREL, M. 1978. Organic sedimentation and genesis of petroleum in Mahakam delta, Borneo. *Bulletin of the American Association of Petroleum Geologists*, **62**, 1684–1695.

DURAND, B. & NICAISE, G. 1980. Procedures for kerogen isolation. In: DURAND, B. (ed.), *Kerogen*, Editions Technip, Paris, 35–53.

HUC, A. Y., DURAND, B. & MONIN, J. C. 1978. Humic compounds and kerogens in cores from Black Sea sediments, Leg 42B—Holes 379 A, B and 380 A. *Initial Reports of the Deep Sea Drilling Project*, United States Government Printing Office, Washington, **XLII**, 737–748.

PELET, R. 1981. Géochimie organique des sédiments marins profonds du golfe d'Aden et de la mer d'Oman—vue d'ensemble. In: *ORGON IV, Golfe d'Aden, Mer d'Oman*, Editions du CNRS, Paris, 529–547.

PELET, R., BEHAR, F. & MONIN, J. C. 1986. Resins and asphaltenes in the generation and migration of petroleum. In: LEYTHAEUSER, D. & RULLKÖTTER, J. (eds), *Advances in Organic Geochemistry 1985*, Pergamon Press, Oxford, 481–498.

TIERCELIN, J. J., LE FOURNIER, J., HERBIN, J. P. & RICHERT, J. P. 1980. In: *Geodynamic Evolution of the Afro-Arabian Rift System*, Accademia Nazionale dei Lincei, Roma, **47**, 143–163.

TISSOT, B., DURAND, B., ESPITALIÉ, J. & COMBAZ, A. 1974. Influence of nature and diagenesis of organic matter in formation of petroleum. *Bulletin of the American Association of Petroleum Geologists*, **58**, 499–506.

VERHEYEN, T. V., PANDOLFO, A. G. & JOHNS, R. B. 1985. Structural relationships between a brown coal and its kerogen and humic acid fractions. *Organic Geochemistry*, **8**, 375–388.

M. VANDENBROUCKE & F. BEHAR, Institut Français du Pétrole, BP 311, 92506 Rueil Malmaison Cedex, France.

Biological marker compounds as indicators of the depositional environments of petroleum source rocks

J. K. Volkman

SUMMARY: Recent studies concerned with the use of biological marker compounds for assessing the source of organic matter in sediments and crude oils are reviewed and critically assessed. Compounds which are indicative of inputs from bacteria and phytoplankton are reviewed as well as some compounds for which origins have not been determined. The use of ratios of C_{27} and C_{29} steranes for assessing the depositional environment of petroleum source rocks is shown to be particularly open to misinterpretation. Some inferences based on pristane/phytane ratios also need to be reassessed. Areas of research where more studies would be valuable are highlighted.

Introduction

The organic matter present in sedimentary rocks consists of complex assemblages of organic molecules. These include a wide range of structural types, molecular weights, functionalities and polarities. From this mixture, the organic geochemist is able to identify compounds with specific structures which can be related to known biochemicals produced by organisms present when the sediment was deposited. From the presence, or absence, of these biological markers (commonly used synonyms include chemical fossil, molecular fossil or biomarker) one can identify the major sources of organic matter and hence deduce aspects of the sedimentary depositional environment. In some cases this type of information can be obtained from the geological context of the sample or from a study of the fossils present but often such data do not provide the whole picture. Many organisms leave only a partial fossil record and the classical techniques of geology and palaeontology provide little information concerning the chemistry or microbiology of the depositional environment. However, when one combines these techniques with organic geochemistry one can usually develop a far more detailed picture of the palaeoenvironment than that from any of these disciplines taken in isolation.

In this review, I will mainly discuss those studies of compounds in oils and sediments that have structures based on carbon, hydrogen and oxygen and which are amenable to study by capillary gas chromatography-mass spectrometry (GC-MS). This approach has been adopted since this is still the most commonly used technique in organic geochemistry largely because it allows one to separate and identify minute amounts of compounds present in the complex mixtures found in geological samples. Those compounds which still retain structural features inherited from the lipids (fats and other compounds) of living organisms also tend to be found in this fraction. This is not to say that more polar complex compounds, such as sulphur and nitrogen heterocycles, chlorins and porphyrins do not also contain 'biomarker' information but our understanding of the sources of these compounds is much less developed. Techniques based on pyrolysis and stable isotopes can also be invaluable sources of information but these areas are outside the scope of this paper.

Parameters devised by organic geochemists are now in routine use to assess the maturity of an oil and the extent to which it has been biodegraded. These are usually expressed as molecular ratios calculated from peak areas for the respective compounds in gas chromatograms or GC-MS analyses. Progress has also been made in studying the effects of migration on the composition of an oil, and a large number of biomarker parameters have been used to relate one oil to another and to relate oils with possible source rocks. Similar techniques are also widely used in environmental research for studying the fate of petroleum hydrocarbons released into the environment and for 'fingerprinting' oil spills.

Progress in the use of biomarkers to assess palaeoenvironmental conditions of deposition has been slower than in the areas mentioned above but significant advances are now being achieved. Such techniques could have considerable value to the oil exploration industry since they create the possibility of deducing what type of source rock must have been the source of an oil, simply from a study of the composition of the oil itself. Such data, combined with maturity measurements, reduce the number of potential source rocks that need to be examined and ultimately can enhance the chances of finding further oil deposits in an area. An example illustrating the potential of this approach has recently been published (McKirdy et al. 1986 and this volume).

There is a demonstrable need for more organic geochemistry studies of a range of sediment types. These environments should be well defined in terms of their geological and biological settings so that a reliable inventory of biological markers for different depositional environments can be established. Both Recent and ancient sediments should be studied so that the effects of diagenesis can be ascertained. Unfortunately, many potentially valuable biomarkers do not survive for sufficiently long in the geological record to be of use for petroleum studies. Another area of research that does not receive the attention it deserves is the study of the lipid compositions of organisms likely to contribute organic matter to sediments. Such studies can lead to the identification of new biological markers and help to clarify what compounds are truly characteristic of a particular environment. Several excellent reviews have been published detailing some of the organic compounds found in Recent sediments (e.g., Simoneit 1978; Cranwell 1982; Johns 1986). A bibliography of studies in marine organic geochemistry is also available (Gagosian 1983), and Mackenzie (1984) has written a valuable overview of organic geochemistry techniques applied to petroleum research. In view of this extensive literature, the present review is mainly concerned with more recent developments particularly those related to the use of biomarkers for assessing palaeoenvironmental depositional conditions. Some of the difficulties and ambiguities in the use of biological markers will also be highlighted.

Marine and lacustrine environments

The terms 'marine' and 'lacustrine' clearly signify different depositional types, but they fail to convey the enormous diversity of environments included within these categories. In the broader sense, marine environments include coastal, shelf and open ocean settings, each of which includes recognizably different areas in terms of productivity and depositional environment. Similarly, the term lacustrine could include freshwater and saline lakes of vastly differing areas and ecologies, hypersaline environments such as sabkhas, playa lakes and algal mat communities to name just a few. Each of these environments will leave a different lipid signature in the sedimentary rocks and the task facing the organic geochemist is to determine which compound, or group of compounds, is characteristic of each type. Although some of these environments are unlikely to give rise to petroleum source rocks, and thus they may seem to be of academic interest to the petroleum geologist, it is only from a comparison of organic geochemical data from all these different environments that the validity or usefulness of a biological marker can be established.

Many environments have features in common, so it should come as no surprise that the distribution of compounds in some lipid classes look remarkably similar irrespective of the type of sediment. Pentacyclic triterpenoids of the hopane type provide an excellent example. Derivatives of hopanoids occur widely in bacteria and, hence, they have been found in every sediment studied to date (Ourisson et al. 1979). However, differences in the proportions of compounds within this series have been noted which may be useful as palaeoenvironmental indicators (see later). This illustrates that the amount of a given biomarker may be of more significance than merely its presence or absence. For example, the ratio of steranes to hopanes (Mackenzie et al. 1982), or of methylsteranes to desmethylsteranes (Shi Ji-Yang et al. 1982), can be used to group related source rocks or oils, but the underlying reasons for these variations in terms of different depositional environments are still poorly understood (Mackenzie 1984).

The main sources of organic matter in sediments are phytoplankton, bacteria and vascular plants with (usually) smaller contributions from aquatic fauna and other organisms. Within each group of organisms one can expect that there will be some common features of biochemistry as well as some major differences. Sometimes a single species might contain an unusual lipid distribution, but more commonly a group of taxonomically-related organisms have distinctive and diagnostic lipid compositions. However, it is only when such organisms are restricted to a specific environment that the organic geochemist will find compounds useful as biological markers for a given depositional environment.

Biological markers for vascular plants

It is a relatively straightforward task to recognize compounds derived from vascular plants in immature sediments. Typically one finds high concentrations of long-chain (C_{25}–C_{35}) n-alkanes with a strong predominance of odd chain-lengths, long-chain (C_{22}–C_{32}) n-alkanoic acids and n-alkanols with a high predominance of even chain-lengths, certain hydroxy acids and dicarboxylic acids characteristic of plant cutin and suberin, diterpenoid (resin) acids, long-chain (C_{40}–C_{64}) saturated alkyl (wax) esters and pentacyclic triterpenoids of the oleanane, ursane, friedelane and lupane type. These compounds are discussed in detail by Cranwell (1982).

In more mature samples one finds fewer biological markers (see Fig. 1 for examples), since many of the compounds mentioned above are degraded or converted to other compounds. Over geological time most compounds are broken down but some are simply defunctionalized with, in many cases, concomitant aromatization, ring cleavage or rearrangement. A good example is the formation of the aromatic hydrocarbon retene (II) from abietic acid (I) which occurs quite early in the diagenetic sequence (e.g., LaFlamme & Hites 1978). The clear link established between this product and its precursor permits one to use retene as a biological marker for organic matter from conifers. However, there are many other diagenetically formed compounds for which a clear product–precursor relationship has not yet been established and the information potential of the compound remains unexploited. A good example is perylene (III), the origin of which is still the subject of much debate (e.g. Wakeham et al. 1979).

Pentacyclic triterpanes

Crude oils contain only a few unambiguous biological markers for vascular plants. Pentacyclic triterpanes have proven to be the most useful and of these 18α(H)-oleanane (V) is the most commonly cited (Brassell & Eglinton 1983a). This pentacyclic alkane has been identified in many oils and shales, most of which can be clearly attributed to a terrestrial source and in some cases specifically to a deltaic depositional environment. Examples include oils and shales from the Niger (Hills & Whitehead 1966; Ekweozor et al. 1979a, b) and Mahakham deltas (Hoffman et al. 1984), oils from south-east Hungary (Sajgo 1984) and Brunei, Sabah and Indonesia (Gran-

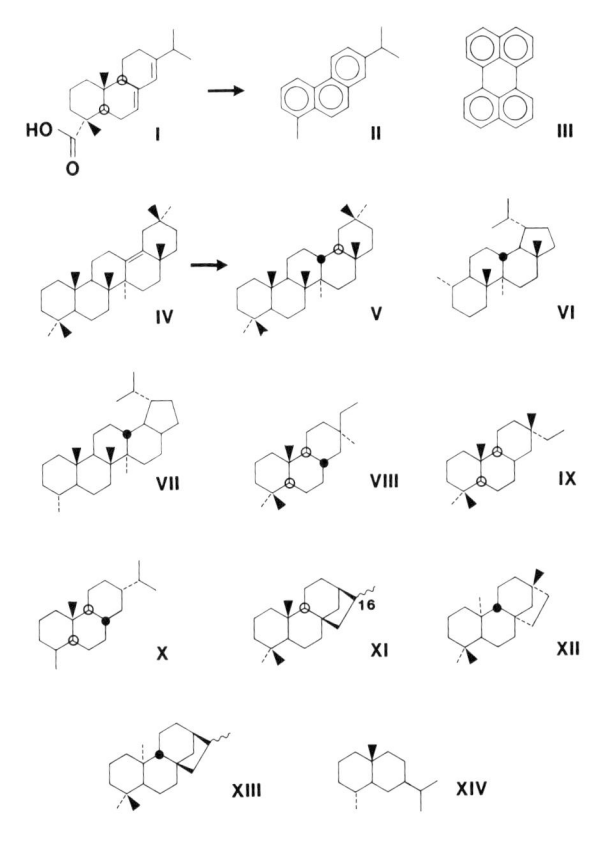

FIG. 1. Structures of compounds proposed as biological markers for vascular plants. I: abietic acid; II: retene (diagenetically formed from abietic acid); III: perylene (origin unknown, doubtful status); IV: olean-13(18)-ene; V: 18α(H)-oleanane; VI: C_{24} cycloalkene (presumed diagenetic product from cleavage of the A-ring of 3-oxytriterpenoids); VII: 28,30-bisnorlupane (status uncertain); VIII: pimarane; IX: isopimarane; X: abietane; XI: phyllocladane; XII: *ent*-beyerane; XIII: *ent*-kaurane; XIV: eudesmane.

tham et al. 1983) and source rocks from China (Brassell et al. 1986a). An unsaturated analogue, olean-13(18)-ene (IV), has been identified in immature shales from the Niger Delta (Ekweozor et al. 1979a), and this compound is thought to be an intermediate in the diagenetic transformation of oxygenated 'oleanoids' (Brassell & Eglinton 1983a) to 18α(H)-oleanane.

It should be noted that the presence of 18α(H)-oleanane, or other triterpenoids of higher plant origin, in an oil is not sufficient evidence that the oil originated from a terrestrial source rock. Large amounts of terrigenous matter also occur in some marine sediments from coastal and deep-sea areas. Brassell et al. (1981a) identified olean-12-en-3-one, urs-12-en-3-one and taraxer-14-en-3-one in immature sediments from the Middle America Trench and a similar range of ketones and alcohols has also been found in sediments from the Japan Trench (Brassell & Eglinton 1983a). Both suites of sediments contained significant amounts of other biological markers for vascular plants. Some deep-sea sediments (Brassell & Eglinton 1983a) also contain a series of C_{24} cycloalkenes and cycloalkanes (e.g., VI) which appear to be derived from cleavage of the A-ring of 3-oxytriterpenoids (Corbet et al. 1980), and thus indirectly indicate the presence of terrigenous organic matter. 28,30-Bisnorlupanes (VII) which are abundant in some Tertiary sediments from West Greenland, may also represent degradation products of oxygenated lupanes of plant origin (Rullkotter et al. 1982a).

Tri- and tetracyclic diterpanes

The isolation and identification of polycyclic diterpanes, and their oxygenated analogues, in coals, lignins, resins, sediments and oils is an active area of organic geochemistry research which is too broad to review here. Tricyclic diterpanes with pimarane (VIII), isopimarane (IX) and abietane (X) skeletons, which are abundant in conifer resins, are commonly reported as are tetracyclic diterpanes having phyllocladane (XI), *ent*-beyerane (XII) and *ent*-kaurane (XIII) skeletons (Noble et al. 1985). Although a few diterpanes are not derived from vascular plants (e.g., Zumberge 1983), most are unambiguous biological markers for organic matter of plant origin and in some cases they can be related to specific types of plants. However, the diagenetic pathways by which these hydrocarbons are produced in sediments are not well understood, and many structures still remain to be elucidated. For leading references, the reader is referred to Simoneit (1977), Livsey et al. (1984) and Mackenzie (1984).

Bicyclic alkanes

C_{14}–C_{16} alkanes are found in most crude oils and these can be readily recognized from m/z 123 mass fragmentograms. Some have drimane or rearranged drimane skeletons which are probably formed from hopanoid precursors during diagenesis (Kagramanova et al. 1976; Alexander et al. 1984) whereas others, such as those with the eudesmane skeleton (XIV; Alexander et al. 1983) are clearly linked to a higher plant source. In crude oils, these lower molecular weight cyclic alkanes are often more abundant than pentacyclic triterpenoids but their potential as source and diagenesis indicators has yet to be exploited fully.

General comment

The above data highlight a general problem facing the organic geochemist. Even though a given compound may have a known source, such as vascular plants in this case, this is usually not sufficient evidence to define the palaeoenvironment. In most cases, information from several biological markers must be obtained before a self-consistent picture can be established. The absolute concentration of the biomarker also provides important information. Clearly, one would only expect high concentrations of biological markers for vascular plants in environments where terrigenous matter accumulates such as in lakes, estuaries and some coastal sediments but moderate amounts have been detected in some deep-sea sediments far from land as a result of inputs from turbidity currents and atmospheric transport. In cases where most of the terrigenous organic matter is from aeolian sources one would expect to find mainly plant wax compounds such as long-chain hydrocarbons, fatty acids and alcohols but not oxygenated triterpenoids or steroids (e.g., Gagosian et al. 1982). Where riverine inputs are important, the organic matter is likely to be highly reworked and degraded but some of the more resistant compounds, such as certain triterpenoids and 24-ethylcholesterol, are likely to be present in significant amounts.

Biological markers for phytoplankton

The lipid compositions of most phytoplankton species are dominated by esterified fatty acids (mainly triacylglycerols and phospholipids) together with smaller amounts of free fatty acids, hydrocarbons and sterols. The use of sterols as biological markers for phytoplankton has been discussed recently (Volkman 1986). Fatty acid distributions are often quite different from one algal class to another and so these compounds are often used to determine sources of organic matter in Recent sediments (e.g., Volkman et al. 1980c). Hydrocarbon distributions tend to be less distinctive and generally of less value as biological markers. A possible exception is the occurrence of 7- and 8-methylheptadecane in cyanobacteria. Also, hydrocarbons are generally not abundant in algae. Both compound classes tend to be degraded in sediments either because the compounds are relatively short-chain (i.e., $<C_{20}$) or contain several double bonds. Consequently, with few exceptions, they tend to be of limited use as palaeoenvironmental indicators for ancient sediments. Fortunately for the organic geochemist a few algae synthesize unusual lipids with distinctive structures which can be related to geologically occurring compounds.

Botryococcane

Botryococcane (XV, Fig. 2) is an unusual C_{34} acyclic isoprenoid alkane which appears to be very restricted in its geological occurrence. It is thought to derive from hydrogenation of C_{34} alkenes, termed botryococcenes, which have the same skeleton and occur in very high concentration in some races of the freshwater green alga *Botryococcus braunii* (Maxwell et al. 1968; Cox et al. 1973; Metzger et al. 1985) but not in other algae. Biopolymers produced by this alga have been shown to be an important source of algal kerogens (Largeau et al. 1986). Recently, a number of related alkenes having the general formula C_nH_{2n-10} (n = 30–37) have also been identified in the same alga (reviewed by Volkman & Maxwell 1986).

Botryococcane, and related alkanes, have been found in oils from Indonesia (Moldowan & Seifert 1980; Seifert & Moldowan 1981), which is consistent with the lacustrine source inferred for these oils. More recently, Brassell et al. (1986a and this volume) have identified C_{31} and C_{33} 'botryococcanes' in the lacustrine Maoming oil shale in China. It is interesting that this shale did not contain the C_{34} botryococcane which could indicate that the algal source was a different race of *B. braunii* to that which is common in lakes today.

Recently, botryococcane was found in a number of coastal bitumens recovered from stranding sites in South Australia and Victoria, Australia (McKirdy et al. 1986 and this volume). Carbon isotope data indicate that these bitumens originate from algal organic matter and the abundance of botryococcane suggested that deposition occurred in a fresh to saline lacustrine environment. Those bitumens containing most botryococcane exhibited pristane to phytane ratios of about 2, sulphur contents of about 0.3%, high hopane to sterane ratios (~ 55) and high concentrations of 4-methyl steranes; the latter probably originating from the 4-methyl sterols of dinoflagellates. From these data, McKirdy et al. (1986) postulated that these oils were derived from extensive organic

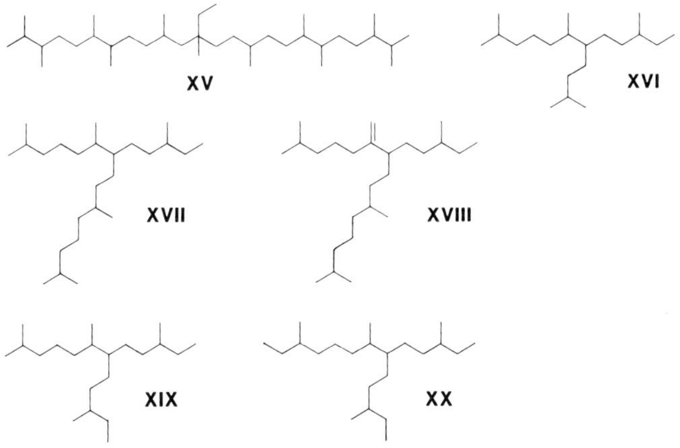

$CH_3(CH_2)_{13}CH=CH(CH_2)_5CH=CH(CH_2)_{12}COCH_3$ XXI

$CH_3(CH_2)_{13}CH=CH(CH_2)_5CH=CH(CH_2)_{12}COCH_2CH_3$ XXII

FIG. 2. Structures of some unusual acyclic lipids. XV: botryococcane; XVI: 2,6,10-trimethyl-7-(3-methylbutyl)-dodecane; XVII: 2,6,10,14-tetramethyl-7-(3-methylpentyl)-pentadecane; XVIII: alkene of XVII; XIX, XX: C_{21} and C_{22} analogues of XVII; XXI: heptatriaconta-15,22-dien-2-one; XXII: octatriaconta-16,23-dien-3-one.

rich source beds deposited in deep meromictic lakes along Australia's southern continental margin. This was an important result since potential source rocks have yet to be penetrated by the drill and the geochemical data conclusively demonstrated the presence of a new class of non-marine crude oils in this region. Another interesting feature of the data was that these bitumens were very waxy (i.e., they contained high concentrations of long-chain alkanes) despite being primarily of algal origin. Waxy crudes are usually attributed to a terrigenous source since long-chain compounds are abundant in vascular plants, but these and other data suggest that waxy crudes can also be produced from organic matter derived from certain algae such as *Botryococcus*.

2,6,10-trimethyl-7-(3-methylbutyl)-dodecane and related compounds

This C_{20} alkane (XVI) has an unusual isoprenoid structure composed of four isoprene units linked in a non-linear fashion. This compound appears to be just one member of a new class of isoprenoid compounds which also includes C_{25} and C_{30} analogues, and perhaps C_{21} and C_{22} hydrocarbons as well (Dunlop & Jeffries 1985). The value of such compounds as palaeoenvironmental indicators is still not clear but their wide occurrence in many Recent sediments is sufficient reason to merit further study.

The C_{20} alkane, and a related monoene, are widely occurring in both marine and lacustrine Recent sediments (reviewed by Rowland et al. 1985), although only rarely are they major constituents. There are few reports of their occurrence in older sediments but this may reflect the fact that the C_{20} alkane is poorly resolved from pristane and hence easily overlooked unless GC-MS techniques are used. Significant amounts of XVI have been found in Plio-Pleistocene marine sediments from the eastern Mediterranean and the alkene occurs in marine sediments from the Cariaco Trench (see Rowland et al. 1985). This alkane does not appear to be abundant, or widespread, in crude oils with the notable exception of the Rozel Point crude oil from Utah where it is the second most abundant alkane present (Yon et al. 1982). Since the source of this oil is not yet known, it is of obvious interest to the geochemist to determine the type of depositional conditions which gave rise to such a high abundance of this unusual alkane. Recent studies indicate that the source rocks could have been deposited in a hypersaline environment (see later).

Recently, Rowland et al. (1985) identified the alkane and a C_{20} monoene in the macroscopic green alga *Enteromorpha prolifera*. It seems unlikely that this particular species is the source of these hydrocarbons in most sediments but this finding does point to an origin from algae rather than from bacteria. Significant amounts of the alkane and alkene have been found in particulate matter from estuarine waters (Bayona et al. 1983; Albaiges et al. 1984b) suggesting a planktonic source. High concentrations have also been found in sediments from a hypersaline basin (Dunlop & Jeffries 1985), and it is of interest that oceanic sediments nearby contained only small amounts.

Many more sediments contain high concentrations of acyclic C_{25} alkenes having two, three and four double bonds (see papers by Albaiges et al. 1984a, b; Dunlop & Jeffries 1985; Rowland et al. 1985 for leading references). On hydrogenation, these are reduced to a single isoprenoid alkane thought to be 2,6,10,14-tetetramethyl-7-(3-methylpentyl)pentadecane (XVII; Bayona et al. 1983; Rowland et al. 1985) which is poorly separated from n-C_{21} alkane on non-polar GC phases. These alkenes do not appear to be present in freshwater lacustrine environments, and the fact that they are abundant in organic rich sediments from the Peru upwelling (Volkman et al. 1983a) is significant since similar environments are thought to have been important in the formation of some marine petroleum source rocks. It seems likely that a planktonic organism produces at least some of these compounds since high concentrations of C_{25} trienes and tetraenes have been found in sediment traps deployed in the waters off Peru (Volkman et al. 1983a).

Enteromorpha prolifera also contains a C_{25} acyclic diene which probably has the same skeleton as the C_{25} alkenes which occur in marine sediments (Rowland et al. 1985), but differs in the positions or geometry of the double bonds. A related monoene, thought to have structure XVIII, has also been identified as the major hydrocarbon in sediments from the hypersaline basin studied by Dunlop & Jeffries (1985). These sediments also contained significant amounts of related C_{21} and C_{22} acyclic isoprenoid alkanes (tentatively assigned as XIX and XX), which dominated the hydrocarbon profiles in deeper sediments attesting to their stability (Dunlop & Jeffries 1985). These hydrocarbons may well prove to be useful markers for this type of hypersaline environment. A branched C_{25} triene, of unknown origin, has also been found as a major constituent in a sapropelic mud from a playa lake (Albaiges et al. 1984a).

Very long-chain alkenones, esters, and alkenes

In 1980, Volkman et al. (1980b) reported that a cosmopolitan marine alga, the coccolithophorid

Emiliania huxleyi (Prymnesiophyceae), synthesizes an unusual suite of straight-chain unsaturated C_{37} and C_{38} methyl ketones (XXI) and alkenes and C_{38} and C_{39} ethyl ketones (XXII). In subsequent work, Volkman *et al.* (1980a) also identified methyl and ethyl esters of a diunsaturated C_{36} n-alkenoic acid and various C_{31} and C_{33} n-alkenes in this alga. The ketones are abundant in a wide variety of marine sediments (Volkman *et al.* 1980b; De Leeuw *et al.* 1980; Marlowe *et al.* 1984a, b and references therein) and for a time it was thought that they might be specific markers for organic matter derived from *E. huxleyi*, at least in Recent sediments. Perhaps not surprisingly, further work (Marlowe *et al.* 1984a, b) established that a few other species of Prymnesiophytes, which unlike *E. huxleyi* do not bear coccoliths, also synthesize these compounds although the distributions are not identical. Subsequently Cranwell (1985) and Volkman *et al.* (1986b) showed that these compounds do occur in some lake sediments, but again the distributions are distinctively different from those which occur in marine sediments. These latter results imply that other species of algae must also be capable of synthesizing these compounds.

The rapid expansion in our knowledge of the sources of these ketones and their applications in organic geochemistry is typical of many of the advances in this field of research. In particular, this topic highlights how dramatic improvements in analytical methodology for sediment lipid analyses have not been matched by a corresponding increase in our knowledge of the occurrence of these newer compound classes in plants, animals and bacteria. These long-chain ketones were first recognized in sediments from Walvis Ridge (Boon *et al.* 1978) using direct insertion mass spectrometry but full details of their structures were not published until 1980 (De Leeuw *et al.* 1980; Volkman *et al.* 1980a), largely due to the difficulty of analysing these high molecular compounds with capillary columns available at that time. Capillary gas chromatography techniques have advanced dramatically since then and it is now possible to study very high molecular weight compounds such as wax esters and triacylglycerols (e.g., Cranwell & Volkman 1981) and even porphyrins (e.g., Marriott *et al.* 1982) directly by gas chromatography.

Another important consideration is that these unusual ketones and hydrocarbons had not been reported in the natural products literature of the time and it was only through the efforts of organic geochemists that plausible biological sources were identified (Volkman *et al.* 1980a; Marlowe *et al.* 1984a) which accounted for many of the known geological occurrences. New compounds are continually being identified in sediments but all too often the biological sources of these compounds have not been identified.

Although the specificity of these compounds as biological markers for marine Prymnesiophycean algae now appears to be doubtful, these studies have nonetheless provided much important information. Firstly, they showed that unicellular algae can be significant contributors of very long (C_{37}–C_{40}) straight-chain lipids to sediments with the logical extension that waxy crude oils could have an algal origin. Secondly, the occurrence of these compounds in sediments over 100 million years old (Brassell *et al.* 1986b) attests to the remarkable geological stability of such long-chain compounds. Thirdly, feeding experiments in which *E. huxleyi* was fed to the copepod *Calanus helgolandicus* (Volkman *et al.* 1980b) demonstrated that these compounds are also resistant to biological degradation and hence they tend to survive in biologically active sediments in which shorter-chain lipids are rapidly degraded (Volkman *et al.* 1986b). This may be due to their long chain-length and/or due to the presence of *trans* double bonds instead of the more biologically common *cis* geometry which has recently been demonstrated from infra-red spectrometry studies (J. K. Volkman, unpublished data, 1985).

Lastly, and perhaps of most interest, is that the ratio of ketones having two and three double bonds appears to correlate with growth temperature which has raised the possibility of using these stable compounds as palaeoclimatic indicators (Brassell *et al.* 1986b).

Biological markers for bacteria

Bacteria are now recognized as important sources of lipids in sediments and petroleums (Ourisson *et al.* 1984) and most of the main bacterial groups have distinctive distributions. Since environmental conditions often determine which bacterial groups predominate in any given sediment there is a real prospect of finding specific bacterial lipids which can be used as markers for certain depositional conditions.

Fatty acids are major constituents of many bacteria, with the notable exception of the Archaebacteria, and specific acids characteristic of different groups are readily recognized in most Recent sediments (e.g., Perry *et al.* 1979; Gillan & Sandstrom 1985). Unfortunately, these fatty acids are readily degraded and thus of limited value for studying the ancient sediment record. Other more stable lipids such as acyclic isoprenoids, ether lipids and triterpenoids (Figs 3 and 4) have a long geological record and are potentially much more useful.

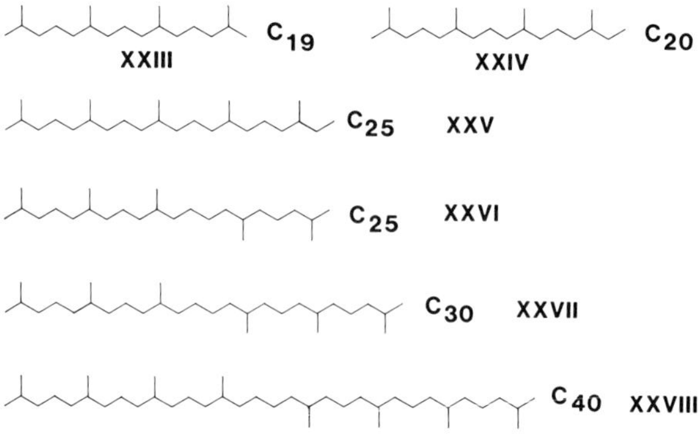

FIG. 3. Structures of acyclic isoprenoid alkanes which may be indicative of bacterial inputs. XXIII: pristane (probably not bacterial); XXIV: phytane; XXV: 2,6,10,14,18-pentamethyleicosane; XXVI: 2,6,10,15,19-pentamethyleicosane; XXVII: squalane; XXVIII: lycopane (status uncertain).

One group of bacteria that is of particular interest to organic geochemists is the Archaebacteria which includes the methanogens, thermoacidophiles and haloalkaliphiles (Woese et al. 1978). Methanogenic bacteria synthesize a variety of lipids based on acyclic isoprenoid skeletons. These include free hydrocarbons and glyceride-analogues where the ester bond to glycerol is replaced by an ether bond. The C_{25} isoprenoid alkane 2,6,10,15,19-pentamethyleicosane (XXVI), which has an irregular isoprenoid skeleton, appears to be a useful marker for methanogenic bacteria (Brassell et al. 1981b).

These bacteria can also be important sources of phytane (XXIV) and C_{30} acyclic isoprenoids such as squalane (XXVII) and partially reduced squalenes in sediments (Brassell et al. 1981b; Albaiges et al. 1984a; Volkman et al. 1986a). C_{40} alkanes, including tail-to-tail linked lycopane (XXVIII), have also been identified in oils and these are also thought to be derived from Archaebacteria (Michaelis & Albrecht 1979; Brassell et al. 1981b; Albaiges et al. 1985) although there are insufficient data to link them specifically to methanogens. These bacteria do not appear to be a significant source of pristane

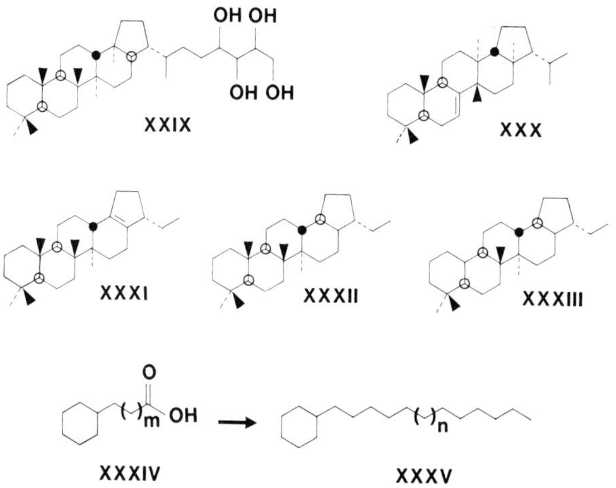

FIG. 4. Structures of cyclic hydrocarbons associated with bacterial inputs. XXIX: C_{35} hopanetetraol (probable precursor of most hopanoids in sediments); XXX: fern-7-ene; XXXI: 28,30-bisnorhop-17(18)-ene; XXXII: 28,30-bisnorhopane; XXXIII: 25,28,30-trisnorhopane; XXXIV: C_{17} ω-cyclohexyl fatty acid; XXXV: alkylcyclohexane (possibly diagenetically derived from XXXIV).

(XXIII; Risatti et al. 1984). For a further discussion of this topic, the reader is referred to the review by Volkman & Maxwell (1986).

The Archaebacteria also includes an interesting group which thrive under harsh saline and alkaline conditions. These halophilic and haloalkaliphilic bacteria synthesize lipids containing acyclic isoprenoid skeletons such as phytane (XXIV) and 2,6,10,14,18-pentamethyleicosane (XXV). Such bacteria are important sources of lipids in present day Dead Sea sediments (Anderson et al. 1977) and ancient hypersaline environments (see later).

Pristane/phytane ratios have been used for many years as indicators of redox potential in sediments (Didyk et al. 1978), with a surprising degree of success when one considers that the underlying premise that both compounds are derived from phytol is clearly not correct in many cases. Archaebacteria can be important sources of phytane and thus low pristane/phytane ratios may simply reflect high populations of methanogenic bacteria in that depositional environment. Phytane can also be contributed from haloalkaliphilic bacteria and so low pristane/phytane ratios are also found in hypersaline environments (ten Haven et al. 1985 and this volume). An additional source of pristane is from tocopherols (Brassell et al. 1983). Clearly, pristane/phytane ratios need to be used with caution but they are still useful provided that the importance of Archaebacterial contributions has been established from other biological markers, such as C_{25}, C_{30} and C_{40} acyclic isoprenoid alkanes, diphytanylglyceryl ethers and dibiphytanylglyceryl ethers (Chappe et al. 1979, 1982; Moldowan & Seifert 1979; Albaigés 1980; Volkman & Maxwell 1986).

Pentacyclic compounds with the hopane skeleton (XXIX) are abundant in many bacteria and consequently hopanes have been found in most mature sediments and crude oils (e.g., Ourisson et al. 1979, 1984). Although the hopanes are particularly useful as indicators of thermal maturity (see Mackenzie 1984, for a review), to date they have been of limited value as palaeoenvironmental indicators. High concentrations of C_{34} and C_{35} hopanes may be indicative of hypersaline conditions (McKirdy et al. 1983), but it is generally true that the distribution of hopanes are remarkably similar in quite unrelated oils. Mackenzie (1984) has noted that a high concentration of hopanes relative to steranes is common in samples derived primarily from Type III organic matter (i.e., terrestrial vascular plants), and that such ratios are useful for correlation purposes. Little use has been made of ratios which compare the concentrations of different lipid classes or compound types, despite the fact that quantitative differences exist between environments (e.g., Comet 1982). Such parameters may well prove to be very useful indicators of palaeoenvironmental conditions. For example, Volkman et al. (1986a) observed high concentrations of acyclic isoprenoid alkanes relative to hopanoid hydrocarbons in sediments from an anoxic meromictic lake which reflected the high proportion of the microbial biomass that was derived from methanogenic bacteria in that environment.

Recent studies have demonstrated that bacteria are possible sources of other pentacyclic triterpenoids such as fernenes, 28,30-bisnorhopane and 25,28,30-trisnorhopane (Fig. 4). Fernenes (XXX) have been identified in several Recent and ancient sediments and it is now thought that they are mainly derived from certain bacteria and not from ferns, despite the abundance of fernenes in the latter (Brassell & Eglinton 1983a; Volkman et al. 1986a). Fern-7-ene and fern-9(11)-ene have been identified as minor lipid constituents of an anaerobic photosynthetic purple bacterium (Howard 1980) but there are still no other reported occurrences in bacteria perhaps because few studies have been carried out. The recent identification of fern-7-ene (XXX) as a major hydrocarbon in anoxic sediments from a saline, meromictic lake in Antarctica (Volkman et al. 1986a), shows that this hydrocarbon is a major constituent in at least one bacterium which is clearly important as a source of sedimentary lipids. Volkman et al. (1986a) also proposed that the presence of fernenes in sediments could indicate that the sediments were deposited under highly reducing conditions, since they found only trace amounts of fernenes in oxic sediments from the same lake. The saturated analogue of fernenes, i.e., fernane, has not yet been identified in ancient sediments or crude oils.

28,30-bisnorhopane (XXXII) and a related C_{27} triterpane, 25,28,30-trisnorhopane (XXXIII) have been identified in a limited number of sediments and petroleums (Seifert et al. 1978; Grantham et al. 1980; Rullkotter et al. 1982a, b; Volkman et al. 1983a; Fowler & Douglas 1984; Moldowan et al. 1984). This bisnorhopane is particularly abundant in the Miocene Monterey Formation (Seifert et al. 1978), and it has been suggested that it may be derived from reduction of 28,20-bisnorhop-17(18)-ene (XXXI) which has been tentatively identified in a sediment from the Gulf of Carpentaria (Rullkotter et al. 1982b). Grantham et al. (1980) and Fowler & Douglas (1984) have suggested that it may be associated with anaerobic microorganisms, but the full significance of these compounds has yet to be established.

Alkylcyclohexanes (XXXV) containing a long

hydrocarbon chain attached to a cyclohexane or methylcyclohexane ring are major constituents of some crude oils. It is possible that they originate from the cyclization of algal fatty acids (Johns *et al.* 1966), but an origin from bacterial lipids is also considered likely (e.g., Fowler & Douglas 1984 and refs. therein) since structurally related C_{17} and $C_{19}\omega$-cyclohexyl fatty acids (XXXIV) have been found in the thermoacidophilic bacterium *Bacillus acidocaldarius* (De Rosa *et al.* 1972). The presence of high concentrations in some organic-rich Ordovician rocks (Fowler & Douglas 1984), suggests that vascular plants are not probable sources of these compounds, at least in those sediments.

The importance of green sulphur bacteria (Chlorobiaceae) as a source of compounds in reef-hosted oils in the Silurian of the Michigan Basin, Canada, and the Devonian of Western Canada has recently been demonstrated by Summons & Powell (1986). These authors identified a suite of 1-alkyl-2,3,6-trimethylbenzenes as major constituents and showed from carbon isotope data and the low abundance of other isomers that these hydrocarbons were probably derived from specific aromatic bacterial carotenoids. These oils also had very low pristane/phytane ratios consistent with the strongly reducing conditions required by anoxygenic photosynthetic sulphur bacteria.

Sterols and steranes

The stability and distinctive structure of the steroid skeleton (XXXVI; Fig. 5) has made it possible to study in great detail the transformation reactions that convert the sterols found in Recent sediments to sterenes in ancient sediments and steranes in crude oils and mature sediments.

These processes have been reviewed elsewhere (Mackenzie *et al.* 1982; Brassell & Eglinton 1983a; Brassell *et al.* 1983, 1984; McEvoy & Maxwell 1983), so only those aspects which relate to the use of steroids as source indicators will be discussed here.

Complex mixtures of sterols occur in all Recent sediments, and more than 90 sterols have been identified in some marine sediments (Brassell & Eglinton 1983a). These differ in the positions of the double bonds, side-chain structure and presence or absence of a methyl group at C-4 (e.g., Smith *et al.* 1982, 1983; Brassell & Eglinton 1983a). The distributions in lacustrine sediments are generally less complex, but this is not always so (e.g., Robinson *et al.* 1984). The sterol distributions found in dissimilar environments do show some significant differences which can be related to the type and abundance of eukaryotic organisms present. As yet there is no convincing evidence that prokaryotic organisms are significant sources of sterols in sediments although small amounts do occur in some cyanobacteria (reviewed by Volkman 1986) and at least one bacterium (Bird *et al.* 1971).

The relative proportions of $C_{26}-C_{30}$ 5α(H)-steranes (XXXVII) in crude oils are often different and such distributions have proved to be extremely useful for correlating one oil with another or for correlating oils with presumed source rocks. A common approach has been to look at the relative abundances of the regular C_{27}, C_{28} and C_{29} 5α(H), 14α(H), 17α(H)-steranes, i.e., 5α(H)-cholestane (XXXVIIa), 5α(H)-24-methylcholestane (XXXVIIb) and 5α(H)-24-ethylcholestane (XXXVIIc) since these are usually fairly prominent in m/z 217 mass fragmentograms. These data are presented either as a set of ratios or in the form of a triangular diagram.

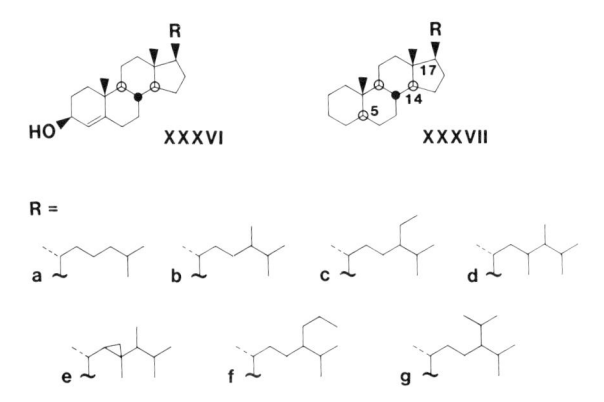

FIG. 5. Examples of steroidal compounds. XXXVI: Δ^5-unsaturated sterol; XXXVII: 5α(H),14α(H),17α(H)-sterane. a–c: common side-chains, a: cholesterol side-chain; b: 24-methyl; c: 24-ethyl. d–g: unusual side-chains; d: 23,24-dimethyl; e: 22,23-methylene-23,24-dimethyl (gorgosterol); f: n-propyl; g: isopropyl.

Huang & Meinschein (1979) suggested that samples from similar environments should have similar ratios of sterols or steranes, and thus such triangular plots would allow one to distinguish between different depositional environments. Unfortunately, while some samples do plot as expected, this approach has proven to be too simplistic for general use. This is perhaps not surprising since the abundances of only three steranes are measured and yet the original sterol distribution from which they were derived might have contained ten times that number of different compounds.

In Fig. 6, I have plotted a few examples from my own studies which do not exhibit the sterol ratios expected by the scheme of Huang & Meinschein (1979). The seawater samples contain far more C_{29} sterols than one would expect if the only major source of such sterols was vascular plants. Clearly marine sources can also be important. The Antarctic sediments are particularly interesting since there are no vascular plants present and yet C_{29} sterols predominate. The sediments from Peru also contain significant amounts of C_{29} sterols despite the fact that this is an area of high algal productivity and the adjacent land is largely desert. In this case, the C_{29} sterols are most likely derived from vascular plants that have been transported into the coastal zone, along with the other terrigenous organic matter, by rivers. Remarkably similar distributions are also found in many lake sediments, such as those from Loch Clair (Cranwell & Volkman 1981), and this raises considerable concern about the usefulness of sterol or sterane ratios as indicators of depositional environment. For a fuller discussion of this topic, the reader is referred to Volkman (1986).

One of the underlying reasons for such ambiguity is that few sterols can be considered as unambiguous markers for a specific group of organisms or environment. In a recent review, Volkman (1986) has shown that few, if any, algal classes can be considered to have a characteristic sterol distribution although certainly, some sterols are more common in particular algal groups than others. For example, 'diatomsterol' (24-methylcholesta-5,22E-dien-3β-ol) is the major sterol of many diatoms but it also predominates in some Prymnesiophytes and occurs in many other algae (Volkman 1986). Only one sterol, dinosterol (XXXVIII: 4α,23,24-trimethyl-5α-cholest-22E-en-3β-ol; Fig. 7), has been shown to be restricted to a single algal class, the dinoflagellates. Further work may establish that other 4-methylsterols are good markers for dinoflagellates, but the possibility of bacterial sources for some of these compounds has also been suggested (Mermoud et al. 1982). Also, there is no reason to expect that the sterol distribution in sediments from a diatom-

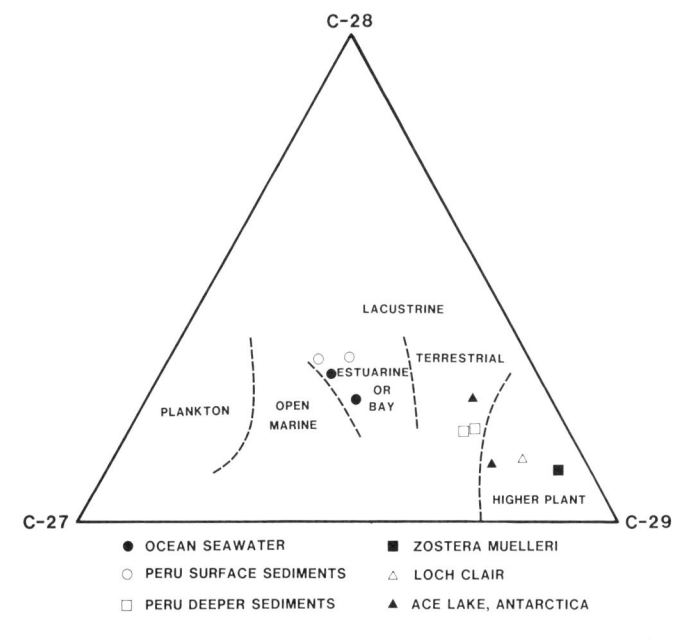

FIG. 6. C_{27}, C_{28}, C_{29} sterol abundances in selected sediments and samples of seawater particles plotted according to the scheme of Huang & Meinschein (1979).

dominated lake environment would be very different from those found in sediments from a diatom-dominated upwelling environment. To further complicate matters, one often finds environments where only a few species of algae dominate and if these should have unusual or atypical sterol distributions then the sediments would also exhibit an unusual sterol distribution. The possibility of selective degradation of specific sterols by bacteria and within food-webs also needs consideration.

Many authors have used the predominance of C_{29} steranes in an oil as evidence that the source rock was rich in terrigenous organic matter. Since it is now clear that algae or cyanobacteria can be important sources of C_{29} sterols in some sediments (Matsumoto et al. 1982; Volkman 1986), one should always check for other indications of higher plants such as the presence of various triterpanes as discussed earlier. Several examples of oils and sediments having a high proportion of C_{29} steranes which are not related to a vascular plant origin (McKirdy et al. 1983, 1986; Fowler & Douglas 1984) are now known and undoubtedly more will be found. Conversely, some lacustrine sediments, such as those from China, have similar ratios of C_{27}, C_{28} and C_{29} steranes (Shi Ji-Yang et al. 1982).

One way to overcome these problems is to look for minor steranes with distinctive side-chain structure such as 23,24-dimethyl (d, Fig. 5) or cyclopropyl (e, Fig. 5) substitution since sterols with these structures appear to be more common in marine environments (e.g., Wardroper et al. 1978). Unfortunately, this is difficult using existing GC-MS techniques. Sterane distributions in crude oils are very complex due to the presence of a variety of isomers differing in their stereochemistry at C-5, C-14, C-17, C-20 and C-24 (XXXVII). In some cases, non-steroidal compounds can also interfere in m/z 217 and m/z 218 mass fragmentograms used to characterize sterane distributions (e.g., Hoffman et al. 1984). The use of better capillary columns for gas chromatography and improved mass spectrometry techniques (e.g., Warburton & Zumberge 1983; Brooks et al. 1984), combined with quantitative measurements (Rullkotter et al. 1984), should improve this situation but it will be some time before these techniques are widely adopted by the oil industry. A step in this direction is the study by Moldowan (1984) who used GC-MS techniques to identify C_{30} (4-desmethyl) steranes in about 40 oils from marine and lacustrine sources. The C_{30} steranes were present in all marine oils younger than the Cambrian but they were not detected in the lacustrine oils suggesting to Moldowan that they can be used as markers for inputs from marine organisms. C_{30} sterols are certainly common in marine environments but they are also abundant in some phytoplankton from the Cryptophyceae and Chrysophyceae (Volkman 1986), which raises the possibility that significant amounts may be found in some lacustrine environments. Further work to establish the side-chain structure of the C_{30} steranes (e.g., n-propyl or isopropyl; f and g, Fig. 5), and more detailed studies of the sterols in lacustrine sediments are needed to establish the usefulness of this parameter. The significance of high concentrations of C_{26} steranes in some oils (e.g., Shi Ji-Yang et al. 1982) also needs to be examined.

It is apparent that the stage has not yet been reached when one can reliably use the distribution of steranes in an oil to deduce what type of source rock must have produced that oil. Indeed, such a goal is unlikely to be reached unless sterane distributions are examined in far more detail than is presently the case.

There are, however, two classes of steranes which do appear to have potential as source markers. These are the 4-methylsteranes (XXXIX, Fig. 7) and the A-norsteranes (XLI). The 4-methylsteranes are believed to derive from the 4-methylsterols (e.g., XXXVIII) produced by dinoflagellates (e.g., De Leeuw et al. 1983; Robinson et al. 1984; McKirdy et al. 1986). Such sterols are not abundant in sediments from

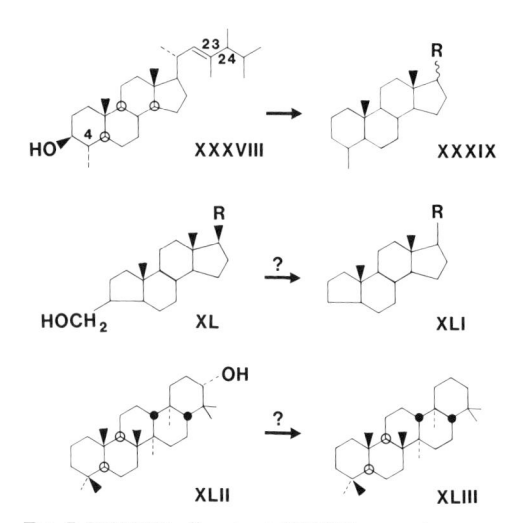

FIG. 7. XXXVIII: dinosterol; XXXIX: general structure of 4-methylsteranes—both 4α and 4β isomers have been found in sediments; XL: 3β-hydroxymethyl-A-norsterane found in sponges; XLI: general structure of A-norsteranes possibly diagenetically formed from XL; XLII: tetrahymanol found in protozoa; XLIII: gammacerane (possibly diagenetically derived from XLII).

upwelling areas (Wardroper et al. 1978; Smith et al. 1982, 1983; Gagosian et al. 1983). Although dinoflagellates are common in both marine and lacustrine environments, most reports of high concentrations of 4-methylsteranes in oils appear to be associated with lacustrine source rocks (Brassell et al. 1986a; McKirdy et al. 1986; Wolff et al. 1986).

C_{26}–C_{28} steranes with modified A-rings (A-norsteranes; XLI) have been found in a Cretaceous black shale (van Graas et al. 1982), and Middle Miocene sediments from the Southern California Bight (McEvoy & Maxwell 1983). It seems likely that these steranes are derived from the 3β-hydroxymethyl-A-nor steranes (XL) which are abundant in some sponges (Minale & Sodano 1974). Since most sponges are found in coastal and continental shelf marine environments these compounds may prove to be useful markers for these environments.

Gammacerane

The pentacyclic triterpenoid gammacerane (XLIII) has attracted considerable attention from geochemists since it is only found in a few oils and it is readily detected using m/z 191 mass fragmentograms. It has been suggested that gammacerane is produced in sediments from degradation of tetrahymanol (XLII) which occurs in protozoa (e.g., Hills et al. 1966; Brassell & Eglinton 1983a) but there may be other sources and until these are determined the value of gammacerane as a palaeoenvironmental indicator will remain limited.

Gammacerane was first identified in the Green River shale (Hills et al. 1966), and since then it has been identified in other 'terrestrial' samples. It occurs in several immature Chinese oils and it could be used to correlate these oils with various potential source rocks (Shi Ji-Yang et al. 1982). Seifert et al. (1984) have detected gammacerane in a number of oil seeps from Greece and they used a 'gammacerane index' (defined as the ratio of the concentrations of gammacerane to 17α(H)-hopane multiplied by 100), to distinguish the various groups of oils. They further showed that gammacerane is highly resistant to biodegradation which is a very useful feature for any biological marker. Gammacerane is also present in oils from Indonesia which are thought to be of 'terrestrial' origin (Seifert & Moldowan 1981), but minor amounts have also been detected in Gulf of Suez oils of presumed 'marine' origin (Rohrback 1983). A very interesting recent finding is the occurrence of gammacerane in sediments from hypersaline environments (see below).

Biological markers for hypersaline environments

Organic rich sediments having the potential for generating oil and gas are produced in some hypersaline environments. Most organic geochemistry studies have focused on contemporary examples but an increasing interest in the study of ancient sediments in the last few years has led to some exciting new results (De Leeuw et al. 1985; ten Haven et al. 1985, this volume; Sinninghe Damsté et al. 1986). These environments are characterized by distinctive microbial communities so one might expect to see a recognizable lipid 'fingerprint' left in these sediments.

Early studies established that a characteristic of many samples from evaporitic facies is an unusual predominance of even carbon C_{20}–C_{34} n-alkanes together with a pristane/phytane ratio less than 1 (ten Haven et al. 1985 and references therein). A predominance of n-C_{22} alkane has also been observed in some sediments (Schenck 1969; Powell & McKirdy 1973; De Leeuw et al. 1985; ten Haven et al. 1985), which may also prove to be a biological marker for this type of environment. The regular C_{25} acyclic isoprenoid alkane 2,6,10,14,18-pentamethyleicosane (XXV) has also been suggested as a marker for saline conditions (Waples et al. 1974), based on its occurrence in Tertiary lagoonal sediments. Polar lipids containing this C_{25} isoprenoid chain are abundant in certain haloalkaliphiles (De Rosa et al. 1983), but this alkane is also found in some methanogenic bacteria (Risatti et al. 1984).

Ten Haven et al. (1985) have identified a number of new compounds in evaporite basin sediments in the Messinian Formation of northern Italy which may prove to be markers for hypersaline environments. These include 4α,24- and 4β,24-dimethylcholestanes (XLIV; Fig. 8), 4-methylspirosterenes (XLV), 4,4-dimethylhomopregnanes (XLVI) and homopregnanes (XLVII). These samples also contained a series of 17α(H),21β(H)-hopanes which exhibited an unusually high abundance of C_{34} and C_{35} components (XLIX) together with a similar distribution of hop-17(21)-enes (XLVIII). From the similarity of the two distributions these authors suggested that the hopanes are formed from reduction of the hop-17(21)-enes, and not from isomerization of the 17β(H),21β(H)-hopanes which is generally considered to be the more common mechanism. High concentrations of C_{35} hopanes have also been found in crude oils derived from pre-Ordovician carbonates from alkaline playa lakes (McKirdy et al. 1983). Although the origin of the C_{35} hopenes and hopanes has not been established, it seems highly probable that they are

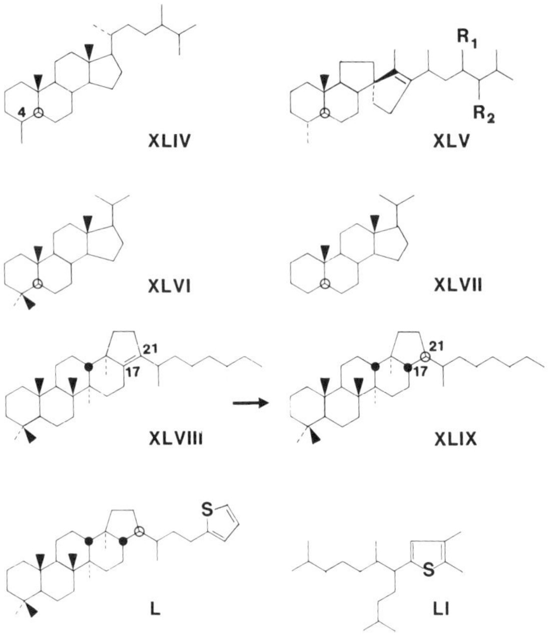

FIG. 8. Structures of unusual lipids proposed as biological markers of hypersaline conditions. XLIV: 4β,24-dimethylcholestane; XLV: 4-methylspirosterene; XLVI: 4,4-dimethyl-5α(H),14β(H),17β(H)-homopregnane; XLVII: homopregnane; XLVIII: pentakishomohop-17(21)-ene; XLIX: 17α(H),21β(H)-pentakishomohopane; L: 30-(2'-methylenethienyl)-17β(H),21β(H)-hopane; LI: 3-methyl-2-(2,6,10-trimethyldodecyl)-thiophene.

derived from bacteria or cyanobacteria which appear to be associated with hypersaline environments.

These Messinian sediments have also been found to contain high concentrations of organic sulphur compounds (Sinninghe Damsté et al. 1986), and indeed crude oils produced from carbonate–evaporite source rocks usually contain high sulphur contents (Tissot & Welte 1978). There is now increasing evidence to suggest that bacterially formed sulphur is incorporated into organic matter at an early stage of diagenesis (see Valisolalao et al. 1984; Brassell et al. 1986c for leading references) but unfortunately this area of organic geochemistry has received little attention until now. It would be unwise given the limited data available to state with any certainty which of these organic sulphur compounds might be markers for hypersaline environments but the predominance of isoprenoid thiophenes, particularly 3-methyl-2-(2,6,10-trimethyldodecyl)-thiophene (LI), is worth noting, and could be related to the predominance of the corresponding alkane (XVI) in Rozel Point crude oil (ten Haven et al. 1985). The relationship, if any, between the occurrence of high concentrations of a C_{35} hopanoid thiophene (L) in some Cretaceous sediments (Valisolalao et al. 1984) and a high abundance in C_{35} hopanes in some oils also warrants further study.

Biological markers for upwelling environments

The upwelling of cold, nutrient-rich waters onto continental shelves produces areas of high productivity and leads to the formation of organic-rich sediments. The potential of biological marker compounds as indicators of such environments in the ancient sediment record was assessed by Brassell & Eglinton (1983b). These authors concluded that no specific markers were known at that time to distinguish upwelling areas from other regions of high productivity. The same statement appears to be true today despite the fact that our understanding of the lipids likely to be found in sediments from upwelling areas has improved considerably.

In general terms, the phytoplankton populations in upwelling environments tend to be dominated by diatoms and one would expect lipids from these algae to be abundant in the sediments. Indeed, diatom-derived sterols are readily recognized in contemporary sediments from the Peru upwelling (Gagosian et al. 1983; Smith et al. 1983) and Walvis Bay (Smith et al. 1982). In productive lakes one often finds high concentrations of small flagellates such as dinoflagellates and green algae which have distinctively different sterol distributions (Robinson et al. 1984; Volkman 1986). In high productivity areas, one would also expect steroids to be considerably more abundant than the bacterial hopanoids leading to higher sterane/hopane ratios in ancient sediments from upwelling systems.

Remineralization of the high concentrations of organic matter produced in upwelling areas leads to oxygen-depleted bottom waters and anoxic surface sediments. Enhanced concentrations of biological markers for methanogenic bacteria such as phytane (XXIV) or 2,6,10,15,19-pentamethyleicosane (XXVI) would be expected in such sediments but not to the extent that they dominate the lipid distributions unless the anoxic conditions are persistent and the activity of other anaerobic bacteria, such as the suphate-reducers, is suppressed. Such conditions are more common in meromictic lakes or enclosed basins.

Since upwelling occurs near the coast this raises the possibility that the sediments might receive significant inputs of terrigenous organic matter. Indeed, Brassell & Eglinton (1983b) found significant concentrations of biological markers for higher plants in sediments from areas of high marine productivity in the Japan Trench, Middle America Trench and Walvis Ridge. There is growing evidence that lipids associated with terrigenous organic matter are less readily degraded than marine lipids (e.g., Prahl et al. 1984; Volkman et al. 1983b). In the case of the Peru upwelling, this is reflected by lipid distributions in deeper sediments which look remarkably similar to those found in some lake sediments (Volkman et al. 1983b; Volkman 1986). Studies of how lipid distributions change with depth in other areas of upwelling, such as Walvis Bay, are needed to determine the importance of this phenomenon.

Perhaps the best possibility of finding biological markers characteristic of upwelling areas is from a study of the bacterial flora. In this regard, the sporadic occurrence of bacterial mats of *Thioploca* sp. in sediments from high productivity areas of Chile, Peru and Walvis Bay (Gallardo 1977), clearly points to one topic needing further examination (Volkman et al. 1983b).

Conclusions

Intensive research over the last decade has led to major advances in our understanding of the types, distributions and abundances of organic compounds likely to be preserved in sediments deposited under different environmental conditions. Organic geochemistry has advanced from simply identifying new compounds in oils and sediments to the stage where compositional data are now routinely used as a tool to assess the maturity, source and extent of biodegradation of a crude oil. In most cases, it is still not possible to infer what type of sediment must have produced an oil simply from the composition of the oil, but many features of the depositional environment can be defined using biological marker techniques.

Many biological markers for vascular plants have been identified, and their presence in many oils, including some from 'marine' environments, testifies to the quantitative importance of this source of organic matter. There is also a growing list of biological markers for bacteria, including markers for specific groups such as methanogens and haloalkaliphiles, which have excellent potential as indicators of palaeoenvironmental conditions. Although phytoplankton are clearly major sources of organic matter in many aquatic environments, we still have few specific compounds to use as markers for contributions from the different algal groups, or for distinguishing between marine and lacustrine species. Botryococcane and 4-methylsteranes stand out as exceptions to this generality.

The increasing analytical sophistication of techniques used to study the lipids in sediments and oils has led to the identification of many new compounds whose origins remain unknown. Some of these, such as 28,30-bisnorhopane, are either major constituents or found in a very restricted range of samples and clearly point to important sources of organic compounds about which we know very little. There is a demonstrable need for more detailed comparative studies of the lipid compositions of well defined Recent and ancient sediments as well as more information on the lipid compositions of organisms likely to contribute organic matter to sediments. Such studies have demonstrated that interpretations based on widely used parameters such as pristane/phytane ratio and ratios of C_{27}, C_{28} and C_{29} steranes have often been overly simplistic and in some cases misleading. However, such data can provide useful information provided that they are used in conjunction with information from other biological markers. New parameters need to be developed, and those which compare the abundances of selected compounds in different compound classes such as acyclic isoprenoids, steroids and triterpenoids appear to have most potential as indicators of depositional environment and relative contributions of organic matter from different organisms.

ACKNOWLEDGEMENTS: I am grateful to the many colleagues cited in the references whose collaboration on joint projects and stimulating discussions have contributed to the ideas and results reported in this paper. Drs S. C. Brassell, J. W. Farrington and A. S. Mackenzie provided very helpful critiques of early drafts of this manuscript. P. Chandler (WHOI) and B. Baker (CSIRO) typed the paper and B. Hansen and J. Nunn contributed the artwork.

References

ALBAIGÉS, J. 1980. Identification and geochemical significance of long chain acyclic isoprenoid hydrocarbons in crude oils. *In*: DOUGLAS, A. G. & MAXWELL, J. R. (eds), *Advances in Organic Geochemistry, 1979*, Pergamon Press, Oxford, 19–28.

——, ALGABA, J. & GRIMALT, J. 1984a. Extractable and bound neutral lipids in some lacustrine sediments. *Organic Geochemistry*, **6**, 223–236.

——, BORBON, J. & WALKER, W. II. 1985. Petroleum isoprenoid hydrocarbons derived from catagenic degradation of Archaebacterial lipids. *Organic Geochemistry*, **8**, 293–297.

——, GRIMALT, J., BAYONA, J. M., RISEBROUGH, R., DELAPPE, B. & WALKER, W. II. 1984b. Dissolved, particulate and sedimentary hydrocarbons in a deltaic environment. *Organic Geochemistry*, **6**, 237–248.

ALEXANDER, R., KAGI, R. I. & NOBLE, R. 1983. Identification of the bicyclic sesquiterpenes drimane and eudesmane in petroleum. *Journal of the Chemical Society, Chemical Communications*, 226–228.

——, ——, & VOLKMAN, J. K. 1984. Identification of some bicyclic alkanes in petroleum. *Organic Geochemistry*, **6**, 63–70.

ANDERSON, R., KATES, M., BAEDECKER, M. J., KAPLAN, I. R. & ACKMAN, R. G. 1977. The stereoisomeric composition of phytanyl chains in lipids of Dead Sea sediments. *Geochimica et Cosmochimica Acta*, **41**, 1381–1390.

BAYONA, J. M., GRIMALT, J., ALBAIGÉS, J., WALKER, W. II, DELAPPE, B. W. & RISEBROUGH, R. W. 1983. Recent contributions of high resolution gas chromatography to the analysis of environmental hydrocarbons. *Journal of High Resolution Chromatography & Chromatography Communications*, **6**, 605–611.

BIRD, C. W., LYNCH, J. M., PIRT, F. J., REID, W. W., BROOKS, C. J. W. & MIDDLEDITCH, B. S. 1971. Steroids and squalene in *Methylococcus capsulatus* growth on methane. *Nature*, **230**, 473–474.

BOON, J. J., VAN DER MEER, F. W., SCHUYL, P. J. W., DE LEEUW, J. W., SCHENCK, P. A. & BURLINGAME, A. L. 1978. Organic geochemical analyses of core samples from site 362 Walvis Ridge, DSDP Leg 40. *In*: BOLLI, H. M., RYAN, W. B. et al. (eds), *Initial Reports of the Deep Sea Drilling Project*, **40**(3), Washington (US Government Printing Office) 627–637.

BRASSELL, S. C. & EGLINTON, G. 1983a. Steroids and triterpenoids in deep sea sediments as environmental and diagenetic indicators. *In*: BJORØY, M. et al. (eds), *Advances in Organic Geochemistry, 1981*, John Wiley & Sons Ltd, Chichester, 684–697.

—— & —— 1983b. The potential of organic geochemical compounds as sedimentary indicators of upwelling. *In*: SUESS, E. & THIEDE, J. (eds), *Coastal Upwelling. Its Sediment Record*, Plenum Press, New York, NATO Conference Series IV, **10A**, 545–571.

——, —— & FU JIA MO. 1986a. Biological marker compounds as indicators of the depositional history of the Maoming oil shale. *In*: LEYTHAEUSER, D. & RÜLLKOTTER, J. (eds), *Advances in Organic Geochemistry, 1985*, Pergamon Press, Oxford, 927–941.

——, & MAXWELL, J. R. 1981a. Preliminary lipid analyses of two Quaternary sediments from the Middle America Trench, Southern Mexico Transect, Deep Sea Drilling Project Leg 66. *In*: WATKINS, J. S., MOORE, J. C. et al. (eds), *Initial Reports of the Deep Sea Drilling Project*, **66**, 557–580.

——, & —— 1983. The geochemistry of terpenoids and steroids. *Biochemical Society Transactions*, **11**, 575–586.

——, ——, MARLOWE, I. T., PFLAUMANN, U. & SARNTHEIN, M. 1986b. Molecular stratigraphy: a new tool for climatic assessment. *Nature*, **320**, 129–133.

——, LEWIS, C. A., DE LEEUW, J. W., DE LANGE, F. & SINNINGHE DAMSTÉ, J. S. 1986c. Isoprenoid thiophenes: novel products of sediment diagenesis? *Nature*, **320**, 160–162.

——, McEVOY, J., HOFFMAN, C. F., LAMB, N. A., PEAKMAN, T. M. & MAXWELL, J. R. 1984. Isomerisation, rearrangement and aromatisation of steroids in distinguishing early stages of diagenesis. *Organic Geochemistry*, **6**, 11–23.

——, WARDROPER, A. M. K., THOMSON, I. D., MAXWELL, J. R. & EGLINTON, G. 1981b. Specific acyclic isoprenoids as biological markers of methanogenic bacteria in marine sediments. *Nature*, **290**, 693–696.

BROOKS, P. W., MEYER, T. & CHRISTIE, O. H. J. 1984. A comparison of high resolution GC-MS and selected metastable ion monitoring of polycyclic steranes and triterpanes. *Organic Geochemistry*, **6**, 813–816.

CHAPPE, B., ALBRECHT, P. & MICHAELIS, W. 1982. Polar lipids of Archaebacteria in sediments and petroleum. *Science*, **217**, 65–67.

——, MICHAELIS, W., ALBRECHT, P. & OURISSON, G. 1979. Fossil evidence for a novel series of archaebacterial lipids. *Naturwissenschaften*, **66**, 522–523.

COMET, P. A. 1982. The use of lipids as facies indicators. Ph.D. thesis, University of Bristol, 387 pp.

CORBET, B., ALBRECHT, P. & OURISSON, G. 1980. Photochemical or photomimetic fossil triterpenoids in sediments and petroleum. *Journal of the American Chemical Society*, **102**, 1171–1173.

COX, R. E., BURLINGAME, A. L., WILSON, D. M., EGLINTON, G. & MAXWELL, J. R. 1973. Botryococcene—a tetramethylated acyclic triterpenoid of algal origin. *Journal of the Chemical Society. Chemical Communications*, 284–285.

CRANWELL, P. A. 1982. Lipids of aquatic sediments and sedimenting particulates. *Progress in Lipid Research*, **21**, 271–308.

—— 1985. Long-chain unsaturated ketones in Recent lacustrine sediments. *Geochimica et Cosmochimica Acta*, **49**, 1545–1551.

—— & VOLKMAN, J. K. 1981. Alkyl and steryl esters in a Recent lacustrine sediment. *Chemical Geology*, **32**, 29–43.

DE LEEUW, J. W., VAN DER MEER, F. W., RIJPSTRA, W. I. C. & SCHENCK, P. A. 1980. On the occurrence and structural identification of long-chain unsaturated ketones and hydrocarbons in Recent and sub-Recent sediments. *In*: DOUGLAS, A. G. & MAXWELL, J. R. (eds), *Advances in Organic Geochemistry, 1979*, Pergamon Press, Oxford, 211–217.

——, RIJPSTRA, W. I. C., SCHENCK, P. A. & VOLKMAN, J. K. 1983. Free, esterified and residual bound sterols in Black Sea Unit 1 sediments. *Geochimica et Cosmochimica Acta*, **47**, 455–465.

——, SINNINGHE DAMSTÉ, J. S., KLOK, J., SCHENCK, P. A. & BOON, J. J. 1985. Biogeochemistry of Gavish Sabkha sediments. 1. Studies on neutral reducing sugars and lipid moieties by gas chromatography-mass spectrometry. *In*: FRIEDMAN, G. M. & KRUMBEIN, W. E. (eds), *Hypersaline Ecosystems—The Gavish Sabkha. Ecological Studies 53*, Springer-Verlag, New York, 350–367.

DE ROSA, M. A., GAMBACORTA, L., MINALE, L. & BU'LOCK, J. D. 1972. The formation of ω-cyclohexyl-fatty acids from shikimate in an acidophilic thermophilic *Bacillus*. *Biochemistry Journal*, **128**, 751–754.

——, GAMBACORTA, A., NICOLAUS, B. & GRANT, W. D. 1983. A C_{25}, C_{25} diether core lipid from Archaebacterial haloalkaliphiles. *Journal of General Microbiology*, **129**, 2333–2337.

DIDYK, B. M., SIMONEIT, B. R. T., BRASSELL, S. C. & EGLINTON, G. 1978. Organic geochemical indicators of palaeoenvironmental conditions of sedimentation. *Nature*, **272**, 216–222.

DUNLOP, R. W. & JEFFRIES, P. R. 1985. Hydrocarbons of the hypersaline basins of Shark Bay, Western Australia. *Organic Geochemistry*, **8**, 313–320.

EKWEOZOR, C. M., OKOGUN, J. I., EKONG, D. E. U. & MAXWELL, J. R. 1979a. Preliminary organic geochemical studies of samples from the Niger Delta (Nigeria). I. Analyses of crude oils for triterpanes. *Chemical Geology*, **27**, 11–28.

——, ——, —— & —— 1979b. Preliminary organic geochemical studies of samples from the Niger Delta (Nigeria). II. Analyses of shale for triterpenoid derivatives. *Chemical Geology*, **27**, 29–37.

FOWLER, M. G. & DOUGLAS, A. G. 1984. Distribution and structure of hydrocarbons in four organic-rich Ordovician rocks. *Organic Geochemistry*, **6**, 105–114.

GAGOSIAN, R. B. 1983. Review of marine organic geochemistry. *Reviews of Geophysics and Space Physics*, **21**, 1245–1258.

——, VOLKMAN, J. K. & NIGRELLI, G. E. 1983. The use of sediment traps to determine sterol sources in coastal sediments off Peru. *In*: BJORØY, M. *et al.* (eds), *Advances in Organic Geochemistry, 1981*, John Wiley & Sons Ltd, Chichester, 369–379.

——, ZAFIRIOU, O. C., PELTZER, E. T. & ALFORD, J. B. 1982. Lipids in aerosols from the tropical North Pacific: temporal variability. *Journal of Geophysical Research*, **87**, 11133–11144.

GALLARDO, V. 1977. Large benthic microbial communities in sulphide biota under Peru–Chile subsurface countercurrent. *Nature*, **288**, 331–332.

GILLAN, F. T. & SANDSTROM, M. W. 1985. Microbial lipids from a nearshore sediment from Bowling Green Bay, North Queensland: the fatty acid composition of intact lipid fractions. *Organic Geochemistry*, **8**, 321–328.

GRANTHAM, P. J., POSTHUMA, J. & BAAK, A. 1983. Triterpanes in a number of far-eastern crude oils. *In*: BJORØY, M. *et al.* (eds), *Advances in Organic Geochemistry, 1981*, John Wiley & Sons Ltd, Chichester, 675–683.

——, —— & DE GROOT, K. 1980. Variation and significance of the C_{27} and C_{28} triterpane content of a North Sea core and various North Sea crude oils. *In*: DOUGLAS, A. G. & MAXWELL, J. R. (eds), *Advances in Organic Geochemistry, 1979*, Pergamon Press, Oxford, 29–38.

HILLS, I. R. & WHITEHEAD, E. V. 1966. Triterpanes in optically active petroleum distillates. *Nature*, **209**, 977–979.

——, ——, ANDERS, D. E., CUMMINS, J. J. & ROBINSON, W. E. 1966. An optically active triterpane, gammacerane, in Green River, Colorado, oil shale bitumen. *Journal of the Chemical Society. Chemical Communications*, 752–754.

HOFFMANN, C. F., MACKENZIE, A. S., LEWIS, C. A., MAXWELL, J. R., OUDIN, J. L., DURAND, B. & VANDENBROUCKE, M. 1984. A biological marker study of coals, shales and oils from the Mahakam Delta, Kalimantan, Indonesia. *Chemical Geology*, **42**, 1–23.

HOWARD, D. L. 1980. Polycyclic triterpenes of the anaerobic photosynthetic bacterium *Rhodomicrobium vannielii*. Ph.D. thesis, University of California, Los Angeles.

HUANG, W. Y. & MEINSCHEIN, W. G. 1979. Sterols as ecological indicators. *Geochimica et Cosmochimica Acta*, **43**, 739–745.

JOHNS, R. B. (ed.), 1986. *Biological Markers in the Sedimentary Record*. Elsevier, Amsterdam.

——, BELSKY, T., MCCARTHY, E. D., BURLINGAME, A. L., HAUG, P., SCHNOES, H. K., RICHTER, W. J. & CALVIN, M. 1966. The organic geochemistry of ancient sediments II. *Geochimica et Cosmochimica Acta*, **30**, 1191–1222.

KAGRAMANOVA, G. R., PUSTIL'NIKOVA, S. D., PEKT, T., DENISOV, Y. V. & PETROV, A. A. 1976. Sesquiterpane hydrocarbons of petroleums. *Neftekhimiya*, **16**, 18–22 (In Russian).

LAFLAMME, R. E. & HITES, R. A. 1978. The global distribution of polycyclic aromatic hydrocarbons in recent sediments. *Geochimica et Cosmochimica Acta*, **42**, 289–303.

LARGEAU, C., KADOURI, A., DERENNE, S. & CASADEVALL, E. 1986. Comparative study of the pyrolysates of Botryococcus-derived kerogens and of related biopolymers. *In*: LEYTHAEUSER, D. & RÜLLKOTTER, J. (eds), *Advances in Organic Geochemistry, 1985*, Pergamon Press, Oxford, 1023–1032.

LIVSEY, A., DOUGLAS, A. G. & CONNAN, J. 1984. Diterpenoid hydrocarbons in sediments from an offshore (Labrador) well. *Organic Geochemistry*, **6**, 73–81.

MACKENZIE, A. S. 1984. Applications of biological markers in petroleum geochemistry. *Advances in Petroleum Geochemistry*, **1**, 115–214.

——, BRASSELL, S. C., EGLINTON, G. & MAXWELL, J. R. 1982. Chemical fossils: the geological fate of steroids. *Science*, **217**, 491–504.

MARLOWE, I. T., BRASSELL, S. C., EGLINTON, G. & GREEN, J. C. 1984a. Long-chain unsaturated ketones and esters in living algae and marine sediments. *Organic Geochemistry*, **6**, 135–141.

——, GREEN, J. C., NEAL, A. C., BRASSELL, S. C., EGLINTON, G. & COURSE, P. A. 1984b. Long-chain (n-C_{37}-n-C_{39}) alkenones in the Prymnesiophyceae. Distribution of alkenones and other lipids and their taxonomic significance. *British Journal of Phycology*, **19**, 203–216.

MARRIOTT, P. J., GILL, J. P., EVERSHED, R. P., EGLINTON, G. & MAXWELL, J. R. 1982. Capillary GC and GC-MS of porphyrins and metalloporphyrins. *Chromatographia*, **16**, 304–308.

MATSUMOTO, G., TORII, T. & HANYA, T. 1982. High abundance of algal 24-ethylcholesterol in an Antarctic lake sediment. *Nature*, **299**, 52–54.

MAXWELL, J. R., DOUGLAS, A. G., EGLINTON, G. & MCCORMICK, A. 1968. The botryococcenes—hydrocarbons of novel structure from the alga *Botryococcus braunii*, Kutzing. *Phytochemistry*, **7**, 2157–2171.

MCEVOY, J. & MAXWELL, J. R. 1983. Diagenesis of steroidal compounds in sediments from the Southern California Bight (DSDP Leg 63, Site 467). *In*: BJORØY, M. *et al.* (eds), *Advances in Organic Geochemistry, 1981*, John Wiley & Sons Ltd, Chichester, 449–464.

MCKIRDY, D. M., ALDRIDGE, A. K. & YMPA, P. J. M. 1983. A geochemical comparison of some crude oils from pre-Ordovician carbonate rocks. *In*: BJORØY, M. *et al.* (eds), *Advances in Organic Geochemistry, 1981*, John Wiley & Sons Ltd, Chichester, 99–107.

——, COX, R. E., VOLKMAN, J. K. & HOWELL, V. J. 1986. Botryococcane in a new class of Australian non-marine crude oils. *Nature*, **320**, 57–59.

MERMOUD, F., GULACAR, F. O., SILES, S., CHASSAING, B. & BUCHS, A. 1982. 4-methylsterols in recent lacustrine sediments: terrestrial, planktonic or some other origin? *Chemosphere*, **11**, 557–567.

METZGER, P., BERKALOFF, C., CASADEVALL, E. & COUTE, A. 1985. Alkadiene and botryococcene-producing races of wild strains of *Botryococcus braunii*. *Phytochemistry*, **24**, 2305–2312.

MICHAELIS, W. & ALBRECHT, P. 1979. Molecular fossils of archaebacteria in kerogen. *Naturwissenschaften*, **66**, 420–422.

MINALE, L. & SODANO, G. 1974. Marine sterols: unique 3β-hydroxymethyl-A-nor-5α-steranes from the sponge *Axinella verrucosa*. *Journal of the Chemical Society. Perkin Transactions I.*, 2380–2384.

MOLDOWAN, J. M. 1984. C_{30}-steranes, novel markers for marine petroleums and sedimentary rocks. *Geochimica et Cosmochimica Acta*, **48**, 2767–2768.

—— & SEIFERT, W. K. 1979. Head to head linked isoprenoid hydrocarbons in petroleum. *Science*, **204**, 169–171.

—— & —— 1980. First discovery of botryococcane in petroleum. *Journal of the Chemical Society. Chemical Communications*, 912–914.

——, ——, ARNOLD, E. & CLARDY, J. 1984. Structure proof and significance of stereoisomeric 28,30-bisnorhopanes in petroleum and petroleum source rocks. *Geochimica et Cosmochimica Acta*, **48**, 1651–1661.

NOBLE, R. A., ALEXANDER, R., KAGI, R. I. & KNOX, J. 1985. Tetracyclic diterpenoid hydrocarbons in some Australian coals, sediments and crude oils. *Geochimica et Cosmochimica Acta*, **49**, 2141–2147.

OURISSON, G., ALBRECHT, P. & ROHMER, M. 1979. The hopanoids. Palaeochemistry and biochemistry of a group of natural products. *Pure and Applied Chemistry*, **51**, 709–729.

——, ALBRECHT, P. & ROHMER, M. 1984. The microbial origin of fossil fuels. *Scientific American*, **251**, 44–51.

PERRY, G. J., VOLKMAN, J. K., JOHNS, R. B. & BAVOR, H. J. JR. 1979. Fatty acids of bacterial origin in contemporary marine sediments. *Geochimica et Cosmochimica Acta*, **43**, 1715–1725.

POWELL, T. G. & MCKIRDY, D. M. 1973. The effect of source material, rock type and diagenesis on the n-alkane content of sediments. *Geochimica et Cosmochimica Acta*, **37**, 623–633.

PRAHL, F. E., EGLINTON, G., CORNER, E. D. S., O'HARA, S. C. M. & FORSBERG, T. E. V. 1984. Changes in plant lipids during passage through the gut of Calanus. *Journal of the Marine Biological Association*, **64**, 317–334.

RISATTI, J. B., ROWLAND, S. J., YON, D. A. & MAXWELL, J. R. 1984. Stereochemical studies of acyclic isoprenoids—XII. Lipids of methanogenic bacteria and possible contributions to sediments. *Organic Geochemistry*, **6**, 93–104.

ROBINSON, N., EGLINTON, G., BRASSELL, S. C. & CRANWELL, P. A. 1984. Dinoflagellate origin for sedimentary 4α-methylsteroids and 5α(H)-stanols in lake sediments. *Nature*, **308**, 439–442.

ROHRBACK, B. G. 1983. Crude oil geochemistry of the Gulf of Suez. *In*: BJORØY, M. *et al.* (eds), *Advances in Organic Geochemistry, 1981*, John Wiley & Sons Ltd, Chichester, 39–48.

ROWLAND, S. J., YON, D. A., LEWIS, C. A. & MAXWELL, J. R. 1985. Occurrence of 2,6,10-trimethyl-7-(3-methylbutyl)-dodecane and related hydrocarbons in the green alga *Enteromorpha prolifera* and sediments. *Organic Geochemistry*, 8, 207-213.

RULLKÖTTER, J., LEYTHAEUSER, D. & WENDISCH, D. 1982a. Novel 23,28-birnorlupanes in Tertiary sediments. Widespread occurrences of nuclear demethylated triterpanes. *Geochimica et Cosmochimica Acta*, 46, 2501-2509.

——, VON DER DICK, H. & WELTE, D. H. 1982b. Organic petrography and extractable hydrocarbons of sediments from the Gulf of California, Deep Sea Drilling Project. *In*: CURRAY, J. R., MOORE, D. G. et al. *Initial Reports of the Deep Sea Drilling Project*, 64(2), Washington (US Government Printing Office) 837-853.

——, MACKENZIE, A. S., WELTE, D. H., LEYTHAEUSER, D. & RADKE, M. 1984. Quantitative gas chromatography-mass spectrometry analysis of geological samples. *Organic Geochemistry*, 6, 817-827.

SAJGÓ, Cs. 1984. Organic geochemistry of crude oils from south-east Hungary. *Organic Geochemistry*, 6, 569-578.

SCHENCK, P. A. 1969. The predominance of C_{22} n-alkane in rock extracts. *In*: SCHENCK, P. A. & HAVENAAR, I. (eds), *Advances in Organic Geochemistry, 1968*, Pergamon Press, Oxford, 261-268.

SEIFERT, W. K. & MOLDOWAN, J. M. 1981. Paleoreconstruction by biological markers. *Geochimica et Cosmochimica Acta*, 45, 783-794.

——, —— & DEMAISON, G. J. 1984. Source correlation of biodegraded oils. *Organic Geochemistry*, 6, 633-643.

——, ——, SMITH, G. W. & WHITEHEAD, E. V. 1978. First proof of structure of a C_{28}-pentacyclic triterpane in petroleum. *Nature*, 271, 436-437.

SHI, JI-YANG, MACKENZIE, A. S., ALEXANDER, R., EGLINTON, G., GOWAR, A. P., WOLFF, G. A. & MAXWELL, J. R. 1982. A biological marker investigation of petroleums and shales from the Shengli oilfield, the People's Republic of China. *Chemical Geology*, 35, 1-31.

SIMONEIT, B. R. T. 1977. Diterpenoid compounds and other lipids in deep-sea sediments and their geochemical significance. *Geochimica et Cosmochimica Acta*, 41, 463-476.

—— 1978. The organic chemistry of marine sediments. *In*: RILEY, J. P. & CHESTER, R. (eds), *Chemical Oceanography, Volume 7*. Academic Press, London, 233-311.

SINNINGHE DAMSTÉ, J. S., TEN HAVEN, H. L., DE LEEUW, J. W. & SCHENCK, P. A. 1986. Organic geochemical studies of a Messinian evaporitic basin, northern Apennines (Italy) II: Isoprenoid and n-alkyl thiophenes and thiolanes. *In*: LEYTHAEUSER, D. & RÜLLKOTTER, J. (eds), *Advances in Organic Geochemistry, 1985*, Pergamon Press, Oxford, 791-805.

SMITH, D. J., EGLINTON, G., MORRIS, R. J. & POUTANEN, E. L. 1982. Aspects of the steroid geochemistry of a recent diatomaceous sediment from the Namibian shelf. *Oceanologica Acta*, 5, 365-378.

——, ——, —— & —— 1983. Aspects of the steroid geochemistry of an interfacial sediment from the Peruvian upwelling. *Oceanologica Acta*, 6, 211-219.

SUMMONS, R. E. & POWELL, T. G. 1986. Chlorobiaceae in Palaeozoic seas revealed by biological markers, isotopes and geology. *Nature*, 319, 763-765.

TEN HAVEN, H. L., DE LEEUW, J. W. & SCHENCK, P. A. 1985. Organic geochemical studies of a Messinian evaporitic basin, northern Apennines (Italy) I. Hydrocarbon biological markers for a hypersaline environment. *Geochimica et Cosmochimica Acta*, 49, 2181-2191.

TISSOT, B. P. & WELTE, D. H. 1978. *Petroleum Formation and Occurrence*. Springer-Verlag, Berlin.

VALISOLALAO, J., PERAKIS, N., CHAPPE, B. & ALBRECHT, P. 1984. A novel sulfur containing C_{35} hopanoid in sediments. *Tetrahedron Letters*, 25, 1183-1186.

VAN GRAAS, G., DELANGE, F., DE LEEUW, J. W. & SCHENCK, P. A. 1982. A-nor steranes, a novel class of sedimentary hydrocarbons. *Nature*, 296, 59-61.

VOLKMAN, J. K. 1986. A review of sterol markers for marine and terrigenous organic matter. *Organic Geochemistry*, 9, 83-99.

—— & MAXWELL, J. R. 1986. Acyclic isoprenoids as biological markers. *In*: JOHNS, R. B. (ed.), *Biological Markers in the Sedimentary Record*, Elsevier, Amsterdam, 1-42.

——, ALEXANDER, R., KAGI, R. I. & RULLKÖTTER, J. 1983a. GC-MS characterization of C_{27} and C_{28} triterpanes in sediments and petroleum. *Geochimica et Cosmochimica Acta*, 47, 1033-1040.

——, ALLEN, D. I., STEVENSON, P. L. & BURTON, H. R. 1986a. Bacterial and algal hydrocarbons in sediments from a saline Antarctic lake, Ace Lake. *In*: LEYTHAEUSER, D. & RÜLLKOTTER, J. (eds), *Advances in Organic Geochemistry*, 1985, Pergamon Press, Oxford, 671-681.

——, BURTON, H. R., EVERITT, D. A. & ALLEN, D. I. 1986b. Pigment and lipid compositions of algal and bacterial communities in Ace Lake, Vestfold Hills, Antarctica. *Developments in Hydrobiology* (In press).

——, EGLINTON, G., CORNER, E. D. S. & FORSBERG, T. E. V. 1980a. Long-chain alkenes and alkenones in the marine coccolithophorid *Emiliania huxleyi*. *Phytochemistry*, 19, 2619-2622.

——, ——, —— & SARGENT, J. R. 1980b. Novel unsaturated straight-chain C_{37}-C_{39} methyl and ethyl ketones in marine sediments and a coccolithophore *Emiliania huxleyi*. *In*: DOUGLAS, A. G. & MAXWELL, J. R. (eds), *Advances in Organic Geochemistry, 1979*, Pergamon Press, Oxford, 219-228.

——, FARRINGTON, J. W., GAGOSIAN, R. B. & WAKEHAM, S. G. 1983b. Lipid composition of coastal marine sediments from the Peru upwelling region. *In*: BJORØY, M. et al. (eds), *Advances in Organic Geochemistry, 1981*, John Wiley & Sons Ltd, Chichester, 228-240.

——, JOHNS, R. B., GILLAN, F. T., PERRY, G. J. & BAVOR, H. J. JR. 1980c. Microbial lipids of an intertidal sediment—1. Fatty acids and hydrocar-

bons. *Geochimica et Cosmochimica Acta*, **44**, 1133–1143.
WAKEHAM, S. G., SCHAFFNER, C., GIGER, W., BOON, J. J. & DE LEEUW, J. W. 1979. Perylene in sediments from the Nambian Shelf. *Geochimica et Cosmochimica Acta*, **43**, 1141–1144.
WAPLES, D. W., HAUG, P. & WELTE, D. H. 1974. Occurrence of a regular C_{25} isoprenoid hydrocarbon in Tertiary sediments representing a lagoonal, saline environment. *Geochimica et Cosmochimica Acta*, **38**, 381–387.
WARBURTON, G. A. & ZUMBERGE, J. E. 1983. Determination of petroleum sterane distributions by mass spectrometry with selected metastable ion monitoring. *Analytical Chemistry*, **55**, 123–126.
WARDROPER, A. M. K., MAXWELL, J. R. & MORRIS, R. J. 1978. Sterols of a diatomaceous ooze from Walvis Bay. *Steroids*, **32**, 203–221.

WOESE, C. R., MAGRUM, L. J. & FOX, G. E. 1978. Archaebacteria. *Journal of Molecular Evolution*, **11**, 245–252.
WOLFF, G. A., LAMB, N. A. & MAXWELL, J. R. 1986. The origin and fate of 4-methyl steroid hydrocarbons. 1—Diagenesis of 4-methyl sterenes. *Geochimica et Cosmochimica Acta*, **50**, 335–342.
YON, D. A., MAXWELL, J. R. & RYBACK, G. 1982. 2,6,10-trimethyl-7-(3-methylbutyl)dodecane, a novel sedimentary biological marker compound. *Tetrahedron Letters*, **23**, 2143–2146.
ZUMBERGE, J. E. 1983. Tricyclic diterpane distributions in correlation of Palaeozoic crude oils from the Williston Basin. *In*: BJORØY, M. *et al*. (eds), *Advances in Organic Geochemistry, 1981*, John Wiley & Sons Ltd, Chichester, 738–745.

J. K. VOLKMAN, CSIRO, Division of Oceanography, GPO Box 1538, Hobart, Tasmania 7001, Australia.

Application of biological markers in the recognition of palaeo-hypersaline environments

H. L. ten Haven, J. W. de Leeuw, J. S. Sinninghe Damsté, P. A. Schenck, S. E. Palmer & J. E. Zumberge

SUMMARY: In this study the saturated and aromatic hydrocarbon fractions of a marl sample from a Messinian (late Miocene) evaporitic basin located in the northern Apennines, and four oils, Rozel Point oil (Utah, USA; Miocene) and three seep oils from Sicily (Messinian), have been studied by GC with simultaneous FID and FPD detection and by GC-MS. All samples show characteristics which might be linked to hypersaline conditions prevailing during the time of deposition. Some of these characteristics are: a very low pristane/phytane ratio (<0.1), a relatively high abundance of docosane (C_{22}) and gammacerane and a series of extended hopanes and/or hop-17(21)-enes maximizing at C_{35}. The aromatic hydrocarbon fraction of all samples is dominated by organic sulphur compounds of which 2,3-dimethyl-5-(2,6,10-trimethylundecyl) thiophene is the most abundant compound. The suggestion of Meissner et al. (1984), that the source rock of Rozel Point oil was deposited under hypersaline conditions in a playa-lake system, is supported by the organic geochemical characteristics of this oil.

Introduction

Biological markers, compounds which originate from specific structures occurring in living organisms, are widely used in petroleum exploration. Mostly, they are applied to discriminate between a continental versus a marine origin, for oil/source rock correlations and for thermal maturity estimates of source rocks (e.g., Tissot & Welte 1984). The different sources of marine input can, at present, be specified most exactly by biological markers. For example, it is believed that 4-methylsteranes are the diagenetically altered imprints of dinoflagellates (Boon et al. 1979; Robinson et al. 1984; de Leeuw 1986).

Recently several characteristic biological markers and their specific distribution patterns have been correlated with hypersaline conditions prevailing during time of deposition (ten Haven et al. 1985; Sinninghe Damsté et al. 1986).

It is the intention of this paper to present data with similar characteristics observed in a marl layer from the northern Apennines (NAM) and seep oils from Sicily (SSO E1 and E5), the depositional conditions of which are known, and to apply this information to reconstruct the palaeoenvironmental facies of the source rock of the Rozel Point oil (RPO). We will, therefore, discuss only those compounds which are relevant.

Detailed descriptions of the geological setting of the northern Apennines marl layer and of the saturated hydrocarbon fraction extracted from this sample are given by ten Haven et al. (1985). The aromatic hydrocarbon fraction is described by Sinninghe Damsté et al. (1986). Some preliminary geochemical results of the Sicily seep oils (SSO) and their geological setting were published by Palmer & Zumberge (1981). The Italian samples are all Messinian (late Miocene) in age. It is thought that during the late Miocene the most geographically widespread anoxic event since the Cretaceous took place (Thunell et al. 1984), resulting in the deposition of organic-rich sediments, such as the Monterey shale and its contemporary deposits from the circum Pacific. In and around the Mediterranean area the late Miocene is characterized by thick evaporatic deposits, interbedded with organic-rich layers (Cita 1982). The origin and age of the Rozel Point oil is not exactly known. Meissner et al. (1984) suggested that this oil is sourced by playa-lake deposits of the Miocene Salt Lake group.

Experimental

The extraction procedure and the separation of the extracts and oils into saturated hydrocarbon, aromatic hydrocarbon and polar fractions are described by Sinninghe Damsté et al. (1986). Prior to gas chromatography the elemental sulphur was removed with activated copper. Gas chromatography of the saturated hydrocarbon fractions was performed using a Carlo Erba 4160 instrument with on-column injection, equipped with a 25 m fused silica column (0.32 mm) coated with CP-sil 5, programmed from 125 to 330°C at 4°/min with H_2 as carrier gas. Gas chromatography of the aromatic hydrocarbon fraction was performed using a Varian 3700 instrument with simultaneous flame ionization detection (FID) and flame photometric detection (FPD), equipped with a 50 m fused silica column (0.22 mm) coated with CP-sil 5. The conditions of the gas chromatography–mass spectrometry analyses are described by Sinninghe Damsté et al. (1986). Identifications of compounds are based on comparison of relative retention times and mass spectra with those of standards and data reported in the literature (e.g., Sinninghe Damsté et al. 1986; Philp 1985 and references cited therein).

Results and discussion

Saturated hydrocarbon fraction

The gas chromatograms of the saturate fraction of four samples are shown in Figure 1. The numbers in this figure indicate n-alkanes and correspond with their number of carbon atoms. The R_{22} index, defined as $2 \times C_{22}/(C_{21}+C_{23})$, is greater than one for all samples (RPO = 1.7; NAM = 1.9; SSO E1 = 3.0; SSO E5 = 3.1). This predominance of docosane is interpreted as a marker for hypersaline environments (ten Haven et al. 1985) and is also observed in Chinese oils, the source rock of which was deposited in saline lakes (Wang et al. this volume). Tetracosane is even more abundant than docosane in the RPO sample, a phenomenon sometimes also observed in Chinese oils (Wang et al. this volume). The alkanes of RPO show an even over odd predominance (CPI$_{(24-34)}$ = 0.82) in contrast with those of the NAM (CPI = 2.20) and SSO E5 (CPI = 1.20). It has been postulated several times that an even over odd predominance of n-alkanes characterizes a hypersaline environment (see ten Haven et al. 1985 and references cited therein). Sometimes, though, such an even over odd predominance can be obscured due to an additional input of continentally-derived alkanes. We believe, therefore, that the R_{22} index is a better criterion for hypersaline environments as heneicosane (C_{21}) and tricosane (C_{23}) are relatively low components among continentally-derived hydrocarbons.

Monomethyl branched alkanes, such as 2-methylpentadecane (A in Fig. 1) and 2-methylhexadecane (C) are relatively abundant and are thought to reflect the original presence of heterotrophic bacteria. Also in the RPO sample 7-methyl- and 8-methylheptadecane were observed and these compounds are thought to be derived from cyanobacteria (Gelpi et al. 1970).

Isoprenoid alkanes are abundant and phytane

TABLE 1. *Selected compounds identified in the saturated hydrocarbon fraction*

A	2-methylpentadecane
B	2,6,10-trimethylpentadecane
C	2-methylhexadecane
D	3-methylhexadecane
E	2,6,10-trimethyl-(3-methylbutyl) dodecane
F	2,6,10,14-tetramethylpentadecane (pristane)
G	2,6,10,14-tetramethylhexadecane (phytane)
H	2,6,10,14-tetramethylheptadecane
I	2,6,10,14-tetramethyloctadecane
J	5α(H),14α(H),17α(H)- + 5α(H),14β(H),17β(H)-pregnane
K	2,6,10,14,18- and/or 2,6,10,14,19-pentamethyleicosane
L	2,6,10,15,19,23-hexamethyltetracosane (squalane)
M	5β(H),14α(H),17α(H)-20R-cholestane + 5α(H),14α(H),17α(H)-20S-cholestane (esp. RPO)
N	5α(H),14β(H),17β(H)-20R-cholestane + 5α(H),14β(H),17β(H)-20S-cholestane
O	5α(H),14α(H),17α(H)-20R-cholestane
P	5α(H),14β(H),17β(H)-20R-24-methylcholestane 5α(H),14β(H),17β(h)-20S-24-methylcholestane
Q	5α(H),14α(H),17α(H)-20R-24-methylcholestane
R	4α,24-dimethyl-5α(H),14β(H),17β(H)-20R-cholestane + 5α(H), 14α(H), 17α(H)-20S-24-ethylcholestane (esp. RPO)
S	4α,24-dimethyl-5α(H),14β(H),17β(H)-20S-cholestane + 5α(H), 14β(H), 17β(H)-20R-24-ethylcholestane + 5α(H), 14β(H), 17β(H)-20S-24-ethylcholestane
T	5α(H), 14α(H),17α(H)-20R-24-ethylcholestane
U	4α-methyl,24-ethyl-5α(H),14α(H),17α(H)-20R-cholestane + 17α(H),21α(H)-hopane
V	gammacerane
W	22R-pentakishomohop-17(21)-ene 22S-pentakishomohop-17(21)-ene
X	17α(H),21β(H)-22R-pentakishomohopane 17α(H),21β(H)-22S-pentakishomohopane

(G) is the most abundant compound in the saturate fraction of all samples. The pristane/phytane ratio is very low (<0.1), which is indicative of hypersaline environments (ten Haven et al. 1985; see also Albaiges & Torradas 1974; Fu Jiamo et al. this volume; Wang et al. this volume). The C_{25} isoprenoid (K), either 2,6,10,14,18- and/or 2,6,10,14,19-pentamethyleicosane, is also present in all samples and it is noteworthy to mention that the 2,6,10,14,18-C_{25} isoprenoid has been suggested as a biological marker for hypersaline environments (Waples et al. 1974). The 2,6,10,15,19-C_{25} isoprenoid is virtually absent. The presence of squalane (L) suggests that halophilic bacteria were present in the original depositional environment, although a contribution from other bacteria such as methanogenic bacteria cannot be precluded. Recently large quantities of highly branched

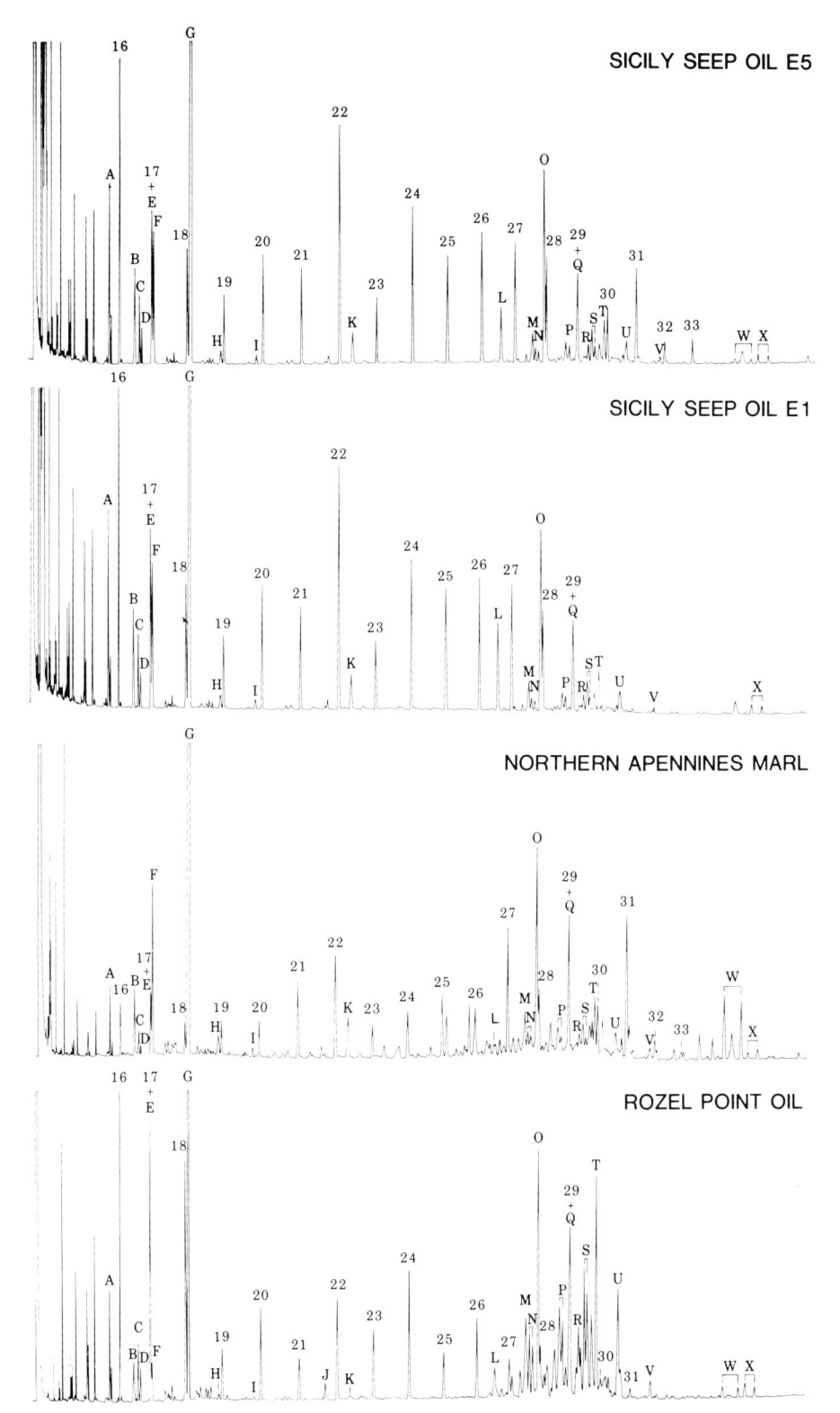

FIG. 1. Gas chromatograms of the saturated hydrocarbon fractions of Sicily seep oils E1 and E5 (E2 and E5 gave identical gas chromatograms), northern Apennines marl sample, and Rozel Point oil. Identifications of letter-labelled compounds are given in Table 1.

alkanes and alkenes were found in sediments from some hypersaline basins of Western Australia (Dunlop & Jefferies 1985). The isoprenoid alkane 2,6,10-trimethyl-7-(3-methylbutyl) dodecane (E) is an important compound occurring in the RPO (Yon et al. 1982) and is also present in small quantities in the other samples. This compound might be a biological marker for *Enteromorpha prolifera* (Rowland et al. 1985), a species which is known to have a salinity tolerance up to 65‰ (Ehrlich & Dor 1985).

Figure 2 shows mass chromatograms of m/z 191 and 367 of the RPO. The distribution patterns of the extended hop-17(21)-ene series, exemplified by the m/z 367 trace, and the extended 17α(H),21β(H)-hopane series are very similar if not identical and maximize at C_{35}. This phenomenon is thought to be very typical for hypersaline environments (ten Haven et al. 1985). Similar distribution patterns of the hopanoids are observed in the other samples (Fig. 1) and also in marl extracts from Sicily evaporitic deposits (Palmer & Zumberge 1981) and in some Chinese oils derived from salt lake evaporitic formations (Fu Jiamo et al. this volume). There are, however, some similar distribution patterns reported from non-hypersaline environments (e.g., McEvoy 1983). A reduction of hopenes to the corresponding hopanes was suggested to explain the similar distribution patterns of these compounds (ten Haven et al. 1985, 1986). Another feature which all samples have in common, is the presence of gammacerane (V in Fig. 1; see also Fig. 2). Gammacerane is also ubiquitous in Chinese oils derived from saline environments (Fu Jiamo et al. this volume; Xie Taijun et al. 1986). In the RPO 5α(H),14α(H),17α(H)- and 5α(H),14β(H),17β(H) pregnane (J), homopregnane and 4-methylpregnane were encountered, but only in small amounts. Sometimes pregnanes and homopregnanes are present as the major steranes in samples from hypersaline environments (ten Haven et al. 1985; see also Fu Jiamo et al. this volume).

The sterane composition and distribution of the Italian samples (NAM, SSOs) are almost identical (Fig. 1), supporting the presumed origin of the seep oils from Messinian formations in Sicily (Palmer & Zumberge 1981). One remarkable characteristic is the almost complete absence of 20S 5α(H),14α(H),17α(H) sterane isomers, whereas the 20S and 20R 5α(H),14β(H),17β(H) steranes (N, P, S) are present in relatively high amounts. This 'maturity' discrepancy can be explained by an alternative diagenetic pathway of steroids, assuming precursor steroids with a Δ^7, Δ^8 and/or $\Delta^{8(14)}$ double bond (ten Haven et al. 1986). Other relatively important steranes are 4-methylsteranes (R, S, U). These compounds may point to an input of dinoflagellates (Boon et al. 1979; Robinson et al. 1984; de Leeuw 1986), which is not surprising considering the wide salinity tolerance of dinoflagellates (Wall & Dale 1974). However, a bacterial origin for the 4-

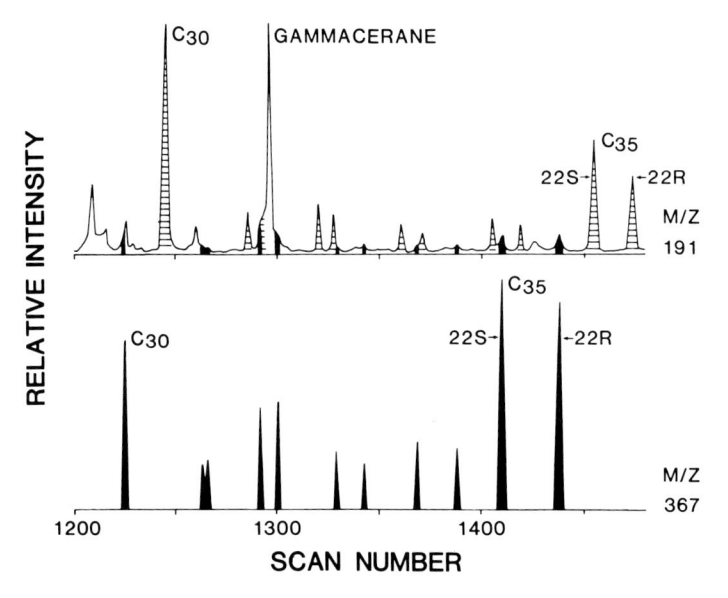

FIG. 2. Mass chromatograms of m/z 191 and 367 in the hopanoid region of the RPO. Hop-17(21)-enes are indicated black and 17α(H),21β(H) hopanes are shaded.

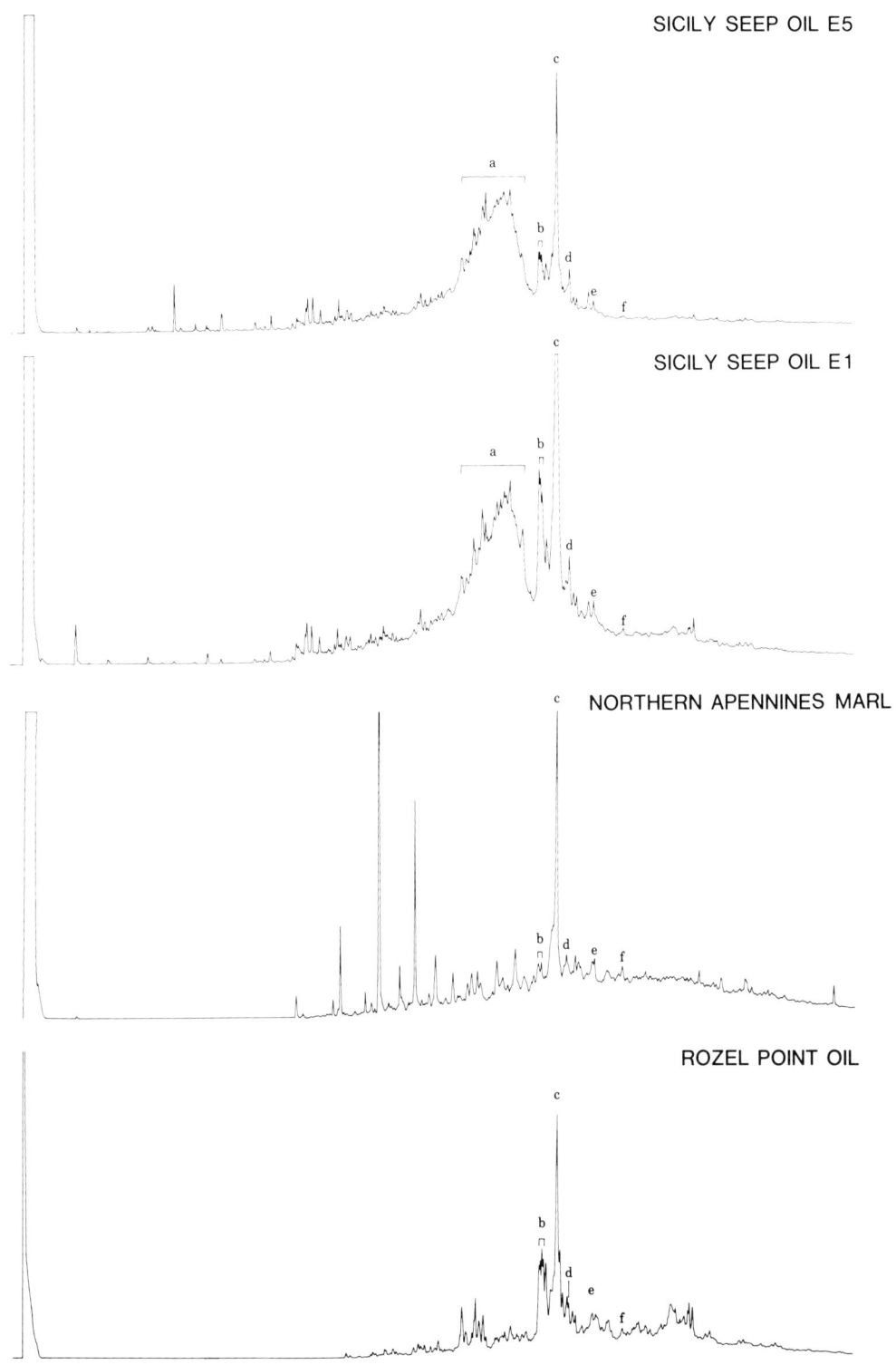

FIG. 3. FPD–gas chromatograms of the aromatic hydrocarbon fraction of Sicily seep oils E1, E5, northern Apennines marl and Rozel Point oil. Identifications of selected compounds are given in Table 2.

TABLE 2. *Selected compounds identified in the aromatic hydrocarbon fraction*

a unresolved hump of a.o. thiolanes
b mixture of mid-chain C_{20} isoprenoid thiophenes
c 2,3-dimethyl-5-(2,6,10-trimethylundecyl) thiophene
d 3-methyl-2-(3,7,11-trimethyldodecyl) thiophene
e 3,5-dimethyl-2-(3,7,11-trimethyldodecyl) thiophene
g 5-ethyl-3-methyl-2-(3,7,11-trimethyldodecyl) thiophene

methyl steranes can at present not totally be precluded, since Bouvier et al. (1976) reported the occurrence of a 4-methyl-$\Delta^{8(14)}$-sterol and a 4-methyl-$\Delta^{8(14),24}$-sterol in the bacterium *Methylococcus capsulatus*.

Aromatic hydrocarbon fraction

The majority of compounds present in the so-called aromatic hydrocarbon fraction consists of organic sulphur compounds (OSC). In view of this it is noteworthy that Thompson (1981) reported a sulphur content of 13.95% for the RPO and Colombo & Sironi (1961) measured up to 10.10% sulphur in Messinian seep oils. Figure 3 shows the gas chromatograms as recorded with an FPD giving a selective response for OSC. A detailed description of the OSC of the NAM sample is reported by Sinninghe Damsté et al. (1986) and OSC of the RPO are reported by de Leeuw (1986) and by Sinninghe Damsté and de Leeuw (1986). The most abundant compound in all samples is identified as 2,3-dimethyl-5-(2,6,10-trimethylundecyl) thiophene (c in Fig. 3). The co-occurrence of this C_{20} isoprenoid thiophene and phytane as the most important compounds in the respective component classes, seems to favour the hypothesis that this isoprenoid thiophene results from an early diagenetic incorporation of sulphur in, for example, archaebacterial phytenes (Brassell et al. 1986). In the RPO a mixture of uncommon isoprenoid thiophenes is observed with the 2,6,10-trimethyl-7-(3-methylbutyl) dodecane carbon skeleton (Sinninghe Damsté et al. 1987). These types of sulphur compounds are relatively abundant in the aromatic fraction. In the saturated fraction the corresponding alkane, 2,6,10-trimethyl-7-(3-methylbutyl) dodecane, is an important compound, which supports the incorporation theory. More information concerning the OSC of the seep oils investigated here is published elsewhere (Sinninghe Damsté et al. 1987; Schmid et al. 1987).

As the precise nature of the OSC is still poorly understood, a direct link with the depositional environment seems rather speculative. However, in view of their dominant presence in all samples, the suggestion of Sinninghe Damsté et al. (1987) that these OSC characterize hypersaline depositional environments, seems to be justified. Moreover the existence of high-sulphur petroleums has been ascribed to sulphur incorporation into organic matter in carbonate–evaporate environments (Tissot & Welte 1984).

Conclusions

All samples show similar biological marker characteristics, which can be attributed to the environment of deposition. Table 3 summarizes these biological markers and their typical distribution patterns.

TABLE 3. *Organic geochemical phenomena related to hypersaline depositional environments*

- Phytane ≫ pristane
- $R_{22} = \dfrac{2 \times \text{n-}C_{22}}{\text{n-}C_{21} + \text{n-}C_{23}} > 1.5$
- High abundance of regular C_{25} isoprenoid
- High abundance of squalane
- High abundance of organic sulphur compounds especially 2,3-dimethyl-5-(2,6,10-trimethylundecyl) thiophene
- 14β(H),17β(H)-sterane concentration relatively high in comparison with the 14α(H),17α(H)-20S steranes
- Relatively high abundance of gammacerane
- Typical distribution patterns of C_{31}–C_{35} hop-17(21)-enes and 17α(H),21β(H)-hopanes, both maximizing at C_{35}

The Rozel Point oil has been suggested to be sourced by playa-lake deposits of the Miocene Salt Lake group (Meissner et al. 1984), and based on the distribution of biological markers and their relative quantities observed in the saturated and aromatic hydrocarbon fractions, we support this suggestion. It seems that the phenomena as described in Table 3, can be used as a key to the past to recognize palaeo-hypersaline environments.

ACKNOWLEDGEMENTS: This study was partly supported by the Netherlands Foundation for Earth Science Research (AWON) with financial aid from the Netherlands Organization for the Advancement of Pure Research (ZWO) (grant 18.23.09). B. C. Schreiber collected the seep oil samples from Sicily. The geological survey of Utah kindly provided the Rozel Point oil. M. A. de Zeeuw gave analytical assistance.

References

ALBAIGES, J. & TORRADAS, J. M. 1974. Significance of the even carbon n-paraffin preference of a Spanish crude oil. *Nature*, **250**, 567–568.

BOUVIER, P., ROHMER, M., VENVENISTE, P. & OURISSON, G. 1976. $\Delta^{8(14)}$-steroids in the bacterium *Methylococcus capsulatus*. *Biochemical Journal*, **159**, 267–271.

BOON, J. J., RIJPSTRA, W. I. C., DE LANGE, F., DE LEEUW, J. W., YOSHIOKA, M. & SHIMIZU, Y. 1979. Black Sea sterol—a molecular fossil for dinoflagellate blooms. *Nature*, **277**, 125–127.

BRASSELL, S. C., LEWIS, C. A., DE LEEUW, J. W., DE LANGE, F. & SINNINGHE DAMSTÉ, J. S. 1986. Isoprenoid thiophenes: novel diagenetic products of sediment diagenesis? *Nature*, **320**, 160–162.

CITA, M. B. 1982. The Messinian salinity crisis in the Mediterranean: a review. *In*: BERCKHEMER, H. & HSÜ, K. J. (eds), *Alpine Mediterranean Geodynamics, Geodynamic Series*, **7**, 113–140.

COLOMBO, U. & SIRONI, G. 1961. Geochemical analyses of Italian oils and asphalts. *Geochimica et Cosmochimica Acta*, **25**, 24–51.

DUNLOP, R. W. & JEFFERIES, P. R. 1985. Hydrocarbons of the hypersaline basins of Shark Bay, Western Australia. *Organic Geochemistry*, **8**, 313–320.

EHRLICH, A. & DOR, I. 1985. Photosynthetic microorganisms of the Gavish Sabkha. *In*: FRIEDMAN, G. M. & KRUMBEIN, W. E. (eds), *Hypersaline Ecosystems—The Gavish Sabkha*, Ecological Studies 53, Springer Verlag, Berlin, 296–321.

GELPI, E., SCHNEIDER, H., MANN, J. & ORO, T. 1970. Hydrocarbons of geochemical significance in microscopic algae. *Phytochemistry*, **9**, 603–612.

TEN HAVEN, H. L., DE LEEUW, J. W. & SCHENCK, P. A. 1985. Organic geochemical studies of a Messinian evaporitic basin, northern Apennines (Italy) I. Hydrocarbon biological markers for a hypersaline environment. *Geochimica et Cosmochimica Acta*, **49**, 2181–2191.

——, ——, PEAKMAN, T. M. & MAXWELL, J. R. 1986. Anomalies in steroid and hopanoid maturity indices. *Geochimica et Cosmochimica Acta*, **50**, 853–855.

DE LEEUW, J. W. 1986. Sedimentary lipids and polysacharides as indicators for sources of input, for microbial activity and short term diagenesis. *In*: SOHN, M. L. (ed.), *Organic Marine Geochemistry, American Chemical Society Symposium Series*, **305**, 33–61.

MCEVOY, J. 1983. The origin and diagenesis of organic lipids in sediments from the San Miguel Gap. Ph.D. dissertation. University of Bristol.

MEISSNER, F. F., WOODWARD, J. & CLAYTON, J. L. 1984. Stratigraphic relationships and distribution of source rocks in the Greater Rocky Mountain Region. *In*: WOODWARD, J., MEISSNER, F. F. & CLAYTON, J. L. (eds), *Hydrocarbon Source Rocks of the Greater Rocky Mountain Region*, Rocky Mountain Association of Geologists, Denver, 1–34.

PALMER, S. E. & ZUMBERGE, J. E. 1981. Organic geochemistry of upper Miocene evaporite deposits in the Sicilian basin, Sicily. *In*: BROOKS, J. (ed.), *Organic Maturation Studies and Fossil Fuel Exploration*, Academic Press, London, 393–426.

PHILP, R. P. 1985. *Fossil Fuel Biomarkers. Applications and Spectra*. Elsevier, Amsterdam.

ROBINSON, N., EGLINTON, G., BRASSELL, S. C. & CRANWELL, P. A. 1984. Dinoflagellate origin for sedimentary 4α-methyl steroids and 5α(H) stanols in lake sediments. *Nature*, **308**, 439–441.

ROWLAND, S. J., YON, D. A., LEWIS, C. A. & MAXWELL, J. R. 1985. Occurrence of 2,6,10-trimethyl-7-(3-methylbutyl) dodecane and related hydrocarbons in the green algae *Enteromorpha prolifera* and sediments. *Organic Geochemistry*, **8**, 207–213.

SCHMID, J., CONNAN, J. & ALBRECHT, P. 1987. Occurrence and geochemical significance of long-chain dialkylthiacyclo-pentanes. *Nature*, **329**, 54–56.

SINNINGHE DAMSTÉ, J. S. & DE LEEUW, J. W. 1986. The origin and fate of isoprenoid C_{20} and C_{15} sulphur compounds in sediments and oils. *International Journal of Environmental Analytical Chemistry*, **23**, 1–19.

——, TEN HAVEN, H. L., DE LEEUW, J. W. & SCHENCK, P. A. 1986. Organic geochemical studies of a Messinian evaporitic basin, northern Apennines (Italy) II: Isoprenoid and n-alkyl thiophenes and thiolanes. *In*: LEYTHAEUSER, D. & RULLKÖTTER, J. (eds), *Advances in Organic Geochemistry, 1985*, Pergamon Press, Oxford, 791–805.

——, DE LEEUW, J. W., KOCK-VAN DALEN, A. C., DE ZEEUW, M. A., DE LANGE, F., RIJPSTRA, W. I. C. & SCHENCK, P. A. 1987. The occurrence and identification of series of organic sulphur compounds in oils and sediment extracts. I. A study of Rozel Point oil (U.S.A.). *Geochimica et Cosmochimica Acta*, **51**, 2369–2397.

THOMPSON, C. J. 1981. Identification of sulfur compounds in petroleum and alternative fossil fuels. *In*: FRIEDLINA, R. KH. & SKOROVA, A. E. (eds), *Organic Sulfur Chemistry*, Pergamon Press, Oxford, 189–200.

THUNELL, R. C., WILLIAMS, D. F. & BELYEA, P. R. 1984. Anoxic events in the Mediterranean in relation to the evolution of late Neogene climates. *Marine Geology*, **59**, 105–134.

TISSOT, B. P. & WELTE, D. H. 1984. *Petroleum Formation and Occurrence*. 2nd edn. Springer-Verlag, Berlin.

WALL, D. & DALE, B. 1974. Dinoflagellates in late Quaternary deep water sediments of Black Sea. *In*: DEGENS, E. T. & ROSS, D. A. (eds), *The Black Sea—Geology, Chemistry and Biology*, Memoir of the American Association of Petroleum Geologists, **20**, 364–380.

WAPLES, D. W., HAUG, P. & WELTE, D. H. 1974. Occurrence of a regular C_{25} isoprenoid hydrocarbon in Tertiary sediments representing a lagoonal, saline environment. *Geochimica et Cosmochimica Acta*, **38**, 381–387.

XIE TAIJUN, WU LIZHEN & JIANG JIGANG, 1986. Formation of oil and gas fields in Jianghan saline lacustrine basin. In: *Proceedings Beijing Petroleum Geology Symposium 1984*. (In press).

YON, D. A., RYBACK, G. & MAXWELL, J. R. 1982. 2,6,10-trimethyl-7-(3-methylbutyl) dodecane, a novel sedimentary biological marker compound. *Tetrahedron Letters*, **23**, 2143–2146.

II. L. TEN HAVEN*, J. W. DE LEEUW, J. S. SINNINGHE DAMSTÉ & P. A. SCHENCK. Delft University of Technology, Department of Chemistry and Chemical Engineering, Organic Geochemistry Unit, De Vries van Heystplantsoen 2, 2628 RZ Delft, The Netherlands.

S. E. PALMER & J. E. ZUMBERGE**. Cities Service Oil and Gas Corporation, P.O. Box 3908, Tulsa, Oklahoma, 74102, USA.

*Present address: Institute of Petroleum and Organic Geochemistry (ICH-5), KFA-Jülich GMBH, P.O. Box 1913, W-5170 Jülich-I, Federal Republic of Germany.

**Present address: Raska Laboratories, 3601 Dunvale, 77063 Houston, Texas, USA.

Interactions of iron, carbon and sulphur in marine and lacustrine sediments

W. Davison

SUMMARY: Sulphur occurs in marine sediments chiefly as pyrite, which can form both directly or diagenetically within the sediment, or directly at the surface of the sediment. Direct formation occurs when reducing and oxidizing conditions co-exist, whereas diagenesis of monosulphides to pyrite involves oxidative transfer of electrons by, for example, decomposition of organic matter. The limited data available indicate that pyrite occurs in low concentrations in freshwater sediments and is formed directly at a redox boundary. Although most of the sulphur is bound within organic compounds, there is relatively little compared with that present in marine sediments. The absolute concentrations of pyrite, and the ratios of carbon to total sulphur, carbon to pyrite sulphur and organic sulphur to pyrite sulphur all have their uses as palaesalinity indicators. Although their use is soundly based due to the elevated concentrations of sulphate in sea compared with fresh waters, their validity is not proven. There are sufficient gaps in our knowledge, particularly about pyrite concentrations and formation processes in freshwater environments, to suggest that at present they should be used cautiously.

Introduction

Once sulphur and iron have been delivered to sediments their pathways and fates are largely determined by the rates of supply and decomposition of organic material. When microbiologically-mediated turnover of organic carbon is high, reducing conditions are often created allowing sulphate to be converted to sulphide. Ferrous iron, which is stable in oxygen-free environments, reacts with sulphide and traps sulphur in the sediment as an iron–sulphur compound. Considerable concentrations of iron sulphides can form in sulphate-rich marine sediments. By contrast iron sulphide formation in freshwater sediments is limited by low concentrations of sulphate. Berner & Raiswell (1984) have exploited this difference in sulphur concentrations by using the ratio of carbon to sulphur to diagnose whether sedimentary rocks were formed in marine or freshwater environments. The lack of sulphur in freshwater sediments has also led to the suggestion that monosulphides, FeS, may be preserved, rather than being converted to pyrite, FeS_2 (Berner et al. 1979). Pyrite formation is not limited by sulphur availability in marine sediments and so the ratio of pyrite to monosulphides may provide an additional indicator of the salinity prevalent during the deposition of sedimentary rocks.

New information, regarding the formation and transformations of iron sulphides within sediments, has recently become available. This paper reviews current ideas of pyrite formation processes in marine and lacustrine systems to provide a basis for appraising the use of elemental ratios as palaeosalinity indicators.

Formation of pyrite

Solubility considerations

Iron sulphides are formed when the ion activity product, calculated from the concentrations of free ferrous and sulphide ions, exceeds the solubility product. Two solubility products are pertinent to this preliminary discussion: that of amorphous FeS and that of pyrite. They can both be expressed as:

$$K_{sp} = a\,Fe^{2+} \times a\,HS^-/a\,H^+,$$

where 'a' represents the activity of an ionic species.

For amorphous FeS the appropriate reaction is:

$$FeS + H^+ = Fe^{2+} + HS^-,$$

whereas for pyrite it is:

$$FeS_2 + H^+ = Fe^{2+} + HS^- + S°.$$

Pyrite is assumed to be in equilibrium with an excess of elemental S which, as a solid phase, has an activity of unity. It is a slightly more oxidized form of sulphur than sulphide.

Amorphous FeS is the most soluble of the iron sulphides, all of which are fairly insoluble. it has a pK_{sp} ($= -\log K_{sp}$) of 3 (Davison 1980), which means that at a concentration of Fe^{2+} of 10^{-5} M, typical of reducing environments, total dissolved

From: FLEET, A. J., KELTS, K. & TALBOT, M. R. (eds), 1988, *Lacustrine Petroleum Source Rocks*, Geological Society Special Publication No. 40, pp. 131–137.

sulphide cannot exceed 2×10^{-5} M at pH 7 and 10^{-6} M at pH 8. Pyrite is much less soluble. Free energy data give a value of pK_{sp} of 13.6 (Goldhaber & Kaplan, 1974), but direct solubility measurements of Tewari et al. (1978) indicate a value of 7. Based on the latter value, if the concentration of Fe^{2+} was 10^{-5} M, total dissolved sulphide would be limited to the exceedingly small concentrations of 2×10^{-9} M at pH 7 and 10^{-10} M at pH 8. Evidently pyrite can form when the sulphide concentration is very small and insufficient to precipitate amorphous FeS.

Marine systems

Marine systems are characterized by high concentrations of sulphate which can be reduced to sulphide. Iron, the second most abundant metal in the earth's crust, is rarely limiting, and so under anoxic conditions bisulphide and iron are usually present in sufficient concentration to exceed the solubility product of both pyrite and amorphous FeS. Amorphous FeS of indeterminate structure precipitates preferentially. It is a high energy form and so may be transformed on aging into lower energy, more crystalline, and hence less soluble analogues, such as mackinawite and greigite (Fig. 1). Transformation to pyrite is not so simple because an oxidation step is required. However, if elemental sulphur is present, the monosulphides may be oxidized to pyrite in a slow process which takes several years. Diagenetic transformations resulting in an increasing pyrite concentration with depth have been observed in, for example, sediments from the Gulf of Mexico (Filipek & Owen 1980).

Pyrite can also form as a primary phase in marine sediments. Howarth (1979) has shown that in tidal salt marshes substantial amounts of pyrite may form within 24 hours. The sediments, which are organic-rich and essentially reducing in character, have an oxidizing component in the form of oxygen which is supplied directly from the roots of the marsh-grass. Elemental sulphur, generated by reaction with sulphides, or from incomplete reduction of sulphate (Fig. 2), reacts with ferrous and sulphide ions to form pyrite.

A similar mechanism may operate at the surface of reducing marine sediments which are overlain by well oxygenated waters (Goldhaber & Kaplan 1974). However, it cannot explain recent observations (Howarth & Jorgenson 1984) of rapid pyrite formation at a depth of 15 cm in sub-tidal sediments. Molecular oxygen is not available under these conditions and it has been suggested (Howarth & Jorgensen 1984) that elemental sulphur may be formed by iron or manganese oxides oxidizing sulphides. A more likely possibility is that energy from the inorganic oxidation of sulphide is used by micro-organisms to synthesize organic material from carbon dioxide and volatile fatty acids in a process known as mixotrophy (Fig. 2). Several organisms including *Beggiatoa* can participate in such reactions (Jones 1985 and pers. comm.). Slow diagenetic formation of pyrite from monosulphides must involve a similar electron transfer mechanism.

Direct pyrite formation occurs most readily in acidic environments (Howarth 1979). As pH decreases larger concentrations of free sulphide and iron are required to achieve the solubility product (see previous section). Thus, the system may become undersaturated with respect to monosulphides, but saturated with respect to pyrite which consequently forms.

Freshwater systems

There have been very few direct determinations of pyrite in freshwater sediments. The only reliable data sets are in two papers published during 1985 (Davison et al. 1985; Nriagu & Soon 1985). Previous estimates (Sugawara et al. 1953, 1954) used doubtful methodology which may have failed to unequivocally measure pyrite (Nriagu & Soon 1985). Because of the lack of data little is known about the mechanism of pyrite formation in freshwater sediments. Current views have been deduced from the variation in the concentration of pyrite with depth (Figs 3–5).

The very high concentrations in the surface sediments from McFarlane and Kelley lakes (Fig. 3) must be viewed cautiously. These two lakes are near Sudbury, Ontario, where there are extensive mining and smelting activities. The lake waters contain elevated concentrations of sulphate, and the sediments are so enriched in nickel and

DIAGENETIC FORMATION OF PYRITE

$$Fe^{2+} + HS^- = FeS\ am. + H^+$$

$$\downarrow (aging)$$

MACKINAWITE, GREIGITE

$$\downarrow (oxidation),\ S°$$

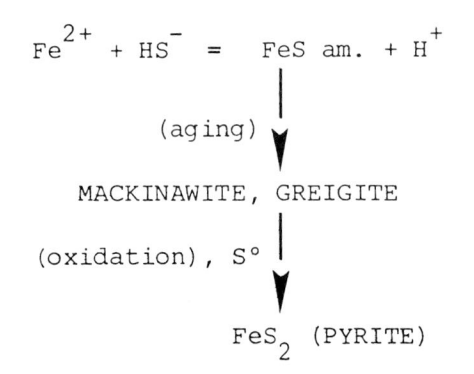

$$FeS_2\ (PYRITE)$$

FIG. 1. Diagenetic formation of pyrite.

Interactions of iron, carbon and sulphur 133

DIRECT FORMATION OF PYRITE

SALT MARSH AND SURFACE SEDIMENTS

$HS^- + \frac{1}{2}O_2 + H^+ = S° + H_2O$ sulphide oxidation

$SO_4^{2-} + 1\frac{1}{2}CH_2O + 2H^+ = S° + 1\frac{1}{2}CO_2 + 2\frac{1}{2}H_2O$ sulphate reduction

$Fe^{2+} + HS^- + S° = FeS_2 + H^+$

BURIED SEDIMENT

$HS^- + 2Fe(OH)_3 + H^+ = S° + 2Fe(OH)_2 + 2H_2O$ iron reduction

$HS^- + CH_2O/\frac{1}{2}CO_2 + H^+ = S° + 1\frac{1}{2}CH_2O + \frac{1}{2}H_2O$ mixotrophy

FIG. 2. Direct formation of pyrite.

copper that a substantial part of the sulphur attributed to pyrite may be bound to these metals (Nriagu & Soon 1985).

The other five lakes for which data are available are in relatively unpolluted regions. Concentrations of pyrite in their sediments may be sensibly compared with values for marine sediments. The latter, with values of pyrite-S in dry weight of sediment of 0.5–2 mg g^{-1} for the Gulf of Mexico (Filipek & Owen 1980), 2–10 mg g^{-1} for a Danish Fjord (Howarth & Jorgensen 1984) and 2 mg g^{-1} for the Santa Catalina Basin (Goldhaber & Kaplan 1974), are much higher than the available freshwater values (Figs 4 and 5).

Pyrite concentrations in the sediments of Turkey and Batchawara Lakes in Ontario, Canada (Fig. 4), are very low, rarely exceeding the detection limit of 30 µg of S g^{-1} dry weight. The waters of these oligotrophic lakes are low in sulphate, ca. 5 mg l^{-1}, and their sediments contain no measurable acid volatile sulphide, the operational definition for the collective monosul-

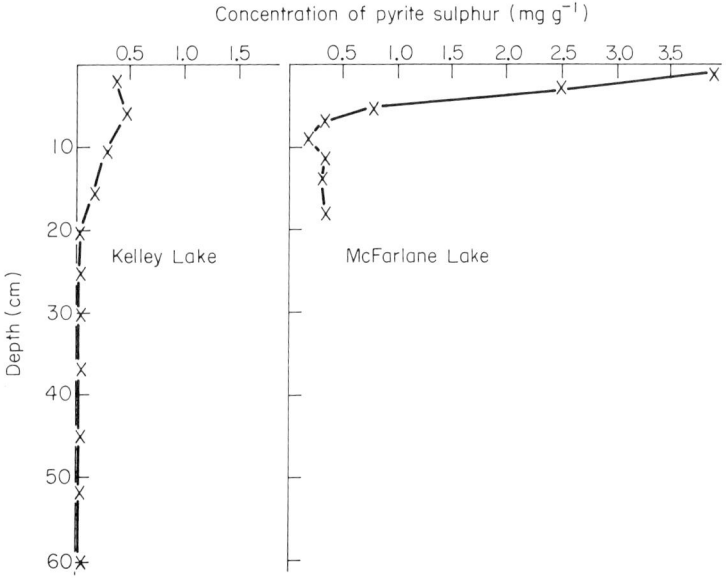

FIG. 3. Variation of pyrite sulphur with depth in sediment from two lakes near Sudbury, Ontario. Data taken from Nriagu & Soon 1985.

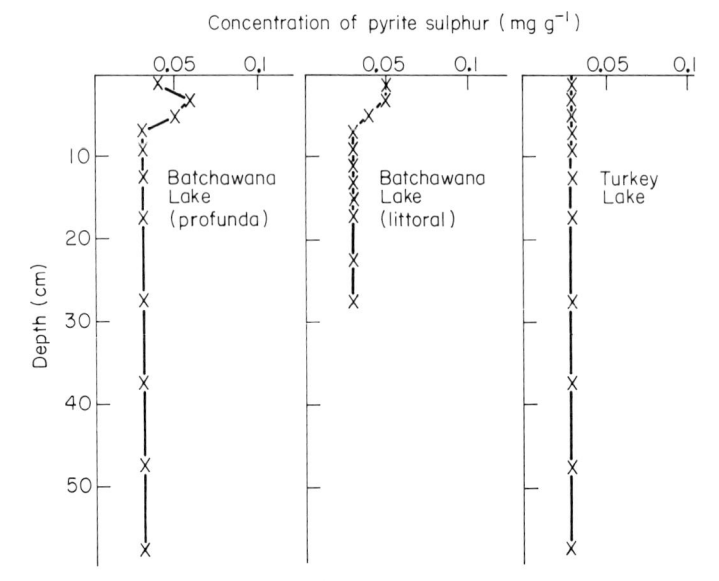

FIG. 4. Variation of pyrite sulphur with depth in sediment from three lakes in an unpolluted region of Ontario. Relatively constant low levels reflect the detection limit of 0.03 mg g^{-1}. Data taken from Nriagu & Soon (1985).

phides. Evidently there is insufficient reduction of sulphate to form iron sulphides, which has led to the suggestion (Nriagu & Soon 1985) that pyrite formation is unlikely to occur in oligotrophic lakes containing less than 5 mg l^{-1} of sulphate.

Independent measurements (Davison et al. 1985) in an oligotrophic lake, Ennerdale Water, in the English Lake District, where the sulphate concentration of the water is also ca. 5 mg l^{-1} (note, concentrations given in a previous paper are now known to be four times too high), show appreciable concentrations of pyrite (Fig. 5). Acid volatile sulphide in these sediments is low, but detectable at 20–100 µg g^{-1} for depths greater than 3 cm, so some reduction of sulphate must occur. The absence of both pyrite and acid volatile sulphide from the top 3 cm of sediment which have some oxidizing characteristics, has prompted the suggestion (Davison et al. 1985) that pyrite is formed directly in the sub-surface sediments. Low values of acid volatile sulphide

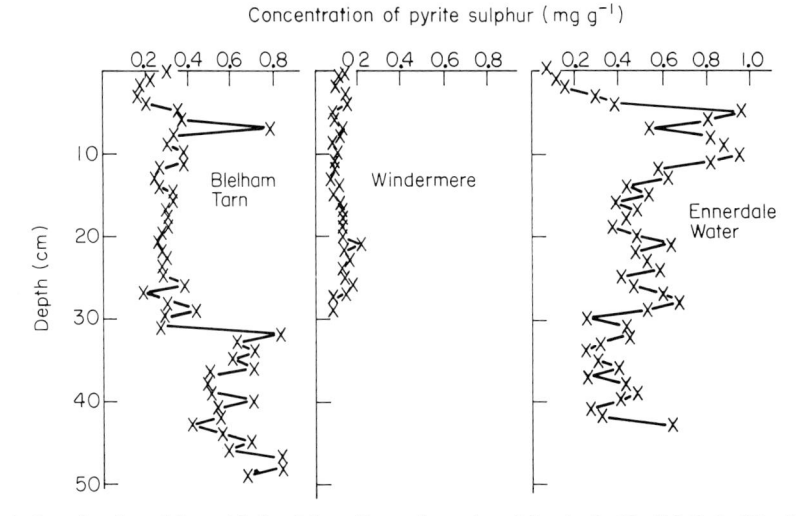

FIG. 5. Variation of pyrite sulphur with depth in sediment from three lakes in the English Lake District.

imply that sulphate reduction is slow, which in the vicinity of the more oxidizing conditions may encourage pyrite formation. The presence of pyrite in Ennerdale Water but not in Turkey and Batchawara lakes may be linked to a difference in pH. The sediments of the Ontario lakes had a much lower pH, 5–5.5, than Ennerdale Water, 6.5.

The two other Lake District lakes (Fig. 5), Windermere and Blelham Tern, are eutrophic according to the definition of Vollenweider (1976). High concentrations of acid volatile sulphur were present in their sediments, presumably due to an ample supply or organic matter and high concentrations of sulphate (ca. 10 mg l^{-1}) in the overlying waters. Pyrite was present at all depths. It did not systematically increase with depth, as observed for marine systems where diagenetic formation from monosulphides has been inferred (Berner 1980). However, the possibility of diagenetic formation cannot be entirely discarded because in the sediments of Blelham Tarn pyrite concentrations are larger at depths greater than 30 cm.

Pyrite is thought (Davison et al. 1985) to form directly at the sediment–water interface. Ferrous and sulphide ions are supplied from the highly reducing sediment, whereas oxygen in the overlying waters provides the necessary oxidizing components. In Blelham Tarn, which becomes seasonally anoxic, there is the additional possibility of direct pyrite formation within the water column.

The authors of the two recent papers on lacustrine systems (Davison et al. 1985; Nriagu & Soon 1985) agree that freshwater pyrite formation does not appear to be directly related to concentrations of acid volatile sulphur, pore water sulphate, elemental sulphur or organic carbon. Pyrite appears to form directly when reducing sediments are subject to oxidizing influences.

Indicators of past salinity

The use of carbon–sulphur ratios to distinguish freshwater from marine sedimentary rocks is based on a simple premise (Berner & Raiswell 1984). In marine systems formation of iron sulphides is not limited by lack of sulphur because there is a considerable concentration of sulphate, ca. 30 mmol l^{-1}, in the water. By contrast, freshwaters have much less sulphate, usually 0.05–0.3 mmol l^{-1}, which may limit iron sulphide formation. The sediments must be suitable for producing iron sulphides and so there should be no other limiting factors. There should be adequate supplies of decomposable organic material to facilitate reduction of sulphate, and of iron, in a reducible form which can react with sulphide.

Implicit, in the use of C/S ratios, is the assumption that most sedimentary sulphur is bound to iron. Thus, sedimentary material containing barite, gypsum or anhydrite will confound the use of C/S ratios. Rocks which are unusually high in carbon, such as coal, have also been recognized as a problem (Berner & Raiswell 1984). The abundance of organic matter may lead to pyrite formation being limited by iron availability. More importantly, however, more sulphur may be present in the organic material than as pyrite.

Recent analyses of freshwater sediments show that if there is a low rate of decomposition of organic carbon they may be virtually devoid of pyrite because there is insufficient reduction of sulphate. Moreover, when there is considerable sulphate reduction, as in productive lakes, pyrite is again a minor component (Davison et al. 1985). Acid volatile sulphides are present in the near surface sediments of productive lakes, but they are drastically depleted at depth. They are not, however, converted diagenetically to pyrite. The elevated levels may represent an increased sulphur burden due to pollution (Nriagu & Coker 1983) or more probably sulphide is lost from the sediment to the overlying water by a recycling mechanism which has yet to be understood (Berner & Westrich 1985). Thus in older sediments most of the sulphur is present within organic compounds. Generally, pyrite formation in freshwater lakes does not appear to be related to the supply of organic carbon (Davison et al. 1985; Nriagu & Soon 1985), and the major form of sulphur in sedimentary rocks with freshwater origins is unlikely to be pyrite. So what of C/S ratios? Table 1 shows the mean ratios of carbon to total sulphur, C/S_t, and carbon to pyrite sulphur, C/S_p, for the near surface and deeper sediments of the freshwater systems previously discussed. Excluding Kelley and McFarlane lakes the ratios for the deeper material are $C/S_t = 40$–120 and $C/S_p = 170$–>6000. The much higher values for C/S_p compared with C/S_t reflect the very low concentrations of pyrite sulphur. However, C/S_t ratios are still sufficiently removed from the ratio of 0.5–5 for marine sediments (Berner & Raiswell 1984), where most sulphur is present as pyrite, for there to be a clear distinction. Apparently the alternative, organic sulphur, pathway in freshwater sediments is small compared with inorganic pyrite formation in marine sediments. Note that even in the polluted Kelley and MacFarlane lakes the ratios in the deeper

TABLE 1. *Carbon and sulphur data* for freshwater sediments*

Lake	Sediment	C %	AVS (mg g^{-1})	S$_p$ (mg g^{-1})	S$_t$ (mg g^{-1})	C/S$_p$	C/S$_t$	FeS$_2$/FeS
Kelley and	Top	2–6	0.1–8	0.4–3	15–20	20–80	2–3	0.2–2
McFarlane	Deeper	3–11	0–2	0.03–0.3	1.2–3	300–1000	20–30	0.3–0.5
Ennerdale	Top	5–10	0–0.2	0.1–1	2–3	300	30	26
	Deeper	6–8	0–0.1	0.4	1.8	170	40	5
Windermere	Top	7–9	0–3	0.1	2–6	700	20	0.9
	Deeper	6–11	0–0.2	0.1	1–2	710	60	1.3
Turkey	Whole	13	<0.03	<0.03	1.5	>4000	90	—
Blelham	Top	13	1–6	0.3	5–12	430	20	0.07
	Deeper	13	0.5–1	0.6	3–4	220	40	0.5
Batchawana	Littoral	14–19	<0.03	0.03	1.4	6000	120	>0.5
	Profundal	22–25	<0.03	0.04	3.6	6000	70	>0.5

*Range or typical values are given for the whole sediment when concentrations are uniform, or for the top and lower sections separately when the top is enriched in AVS or pyrite. Data for Kelley and McFarlane lakes have been combined, and data are given for both littoral and profundal sediments of Batchawana Lake.

sediments were appreciably larger than those for marine sediments (Table 1).

As these conclusions are based on a limited freshwater data set higher concentrations of pyrite in freshwater sediments cannot be ruled out. Concentrations of acid volatile sulphide may reach 10 mg g^{-1} (Davison et al. 1985) implying that low sulphate concentrations do not limit pyrite formation. Moreover, some source rocks, which are believed to have freshwater origins contain pyrite at concentrations of 10 mg g^{-1} (Duncan & Hamilton this volume).

The ratio of pyrite sulphur to acid volatile sulphur has also been suggested as a salinity indicator (Berner et al. 1979). It was based on the idea that in freshwater systems diagenetic transformations of FeS to FeS$_2$ will be limited by lack of sulphur. However, there is little evidence for either FeS preservation or of diagenetic pyrite formation (Davison et al. 1985), so not surprisingly the FeS$_2$/FeS ratios for older lake sediments (Table 1), 0.5–5, lie within the range of 0.2–100 reported for estuarine samples (Table 2).

There are other potential palaeosalinity indicators (Table 2). As many marine sediments are generally rich in pyrite and freshwater sediments deficient, high concentration of pyrite can almost certainly be attributed to marine systems or closed-basin lacustrine systems which are rich in sulphide. A more refined approach may be to consider the ratio of organic to pyrite sulphur. Organic sulphur is usually the dominant fraction in freshwater sediments, giving ratios >1. In marine systems the large concentrations of pyrite sulphur produce ratios <1. In the past, methodological difficulties would have prevented the use

TABLE 2. *Typical values for palaeosalinity indicators*

Indicator	Marine	Freshwater
C/S$_t$	0.5–5[1]	40–120[2]
C/S$_p$	0.5–5[1]	170–6000[2]
FeS$_2$/FeS	0.2–100[3]	0.5–5[4]
Pyrite S (mg g^{-1})	1–20[5]*	0–0.9[4,6]
Org. S./Pyrite S	<0.1[5]	3–100[4,6]

Data taken from (1) Berner & Raiswell (1984); (2) Table 1, this paper; (3) Berner et al. (1979); (4) Davison et al. (1985); (5) Goldhaber & Kaplan (1974); (6) Nriagu & Soon (1985).
*Exceptional data for Pacific Trench sediments, where pyrite was undetectable, was not included.

of this ratio, but the various sulphur fractions are now more amenable to measurement (Davison et al. 1985; Nriagu & Soon 1985). However, until more is known about the diagenesis of organic sulphur (Guerin & Braman 1985), especially the possibility of its conversion to pyrite, such a ratio must be regarded cautiously. All these palaeosalinity tests, exploit the difference in concentration between sulphate in marine and freshwater environments. Their general validity is therefore assured, but only through the appraisal of numerous data sets for recent sediments and source rocks can the range of values, and therefore the confidence of the test, be established.

ACKNOWLEDGEMENTS: I thank John Hilton, Gwyn Jones, David Kinsman and Colin Reynolds for their constructive comments, Joyce Hawksford for typing and the NERC for financial support.

References

BERNER, R. A. 1980. *Early Diagenesis.* Princeton University Press, Princeton, NJ.

—— & RAISWELL, R. 1984. C/S method for distinguishing freshwater from marine sedimentary rocks. *Geology*, **12**, 365–368.

—— & WESTRICH, J. T. 1985. Bioturbation and the early diagenesis of carbon and sulfur. *American Journal of Science*, **285**, 193–206.

——, BALDWIN, T. & HOLDREN, G. R. 1979. Authigenic iron sulphides as palaeosalinity indicators. *Journal of Sedimentary Petrology*, **49**, 1345–1350.

DAVISON, W. 1980. A critical comparison of the measured solubilities of ferrous sulphide in natural waters. *Geochimica et Cosmochimica Acta*, **44**, 803–808.

——, LISHMAN, J. P. & HILTON, J. 1985. Formation of pyrite in freshwater sediments: implications for C/S ratios. *Geochemica et Cosmochimica Acta*, **49**, 1615–1620.

FILIPEK, L. H. & OWEN, R. M. 1980. Early diagenesis of organic carbon and sulphur in outer shelf sediments from the Gulf of Mexico. *American Journal of Science*, **280**, 1097–1112.

GOLDHABER, M. B. & KAPLAN, I. R. 1974. The sulphur cycle. *In*: Goldberg, E. D. (ed.), *The Sea Vol. 5, Marine Chemistry.* Wiley, New York, 303–336.

GUERIN, W. F. & BRAMAN, R. S. 1985. Patterns of organic and inorganic sulfur transformations in sediments. *Organic Geochemistry*, **8**, 259–268.

HOWARTH, R. W. 1979. Pyrite: its rapid formation in a salt marsh and its importance in ecosystem metabolism. *Science*, **203**, 49–51.

—— & JORGENSEN, B. B. 1984. Formation of ^{35}S-labelled elemental sulphur and pyrite in coastal marine sediments (Limfjorden and Kysing Fjord, Denmark) during short-term $^{35}SO_4^{2-}$ reduction measurements. *Geochimica et Cosmochimica Acta*, **48**, 1807–1818.

JONES, J. G. 1985. Microbes and microbial processes in sediments. *Philosophical Transactions of the Royal Society, London, Series A*, **315**, 3–17.

NRIAGU, J. O. & COKER, R. D. 1983. Sulphur in sediments chronicles past changes in lake acidification. *Nature*, **303**, 692–694.

—— & SOON, Y. K. 1985. Distribution and isotopic composition of sulfur in lake sediments of northern Ontario. *Geochimica et Cosmochimica Acta*, **49**, 823–834.

SUGAWARA, K., KOYAMA, T. & KOZAWA, A. 1953. Distribution of various forms of sulphur in lake-, river- and sea-muds. *Journal of Earth Sciences (Nagoya University)*, **1**, 17–23.

——, —— & —— 1954. Distribution of various forms of sulphur in lake-, river- and sea-muds. *Journal of Earth Sciences (Nagoya University)*, **2**, 1–4.

TEWARI, P. H., WALLACE, G. & CAMPBELL, A. B. 1978. The solubility of iron sulphides and their role in mass transport in Girdler-Sulphide heavy water plants. *Report of the Atomic Energy of Canada Limited*, **AECL-5960**, 1–34.

VOLLENWEIDER, R. A. 1976. Advances in defining critical loading levels for phosphorus in lake eutrophication. *Memorie dell Istituto italiano di Idrobiologia*, **33**, 53–83.

W. DAVISON, Freshwater Biological Association, The Ferry House, Ambleside, Cumbria, LA22 0LP, UK.

Possible relationships of stratigraphy and clay mineralogy to source rock potential in lacustrine sequences

R. F. Yuretich

SUMMARY: Petroleum accumulations in lacustrine environments often possess characteristics different from those in marine settings. In general, lacustrine petroleum fields are smaller, the oil has often not migrated far from the source, and oil shales are common. Some of these differences can be attributed to the relative youthfulness of lacustrine basins and to differences in the type of organic matter present. However, additional influences may arise from the stratigraphic framework and clay composition of the source rocks. The scale of the environment may be a major controlling factor. Gradations between large lakes and marginal marine basins exist, but lake basins will more often be areally small, relatively shallow and subject to frequent lateral migrations in depositional environments. These characteristics make the resulting sedimentary sequences much more heterogeneous, both in stratigraphic relationships and in sediment composition. Stratigraphy will ultimately influence the migration and accumulation of the oil; the sediment composition—particularly clay mineralogy—will affect the generating potential of the source rock. Shifting sediment source areas coupled with lake level fluctuations and early diagenetic reactions may produce thin mudstones containing very heterogeneous clay mineral assemblages, which, accordingly, may not undergo significant transformations during burial diagenesis.

Introduction

The potential for petroleum production from lacustrine rocks has not yet received a thorough evaluation from explorationists. Until very recently, the popular wisdom assumed that lake sediments comprised only an insignificant portion of the geologic record. Since the world's largest petroleum accumulations were discovered in marine-shelf environments (Moody 1975), most petroleum geologists paid scant attention to lacustrine environments and processes. On the other hand, the extensive research on the Green River Oil Shale in Wyoming and Utah (Bradley 1929, 1970, 1973; Eugster & Surdam 1973; Desborough 1978) kindled an awareness that lake deposits could be abundantly petroliferous. Subsequent exploration and research has documented the importance of lacustrine hydrocarbons. The Green River Formation in Utah is also an active petroleum province with the potential for 130 million barrels of recoverable petroleum (Ritzma 1972). Research in the modern rift lakes of East Africa has demonstrated that many of these large lakes are actively generating hydrocarbons (Degens et al. 1973) and this has prompted further investigations of ancient tectonic rift settings such as the Mesozoic rifts of eastern North America. Finally, the re-establishment of scientific ties with the People's Republic of China has opened the eyes of western explorationists since a large fraction of their active production comes from lacustrine and related non-marine sedimentary rocks (Yang 1985).

Characteristics of lacustrine petroleum accumulations

From these renewed investigations, several general qualities of lacustrine oil fields can be ascertained. In the first place, they are generally smaller than their marine-derived counterparts. Klemme (1980) classified petroleum basins into eight different varieties. Of these, the intracratonic and rift basins (Klemme's Type 2 and 3) are those in which significant lake deposits are likely to occur. *In toto*, these basins comprise some 32% of the area of continental and offshore sedimentary basins world-wide, and account for about 37% of known recoverable reserves. Secondly, petroleum derived from lacustrine sources is often trapped relatively close to the source rock. Much of the production from the Tertiary Uinta Basin in Utah is from small sand bodies surrounded by hydrocarbon-rich shale (Hunt et al. 1954; Pitman et al. 1982). Finally, extensive regions of oil shale are a notable lacustrine phenomenon: approximately 70% of potential hydrocarbon recovery from oil shales occurs in lake deposits (Tissot & Welte 1978). Oil shales are known from numerous lake basins on several continents, with the largest being in North America (Green River) and Brazil (Paraiba Valley).

There can be several explanations for this state

of affairs. Many lacustrine petroleum fields are producing from Tertiary formations. Accordingly, they may simply not have had enough time to mature into major accumulations. This could account for the relatively small size, short migration distances and large amounts of oil shale. Age, of course, cannot be the only factor here. The Uinta Basin oil fields, despite their relative youth, have been heated to the temperatures required for abundant petroleum genesis (Pitman et al. 1982) yet all of the typical lacustrine characteristics are present. Indeed, about 31% of all recoverable petroleum is in Tertiary reservoirs, and several of the largest giant fields (including Bolivar Coastal in Venezuela and Ahwaz in Iran) are Tertiary in age (Tiratsoo 1984). So it seems likely that some fundamental characteristics of lacustrine sedimentary sequences may be responsible for the observed features of associated petroleum fields. As one possible hypothesis, I propose that stratigraphic relationships in conjunction with the clay mineralogy of the source rock may be important influences upon the size, distribution and maturity of lacustrine oil.

Clay mineral distribution in sedimentary basins

The sedimentary processes which control deposition in lacustrine and shallow-marine environments are basically the same. The analysis of Picard & High (1972) demonstrated that there are few individual characteristics which could unambiguously distinguish between the two. However, it is quite clear that the sum total of several indices can separate lacustrine stratigraphic sequences from those of marine origin (Picard & High 1981). In general, the scale of the original environment imparts most of the diagnostic features to the resultant deposits. Larger-scale basins usually exhibit a greater homogeneity in the sediments deposited within them. Although there is a gradation between large lakes and small, restricted marine basins, lacustrine environments most often occupy the lower end of the size spectrum.

Small basins will display more lateral and vertical variability in the distribution of sediments. This can be displayed in two principal ways: (1) a more restricted development of sandstones and correspondingly less lateral continuity of lithotypes; (2) extensive compositional heterogeneities in shale and mudstones. My primary focus in this review is on the latter aspect.

Marine environments

Clay minerals in both near-shore and profundal parts of the modern ocean basins have noticeable geographic variations. In general, it seems that the heterogeneity increases as the size of the basin decreases. Clay minerals in the open ocean usually consist of extensive mappable areas where a particular clay assemblage dominates. Frequently, these areas comprise an entire ocean basin (Biscaye 1965; Griffin et al. 1968). For example, smectites make up 53% of the clay minerals in the South Pacific and 41% in the Indian Ocean.

Along continental margins the variability in clay mineral assemblages increases. Hathaway (1972) and Pevear (1972) studied the clay minerals of the continental shelf sediments of the eastern United States and were able to group the assemblages into broad geographic zones which reflected the characteristics of the sediment source areas. These zones are still on the order of several hundred square kilometres, and the transitions between the groups are often quite subtle with smectite being the dominant mineral. Such patterns emphasize the importance of long-distance transport and redistribution of clay minerals from fluvial sources by wave and current activity.

In near-shore, restricted basins, lateral changes in clay mineral composition can be quite large. For example, Hiscott (1984) noted that the clay minerals in Cretaceous marine strata of off-shore eastern Canada display pronounced variations among wells in close proximity to one another. Smectite, in particular, exhibited large lateral changes. Hiscott attributed this to the preponderance of small basins and local sediment source areas prevalent at that time. Areas of exposed plateau basalts contributed the smectite to the muds. The absence of redistribution by bottom currents also likely contributes to the heterogeneity.

This inverse relationship between basin size and clay mineral variability is preserved in ancient marine deposits as well. Hiscott's (1984) study of eastern Canada showed pronounced vertical variability in the clay mineral assemblages from oil wells and DSDP cores. In particular, the clays become very homogeneous in the Tertiary as major continent-wide rivers developed and spread detritus over a wide area.

The persistence of uniform clay assemblages has also been documented for many other marine petroliferous basins. The Niger Delta Region, dominated as it is by a very large river system, has a consistent clay mineral composition in shales ranging in age from Eocene to the present, representing thicknesses in excess of 3000 m (Lambert-Aikhionbare & Shaw 1982). Karlsson et al. (1979) performed detailed analyses of the clay mineralogy from a core in the North Sea, generally above the temperature zone in which burial diagenetic reactions are expected to occur. Although there are some variations in clay mineral ratios from sample to sample, the mineralogy on the whole is remarkably constant. There is a gradual vertical change in mineral abundances with depth, reflecting the lateral migration of sedimentary facies with time during the deposition of this 2500 m of sediment.

Data from a 5000-m thick Miocene shale sequence in Lousiana show similar consistencies (Hinch 1980). Progressive diagenesis decreases the relative proportion of expandable clays with depth in the well, but otherwise the overall clay assemblages remain remarkably constant. Obviously, changes in the tectonic and geochemical conditions will influence the stability of clay mineral types in any basin, yet it seems apparent that basin scale will be a primary control on the initial variability of extant clays.

Lacustrine environments

In lacustrine sequences, long sections of relative constancy are rarely, if ever, achieved and data for these environments are much less abundant. In fact, thick stratigraphic sections of any kind are very rare in lacustrine sedimentary rocks. The mean thickness of 106 units ranging from Precambrian to Pleistocene age in the western United States is about 300 m, with only a few exhibiting thicknesses of greater than 1000 m (Picard & High 1972). Even within these relatively short intervals, lake environments show more stratigraphic variability, reflecting a greater sensitivity to small-scale facies changes than most marine deposits. Diversity in clay mineralogy is a hallmark of most lacustrine sediments, reflecting these numerous facies changes (Picard & High 1972).

One reason for such variability in lake sediments is the ubiquitous influence of local source areas on lake sediment mineralogy. In the open marine realm, the prevalence of large river systems contributing to the sediment budget on continental shelves serves to homogenize the sediment composition brought to the depositional site. In addition, long-shore or on-shore currents can often re-work the original accumulation into a more uniform assemblage (Pevear 1972). These secondary processes are often less vigorous in most lakes. Even in very large lake basins, local source areas can leave their distinctive imprint upon sediment composition and subsequent re-working is usually insufficient to obliterate these signals.

A case in point is Lake Turkana in Kenya (Yuretich 1979, 1986). There are three major sediment sources for this lake (Fig. 1). To the north is an extensive basaltic volcanic plateau with a humid tropical climate. The perennial Omo River delivers the weathering products from this region to the northern part of the lake. In the western and south-western parts of the drainage basin, the Turkwel and Kerio Rivers drain a predominantly metamorphic source area in the Loriu hills and along the rift valley escarpment. The climate here is generally arid, and most of the detritus is introduced into the middle section of the lake. Along the eastern and western shores, very small ephemeral streams provide sediment to the lake from very localized sources.

These various lines of sediment supply have left their distinctive imprints upon the clay minerals in the sediments of Lake Turkana (Fig. 2). Kaolinite is more prevalent in the extreme north and illite is abundant in the central portions of the lake. Although smectite is ubiquitous, the highest concentrations are found along the eastern shore. Circulation in the lake is quite vigorous (Ferguson & Harbott 1982), but is still not sufficient to obliterate the source area influences.

Such facies relationships become more common when examined in three dimensions. Lakes are notable for rapid fluctuations in size and shape; many modern lakes have undergone substantial changes at several times during the Holocene Epoch, usually in response to climatic variations (Kutzbach & Street-Perrot 1985). Clay mineral suites can shift dramatically during climatic excursions in response to (a) migrating sedimentary facies; (b) changes in weathering processes and sediment supply to the lake basin and (c) variations in lake water chemistry. This latter variable will affect the nature and degree of early diagenetic reactions occurring in the sediment mass, and is particularly well illustrated in cores collected from East African lakes. Stoffers & Singer (1979) observed radical changes in clay mineral composition over very short intervals in cores from Lake Albert (Mobutu Sese Seko). The principal feature is the alternating presence of smectite, interstratified illite–smectite and illite (Fig. 3). On the basis of diatom assemblages plus the presence of other salinity-sensitive minerals, the authors interpreted the clay minerals as responding to changes in water chemistry induced by climatic fluctuations. Other studies of various lacustrine settings (Gac *et al.* 1977; Jones & Weir 1983) have confirmed the general picture: kaolinites are more prevalent in freshwater environments, whereas smectite, interstratified smectite–illite and illite can form under increasing lake water alkalinity and salinity.

Although the foregoing observations are based on modern lakes, comparable patterns have been

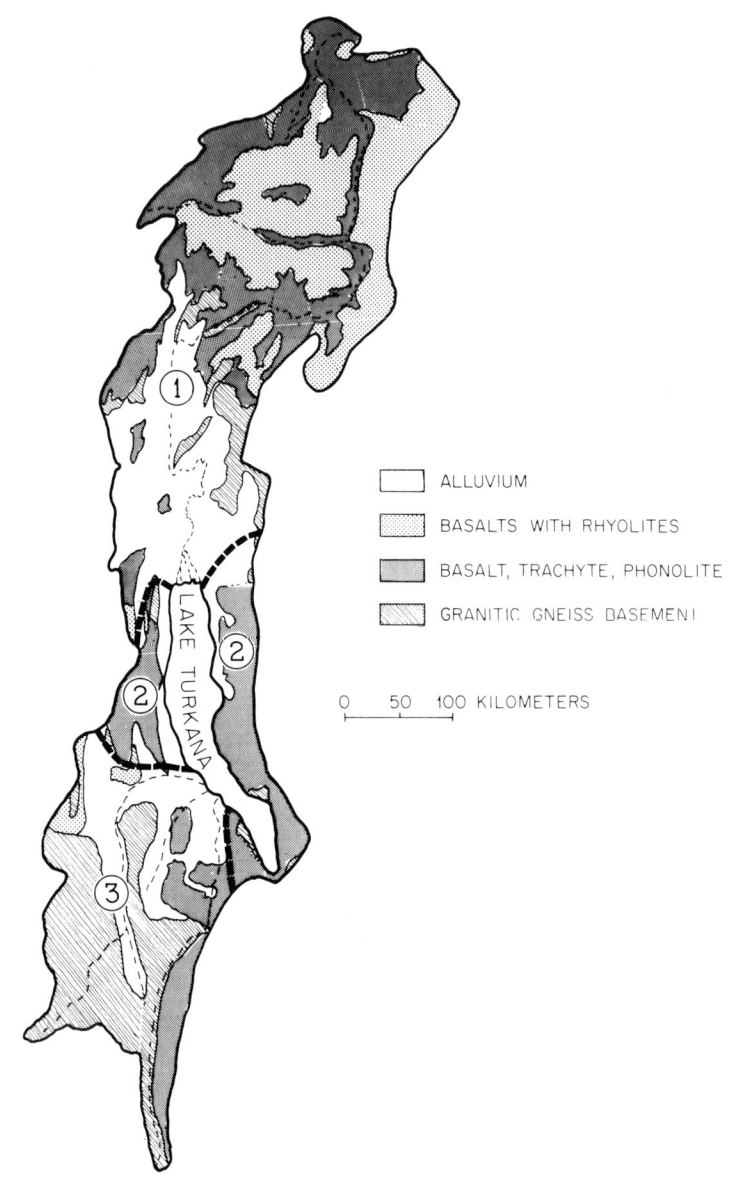

FIG 1. Geology and drainage network of principal sediment source areas around Lake Turkana, Kenya (from Yuretich 1986).

identified in ancient lacustrine sequences. While analysing the clay mineralogy in the Eocene Green River Formation of Utah, Dyni (1976) observed a pronounced facies dependency on their distribution (Fig. 4). This is most likely caused by both physical processes of sedimentation (hydrodynamic sorting, cf. Gibbs 1977) as well as post-depositional reactions with lake water. During low lake levels, the water was restricted to the basin centre and highly concentrated; clays would have been entirely converted to more crystalline, stable minerals (zeolites and feldspars). When the lake was larger, it was also fresher and the clays reflect this as well.

Such diagenetic processes can lead to even further complications in clay mineral stratigraphy. In the Mesozoic rift lakes of the Connecticut River valley, the clay mineral relationships are

CLAY MINERAL PROVINCES

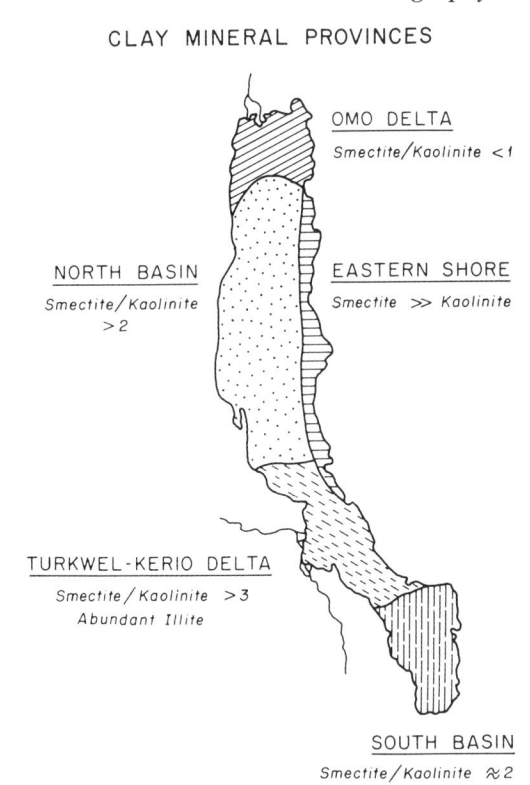

FIG. 2. Clay minerals in the surface sediments of Lake Turkana which reflect the source rock types and climate in the various sediment source areas (from Yuretich 1986).

considerably diverse (April 1981). In addition to the changes caused by lake level fluctuations and sediment–water interactions, the expulsion of highly concentrated pore waters during compaction altered the clay mineral assemblages of overlying lake beds (Fig. 5) giving rise to a mixed-layer chlorite–smectite. Such complications emphasize that the natural state of affairs in lacustrine environments promotes extensive variability in size, shape, lake water chemistry, sediment source areas and diagenetic processes. Therefore it is more difficult than in marine situations to accumulate thick sequences of mudstone of reasonably uniform clay mineral composition.

Burial diagenesis

The burial diagenesis of clay minerals in shales has been documented for numerous marine sequences (Dunoyer de Segonzac 1970; Perry & Hower 1970; Pearson et al. 1981). Most researchers now agree that progressive compaction and concomitant temperature increase cause the prevailing clay mineral assemblage to undergo a predictable and regular change. The general pattern is for expandable smectites and mixed-layer clays to gradually increase the relative proportion of illite layers in the crystal lattice (Fig. 6). Contemporaneously, kaolinite usually decreases with depth, and chlorite often makes an initial appearance at greater depths within the sedimentary sequence. The overall reaction for these mineralogical alterations is believed to be (Hower et al. 1976):

$$\text{smectite} + \text{K-feldspar} + \text{mica} = \text{illite} + \text{chlorite} + \text{quartz}.$$

Many authors have subsequently believed that the smectite-to-illite conversion, in particular, is of importance to processes of primary migration from shale source rocks because it occurs over the same depth and temperature interval as hydrocarbon generation (Foscolos & Powell 1979; Tissot & Welte 1978; Bruce 1984). During the transition of smectite to illite, the clays lose water in two principal stages (Fig. 7). First, at relatively low temperatures interlayer water is lost; at greater depths, roughly coinciding with the maximum stage of hydrocarbon production, bound water is removed from the clay structure. As Burst (1969) has observed, this latter process probably occurs gradually over a long interval of time.

This water loss is viewed by many to be of potential importance to petroleum migration. A net increase in the amount of liquid water in the shale source rock could assist in the movement of petroleum hydrocarbons into reservoirs. In addition, a slight porosity increase usually accompanies this de-watering (Fig. 7) and is often cited as a means of providing some 'breathing room' in an otherwise suffocating environment. A third effect is that loss of mineral water can often cause severe undercompaction in clay-rich sequences.

Considerable disagreement exists about the relative importance of any of these factors. The need for water in petroleum migration is debatable, since many authors argue for movement of hydrocarbons in an oil phase (McAuliffe 1979). However, it seems likely that there are several mechanisms of petroleum migration operating coincidentally and the addition of water into the system could enhance these other processes (Barker 1980). The effectiveness of the porosity increase is also open to question. Many people feel that the relative magnitude of the increase is too small to provide additional significant pore space for petroleum migration and others feel it may actually inhibit petroleum movement because shale permeability remains low (Magara

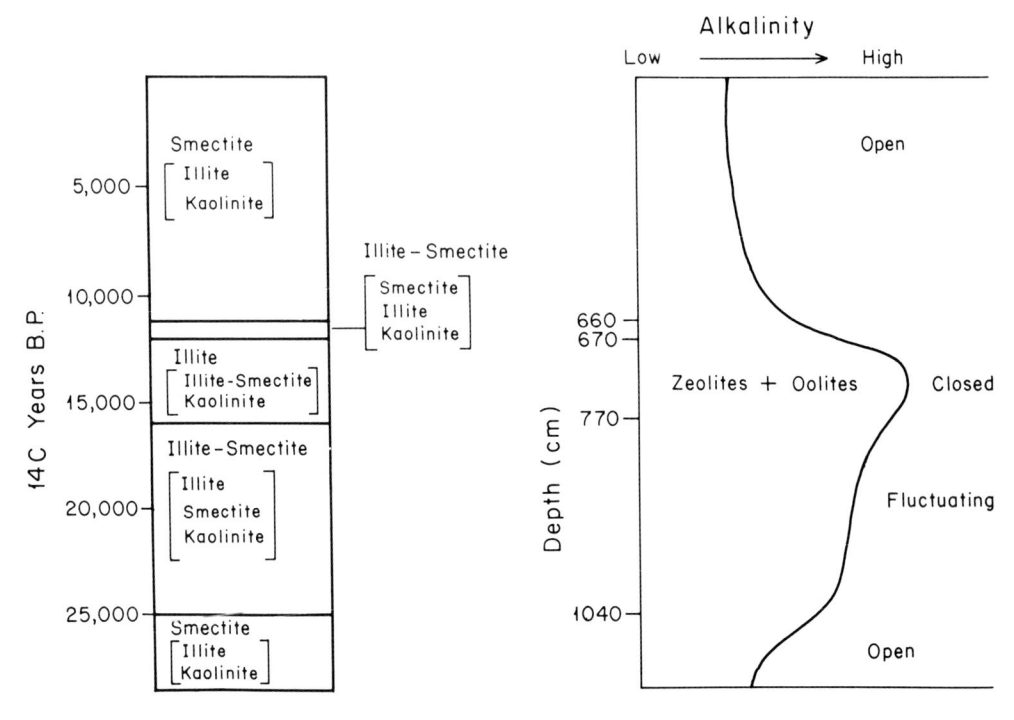

FIG. 3. Changes in clay mineral assemblages with depth and age in cores from Lake Albert (Mobutu Sese Seko). These changes are most likely a function of fluctuating hydrology and salinity, shown at right.

FIG. 4. Inferred palaeoenvironments and clay mineral assemblages in the Eocene Green River Formation, Utah. Changes in clays result from both hydrodynamic phenomena and post-depositional diagenetic reactions.

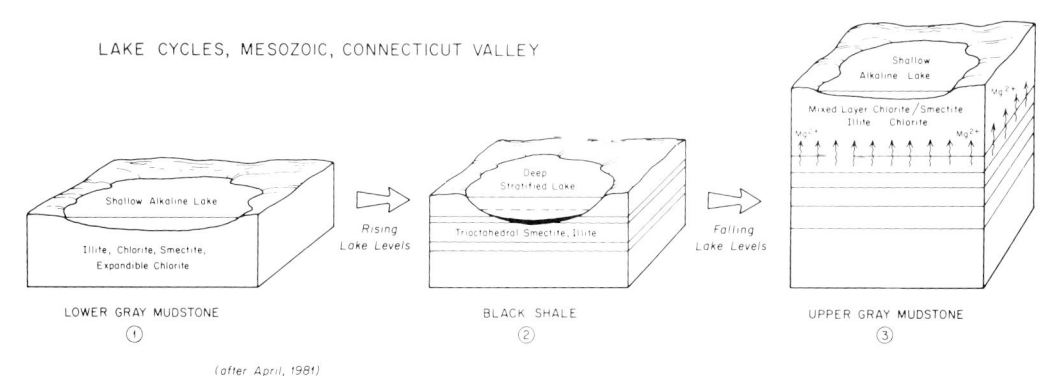

LAKE CYCLES, MESOZOIC, CONNECTICUT VALLEY

FIG. 5. Interpretation of clay mineral changes in the lake deposits of a portion of the Jurassic strata, Connecticut rift valley. Upward migration of magnesium from black shale during compaction promoted formation of unusual chlorite/smectite as diagenetic product.

1975). To counter this, the role of clay de-watering in maintaining undercompacted zones for long periods of time can keep potential migration pathways open (Bruce 1984). However, once again there is a diversity of opinion on this subject.

Regardless of the exact mechanisms, similar burial diagenetic processes and petroleum generation should occur in lacustrine rocks as well. Information on these is much scantier at present. Heling (1974) has studied the smectite-to-illite conversion in several cores from drill holes in the Rhine Graben of Germany which intersect some lacustrine sedimentary rocks. The conversion to illite does occur in both marine and non-marine formations, but the apparent temperature (i.e., depth) of the change is more variable in the lacustrine rocks, apparently reflecting the effects of interstitial water chemistry on the reaction. Dunoyer de Segonzac (1970) recognized the importance of interstitial water composition on the clay mineral transformation. He concluded that smectite was stable at temperatures greater than those normally encountered during diagenesis, but that the presence of sufficient magnesium, sodium and, particularly, potassium was most important for the formation of mixed-layer clays. In many freshwater lacustrine settings, these ions may be in short supply, thereby inhibiting the conversion of smectite. Furthermore Heling (1974) recognized an intrinsic clay-assemblage variability in the non-marine deposits of the Rhine Graben which is a function of fluctuations in the environment at the time of deposition.

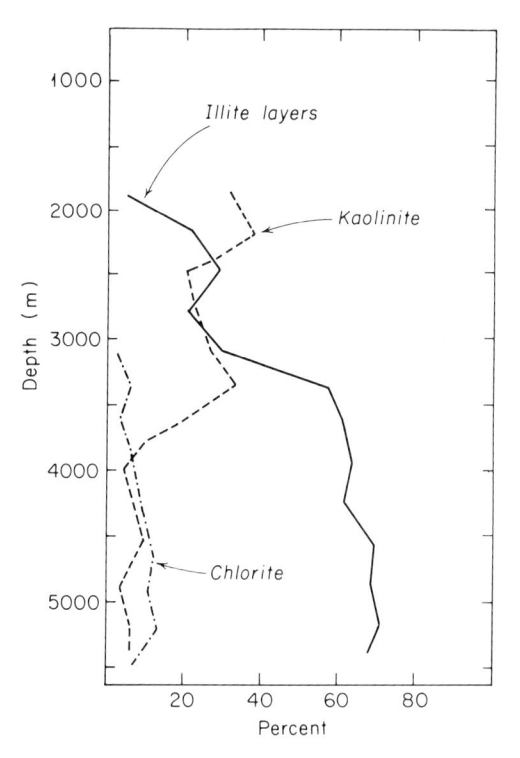

The Belfry lake beds

A good example of clay mineral variability within a lacustrine setting is provided by the Paleocene Fort Union Formation in the Bighorn Basin of Montana and Wyoming (Yuretich *et al.* 1984). The northern part, or Clarks Fort Basin, contains a well developed non-marine sequence which centres around both swamp and lake deposits (Fig. 8). The lake deposits are contained within the Belfry Member (Yuretich *et al.* 1984) in which I have recently identified three sub-facies. The

FIG. 6. Clay mineral changes in Tertiary shale sequences, offshore Gulf of Mexico.

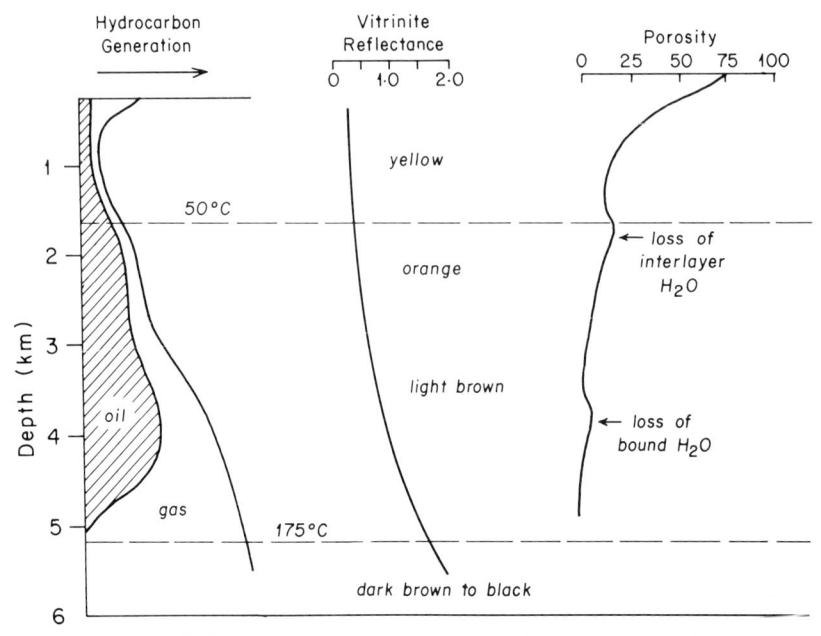

FIG. 7. Water loss from expandable clays and corresponding porosity increase in potential source rocks approximately coincides with maximum production of petroleum hydrocarbons (compiled from various sources).

lowermost unit is a mudstone-dominated facies which may be representative of an on-shore deltaic backswamp. The middle unit is the most distinctive part, characterized by numerous laterally continuous sandstones. These are probably either crevasse-splay or subaqueous delta-front sandstones right along the lake margin. The third and topmost lithology consists of mudstone and limestone which may represent a return to a meandering-stream environment. In general, I view this as a shallow lake (about 15 m[?] maximum depth) with a low energy muddy shoreline. The size and water chemistry of the lake fluctuated repeatedly in response to tectonic pulses from the uplands to the west.

The resultant stratigraphy is, indeed, complex. Of particular interest to this analysis are the clay minerals in the mudstones. In our initial survey (Yuretich et al. 1984), four principal clay mineral assemblages were identified. Type 1 contains an expandable mixed-layer clay (perhaps a smectite–vermiculite) as its most abundant component, with subordinate illite and kaolinite. In Type 2, illite becomes more prominent and chlorite is also present. Type 3 has the expandable phase looking more like a pure smectite and kaolinite shows increasing importance. Finally, Type 4 is the most distinctive assemblage with no expandable clays: illite is dominant with chlorite and kaolinite also present.

The available palaeotectonic evidence suggests that these sedimentary rocks have never been buried very deeply and no progressive diagenesis of the clays is expected. The abundance of smectite and expandable clays throughout the sections corroborates the lack of elevated temperatures. Since these are outcrop samples, there was some concern that the clay assemblages might be controlled by the weathering of the exposed mudstones. However, excavations and detailed samplings at several sites demonstrates that these are primary clay assemblages reflecting original depositional conditions.

Our initial investigation also showed that there was an extraordinary degree of variability in the clay mineral assemblages as seen in two vertical sections (Yuretich et al. 1984). Subsequent analyses coupled with precise lithostratigraphic correlation show that this variability is maintained over a wide geographic area (Fig. 9). However, at least one pattern is emerging: the upper parts of the lacustrine facies in the northeastern part of the study area contain fewer expandable clays. The initial lithologic and palaeocurrent studies place this region in the deeper, basinward part of the lake environment. Exactly why expandable clays should be absent from this facies is still under investigation, but whether it is a function of source area changes, hydrodynamic segregation or early diagenesis, the net result is the same: the variation in initial expandable clay content would limit the degree

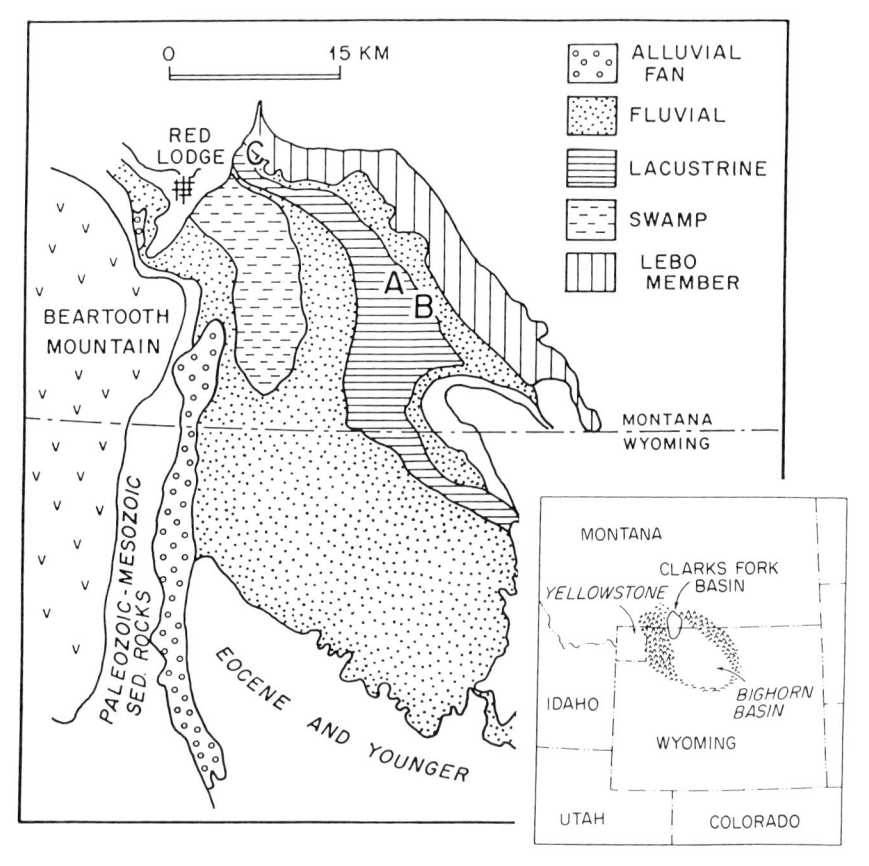

FIG. 8. Outcrop map of sedimentary facies in the Paleocene Fort Union Formation of the Clarks Fork Basin, Montana and Wyoming. Belfry Member coincides approximately with the mapped distribution of lacustrine facies. Inset: regional location map (after Yuretich et al. 1984).

of progressive transformation of the clays during burial.

Discussion and summary

The scenario developed here implies that basin size, environmental fluctuations and interstitial water chemistry may cause the clay minerals in lacustrine sedimentary sequences to behave differently under compaction and burial than those in marine settings. Many marine sequences, particularly those on continental shelves or in deltaic areas with good petroleum potential, will contain thick shale formations. Because of the ability of large source areas, well-developed transport systems and submarine currents to homogenize sediment composition, the shales can frequently possess a consistent clay mineral assemblage with smectites and mixed-layer clays an important fraction. Marine pore waters will contain sufficient dissolved cations for the conversion of smectite to illite to occur during burial, and the release of water from this thermal transformation may assist in petroleum migration by either promoting undercompaction of the shales or assisting in the migration of hydrocarbon fluids. Many sandstones within such sequences will also have a large lateral extent and petroleum forced out of the shales into these sandstones can migrate fairly long distances (Fig. 10a).

In lacustrine sequences, the smaller scale of the stratigraphic relationships tends to restrict many aspects of petroleum generation and migration. Shales are usually thinner and the clay mineral assemblages they contain may vary greatly over short vertical and horizontal distances. In addition, fluctuations in lake water chemistry will also affect the timing of smectite-to-illite conversion. During freshwater episodes, there will be insufficient cations in the pore waters to effect burial transformation at the usual temperatures.

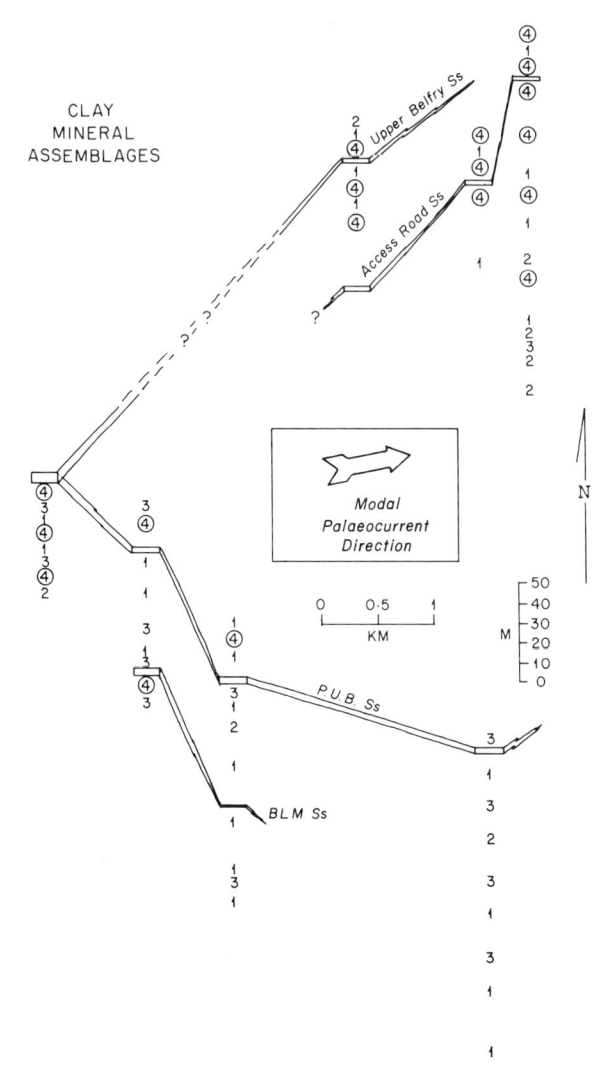

FIG. 9. Correlated sections of Belfry Member in Clarks Fork Basin. Numbers along sections refer to different clay mineral assemblages (see text). Palaeocurrent direction based on mean of eighty-seven measurements in sandstones across the map area.

On the other extreme, high salinity and alkalinity can cause diagenesis to occur very early, long before the breakdown of kerogen into petroleum hydrocarbons can occur. Consequently, transformation of smectites in this environment would be a localized and somewhat erratic process. Subsequent conversion of kerogen to liquid hydrocarbons could be less efficient owing to a lower catalytic effect in smectite-poor locales as well as a proportionately lower volume of water released during compaction. This may help explain why oil shales seem so common in palaeolake settings. Furthermore, the variability in stratigraphy with fewer laterally extensive sandstone beds could help ensure that primary migration is restricted to regions proximal to the source rock (Fig. 10b).

As in all things geological, gradations will exist between marine and lacustrine characteristics. Small, marginal marine basins can exhibit a facies variability akin to lakes; on the other hand, large lakes are more resistant to change than their smaller counterparts and can possess many marine-like characteristics, particularly if they are saline. However, my (outrageous?) hypothesis presented here helps explain some very noticeable

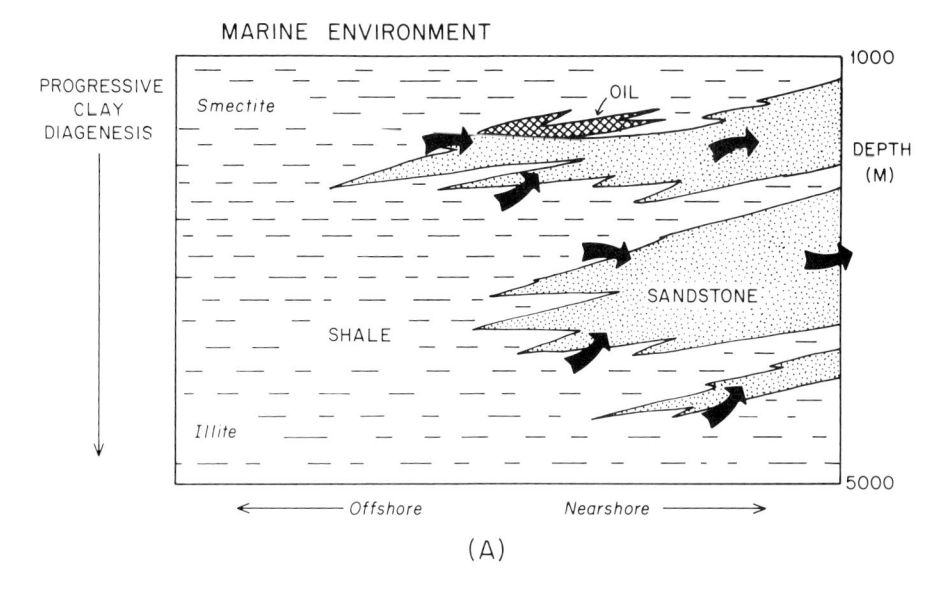

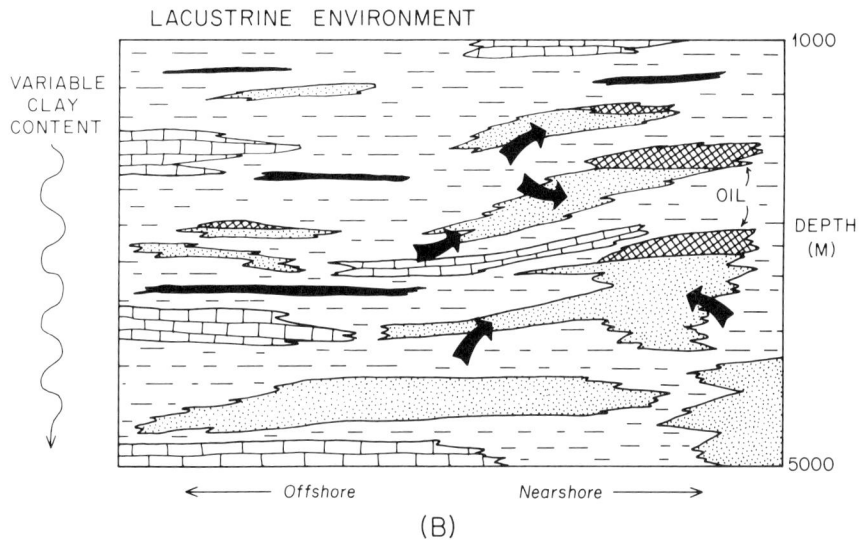

FIG. 10. Possible scenarios for petroleum generation and migration in marine and lacustrine environments. (A) In marine settings, abundant water released during smectite diagenesis promotes fluid movement. Laterally extensive sandstones can receive fluids and allow for long-distance migration. (B) In lacustrine situations, there is less fluid released because of highly variable clay mineral content. Stratigraphic complexities, particularly small sandstone bodies, restrict migration and result in localized petroleum accumulations.

distinctions between marine and lacustrine petroleum accumulations. Our present knowledge of lacustrine clay minerals is, admittedly, very limited, and further research on lacustrine source rocks and their behaviour during burial diagenesis is needed to see how widespread these patterns are.

ACKNOWLEDGEMENTS: The thorough review by H. Chamley of an earlier version of this paper raised several thought-provoking questions some (but, unfortunately, not all) of which I have tried to address in the revision. Sincere thanks also go to Marie Litterer, Technical Illustrator, University of Massachusetts, for her expert drafting of the diagrams contained in this paper. The studies of the Fort Union Formation are being supported by grant EAR83-06153 from the U.S. National Science Foundation.

References

APRIL, R. A. 1981. Trioctahedral smectite and interstratified chlorite/smectite in Jurassic strata of the Connecticut Valley. *Clays and Clay Minerals*, **29**, 31–39.

BARKER, C. 1980. Primary migration: the importance of water–mineral–organic matter interaction in the source rock. *In*: ROBERTS, W. H. & CORDELL, R. J. (eds), *Problems of Petroleum Migration*, American Association of Petroleum Geologists, Studies in Geology, **10**, Tulsa, OK, 19–31.

BISCAYE, P. E. 1965. Mineralogy and sedimentation of Recent deep-sea clay in the Atlantic Ocean and adjacent seas and oceans. *Bulletin of the Geological Society of America*, **76**, 803–832.

BRADLEY, W. H. 1929. Shore phases of the Green River Formation in northern Sweetwater County, Wyoming. *United States Geological Survey Professional Paper 140-D*, 121–131.

—— 1970. Green River oil shale—concept of origin extended. *Bulletin of the Geological Society of America*, **81**, 985–1000.

—— 1973. Oil shale formed in a desert environment: Green River Formation, Wyoming. *Bulletin of the Geological Society of America*, **84**, 1121–1124.

BRUCE, C. H. 1984. Smectite dehydration—its relation to structural development and hydrocarbon accumulation in northern Gulf of Mexico Basin. *Bulletin of the American Association of Petroleum Geologists*, **68**, 673–683.

BURST, J. F. 1969. Diagenesis of Gulf Coast clayey sediments and its possible relation to petroleum migration. *Bulletin of the American Association of Petroleum Geologists*, **53**, 73–93.

DEGENS, E. T., VON HERZEN, R. P., WONG HOW-KIN, DEUSER, W. G. & JANNASCH, H. W. 1973. Lake Kivu: structure, chemistry and biology of an East African rift lake. *Geologische Rundschau*, **62**, 245–277.

DESBOROUGH, G. A. 1978. A biogenic-chemical stratified lake model for the origin of oil shale of the Green River Formation: an alternative to the playa-lake model. *Bulletin of the Geological Society of America*, **89**, 961–971.

DUNOYER DE SEGONZAC, G. 1970. The transformation of clay minerals during diagenesis and low-grade metamorphism: a review. *Sedimentology*, **15**, 281–346.

DYNI, J. R. 1976. Trioctahedral smectite in the Green River Formation, Duchesne County, Utah. *United States Geological Survey Professional Paper 967*, 14 pp.

EUGSTER, H. P. & SURDAM, R. C. 1973. Depositional environment of the Green River Formation of Wyoming, a preliminary report. *Bulletin of the Geological Society of America*, **84**, 1115–1120.

FERGUSON, A. J. D. & HARBOTT, B. J. 1982. Geographical, physical and chemical aspects of Lake Turkana. *In*: HOPSON, A. J. (ed.), *Lake Turkana: a Report on the Findings of the Lake Turkana Project, 1972–1975*, Overseas Development Administration, London, 1–107.

FOSCOLOS, A. E. & POWELL, T. G. 1979. Mineralogical and geochemical transformation of clays during burial-diagenesis (catagenesis): relation to oil generation. *In*: MORTLAND, M. M. & FARMER, V. C. (eds), *International Clay Conference 1978, Developments in Sedimentology*, **27**, Elsevier, Amsterdam, 261–270.

GAC, J., DROUBI, A., FRITZ, B. & TARDY, Y. 1977. Geochemical behaviour of silica and magnesium during the evaporation of waters in Chad. *Chemical Geology*, **19**, 215–228.

GIBBS, R. J. 1977. Clay mineral segregation in the marine environment. *Journal of Sedimentary Petrology*, **47**, 237–243.

GRIFFIN, J. J., WINDOM, H. & GOLDBERG, E. D. 1968. The distribution of clay minerals in the world ocean. *Deep-Sea Research*, **15**, 433–459.

HATHAWAY, J. C. 1972. Regional clay-mineral facies in estuaries and continental margin of the United States east coast. *In*: NELSON, B. (ed.), Environmental Framework of Coastal Plain Estuaries, *Memior of the Geological Society of America*, **133**, 293–316.

HELING, D. 1974. Diagenetic alteration of smectite in argillaceous sediments of the Rhinegraben (SW Germany). *Sedimentology*, **21**, 463–472.

HINCH, H. H. 1980. The nature of shales and the dynamics of hydrocarbon expulsion in the Gulf Coast Tertiary section. *In*: ROBERTS, W. H. & CORDELL, R. J. (eds), *Problems of Petroleum Migration*, American Association of Petroleum Geologists, Studies in Geology, **10**, Tulsa, OK, 1–18.

HISCOTT, R. N. 1984. Clay mineralogy and clay-mineral provenance of Cretaceous and Paleogene strata, Labrador and Baffin shelves. *Bulletin of Candian Petroleum Geology*, **32**, 272–280.

HOWER, J., ESLINGER, E. V., HOWER, M. E. & PERRY, E. A. 1976. Mechanism of burial metamorphism of argillaceous sediment: 1. mineralogical and chemical evidence. *Bulletin of the Geological Society of America*, **87**, 725–737.

HUNT, J. M., STEWART, F. & DICKEY, P. A. 1954. Origin of hydrocarbons of Uinta Basin, Utah. *Bulletin of the American Association of Petroleum Geologists*, **38**, 1671–1698.

JONES, B. F. & WEIR, A. H. 1983. Clay minerals of Lake Albert, an alkaline, saline lake. *Clays and Clay Minerals*, **31**, 161–172.

KARLSSON, W., VOLLSET, J., BJORLYKKE, K. & JORGENSEN, P. 1979. Changes in mineralogical composition of Tertiary sediments from North Sea wells. *In*: MORTLAND, M. M. & FARMER, V. C. (eds), *International Clay Conference 1978, Developments in Sedimentology*, **27**, Elsevier, Amsterdam, 281–290.

KLEMME, H. D. 1980. Petroleum basins—classification and characteristics. *Journal of Petroleum Geology*, **3**, 187–207.

KUTZBACH, J. E. & STREET-PERROT, F. A. 1985. Milankovitch forcing of fluctuations in the level of tropical lakes from 18 to 0 kyr BP. *Nature*, **317**, 130–134.

LAMBERT-AIKHIONBARE, D. O. & SHAW, H. F. 1982. Significance of clays in the petroleum geology of the Niger Delta. *Clay Minerals*, **17**, 91–103.

MAGARA, K. 1975. Reevaluation of montmorillonite dehydration as cause of abnormal pressure and hydrocarbon migration. *Bulletin of the American Association of Petroleum Geologists*, **59**, 292–302.

MCAULIFFE, C. D. 1979. Oil and gas migration: chemical and physical constraints. *Bulletin of the American Association of Petroleum Geologists*, **63**, 761–781.

MOODY, J. D. 1975. Distribution and geological characteristics of giant oil fields. *In*: FISCHER, A. G. & JUDSON, S. J. (eds), *Petroleum and Global Tectonics*, Princeton University Press, Princeton, NJ, 307–320.

PEARSON, M. J., WATKINS, D. & SMALL, J. S. 1982. Clay diagenesis and organic maturation in northern North Sea sediments. *In*: VAN OLPHEN, H. & VENIALE, F. (eds), *International Clay Conference 1981, Developments in Sedimentology*, **35**, Elsevier, Amsterdam, 665–675.

PERRY, E. A. & HOWER, J. 1970. Burial diagenesis in Gulf Coast pelitic sediments. *Clays and Clay Minerals*, **18**, 165–177.

PEVEAR, D. R. 1972. Source of recent nearshore marine clay, southeastern United States. *In*: NELSON, B. (ed.), Environmental Framework of Coastal Plain Estuaries, *Memoir of the Geological Society of America*, **133**, 317–336.

PICARD, M. D. 1981. Physical stratigraphy of ancient lacustrine deposits. *In*: ETHRIDGE, F. G. & FLORES, R. M. (eds), *Recent and Ancient Nonmarine Depositional Environments: Models for Exploitation*. Society of Economic Paleontologists and Mineralogists, Special Publication, **31**, Tulsa, OK, 233–260.

—— & HIGH, L. R. 1972. Criteria for recognizing lacustrine rocks. *In*: RIGBY, J. K. & HAMBLIN, W. K. (eds), *Recognition of Ancient Sedimentary Environments*. Society of Economic Paleontologists and Mineralogists, Special Publication, **16**, Tulsa, OK, 108–145.

PITMAN, J. K., FOUCH, T. D. & GOLDHABER, M. B. 1982. Depositional setting and diagenetic evolution of some Tertiary unconventional reservoir rocks, Uinta Basin, Utah. *Bulletin of the American Association of Petroleum Geologists*, **66**, 1581–1596.

RITZMA, H. R. 1972. The Uinta Basin. *In*: *Geologic Atlas of the Rocky Mountain Region*. Rocky Mountain Association of Geologists, Denver, CO, 276–278.

STOFFERS, P. & SINGER, A. 1979. Clay minerals in Lake Mobutu Sese Seko (Lake Albert)—their diagenetic changes as an indicator of the paleoclimate. *Geologische Rundschau*, **68**, 1009–1024.

TIRATSOO, E. N. 1984. *Oilfields of the World*, Scientific Press, Beaconsfield, England, 392 pp.

TISSOT, B. & WELTE, D. H. 1978. *Petroleum Formation and Occurrence*, Springer-Verlag, New York, 538 pp.

YANG W. 1985. Daqing oil field, People's Republic of China: a giant field with oil of nonmarine origin. *Bulletin of the American Association of Petroleum Geologists*, **69**, 1101–1141.

YURETICH, R. F. 1979. Modern sediments and sedimentary processes in Lake Rudolf (Lake Turkana), eastern rift valley, Kenya. *Sedimentology*, **26**, 313–331.

—— 1986. Controls on the composition of modern sediments, Lake Turkana, Kenya. *In*: FROSTICK, L., RENAUT, R., REID, I. & TIERCELIN, J. (eds), *Sedimentation in the African Rift System*, Geological Society Special Publication, **23**, Blackwell Scientific Publications, Oxford, 135–146.

——, HICKEY, L. J., GREGSON, B. P. & HSIA, Y. 1984. Lacustrine deposits in the Paleocene Fort Union Formation, northern Bighorn Basin, Montana. *Journal of Sedimentary Petrology*, **54**, 836–852.

R. F. YURETICH, Department of Geology and Geography, University of Massachusetts, Amherst, Massachusetts 01003, USA.

Palaeo-environment information from deep water siderite (Lake of Laach, West Germany)

B. Bahrig

SUMMARY: A comparative study from modern and ancient sediments from the Lake of Laach was performed by means of chemical and stable isotope analyses. In the modern environment siderite is precipitated around CO_2-spring vents. In contrast, siderite from the Pleistocene/Holocene boundary occurs as small layers in a varved sequence. The stable isotope data provide strong evidence for an analogous formation of both types of siderite. The different CO_2 sources of the siderites allow reconstruction of a distinct rise of productivity at the Pleistocene/Holocene transition. It is proposed that siderite may be developed as a useful tool for tracing both palaeo-environmental and early diagenetic processes.

Siderite is a common mineral phase in oil shales (Weber & Hofmann 1982; Negendank et al., 1982), coal-bearing sequences (Curtis 1967) and other potential petroleum source rocks. Up to now, stable isotope analysis of siderite has only been applied in order to solve diagenetic problems (e.g., Gautier 1982). As such it may be used to identify marine or fresh pore waters, and often to trace certain diagenetic sources of CO_2. Recent work on siderites from the German coal measures and from different Eocene oil shales (Bahrig & Conze 1986) gives evidence for the value of stable isotope and chemical investigations on siderite for the reconstruction of the conditions of accumulation and preservation of different types of organic matter.

Isotopic ratios of the siderite presented in this study stress the idea that palaeo-environmental information also may be preserved in the stable isotope composition of this mineral, because of the strong linkage of early diagenetic and sedimentary conditions.

Setting and geology

The Lake of Laach is situated in the western part of the Rhenish Massif. It occupies an 11 000-year old caldera structure that was formed during a short eruption phase at the end of the Late Glacial period (Firbas 1953). Today the lake is 54 m deep; it was situated in a hydrologically closed basin, until an artificial outlet was built in 1164 A.D. The water from the small catchment area (tuffs and older scoria cones) enters the lake mostly by sub-surface percolation (Henning 1965).

The lake is of holomictic–dimictic type, i.e., the water mass is completely mixed twice a year. During the summer period two separate water layers develop. In the warmer, upper water mass algal blooms induce an open water precipitation of low-Mg-calcite. The lower water mass is characterized by O_2-depletion (Matterne & Scharf 1977; Bahrig 1985).

The most important features of the modern lake are CO_2 springs at the lake margins and on the lake bottom, where volcanic gas bubbles rise (Fig. 1).

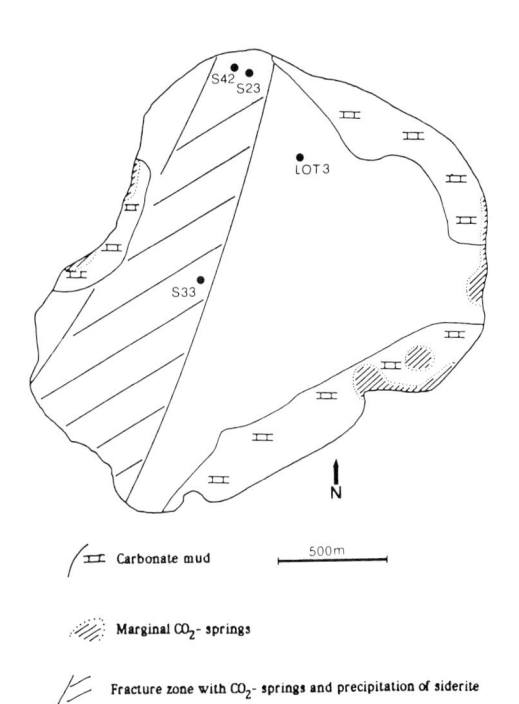

FIG. 1. Sketch map of the Lake of Laach and sample localities. The distribution of different types of CO_2 springs at the lake bottom is indicated. Siderite occurs around spring vents along a N–S trending fracture zone.

From: FLEET, A. J., KELTS, K. & TALBOT, M. R. (eds), 1988, Lacustrine Petroleum Source Rocks, Geological Society Special Publication No. 40, pp. 153–158.

Data and discussion

Siderite from the modern lake environment

The surface sediments show a distinct zonation of the lake bottom into: (1) a marginal zone of clastics; (2) a carbonate zone; and (3) a basinal zone with diatomaceous and organic-rich muds. In the western part of the lake this pattern is disturbed by a tectonic fracture zone, where acidic, CO_2-rich springs debouch into the lake water, and any calcite is dissolved (Fig. 1). Individual spring vents exhibit red aureoles of iron minerals, mostly siderite. In these aureoles up to 50% of the sediment is composed of 2–10 micron-sized, round-shaped siderite crystals.

The model in Fig. 2 is proposed to explain the precipitation of the siderite.

Acidic spring water enriched in CO_2 and Fe^{2+}, rising in the spring vents, mixes with the anoxic, more alkaline pore solutions and lake water. Both the precipitated calcite and shell debris are dissolved. The resulting removal of CO_2 and rising pH lead to an oversaturation of the solution with respect to siderite, which is precipitated at the sediment/water contact and in the sediment.

Stable isotope data

Stable isotope analysis was performed following the usual phosphoric acid method, the reaction temperature was 60°C, a temperature correction was made to 25°C after Bahrig & Oberhänsli (in

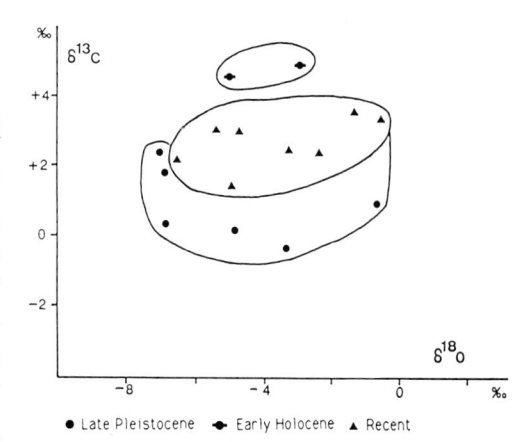

FIG. 3. Stable isotope data of ancient and recent siderites from the Lake of Laach. The similarity of the data indicates that no substantial change in the precipitation mechanism has taken place. However, the composition of the early Holocene siderites seems to indicate a distinct change in at least one CO_2 component during the lake history. For further discussion, see text.

prep.); samples were cleaned of organic matter by heating to 400°C for half an hour in a vacuum. The $\delta^{18}O$ of siderite from different sites (Fig. 3) is within the normal range of a freshwater carbonate (Hoefs 1980). The scattering of the data is due to different water depths, and resulting differences of the precipitation temperatures, and local fluctuations of the proportions of different solutions.

The $\delta^{13}C$ is relatively heavy and indicates an additional source of ^{13}C besides the lake water ($\delta^{13}C = -11‰$). Such positive $\delta^{13}C$ might result from microbial methanogenesis, but analyses of the gas rising from the lake bottom showed no trace of methane. Additionally, the concentration of the siderite around spring vents points to a primary source of the isotopically heavy CO_2. There are two possible sources for this:

(1) *Volcanic CO_2*. The $\delta^{13}C$ is in the region of $-2‰$ PDB. Equilibration with pure water would result in a $\delta^{13}C$ of about $+8‰$ (fractionation factors from Friedman & O'Neil 1977).

(2) *Carbonate from dissolution of calcite*. The very common charophyte remains exhibit a $\delta^{13}C$ of $+8‰$, shells from snails of $+3‰$, precipitated calcite shows a $\delta^{13}C$ in the same order. Bicarbonate resulting from dissolution of these carbonates would show a $\delta^{13}C$ of $+1$ to $+6‰$. Recycling of this calcite would also offer an explanation for the high Ca content of the siderite, which is in the range of 15% $CaCO_3$.

As a result, three different sources of CO_2 can be distinguished in the modern environment: (1)

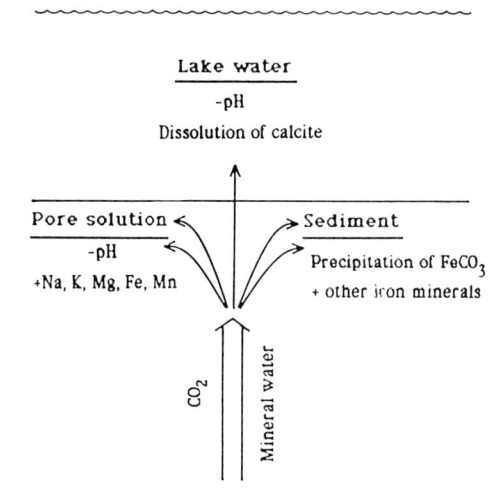

FIG. 2. Chemical model for the precipitation of siderite around spring vents. The Fe carbonate is thought to be precipitated because of a lowering of the pH of the rising mineral-enriched waters, produced by degassing of CO_2 and mixing with the more alkaline pore solutions.

TABLE 1. *Stable isotope data from Lake of Laach sediments*

Sample	Depth (cm)	$\delta^{13}C$ ‰ PDB	$\delta^{18}O$ ‰ PDB	Carbonate
S 23	15–20	+1.40	−4.92	siderite
	25–30	+3.05	−5.33	siderite
	40–45	+2.13	−6.53	siderite
	55–60	+2.99	−4.70	siderite
S 33	0–5	+3.57	−1.29	siderite
	10–15	+2.43	−2.27	siderite
S 42	0–15	+2.49	−3.19	siderite
	23–26	+3.33	−0.50	siderite
Lot 3	223–225	+7.37	−3.48	calcite
		+4.59	−4.93	siderite
	231–233	+6.54	−5.44	calcite
		+4.97	−2.85	siderite
	247–249	+8.33	−5.34	calcite
		+2.40	−6.99	siderite
	260–263	+6.23	−4.59	calcite
		+0.90	−0.62	siderite
	265–267	+1.85	−6.87	siderite
	276–278	−0.33	−3.37	siderite
	286–287	+0.35	−6.84	siderite
	297–299	+0.18	−4.17	siderite

bicarbonate from lake water, with components of biodegradational and atmospheric CO_2; (2) volcanic gas; and (3) bicarbonate from dissolution of Ca carbonate.

Siderite from ancient, varved sediments

The profile of the piston core Lot 3 (for location see Fig. 1) shows a complete sedimentary sequence from the deep water facies area (Fig. 4). It is composed of a younger part, exhibiting different mud types, and an older part composed of varved clastic sediments. The latter belong to the Dryas 3 to Preboreal stages (11 000–9000 years B.P.). At the time of their deposition the lake was about 90 m deep, the scarce vegetation cover and the steep slopes of the crater basin promoted a high input of clastics (Bahrig 1985).

A single varve is composed of a coarse-grained silt layer that accumulated during snow melt, overlain by a medium- to fine-grained silt representing summer to winter sedimentation. Diatom distribution supports this interpretation. The siderite grains occur in the part of the varve sublaminae that corresponds to late spring to autumn deposition. In most cases the maximum concentration of up to 70% is reached in the transition zone from coarse- to fine-grained sediment. The 2–4 micron-sized crystals exhibit a xenomorphic shape. As no characteristic features of diagenetic formation can be observed, precipitation at or near the sediment/water contact is presumed.

Stable isotope data

The stable isotope data of these late- to post-glacial siderites are in the same range as those of the modern data (Table 1, Fig. 3). The slightly higher $\delta^{13}C$ of the latter may be due to the higher productivity of the lake and the resulting ^{12}C depletion of the lake waters (McKenzie 1982), or to a recent increase in the production of volcanic CO_2. Thus, the data indicate a similar formation process in the modern and ancient environments, but the periodic and stratiform occurrence points to a laterally more extensive precipitation in the ancient lake.

Siderite can only form in a restricted, oxygen-depleted environment with a high CO_2 partial pressure (Huber 1958; Curtis & Spears 1968; Stumm & Morgan 1981). Taking into account the great water depth of the lake—about 90 m—at that time (Bahrig 1985), the connection between siderite formation and warmer seasons, as shown by the siderite distribution in the varves, indicates that a periodic stratification of the water mass took place.

Two models for a formation under these conditions are possible:

(1) *Static model:* the mineral was precipitated at the contact of the sediment to an overlying stagnant, O_2-poor, water mass which was provided with excess CO_2 and Fe^{2+} from volcanic springs.

(2) *Dynamic model:* a situation similar to that of model 1 developed during summer stratification. At times of turnover, oxygen was added to the hypolimnion by mixing of the water masses. The high amount of Fe^{2+} became unstable, Fe was precipitated as hydroxide. By this mechanism, it accumulated at the sediment/water contact. Remobilization as Fe(II) occurred in this still anoxic boundary layer, resulting in a supersaturation and precipitation of siderite.

The stratigraphic distribution of the $\delta^{13}C$ data (Fig. 4) shows a marked shift from the early Dryas 3 siderites to those at the Pleistocene/Holocene boundary. At the same depth the organic carbon and diatom contents of the sediment are rising and a calcite occurs that is characterized by an extremely heavy $\delta^{13}C$. From the crystal shape, rhombohedra with a step-like surface structure, these calcites seem to have been precipitated in the open water (Otsuki & Wetzel 1974; Kelts & Hsü 1978). The extreme $\delta^{13}C$ can only be explained by a strong kinetic (biogenic) effect, and their formation seems to be coupled to

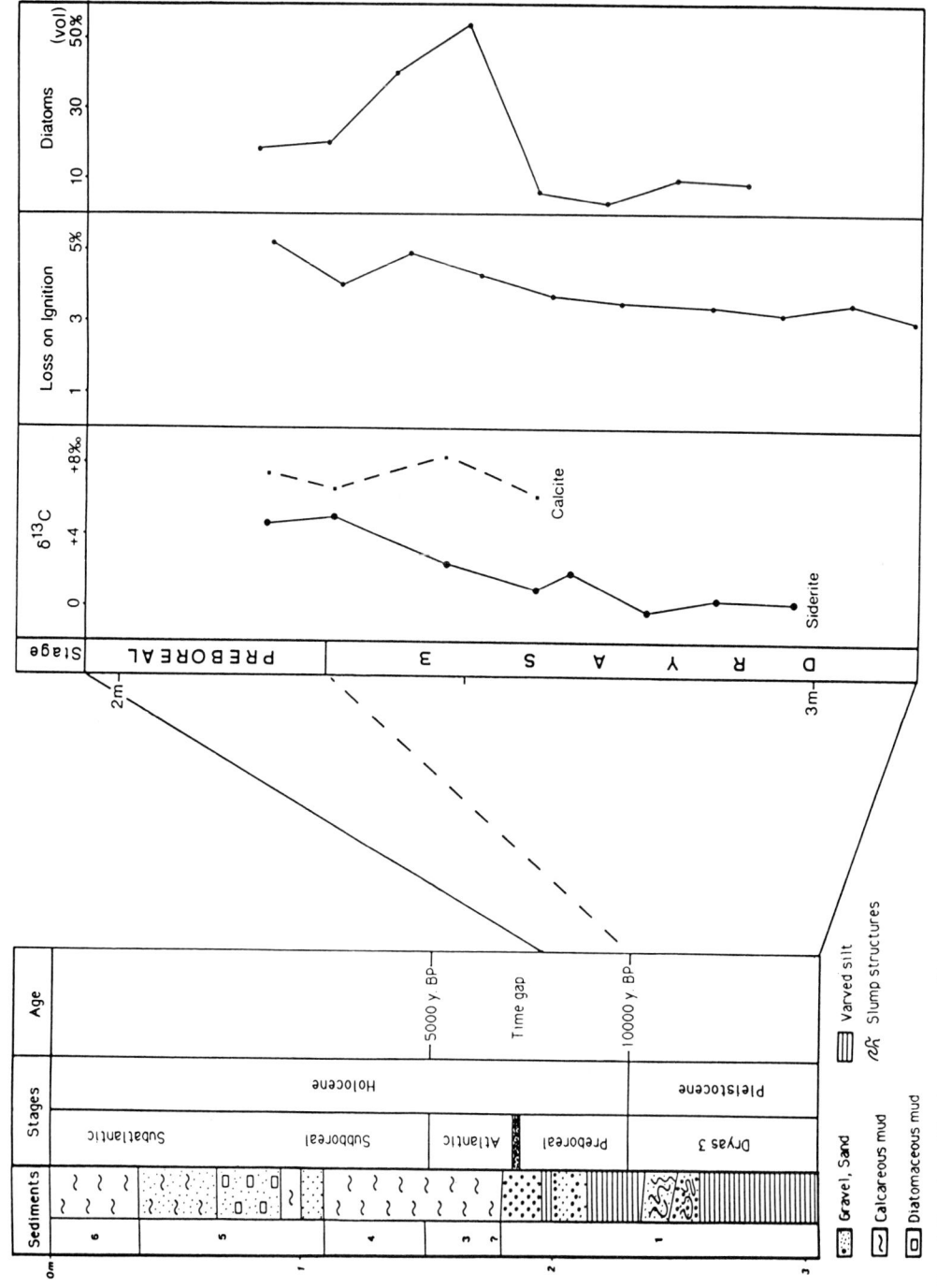

FIG. 4. Stratigraphic distribution of ^{13}C of carbonates, organic matter (as loss on ignition at 550°C), and diatom content in the varved silt sequence from the Lake of Laach. Both the rise of biogeneous components and the occurrence of an open water precipitated calcite indicate a marked rise in the productivity of the lake's ecosystem at the end of the late glacial period.

the rise of the other productivity-dependent parameters like diatom content and organic matter.

How could the signal of this ecologic shift near the Pleistocene/Holocene boundary reach a benthic carbonate? To solve this problem, one has to consider the three reservoirs of the modern siderite-CO_2 (lake water, dissolution of calcite, volcanic springs).

(1) *Lake water CO_2*: This could have changed its composition due to increased productivity. Planktonic algae, as well as macrophytes, selectively extract ^{12}C for their tissues. The positive shift of diatom content and organic carbon fraction suggests a marked rise in net productivity and in sedimentation of plant material. Thus, a continuous depletion of ^{12}C from the lake water may have occurred, leaving the residual carbon pool enriched in ^{13}C.

(2) *CO_2 from dissolution of calcite*: From Fig. 4, it is evident that at the same time as the enrichment in ^{13}C in the siderite started, the first calcite appeared in the sediments. If this calcite was precipitated higher in the water column, e.g., in the epilimnion, dissolution might have taken place in a more acid, CO_2-enriched hypolimnion (lower water mass) or at the sediment surface. By this mechanism ^{13}C-enriched HCO_3^- could be provided to the sediment surface without markedly changing the isotopic composition of the whole water mass, resulting in a strong effect on the $\delta^{13}C$ of the bottom water.

According to Fig. 4, a link may exist between the climate-dependent rise in productivity and precipitation of the isotopically heavy calcite. This offers a possibility of an indirect transfer of the ecologic signal to the benthic water mass.

(3) *Volcanic springs*: The third CO_2 source, the volcanic springs, only allows discussion of a more speculative alternative. A higher level of activity or stronger gas evasion due to a drop of the lake level would have provided additional ^{13}C-enriched CO_2 to a hypolimnetic water mass.

Conclusions

All these explanations include transfer of climatic and ecologic data to the benthic environment, thus demonstrating the strong connection of the early diagenetic environment of lakes to the climate-dependent chemical and biological parameters of sedimentation.

Dispersed and layered siderite types are mostly restricted to very early diagenetic phases of organic-rich sediments and may thus show not only a signal of microbial decomposition processes, as in oil shales (Bahrig & Conze 1986), but also give information about the climatic and limnologic reasons for the accumulation of organic matter. The example from the Lake of Laach demonstrates that an analysis of the different sources of CO_2 of early diagenetic/ synsedimentary carbonates provides data for a more detailed reconstruction of an ancient lacustrine environment and may thus help to trace potential petroleum source rocks.

The special value of siderite for these purposes lies in its restriction to anoxic environments that are characterized by a high potential for preservation of organic material.

ACKNOWLEDGEMENTS: This project was supported by the Deutsche Forschungsgemeinschaft (Nr. Fü 66/29).

References

BAHRIG, B. 1985. Sedimentation und Diagenese im Laacher Seebecken (Osteifel). *Bochumer geologische und geotechnische Arbeiten*, **19**, 231 pp.

—— & CONZE, R. 1986. Siderit als Tracer diagenetischer Prozesse. *1. Treffen deutschsprachiger Sedimentologen*, Freiburg 1986, 8–11.

CURTIS, C. D. 1967. Diagenetic iron minerals in some British Carboniferous sediments. *Geochimica et Cosmochimica Acta*, **31**, 2109–2123.

—— & SPEARS, C. D. 1968. The formation of sedimentary iron minerals. *Economic Geology*, **63**, 257–270.

FIRBAS, F. 1953. Das absolute Alter der jüngsten vulkanische Eruption im Bereich des Laacher Sees. *Die Naturwissenschaften*, **40**, 54–55.

FRIEDMAN, I. & O'NEIL, J. R. 1977. Compilation of stable isotope fractionation factors of geochemical interest. *In*: FLEISCHER, M. (ed.), *Data of Geochemistry*. 6th edn. United States Geological Survey Professional Paper 440–K, 12+98 pp.

GAUTIER, D. L. 1982. Siderite concretions: Indicators of early diagenesis in the Gammon Shale (Cretaceous). *Journal of Sedimentary Petrology*, **52**, 859–871.

HENNING, I. 1965. Das Laacher-See-Gebiet. Eine Studie zur Hydrologie und Klimatologie. *Arbeiten zur Rheinischen Landeskunde*, **22**, 135+25 pp.

HOEFS, J. 1980. *Stable Isotope Geochemistry*, Springer-Verlag, Berlin, 208 pp.

HUBER, N. K. 1958. The environmental control of sedimentary iron minerals. *Economic Geology*, **53**, 123–140.

KELTS, K. & HSÜ, K. J. 1978. Freshwater carbonate sedimentation. *In:* LERMAN, A. (ed.), *Lakes:*

Chemistry, Geology, Physics, Springer-Verlag, New York, 295–323.
MATTERNE, M. & SCHARF, B. W. 1977. Zur Eutrophierung und Restaurierung de Laacher Sees. *Archiv für Hydrobiologie,* **80**, 506–518.
MCKENZIE, J. A. 1982, Carbon-13 cycle in Lake Greifen: a model for restricted ocean basins. *In*: SCHLANGER, S. O. & CITA, M. B. (eds), *Nature and Origin of Cretaceous Carbon-rich Facies*, Academic Press, London, 197–207.
NEGENDANK, J. F. W., IRION, G. & LINDEN, J. 1982. Ein eozänes Maar bei Eckfeld nordöstlich Manderscheid (SW-Eifel). *Mainzer Geowissenschaftliche Mitteilungen,* **11**, 157–172.
OTSUKI, A. & WETZEL, R. G. 1974. Calcium and total alkalinity budgets and calcium carbonate precipitation of a small hardwater lake. *Archiv für Hydrobiologie,* **73**, 14–30.
STUMM, W. & MORGAN, J. J. 1981. *Aquatic Chemistry.* 2nd edn. Wiley, New York, 780 pp.
WEBER, J. & HOFMANN, U. 1982. Kernbohrungen in der eozänen Fossillagerstätte Grube Messel bei Darmstadt. *Geologische Abhandlungen Hessen,* **83**, 58 pp.

B. BAHRIG, Institut für Wasserforschung, Zum Kellerbach 52, D-5840 Schwerte—Geisecke, West Germany.

Spores and pollen in oils as indicators of lacustrine source rocks

Jiang De-xin

SUMMARY: More than 200 species of fossil spores and pollen extracted from 68 crude oil samples collected from the lacustrine sedimentary basins of China, including the Talimu Basin, the Zhungeer Basin, the Jiuquan Basin, the Chaidamu Basin, the Eerduosi Basin, the Xialiaohe Depression, etc., have been investigated. The spores and pollen found in the crude oil are interpreted as a kind of 'trace element' indicating the pathway of the petroleum migration from the source rock to the reservoir. The principle of why some of the spores and pollen found in the crude oil can act as a kind of indicator of source rock, and the methods used to find and recognize this kind of indicator, are explained in this paper. Based on the investigations of the spores and pollen from the crude oils, the spore and pollen indicators of lacustrine petroleum source rocks of the continental sedimentary basins are proposed.

The spores and pollen acting as the indicators of the lacustrine petroleum source rocks usually indicate a tropical or subtropical wet depositional climate. It is therefore, concluded that a wet and hot climate is favourable for the deposition of lacustrine petroleum source rocks.

As long ago as the year 1923, when White found the spore sacs of *Protosalvinia* and some megaspores of Lycopod from the Upper Devonian black shales of Kentucky in the United States of America, it was suggested that these plants could be a mother substance of natural petroleum (White & Stadnichenko 1923). In 1937, Sanders (1937) found some fossil spores and algae in Mexican and Romanian crude oil samples so further supporting the theory of the organic genesis of petroleum. Then Waldschmidt (1941), Timofeev & Karimov (1953), Banerjee (1965) and Dejersey (1965) all found plant microfossils in crude oils from America, Russia, Australia and India. Subsequently some experiments on petroleum migration were made by Chepikov and others in the Soviet Union (Chepikov & Medvedeva 1971).

Since 1974, I have found more than 200 species of fossil spores and pollen in crude oil samples collected from the petroleum provinces of the continental sedimentary basins in China. Based on the palynological investigations, it was considered that these fossil spores and pollen can serve not only as a witness of the organic genesis of petroleum, but also as indicators of the nature of the source rock. The source rocks of some petroleum provinces of China were assessed using this approach (Jiang De-xin & Yang Huiqiu 1980, 1981, 1982, 1983) and are discussed in this paper. Why the fossil spores and pollen in crude oils can act as indicators of lacustrine petroleum source rocks, and how to find and recognize them, is also explained in detail.

Materials and methods

Seventy-five crude oil samples were collected from 17 oil-fields of the continental petroliferous basins of China, including the Talimu Basin, the Zhungeer Basin, the Chaidamu Basin, the Jiuquan Basin, the Eerduosi Basin, the Sanshui Basin and the Xialiaohe Depression (Fig. 1). More than 200 species of fossil spores and pollen extracted from the crude oil samples were examined, and some of them were found to be indicators of the lacustrine petroleum source rocks.

All the crude oil samples were prepared for examination using a filtering method (Jiang De-xin *et al.* 1974) (Fig. 2):

1. The samples were diluted with benzene or gasoline for filtration. Three to five litres of crude oil was used for each sample.
2. The diluted samples were filtered through the filter papers under temperature of 70°–75°C to remove the organic solvents.
3. The filter papers together with the remaining residues were extracted with benzene, ether, ketone, alcohol and other appropriate organic solvents to remove the oily, waxy and bituminous matter using a Soxhlet apparatus.
4. The residues including insoluble organic matter (kerogen) and mineral matter were concentrated and water washed by centrifuging, and in some cases separated by heavy liquid flotation.
5. The organic residues were mounted in glycerine jelly for study by light microscopy.

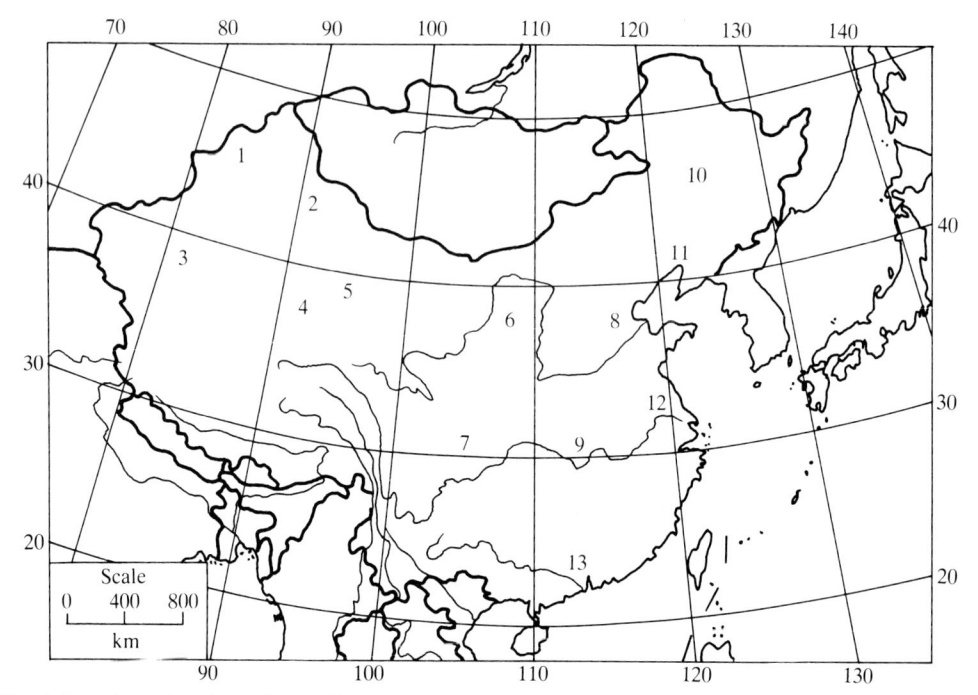

FIG. 1. Locations of continental petroliferous basins of China. 1 Zhungeer (Junggar) Basin, 2 Tulufan (Turpan) Basin, 3 Talimu (Tarim) Basin, 4 Chaidamu (Qaidam) Basin, 5 Jiuquan Basin, 6 Eerduosi Basin, 7 Sichuan Basin, 8 Huabei Basin, 9 Jianghan Basin, 10 Songliao Basin, 11 Xialiaohe Depression, 12 Subei Basin, 13 Sanshui Basin.

Most of the crude oil samples were found to contain fossil spores and pollen, but some of them contain few or no microfossils.

Principles

The fossil spores and pollen of petroleum source rocks can be carried to petroleum accumulations during the course of petroleum migration, because the spores and pollen are minute, light and have strong walls (exine). Also the microfossils from the formations through which the oil and gas migrated can be picked up by the petroleum fluid. In addition, the spores and pollen from the reservoir rocks may be found in the petroleum. Consequently, the spores and pollen extracted from a crude oil sample are usually a three-part assemblage. For the sake of convenience, it is called the petroleum sporo-pollen assemblage (Jiang De-xin & Yang Huiqiu 1980).

The spores and pollen indicate the geological ages of the original rock formations in which the spores and pollen were originally buried. This is especially so for the index species. When you find them in crude oils, you can know which strata a petroleum has migrated through. Therefore, the spores and pollen extracted from crude oil samples can be used as a kind of 'trace element' indicating the pathway of the petroleum migration from the source rocks to the petroleum pool. The laboratory experiments on petroleum migration conducted by Chepikov demonstrated that the microfossils entrapped by oil and gas from source and carrier rocks can migrate not only together with the oil, but also with the gas, and they can also be separated from the oil and gas without any appreciable change in structural features (Chepikov & Medvedeva 1971). It was suggested that the spores and pollen should serve as a kind of reliable indicator of the petroleum migration route. The above mentioned idea that the spores and pollen can act as the trace elements of the petroleum migration track might be corroborated by the results of experiments.

The spores and pollen involved in a petroleum sporo-pollen assemblage are usually different in geological age, because the source rocks, the rock formation through which oil and gas passed and the reservoir rocks of a petroleum province are usually of different ages. The reservoir rocks of an oil-field or a petroliferous province are always known, so the spores and pollen of the reservoir rocks can also be identified. If we subtract these

Spores and pollen in oils 161

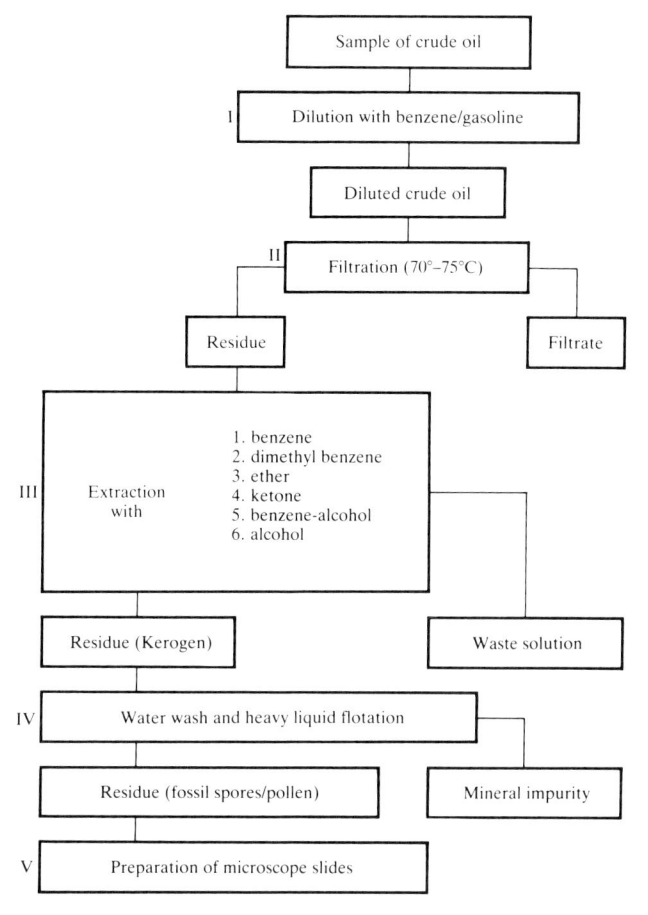

```
                    ┌─────────────────────┐
                    │ Sample of crude oil │
                    └──────────┬──────────┘
                               │
              I   ┌─────────────────────────────┐
                  │ Dilution with benzene/gasoline │
                  └──────────────┬──────────────┘
                                 │
                      ┌──────────────────┐
                      │ Diluted crude oil │
                      └─────────┬────────┘
                                │
             II    ┌──────────────────────┐
                   │ Filtration (70°–75°C) │
                   └────┬────────────┬────┘
                        │            │
                   ┌─────────┐   ┌─────────┐
                   │ Residue │   │ Filtrate │
                   └────┬────┘   └─────────┘
                        │
           ┌──────────────────────────────────────┐
           │                  1. benzene          │
           │                  2. dimethyl benzene │
      III  │  Extraction      3. ether            │
           │     with         4. ketone           │
           │                  5. benzene-alcohol  │
           │                  6. alcohol          │
           └───────┬───────────────────────┬──────┘
                   │                       │
          ┌────────────────┐       ┌───────────────┐
          │ Residue (Kerogen) │    │ Waste solution │
          └────────┬────────┘       └───────────────┘
                   │
       IV  ┌────────────────────────────────────┐
           │ Water wash and heavy liquid flotation │
           └─────────┬──────────────────┬───────┘
                     │                  │
      ┌─────────────────────────┐  ┌─────────────────┐
      │ Residue (fossil spores/pollen) │  │ Mineral impurity │
      └────────────┬────────────┘  └─────────────────┘
                   │
        V  ┌───────────────────────────────┐
           │ Preparation of microscope slides │
           └───────────────────────────────┘
```

FIG. 2. Method of extraction of spores/pollen from crude oil samples.

spores and pollen from the petroleum sporo-pollen assemblage, we will find that the remainder indicate the source rocks and the carrier beds. In the continental lacustrine petroliferous basins, anticlines are an important type of trap. With this type of trap, the source rocks underlie the carrier beds which the reservoir rocks, in turn, overlie, because oil and gas always migrate upwards along fissures, faults and unconformities. The source rocks are, therefore, the oldest beds present and the reservoir rocks the youngest. The source rocks of a continental lacustrine petroliferous basin can be deduced, therefore, from the petroleum sporo-pollen assemblage. The most ancient spores and pollen in the assemblage usually indicate the source rocks, those of intermediate age the age of the carrier beds and the youngest the reservoir rocks. Sometimes the petroleum sporo-pollen assemblage is of one age indicating that the source rocks, reservoir rocks, and carrier beds belong to the same formation. For peculiar types of the petroleum pools, such as palaeocryptodome reservoirs, metamorphic-rock fissure reservoirs, etc., the microfossils extracted from the crude oils can also indicate the age of the source rocks. Under these circumstances, the source rocks are younger than the reservoir rocks.

Results

Eerduosi Basin (Table 1)

The Eerduosi Basin is situated in the centre of North China (6, Fig. 1). The Mesozoic arched stratified petroleum pools are widespread in the basin. Some Late Triassic species of spores and pollen, such as *Punctatisporites triassicus, Punctatisporites textatus, Calamospora impexa, Marattisporites scabratus, Pityosporites divulgatus*, etc.,

TABLE 1. *Spore/pollen indicators of petroleum source rocks of Eerduosi Basin*

Spore/pollen indicators	Localities of crude oil samples	Reservoir rocks	Source rocks
Cyathidites minor Gleicheniidites senonicus Cycadopites nitidus C. typicus C. minimus Classopollis classoides C. annulatus	Zhangqing Loc. No. Z1–7	Yanan Formation (J_{1-2})	Yanan Formation (J_{1-2})
Punctatisporites triassicus P. textatus Calamospora impexa Marattisporites scabratus Pityosporites divulgatus	Zhangqing Loc. No. Z2–5	Yanan Formation (J_{1-2})	Yanzhang Formation (T_3)

have been found from the crude oil samples taken from the Jurassic reservoir of the basin. Jurassic species as *Cyathidites minor, Gleicheniidites senonicus, Cycadopites nitidus, Cycadopites typicus, Cycadopites minimus, Classopollis classoides*, etc. have also been found in the petroleum sporo-pollen assemblage. All of the species are widely distributed in the lacustrine black shales and grey siltstones of the Upper Triassic Yanzhang Formation or the Lower–Middle Jurassic Yanan Formation of the region. They must have come from these two formations. According to the above reasoning, the Late Triassic species indicate that the Yanzhang Formation is the source rock of the basin. The Yanan Formation might be another source rock. Consequently, the above species may be regarded as the spore or pollen indicators of the lacustrine source rocks in the region.

Talimu Basin (Table 2)

The Talimu Basin of South Xinjiang is the largest continental basin in China (3, Fig. 1). It is known as the heart of Central Asia. Arched stratified petroleum pools are common there. Two petroleum provinces of the basin have been studied here.

The following Jurassic species of spores and pollen have been found in crude oil samples from the Neogene reservoir of the Yecheng Petroleum Province of the Talimu Basin: *Deltoidospora perpusilla, Deltoidospora gradata, Cyathidites australis, Cyathidites minor, Cibotiumspora paradoxa, Dictyophyllidites harrisii, Dictyophyllum rugosum, Gleicheniidites senonicus, Gleicheniidites rousei, Undulatisporites concave, Osmundacidites well-manii, Leptolepidites major, Apiculatisporis ovalis, Bennettiteaepollenites lucifer, Cycadopites nitidus, Cycadopites typicus, Cycadopites minimus, Cycadopites carpentieri, Chasmatosporites elegans, Classopollis classoides, Classopollis annulatus, Podocarpidites major, Podocarpidites multesimus, Parvisaccites enigmatus, Quadraeculina limbata, Caytonipollenites pallidus, Pteruchipollenites thomasii, Alisporites grandis, Alisporites bilateralis, Eucommiidites troedssonii*, etc. ... All of the species have been found in the Jurassic deposits of the Eurasian continent and the North American continent. And most of the species are the index species of the Middle Jurassic. They are widespread in the Middle Jurassic lacustrine deposits in this and adjacent regions. Many Neogene species of spores and pollen have also been found in the crude oil samples, but the Jurassic species are the most ancient species in the petroleum sporo-pollen assemblage. They are, therefore, considered to have been brought from the source rocks into the petroleum pool by oil and gas in the course of the petroleum migration. We have good reason to say that the Middle Jurassic species indicate that the Middle Jurassic Yangye-Taerga Formation should be the source rock of the region. The above Jurassic species, therefore, can serve as the spore and pollen indicators of the lacustrine source rocks of the region.

The spores and pollen extracted from the crude oil samples taken from the Jurassic reservoir of the Kuche Petroleum Province of the Talimu Basin (Table 3) are as follows: *Deltoidospora perpusilla, Deltoidospora gradata, Cyathidites minor, Hymenophyllumsporites deltoidus, Cingulatisporites problematicus, Cibotiumspora paradoxa, Osmundacidites wellmanii, Cycadopites nitidus, Cycadopites typicus, Podocarpidites multesimus,*

TABLE 2. *Spore/pollen indicators of petroleum source rocks of Yecheng Petroleum Province of Talimu Basin*

Spore/pollen indicators	Localities of crude oil samples	Reservoir rocks	Source rocks
Deltoidospora perpusilla	Kekeya	Ashi	Yangye-
D. gradata	Locs. No. K6	Formation	Taerga
Cyathidites australis	No. K10	(N_2)	Formation
C. minor			(J_2)
Cibotiumspora paradoxa		Pakabulake	
Dictyophyllidites harrisii		Formation	
Dictyophyllum rugosum		(N_1)	
Gleicheniidites senonicus			
G. rousei			
Undulatisporites concave			
Osmundacidites wellmanii			
Leptolepidites major			
Apiculatisporis ovalis			
Bennettiteaepollenites lucifer			
Cycadopites nitidus			
C. typicus			
C. minimus			
C. carpentieri			
Chasmatosporites elegans			
Classopollis classoides			
C. annulatus			
Podocarpidites major			
P. multesimus			
Parvisaccites enigmatus			
Quadraeculina limbata			
Caytonipollenites pallidus			
Pteruchipollenites thomasii			
Alisporites grandis			
A. bilateralis			
Eucommiidites troedssonii			

TABLE 3. *Spore/pollen indicators of petroleum source rocks of Kuche (Kuga) Petroleum Province of Talimu Basin*

Spore/pollen indicators	Localities of crude oil samples	Reservoir rocks	Source rocks
Deltoidospora perpusilla	Kangcun	Kangcun	Kuzilenuer
D. gradata	Jilishi	Formation	Formation
Cyathidites minor	Heiyingshan	(N_1)	(J_2)
Hymenophyllumsporites deltoidus	Yiqikelike		
Cingulatisporites problematicus			Yangxia
Cibotiumspora paradoxa			Formation
Osmundacidites wellmanii			(J_1)
Cycadopites nitidus			
C. typicus		Kuzilenuer	
Podocarpidites multesimus		Formation	
Alisporites grandis		(J_2)	
A. bilateralis			
Parvisaccites enigmatus			
Abietineaepollenites microalatus			
A. minimus			

Alisporites grandis, Alisporites bilateralis, Parvisaccites enigmatus, etc. . . . All of these species have been found in the lacustrine deposits of the Lower Jurassic Yangxia Formation or the Middle Jurassic Kuzilenuer Formation in this region. Neither older nor younger spores and pollen are found in the crude oil samples. The petroleum sporo-pollen assemblage contains only the contemporaneous spores and pollen. This suggests that the source rocks are much the same age as the reservoir rocks. Consequently, the Jurassic species indicate both the reservoir rocks and the source rocks, they qualify as the spore and pollen indicators of the lacustrine source rocks in the region. In addition, some of the above Jurassic species, such as *Deltoidospora perpusilla, Cyathidites minor, Cibotiumspora paradoxa, Cycadopites nitidus, Cycadopites typicus, Parvisaccites enigmatus, Abietineaepollenites minimus*, etc., have also been found in the crude oil samples taken from both the Cretaceous and Neogene reservoirs of the present petroliferous province. Therefore, they are also taken to be the petroleum source indicators.

Zhungeer Basin (Table 4)

The Zhungeer Basin of North Xinjiang (1, Fig. 1) is a well known continental petroliferous basin in China. The following Jurassic species of spores and pollen have been obtained from the oils of the Tertiary reservoir of the southern petroliferous province of the basin: *Cyathidites minor, Dictyophyllidites harrisii, Cycadopites nitidus, Cycadopites typicus, Cycadopites minimus, Callialasporites dampieri, Araucariacites australis, Podocarpidites canadensis, Podocarpidites multesimus, Parvisaccites enigmatus, Caytonipollenites pallidus, Alisporites grandis, Alisporites bilateralis, Pteruchipollenites thomasii, Pteruchipollenites microsaccus, Abietineaepollenites dunrobinensis, Abietineaepollenites minimus, Eucommiidites troedssonii*, etc. . . . A lot of Tertiary species of spores and pollen have also been found in the crude oil samples, but the Jurassic species are the most ancient elements in the petroleum sporo-pollen assemblage. The Jurassic species, therefore, indicate that the Jurassic lacustrine deposits should be the source rocks of the southern petroliferous province of the basin, thus the above mentioned species should be the spore and pollen indicators of the lacustrine source rocks in this region.

Jiuquan Basin (Table 5)

I have found Early Cretaceous species of spores and pollen such as *Schizaeoisporites zizyphinus, Cibotiumspora paradoxa, Monosulcites minimus, Psophosphaera grandis*, etc. in the crude oil samples taken from both the Early Cretaceous reservoir and the Tertiary reservoir of the Yumen Petroleum Province of the Jiuquan Basin (5, Fig. 1) in Gansu Province of China. In addition, some

TABLE 4. *Spore/pollen indicators of petroleum source rocks of southern petroliferous province of Zhungeer Basin*

Spore/pollen indicators	Localities of crude oil samples	Reservoir rocks	Source rocks
Cyathidites minor	Dushanzi	Shawan	Xishanyao–
Dictyophyllidites harrisii	Locs.	Formation	Toudenhe
Cycadopites nitidus	No. D58	(E_3–N)	Formation
C. typicus	No. D63		(J_2)
C. minimus	No. D68		
Callialasporites dampieri	No. D201		Badaowan–
Araucariacites australis			Sangonghe
Podocarpidites canadensis			Formation
P. multesimus			(J_1)
Parvisaccites enigmatus			
Quadraeculina limbata			
Q. minor			
Caytonipollenites pallidus			
Pteruchipollenites thomasii			
P. microsaccus			
Alisporites grandis			
A. bilateralis			
Abietineaepollenites dunrobinensis			
A. minimus			
Eucommiidites troedssonii			

TABLE 5. *Spore/pollen indicators of petroleum source rocks of Jiuquan Basin*

Spore/pollen indicators	Localities of crude oil samples	Reservoir rocks	Source rocks
Schizaeoisporites zizyphinus *S.* cf. *cretacius* *Cibotiumspora paradoxa* *Monosulcites minimus* *Bennetiteaepollenites* sp. *Psophosphaera grandis*	Laojunmiao Yaerxia Shiyougou Baiyanghe	Baiyanghe Formation (N) Huoshaogou Formation (E) Xinminbu Formation (K)	L. Xinminbu Formation (K_1)

Early Cretaceous species have been obtained, even from the fissured Silurian metamorphic reservoir of the region. The above Early Cretaceous species of spores and pollen are common in the lacustrine black shales of the Lower Xinminbu Formation. The fact that the spores and pollen from the Early Cretaceous black shales appear both in the Tertiary reservoir and in the Silurian reservoir results from the petroleum migration. It shows that the black shales of the Lower Xinminbu Formation may be considered to be the source rocks of the region and the above Early

TABLE 6. *Spore/pollen indicators of petroleum source rocks of Chaidamu Basin*

Spore/pollen indicators	Localities of crude oil samples	Reservoir rocks	Source rocks
Deltoidospora regularis *Cyatheacidites annulata* *Polypodiaceaesporites haardti* *P. nutidus* *Polypodiisporites favus* *Cycadopites giganteus* *Ginkgopites dubia* *Podocarpidites nageiaformis* *P. paranageiformis* *Abietineaepollenites cembraeformis* *Pinuspollenites labdacus* *P. banksianaeformis* *Piceaepollenites alatus* *P. tobolicus* *Cedripites deodariformis* *Tsugaepollenites igniculus* *T. viridifluminipites* *Quercoidites henrici*	Youquanzi Daifengshan Xianshuiquan Huatugou	Youshashan Formation (N) U. Ganchaigou Formation (N_1)	Ganchaigou Formation (E_3–N_1)
Cyathidites minor *Dictyophyllidites harrisii* *Osmundacidites wellmanii* *Duplexisporites problematicus* *D. gyratus* *Cycadopites nitidus* *C. typicus* *C. carpentieri* *Callialasporites dampieri* *Podocarpidites major* *P. multesimus* *Parvisaccites enigmatus* *Pteruchipollenites thomasii* *Alisporites bilateralis*	Yuka Lenghu	Hongshuigou Formation (J_3) Daimeigou Formation (J_2)	Daimeigou Formation (J_2)

Cretaceous species of spores and pollen may be referred to as the spore and pollen indicators of the lacustrine source rocks in the basin.

Chaidamu Basin (Table 6)

The Chaidamu Basin (4, Fig. 1) is a large continental petroliferous basin in Qinghai Province of China. I have found some Jurassic species of spores and pollen, such as *Cyathidites minor, Dictyophyllidites harrisii, Osmundacidites wellmanii, Duplexisporites problematicus, Duplexisporites gyratus, Cycadopites nitidus, Cycadopites typicus, Cycadopites carpentieri, Callialasporites dampieri, Podocarpidites major, Podocarpidites multesimus, Parvisaccites enigmatus, Alisporites bilateralis, Pteruchipollenites thomasii*, etc., in the crude oil samples taken from the Jurassic reservoir of the northern edge of the Chaidamu Basin. I have also found some Tertiary species of spores and pollen in the crude oil samples taken from the Neogene reservoir of the basin. The Tertiary species of spores and pollen obtained from the Neogene reservoir are of *Deltoidospora regularis, Cyatheacidites annulata, Polypodiaceaesporites haardti, Polypodiaceaesporites nutidus, Polypodiisporites favus, Cycadopites giganteus, Ginkgopites dubia, Podocarpidites nageiaformis, Podocarpidites paranageiformis, Abietineaepollenites cembraeformis, Pinuspollenites labdacus, Pinuspollenites banksianaeformis, Piceaepollenites alatus, Piceaepollenites tobolicus, Cedripites deodariformis, Tsugaepollenites igniculus, Tsugaepollenites viridifluminipites, Quercoidites henrici*, etc.... Most of the species have been found in the Oligocene deposits and some of them have been found in the Miocene deposits of this region and the adjacent regions. Based on the investigation of the petroleum sporo-pollen assemblages, both the Lower–Middle Jurassic Formation and the Oligocene–Miocene Formation may be considered to be the favourable source rocks, and the above species may be regarded as the spore and pollen indicators of the lacustrine source rocks in this basin.

Xialiaohe Depression (Table 7)

The Xialiaohe Depression (11, Fig. 1) is a petroleum province in Northeastern China. The following Tertiary species of spores and pollen have been found from the crude oil samples taken from the Oligocene reservoir of the Liaohe oilfield of the depression: *Stridiporosporites bistriatus, Multicellaesporites margaritus, Multicellaesporites desmodes, Multicellaesporites tenerus, Multicellaesporites ovatus, Dicellaesporites popovii, Inapertisporites rotundus, Pterisisporites undulatus, Plicifera decora, Abietineaepollenites cembraeformis, Abietineaepollenites microsibiricus, Cedripites diversus, Keteleeria dubia, Ephedripites cheganicus, Quercoidites asper, Chenopodipollis microporatus, Chenopodipollis multiporatus*, etc.... All species distributed in the Eocene to Oligocene Shahejie Formation in this and adjacent regions. Neither earlier nor later species are found in the petroleum sporo-pollen assemblage, indicating that the Shahejie Formation should be the regional source

TABLE 7. *Spore/pollen indicators of petroleum source rocks of Xialiaohe Depression*

Spore/pollen indicators	Localities of crude oil samples	Reservoir rocks	Source rocks
Stridiporosporites bistriatus	Liaohe	Dongying	Shahejie
Multicellaesporites margaritus	Locs.	Formation	Formation
M. desmodes	No. L7-5	(E_3)	(E_{2-3})
M. tenerus	No. L1-1		
M. ovatus	No. L5-2	Shahejie	
Dicellaesporites popovii	No. L8-11	Formaton	
Inapertisporites rotundus	No. L213	(E_{2-3})	
Pterisisporites undulatus			
Plicifera decora			
Abietineaepollenites cembraeformis			
A. Microsibiricus			
Pinuspollenites labdacus			
Cedripites diversus			
Keteleeria dubia			
Ephedripites cheganicus			
Quercoidites asper			
Chenopodipollis microporatus			
C. multiporatus			

TABLE 8. *Spore/pollen indicators of petroleum source rocks of Sanshui Basin*

Spore/pollen indicators	Localities of crude oil samples	Reservoir rocks	Source rocks
Sphagnumsporites antiquasporites *Cyathidites* cf. *minor* *Plicifera decora* *Polypodiaceaesporites haardti* *Dacrydiumpollenites rarus* *Cedripites eocenicus* *C. deodariformis* *Pinuspollenites labdacus* *P. minutus* *P. bankianaeformis* *Momipites coryloides* *Quercoidites minutus* *Q. asper*	Sanshui Locs. No. S1 No. SS12	Buxin Formation (E)	Buxin Formation (E)

rock. The above species, therefore, may be taken as the spore and pollen indicators of the source rocks in the region.

Sanshui Basin (Table 8)

The Sanshui Basin (13, Fig. 1) is a small continental petroliferous basin in Guangdong Province of China. The petroleum sporo-pollen assemblage from the Eogene reservoir of this petroleum province consists mainly of *Sphagnumsporites antiquasporites*, *Cyathidites* cf. *minor*, *Plicifera decora*, *Polypodiaceaesporites haardti*, *Dacrydiumpollenites rarus*, *Cedripites eocenicus*, *Cedripites deodariformis*, *Pinuspollenites labdacus*, *Pinuspollenites minutus*, *Pinuspollenites bankianaeformis*, *Momipites coryloides*, *Quercoidites minutus*, *Quercoidites asper*, etc.... All of the species have been found in the Eogene lacustrine deposits in South China. It implies that the source rocks of the region should belong to the Eogene System and the above species may serve as the spore and pollen indicators of the lacustrine source rocks in the basin.

Discussion

According to the theories about the origin, migration and accumulation of petroleum, the fossil spores and pollen in the crude oils must have come from the source rocks, carrier beds and reservoir rocks. Judging from the results of this study, fossil spores and pollen can be expelled from source rocks and carried to petroleum pools. Most species of the spores and pollen can pass through the pathway of petroleum migration, because they have minute sizes (< 100 μm). If the distance of petroleum migration is comparatively short, it is possible to find numerous spores and pollen in the crude oils. I have found 506 specimens of 67 species of spores and pollen referred to 39 genera in crude oil from the Qiketai oil-field of the Tulufan Basin, Xinjiang (Fig. 1, Table 9). The abundances of the spores and pollen in the crude oils are relevant to their original contents in the source rocks, their sizes and the distance of petroleum migration.

Conclusions

The results of this study come to the conclusion that the Late Triassic, the Middle Jurassic, the Early Cretaceous and the Oligocene are the principal times of formation of lacustrine petroleum source rocks in the continental petroliferous basins in China. The influence of palaeoclimate and palaeoenvironment can be seen on the palynofloras of these periods. The palynofloras of the Late Triassic, the Middle Jurassic, the Early Cretaceous and the Oligocene species, all usually indicate a tropical or subtropical wet climate. It may, therefore, be considered that the lacustrine deposition under hot or warm and wet climate conditions was most favourable for the formation of the lacustrine petroleum source rocks in continental facies.

ACKNOWLEDGMENTS: The writer wishes to thank Professor Hsü Jen, Professor A. Traverse and Professor Zhu Xia for their invaluable help and advice. The writer is grateful to Mr Hu Boliang and Mr He Zhuosheng for their help in discussions of geochemical and stratigraphical problems and in providing many samples. Thanks are also due to Ms Yang Huiqiu, Ms Du Jine and Mr Wu Ping for technical assistance.

TABLE 9. *Fossil spores and pollen in crude oil from Qiketai oil-field of Tulufan Basin*

Spores/pollen	Number of specimens (abundance of spores/pollen)	Spores/pollen	Number of specimens (abundance of spores/pollen)
Deltoidospora magna	4	*Pseudowalchia ovalis*	4
D. gradata	2	*P. landesii*	3
Cyathidites australis	9	*P. biangulina*	2
C. minor	19	*Podocarpidites unicus*	12
Dictyophyllidites harrisii	18	*P. major*	22
Cibotiumspora paradoxa	4	*P. multicina*	14
C. jurienensis	3	*P. multesimus*	12
Gleicheniidites rousei	5	*P. rousei*	6
G. conflexus	4	*P. wapellaensis*	3
Concavisporites sp.	7	*Platysaccus lopsinensis*	4
Biretisporites sp.	2	*Parvisaccites enigmatus*	26
Undulatisporites concavus	4	*Ovalipollis minor*	11
Granulatisporites arenaster	2	*O. canadensis*	6
G. minor	3	*Alisporites grandis*	12
Osmundacidites wellmanii	7	*A. bilateralis*	14
Todisporites minor	7	*A. thomasii*	7
Apiculatisporis globasus	2	*Pityosporites similis*	4
Converrucosisporites venitus	2	*P. divulgatus*	3
Leptolepidites major	4	*Piceites expositus*	5
Aratrisporites cf. *scabratus*	3	*P. podocarpoides*	3
Cycadopites nitidus	6	*P. latens*	2
C. typicus	3	*Protopicea exilioides*	8
C. subgranulosus	2	*Pseudopicea variabiliformis*	2
Chasmatosporites elegans	8	*Piceaepollenites complanatiformus*	8
Classopollis classoides	14	*Abietineaepollenites dunrobinensis*	11
C. itunensis	37	*A. microalatus*	4
C. annulatus	10	*A. minimus*	6
C. qiyangensis	6	*Cedripites minor*	34
Callialasporites dampieri	11	*C.* sp.	12
C. radius	4	*Eucommiidites troedssonii*	5
Araucariacites australis	2	Total	506
Inaperturopollenites dettmannii	3		
Protoconiferus funarius	3		
Paleoconiferus asaccatus	2	Reservoir rocks	Toudenhe Formation (J_2)
Caytonipollenites pallidus	3		
Vitreisporites jurassicus	14	Source rocks	Xishanyao–Toudenhe Formation (J_2)
V. jansonii	2		Badaowan–Sangonghe Formation (J_1)

References

BANERJEE, D. 1965. Microflora in crude petroleum from Assam, India. *Transactions of Mining Geology and Metal Institute, India*, **62**, 51–66.

CHEPIKOV, K. R. & MEDVEDEVA, A. M. 1971. Spores and pollen in oil and gas as migration indices. *Journal of Palynology*, **7**, 56–59.

DEJERSEY, N. J. 1965. Plant microfossils in some Queensland crude oil samples. *Publication of the Geological Survey of Queensland*, **329**, 1–9.

JIANG DE-XIN & YANG HUIQIU 1980. Petroleum sporo-pollen assemblages and oil source rock of Yumen oil-bearing region in Gansu. *Acta Botanica Sinica*, **22**, 280–285 (in Chinese).

—— & —— 1981. Studies of source rocks judging from spores and pollen. *Memoirs of Lanzhou Institute of Geology, Academia Sinica*, **1**, 99–110 (in Chinese).

—— & —— 1982. Sporo-pollen and oil source. *Proceedings of the Symposium on Petroleum Geoscience, Academia Sinica*, Science Press, Beijing, 157–162 (in Chinese).

—— & —— 1983. Petroleum sporo-pollen assemblage and oil source rock of Kuche sag in Xinjiang. *Acta Botanica Sinica*, **25**, 179–186 (in Chinese).

——, —— & DU JINE 1974. Significance and method of sporo-pollen analysis of crude oil. *Journal of Botany*, **1**, 31–32 (in Chinese).

SANDERS, J. M. 1937. The microscopical examination

of crude petroleum. *Journal of the Institution of Petroleum Technologists*, **23**, 525–573.

TIMOFEEV, B. V. & KARIMOV, A. K. 1953. Spores and pollen in mineral oil. *Reports of Academy of Sciences of USSR*, **92**, 151–152 (in Russian).

WALDSCHMIDT, W. A. 1941. Progress report on microscopic examination of Permian crude oils. *Bulletin of the American Association of Petroleum Geologists*, **25**, 934.

WHITE, D. & STADNICHENKO, T. 1923. Some mother plants of petroleum in the Devonian black shales. *Economic Geology*, **18**, 238–252.

JIANG DE-XIN, Lanzhou Institute of Geology, Academia Sinica, Lanzhou, Gansu, China.

Part III
Case Studies

Palaeolimnology and organic geochemistry of the Middle Devonian in the Orcadian Basin

A. D. Duncan & R. F. M. Hamilton

SUMMARY: An interdisciplinary study of the Middle Devonian lacustrine laminites from the Orcadian Basin of NE Scotland has enabled the construction of a predictive source-rock model which we have applied to the Moray Firth area.

Parameters characterizing areal source-rock extent, primary organic input, and the depths, salinity levels and redox potential of the depositional water column are discussed, together with the quantity and maturity of the preserved sedimentary organic matter.

On the basis of this model, we suggest that a Middle Devonian component is present in the Beatrice crude (an atypical North Sea oil) in the Moray Firth Basin. This proposal is largely supported by molecular geochemical evidence.

Context of the Orcadian Basin outcrop area

Recent source-rock studies using mostly bulk organic geochemical techniques (Hamilton & Trewin 1985; Marshall et al. 1985; Parnell 1985) have demonstrated that the Middle Devonian 'fish bed' laminites from the Orcadian Basin outcrop area have good petroleum source-rock potential. The inference from palaeocurrent vectors (Mykura 1983), however, is that the present outcrop represents only the southern and western flanks of a much larger basin which, accepting the extrapolation of Ziegler (1982), extended some 600 km from the present day southern shores of the Moray Firth across the North Sea to the Hornelen area on the west coast of Norway. The aims of this study were to undertake detailed source-rock characterization and an assessment of palaeolimnological conditions, in order to provide a basis for predictive source-rock modelling in the Orcadian Basin.

'Fish bed' laminites

The most common rock types in the Orcadian Basin to have been considered as potential source rocks are the dark, organic-rich microlaminated units commonly known as 'fish beds', due to the beautifully preserved fossil fish they often contain. Their lithology usually comprises a tripartite association of micritic carbonate (calcite and/or dolomite), detrital material, and organic layers. The total thickness of the triplets averages from 0.01–1.0 mm. Modal proportions of individual components vary, but 'carbonate laminite' and 'detrital laminite' represent convenient end members. Although certain examples exhibit the classic, seasonally related tripartite 'varve' sequence (e.g. Rayner 1963), the general sequential arrangement and individual layer definition can only be described as variable.

Organic contents as plotted on the simple three component system (Fig. 1) range up to 5% with no obvious relationship linking variance with a carbonate or detrital dominated host. Micronodular textures with accompanying stylolitization are commonly developed in carbonate-rich laminites and are related to diagenetic pressure-solution effects. Intrastratal deformation features such as microfaults, slumps and complete dislocation planes are common.

Typically, laminite rock units vary up to 2 m in thickness and grade through synaeresis-cracked siltstone/shale laminites into coarser siltstone and fine sandstone beds bearing desiccation-cracked surfaces and current oscillation ripples. This lithological gradation is commonly repetitive and forms the rhythms or cycles (Crampton & Carruthers 1914; Donovan 1980) so characteristic of the Orcadian Basin and indeed many other ancient lacustrine sequences. Individual cycle thickness varies from 3–10 m with a mean at ca. 6 m. The sequential arrangement, thickness variations, and inferred periodicities of stacked cyclic sequences in Caithness have been related to long-term climatic variations (Donovan 1980). Fourier analysis of several sequences from the Lower and Upper Caithness Flagstone Group has suggested that a Milankovitch-type climatic forcing model was the dominant control over cycle development (Hamilton 1986).

Depositional models

General models

The two general models proposed for the production of the distinctive organic-rich microlami-

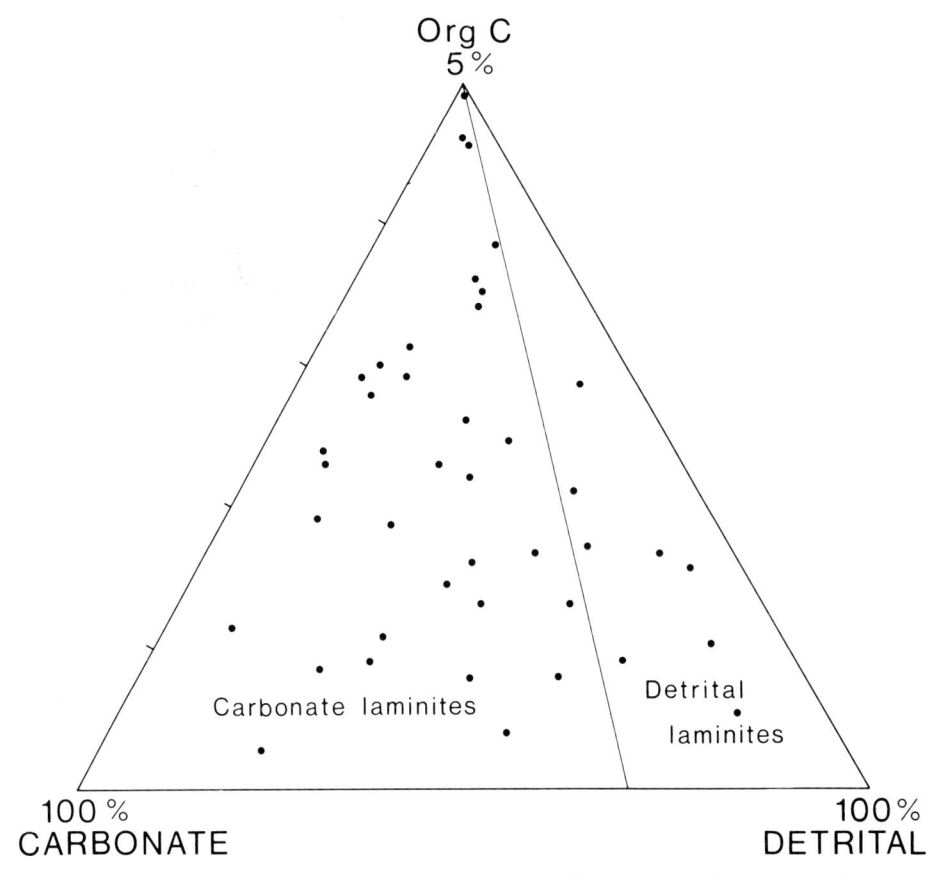

FIG. 1. Compositional variation within Orcadian Basin fish bed laminites expressed in terms of a simple carbonate–organic carbon–detrital ternary system.

nated lake sediments are the shallow algal mat lake and the deep stratified lake models. The shallow algal mat lake or 'playa lake model' (Eugster & Surdam 1973) produces organic microlamination through the colonization of the sediment surface by epipelic algae and bacteria in continuous mats. Alternating layers of allochthonous carbonate or clastic material are swept in as intermittent influxes from the lake margins. Analogy is made with modern examples of high productivity water bodies in arid areas, such as the lakes and lagoons in South and Western Australia (Logan et al. 1974; von der Borch et al. 1977) and the Red Sea (Friedman et al. 1973). Similarly, the occurrence of mats is constrained by water clarity and depth to satisfy the photosynthesis levels necessary for growth. Typically, flat algal mats are restricted to very shallow and fluctuating water depths (0–4 m) with associated stromatolites and desiccated surfaces.

The playa lake model has been applied to many rock sequences within the Green River Formation (Eugster & Surdam 1973; Eugster & Hardie 1975; Lundell & Surdam 1975) including, more recently, the organic laminites of the Laney Member (Buccheim & Surdam 1977; Surdam & Wolfbauer 1975) which are similar in lithological character and setting to the Orcadian Basin 'fish bed' laminites.

According to the stratified lake model, originally proposed by Bradley (1929), the delicate organic microlamination is produced by seasonal variations in the type and quantity of material settling out of a stratified water column. Accumulation occurs below wave base within the anoxic hypolimnion of a perennially stratified (meromictic) lake. Organic input is derived from planktonic algal 'blooms' in the photic epilimnion and allochthonous organic detritus transported from the land and the shallow margins. Laminites pass landward into a zone of non-microlaminated sediment and marginal deposits which may contain epipelic algal colonization features, such as stromatolite domes and mats, and are subject to intermittent desiccation.

In terms of source-rock modelling for an ancient lacustrine basin, the differences between the two models carry important implications. The principal differences are: original water depth, types of biotic communities, lateral morphology of the organic laminite (source rock) lithosome, and the types of vertical sedimentary sequences developed in one area.

Although the Orcadian Basin fish bed laminites and associated sediments have been interpreted within the context of a stratified lake model (Rayner 1963; Donovan 1975, 1980; Mykura 1976), many of the sedimentary features resemble those of playa lake systems (Eugster & Surdam 1973). This, augmented by our expanding knowledge of organic production and preservation within shallow algal mat lakes (e.g. De Deckker this volume) validates a reappraisal of these sediments.

As modern algal mats are most commonly limited to relatively shallow water depths of less than 4 m, any estimate of depositional water depth for a particular organic laminite horizon could be used to discriminate between the two depositional models. Although extensive algal mats form today only in specialized environments which are effectively hostile to grazing animals, it is not reasonable to apply this limitation to ancient sequences due to uncertainties regarding the adaptation and effectiveness within groups of grazing animals through geological time. However, it is perhaps fair to assume that the principal limitation on photosynthesis, namely depth of light penetration, has remained the same since the appearance of algae in the Precambrian.

Water depth estimates

Several methods have been used to estimate the water depths in ancient lakes. One assumes that a lake basin, once formed, fills up with sediment to the level of its outlet, such that the depth of water is equal to the compaction-corrected rock interval between the deepest lake phase sediments and beds containing the first evidence of emergence (Van Houten 1977; Fannin 1970). This model is not reliable, however, in view of documented examples such as Lake Bonneville (Morrison 1966) or the East African Rift lakes (Beadle 1981), where the maximum water depths inferred from fossil outlet points bear no fixed or necessary relationship to the corresponding sedimentary sequences deposited during the last deep lake cycle. Indeed the thickness of the regressive sedimentary sequence may be anything approaching two orders of magnitude smaller than the maximum water depth.

Another approach is to trigonometrically calculate water depth by extrapolating an inferred lake marginal slope angle across a single lacustrine sedimentary unit (Hubert et al. 1976). However due to the small angles involved and morphometrical variations possible over a lake basin, this method may also be discounted.

The differential relief of a lake floor which can be proven to have been completely submerged provides a minimum value for maximum water depth during that particular lake phase. At several localities near the western margins of the Orcadian Basin outcrop area, the irregular Caledonian basement topography is exhumed below a complex of coarse immature clastics and marginal carbonates (Donovan 1975). In at least two examples, individual limestone units may be traced over an original depositional surface of some 10 m differential relief.

Another method uses the delicate microlamination of the sediment itself, and the fact that whole undisturbed fish carcasses are commonly preserved through several bedding planes implying that the lacustrine laminite units accumulated below wave base. The depth of wave base in modern low latitude lakes is directly influenced by wind-induced wave action within the lake body (Beadle 1981; Serruya & Pollingher 1983), except in hypersaline examples where the propagation of wave energy down through a water column is restricted through attenuation in the commonly established hypersaline density stratification.

Smith & Sinclair (1972) have produced an empirically derived formula relating wave period (T), wind speed (w) and fetch (Lf) which may be applied to non-hypersaline lakes,

$$g \cdot T/w = 0.46 \cdot (g \cdot Lf/w^2)^{0.28} \quad \text{(I)}$$

By assuming wave base (WB) = $\lambda/2$ and wavelength (λ) = $1.56T^2$ (Håkanson & Jansson 1983), and gravity (g) = 9.81 m s^{-2}, the original equation (I) may be restated in terms of wave base, fetch and wind speed:

$$WB = 0.00616 \, Lf^{0.56} w^{0.88} \quad \text{(II)}$$

For ancient lake examples, a minimum value for potential fetch may be reasonably regarded as the lateral extent of an undisturbed microlaminated horizon (Fig. 2). Palaeowind speeds and directions which govern the effective fetch may be derived from direct measurement of aeolian features within the lake basin or inferred from global palaeoclimate models (e.g. Parrish 1982; Summerhayes this volume) based on palaeogeo-

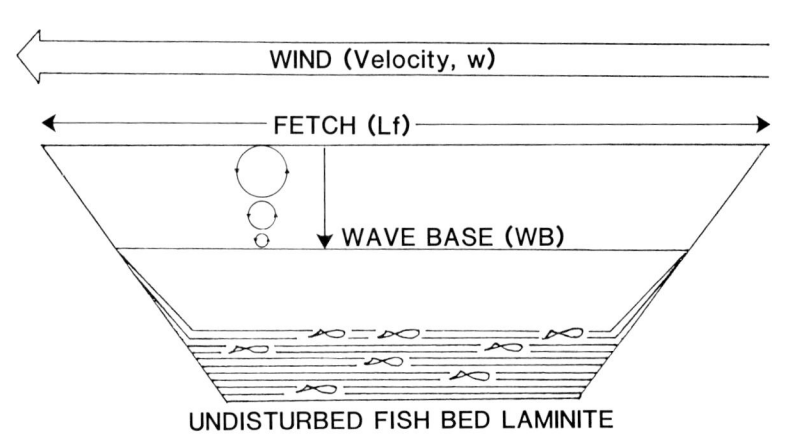

FIG. 2. Schematic model illustrating empirical relationship between wave fetch (Lf), wave base (WB), and wind speed (w).

graphic reconstructions for the area (e.g. Smith et al. 1973).

Within the Orcadian Basin, the generally poor level of exposure, structural complexity and an unsatisfactory biostratigraphic control (Richardson 1965; Donovan et al. 1974; J. Marshall, personal communication 1985) serve to hinder lateral correlation of individual fish bed horizons. One fish bed horizon, however which contains the distinctive Achanarras fish fauna, has been traced across the entire outcrop area (Donovan et al. 1974; Westoll 1977; Trewin 1986; Hamilton, unpublished data). This correlation is due more to the relatively rich and diverse fauna than any lithological peculiarity. Essentially identical laminite horizons occur within cyclic sequences above and below each of the Achanarras correlatives.

Potential wave fetch may be derived from the lateral extent of undisturbed microlaminated lithosome delineated by the Achanarras horizon correlatives. Estimates of reasonable maximum (storm) wind speeds and directions are derived from the contemporaneous draas and some coarse grained barchanoid dune sets fringing the basin margin. These data (kindly supplied by David Rogers of Cambridge), when augmented with the palaeogeographic reconstructions and associated palaeoclimate modelling, are consistent with the strongest winds blowing along the axis of the basin from the south. Analogy is drawn with African lake basins today where strong winds are funnelled down the rift valleys and directly influence the depth of wave mixing (Beadle 1981). For the purpose of this calculation, a conservative storm wind speed of 30 m s^{-1} is proposed for the Middle Devonian in the Orcadian Basin.

Using two correlations, to which different levels of confidence can be attached, depths of 75 m or 122 m can be calculated for wave base during Achanarras time (Fig. 3). The first correlation involves only the Caithness–Orkney area, where the horizon can be traced laterally to give a potential fetch of some 95 km which produces a wave base depth of 75 m. Detailed sampling at four localities, Niandt (N), Achanarras (A), Baligill (B) and Sandwick (S), produces an essentially identical vertical distribution of at least 8 fish genera (all at species level) and a strongly similar lithological succession (Trewin 1976, 1986; Hamilton, unpublished data). A more tenuous correlation, which involves an interpretational restoration of the post-Middle Devonian movements on major transcurrent faults (Mykura 1983) such as the Great Glen Fault, links the fish beds on the south side of the Moray Firth with the Melby fish bed (M) in Shetland. This extrapolation gives a potential fetch of some 225 km with a calculated wave base of 122 m.

This method takes no account of other significant current sources such as seiches, turbulent inflow, or long period internal waves which all propagate energy down through the water column (Sly 1978; Beadle 1981). However, the wave base approximations generated clearly favour the deep stratified lake model over shallow algal mat playas. Accepting the water depth estimates, the discrete siltstone units which occur interbedded with laminite through some 300 m of strata above and below the Achanarras horizon in Caithness and often contain basal rip-up clasts and sole structures, may be interpreted as turbidites accumulating in a deep lake environment. Current directions from flute and groove casts on the turbidite bases are from the NW suggesting that water depth increased down a palaeoslope to the

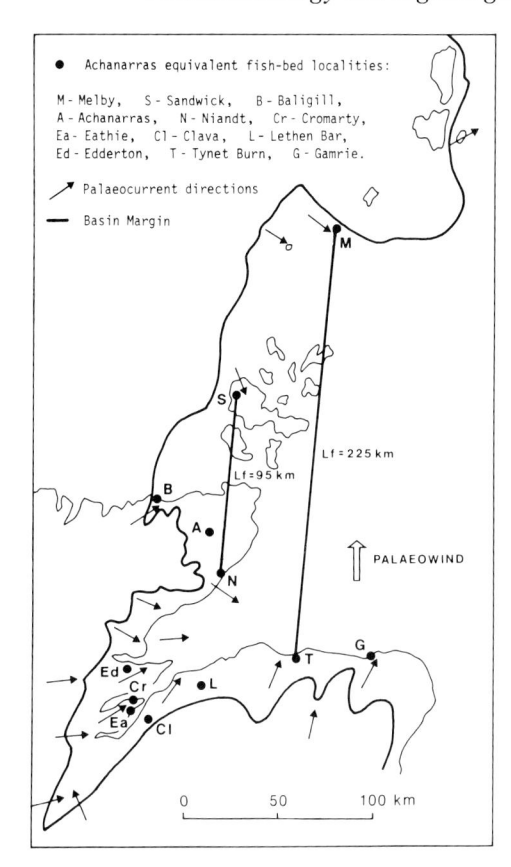

FIG. 3. Middle ORS palaeogeography of northern Scotland (after Mykura 1983) showing the position of the Achanarras equivalent fish-bed laminite horizons used to delineate the area of undisturbed microlaminated lithosome.

SE and greater development of deep lake facies may be reasonably predicted in this direction.

Palaeosalinity determinations

Due to the importance of water salinity levels in influencing lacustrine bioproduction and subsequent organic preservation levels (Kelts this volume), estimates of palaeosalinity were made for the Orcadian Basin fish bed laminites using general lithological features and geochemical parameters.

Carbonate chemistry

Dolomite and calcite, in ferroan and less common non-ferroan species, are the predominant carbonate types within fish beds. Studies of microtextures using stained thin section, cathodoluminescence (CL) and scanning electron microscopy (SEM) (Hamilton, unpublished data), provide evidence for a primary and/or early diagenetic origin for the dolomite. Examples from laminites, where the fine interlayer distribution of dolomite concentrations and the dolomite/calcite ratio are preserved within 10% variance over distances of 50–100 m, and in one case 7 km (Donovan 1971; Hamilton, unpublished data), also support an early origin.

Stable isotope analyses of over a hundred samples from the fish beds which contain coexisting calcite–dolomite pairs reveal a consistent $\delta^{18}O$ enrichment within the dolomite fractions. Overall, the calcites range from around -9.0 to $-2.0‰$ and the dolomites from -6.0 to $+1.0‰$. Additionally, the $\delta^{18}O$ dolomite–calcite lies predominantly within the range of $+3-4‰$. Land (1980) found that the value of $\delta^{18}O$ dolomite–calcite formed in isotopic equilibrium under sedimentary temperatures should be $+3\pm1‰$ and McKenzie (1981), showed that for naturally occurring dolomite–calcite mixtures the value is $+3.2‰$. This overall isotopic evidence is supported by vertical isotopic profiling through a carbonate laminite fish bed using a fine 5 mm or 10 mm sampling interval (Fig. 4). In this example $\delta^{18}O$ and $\delta^{13}C$ enrichment cycles are sympathetically related to discrete mass fish mortality horizons and primary/early diagenetic chert nodule development.

The early dolomite implies that Mg concentrations within the depositional waters were at, or approaching, saturation with Mg/Ca > 2 (Muller et al. 1972) and they thus forced a primary dolomite precipitation. Given the limited documentation of this phenomenon in 'normal' modern lacustrine environments, any primary precipitate was more probably a high-Mg calcite or aragonitic precursor which was rapidly converted to dolomite during early diagenesis. Any additional Mg would be provided through clay breakdown reactions (McHargue & Price 1982) or decomposition of chlorophyll-bearing organic compounds (Desborough 1979). Such situations exist today only in lakes with high salinities (Callander 1968; Muller 1971; Degens et al. 1973; Botz & von der Borch 1984).

In the context of typical $\delta^{18}O$ values for carbonates and cherts from sedimentary rocks (Fig. 5), the Orcadian Basin carbonates, which lie predominantly within the range -7.5 to $0‰$, overlap both continental and marine carbonates. This is consistent with carbonate evolution from evaporitically-modified (saline?) continental groundwaters.

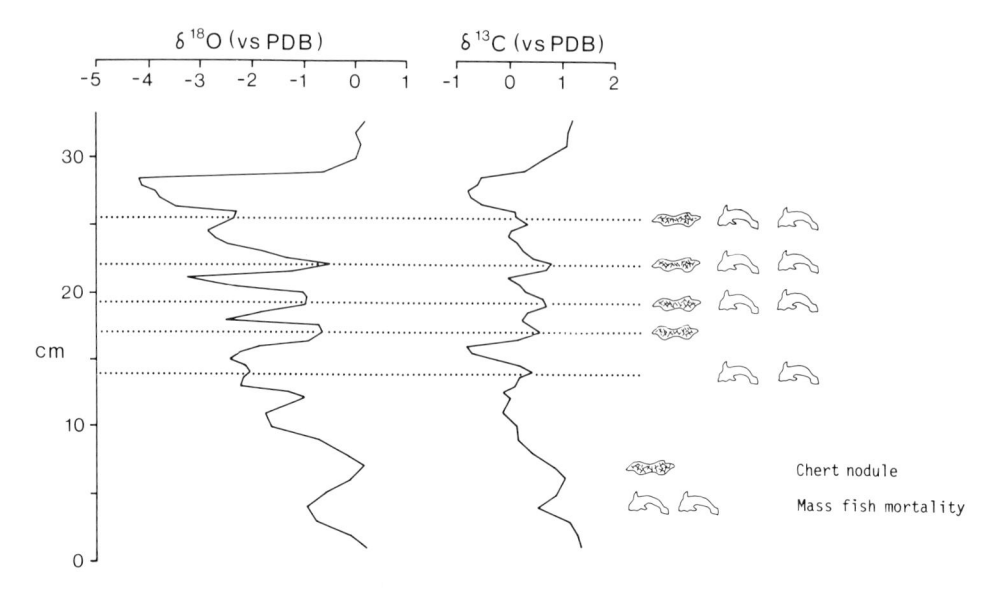

FIG. 4. $\delta^{18}O$ and $\delta^{13}O$ carbonate isotope profile through an Upper Caithness Flagstone Group fish bed demonstrating the sympathetic relationship between carbonate composition and primary depositional features; mass fish mortality horizons and chert nodule development.

Synaeresis cracks

Subaqueous shrinkage cracks, which upon compaction produce fang structures (Donovan & Foster 1972), are ubiquitous within the flagstone successions of the Orcadian Basin. 'Fangs' commonly occur within the shale and laminated siltstone beds of the B and C divisions of Donovan (1980) and rarely within fish bed laminates with which they exhibit gradational contact. The production of synaeresis cracks is dependent upon the presence of a swelling clay component within the sediment layer being cracked, and most importantly upon saline water influx (Jungst 1934; Burst 1965; Donovan & Foster 1972).

Chert nodules

Nodular chert, in the form of bedding parallel lenticles and discrete oblate nodules, is sporadically developed within laminite horizons in the Orcadian Basin. Textural relationships show that the host sediment has commonly been squeezed and deformed plastically around the nodule margins suggesting a primary/early diagenetic chert origin. A Magadi-type model has been proposed for the chert (Parnell this volume); essential to this model are elevated Na concentrations within the depositional waters to allow the formation of a Na-silicate precursor mineral.

Other examples of inorganic chert formation under sedimentary conditions (e.g. Peterson & von der Borch 1965) are recorded mostly from brackish–saline environments with alkaline affinities.

C/S ratios

The ratio of organic carbon to reduced sulphur can be used to distinguish freshwater from marine sediments and also saline (high sulphate) from non-saline lacustrine phases (Berner & Raiswell 1984). This method relies upon the fact that much lower concentrations of dissolved sulphate characterize fresh waters compared to marine or saline lake environments. It assumes that all available sulphate within a particular salinity environment is reduced bacterially to produce an amount of sedimentary pyrite which is environment-specific for the corresponding organic enrichment. In this context, Davison (this volume) has demonstrated experimentally that the formation and accumulation of sedimentary pyrite, although far more complex than originally postulated by Berner and Raiswell (1984), is essentially controlled by the sulphate concentration of the primary depositional waters.

C/S ratios were obtained from fish bed laminite samples which contained a ferroan carbonate species and pyrite as the sole sulphide mineral, most commonly occurring as microscopic framboidal aggregates within organic layers. For

OXYGEN ISOTOPES

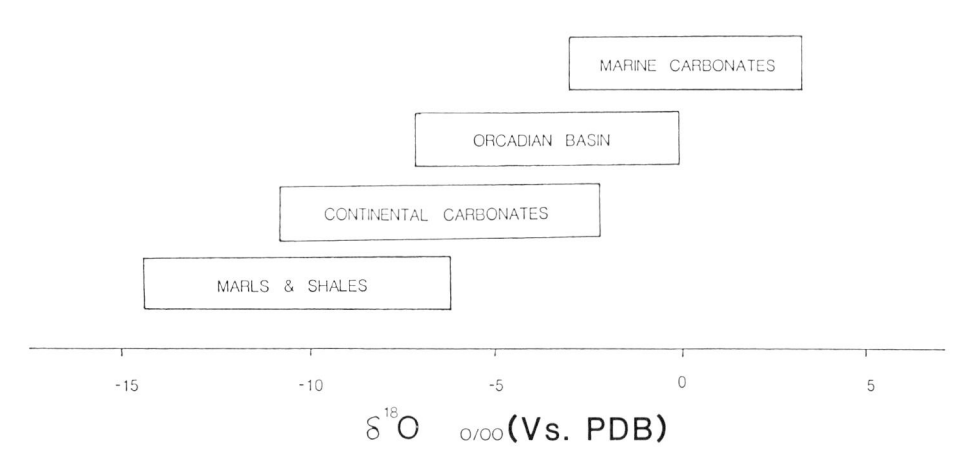

FIG. 5. Typical range of $\delta^{18}O$ values from sedimentary rocks including Orcadian Basin carbonates (Keith & Weber 1964; Hudson 1977; Margaritz & Gat 1981).

comparison C/S data sets are presented for modern freshwater lakes (Berner & Raiswell 1984) and two other ancient lacustrine sequences from similar settings to the Orcadian Basin (Fig. 6); the Upper Triassic Lockatong Formation of the Newark Supergroup (supplementary Newark data kindly supplied by Paul Olsen) and the Laney Shale Member of the Eocene Green River Formation.

Compared to the trend of C/S ratios obtained for modern freshwater sediments, the Orcadian Basin laminites exhibit marked S enrichment (up to 1%) particularly in the 1-2% TOC range. The Orcadian data is distinct from those for the Green River Laney which, despite the relatively small data set, are all consistent with a low sulphate freshwater origin. They are more similar to those for the Newark Basin data. Indeed the Lockatong sequence of the Newark Basin also contains synaeresis cracks, dolomite, gypsum, glauberite, halite pseudomorphs and abundant syngenetic analcime, the sodium-rich zeolite (Van Houten 1964).

The Orcadian Basin laminite trend is therefore interpreted as resulting from deposition within an environment where sulphate levels within the water column were consistently elevated compared to normal freshwater. Leventhal (1983) has interpreted 'excess' sulphide trends as a product of water column iron-sulphide formation supplementing sediment pyrite formation. We suggest that the limitation of organic enrichment to values of less than ca. 5.0% organic carbon could be due to a restricted allochthonous organic input related to a sparse, poorly diversified Devonian land flora, and/or high rates of bacterial degradation of organic material within near-surface sediments thus reducing the preservable organic enrichment. This may be further supported by the low C/S values within the 1-2% TOC range which perhaps reflect a tendency towards an end point of bacterial turnover reactions due to some physical limitation such as top sediment permeability.

Discussion

The complementary results from the palaeosalinity indicators used here support the model of a saline-brackish water closed lake system. The fact that the data set was gathered from all stratigraphic levels within the Middle Devonian lacustrine sequence suggests that this general depositional regime prevailed over a long time

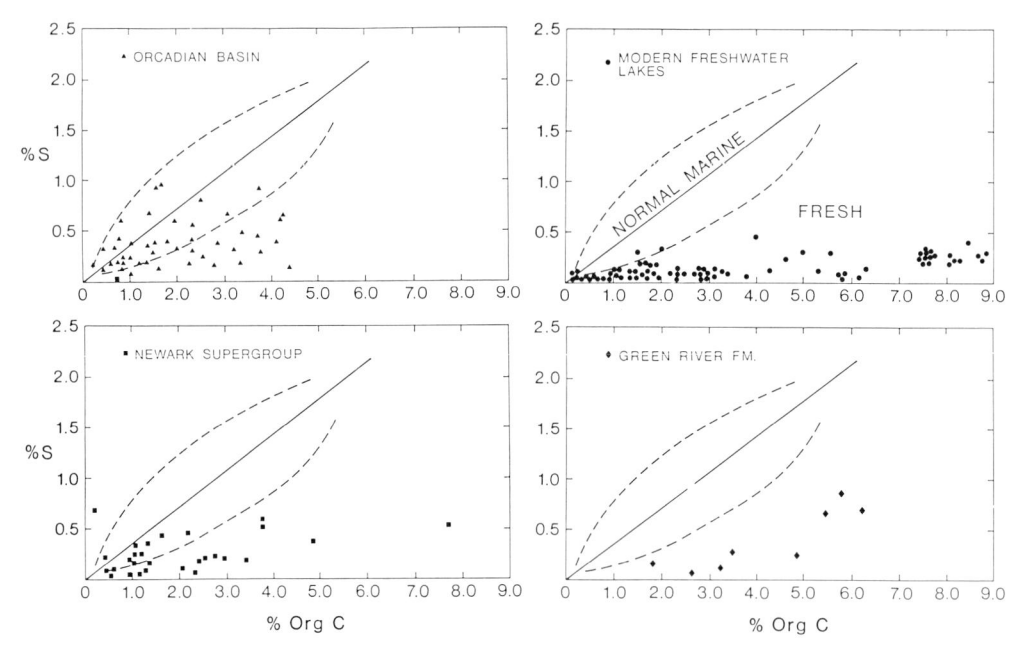

FIG. 6. C/S ratios from the Orcadian Basin, Newark Supergroup of the Newark Basin, Laney Member of the Green River Formation, and modern freshwater lakes (sources as in text).

period. To sustain these levels of ionic concentration a steady solute inflow must have occurred. Given the inferred relative ionic abundances it seems most probable that the inflow was strongly influenced by the weathering of the thick Cambro–Ordovician shelf carbonate succession presently cropping out in the NW Highlands. Direct chemical weathering of dolomitic limestones and evaporite facies beds (Allison & Russell 1985) from the Durness Group and the Fucoid Beds could provide inflow solutions with $Ca \simeq Mg \gg Na, K$ and both bicarbonate and sulphate anions.

Solutes may have also been periodically acquired through volcanic activity from direct fallout of airborne ejecta i.e. ash and gases. Groundwaters from the weathering of lava piles around the Orkney and Shetland volcanic centres may also have contributed to the inflow. As a consequence of the limited Devonian terrestrial vegetation cover capable of binding and fixing sediment, the quantity of dust held in the atmosphere must have been greatly increased. As such, the process of solute acquisition from rainfall after atmospheric dust dissolution should also be considered.

The evolution of Orcadian Basin inflow waters would have been first influenced by the early precipitation of insoluble carbonates, as is supported by the basin margin limestones and openwork cements of marginal alluvial screes which are predominantly low-Mg, non-ferroan calcite. Subsequent precipitation of high Mg calcite and dolomite formation, as already described, would have left brines enriched in more soluble salts such as Na^+ and K^+, which require, by modern analogue further concentration factors in the order of 25 to 250 (Hardie 1968; Jones et al. 1977) before direct precipitation of evaporite minerals such as halite or trona will occur. This is consistent with the extremely rare occurrence of 'evaporite' minerals in the Orcadian Basin deep lake laminites.

The absence of gypsum is noteworthy, especially considering the relatively high (ca. 1%) S concentrations as pyrite in the anoxic 'fish bed' sediments. By analogy to the explanation given by Neev and Emery (1967) for the Dead Sea sediments, the absence of gypsum is attributed to bacterial reduction.

The general absence of evaporite minerals in the playa lake facies desiccated beds which occur in deep stratified lake–playa lake sedimentary cycles throughout the Flagstone Group, requires a more involved model. It is possible that any evaporite mineral accumulation during a desiccation phase took the form of thin crusts disseminated through the top sediment; subsequent lake transgression and re-wetting then caused evaporite dissolution thus removing any trace. The rarity of pseudomorphs and sediment collapse structures in these playa lake beds negates the importance of this process.

It is also possible that lake system salinity levels remained critically buffered by ionic discharge via runoff to a possible marine connection to the SE (Pennington 1975), which would have made the Orcadian Basin a restricted open lake system. This model however requires a careful balancing of lake levels and corresponding outlet point datum levels.

We therefore postulate that an evaporite facies, forming from brine concentration during desiccation phases, was developed in parts of the Orcadian Basin central area (Fig. 7), presently located in the offshore northern North Sea area. Such concentric facies zonations have been recorded from numerous other modern and ancient saline lake examples (Eugster & Hardie 1978). This model is consistent with the sedimentological data which suggest that the deepest area of the Orcadian Basin existed to the east of the present outcrop. The composition and effect of such basin centre brines on organic preservation may be important; a sulphate-rich brine forced down through the sediment pile by capillary action pumping would allow bacterial degradation to occur at greater depth within the sediment, thus reducing organic enrichment in all but the most impermeable source rock units.

The organic geochemical approach

In recent years, organic geochemistry has become an important exploration tool, with source-rock characterization of samples cropping out on the flanks of a sedimentary basin representing a particularly useful approach. This section of the study therefore addresses the questions of the type, amount and maturity of the organic matter (OM) present in the Middle Devonian laminites of the Orcadian Basin. A variety of complementary analytical techniques have been used with considerable emphasis being placed on biological marker data. Mackenzie (1984) indicated that specific molecular compound classes can be used to assess:

(a) the maturity of a crude oil or potential source rock,
(b) the extent of biodegradation and migration of subsurface hydrocarbon accumulations, and
(c) genetic relationships between families of crude oils and potential source beds.

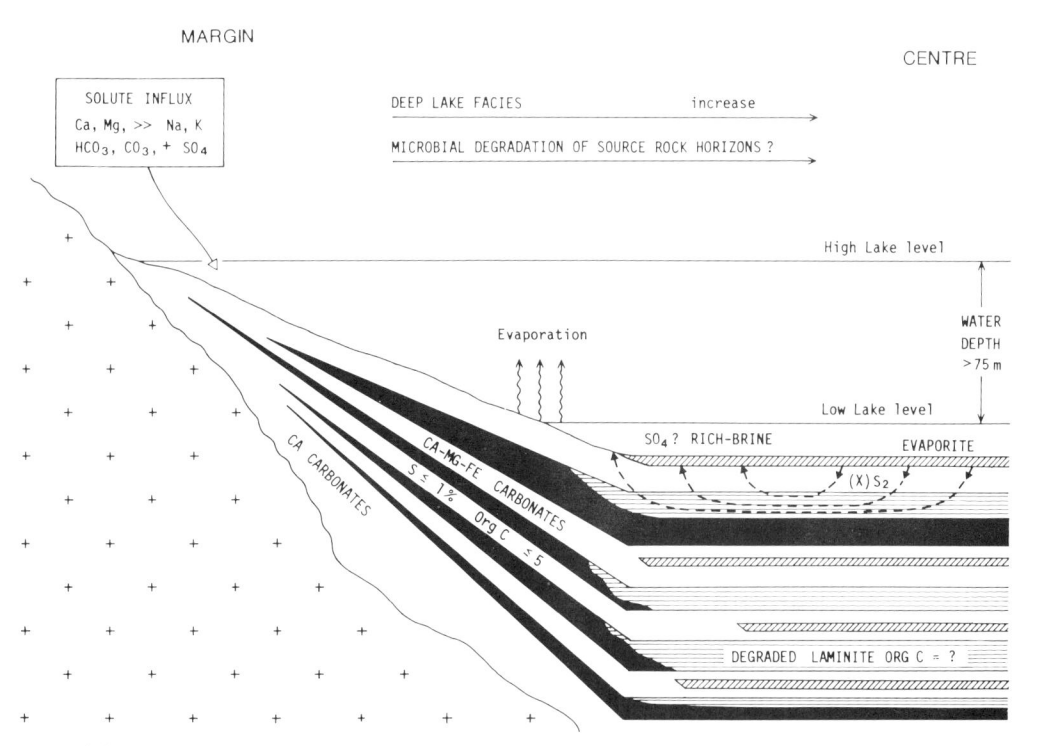

FIG. 7. Schematic model which shows the depositional and preservational features of fish bed laminite (source rock) horizons within the Orcadian Basin.

Analytical procedures

A comprehensive account of experimental methods is given in Duncan (1986). However, preparation, fractionation and analytical techniques employed throughout are directly analogous to those referenced in e.g. Mackenzie *et al.* (1984).

Bulk geochemistry

Ten samples of Eifelian to Givetian age, representative of the Middle Devonian lacustrine laminites, were taken from five stratigraphic horizons around the Orcadian Basin (Fig. 8). The bulk geochemical and lithological data presented in Table 1 show that the lithological variation within the sample set is slight (with the exception of RH) and suggests only limited variation in kerogen type. The present day relative stratigraphic positions of each sample are indicated by a depth index (1=highest, 5=lowest). Both marginal and more distal basin sediments were analysed. Despite the variations noted in Table 1, relationships between lithology, organic content and hydrocarbon yield are not apparent, and

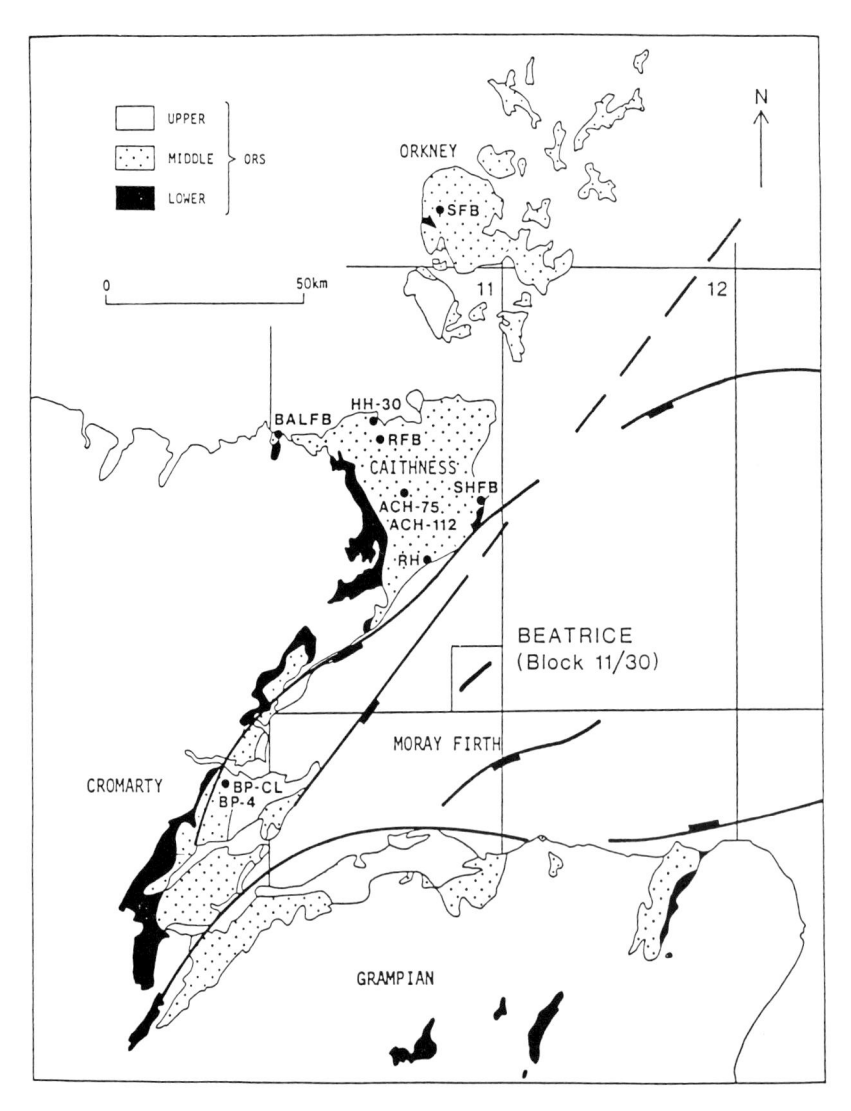

FIG. 8. Distribution of Lower, Middle and Upper Devonian sediments in northern Scotland, and the location of the Beatrice oilfield.

TABLE 1. *Lithological and bulk geochemical parameters*

Depth index and locality	Sample	Lithology	Carbonate (wt %)	TOC[1] (wt %)	TSE[2] yield (ppm)	Bitumen ratio[3] (mg/g)	Asph.[4] (wt %)	Spore[5] colour (7-point scale)	R_0 equivalent (calculated)
1. Robbie's fish bed	RFB	Carbonate laminite* micronodular fabric	46	3.60	3148	87.4	<1	4–5	0.50–0.75[+]
2. Holburn Head	HH-30	Organic laminite	34	3.82	3536	92.6	<1	3/4–4	0.40–0.55[+]
Black Park (Edderton)	BP-CL	Carbonate laminite micronodular	64	1.03	311	30.2	3.1	3	0.4
	BP-4	Detrital laminite, micaceous	12	1.00	362	3.6	1.9	3	0.4
Baligill	BAL FB	Carbonate laminite micronodular	73	0.85	101	12.0	3.2	3–3/4	0.55
3. Sandwick	SFB	Carbonate laminite micronodular	38	1.56	1586	102.0	1.7	—	—
Achanarras	ACH 75/7	Detrital laminite	27	1.58	602	38.1	6.2	4/5–5	—
	ACH 112-4	Carbonate laminite	33	1.30	414	31.8	10.8	4/5–5	0.75[+]
4. Robbery Head	RH	Micaceous bituminous laminite, slickensided	27	2.11	628	29.7	3.0	—	—
5. South Head	SHFB	Carbonate laminite	65	0.88	26	2.9	5.3	—	—

* Laminite refers to dark fine-grained lacustrine sediments with microlaminated (0.1–1 mm) fabric, typically comprising a tripartite association of carbonate, detrital and organic matter layers.
[1] Total organic carbon. [2] Total soluble extract. [3] TSE/TOC. [4] Asphaltene yield. [5] Batten (1981).

the bitumen ratio does not conform to an increasing, depth-related trend.

Transmitted light microscopy reveals limited amounts of terrestrially-derived cuticular material and palynomorphs, consistent with the restricted vegetation cover during Middle Devonian times. The bulk of each of the preparations comprises amorphous organic matter (AOM) of probable algal origin, thus constraining the kerogen to Types I or II. Rock-Eval data were not available for this sample set. However, hydrogen indices from the data of Parnell (1985), display non-maturity related variations suggesting changes in kerogen type for contemporaneous sediments. A similar scatter of data points was also noted by Marshall et al. (1985) for Middle Devonian Orcadian Basin laminites and confirms the limitations of Rock-Eval pyrolysis for typing organic matter, as noted by Katz (1983). The conclusion of Parnell (1985), that a number of samples show good source potential, is confirmed by the bitumen extract and TOC data presented in Table 1. Samples RFB and HH-30 represent particularly good source rocks (high yield, high TOC), with the Achanarras and SFB samples also showing good source potential. Generally low asphaltene yields are consistent with maturity estimates derived from molecular geochemistry. Spore colouration data for selected samples places the Achanarras material and RFB (Table 1) close to the principal zone of oil generation (e.g. Batten 1981; Tissot & Welte 1984) although, again, no consistent depth-related trend is apparent. Hillier & Marshall (this volume) show that conflicting indices, hence of doubtful validity, may be obtained using this technique. We ascribe these variations to matrix effects.

Molecular geochemistry I

The results of gas chromatographic analyses of the ten samples are summarized in Table 2, whilst representative gas chromatograms are shown in Figs 9 and 10. An arbitrary subdivision into types A_1 and B_1 was made according to the principal features of the resulting C_{15+} distributions. Both categories describe unimodal, low-molecular-weight skewed aliphatic hydrocarbon profiles, whilst the essential differences are listed in Table 3. Minor internal variations in these distributions were designated A_{11} and B_{11}. For example, the high molecular weight carotenoid components characteristic of type B distributions represent the fully saturated, geologically-stable equivalents of the ubiquitous naturally occurring pigments. β-carotane is most commonly attributed to an algal source (Tissot & Welte 1984) and has been reported from a variety of lacustrine sediments including the Middle Devonian of the Orcadian Basin (see Hall & Douglas 1983; Jiang Zusheng & Fowler 1985 and references therein).

The smaller peak eluting before β-carotane has an identical mass spectrum with that for γ-carotane as indicated by Jiang Zusheng & Fowler (1985). However, maturity ratios calculated using the γ/β-carotane index proposed by these authors do not parallel the data from more widely used maturation parameters (e.g. Mackenzie et al. 1980). All distributions are consistent with a substantial algal/bacterial input as indicated by common maxima in the n-C_{17} to n-C_{19} region, although further differentiation is not possible owing to the lack of substantial n-alkane predominance in this region.

The low concentration of alkanes above n-C_{25}

TABLE 2. *Molecular geochemical parameters I*

depth index	Sample	Distribution type	n-alkane maximum	Pristane[a] / Phytane	Pristane[b] / n-C_{17}	$^{CPI}(C_{25-33})$, $(C_{17-33})^a$	β-carotane	$\frac{\gamma}{\beta}$carotane[a]
1*	RFB	B_1	C_{19}	1.39	0.80	1.55, 1.03	++	0.25
2	HH-30	B_{11}	C_{20}	1.49	0.57	1.08, 1.12	+++	0.34
	BP-CL	A_{11}	C_{21}	2.08	0.30	1.08, 1.12	—	—
	BP-4	A_{11}	C_{21}	2.00	0.26	1.12, 1.25	—	—
	BAL FB	A_1	C_{19}	0.94	0.17	1.01, 1.02	—	—
3	SFB	B_{11}	$C_{17}=C_{19}$	1.44	0.46	0.97, 1.02	+	0.42
	ACH 75/7	B_1	C_{17}	1.08	0.70	1.13, 1.08	++++	0.23
	ACH 112-4	B_1	$C_{17}=C_{19}$	1.04	0.91	1.10, 1.09	++	0.27
4	RH	A_1	C_{17}	1.32	0.24	1.03, 1.01	—	—
5	SHFB	A_1	C_{19}	0.70	0.32	1.03, 1.05	—	—

*Indicates relative stratigraphic position (depth index).
[a] Calculated from peak heights on gas chromatograms.
[b] Calculated from m/z 85 traces.
+ Indicates relative abundance.

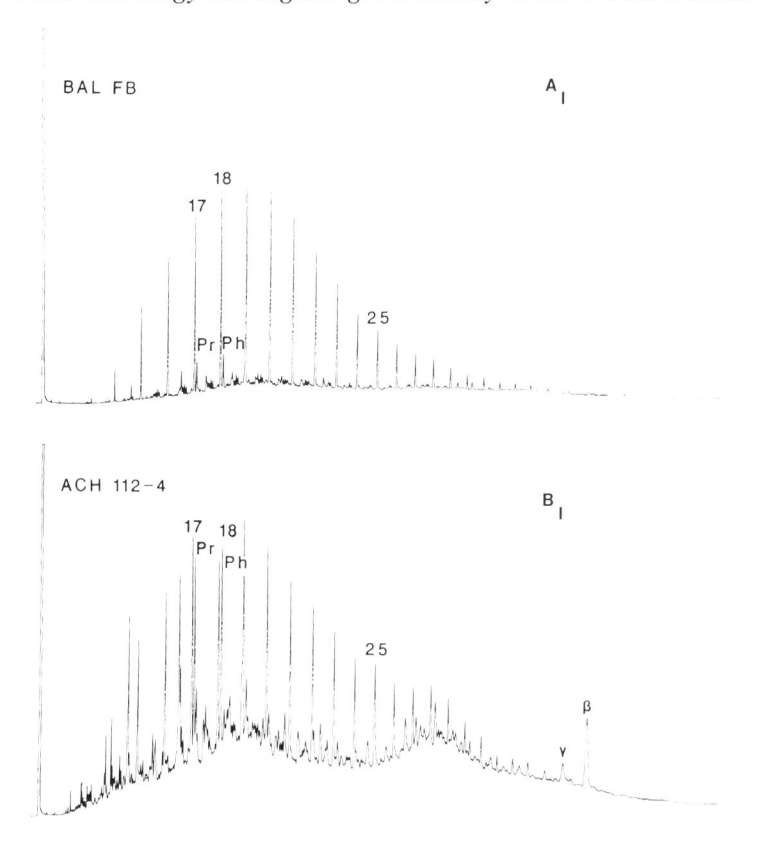

FIG. 9. Gas chromatograms of saturates distributions A_1 and B_1. Peaks labelled according to carbon number: Pr = pristane, Ph = phytane, γ, β = γ- and β-carotane, respectively.

and lack of odd-over-even predominance (OEP) in most samples is considered a function of maturity and the lack of terrigenous input available for autochthonous microbial reworking. The n-C_{21} maximum and 'stepped' n-alkane profile common to the Blackpark samples may reflect a specific organic matter input from an unknown source, or possibly bacterial reworking. Despite their fresh petrographic appearance and collection from quarried rock section at up to 5 m from the original outcrop surface, organic weathering of these samples cannot be discounted. However, chromatographic evidence suggests that it has not been pervasive.

Pristane/phytane (Pr/Ph) ratios are subject to a variety of interpretations (e.g. Philp 1985), on account of their susceptibility to a variety of geological processes. Although both of these lipids are now known not to be exclusively derived from chlorophyll (e.g. Goossens et al. 1984), Pr/Ph ratios remain most commonly applied as environmental parameters (e.g. Didyk et al. 1978). Values for these samples (Table 2) indicate deposition in an anaerobic environment, below a fluctuating oxic/anoxic interface in the water column (e.g. Didyk et al. 1978). Differences in the isoprenoid/n-alkane ratios, expressed by Pr/n-C_{17} and Ph/n-C_{18} ratios (Tables 2 and 3), can also assist in the classification of aliphatic hydrocarbon distributions. Possible interpretations of these are described by Philp (1985; see also Leythaeuser et al. 1984).

Variations in the Pr/n-C_{17} ratio observed in this sample set are attributed to non-specific source effects, although all values are consistent with deposition in open water conditions (Didyk et al. 1978). It is of interest to note that the presence of β- and γ-carotane is concurrent with the higher Pr/n-C_{17} and Ph/n-C_{18} ratios, although a direct reason for this empirical relationship is not known. However, these B_1- and B_{11}-type distributions also contain prominent unresolved complex mixtures (UCMs) of naphthenic compounds which may imply impregnation with

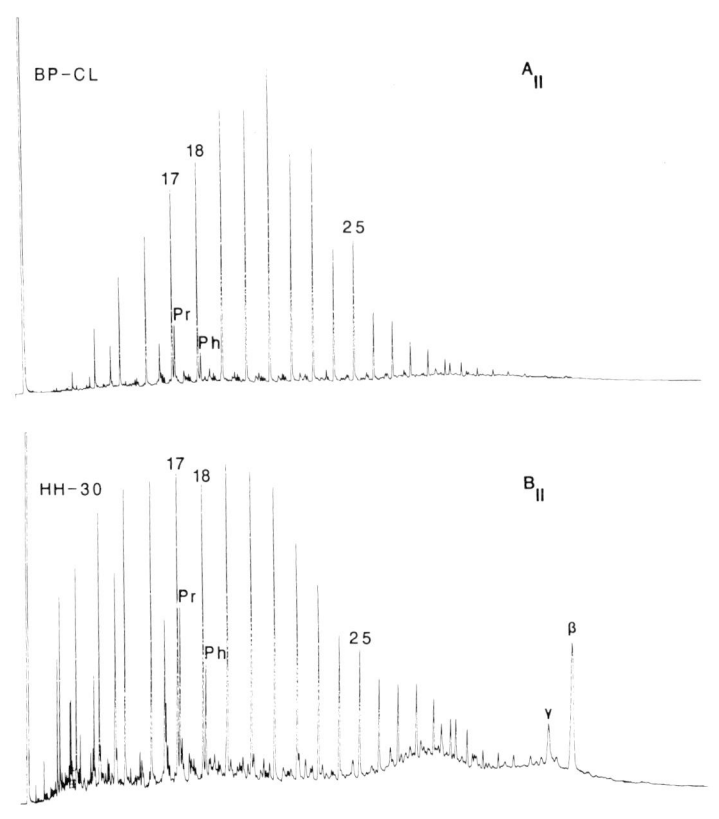

FIG. 10. Gas chromatograms of saturates distributions A_{11} and B_{11}; annotation as for Fig. 9.

migrated petroleum (e.g. Simoneit 1978); these samples also generally display the highest bitumen ratios.

The isoprenoid distributions for classes A and B are also in marked contrast to each other.

Retention time and mass spectral data suggested that the branched acyclic compounds most prominent in type B_1 samples, belonged to the regular head-to-tail linked series (e.g. Holzer *et al.* 1979), although lack of molecular ions pre-

TABLE 3. *Parameters used for differentiation of chromatogram types*

	Type		Type	
	A_1	A_{11}	B_1	B_{11}
β-,γ-carotane	—	—	+	+
Pristane/n-C_{17}	<0.40	<0.40	<0.65	0.40–0.65
Phytane/n-C_{18}	0.10–0.14		0.65–1.00	0.30–0.40
Regular acyclic isoprenoids	absent or in very low abundance		abundant	
UCM*	absent		prominent low and high mol. wt. naphthenic humps	high mol. wt. hump only
Distribution	C_{17}, C_{19} max; smooth profile	C_{21} max; stepped profile	ramped, low mol. wt. profile C_{17}–C_{20} max	more abundant low mol. wt. alkanes

*Unresolved complex mixture.

cluded their unequivocal assignment. These compounds, which signify an origin from methanogenic archaebacteria, have been previously noted from the Officer Basin carbonates, South Australia, by McKirdy et al. (1985), and from certain lacustrine sediments in China (Fu Jiamo et al. this volume).

Molecular geochemistry II

Biological marker compounds, i.e. those reflecting an unambiguous link with a natural product or precursor, were present in most cases in rather low abundance; this supports other maturity determinations made on these samples. A number of conventional biomarker ratios which reflect maturation and source differences (e.g. Mackenzie 1984 and references therein) were calculated from the sterane and triterpane distributions. The results of these molecular geochemical parameters are tabulated in Table 4.

Configurational changes at C-22 and stereochemical changes at C-17 and C-21 for the C_{32} and C_{30} pentacyclic hopanes respectively, indicate that in both cases, isomerization reactions involving these molecules, are complete (e.g. Seifert & Moldowan 1980). On this basis, the maturity of the samples can be classed as, at least, early catagenetic. (It is not known if some of the higher values in Table 4 for the first ratio are distorted by coelution, or if they are artefacts of the quantitation programme.)

Although McKirdy et al. (1983, 1985) and Rullkotter et al. (1985) noted that low concentrations of the 18α(H)-trisnorneohopane (T_s) appeared characteristic of carbonate source rocks, a rise in the T_s/T_s+T_m ratio with increasing maturity may still be expected (e.g. Seifert & Moldowan 1978). The carbonate content of these samples is extremely variable (12-73%; mean 42%), and the uniformly low values for this ratio (<27%) obtained by the above authors were exceeded in each case for the Orcadian Basin samples. This feature and the lack of a maturity trend, may imply that source/environmental control in this case is more important than the nature of the inorganic matrix.

A representative distribution of the bacterially-derived hopanes is shown in Fig. 11. In all but one example C_{29} (αβ)-norhopane was subordinate to C_{30} (αβ)-hopane, and the $C_{32}+$ homologues are present in extremely low abundance. Fragmentograms for m/z 191 and m/z 177 (not shown) indicate that the C_{28} (αβ)-bisnorhopane, a characteristic component of many North Sea crude oils (Mackenzie et al. 1984) and also prominent in many Upper Jurassic source rocks from the Moray Firth Basin (Duncan, 1986), is absent from these Devonian sediments. It is thought that peaks eluting in the C_{27} and C_{28} region may represent higher molecular weight members of the tricyclic and tetracyclic terpane series (e.g. Hall & Douglas 1983).

A substantial variation in both the internal distribution and relative concentration of tricyclic triterpanes with respect to the pentacyclic triterpanes was observed, and is shown in Fig. 12. These compounds are also thought to have a microbial or possibly algal origin. Seifert & Moldowan (1978, 1981) and Palacas (1984) considered the tricyclic components as powerful source-specific parameters, and based on both visual inspection of m/z 191 fragmentograms and quantitation of their abundance relative to the hopanes, were able to distinguish oil families and establish oil-source correlation. However, Aquino Neto et al. (1983) showed that the relative abundance of tri- and tetracyclic triterpanes increased with increasing thermal maturation.

In this case, source differences are believed to exert the principal control over the qualitative and quantitative variations noted, with the possible exception of RH (localized heating due to slickensiding?—see Table 1), as there appears to be no correlation between both the ratio of tricyclic triterpanes to pentacyclic hopanes, the internal C_{23}/C_{21} ratio proposed by Ekweozor & Strauss (1983), and other molecular geochemical maturation parameters. The rather limited lithological variations and algal-dominated chromatograms documented in Tables 1 and 2 indicate that the source differences may be secondary, i.e. related to depositional environment, such as salinity, water temperature and pH, rather than a reflection of variation in the primary organic input. Figure 14 shows the distributional variation recorded from the ten samples. Investigation of better characterized, predominantly marine Jurassic source rocks from the Inner Moray Firth Basin shows (Duncan 1986) that tricyclic triterpane distributions are dominated, for the greatest part, by the C_{23} component, and resemble the SHFB example (Fig. 13).

Further source and maturity information is provided by an examination of the sterane distributions using m/z 217 and m/z 218 fragmentograms (Figs 14 and 15).

Primary sterol input is of predominantly algal origin (e.g. Brassell et al. 1983); isomerization reactions within the most thermodynamically stable of these compounds, the steroidal alkanes, reflect their response to thermal stress. These include the reactions involving configurational changes at the C-20 chiral centre and stereochemical changes at positions C-14 and C-17. These record the thermal transformations of the mole-

TABLE 4. Molecular geochemical parameters II

Sample	Hopanes[1] $C_{32}\alpha\beta \dfrac{22S}{22S+22R}$[a]	$C_{30} \dfrac{\alpha\beta}{\alpha\beta+\beta\alpha}$[b]	$\dfrac{T_s}{T_s+T_m}$[c]	$\dfrac{C_{29}}{C_{30}}$[d]	Tricyclics[2] $\dfrac{C_{23}}{C_{21}}$[e]	$\dfrac{C_{21}+C_{23}}{C_{30}}$[f]	$C_{29}\alpha\alpha\alpha \dfrac{20S}{20S+20R}$[g]	Steranes[3] $C_{29}\dfrac{\alpha\beta\beta}{\alpha\alpha\alpha+\alpha\beta\beta}$[h]	$C_{27}:C_{28}:C_{29}$[i] ($\alpha\beta\beta 20R+20S$)	Hopane[j]/Sterane
RFB	61	84	31	0.73	0.80	0.70	53	54	9:52:39	2
HH-30	64	86	30	0.65	0.82	0.45	57	50	10:48:42	4
BP-CL	62	85	33	0.57	0.89	0.56	39	nd₁	†32:30:38	13
BP-4	64	81	45	0.51	0.88	0.58	nd	nd	†35:37:28	18
BAL FB	60	90	34	0.83	0.80	1.07	36	63	37:28:35	7
SFB	71	93	42	0.67	1.14	1.85	40	71	8:54:38	2
ACH 75/7	67	94	47	0.72	1.17	1.02	56	66	9:51:40	1
ACH 112-4	68	90	51	0.69	1.05	1.61	57	69	7:55:38	1
RH	nd	nd	nd	nd	0.72	nd	30	66	9:59:32	nd
SHFB	65	86	41	1.24*	2.21	2.75	nd	nd	nd	nd

Measurements a–c, g–i expressed as percentages.
nd—not determined, insufficient concentration of components.
nd₁—coelution problems.
[1,2]—measured from assigned peak areas in m/z 191 fragmentograms.
[3]—measured from assigned peak areas in m/z 217 and m/z 218 fragmentograms.
*—possible coelution.
†—diasteranes present.

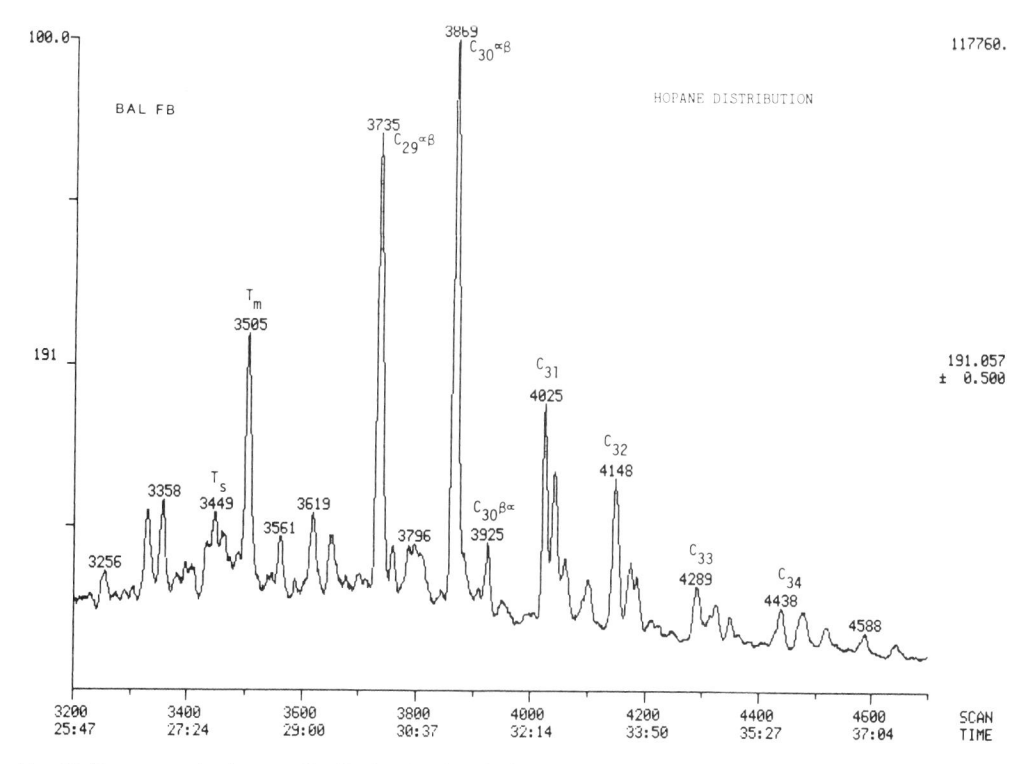

FIG. 11. Representative hopane distribution (peaks labelled as in text; C_{32} isomers labelled S and R) from m/z 191 mass fragmentogram.

cules, from bio- to geolipids. The first of these, recorded by the C_{29} $\alpha\alpha\alpha$ 20S/20S+20R sterane ratio (e.g. Mackenzie et al. 1984), is the most maturation-specific and kinetically ideal of the aliphatic biomarker ratios, the reaction end point ($\sim 55\%$), being coincident with peak generation (e.g. Mackenzie 1984). The spread in maturities recorded here is inconsistent with their present day stratigraphic level, but the Achanarras, HH-30 and RFB samples may be ranked as extremely mature.

The $\alpha\beta\beta/\alpha\alpha\alpha + \alpha\beta\beta$ ratio, measured for C_{29} steranes also shows considerable variation (Table 4). Although the measurements signify that all samples have obtained a substantial degree of maturity (reaction end point $\sim 80-85\%$), the general lack of parallelism with the 20S/20S+20R ratio indicates that this interpretation is oversimplistic. The problematic nature of this measurement is discussed by Mackenzie (1984).

Rullkotter et al. (1985) considered investigation of steroid aromatization essential in corroborating maturity assessments of carbonates based on saturated biomarker ratios. However, the extents of hopane and sterane isomerization recorded here (Table 4) are consistent not only with other molecular and bulk geochemical parameters listed in Tables 1, 2 and 4, but also the age and inferred burial history of these sediments (Hamilton & Trewin 1985); together they confirm that several of the onshore samples lie within the oil window.

The hopane/sterane ratio is largely maturity independent, ostensibly reflecting the relative contribution of bacterial and algal lipid material incorporated into the sediment. However, the sterane concentration is a measure of the autochthonous eukaryotic algal production in the water column (Mackenzie et al. 1984), whilst the hopane concentration reflects not only the bacterial input to the sediment, but also the extent of microbial reworking of e.g. higher plant material. The effects of both primary and secondary microbiological processes are therefore inherent in this ratio. High values are believed to indicate a substantial terrestrially-derived component (e.g. Mackenzie 1984). In this sample set, the ratios recorded fall into three categories, with most samples having values of 4 or less. Values of 7.2, 12.9 and 17.5 were recorded for the Blackpark and Baligill samples (Table 4), while the extremely low concentration of hopanes and ster-

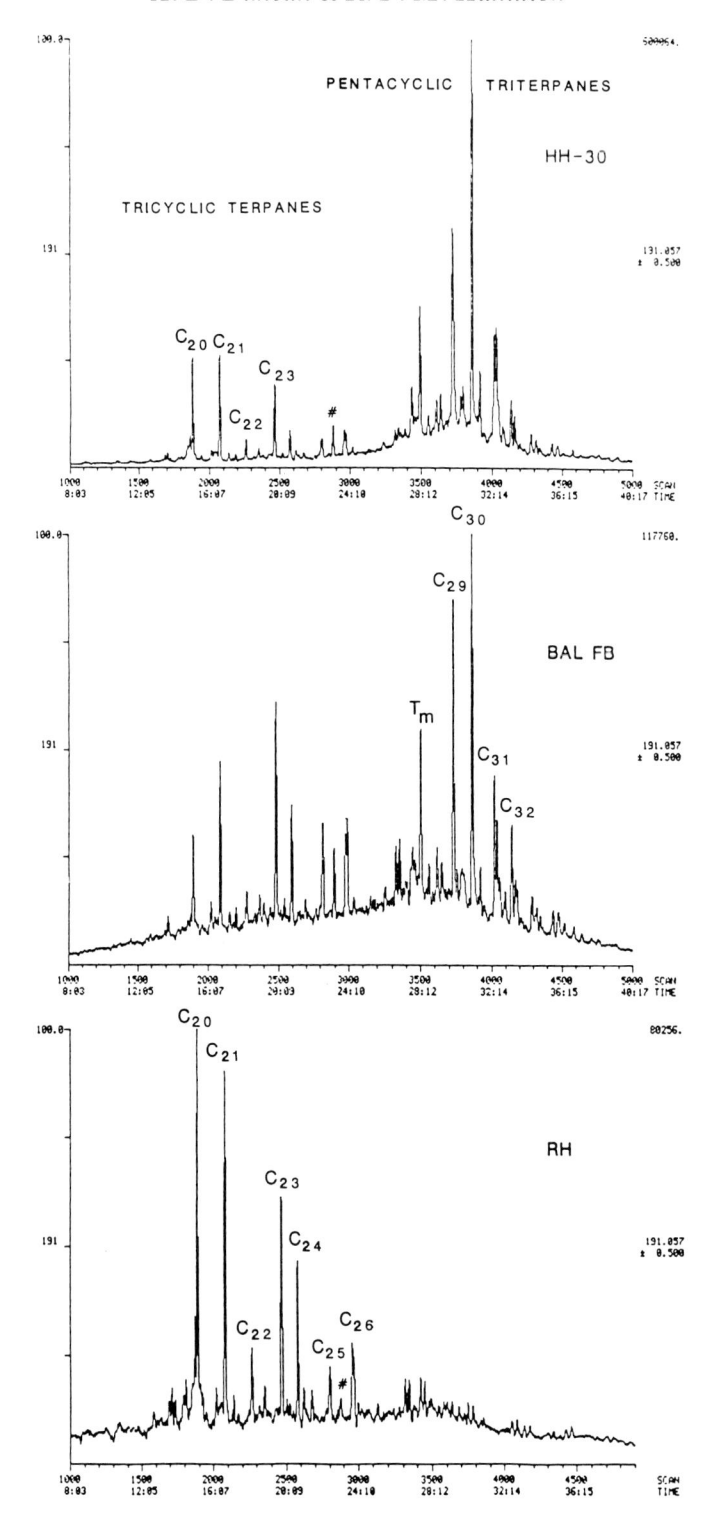

FIG. 12. m/z 191 mass fragmentograms documenting the extent of variation in the relative abundance of tricyclic to pentacyclic triterpanes for Middle Devonian samples; peaks labelled as in text.

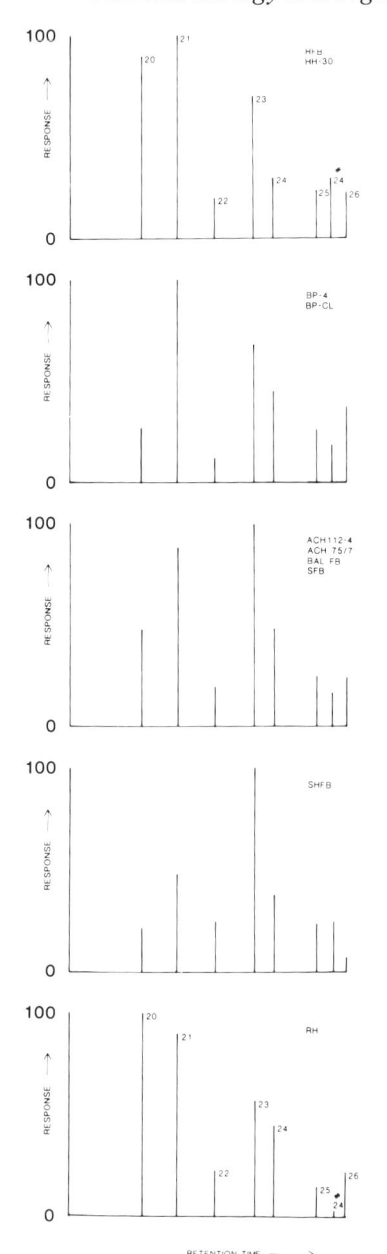

samples, consistent with a higher degree of terrigenous influx. Lower values for this ratio are restricted to those samples considered to reflect the deeper, more distal lacustrine facies. Several of the lower values appear to have marine affinities, based on the distributions presented by Mackenzie et al. (1984). However, McKirdy et al. (and reference therein, this volume) show a considerable spread in values for Australian lacustrine sequences of variable age and salinity, and populated by very different biotas.

Plotting the relative abundances of the regular C_{27}, C_{28} and C_{29} sterane components on a triangular diagram highlights source variation within a sample suite (e.g. Rullkotter et al. 1985; Sajgo 1984). By restricting the assessment to only one, or a pair of diastereomers, maturity effects resulting from isomerization of the complex mixture of components can be minimized (although see Shi Ji-Yang et al. 1982). The $\alpha\beta\beta$ R and S isomers plotted in Fig. 17 indicate that six of the nine samples for which the measurement was made, plot in a tight cluster, close to the C_{28}–C_{29} baseline, while three plot more centrally, thus indicating genetic differences between the Baligill/Blackpark group and the other samples. The unusual depletion in C_{27} sterane components (Fig. 15) was also recorded by Hall & Douglas (1983) for Orcadian Basin Devonian samples, & by McKirdy et al. (1985) for Australian carbonate-sourced oils.

Further internal variations in the sterane distribution are apparent from a consideration of the diasterane content of the samples (Table 4; e.g. Fig. 14). Diagenetic formation of these backbone rearranged steranes from regular steradiene precursors is well documented (e.g. Seiskind et al. 1979). The clay-catalysed protonation mechanism responsible, results in their ubiquitous distribution in clastic and argilaceous sediments (Rullkotter et al. 1985), while their absence is usually indicative of a carbonate-rich lithology (e.g. McKirdy et al. 1985; Palacas et al. 1985).

FIG. 13. Internal variation recorded in tricyclic triterpane distribution; peaks normalized to the most abundant component. (# : C_{24} tetracyclic triterpane.)

anes, in the Robbery Head and South Head samples precluded measurement of the ratio. As can be noted from Fig. 16 the higher hopane/sterane ratios are recorded for the more marginal

Discussion

The suggested source/environmental specificity of the tricyclic triterpane distribution does not appear to correlate with any other major source-related feature—overlapping trends are recorded. These sediments may perhaps be regarded as atypical in that only five of the samples contain a predominance of the C_{23} tricyclic component, the most common major peak (e.g. Aquino Neto et al. 1983; Palacas et al. 1985).

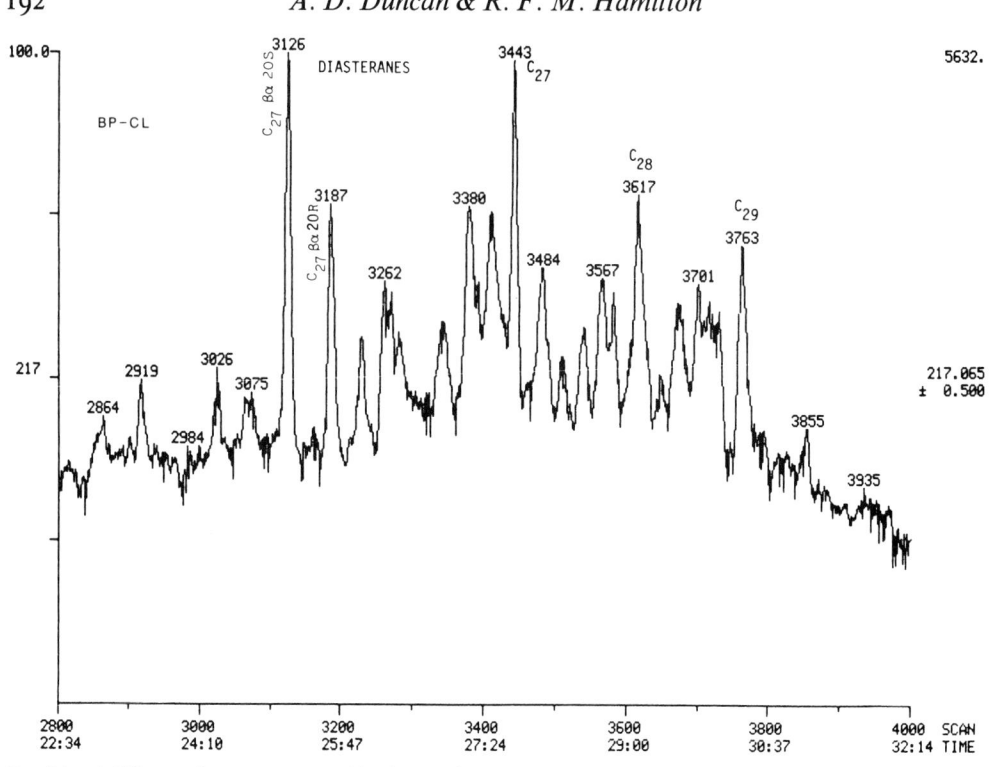

FIG. 14. m/z 217 mass fragmentogram of basin margin sample, recording presence of diasteranes (peaks labelled by carbon number, C_{29} isomers labelled S and R).

McKirdy et al. (1985) attributed the distinctive C_{28}- and C_{29}-dominated sterane distributions recorded from Australian sediments to a cyanobacterial and archaebacterial lipid input, which included a halophilic component. This environmental assessment was consistent with:

(i) predictions from the sedimentology of the Officer Basin carbonates; and
(ii) the occurrence of squalane and other isoprenoid alkanes, common in procaryotes tolerant of such environmental extremes, in the Officer Basin oils and source rocks.

Although the same sterane signature was obtained for the Orcadian Basin samples, and similarities in the isoprenoid distribution are also apparent, the consistent lack of evidence for an extensive evaporitic facies (e.g. gypsum, anhydrite, trona) in the Caithness succession indicates that a hypersaline depositional environment need not necessarily be invoked. However conditions of elevated salinity almost certainly prevailed as recorded by the inorganic geochemical palaeosalinity indicators (see above).

It was noted in this study that fluctuation in the hopane/sterane ratio was controlled by the relative abundance of the steranes. Uniformly low hopane/sterane ratios recorded for the less marginal samples and the consistent lack of a marked $n-C_{15}-n-C_{17}$ predominance indicate dilution of the prokaryotic input by a eucaryotic algal source. This is supported by the considerably lower relative abundances of acyclic isoprenoid alkanes present in the Devonian laminites compared with the Officer Basin carbonates (McKirdy et al. 1985), implying only limited microbial reworking by, for example, methanogens. The C_{27}-depleted steranes distribution within this sample suite is a generally consistent feature; it exhibits no sympathetic variation with other more subtle genetic (e.g. tricyclic triterpane profile, chromatogram type) and maturity changes. Such a pattern is itself considered representative of lacustrine sequences by Jiang Zusheng & Fowler (1985).

The low to marginal predominance of hopanes over steranes (with the exception of the marginal samples) suggests that the abundance of bacterial heterotrophs kept pace with that of the primary algal producers, possibly augmented by limited

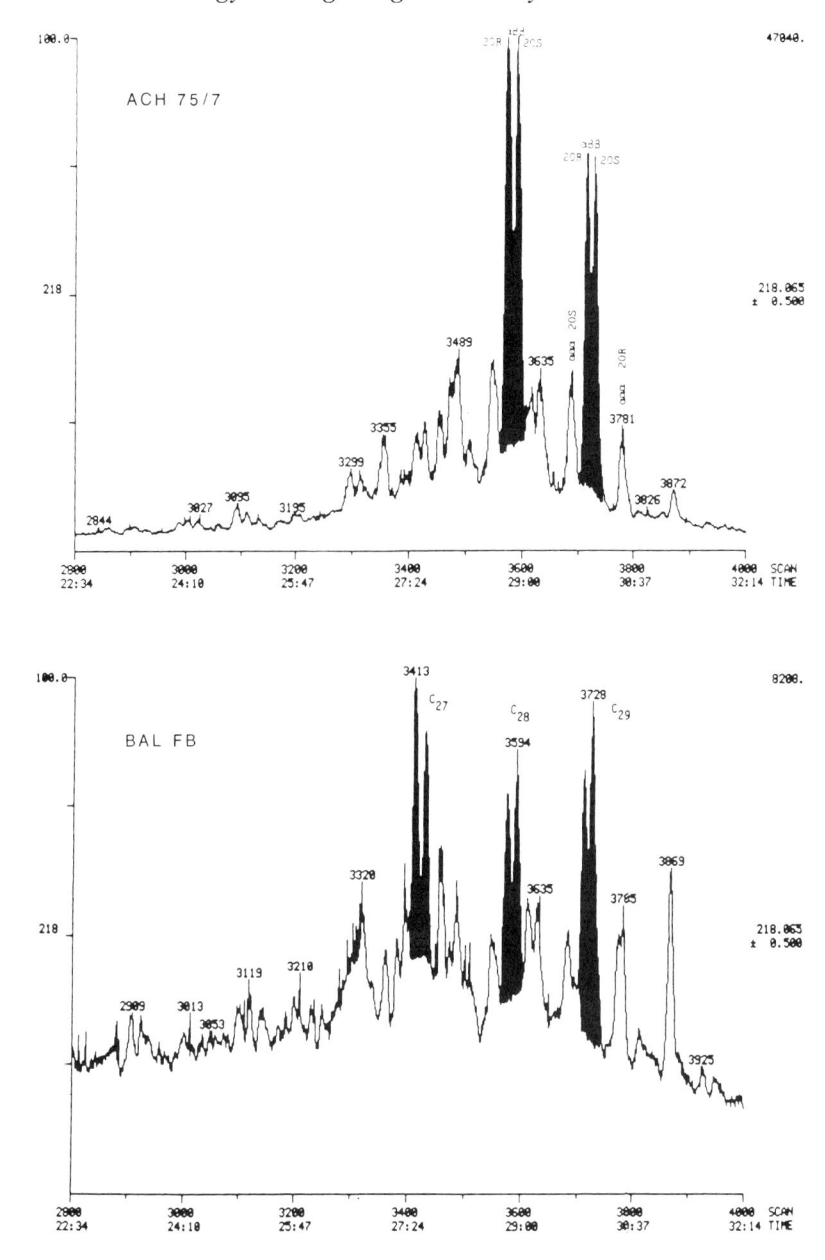

FIG. 15. m/z 218 mass fragmentograms showing compositional variation between offshore lacustrine laminites (upper diagram) and proximal samples (lower diagram); $\alpha\beta\beta$ isomers shaded.

fluctuations in a nutrient supply of fluviatile origin. The principal control on the extent of the photoautotrophic cyanobacterial and algal population supported by the lake would have been the depth of the euphotic zone.

Pr/Ph ratios substantially less than unity, believed to be a common feature of carbonate source rocks (Palacas 1984) were not recorded here. Sulphur concentration in the analysed samples ($<1\%$ by weight) are high by non-marine standards (Berner & Raiswell 1984) and imply reducing bottom water conditions.

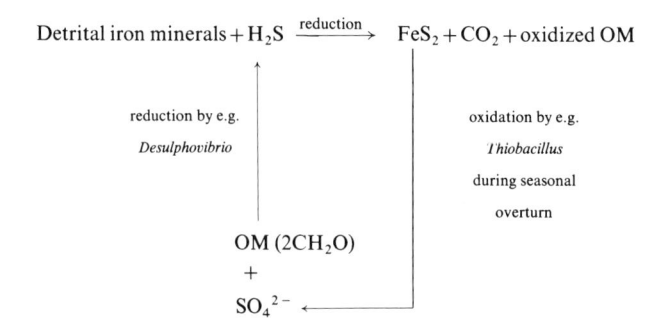

$$\text{Detrital iron minerals} + H_2S \xrightarrow{\text{reduction}} FeS_2 + CO_2 + \text{oxidized OM}$$

reduction by e.g. *Desulphovibrio*

oxidation by e.g. *Thiobacillus* during seasonal overturn

$$OM\ (2CH_2O) + SO_4^{2-}$$

The restricted occurrence, often in high concentration, of carotenoid-derived alkanes is consistent with deposition in both anaerobic and elevated salinity conditions (e.g. Jiang Zusheng & Fowler 1985). The sporadic occurrence of β- and γ-carotane in these Devonian laminites, although associated with higher sterane maturity ratios, most probably reflects such environmental fluctuations, and may furthermore represent a sensitive monitor of the primary algal input.

The organic geochemistry is therefore supportive of a deep stratified lake origin for these sediments. Free-floating cyanobacteria and algal phytoplankton represent the main OM input, with preservation occurring below the chemocline in anoxic waters after percolation down through the water column. The relatively modest extent of reworking by methanogenic, sulphate-reducing and halophilic bacteria in most cases does not favour an alternative scenario whereby

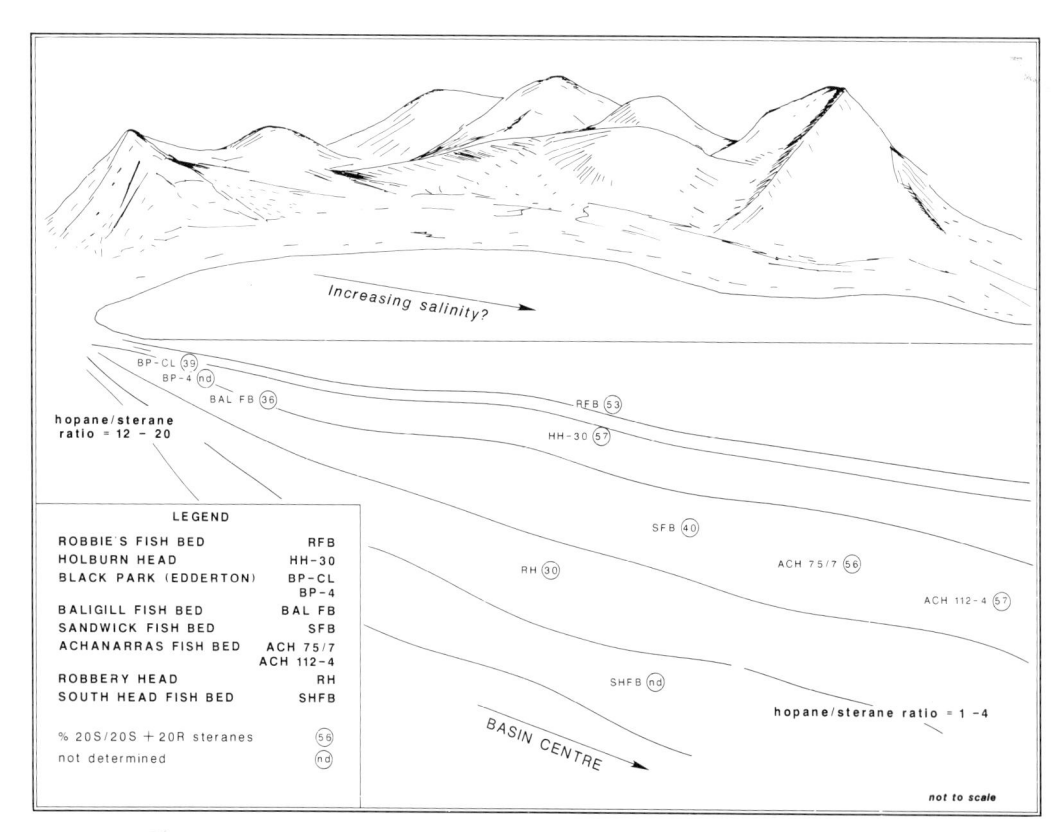

FIG. 16. Schematic illustration of stratigraphic, geographic and maturity relationships for Middle Devonian sample suite.

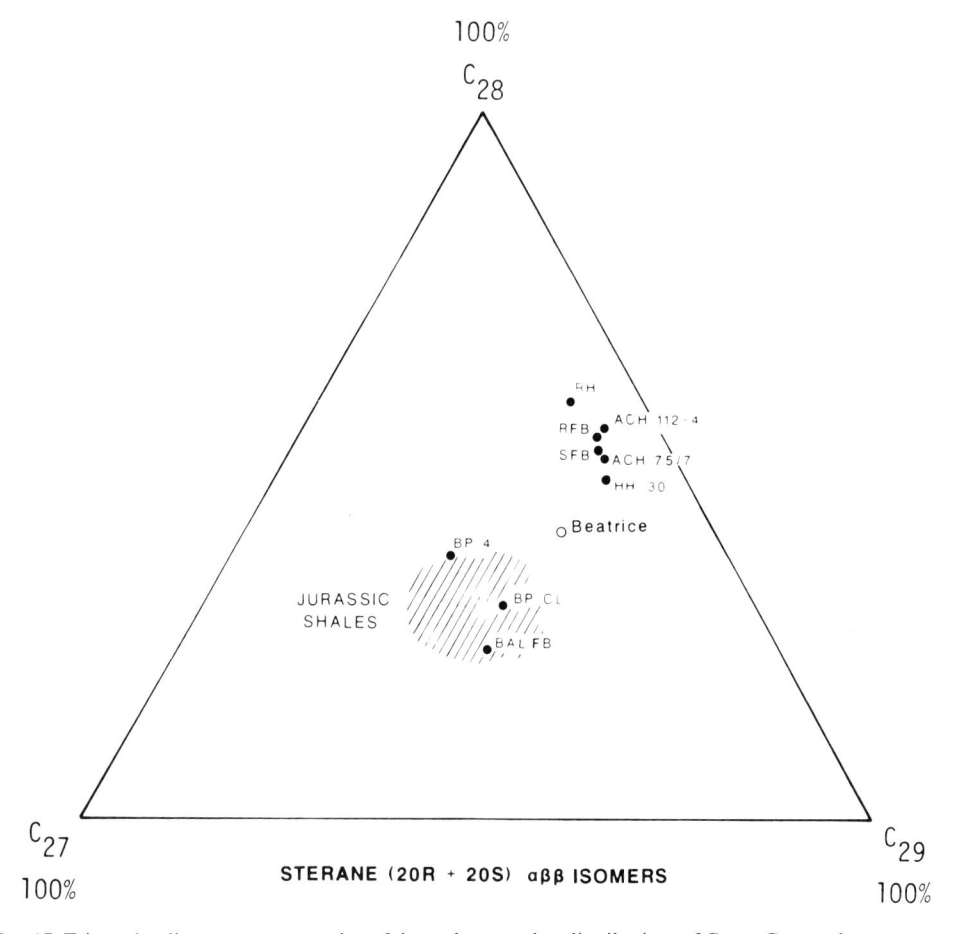

FIG. 17. Triangular diagram representation of the carbon number distributions of C_{27} to C_{29} regular steranes (from m/z 218 fragmentograms).

epipelic cyanobacterial communities undergo suspension and transportation from the shallow lake margins to a deeper water site.

Algal stromatolites are very rare in the Caithness succession but, where found in the Middle Devonian, are confined to basement margin outcrops (Donovan 1975). Given that the hopanes are also constituents of blue-green algae (Ourisson et al. 1979), a greater variation in hopanoid concentration than that recorded, might be expected for the deeper water samples as a result of mobilization of algal mat communities.

Of the other proposed carbonate-specific biomarker features proposed by McKirdy et al. (1983, 1985), i.e.

(i) a predominance of the C_{35} hopane over its C_{34} homologue, and
(ii) the absence of diasteranes,

only the second was observed in these samples (Table 4). The two marginal Blackpark samples may be distinguished from the others by the presence of these compounds. Although both represent marginal lithologies, only BP-4 contained appreciable clay. (20S/20S+20R) ratios for the 13β(H), 17α(H)-C_{27} components indicate complete isomerization.)

Beatrice—a test of the model

In a geochemical investigation of North Sea crude oils and potential source rocks, Mackenzie et al. (1984) demonstrated compositional dissimilarities between the Beatrice oil (Inner Moray Firth; Fig. 8) and the remainder of the sampled Central, Viking and Witchground Graben oils.

The latter are all considered to be sourced by the 'hot' Kimmeridge Clay Formation (Mackenzie et al. 1984; Cornford 1984) and can be correlated with a generally uniform kerogen type.

Jurassic source problems

The Beatrice oil differs from other North Sea oils on the basis of bulk geochemistry, isotopic signature and biomarker distributions. It has long been thought to originate, in whole or in part, from a source other than the oil-prone Kimmeridgian facies. Barnard & Cooper (1981) cited the Bathonian algal oil shales as a possible source of the high-wax Beatrice crude. However, Mackenzie et al. (1984) recognized the volumetric constraints on locally-drained Middle Jurassic source beds, extensive lateral migration being rare in the formation of North Sea hydrocarbon accumulations (Parsley 1984).

The majority of the Upper Jurassic material around the Beatrice structure attains a maximum depth of only around 2000 m (e.g. Linsley et al. 1980). A phase of early Tertiary uplift, variously estimated as 500–700 m by Pearson & Watkins (1983), 1000 m (McQuillin et al. 1982; Mackenzie et al. 1984) and 1500 m (Chesher & Bacon 1975), is recognized within the Inner Moray Firth Basin. However, a detailed study of the Upper Jurassic across blocks 11 and 12 (Duncan 1986) confirms that it is largely immature, with basal Oxfordian shales only locally achieving marginally mature status.

Interpreting the compositional differences in the oils as a response to a facies change, Mackenzie et al. (1984) tentatively proposed a deeply buried (>3 km) Upper Jurassic source for Beatrice, where the prerequisite high input of terrestrial lipids could be provided via sediments shed SE across the Great Glen Fault. Figure 17 includes both sterane distributions of the Beatrice crude (taken from Mackenzie et al. 1984), and the range of compositions obtained from analyses of Upper Jurassic shales, both onshore and offshore, from the Inner Moray Firth Basin. It should be noted that the sterane distributions of Middle Jurassic material cropping out on the Helmsdale coast are virtually identical with their Upper Jurassic analogues.

The Devonian contribution

There is an obvious depletion in the C_{27} sterane component for Beatrice and, although not as marked as that obtained for the bulk of the Devonian samples, suggests a hybrid origin for the oil, with contributions from both Devonian and Jurassic source rocks (Fig. 17).

The extent of Old Red Sandstone subcrop preserved within the limits of the Mesozoic Inner Moray Firth Basin is largely unknown. From the IGS deep seismic survey of the Moray Firth Basin, Chesher & Lawson (1983) considered Old Red Sandstone to be preserved throughout the bounds of the basin; acoustic attenuation precluded any comprehensive account of depths and thicknesses over the area. On the adjacent coastline 22 km to the NW of Beatrice, some 2.5 km of the Caithness Flagstone Group sequence is presently exposed. Also, large blocks of Old Red Sandstone, including fish bed laminite lithologies, occur in abundance within the Upper Jurassic Helmsdale Boulder Beds confirming the presence of Orcadian Basin source-rock facies on the upthrown side of the Helmsdale Fault during Mesozoic times. Recent publications by Richards (1985a, b) verify the preservation of rocks of Lower and Upper Devonian age, both argillaceous and predominantly arenaceous sequences being represented, several of the latter being of reservoir quality. Extension of the relatively shallow Lower and Middle Jurassic drilling objectives will be needed to delineate the offshore succession. However, biomarker ratios derived from these upfaulted Middle Devonian outcrops do not imply overmaturity of the offshore succession where present in the Inner Moray Firth Basin.

Comparison of maturity measurements for most of the deep Orcadian lacustrine samples and the Beatrice crude (Mackenzie et al. 1984) support a contribution from a Devonian source rock to the oil. Although the Devonian bitumens cannot be classed as waxy this is possibly a function of maturity. The high n-alkane content of Beatrice could conceivably be derived from the in-reservoir mixing of two non-biodegraded oils.

Disparity is observed between the hopane/sterane ratios for the majority of the Devonian material and that for Beatrice oil (Mackenzie et al. 1984). However, the very high degree (~70%) of aromatization (Mackenzie et al. 1984) observed for the oil is indicative of high maturity. Thermal degradation of kerogen is known to release high concentrations of hopanes from the bound lipid fraction (S. Brassell personal communication); this may perhaps explain the observed differences.

Despite the age-dependence demonstrated by Degens (1969) for oil-isotopic compositions, they are predominantly source-controlled (Schoell 1984). Sofer (1984) however, stated that the isotopic value of a single oil component (as measured by Mackenzie et al. 1984) is not unique to its depositional environment, advocating analyses of both saturates and aromatics fractions to infer terrigenous-sourced input. We anticipate that detailed carbon isotope and aromatic biomarker measurements would confirm our proposal of a Devonian source contribution.

Barnard and Cooper (1981) proposed that good quality Devonian source rocks from the Orcadian Basin were largely attributable to bitumen impregnation from younger overlying sediments, subsequently removed by erosion. Hydrocarbons of Jurassic age cannot be responsible, as evidenced by both source (e.g. sterane distributions) and maturity parameters. Where impregnation is suspected, intraformational oil staining seems more likely (see also Parnell 1985). Biomarker signatures of the Liassic Lady's Walk Shale do not differ significantly from those of their later Jurassic counterparts; this unit is however, notably organic poor (Duncan 1986). The available evidence suggests that oil derived from mature Middle Devonian lacustrine laminites augments the Jurassic input, although the actual stratigraphic age of the Mesozoic component may not be differentiated from the available data.

From this test, it is clear that a similar approach would be useful in exploration play development for other areas of the Moray Firth where suitable structures have been geophysically identified, but the timing and nature of petroleum migration and entrapment from 'traditional' Mesozoic source rocks precludes hydrocarbon accumulation.

Conclusions

1. Orcadian Basin fish bed laminites were deposited below wave base in the anoxic bottom waters of deep stratified lakes.
2. Onshore sedimentary facies zonation and palaeocurrent data predict a greater development of deep lake (source rock) facies towards the original basin centre; presently the Moray Firth and North Sea areas.
3. The abundant dolomite carbonate found in many fish bed laminites has a primary or very early diagenetic origin.
4. Organic and inorganic geochemistry indicates that Orcadian Basin lake waters were saline–brackish during deep lake phases encouraging high levels of bioproductivity.
5. The development of evaporite facies is postulated for the basin centre during low lake stand desiccation phases.
6. Cyanobacteria and algal lipids represent the primary organic input to the basin.
7. Distinction between marginal and offshore lacustrine samples on the basis of their molecular distributions is indicated; carbonate and clay-rich lithologies may be differentiated using similar criteria.
8. Biomarker maturity ratios, based on analysis of aliphatic hydrocarbons, indicate that several of the onshore Middle Devonian laminites have achieved oil-generation status.
9. A Middle Devonian component to the Beatrice oil is suggested.

ACKNOWLEDGEMENTS: Valuable criticism of the manuscript was made by Dr B. Kneller and other Aberdeen Departmental staff members. Helpful discussions with Dr S. Grigson and Dr T. Fallick are also acknowledged. We are grateful for all technical services rendered in the preparation of this manuscript. All carbonate isotope work was carried out at SURRC, East Kilbride. The authors acknowledge financial support for this study from Exploration and Production Services (UK) North Sea Ltd., and the Department of Geology and Mineralogy, University of Aberdeen, whilst in the tenureship of NERC studentships. Dr R. Archer and Dr G. Speers who reviewed the manuscript are also thanked for their comments.

References

ALLISON, I. & RUSSELL, M. J. 1985. Anhydrite discovered in the Fucoid Beds of NW Scotland. *Journal of Sedimentary Petrology*, 55, 917–918.

AQUINO NETO, F. R., TRENDEL, J. M., RESTLE, A., CONNAN, J. & ALBRECHT, P. 1983. Occurrence and formation of tricyclic and tetracyclic terpanes in sediments and petroleums. *In*: BJORØY, M. *et al.* (eds), *Advances in Organic Chemistry 1981*. John Wiley & Sons Ltd., Chichester, 659–667.

BATTEN, D. J. 1981. Palynofacies, organic maturation and source potential for petroleum. *In*: BROOKS, J. (ed.), *Organic Maturation Studies and Fossil Fuel Exploration*. Academic Press, London, 201–223.

BARNARD, P. C. & COOPER, B. S. 1981. Oils and source rocks of the North Sea area. *In*: ILLING, L. V. & HOBSON, G. D. (eds), *Petroleum Geology of the Continental Shelf of North-west Europe*. Institute of Petroleum, London, 169–175.

BEADLE, L. C. 1981. *The Inland Waters of Tropical Africa*, 2nd edn. Longman, London.

BERNER, R. A. & RAISWELL, R. 1984. C/S method for distinguishing freshwater from marine sedimentary rocks. *Geology*, 12, 365–368.

BOTZ, R. W. & VON DER BORCH, C. C. 1984. Stable isotope study of carbonate sediments from the Coorong Area, South Australia. *Sedimentology*, 31, 837–849.

BRADLEY, W. C. 1929. The varves and climate of the

Green River epoch. *Professional Paper. United States Geological Survey*, **158E**, 86–110.

BRASSELL, S. C., EGLINTON, G. & MAXWELL, J. R. 1983. The geochemistry of terpenoids and steroids. *Biochemical Society Transactions*, **11**, 575–586.

BUCCHEIM, P. H. & SURDAM, R. C. 1977. Fossil catfish and the depositional environment of the Green River Formation, Wyoming. *Geology*, **5**, 196–198.

BURST, J. F. 1965. Subaqueously formed shrinkage cracks in clay. *Journal of Sedimentary Petrology*, **35**, 348–353.

CALLANDER, E. 1968. The post glacial sedimentology of Cevils Lake, North Dakota. Ph.D. thesis, University of North Dakota.

CHESHER, J. A. & BACON, M. 1975. A deep seismic survey in the Moray Firth. *Report of the Institute of Geological Sciences*, **75/11**.

—— & LAWSON, D. 1983. The geology of the Moray Firth. *Report of the Institute of Geological Sciences*, **85/5**, 32 pp.

CORNFORD, C. 1984. Source rocks and hydrocarbons of the North Sea. *In*: GLENNIE, K. W. (ed.), *Introduction to the Petroleum Geology of the North Sea*. Blackwell Scientific Publications, Oxford, 171–204.

CRAMPTON, C. B. & CARRUTHERS, R. G. 1914. The Geology of Caithness. *Memoirs of the Geological Survey*.

DEGENS, E. T. 1969. Biogeochemistry of stable carbon isotopes. *In*: EGLINTON, G. & MURPHY, M. T. J. (eds), *Organic Geochemistry, Methods and Results*, Springer-Verlag, New York, 306–329.

——, VON HERZEN, R. P., WONG, H. K., DEUSRER, W. G. & JANNASCH, H. W. 1973. Lake Kivu: Structure, chemistry and biology of an East African rift lake. *Geologische Rundschau*, **62**, 245–277.

DESBOROUGH, G. A. 1978. A biogenic-chemical stratified lake model for the origin of oil shale of the Green River Formation: an alternative to the playa-lake model. *Bulletin of the Geological Society of America*, **89**, 961–971.

DIDYK, B. M., SIMONEIT, B. R. T., BRASSELL, S. C. & EGLINTON, G. 1978. Organic geochemical indicators of palaeoenvironmental conditions of sedimentation. *Nature*, **272**, 216–221.

DONOVAN, R. N. 1971. The geology of the coastal tract near Wick, Caithness. Ph.D. thesis, University of Newcastle-upon-Tyne (unpublished).

—— 1975. Devonian lacustrine limestones at the margin of the Orcadian Basin, Scotland. *Journal of the Geological Society*, **131**, 489–510.

—— 1980. Lacustrine cycles, fish ecology and stratigraphic zonation in the Middle Devonian of Caithness. *Scottish Journal of Geology*, **16**, 35–50.

—— & FOSTER, R. J. 1972. Subaqueous shrinkage cracks from the Caithness Flagstone Series (Middle Devonian) of North-east Scotland. *Journal of Sedimentary Petrology*, **42**, 309–317.

——, —— & WESTOLL, T. S. 1974. A stratigraphical revision of the Old Red Sandstone of North-eastern Caithness. *Transactions of the Royal Society of Edinburgh*, **69**, 167–201.

DUNCAN, A. 1986. Organic geochemistry applied to source potential and tectonic history of the Inner Moray Firth basin. Ph.D. thesis, University of Aberdeen.

EKWEOZOR, C. M. & STRAUSS, O. P. 1983. Tricyclic terpanes in the Athabasca oil sands: their geochemistry. *In*: BJORØY, M. et al. (eds), *Advances in Organic Geochemistry 1981*. John Wiley & Sons Ltd., Chichester, 746–766.

EUGSTER, H. P. & HARDIE, L. A. 1975. Sedimentation in an ancient playa-lake complex: the Wilkins Peak Member of the Green River Formation of Wyoming. *Bulletin of the Geological Society of America*, **86**, 319–334.

—— & —— 1978. Saline lakes. *In*: LERMAN, A. (ed.), *Lakes: Chemistry, Geology, Physics*. Springer-Verlag, London.

—— & SURDAM, R. C. 1973. Depositional environment of the Green River Formation of Wyoming: a preliminary report. *Bulletin of the Geological Society of America*, **84**, 1115–1120.

FANNIN, N. G. T. 1970. The sedimentary environment of the Old Red Sandstone of Western Orkney. Ph.D. thesis, University of Reading (unpublished).

FRIEDMAN, G. M., AMIEL, A. J., BROWN, M. & MILLER, D. S. 1973. Generation of carbonate particles and laminites in algal mat—example from sea marginal hypersaline pool, Gulf of Aqaba, Red Sea. *Bulletin of the American Association of Petroleum Geologists*, **57**, 541–557.

GOOSSENS, H., DE LEEUW, J. W., SCHENCK, P. A. & BRASSELL, S. C. 1984. Tocopherols as likely precursors of pristane in ancient sediments and crude oils. *Nature*, **312**, 440–442.

HÅKANSON, L. & JANSSON, M. 1983. *Principles of Lake Sedimentology*. Springer-Verlag, London.

HALL, P. B. & DOUGLAS, A. G. 1983. The distribution of cyclic alkanes in two lacustrine deposits. *In*: BJORØY, M. et al. (eds), *Advances in Organic Geochemistry 1981*. John Wiley & Sons Ltd., Chichester, 575–581.

HAMILTON, R. F. M. 1986. Comparative palaeolimnology of the Middle Devonian Orcadian Basin. Ph.D. thesis, University of Aberdeen.

—— & TREWIN, N. H. 1985. *The Petroleum Geology of the Orcadian Basin*, Petroleum Exploration Society of Great Britain Field Guide, PESGB, London.

HARDIE, L. A. 1968. The origin of the recent non-marine evaporite deposit of Saline Valley, Inyo County, California. *Geochimica et Cosmochimica Acta*, **32**, 1279–1301.

HOLZER, G., ORO, J. & TORNABENE, T. G. 1979. Gas chromatographic–mass spectrometric analysis of neutral lipids from methanogenic and thermoacidophilic bacteria. *Journal of Chromatography*, **186**, 795–809.

HUBERT, J. H., REED, A. A. & CAREY, P. J. 1976. Paleogeography of the East Berlin Formation, Newark Group, Connecticut Valley. *American Journal of Science*, **276**, 1183–1207.

HUDSON, J. D. 1977. Stable isotopes and limestone lithification. *Journal of the Geological Society of London*, **133**, 637–660.

JIANG ZUSHENG & FOWLER, M. G. 1985. Carotenoid-derived alkanes in oils from north western China. *In*: SCHENCK, P. A., DE LEEUW, J. W. & LIJMBACH,

JONES, B. F., EUGSTER, H. P. & RETTIG, S. L. 1977. Hydrochemistry of the Lake Magadi Basin, Kenya. *Geochimica et Cosmochimica Acta*, **41**, 53–72.

JUNGST, H. 1934. Sur Geologischen Bedeutung der Synareses. *Geologische Rundschau*, **25**, 312–325.

KATZ, B. J. 1983. Limitations of 'Rock-Eval' pyrolysis for typing organic matter. *Organic Geochemistry*, **4**, 195–199.

KEITH, M. L. & WEBER, J. N. 1964. Carbon and oxygen isotopic composition of selected limestones and fossils. *Geochimica et Cosmochimica Acta*, **28**, 1787–1816.

LAND, L. S. 1980. The isotopic and trace element geochemistry of dolomite: the state of the art. *Special Publication. Society of Economic Palaeontologists and Mineralologists*, **28**, 87–110.

LEVENTHAL, J. S. 1983. An interpretation of carbon and sulphur relationships in the Black Sea sediments as indicators of environments of deposition. *Geochimica et Cosmochimica Acta*, **47**, 133–137.

LEYTHAEUSER, D., MACKENZIE, A. S., SCHAEFER, R. G. & BJØRØY, M. 1984. A novel approach for recognition and quantification of hydrocarbon migration effects in shale–sandstone sequences. *Bulletin of the American Association of Petroleum Geologists*, **68(2)**, 196–219.

LINSLEY, P. N., POTTER, H. C., MCNAB, G. & RACHER, D. 1980. The Beatrice Field, Inner Moray Firth, UK, North Sea. *In*: HALBOUTY, M. T. (ed.), *Giant Oil and Gas Fields of the Decade: 1968–1978*. *Memoir of the American Association of Petroleum Geologists*, **30**, 117–129.

LOGAN, B. W., HOFFMAN, P. & GEBELEIN, C. F. 1974. Algal mats, cryptalgal fabrics and structures, Hamelin Pool, Western Australia. *Memoir of the American Association of Petroleum Geologists*, **22**, 140–194.

LUNDELL, L. & SURDAM, R. C. 1975. Playa-like deposition: Green River Formation, Piceance Creek Basin, Colorado. *Geology*, **3**, 493–497.

MACKENZIE, A. S. 1984. Applications of biological markers in petroleum geochemistry. *In*: BROOKS, J. & WELTE, D. H. (eds), *Advances in Petroleum Geochemistry Vol. 1*. Academic Press, London, 115–214.

——, PATIENCE, R. L., MAXWELL, J. R., VANDENBROUCKE, M. & DURAND, B. 1980. Molecular parameters of maturation in the Toarcian shales, Paris Basin, France—I. Changes in the configurations of acyclic isoprenoid alkanes, steranes and triperpanes. *Geochimica et Cosmochimica Acta*, **44**, 1709–1721.

——, MAXWELL, J. R., COLEMAN, M. L. & DEEGAN, C. E. 1984. Biological markers and isotope studies of North Sea crude oils and sediments. *In: Proceedings of the Eleventh World Petroleum Congress, London*, **2**. John Wiley and Sons Ltd., Chichester, 45–56.

MAGARITZ, M. & GAT, J. R. 1981. Review of the natural abundance of hydrogen and oxygen isotopes. *In*: GAT, J. R. & GONFIANTINI, R. (eds), *Stable Isotope Hydrology: Deuterium and Oxygen-18 in the Water Cycle*. International Atomic Energy Agency, Vienna.

MARSHALL, J. E. A., BROWN, J. F. & HINDMARSH, S. 1985. Hydrocarbon source rock potential of the Devonian rocks of the Orcadian basin. *Scottish Journal of Geology*, **21**, 301–320.

MCHARGUE, T. R. & PRICE, R. C. 1982. Dolomite from clay in argillaceous or shale-associated marine carbonates. *Journal of Sedimentary Petrology*, **52**, 873–885.

MCKENZIE, J. A. 1981. Holocene dolomitization of calcium carbonate sediments from the coastal sabkhas of Abu Dhabi, UAE: a stable isotope study. *Journal of Geology*, **89**, 185–198.

MCKIRDY, D. M., ALDRIDGE, A. K. & YPMA, P. J. M. 1983. A geochemical comparison of some crude oils from Pre-Ordovician Carbonate Rocks. *In*: BJØRØY, M. et al. (eds), *Advances in Organic Geochemistry 1981*. John Wiley & Sons Ltd., Chichester, 99–107.

——, KANTSLER, A. J., EMMETT, J. K. & ALDRIDGE, A. K. 1985. Hydrocarbon genesis and organic facies in Cambrian carbonates of the Eastern Officer Basin, South Australia. *In*: PALACAS, J. G. (ed.), *Petroleum Geochemistry and Source Rock Potential of Carbonate Rocks*. Studies in Geology, **18**. American Association of Petroleum Geologists, Tulsa, 13–32.

MCQUILLIN, R., DONATO, J. A. & TULSTRUP, J. 1982. Dextral displacement of the Great Glen Fault as a factor in the development of the Inner Moray Firth Basin. *Earth and Planetary Science Letters*, **60**, 127–139.

MORRISON, R. B. 1966. Predecessors of Great Salt Lake. *In*: STOKES, W. L. (ed.), *Guidebook to the Geology of Utah*, **20**, Utah Geological Society.

MULLER, G. 1971. Aragonite: inorganic precipitation in a freshwater lake. *Nature Physical Sciences*, **229**, 18.

——, IRION, G. & FORSTNER, U. 1972. Formation and diagenesis of inorganic Ca–Mg carbonates in the lacustrine environment. *Naturwissenschaften*, **59(4)**, 158–164.

MYKURA, W. 1976. *Orkney and Shetland*, British Regional Geology, Institute of Geological Sciences, HMSO, Edinburgh.

—— 1983. The Old Red Sandstone. *In*: CRAIG, G. Y. (ed.), *The Geology of Scotland*, 2nd edn. Scottish Academic Press, Edinburgh, 205–251.

NEEV, D. & EMERY, K. D. 1967. The Dead Sea. Depositional processes and environments of evaporites. *Israel Geological Survey Bulletin*, **41**, 147.

OURISSON, G., ALBRECHT, P. & ROHMER, M. 1979. The hopanoids: palaeochemistry and biochemistry of a group of natural products. *Pure and Applied Chemistry*, **51**, 709–729.

PALACAS, J. G. 1984. Carbonate rocks as source rocks of petroleum: geological and chemical characteristics and oil-source correlations. *In: Proceedings of the Eleventh World Petroleum Congress, London*, **2**. John Wiley and Sons Ltd., Chichester, 31–43.

——, ANDERS, D. E. & KING, J. D. 1985. South Florida Basin—a prime example of carbonate source rocks of petroleum. *In*: PALACAS, J. G. (ed.) *Petroleum Geochemistry and Source Rock Potential of Carbonate Rocks*. Studies in Geology, **18**. American Association of Petroleum Geologists, Tulsa, 71–86.

PARNELL, J. 1985. Hydrocarbon source rocks, reservoir rocks and migration in the Orcadian Basin. *Scottish Journal of Geology*, **21**, 321–335.

PARRISH, J. T. 1982. Upwelling and petroleum source beds, with reference to the Paleozoic. *Bulletin of the American Association of Petroleum Geologists*, **66(6)**, 750–774.

PARSLEY, A. J. 1984. North Sea hydrocarbon plays. *In*: GLENNIE, K. W. (ed.), *Introduction to the Petroleum Geology of the North Sea*. Blackwell Scientific Publications, Oxford, 205–230.

PEARSON, M. J. & WATKINS, D. 1983. Organofacies and early maturation effects in Upper Jurassic sediments from the Inner Moray Firth Basin, North Sea. *In*: BROOKS, J. (ed.), *Petroleum Geochemistry and the Exploration of Europe*. Blackwell Scientific Publications, Oxford, 147–160.

PENNINGTON, J. J. 1975. The geology of the Argyll Field. *In*: WOODLAND, A. W. (ed.), *Petroleum and the Continental Shelf of Northwest Europe*. Applied Science Publishers, Barking, 285–291.

PETERSON, M. N. A. & VON DER BORCH, C. C. 1965. Chert: modern inorganic deposition in a carbonate precipitating locality. *Science*, **149**, 1501–1503.

PHILP, R. P. 1985. Biological markers in fossil-fuel production. *Mass Spectrometry Reviews*, **4**, 1–54.

RAYNER, D. H. 1963. The Achanarras Limestone of the Middle Old Red Sandstone, Caithness, Scotland. *Proceedings of the Yorkshire Geological Society*, **34**, 117–138.

RICHARDS, P. C. 1985a. A Lower Old Red Sandstone lake in the offshore Orcadian Basin. *Scottish Journal of Geology*, **21**, 101–105.

—— 1985b. Upper Old Red Sandstone sedimentation in the Buchan oilfield, North Sea. *Scottish Journal of Geology*, **21**, 227–237.

RICHARDSON, J. B. 1965. Middle Old Red Sandstone spore assemblages from the Orcadian Basin, North-east Scotland. *Palaeontology*, **7**, 559–605.

RULLKOTTER, J., SPIRO, B. & NISSENBAUM, A. 1985. Biological marker characteristics of oils and asphalts from carbonate source rocks in a rapidly subsiding graben, Dead Sea, Israel. *Geochimica et Cosmochimica Acta*, **49**, 1357–1370.

SAJGO, C. S. 1984. Organic geochemistry of crude oils from South-east Hungary. *In*: SCHENCK, P. A., DE LEEUW, J. W. & LIJMBACK, G. W. M. (eds), *Advances in Organic Geochemistry 1983*. Pergamon Press, Oxford, 569–678.

SCHOELL, M. 1984. Recent advances in petroleum isotope geochemistry. *In*: SCHENCK, P. A., DE LEEUW, J. W. & LIJMBACH, G. W. M. (eds) *Advances in Organic Geochemistry 1983*. Pergamon Press, Oxford, 645–663.

SEIFERT, W. K. & MOLDOWAN, J. M. 1978. Applications of steranes, terpanes and mono-aromatics to the maturation, migration and source of crude oils. *Geochimica et Cosmochimica Acta*, **42**, 77–95.

—— & —— 1980. The effect of thermal stress on source rock quality as measured by hopane stereochemistry. *In*: DOUGLAS, A. G. & MAXWELL, J. R. (eds), *Advances in Organic Geochemistry 1979*. Pergamon Press, Oxford, 229–237.

—— & —— 1981. Paleoreconstruction by biological markers. *Geochimica et Cosmochimica Acta*, **45**, 783–794.

SERRUYA, C. & POLLINGHER, C. 1983. *Lakes of the Warm Belt*. Cambridge University Press, Cambridge.

SHI JI-YANG, MACKENZIE, A. S., ALEXANDER, R., EGLINTON, G., GOWER, A. P., WOLFF, G. A. & MAXWELL, J. R. 1982. A biological marker investigation of petroleums and shales from the Shengli Oilfield, China. *Chemical Geology*, **35**, 1–31.

SISEKIND, O., JOLY, G. & ALBRECHT, P. 1979. Simulation of the geochemical transformation of sterols; superacid effect of clay minerals. *Geochimica et Cosmochimica Acta*, **43**, 1675–1679.

SIMONEIT, B. R. T. 1978. The organic chemistry of marine sediments. *In*: RILEY, J. P. & CHESTER, R. (eds), *Chemical Oceanography*, **7**. Academic Press, London, 233–311.

SLY, P. J. 1978. Sedimentary processes in lakes. *In*: LERMAN, A. (ed.), *Lakes: Chemistry, Geology, Physics*. Springer-Verlag, London.

SMITH, A. G., BRIDEN, J. C. & DREWRY, G. E. 1973. Phanerozoic world maps. *In*: HUGHES, N. F. (ed.), *Organisms and Continents Through Time. Special Papers in Palaeontology*, **12**, 1–42.

SMITH, I. R. & SINCLAIR, I. J. 1972. Deep water waves in lakes. *Freshwater Biology*, **2**, 387–399.

SOFER, Z. 1984. Stable carbon isotope compositions of crude oils: application to source depositional environments and petroleum alteration. *Bulletin of the American Association of Petroleum Geologists*, **68**, 31–49.

SURDAM, R. R. & WOLFBAUER, C. A. 1975. Green River Formation, Wyoming: a playa-lake complex. *Bulletin of the Geological Society of America*, **86**, 335–345.

TISSOT, B. P. & WELTE, D. H. 1984. *Petroleum Formation and Occurrence*, 2nd edn. Springer-Verlag, New York.

TREWIN, N. H. 1976. Correlation of the Achanarras and Sandwick Fish Beds, Middle Old Red Sandstone, Scotland. *Scottish Journal of Geology*, **12**, 205–208.

—— 1986. Palaeoecology and sedimentology of the Achanarras Fish Bed of the Middle Old Red Sandstone. *Transactions of the Royal Society of Edinburgh. Earth Sciences*, **77**, 21–46.

VAN HOUTEN, F. B. 1964. Cyclic lacustrine sedimentation, Upper Triassic Lockatong Formation, central New Jersey and adjacent areas. *Bulletin. State Geological Survey, Kansas*, **169**, 497–531.

—— 1977. Triassic–Liassic deposits in Morocco and eastern North America. *Bulletin of the American Association of Petroleum Geologists*, **61**, 79–99.

VON DER BORCH, C. C., BOLTON, B. & WARREN, J. K. 1977. Environmental setting and microstructure of subfossil lithified stromatolites associated with evaporites, Marion Lake, South Australia. *Sedimentology*, **24**, 693–708.

WESTOLL, T. S. 1977. Northern Britain. *In*: HOUSE, M. J. et al. *A Correlation of Devonian Rocks of the British Isles*. Geological Society Special Publication, **7**, Blackwell Scientific Publications, Oxford, 110.

ZIEGLER, P. A. 1982. *Geological Atlas of Western and Central Europe*. Elsevier, Amsterdam.

A. D. DUNCAN* & R. F. M. HAMILTON†, Department of Geology and Mineralogy, University of Aberdeen, Marischal College, Aberdeen, AB9 1AS.

*Present address: The British Petroleum plc, BP Research Centre, Chertsey Road, Sunbury-on-Thames, Middlesex, TW16 7LN.
†Present address: Shell Exploration and Production Company, PO Box 162, 2501 AN The Hague, The Netherlands.

Hydrocarbon source rocks, thermal maturity and burial history of the Orcadian Basin, Scotland

S. J. Hillier & J. E. A. Marshall

Potential hydrocarbon source rocks of lacustrine origin from the Devonian Orcadian Basin have been evaluated for organic matter richness, type and maturity. Techniques used involve pyrolysis, TOC, visual kerogen typing, spore colour and vitrinite reflectance. A widespread anomaly in assessing maturity is that a significant variation in spore colour from the AOM rich lacustrine sediments is observed. These give lower maturity values by up to $1\frac{1}{2}$ points (on a 1–10 scale) than spores from corresponding woody dominated kerogens. This discrepancy is reduced at higher maturity levels, disappearing at a spore colour of about 8. Small scale surveys of the Orcadian Basin thus give a disproportionate content of lower rank kerogens than in fact exist.

The results of a regional survey across the Orcadian Basin show that the sequences of the Walls Sandstone, Fair Isle and South East Shetland are all significantly overmature and contain TOC contents of only minor richness. The main Caithness and Orkney development of the Orcadian Basin does contain significant hydrocarbon source rocks rich in oil-prone kerogen and with maturities from the mid-oil window to into the dry gas zone. The survival of the hydrocarbons which must have been generated from this sequence is in doubt with its shallow burial since the end Devonian. A major deep oxidation episode almost certainly occurred in the Permo-Carboniferous. Some hydrocarbons may have survived in sandstone reservoirs formed by packets of aeolian sediment deposited on the margins of major lake units. Offshore accumulations may also be preserved in half-graben structures filled with coarse Devonian clastics and surrounded by distal organic-rich lacustrine sediments.

S. J. HILLIER & J. E. A. MARSHALL, Dept. of Geology, The University, Highfield, Southampton, SO9 5NH, UK.

Significance of lacustrine cherts for the environment of source-rock deposition in the Orcadian Basin, Scotland

J. Parnell

SUMMARY: Middle Old Red Sandstone lacustrine sequences in the Orcadian Basin include rocks of petroleum source-rock potential, most extensively exposed onshore in Orkney and Caithness. The richest source rocks are carbonate laminites, in which carbonate and carbonaceous laminae are believed to be the deposits of algal blooming and mortality in a stratified lake (Donovan 1980). Many carbonate laminites and some other organic-rich laminites contain chert nodules, interpreted as magadi-type chert enclosing pseudomorphs after (?) trona. The cherts and evaporites were precipitated in sub-lacustrine sediment from groundwaters rich in dissolved sodium carbonate/bicarbonate. This implies that the source rocks were deposited from saline, alkaline lake waters.

Introduction

The source-rock potential of the cyclic lacustrine sequences of Middle Old Red Sandstone (ORS) in the Orcadian Basin is a subject of much recent interest. Analyses by gas chromatography and gas chromatography–mass spectrometry (Douglas et al. 1983), Rock-Eval pyrolysis (Parnell 1985a) and spore colour/vitrinite reflectance (Marshall et al. 1985) have been published, and it is clear that the sequence contains viable source rocks. Some of the rocks with the richest source-rock potential contain cherts (Parnell 1986a). This contribution briefly reviews the distribution of organic carbon in the cyclic sequences, and discusses the significance of the cherts for the conditions of source-rock deposition.

The cyclic sediments of the Middle ORS, exposed principally in Orkney and Caithness, have long been appreciated as lacustrine (Godwin-Austen 1856; Geikie 1879) and have been described by Donovan (1975, 1980) using a widely-quoted model of lake stratification and lake-margin sedimentation. The Middle ORS stratigraphy of Orkney and Caithness is described by Mykura (1976) and Donovan et al. (1974) respectively.

Cyclic sedimentation and distribution of organic carbon

Middle ORS sediments in Orkney and Caithness comprise cycles (typically about 10 m thick) from black mudrocks at the base through laminated siltstones to sandy beds with desiccation cracks (Fig. 1). The cycles represent lacustrine regression and transgression, with fluviatile incursions at low lake level (Mykura 1976). Sediment accumulation was generally during the regressive episode, resulting in asymmetric cycles as shown in Fig. 1A.

There are however cyclic sequences which show no evidence of fluviatile sedimentation where the various lithologies in each cycle represent different lake levels in an enclosed basin. The symmetrical cycle in Fig. 1B (modified from Donovan 1980) is composed of four lithological associations which represent the basic lacustrine facies. The descriptions and interpretations of these lithologies are summarized here from the work of Donovan (1980) to which the reader is referred for further details:

(A) grey-black carbonate laminite, composed of 0.5 mm triplets of micritic carbonate, organic carbon and siliciclastic laminae, the deposits of a thermally stratified lake (see below). Some carbonate laminites lack siliciclastic matter and are petrographically limestones;

(B) grey-black alternating carbon-rich shale and grey coarse siltstone (pair up to 7 mm), deposits of a shallow, sediment-starved, permanent lake;

(C) alternating dark grey carbon-rich shale and grey coarse siltstone (pair up to 10 mm); silts are lensoid and often current-rippled, also deposits of a permanent lake;

(D) alternating green-grey shale and siltstone/fine sandstone, deposits of an impermanent lacustrine environment. Current ripples, wave ripples and desiccation cracks are abundant, and pseudomorphs after gypsum may occur.

Carbonate-rich beds

Carbonate-rich beds are common within the dark, carbon-rich sediments. Two principal types may be distinguished; the carbonate laminites

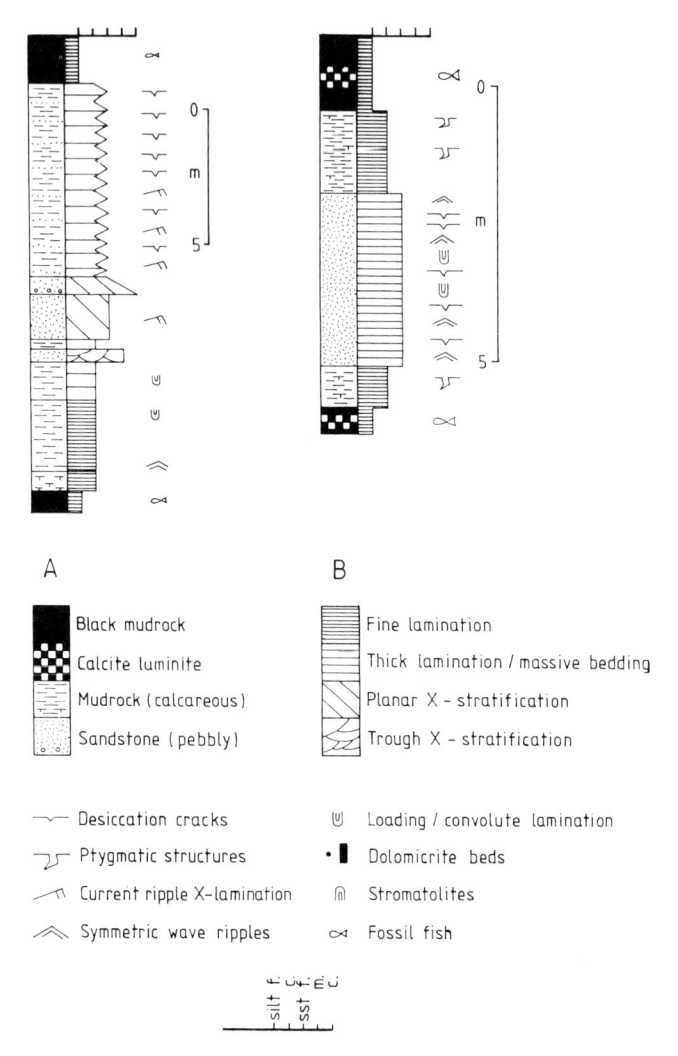

FIG. 1. Fluviolacustrine (A) and lacustrine (B) Middle ORS cycles. Log A from Lower Stromness Group, Muckle Glen, Orkney (HY 225055); log B from Latheron Subgroup, Caithness (after Donovan 1980). Note asymmetry of cycle A due to sediment accumulation during lacustrine regression.

(lithology A above) and thin dolomicrite beds which occur in lithologies B, C and D. The stratigraphic distribution of the two types of carbonate is recorded by Donovan et al. (1974). Cherts occur in the carbonate laminites and in other organic-rich laminites which lack dolomicrite beds (some lithology B). Because of this apparent mutual exclusion of two types of chemical sedimentation, it is important to review briefly the origin of the carbonates.

The carbonate laminites are composed of couplets of micritic carbonate (calcite or dolomite) and organic matter (Fig. 2a); the organic matter may be associated with phosphate (collophanite) and rare collophanite laminae are recorded. Alternatively triplets of carbonate, organic carbon and siliciclastic laminae may be preserved. Diagenetic modification of the carbonate laminae resulted in local recrystallization to microspar and fibrous calcite. Many other beds of organic-rich laminites (lithology B) may lack the carbonate laminae, and consist of couplets of siliciclastic grains and organic matter/clay, both of which may nevertheless contain authigenic carbonate. Donovan (1975, 1980 for detailed accounts) envisages the carbonate and organic laminae as the products of seasonal algal blooming and mortality in a thermally stratified lake,

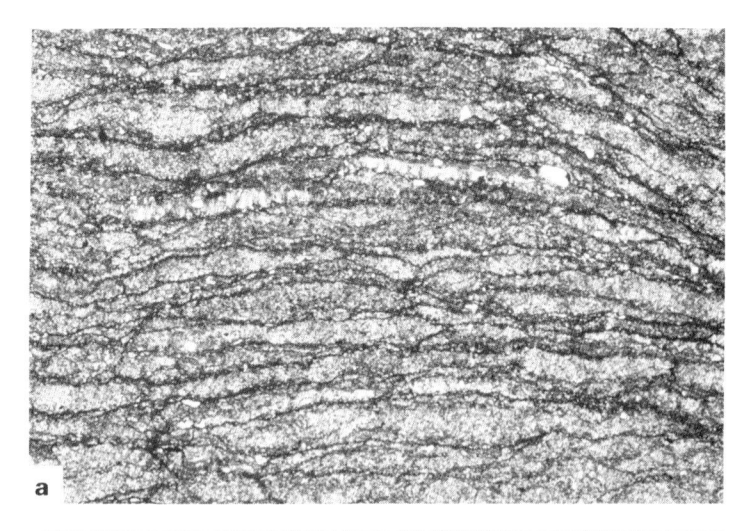

FIG. 2. (a) Photomicrograph of carbonate laminite, consisting of calcite laminae (thick, light) and kerogenous laminae (dark). Calcite is recrystallized to fine sparite, Papa Westray, Orkney (HY 501558) (× 14). (b) Domes of dolomicrite (above hammer) reminiscent of domal stromatolites, overlying beds containing desiccation cracks and overlain by bed of lacustrine sandstone, Harra Ebb, Orkney (HY 216148).

but the more abundant carbon/siliciclastic laminites were probably deposited in non-/intermittently-stratified waters.

The dolomicrite beds are up to 5 cm thick; the thickness varies laterally but beds can be traced over at least several hundred square metres. Sequences containing up to several dozen beds occur in sections of a few metres (Fig. 5A). In some cases the beds are seen to incorporate, and partially replace, siltstone/sandstone laminae. Other beds are pure dolomicrite; the carbonate in these beds may penetrate desiccation cracks in the substrate and appears to have formed 'crusts' on the sediment surface. The beds may have well-defined tops whilst showing gradational bases. These crusts sometimes occur as low-amplitude domes (Fig. 2b), particularly on desiccated surfaces where the domes may be separated by desiccation cracks. Although the dolomicrites generally weather to a distinctive orange colour, some beds incorporate sufficient organic matter (up to 1%) to remain grey/black. Donovan (1978) suggests that the dolomicrites may be due to a seepage reflux of brines during low lake level. However their morphology is very similar to that of stromatolitic deposits, also dolomitic, which occur in the same lithologies (including black organic-rich laminites) and in the same stratigraphic intervals. Indeed some dolomicrite domes have a stromatolitic coating to them. In many cases it is difficult to determine whether a dolomitic deposit contains algal sediment, and I have noted dolomicrites with a texture comparable to the thrombolites of Aitken (1967). There is thus a distinct possibility that algae had a role in the genesis of the dolomicrites. Magnesium preferentially concentrated by algae could be released during degradation and made available for dolomite precipitation (Gebelein & Hoffman 1973). Some etched samples of dolomicrite show sparite/pyrite pseudomorphs after gypsum. The precipitation of dolomite would have lowered the Mg/Ca ratio of the lake water/pore water and made calcium ions available for gypsum precipitation.

Organic carbon contents

The organic carbon contents of the lithologies A–D range up to 4.5% (average 3%), 2.5% (1.5%), 1.5% (0.8%) and 0.8% (0.3%) respectively (based on 100 samples, including those reported in Parnell [1985a]). The carbonate laminites are clearly the richest in organic matter. All carbonate laminites contain traces of hydrocarbon. Oil bleeds from freshly opened joints, tension fractures are infilled with solid hydrocarbon, and fossil fish are partly replaced by hydrocarbon. In a 162 m section figured by Donovan (1980), the lithologies A–D account for 6.5, 11.0, 38.5 and 44.0% of the whole respectively, equivalent to an average organic carbon content of 0.8% over the 162 m. Lithology C, deposits of a shallow impermanent lake, accounts for the highest proportion of organic carbon in the total section. It should be emphasized that some cycles are incomplete and do not include the carbonate laminite. The symmetrical lacustrine cycles represent sedimentation in the basin centre. Mixed lacustrine and fluviatile cycles are more characteristic of sedimentation at/near the basin margins. The organic-rich carbonate laminites and black mudrocks may occur in cycles proximal or distal to the basin margin as they represent high lake levels, but the average organic carbon content over a whole cycle is higher in the distal setting.

The difference in organic carbon content between the carbonate laminites and other dark mudrocks is indicated in Fig. 3 (original Rock-Eval pyrolysis data in Parnell 1985a). The carbonate laminites are generally richer in carbon but a cross-plot against hydrogen index (HI, an expression of hydrocarbon yield which takes account of TOC) shows that the hydrocarbon yield per organic carbon content in these samples is also richer. The cross-plot shows that the total

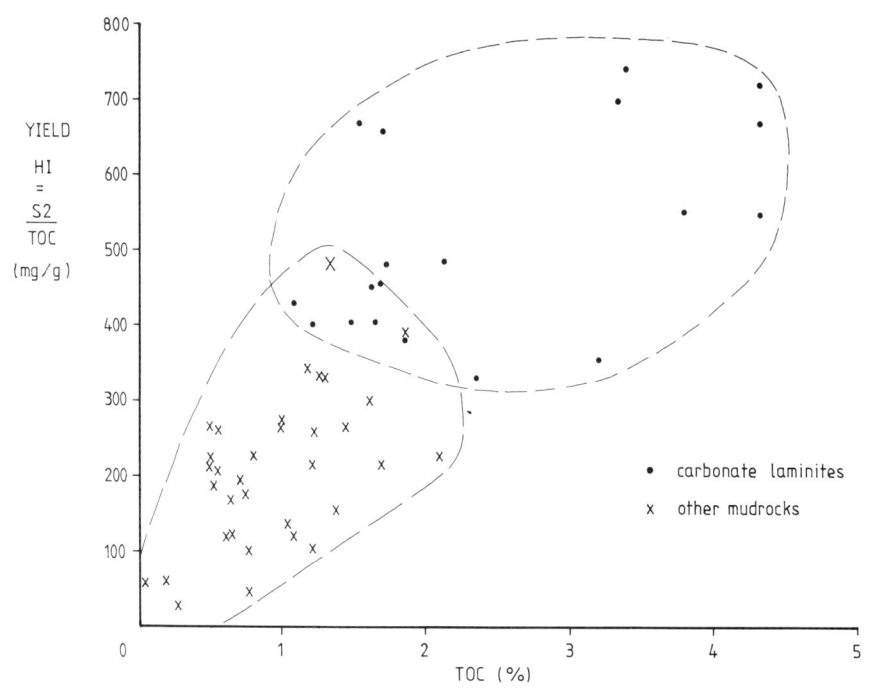

FIG. 3. Cross-plot of hydrogen index against total organic carbon for Middle ORS samples. Note increase in yield with TOC content and greater yield/TOC of carbonate laminites. Data from Parnell (1985a).

yield does not increase proportionally with TOC content, and the yield per organic carbon content is greater in the richer samples than in the lean samples. The data could indicate that in the leaner samples the proportion of terrigenous (land plant) or inert organic matter is greater. This would be acceptable geologically because the leaner samples represent shallower water environments with evidence for sediment transport, and geochemically because the proportion of migrated bitumen in the TOC content is greater in the leaner samples (Parnell 1985a, and unpublished petrographic data of A. C. Hutton). Rock-Eval pyrolysis experiments by Katz (1981) have yielded similar data which show a greater yield from samples with a carbonate matrix as compared to an argillaceous matrix, and an increasing yield per carbon content in richer samples. It is therefore possible that the pyrolysis technique exaggerates the yield from carbon-rich samples relative to the others.

Cherts

Stratigraphic and geographic distribution

Beds of replacement chert nodules and laminae occur in the organic-rich lithologies A and B, which are believed to represent permanent lacustrine sedimentation. Some carbonate laminites (lithology A) include thin (up to 2 cm) beds of laminite poor in carbonate (lithology B). Although chert occurs in both phases, nodules are concentrated in the carbonate-poor beds (Figs 6a and b).

Cherts are recorded in carbonate laminites at localities throughout the main outcrop of lacustrine deposits in Orkney and Caithness (Fig. 4). The relative importance of different lithologies at each stratigraphic level in Caithness has been assessed by Donovan et al. (1974). Their data show that carbonate laminites are an important (>5%) component of two particular stratigraphic intervals, the Robbery Head/Lybster Subgroups and the Mey Subgroup (see below). A similar distribution of carbonate laminites is found in Orkney. Cherts occur in the carbonate laminites deposited during both intervals. The several published records of chert (Steavenson 1928; Donovan et al. 1974; Mykura 1976; Donovan 1978; Fortey & Michie 1978) can be attributed to beds belonging to these intervals. A recent examination of the main sections in southern Orkney and Caithness has shown that cherts are more important than hitherto has been recognized. Carbonate laminites contain chert more often than not. In many cases the chert is difficult to detect except where highlighted by differential weathering. Because the carbonate laminites which host the cherts have the highest hydrocarbon source-rock potential, both in terms of total yield and yield per volume of organic matter, information which can be gleaned from the cherts about the lacustrine environment is valuable.

Organic-rich carbonate laminites are prominent in the Lower Stromness Group of Orkney (Robbery Head and Lybster Subgroups of Caithness). This Group is followed by an unusually thick (60 m) cycle containing the Sandwick Fish Bed (Niandt Limestone of Caithness). The cycle consists predominantly of laminite and represents the period of maximum expansion of lacustrine conditions (Mykura 1983). Chert occurs in each of the carbonate laminites in the Lower Stromness Group succession exposed in Orkney. Two prominent beds of nodular chert even form marker horizons (Mykura 1976). The succession is exposed in at least eight sections in south-west Orkney; chert beds can be identified in all of these sections (Parnell 1986a). Carbonate laminites in equivalent sections in Caithness, beneath the Niandt Limestone at Niandt and Halkirk, similarly yield chert (e.g., in the Robbery Head Limestone at Niandt, ND 222333). The carbonate laminite and other laminites within the Sandwick Fish Bed cycle near Stromness, Orkney (HY 243080), contain numerous thin (<2 mm) chert lenses.

Overlying the Sandwick Fish Bed cycle, the Upper Stromness Group and lower Rousay Group in Orkney (Latheron and Ham-Scarfskerry Subgroups of Caithness) contain a few carbonate laminites. In the upper Rousay Group (Mey Subgroup of Caithness) red beds become increasingly important, but some prominent carbonate laminites occur, commonly containing chert. Examples occur in each of the three Mey Subgroup outcrops in Caithness, at Pennyland (ND 108690), Harrow (ND 288743) and Ackergill (ND 358545).

In both Orkney and Caithness the cherts occur in sequences which also include tuffaceous sediments. There are several tuffaceous horizons in the Lower Stromness Group/Sandwick Fish Bed sequence of Orkney (Mykura 1976), and in the equivalent sequences at Halkirk and Niandt, Caithness (Fortey & Michie 1978).

Magadi-type chert model

A detailed petrology of cherts in the basin is given elsewhere (Parnell 1986a). In summary, the cherts are believed to be early-diagenetic magadi-type cherts, i.e., cherts converted from a hydrated sodium silicate precursor (magadiite) by loss of

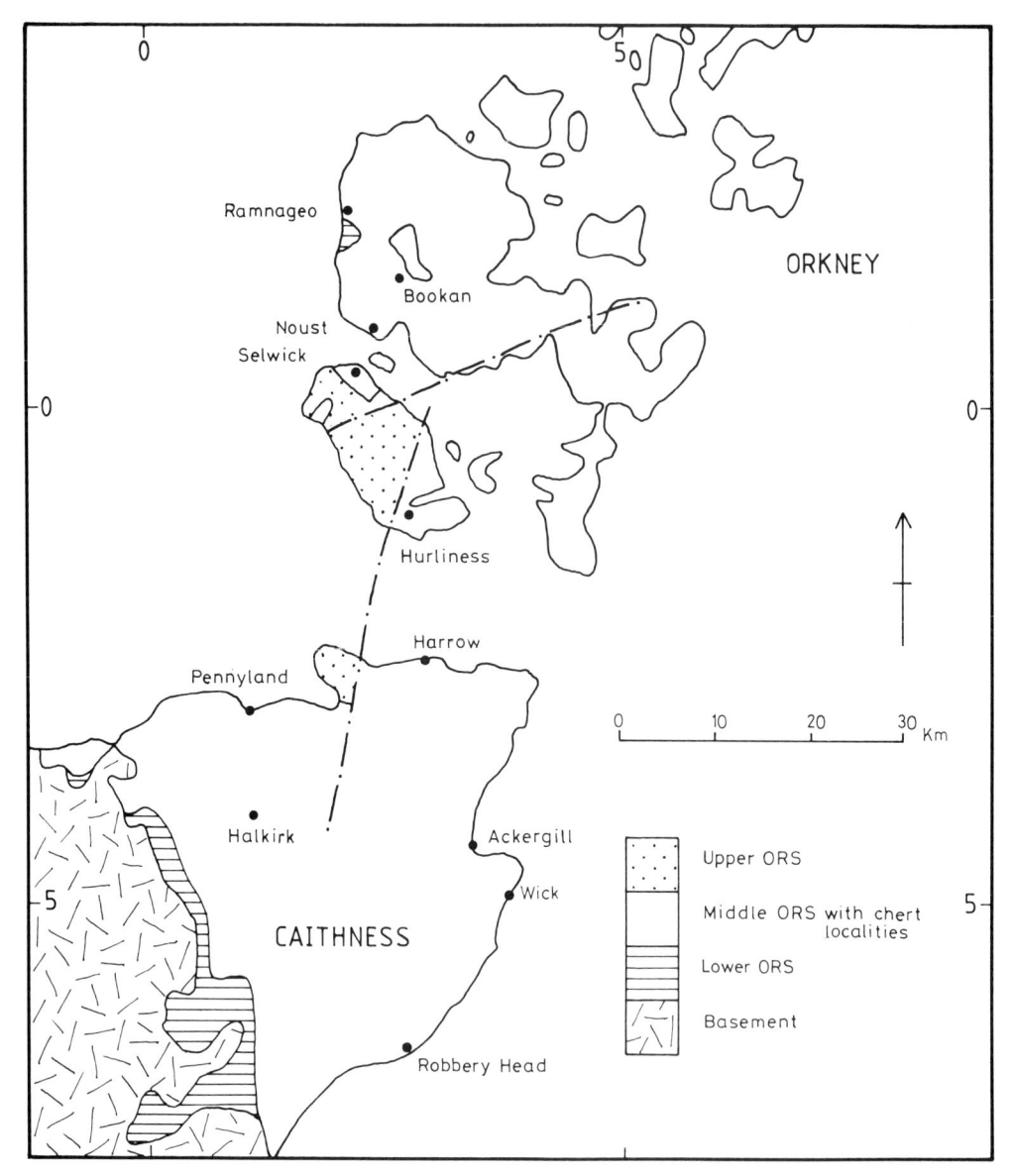

FIG. 4. Location of chert-bearing sections in Middle ORS of Orkney and Caithness. Further details of sections in south-west Orkney in Parnell (1986a). Cherts occur in organic-rich laminites in all sections.

water and release of sodium into solution. They exhibit salt-casts, shrinkage and soft-sediment deformation features which characterize this type of chert (Eugster 1969; 1985). The chert includes length-slow chalcedony and zebraic chalcedony, which indicates alkaline conditions of formation (Folk & Pittman 1971). Lamination within the cherts is continuous with lamination in the host rock, suggesting that the precursor magadiite was precipitated by groundwater reactions rather than directly in the lake water column (see Rooney et al. 1969 for a modern analogue in Oregon). The chert occurs as laminae and as tabular to vertically elongate nodules (Fig. 6b). Elongate nodules are linear/polygonal in plan (Fig. 6c) with central fractures up which sediment has been injected (Fig. 6b). The fractures, with injected sediment, continue outside the chert nodules. Those fractures were interpreted by Parnell (1983a) as desiccation cracks, but they do not contain sediment infillings from above and they occur in carbonate laminites which are thought to be the deposits of a permanent stratified lake. They may alternatively be due to syneresis during the magadiite–chert transformation or induced by sediment injection during compaction. The chert laminae and nodules contain radiating pseudomorphs after an evaporite mineral.

Pseudomorphs within cherts

The pseudomorphs are isolated crystals, sheaves of crystals and radiating crystal clusters up to 1 mm diameter (Fig. 6d). Individual crystals have an acicular habit, with parallel or slightly divergent lengthwise faces and occasionally well-preserved terminal faces. The best preserved examples show that the primary mineral was monoclinic. They cross-cut depositional lamination and sediment is incorporated between individual blades. Pseudomorphs occur particularly adjacent to the central fractures in the chert nodules. The pseudomorphs are presently quartz (single crystals or mosaic quartz), chalcedony, or carbonate replacive after quartz.

Pseudomorph-bearing chert laminae are abundant and show good preservation in the lower part of the Sandwick Fish Bed cycle at Noust (HY 242080), including a carbonate laminite (Fig. 5B). This section shows no evidence for desiccation, and pseudomorphs occur in the fish bed *sensu stricto* (see Mykura 1976), in which fossil fish are particularly abundant and relatively complete. Fossil fish even occur within the pseudomorph-bearing layers. The pseudomorph-bearing section (base of Fig. 5B) immediately overlies a 0.5 m section which shows an upward transition from beds containing desiccation cracks through beds with ptygmatic sedimentary structures (see Donovan & Foster 1972) to a bed of mud-flake breccia. Another chert-bearing bed, 14 m below the Sandwick Fish Bed, occurs above and below beds containing stromatolites/stromatolite debris (HY 243080, 227092). However there is no evidence for sediment transport (current and wave ripples, coarse silt lenses, scour surfaces, mud-flake breccias) in the chert beds. The chert laminae and nodules are rarely laterally continuous. They show an intermittent lateral development, and thin laminae cannot be traced for more than a few metres (compare two sections through the same bed in Fig. 5B, only 100 m apart but with chert laminae preserved at different horizons. All laminae contain pseudomorphs).

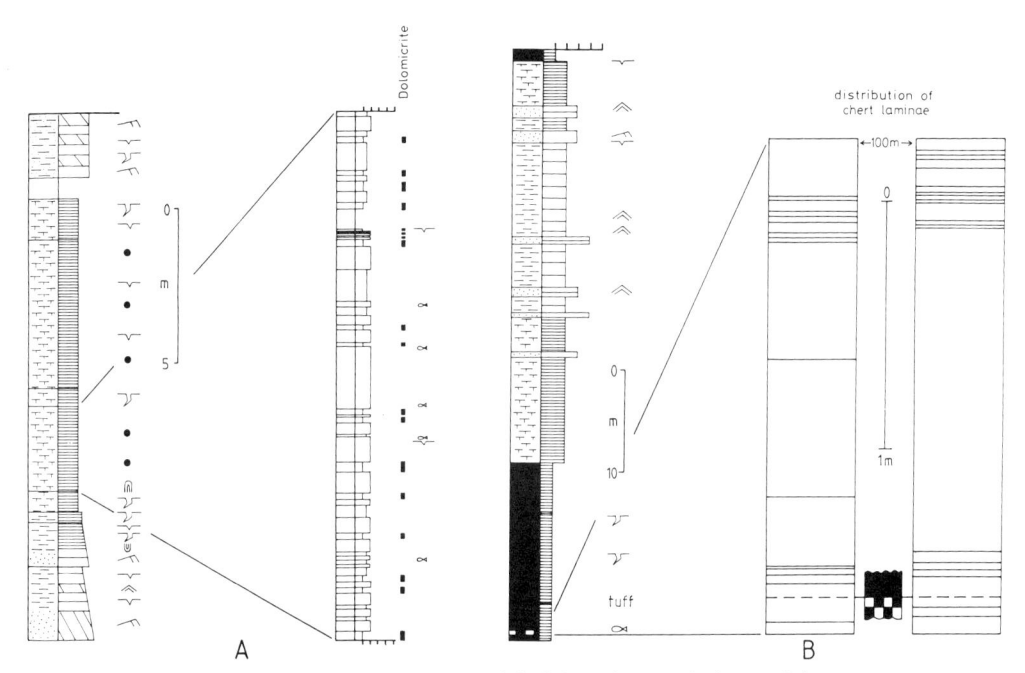

FIG. 5. (A) Sequence in Upper Stromness Group, Burwick, Orkney (ND 442835), containing numerous dolomicrite beds. Dolomicrites occur with and without evidence for exposure, and locally contain pseudomorphs after gypsum. (B) Log through Sandwick Fish Bed cycle, Noust (HY 242080), a thick lacustrine unit. Lower part of bed is magnified from two localities 100 m apart, repeated by faulting, to show rapid lateral variation in distribution of chert laminae. Key as in Fig. 1.

FIG. 6. (a) Chert nodules in organic-rich laminites, developed preferentially in carbonate-poor layers (light) rather than carbonate-rich phases (grey), Ramnageo (HY 225173). (b) Section through chert nodule in organic-rich laminite, showing continuation of laminae between nodule and host, differential compaction of nodule and host, and vertical fracture within nodule, Noust (HY 244079) (×1.5). (c) Chert nodules exhibiting polygonal pattern, in laminites, Ramnageo. (d) Pseudomorphs after (?) trona within chert nodule, Sandwick Fish Bed, Noust (NY 242080) (×26).

Other features relating to evaporite deposits include the occurrence of pseudomorphs after borates (Westoll 1979) and authigenic aegirine in Caithness. The aegirine occurs in proximity to cherts within organic-rich laminite (Fortey & Michie 1978). Pseudomorphs after gypsum occur in the Orcadian Basin Middle ORS, in dolomicrites, siltstones and sandstones (Parnell 1985b), but I have not observed them in organic-rich sediments. They have a rhomboid morphology and are commonly pyritized. The pseudomorphs within the cherts have a distinct morphology and are not pyritized.

Discussion

Chemistry of lake waters

Occurrences of magadiite and magadi-type chert are the deposits of saline, alkaline lakes rich in dissolved sodium carbonate/bicarbonate (Hay 1968). Sodium carbonate brines have a high pH (generally > 9.5; Hay 1968), and the high solubility of silica in such brines is important to the precipitation of magadiite. Jones et al. (1967) record 2700 ppm dissolved silica in the type locality at Lake Magadi, Kenya, a consequence of evaporative concentration. Magadiite can be precipitated from silica-rich sodium carbonate brines as a result of mixing with fresh water which lowers the pH (Eugster et al. 1968). In a stratified lake with fresh water overlying a sodium carbonate brine, the interface between waters of different composition is potentially a zone of magadiite precipitation (Hay 1968). The Orcadian cherts appear to have been precipitated by groundwaters (see above). However, the chemistry of the groundwaters is likely to reflect the chemistry of the lake, and the occurrence of magadi-type cherts does suggest that the lake waters were alkaline.

Identity of pseudomorphs

The most plausible monoclinic precursors for the pseudomorphs are gypsum and trona ($Na_2CO_3 \cdot NaHCO_3 \cdot 2H_2O$). Pseudomorphs after gypsum have not been recorded in organic-rich sediments in the basin. The lack of pyritization of radiating pseudomorphs, which might be expected in anoxic sediments, suggests that they may not be a sulphate mineral. If the cherts are magadi-type then gypsum is a very unlikely precursor. Calcium supplies are rapidly exhausted in an alkaline basin so that gypsum saturation is never achieved (Hardie & Eugster 1970). Trona is much more likely to be associated with magadi-type cherts (Eugster 1985) which, as noted above, are deposits of saline lakes rich in dissolved sodium carbonate/bicarbonate. The magadiite–chert transformation involves the release of sodium ions which become available for the precipitation of trona and other sodium carbonates.

Radiating clusters of trona crystals with similar crystal morphology are figured from the Eocene Green River Formation by Bradley & Eugster (1969), Deardorff & Mannion (1971) and Birnbaum & Radlick (1982), although the Eocene examples are generally larger. There is supporting evidence for the importance of sodium in the brine chemistry, in the occurrences of authigenic aegirine and pseudomorphs after borates, and sodic plagioclase overgrowths in some dolomitic sediments. The only other documented occurrence of authigenic aegirine in a sedimentary sequence is in the trona-bearing sediments of the Green River Formation (Milton et al. 1960). The association of magadi-type cherts with tuffaceous sediments has also been noted in sequences elsewhere (e.g., Eugster 1985). The association is significant to the chemistry of the lake waters. Fortey & Michie (1978) suggest that appreciable quantities of volcanic material rich in sodium would have been available to the Orcadian Basin during these times.

Electron microscopical examination of the cherts has failed to identify any surviving trona. This is hardly surprising. Even in the Eocene Green River Formation, many soluble sodium minerals are recorded only in well cuttings. Trona is highly soluble and therefore a transient mineral and we are fortunate to have any evidence for its existence in sediments of Devonian age.

Conditions of trona precipitation

Chemistry. Trona and borates are unique to saline lake deposits (Picard & High 1972). For trona to precipitate the parent brine must have been rich in bicarbonate ions (HCO_3^-), the $HCO_3^-/Ca^{2+} + Mg^{2+}$ molar ratio must have exceeded unity and other anions (Cl^-, SO_4^{2-}) must be insignificant (Eugster & Hardie 1975). Alkaline earth carbonates must have already precipitated such that the brine was dominated by Na^+ and carbonate species (Eugster 1970). The trona may grow *within* the carbonates. Several of the Orcadian occurrences are in carbonate laminite (calcite or dolomite), and trona in the Green River Formation occurs in dolomitic mudstone (Smoot 1983).

Trona deposits in the Green River Formation have been the subject of much discussion by Eugster and other workers (various papers cited above) who have debated, without consensus, whether the trona was deposited from the saline hypolimnion of a stratified lake or as a result of direct evaporation of a very shallow lake. Lakes which precipitate trona today are not permanently stratified: Lake Magadi becomes seasonally stratified when pluvial run-off covers sodium-rich brines, but the epilimnion then evaporates, stratification breaks down and ionic concentrations eventually reach the trona-precipitating stage (Eugster 1970). Eugster (1985) prefers a playa-lake model for trona precipitation in the Green River Formation. However I believe that in the Orcadian Basin, trona is precipitated in the sediment from connate waters inherited from the saline hypolimnion. The cross-cutting relationship with lamination suggests that they grew displacively during early diagenesis. The host sediments do not exhibit sedimentary structures like ripple marks, cross-bedded channel sands and desiccation features shown by some trona-bearing sediments of the Green River Formation (Birnbaum & Radlick 1982). It is possible that trona, along with other minerals, *was* deposited in efflorescent crusts on lake-margin flats due to intense evaporation, and was then dissolved during lacustrine transgression. This process would have significantly increased the salinity of the lake waters (Jones *et al.* 1977). As trona is highly soluble, selective dissolution of trona would have created a brine more enriched in sodium than the waters flowing into the basin. Although there is evidence for gypsum in the lake flat sediments (see above) and gypsum would not be cogenetic with trona, some modern playas show frequent fluctuations in brine chemistry and alternate between gypsum and trona precipitation.

Sub-lacustrine precipitation. The mode of occurrence of the cherts and their enclosed pseudomorphs suggests a sub-lacustrine precipitation of trona rather than precipitation in a salt-pan. The section below the cherts at Noust, containing desiccation cracks and mud-flake breccias, represents a lacustrine transgression during which erosion of lake-margin flats occurred. The fact that trona was precipitated at times when the lake was permanent, and hence at relatively high level and low salinity, is an argument for stratification.

The intermittent nature of the chert laminae, exemplified by their different distribution through two adjacent sections through the same sequence (Fig. 5B) suggests that the conditions in which chert/trona was precipitated may have been more prevalent than is now apparent. It is likely that trona would have been precipitated both within and without the chert laminae/ nodules, but has only been preserved within them. Despite the high solubility of trona, a sub-lacustrine setting for trona precipitation is not an unrealistic model. There are several other instances of trona precipitation in the organic-rich sediments of alkaline lake deposits.

A comparable example of pseudomorphed sodium carbonate minerals in an alkaline playa-lake sequence is in the Cambrian Officer Basin, South Australia. As in the Orcadian Basin, the sequence includes organic-rich mudrocks with good source potential, within which are magadi-type chert nodules. The cherts are cross-cut by rosettes of bladed crystals, thought to be after trona (White & Youngs 1980). Oil bleeds from cavities in both the Cambrian cherts (White & Youngs 1980) and the Orcadian cherts (Parnell 1983b). Radiating crystal moulds after (?) trona similarly occur in laminated black mudstones, bearing fossil fish, in the Jurassic East Berlin Formation, Connecticut (Demicco & Gierlowski Kordesch 1986). The black mudstones are regarded as the deposits of a perennial lake with anoxic bottom waters (Hubert *et al.* 1976) and are a source of local hydrocarbon shows (Parnell 1986b). A recent analogue for the occurrence of magadiite and acicular crystals of trona in basinal facies organic-rich laminites occurs in the Holocene sediments of Lake Bogoria, northern Kenya rift valley (Tiercelin *et al.* 1982). At present the lake is a closed-basin saline lake with concentrated sodium carbonate/bicarbonate brines (Tiercelin *et al.* 1981). The lake is shallow (<15 m) but meromictic, and shows evidence for fluctuations in level due to short-term climatic cycles. A comparison between regressive–transgressive cycles in the Orcadian lake with those in Quaternary African deposits has already been made by Donovan (1980).

Importance of organic-rich sediment

Magadiite and trona were precipitated in a sub-lacustrine setting because (1) the waters in the sediment inherited from the hypolimnion were sodium carbonate brines, and (2) a supply of CO_2 was available in the sediments.

Sodium carbonate-rich bottom waters in the Orcadian lake would have enhanced density/ chemical stratification of the lake. The alkaline and highly anoxic nature of the hypolimnion would have prevented macrofaunal activity, explaining the lack of bioturbation and the fine lamination of the laminites. The epilimnion was by contrast oxic and supported planktonic algae, whose deposits were preserved in the hypolimnion. Anaerobic decay in the organic-rich sediment deposited during permanent lacustrine intervals could have played an important role in magadiite precipitation by increasing the P_{CO_2} and lowering the pH of the groundwaters. The pH could also be reduced by dilution with run-off but this is considered unlikely in the case of cherts at the base of a 20 m (at least) section of organic-rich permanent lake deposits (Fig. 5B).

Trona precipitation may similarly have been enhanced by the release of biogenically derived CO_2. CO_2 can induce precipitation in a brine undersaturated with respect to trona (Bradley & Eugster 1969; Abd-El-Malek & Rizk 1963). Bacterial generation of CO_2 (e.g., via fermentation reactions) continues within organic-rich mud during burial, and trona crystals grow within the sediment (Eugster & Smith 1965). As P_{CO_2} increases, the solubility of trona decreases markedly (Eugster 1966) and trona precipitates. Where pseudomorphs occur adjacent to fractures in chert nodules, trona appears to have been precipitated by waters escaping from the sediment during syneresis or compaction.

The occurrence of magadi-type cherts and pseudomorphs after trona in rocks which have relatively good source-rock potential is probably not coincidental. The high P_{CO_2} that inevitably results from anaerobic decay of organic matter in the hypolimnion may have been essential for their formation.

Summary

Lacustrine rocks with petroleum source-rock potential occur in the Middle ORS of the Orcadian Basin. Middle ORS sequences in Orkney and Caithness consist of fluviolacustrine cycles, and lacustrine cycles showing evolution from a stratified lake through shallow permanent to impermanent lake conditions. The lithologies representing quiescent conditions show the highest concentration of organic carbon, but the bulk of the organic carbon over the whole sequence is preserved in the shallow water lithologies.

Many dark carbon-rich beds are also carbonate-rich. Sediments of shallow water lithologies contain dolomicrite beds, whilst the deposits of a stratified lake, with the highest source-rock potential, are carbonate laminites. The carbonate laminites, and some other organic-rich laminites, contain cherts which are more widespread than has previously been recognized. The cherts have been interpreted as magadi-type cherts (Parnell 1986a), which are deposited today in saline, alkaline lakes rich in dissolved sodium carbonate/bicarbonate. They contain pseudomorphs after a monoclinic mineral which may have been the sodium carbonate–bicarbonate mineral trona. Trona is also a deposit of saline, alkaline lakes. Both chert and (?) trona show evidence of growth within the organic-rich sediment at the bottom of the Orcadian lake rather than direct precipitation from the lake waters. The chemistry of the pore waters in the sub-lacustrine sediment would to a large degree have reflected the chemistry of the overlying lake waters. The epilimnion, where organic productivity was high, may have been comparatively fresh even during the periods when chert was precipitated. However, the hypolimnion was alkaline and anoxic during those periods, hence the preservation of organic matter was good and the deposits of the Orcadian lake in a stratified state are the best quality source rocks in the basin.

ACKNOWLEDGEMENTS: I am grateful to E. Lawson and S. Watters for cartographic and photographic assistance. T. Janaway kindly supplied Fig. 2b and helped record Fig. 5A.

References

ABD-EL-MALEK, Y. & RIZK, S. G. 1963. Bacterial sulphate reduction and the development of alkalinity. III. Experiments under natural conditions in the Wadi Natrun. *Journal of Applied Bacteriology*, **26**, 20–26.

AITKEN, J. D. 1967. Classification and environmental significance of cryptalgal limestones and dolomites, with illustrations from the Cambrian and Ordovician of southwestern Alberta. *Journal of Sedimentary Petrology*, **37**, 1163–1178.

BIRNBAUM, S. J. & RADLICK, T. M. 1982. A textural analysis of trona and associated lithologies, Wilkins Peak Member, Eocene Green River Formation, Southwestern Wyoming. In: HANDFORD, C. R., LOUCKS, R. G. & DAVIS, G. R. (eds), *Depositional and Diagenetic Spectra of Evaporites—a Core Workshop. Society of Economic Paleontologists and Mineralogists Core Workshop*, **3**, 75–99.

BRADLEY, W. H. & EUGSTER, H. P. 1969. Geochemistry and paleolimnology of the trona deposits and associated authigenic minerals of the Green River Formation of Wyoming. *Professional Paper. United States Geological Survey*, **496–B**, 1–71.

DEARDORFF, D. L. & MANNION, L. E. 1971. Wyoming trona deposits. *Contributions to Geology, University of Wyoming*, **10**, 25–37.

DEMICCO, R. V. & GIERLOWSKI KORDESCH, E. 1986. Facies sequences of a semi-arid closed basin: the Lower Jurassic East Berlin Formation of the Hartford Basin, New England, U.S.A. *Sedimentology*, **33**, 107–118.

DONOVAN, R. N. 1975. Devonian lacustrine limestones

at the margin of the Orcadian Basin, Scotland. *Journal of the Geological Society*, **131**, 489–510.
—— 1978. The Orcadian Basin. *In*: FRIEND, P. F. & WILLIAMS, B. P. J. (eds), *Field Guide to Devonian of Scotland, the Welsh Borderlands and South Wales*, Palaentological Association.
—— 1980. Lacustrine cycles, fish ecology and stratigraphic zonation in the Middle Devonian of Caithness. *Scottish Journal of Geology*, **16**, 35–50.
—— & FOSTER, R. J. 1972. Subaqueous shrinkage cracks from Caithness Flagstone Series (middle Devonian) of northeast Scotland. *Journal of Sedimentary Petrology*, **42**, 309–317.
——, —— & WESTOLL, T. S. 1974. A stratigraphical revision of the Old Red Sandstone of Northeastern Caithness. *Transactions of the Royal Society of Edinburgh*, **69**, 167–201.
DOUGLAS, A. G., HALL, P. B. & SOLLI, H. 1983. Comparative organic geochemistry of some European oil shales. *In*: MIKNIS, F. P. & MCKAY, J. F. (eds), *Geochemistry and Chemistry of Oil Shales*, American Chemical Society, 59–84.
EUGSTER, H. P. 1966. Sodium carbonate–bicarbonate minerals as indicators of P_{CO_2}. *Journal of Geophysical Research*, **71**, 3369–3377.
—— 1969. Inorganic bedded cherts from the Magadi area, Kenya. *Contributions in Mineralogy and Petrology*, **22**, 1–31.
—— 1970. Chemistry and origin of the brines of Lake Magadi, Kenya. *Special Paper. Mineralogical Society of America*, **3**, 215–235.
—— 1985. Oil shales, evaporites and ore deposits. *Geochimica et Cosmochimica Acta*, **49**, 619–635.
—— & HARDIE, L. A. 1975. Sedimentation in an ancient playa-lake complex: the Wilkins Peak Member of the Green River Formation of Wyoming. *Bulletin of the Geological Society of America*, **86**, 319–334.
—— & SMITH, G. I. 1965. Mineral equilibria in the Searles Lake evaporites, California. *Journal of Petrology*, **6**, 473–522.
—— JONES, B. F. & SHEPPARD, R. A. 1968. New hydrous sodium silicates from Kenya, Oregon and California: possible precursors of chert (abstr.). *Special Publication. Geological Society of America*, **115**, 60–61.
FOLK, R. L. & PITTMAN, J. S. 1971. Length-slow chalcedony: A new testament for vanished evaporites. *Journal of Sedimentary Petrology*, **41**, 1045–1058.
FORTEY, N. J. & MICHIE, U. McL. 1978. Aegirine of possible authigenic origin in Middle Devonian sediments in Caithness, Scotland. *Mineralogical Magazine*, **42**, 439–442.
GEBELEIN, C. D. & HOFFMAN, P. 1973. Algal origin of dolomite laminations in stromatolitic limestone. *Journal of Sedimentary Petrology*, **43**, 603–613.
GEIKIE, A. 1879. On the Old Red Sandstone of Western Europe. *Transactions of the Royal Society of Edinburgh*, **28**, 345–452.
GODWIN-AUSTEN, R. 1856. On the possible extension of the Coal-Measures beneath the south-eastern part of England. *Quarterly Journal of the Geological Society*, **12**, 38–73.

HARDIE, L. A. & EUGSTER, H. P. 1970. The evolution of closed-basin brines. *Special Publication. Mineralogical Society of America*, **3**, 273–290.
HAY, R. L. 1968. Chert and its sodium-silicate precursors in sodium-carbonate lakes of East Africa. *Contributions in Mineralogy and Petrology*, **17**, 255–274.
HUBERT, J. F., REED, A. A. & CAREY, P. J. 1976. Paleogeography of the East Berlin Formation, Newark Group, Connecticut Valley. *American Journal of Science*, **276**, 1183–1207.
JONES, B. F., EUGSTER, H. P. & RETTIG, S. L. 1977. Hydrochemistry of the Lake Magadi Basin, Kenya. *Geochimica et Cosmochimica Acta*, **41**, 53–72.
——, RETTIG, S. L. & EUGSTER, H. P. 1967. Silica in alkaline brines. *Science*, **158**, 1310–1314.
KATZ, B. J. 1981. Limitations of Rock-Eval pyrolysis for typing organic matter (abstr.). *Bulletin of the American Association of Petroleum Geologists*, **65**, 944.
MARSHALL, J. E. A., BROWN, J. F. & HINDMARSH, S. 1985. Hydrocarbon source rock potential of the Devonian of the Orcadian Basin. *Scottish Journal of Geology*, **21**, 301–320.
MILTON, C., CHAO, E. C. T., FAHEY, J. J. & MROSE, N. E. 1960. Silicate mineralogy of the Green River Formation of Wyoming, Utah. *Report of the Twenty-first Session of the International Geological Congress*, **21**, 171–184.
MYKURA, W. 1976. *British Regional Geology: Orkney and Shetland*. Institute of Geological Sciences, Edinburgh.
—— 1983. Old Red Sandstone. *In*: CRAIG, G. Y. (ed.), *Geology of Scotland*, Scottish Academic Press, Edinburgh, 205–251.
PARNELL, J. 1983a. Ancient duricrusts and related rocks in perspective: a contribution from the Old Red Sandstone. *Special Publication. Geological Society of London*, **11**, 197–209.
—— 1983b. The distribution of hydrocarbon minerals in the Orcadian Basin. *Scottish Journal of Geology*, **19**, 205–213.
—— 1985a. Hydrocarbon source rocks, reservoir rocks and migration in the Orcadian Basin. *Scottish Journal of Geology*, **21**, 321–336.
—— 1985b. Evidence for evaporites in the O.R.S. of Northern Scotland: replaced gypsum horizons in Easter Ross. *Scottish Journal of Geology*, **21**, 377–380.
—— 1986a. Devonian magadi-type cherts in the Orcadian Basin. *Journal of Sedimentary Petrology*, **56**, 495–500.
—— 1986b. Hydrocarbons and metalliferous mineralization in a lacustrine rift basin: the Hartford–Deerfield Basin, Connecticut Valley. *Neues Jahrbuch fur Mineralogie Abhandlungen*, **154**, 93–110.
PICARD, M. D. & HIGH, L. R. 1972. Criteria for recognizing lacustrine rocks. *Special Publication. Society of Economic Paleontologists and Mineralogists*, **16**, 108–145.
ROONEY, T. P., JONES, B. F. & NEAL, J. T. 1969. Magadiite from Alkaline Lake, Oregon. *American Mineralogist*, **54**, 1034–1043.
SMOOT, J. P. 1983. Depositional subenvironments in an

arid closed basin; the Wilkins Peak Member of the Green River Formation (Eocene), Wyoming, U.S.A. *Sedimentology*, **30**, 801–827.

STEAVENSON, A. G. 1928. Some geological notes on three districts in northern Scotland. *Transactions of the Geological Society of Glasgow*, **18**, 193–233.

TIERCELIN, J. J., PERINET, G., LE FOURNIER, J., BIEDA, S. & ROBERT, P. 1982. Lacs du rift est-Africain, exemples de transition eaux douces-eaux salées: le lac Bogoria, rift Gregory, Kenya. *Memoiré Societé Géologique de France*, **144**, 217–230.

——, RENAULT, R. W., DELIBRIAS, G., LE FOURNIER, J. & BIEDA, S. 1981. Late Pleistocene and Holocene lake level fluctuations in the Lake Bogoria basin, northern Kenya rift valley. *Palaeoecology of Africa*, **13**, 105–120.

WESTOLL, T. S. 1979. Devonian fish biostratigraphy. *Special Papers in Palaeontology*, **23**, 341–353.

WHITE, A. H. & YOUNGS, B. C. 1980. Cambrian alkali playa-lacustrine sequences in the northeastern Officer Basin, South Australia. *Journal of Sedimentary Petrology*, **50**, 1279–1286.

J. PARNELL, Department of Geology, Queen's University, Belfast, BT7 1NN, UK.

The lacustrine Burdiehouse Limestone Formation—a key to the deposition of the Dinantian Oil Shales of Scotland

G. W. F. Loftus & J. T. Greensmith

SUMMARY: Oil shales, and the less common limestones associated with them, represent freshwater lacustrine phases within the Dinantian Oil Shale Group at the eastern end of the Midland Valley of Scotland. The understanding of the areally extensive Burdiehouse Limestone Formation is fundamental to an appreciation of the palaeolimnology of the lake in which algal oozes (consisting mainly of planktonic algae analogous to the modern form *Botryococcus braunii*), later to become oil shales, accumulated. The lake, referred to as Lake Cadell, was shallow (<100 m) with gently sloping margins. It was meromictic and formed a closed system during deposition of the limestones, draining eastwards to the sea during oil shale deposition. The climate was wet tropical ensuring ideal conditions for a prolific flora of algae and vascular plants both around and within the lake, and providing a regular freshwater input. The lacustrine phase forms part of the cyclical Oil Shale Group succession in which each sedimentary cycle can be resolved into four phases indicating long-term lake expansion and contraction.

Geological setting

In the Midland Valley of Scotland the lowest major lithostratigraphic division of the Carboniferous is the Calciferous Sandstone Measures (MacGregor 1960) of Dinantian age. Within the Edinburgh area at the eastern end of the Midland Valley the Calciferous Sandstone succession can be lithologically subdivided into the 1100 m thick Cementstone Group overlain by the up to 2000 m thick Oil Shale Group (Francis 1983). The main outcrops of the latter occur around the Firth of Forth in West Lothian, Midlothian and southern Fife.

The Oil Shale Group sediments accumulated in a series of north–south trending tectonically and volcanically active basins (George 1958) and were for most of the time land-locked to the north, south and south-west. They periodically also became barred in the east cutting off the sea and forming a group of standing bodies of water, collectively referred to as Lake Cadell (Greensmith 1968) (Fig. 1). The relief, geological make-up, climate and drainage of the surrounding land areas determined the nature of the sediments deposited within the lakes and resulted in the complex interaction between freshwater lacustrine, transitional lagoonal and fully marine conditions recorded in the succession (Loftus 1985).

Fluvio-deltaic sandstones, siltstones and shales dominate the Oil Shale Group, forming some 90% by thickness. These were introduced into the lake system from nearby land sources and more distant northeasterly sources (Greensmith 1962, 1966) (Fig. 1). Sedimentation was also distinguished by the accumulation, at intervals, both of thin limestones, of which the Burdiehouse limestones are the best known, and kerogen-rich oil shales. Oil shales and Burdiehouse limestones are the most laterally persistent horizons in the Group. Ashes, marls, cementstones, ironstones, coals, seat earths and marine shales are more sporadic in appearance and constitute about 10% of aggregate thickness (Mitchell & Mykura 1962). Lateral facies changes are rapid in the Group, more especially among the siliciclastic terrigenes. The Burdiehouse limestones and the oil shales vary in thickness, but both tend to be laterally persistent, probably reflecting periods of lake expansion and diminished clastic input into depocentres.

Oil shales

The discovery in 1885 of oil shales at Broxburn in West Lothian stimulated economic interest in the Lothian succession and dominated geological interest in the region for the next century. Oil shale mining began in the same year, rapidly accelerated to 1913, virtually collapsed in 1919, and finally ceased in 1962. About half of the probable oil shale resources remain unexploited (Cameron & McAdam 1978). Industrial interest today is focused upon mineral oil believed to have originated from the kerogenous source rocks represented by the Oil Shale Group (Parnell 1984).

Composition and lithology

Oil shales form about 3% by thickness of the succession and consist of highly kerogenous

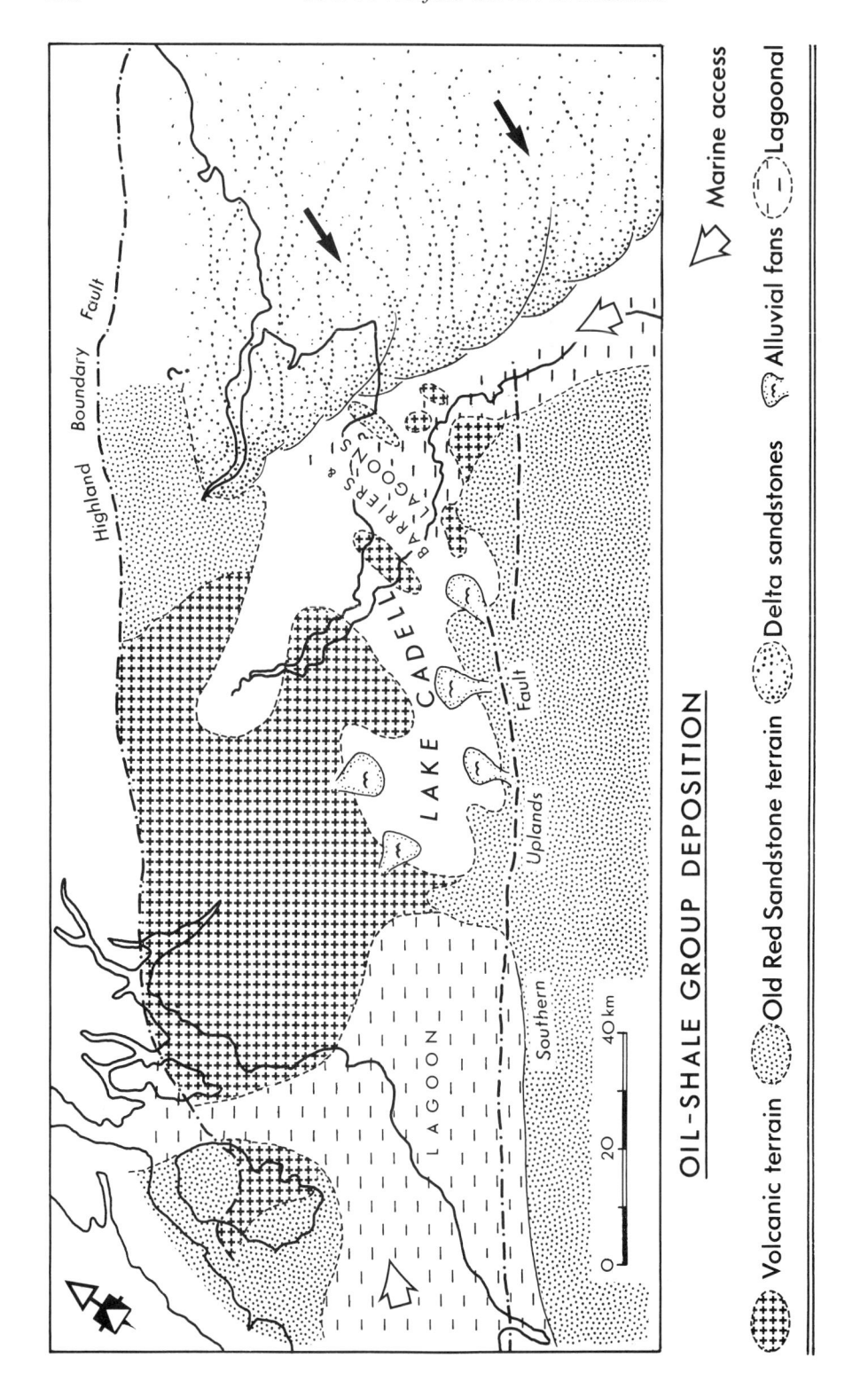

FIG. 1. Generalized Oil Shale Group geography at Burdiehouse Limestone Formation times. The eastern and north-eastern margins of Lake Cadell merged into a lagoonal–barrier complex with intermittent access to the open sea.

sediments ranging in thickness from a few centimetres to 5 m in a single seam. Some are developed as multiple seams e.g., the Broxburn Shale, where up to five seams constitute as much as 18 m aggregate thickness productive shale in 42 m of strata. The oil shales have total organic carbon contents of up to 35%, are mostly immature, and in the past have yielded up to 200 litres of petroleum spirit per tonne on retorting. There are 11 workable oil shale horizons that are widely distributed and can be traced in West Lothian, Midlothian and to a lesser extent, Fife. Certain horizons may also be present beneath the Central Coalfield to the west.

Thin-sections of the oil shales reveal a groundmass consisting of thinly laminated (20–30 μm) brown structureless organic material. The laminations are lenticular probably as a consequence of compaction and rarely extend laterally for more than a few millimetres. Anastamosing bituminous films resembling alginite B are present. The groundmass supports a variety of allochthonous grains, including translucent, orange to yellow resinous globular bodies (resembling alginite A, *ibid.*) that are considered to be of algal origin (Conacher 1917, 1938; Blackburn & Temperley 1936; MacGregor 1938; Greensmith 1968; Cane 1969; Parnell 1984), spore and plant detritus, fish remains, calcareous algae, a few ostracods and accessory quantities of detrital quartz, with detrital and autochthonous pyrite and carbonate.

Environments of deposition

A variety of depositional settings have been proposed for the oil shales. These range from marine (Robinson & Dineen 1967) to fresh water (Greensmith 1968) in either a lagoonal (Greensmith 1962), stratified lake (Moore 1968), paludal (Miknis *et al.* 1979) or playa context (Parnell 1984). As the oil shales are very fine grained and originally appear to have been very finely and evenly laminated (though the latter is now largely lost), they were probably deposited in a body of water unaffected by bottom turbulence and temporarily isolated from siliciclastic input. Moreover the absence of a marine fauna suggests that they were laid down in an environment protected from marine influence. In East Fife, the facies equivalents are predominantly sandstone bodies of fluvio-deltaic aspect. It is tongues of these sandstones in the form of barriers which appear to have intermittently blocked the entry of the sea into the Edinburgh Basin (Greensmith 1966). The lateral passage of oil shales into areally restricted coals (Anderson 1942; Richey 1942; Kennedy 1943) suggests that the body of water had relatively stable shorelines during each phase of oil shale accumulation, a view supported by an absence of evaporites. In fact there are no evaporites observed in the Oil Shale Group (contrast with Green River Formation of Wyoming; Bradley & Eugster 1969). The overall implication is that within the water body when the oil shales were forming there was a relative balance between recharge, evaporation and outflow to the east, maintaining a relatively steady level with only minor fluctuations (*cf.*, Hutchinson 1957).

The kerogenous components of the oil shales are generally argued to have been deposited as a result of the mass mortality of algal blooms at or near the oxygenated surface of a water body. Though difficult to identify, the species are analogous to the modern form *Botryococcus braunii* (Blackburn & Temperley 1936). The areas adjacent to the shorelines were colonized by a rich land flora (Cadell 1925), as evidenced by thin *in situ* coal seams, which provided a ready source for detrital plant remains. The transport of siliciclastic debris into the water body may have been partially inhibited by such vegetation (Moore 1968). In general, however, the oil shales show little post-depositional disturbance other than compaction. By analogy with modern lakes (e.g., Treese & Wilkinson 1982; Tiercelin *et al.* 1981) the oil shales probably collected as flocculent phytoplanktonic oozes in deeper water protected from decomposition in an hypolimnion by anoxic conditions, a view supported by the occurrence of framboidal pyrite believed to have precipitated as an autochthonous early diagenetic phase (Love & Amstutz 1966) and the absence of bioturbation. In this situation the algae were progressively transformed, perhaps by fungal and bacterial decay, into a denser, homogenized organic mass (Moore 1968).

The uniform constitution of the persistent oil shale levels suggests a low bottom relief of the water body. Depth is unknown but the water column must have been sufficient to accommodate a hypolimnion undisturbed by surface turbulence and an epilimnion capable of supporting large carnivorous fish. As slope deformation features and turbidites are extremely rare the implication is that bottom slopes were gentle.

Cyclic sedimentation

Recognition of the position occupied by oil shale seams in the context of depositional cycles is relevant to appreciating their depositional environment. This cyclicity has been evaluated and refined on many occasions with broadly consistent agreement (Richey 1935; Robertson 1948;

Greensmith 1962, 1968; Belt et al. 1975; Francis 1983). The ideal cycle illustrated in Fig. 2 is summarized as follows.

(1) A transgressional phase represented by shales interleaved by thin limestones or cementstone ribs and including rare marine bands.

(2) A lagoonal/lacustrine phase represented by the accumulation of oil shales and including occasional ostracodal limestones. Oil shales sometimes overlie marine bands. This phase frequently concluded with the development of algal, often stromatolitic limestones that are frequently sun-cracked and brecciated, indicating a contraction and lowering of the lagoon or lake levels.

(3) A progradational phase represented by broadly coarsening upward sequences of deltaic aspect. These deltaics are recognized in the Group and its contiguous equivalents in East Fife (Greensmith 1966), although the geometry of many of the sandstone bodies requires detailed investigation. The distal location of a cyclical sequence relative to delta source may exclude the development of the more proximal delta facies, hence cycles in the depocentres lack sandstone units (Fig. 3).

(4) An aggradational phase representing the final stages of delta progradation, confirmed by the presence of fining up, cross-bedded channel, bar, crevasse splay and washover sandstones and siltstones (Belt et al. 1975) laid down on the fluviodeltaic plain, but more significantly including fireclays and coals. The coals are areally restricted, suggestive of extensive braiding of the river systems and impersistence of peat producing conditions.

Causes of delta abandonment and retreat are various and include delta switching, phases of decreased subsidence in the receiving basin and periodic climatic change. The latter is capable of achieving a rise in water level resulting in delta retreat, as demonstrated in Quaternary East African lakes (I. Reid, pers. comm. 1984). A rise in the water level of Lake Cadell may well have been the dominant reason for delta retreat in the Oil Shale Group. The consequences of this retreat are expressed by the presence of ostracodal limestones such as the Burdiehouse limestones and oil shales, both of which are widespread over the basin and probably represent the maximum extent of the water body.

The Burdiehouse Limestone Formation

The Burdiehouse Limestone Formation occupies the same relative position in the sedimentary cycles as do the oil shales (Fig. 2). Moreover, the formation is closely associated with areally restricted oil shales, such as the Camps Shale in West Lothian, and is itself represented by calcite supported oil shale in depocentres (i.e., the profundal deep water facies). Limestones lend

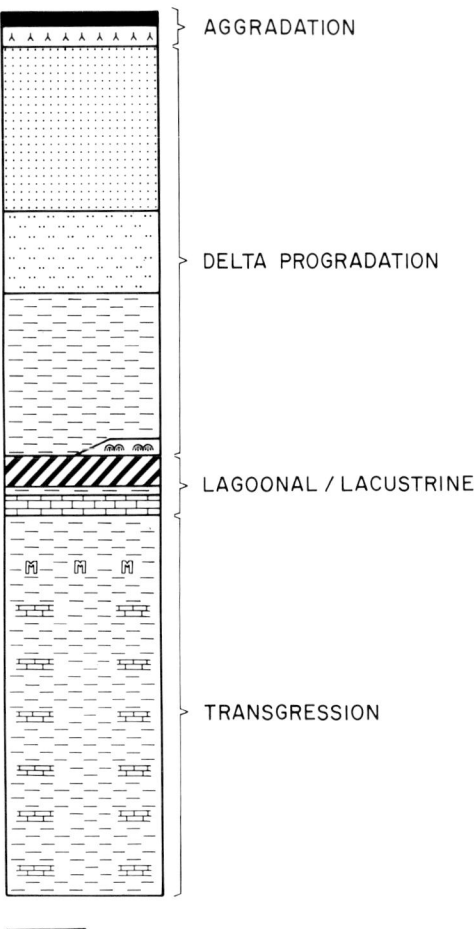

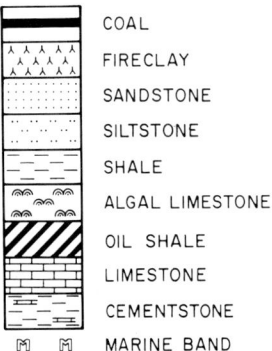

FIG. 2. Diagrammatic section of one complete ideal cycle in the Oil Shale Group.

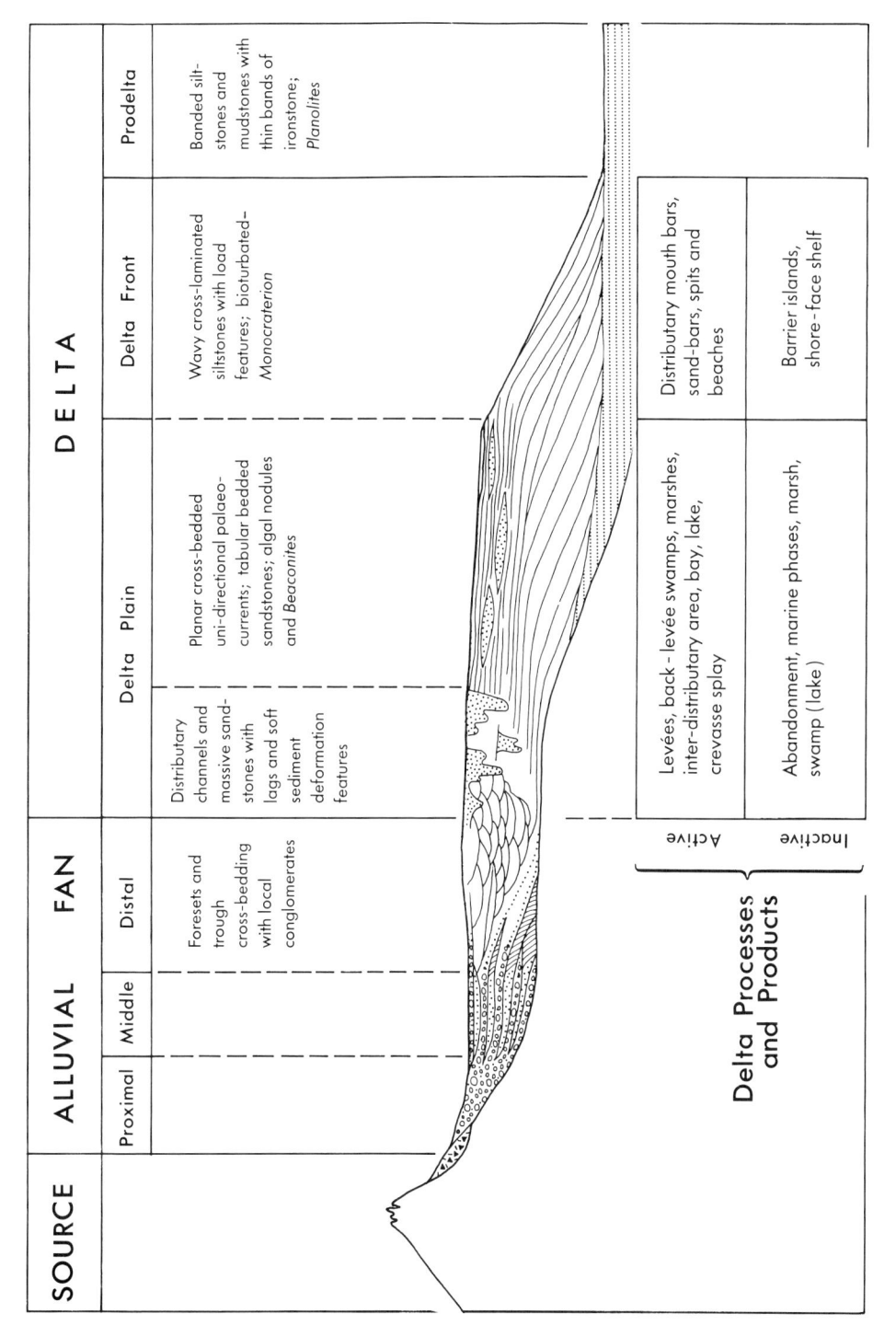

FIG. 3. Schematic cross-section through the alluvial fan-delta complex, illustrating the characteristic features of the sand bodies in the Oil Shale Group.

themselves more readily than the oil shales to environmental analysis and on this basis the Burdiehouse Limestone Formation has been investigated in more detail and used as an analogue for the oil shales. The Burdiehouse Limestone Formation is a major stratigraphic horizon in the Oil Shale Group, marking the boundary between the lower and upper divisions. It is laterally persistent being traceable throughout the Edinburgh area and into southern Fife.

Lithology

The formation consists of between one and six limestone members, from 16 m to 20 cm thick, interbedded variably with bituminous and calcareous shales, sandstone, seat earths and coals, depending upon its geographical position within the basin. The limestones typically reveal regular couplets (Fig. 4) consisting of alternating mudstone–wackestone and packstone–grainstone layers ranging from 1 mm to 10 cm in thickness. These are the dark bands of earlier workers (Carruthers *et al.* 1927).

The limestones are made up of varying proportions of calcite, organic detritus, silica (chert) and siliciclastic material. Detrital clay minerals and clay grade siliciclasts are very rare. The limestones range from bituminous mudstones to algalclastic grainstones often with a rich fauna and flora characteristically restricted in diversity. Burdiehouse limestones are predominantly calcitic; any other carbonate minerals are exceptional and evaporites are absent. Detrital inorganic grains of calcite ranging from 20–350 µm in length are common and occur in a variety of shapes and sizes (Fig. 5a and b). They are observed as aggregates encrusting stems and roots of hydrophytes (epigenic), as single, isolated, hexagonal, intricately zoned grains believed to have formed by accretion during suspension in the water column (endogenic). Endogenic grains are frequently poorly preserved having been abraded before settling or corroded on the sediment/water interface after settling. The largest and best preserved grains are found back-filling rare biogenic burrows. The settling velocity of the largest recorded hexagonal prism (240 µm diameter when approximated to a sphere) was calculated using Stokes's Law and found to be 0.18 m/s, assuming a constant water temperature of 10°C. An upcurrent of similar magnitude would be necessary to allow the grain to have remained in suspension long enough to achieve zonation and is consistent with those observed in modern hydrological systems (Smith 1975). These grains are termed calcimorphs (Loftus 1985) and are often observed in both the

FIG. 4. A cut and polished core from the top of the Burdiehouse Limestone Formation at Brucefield, West Calder displaying regular couplets of wackestone and grainstone. The specimen illustrates the only known occurrence of slope deformation features in the formation. Scale bar 40 mm.

FIG. 5. (a) Epigenic calcimorphs (C) encrusting tubes (probably root traces) filled with authigenic sparry calcite (T). Encrustation along sediment interface also probably occurred (B). The calcimorphs are not zoned. Scale bar 100 µm; (b) intricately zoned hexagonal endogenic calcimorph with diffuse centre (? nucleus). Scale bar 50 µm.

modern and ancient freshwater deposits (Kelts & Hsü 1978; Talbot *et al.* 1984).

Fauna and flora

The fauna is dominated by ostracods forming microcoquinoidal limestones in certain parts of the basin. Of the non-marine genera recognized, only one, *Paraparchites*, is found in any great numbers. Ostracods are rarely absent. Spirorbids are less widespread, and even more restricted in diversity. Two forms only are recognized, one adnate or fixed (*Spirorbis pusillus*) and one free-living (*Spirorbis* sp.). Gastropods and thin-shelled bivalves are few and sporadically distributed. A single species of each was identified. Sponge spicules were recognized from a single outcrop, but not identified to generic level.

Vertebrate remains are common throughout the Burdiehouse Limestone Formation ranging from skeletal detritus (scales, bone fragments, spines and teeth) to complete body fossils and coprolites. A variety of vertebrates have been identified, including amphibians, sharks and carnivorous fish up to 7 m in length (*Rhizodus hibberti*; Loftus 1985), all of which are considered to have been freshwater dwellers (R. Paton, pers. comm. 1984).

Comminuted plant remains are common in central parts of the basin. Twenty-two species of vascular plants have been determined in the limestones (Loftus 1985). The basin margins display *in situ* rootlets where the original lime muds show evidence of pedogenic reworking. Chert lenses are sometimes associated with these disturbed muds. However, calcified algae have contributed by far the largest volumetric proportion of detritus to the Burdiehouse limestones. Eight genera of cyanophytic algae, three chlorophytic algae and a number of indeterminate forms have been identified directly (*ibid.*) (Fig. 10). Their possible influence on the nature of sedimentation in Burdiehouse limestones is discussed later.

Lithological distribution

A number of distinct lithological trends are observed in the limestones and the evidence suggests that they represent changes in the style of sedimentation across the basin from the depocentres to the shoreline. The trends are: (1) increase in bedding thickness and variability in thickness of individual beds; (2) increase in grain size and clastic content; (3) increase in quantity of epigenic calcimorphs; (4) increase in algal diversity and influence; (5) increase in the occurrence of *in situ* floras, pedogenic character and in occurrence of desiccation features; (6) increase in bioturbation and in the degree of sedimentary reworking; (7) increase in the degree of compaction; (8) decrease in the occurrence of displacive carbonate; (9) decrease in quantity of vertebrate remains and endogenic calcimorphs; (10) decrease in overall kerogen content; and (11) decrease in pyrite content.

In summary, thick (up to 16 m), pale yellow, bioturbated, ostracodal, algal limestones occur at the margins and thin (<1 m) finely laminated, black, kerogenous, pyritous limestones with copious vertebrate remains occur towards the basin centre (Fig. 6). These changes enable four main facies to be recognized: paludine, littoral, sublittoral and profundal. A distinct littoral mudflat facies may have existed at the north-eastern margin of the body of water (Fig. 7).

The littoral facies is critical to the understanding of water level fluctuations in Burdiehouse times. It invariably comprises minor cycles of shoreline mudstones passing up into stromatolitic and oncolitic grainstones and capped by shales (Fig. 8). The implication is that each cycle represents a progressive deepening and expansion of the water body and clearly this occurred several times. The fluctuating stands in water level may also account for the rhythmic couplets observed in Burdiehouse limestones.

Diagenesis

The diagenetic history of Burdiehouse limestones varies with depositional environment (Table 1). Vadose fabrics such as dissolution cavities and collapse breccias (Fig. 9a) confirms that the basin margins were subjected to periods of emergence. With the exception of limestones sampled from the profundal facies (Fig. 6), all limestones show some degree of biogenic reworking. Bioturbation is seen as laminar discordancy or destruction and by grain fabric disorientation. Few actual backfilled burrows are preserved. In exceptional cases, pedogenesis occurred, characterized by features such as needle fibre calcite, rhizoliths, clay cutans, circumgranular cracking and alveolar textures (Fig. 9b). However, the majority of the limestones are free of vadose fabrics and were probably unaffected by meteoric pore waters, remaining permanently submerged beneath the standing waters of the basin. These followed a phreatic diagenetic history involving no more than two phases of calcite spar authigenesis with less common cavity infillings of authigenic kaolinite, dolomite, silica (Fig. 9a) or chlorite.

Algal influence

Algae had a dominant influence upon sedimentation during the formation of Burdiehouse limestones, assuming a complex role as rock building organisms by providing both the sediment and by modifying that sediment following deposition. Detrital algae constitute up to 85% of the limestones in the sublittoral facies. Many are neomorphosed, micritized or have been bored by other algae.

The recognition of algal fabrics commonly depends on indirect evidence such as the presence of stromatolites in the littoral facies. Within the limestones there are also preserved other indirect lines of evidence such as cryptalgal clasts, encrustations, algae fenestrae, thrombolites and laminites. However, the direct evidence is that algalclasts are by far the most important contributors of detritus to the limestones (Fig. 9c). They signify a constant reworking of algal growths at the margins of the water body by gentle wave and current activity. Algae from Burdiehouse limestones provide a useful tool for comparative environmental analysis, being closely allied to modern freshwater forms (Loftus 1985). The assemblages are restricted in diversity and contrast strongly with those described from Carbon-

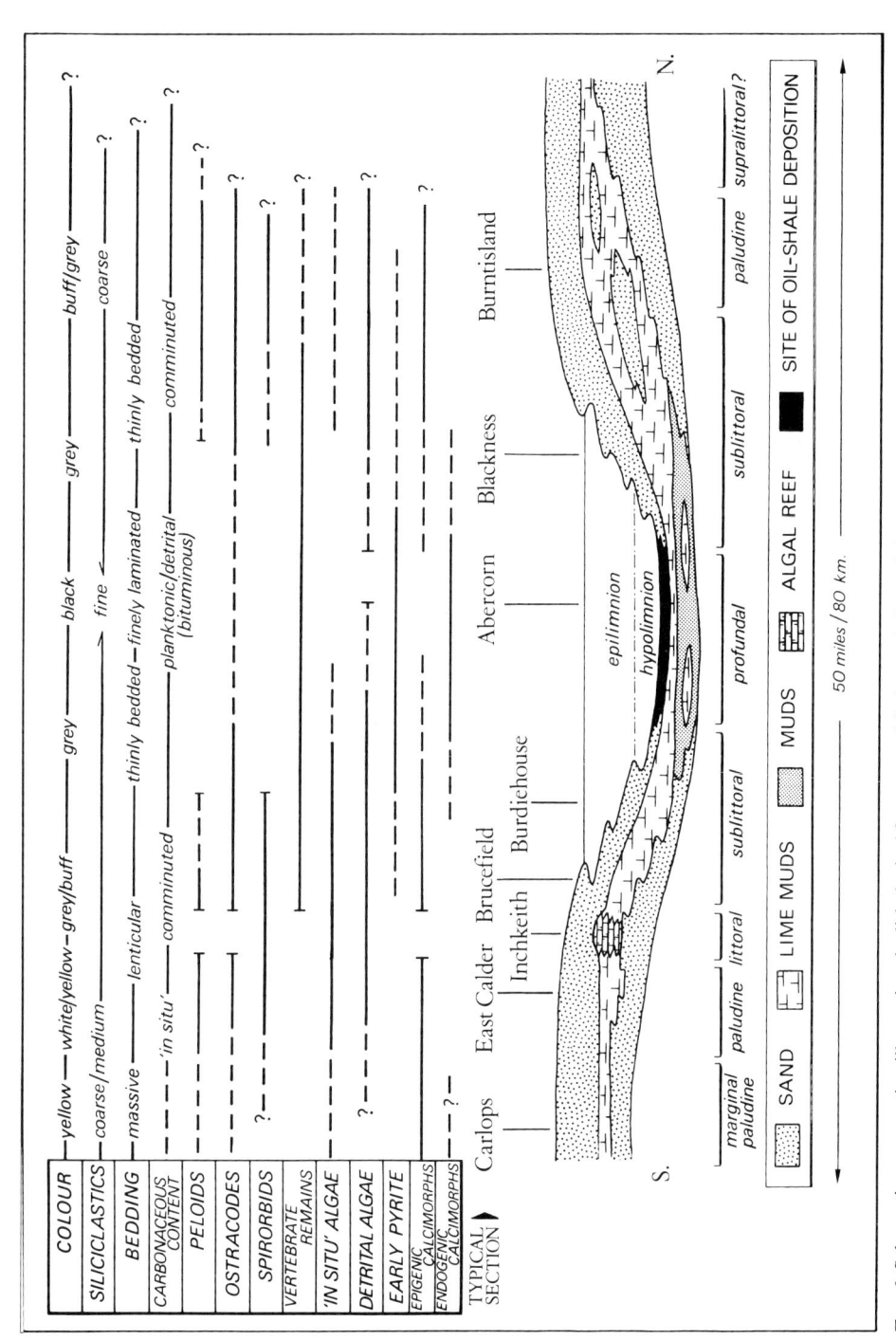

FIG. 6. Schematic cross-section illustrating the lithological features and facies variations of the Burdiehouse Limestone Formation and immediately associated beds across the Lake Cadell basin. Not to scale.

FIG. 7. Lithofacies distribution and inferred extent of Lake Cadell during Burdiehouse Limestone Formation times. The position of the depocentres appears to have been controlled by contemporary tectonic flexuring. The succession thins over the axes, which were the sites of intermittent vulcanicity. Land is indicated by the non-shaded areas.

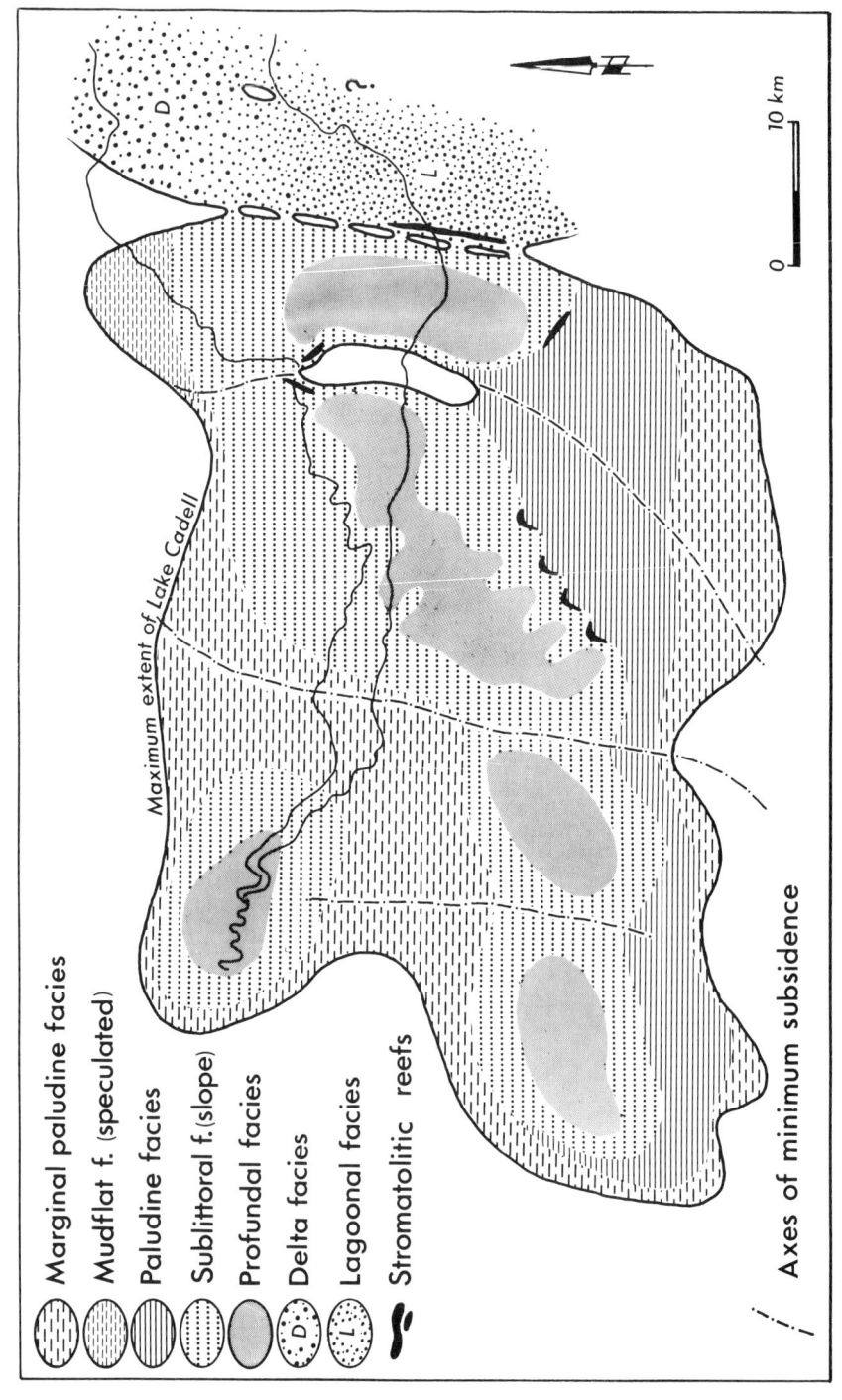

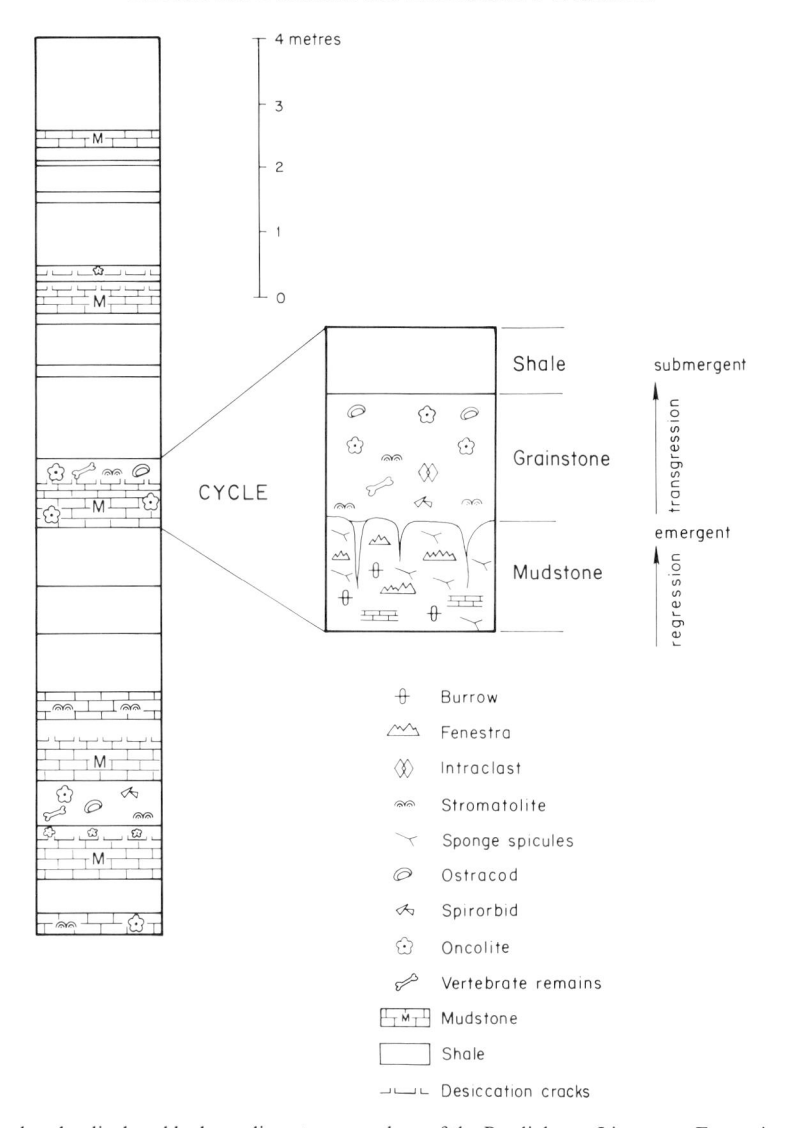

FIG. 8. Littoral cycles displayed by lower limestone members of the Burdiehouse Limestone Formation at Inchkeith.

iferous marine environments (e.g., Chuvashov & Riding 1984).

The algae may have played another role in sedimentation in raising the alkalinity of the surface waters by assimilation of CO_2 and HCO_3^- during photosynthesis (Schneider 1977), perhaps sufficiently to effect the direct chemical precipitation of calcite in the form of calcimorphs (Loftus 1985). Such conditions are most readily achieved in a closed hydrological system.

Palaeolimnology of Lake Cadell

Since Hibbert's original detailed description in 1836, workers have described the Burdiehouse limestones as being lacustrine in origin. This assertion was based upon negative evidence of marine conditions rather than on positive evidence of a non-marine environment. The nature and distribution of cyanophytic algae, vertebrate and invertebrate assemblages, algal fabrics, the predominance of calcite deposition and the styles

TABLE 1. *Cement stratigraphy of the limestone members of the Burdiehouse Limestone Formation*

Diagenetic history	Cement stratigraphy		
	Basinal limestones	Marginal limestones	Pedogenic limestones
Meteoric phreatic diagenesis	Drusy autochthonous calcite Scalenohedral autochthonous calcite Minor neomorphism and grain enlargement Brittle fracture Limpid dolomite Kaolinite cement	Drusy autochthonous calcite Scalenohedral autochthonous calcite Brittle fracture Kaolinite cement Chlorite?	Drusy autochthonous calcite Scalenohedral autochthonous calcite Kaolinite Brittle fracture
Subaerial vadose diagenesis		Microquartz chalcedony and megaquartz Bioclastic moulds Vadose silts and cavity breccias	Needle fibre calcite Rhombohedral brown stained carbonate Clay cutans Micritization (chert)
Emergence		Cemented stromatolites oncolites, intraclasts and desiccation cracks	Algal laminates and desiccation cracks

of limestone diagenesis (outlined in Table 1) indicate that they were deposited in an ancient fresh water lake (Lake Cadell). Lake Cadell covered an area of between 2000 and 3000 km² during the deposition of the formation (Fig. 1). The growth of dense vegetation around the margins and the scarcity of marginal desiccation suggests that the lake was sustained by regular runoff perpetuated by a wet tropical climate from the surrounding, often volcanic, terrain. The physical characteristics, quality of organic preservation and the absence of bioturbation in deeper parts of the lake indicates permanent meromictic stratification of the waters, achieving a depth great enough to both support large free swimming vertebrates and upwelling currents sometimes of the order of 0.18 m/s. Upper limits on the depth of Lake Cadell are imposed by its restricted size, the extent of carbonate shelves, and the scarcity of slope deformation features and turbidites implying the gentlest of bottom relief. Comparison with modern lakes suggests that it was 50–100 m deep at the depocentres (*cf.* Bradley 1966; Murphy & Wilkinson 1980). Short-term fluctuations in the depth are probably reflected partly by interbedded sediments and partly by the limestone couplets and littoral cycles recorded at the margins of the lake. The latter may have been seasonal, reflecting annual partial overturn of the nutrient charged waters from the hypolimnion into the epilimnion and short-term contractions with expansions of the lake margins.

Discussion

The Burdiehouse Limestone Formation and the oil shales in the Oil Shale Group almost certainly accumulated under similar environmental conditions. However, the former is calcitic while the oil shales are not, suggesting physico-chemical differences in the depositional waters. The Burdiehouse limestones record evidence for fluctuating stands in lake level (Fig. 8). In contrast, the lakes in which the oil shales accumulated seem to have possessed relatively stable shorelines. Open lakes (i.e. lakes with an outlet) are characterized by stable shorelines and chemically dilute waters (Eugster & Kelts 1983) so it is possible that this is the model which is most appropriate for the oil shale phases. In contrast during Burdiehouse times, sand barriers, some locked onto a volcanic archipelago, probably blocked the eastern outlet that normally drained the lake, changing its hydrological character from that of an open lake to that of a closed lake (Loftus 1985) (Fig. 1). Removal of the outlet may account for the apparent presence of fluctuating shorelines preserved in Burdiehouse limestones. A progressive saturation of the waters with respect to Ca^{2+} and CO_3^{2-} would have created conditions suitable for the preservation of algae, especially at the lake margins.

The Camps Shale, though one of the least laterally persistent of all the workable oil shales, is intimately associated with the Burdiehouse

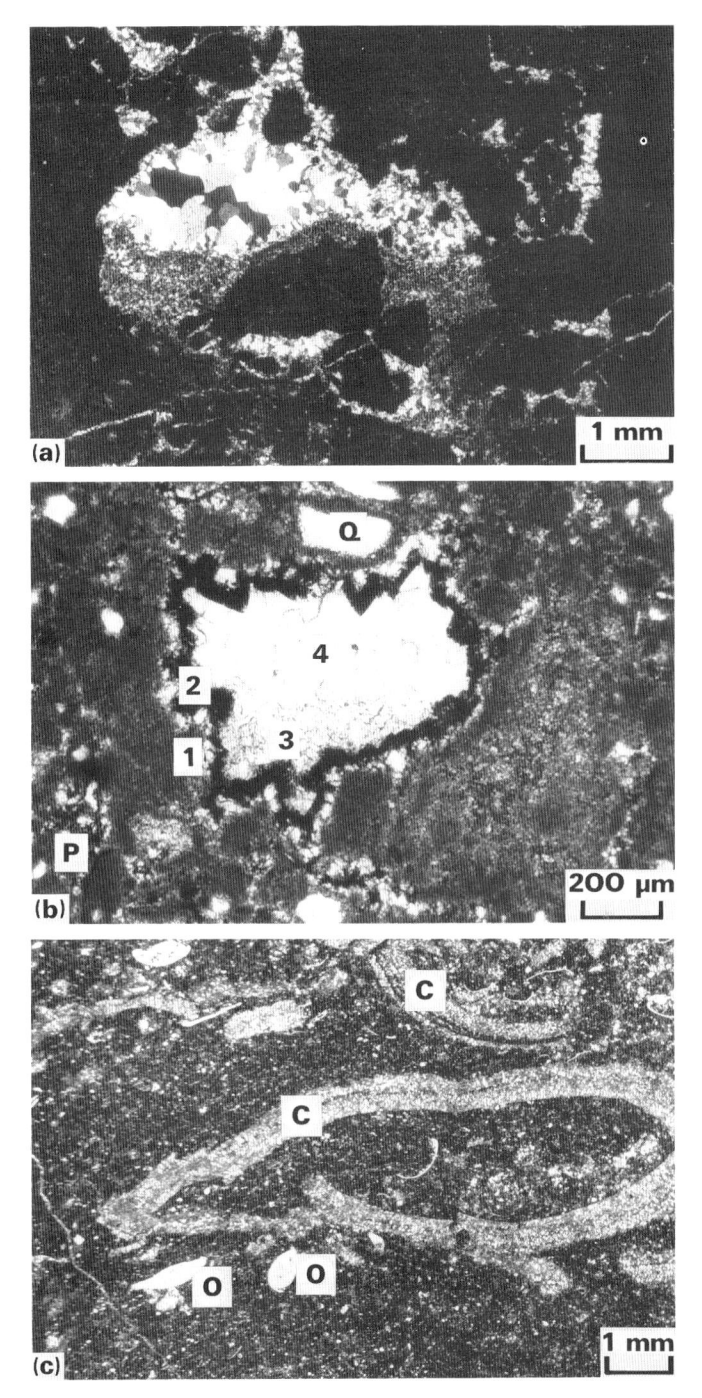

FIG. 9. (a) Dissolution fabric in silicified micritic mudstones, East Calder. The paludine facies displays collapse brecciation and cavity generation with the sedimentation of vadose silt and subsequent fill by megaquartz of the cavities. Scale bar 1 mm. (b) Pedogenic micritic mudstone, East Calder showing peloids (P), quartz grains with circumcrusts (Q) and a rhizoid (root structure) with a four-stage infilling: (1) clear sparry calcite, (2) a dark limonite (after siderite?) coating, (3) a geopetal vadose silt and (4) clear drusy calcite. Scale bar 200 µm. (c) Spalled or collapsed periphytonic algae (C), *Cayeuxia* spp., found in the sublittoral wackestones and packstones. Each of the bright chambers are thalli filled with early calcite cement, as are the articulated ostracods (O). The bioclasts show little evidence of transport. Scale bar 1 mm.

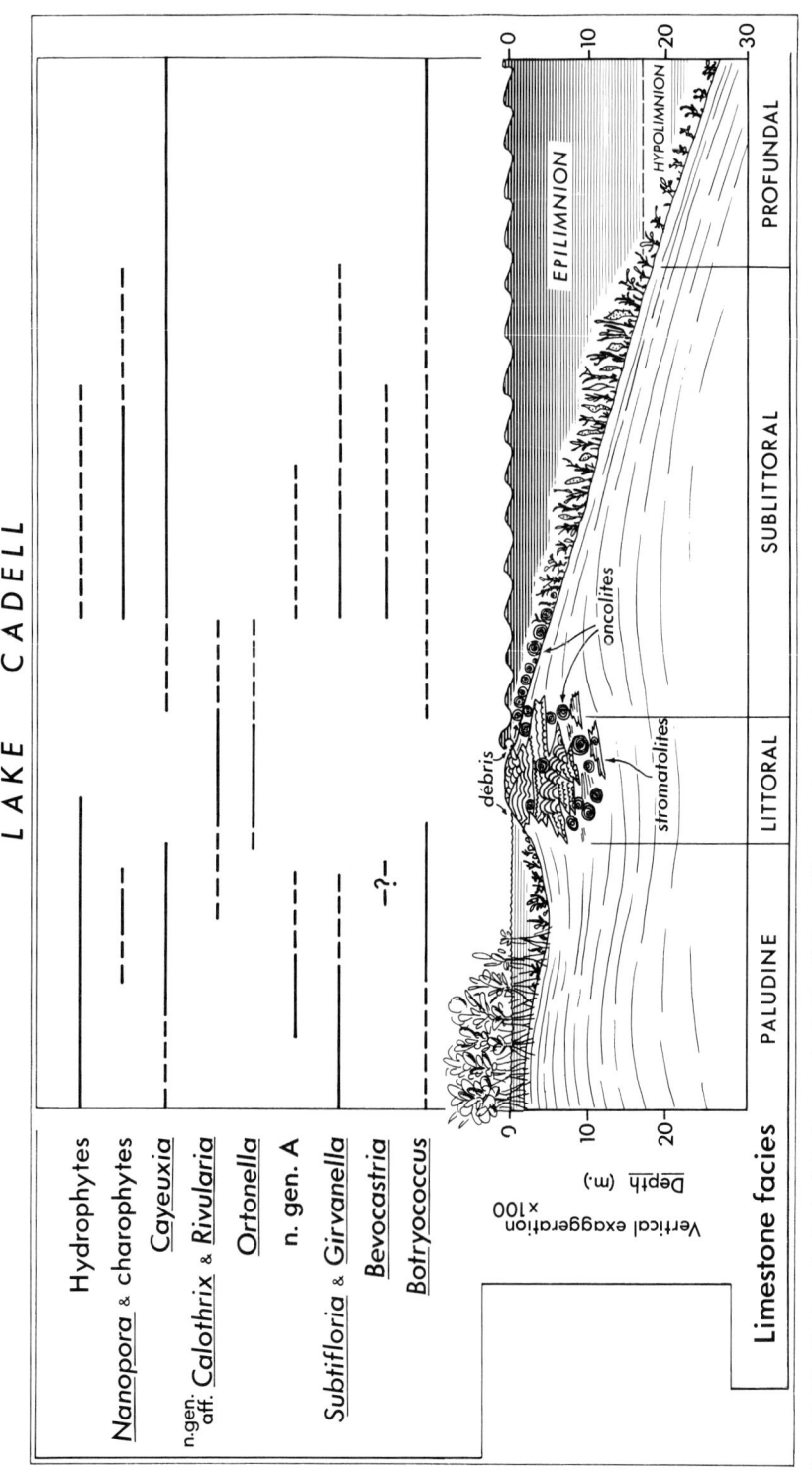

FIG. 10. Distribution of principal algal genera in the Burdiehouse Limestone Formation across Lake Cadell.

Limestone Formation. The profundal deep water facies of the limestone formation shows similar characteristics to oil shales (fine laminations, bituminous, black, pyritous, lacking bioturbation). The oil shales, however, have a much broader areal distribution than the profundal facies of the Burdiehouse Limestone Formation (Fig. 7) (compare with Plate III from Mitchell & Mykura 1962), suggesting that they correspond to deeper water phases of an open Lake Cadell i.e., with depths of about 100 m. The dominance of planktonic algae in the oil-shales, contrasting with the dominance of benthic algae in the Burdiehouse Limestone Formation, suggests that conditions were unsuitable for benthic algae during oil shale deposition. It is suggested here that the shelf areas around the lake, which provided a substrate for benthic algae during Burdiehouse times (Fig. 10), were either reduced in width or were at a further remove from the depocentres while the oil shales were accumulating. Whatever the actual circumstances, benthic algae do not appear to have been positive contributors to the algal oozes (oil shales) accumulating in the depocentres.

Conclusions

Lake Cadell was a complex of freshwater meromictic lakes, set in a wet tropical climate and landlocked on three sides. The lake was subject to long-term fluctuations in lake stand expressed as sedimentary cycles, intermittently opening via lagoons into the sea at its distal eastern end. It was variously bounded by sandy beaches, deltas and barriers. Lake closure led to the deposition of a range of algal-dominated limestones of which the Burdiehouse limestones were the most important with deeper water facies resembling oil shales. Chemical changes and deepening of the lake waters, with a degree of reconnection to the open sea, saw the periodic mass accumulation of planktonic algae analogous to the modern form *Botryococcus braunii*. These collected as flocculent oozes on the low relief bottom of the lake and progressively converted into oil shales.

ACKNOWLEDGEMENTS: We are grateful to many individuals and organizations for assisting in this work, more especially R. Riding, G. F. Elliott and S. Monro and his colleagues in the British Geological Survey. Also LASMO, Tricentrol, BP, Moray Petroleum, Anvil Petroleum, Taylor Woodrow, Livingston New Town Development Corporation, the Royal Scottish Museum and the Hunterian Museum. NERC supported GWFL for 3 years. The bulk of the figures were drawn by Colin Stuart of University College.

References

ANDERSON, J. G. C. 1942. The oil-shales of the Lothians. Area II Pumpherston. *Geological Survey Wartime Pamphlet No. 27.*

BELT, E. S. 1975. Scottish Carboniferous patterns and their palaeoenvironmental significance. *In*: BROUSSARD, M. L. (ed.), *Deltas. Models for Exploration*, Houston Geological Society, Houston, 427–449.

BLACKBURN, K. B. & TEMPERLEY, B. N. 1936. Botryococcus and the algal coals. *Transactions of the Royal Society of Edinburgh*, **58**, 841–868.

BRADLEY, W. H. 1966. Tropical lakes, copropel and oil-shale. *Bulletin of the Geological Society America*, **77**, 1333–1338.

—— & EUGSTER, H. P. 1969. Geochemistry and paleolimnology of the trona deposits and associated authigenic minerals of the Green River Formation of Wyoming. *Professional Paper. United States Geological Survey*, **496-B**, 1–71.

CADELL, H. M. 1925. *The Rocks of West Lothian*, Oliver & Boyd, Edinburgh.

CAMERON, I. & MCADAM, A. D. 1978. The oil-shales of the Lothians, Scotland: present resources and former workings. *Report of the Institute of Geological Sciences, No. 78.*

CANE, R. F. 1969. Coorongite and the genesis of oil-shale. *Geochimica et Cosmochimica Acta*, **33**, 257–265.

CARRUTHERS, R. G., CALDWELL, W., BAILEY, E. M. & CONACHER, H. R. J. 1927. The oil-shales of the Lothians, 3rd edn, *Memoir of the Geological Survey of Scotland*, Her Majesty's Stationery Office, Edinburgh.

CHUVASHOV, B. & RIDING, R. 1984. Principal floras of Palaeozoic marine calcareous algae. *Palaeontology*, **27**, 487–501.

CONACHER, H. R. J. 1917. A study of oil-shales and torbanites. *Transactions of the Geological Society of Edinburgh*, **16**, 164–192.

—— 1938. Coorongite and its occurrence. *In*: *Oil-Shales and Cannel Coals, Part 1*, Institute of Petroleum, London, 42–49.

EUGSTER, H. P. & KELTS, K. 1983. Lacustrine chemical sediments. *In*: GOUDIE, A. S. & PYE, K. (eds), *Chemical Sediments and Geomorphology*, Academic Press, London.

FRANCIS, E. H. 1983. Carboniferous. *In*: CRAIG, G. Y. (ed.), *Geology of Scotland*, 2nd edn, Scottish Academic Press, Edinburgh, 253–296.

GEORGE, T. N. 1958. Lower Carboniferous palaeogeography of the British Isles. *Proceedings of the Yorkshire Geological Society*, **31**, 227–318.

GREENSMITH, J. T. 1962. Rhythmic deposition in the Carboniferous Oil-Shale Group of Scotland. *Journal of Geology*, **70**, 355–364.

—— 1966. Carboniferous deltaic sedimentation in east Scotland: a review and re-appraisal. *In*: SHIRLEY, M. L. & RAGSDALE, J. A. (eds), *Deltas in Their Geologic Framework*, Houston Geological Society, Houston, 143–166.

—— 1968. Palaeogeography and rhythmic deposition in the Scottish Oil-Shale Group. *United Nations Symposium on the Development and Utilisation of Oil Shale Resources*. Section B, Tallinn, 1–16.

HUTCHINSON, G. S. 1957. *A Treatise on Limnology. Volume 1. Geography, Physics and Chemistry*, John Wiley & Sons, New York.

KELTS, K. & HSÜ, K. J. 1978. Freshwater carbonate sedimentation. *In*: LERMAN, A. (ed.), *Lakes: Chemistry, Geology, Physics*, Springer-Verlag, Berlin, 293–323.

KENNEDY, W. Q. 1943. The oil-shales of the Lothians: structure: Area IV (Philpstoun). *Geological Survey Wartime Pamphlet, No. 27*.

LOFTUS, G. W. L. 1985. The petrology and depositional environments of the Dinantian Burdiehouse Limestone Formation of Scotland. Ph.D. thesis, University of London.

LOVE, L. G. & AMSTUTZ, G. C. 1966. Review of microscopic pyrite. *Fortschritte Mineralogie*, **43**, 273–309.

MACGREGOR, A. G. 1960. Divisions of the Carboniferous on the Geological Survey Scottish Maps. *Bulletin. Geological Survey of the United Kingdom*, **16**, 127–130.

MACGREGOR, M. 1938. Conditions of deposition of the oil-shales and cannel coals of Scotland. *In: Oil-Shales and Cannel Coals, Part 1*, Institute of Petroleum, London, 6–18.

MIKNIS, F. P., MACIEL, G. E. & BARTUSKA, V. J. 1979. Cross-polarisation, magic angle spinning ^{13}C NMR spectra of oil-shales. *Organic Geochemistry*, **1**, 169–176.

MITCHELL, G. H. & MYKURA, W. 1962. The geology of the neighbourhood of Edinburgh, 3rd edn, *Memoir. Geological Survey of Scotland*, Her Majesty's Stationery Office, Edinburgh.

MOORE, L. R. 1968. Cannel coals, bogheads and oil shales. *In*: MURCHISON, D. & WESTOLL, T. S. (eds), *Coal and Coal-Bearing Strata*, Oliver & Boyd, London, 19–29.

MURPHY, D. H. & WILKINSON, B. H. 1980. Carbonate deposition and facies distribution in a central Michigan marl lake. *Sedimentology*, **27**, 123–135.

PARNELL, J. 1984. Hydrocarbon minerals in the Midland Valley of Scotland with particular reference to the Oil-Shale Group. *Proceedings of the Geologists Association*, **95**, 275–285.

RICHEY, J. E. 1935. Areas of sedimentation of Lower Carboniferous in the Midland Valley of Scotland. *Summary of Progress, Geological Survey of Great Britain*, No. 2, 13–110.

—— 1942. The oil-shales of the Lothians: structure: Area I (West Calder). *Geological Survey Wartime Pamphlet No. 27*.

ROBERTSON, T. 1948. Rhythm in sedimentation and its interpretation: with particular reference to the Carboniferous sequence. *Transactions of the Edinburgh Geological Society*, **14**, 141–175.

ROBINSON, W. E. & DINEEN, G. V. 1967. Constitutional aspects of oil-shale kerogen. *Proceedings of the 7th World Petroleum Congress*, **3**, 669–680.

SCHNEIDER, J. 1977. Carbonate construction and decomposition by epilithic and endolithic microorganisms in salt and freshwater. *In*: FLUGEL, E. (ed.), *Fossil Algae*, Springer-Verlag, Berlin, 248–269.

SMITH, I. R. 1975. Turbulence in lakes and rivers. *Freshwater Biological Association, Scientific Publication*, No. 291.

TALBOT, M. R., LIVINGSTONE, D. A., PALMER, P. G., MALEY, J., MELACK, J. M., DELIBRIAS, G. & GULLIKSEN, S. 1984. Preliminary results from sediment core from Lake Bosumtwi, Ghana. *Palaeoecology of Africa*, **16**, 173–192.

TIERCELIN, J. J., RENAUT, R. W., DELIBRIAS, G., LE FOURNIER, J. & BIEDA, S. 1981. Late Pleistocene and Holocene lake level fluctuations in the Lake Bogoria Basin, north Kenya Rift Valley. *Palaeoecology of Africa*, **13**, 105–120.

TREESE, K. L. & WILKINSON, B. H. 1982. Peat–marl deposition in a Holocene paludal–lacustrine basin—Sucker Lake Michigan. *Sedimentology*, **29**, 375–390.

G. W. F. LOFTUS, EP/13 Shell Internationale Petroleum Maatschapp, Oostduinlaan 75, The Hague, The Netherlands.

J. T. GREENSMITH, Department of Geological Sciences, University College, Gower Street, London WC1E 6BT, UK.

Lacustrine petroleum source rocks in the Dinantian Oil Shale Group, Scotland: a review

J. Parnell

SUMMARY: The Lower Carboniferous Oil Shale Group includes lacustrine rocks with petroleum source-rock potential. Existing data is reviewed, and new pyrolysis data is presented which shows that the oil shales *sensu stricto* are excellent source rocks with up to 30% TOC and yields of up to 180 mg/g. Other shales and dark limestones in the group also have some source-rock potential. The oil shales contain mixtures of laminar algal matter (lamalginite) and discrete algal bodies (telalginite) which represent planktonic and benthonic algal deposits. The algal bodies are *Botryococcus*-type and referable to *Pila*. There is a general correlation between % alginite and hydrocarbon yield.

Introduction

Oil shales were discovered in the Lower Carboniferous of West Lothian in 1858 and were distilled for oil over a period of 100 years. The shale mining industry in that region is now of only historical interest. Nevertheless, the high oil potential of the oil shales may make them a viable resource once more in the future, and a recent assessment has been made of their reserves (Cameron & McAdam 1978). The oil shales *sensu stricto* (those once mined for oil distillation) are excellent lacustrine source rocks: Barnard & Cooper (1981) have inferred from distillation yields that 'as potential source rocks they are ten to a hundred times better than an average good source rock'. The oil shales were probably responsible for many oil shows in the region (Parnell 1984a), some of which have been commercially exploited. In addition, other shales and limestones within the Oil Shale Group may contain significant quantities of organic carbon. This review summarizes those lithologies in the Oil Shale Group which are organic-rich and may have source-rock potential, lists scattered existing information about the bulk organic geochemistry of the oil shales and presents pyrolysis data. The nature of the organic matter in the sediments is discussed, but I have not attempted a detailed sedimentological analysis of Oil Shale Group sedimentation. For this the reader is referred to papers by Greensmith (1968) and Loftus & Greensmith (this volume), which this review is intended to complement.

The Dinantian Oil Shale Group forms the upper part of the Scottish Calciferous Sandstone Series in the eastern part of the Midland Valley of Scotland (Figs 1 and 2). Basin subsidence, with syndepositional faulting, was probably an important control on sedimentation: the region of oil shale deposition coincides with that of maximum subsidence (Greensmith 1968). The maximum thickness of the Oil Shale Group is about 2000 m, of which 2% is oil shale.

Oil Shale Group sediments

Oil Shale Group sedimentation was rhythmic, reflecting episodes of delta progradation and retreat relative to a lake of several thousand km^2 size. Shales, including oil shales, and limestones were deposited in the lake basin. Periodic emergence is indicated by desiccation cracks in limestones and some shales. Thin marine limestones and shales are also recorded (Greensmith 1968).

Moore (1968) ascribed the oil shales to deposition in a stratified lake in which algae and associated fungal/bacterial organic matter were preserved in the hypolimnion. Dick (1981) further suggested permanent stratification, i.e., with no seasonal overturn of the hypolimnion. Although the oil shales have been generally attributed an 'algal' origin, like many other laminated kerogenous lacustrine deposits, detailed descriptions of the oil shales are lacking and the nature of the algal component has not been clarified. The nature of the algae is discussed below.

Several lithologies within the Oil Shale Group deserve consideration as source rocks. In addition to the oil shales *sensu stricto*, of which there are twelve principal beds or groups of beds (Cameron & McAdam 1978), there are other shales in the sequence which contain significant quantities of organic matter. These shales are predominantly lacustrine but include horizons which yield marine fossils. Some dark-coloured limestones show a lateral and/or vertical gradation to kerogenous shales and themselves may be viable source-rocks.

Figure 3 shows two examples of Oil Shale Group sequences at Society (NT 094793) and South Queensferry (NT 142785), adapted from the records of Cadell *et al.* (1906) and MacGregor

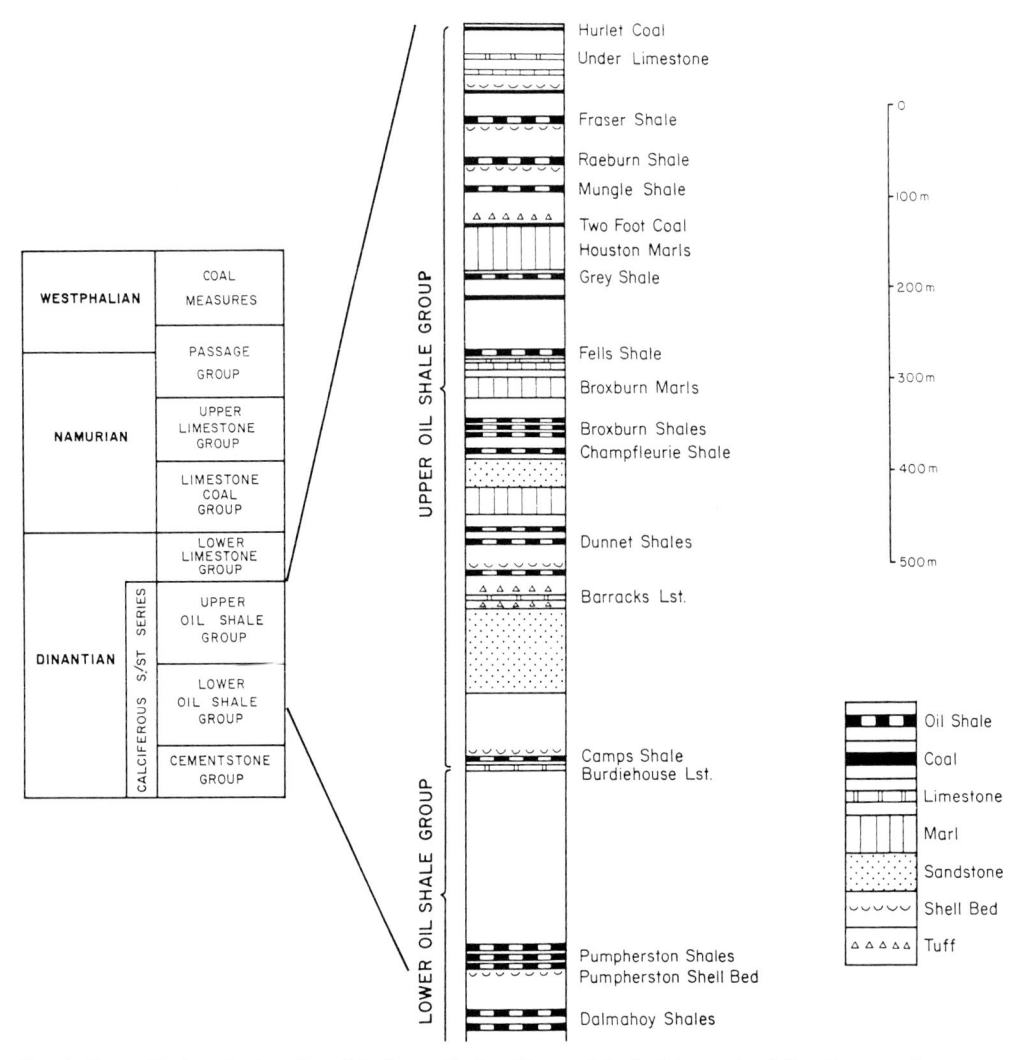

FIG. 1. Generalized vertical section of the Lower Carboniferous of the Lothian region (after Cameron & McAdam 1978).

(1973) respectively. The limestones are kerogenous and the shales are calcareous, emphasizing the transitional nature of the lithologies. In both sequences, shales and limestones contain desiccation cracks, and some limestones are stromatolitic. The Pumpherston Shell Bed is a marine shale (see below) which shows no discontinuities with the rocks above or below. TOC data are given for many beds, and pyrolysis determinations are recorded in Table 1.

Petrography of oil shales and related sediments

The kerogenous sediments of the Oil Shale Group consist of variable mixtures of (i) detrital siliciclastic grains, generally 'floating' in an organic matrix, (ii) plant remains, (iii) yellow algal bodies (Fig. 4), equivalent to the alginite A (telalginite) of Hutton et al. (1980), (iv) very thin (10–20 μm) coalesced laminae of organic matter which have an intimate anastomosing relationship with the mineral matter of the rock (Fig. 5), equivalent to the alginite B (lamalginite) of Hutton et al. (1980), (v) phosphatic fish remains and coprolites, (vi) shell fragments, particularly ostracods, common at certain horizons, (vii) penecontemporaneous/early diagenetic carbonate minerals, (viii) authigenic minerals, e.g., pyrite, albite and chalcedony. The relative proportions of these components can vary very markedly within a

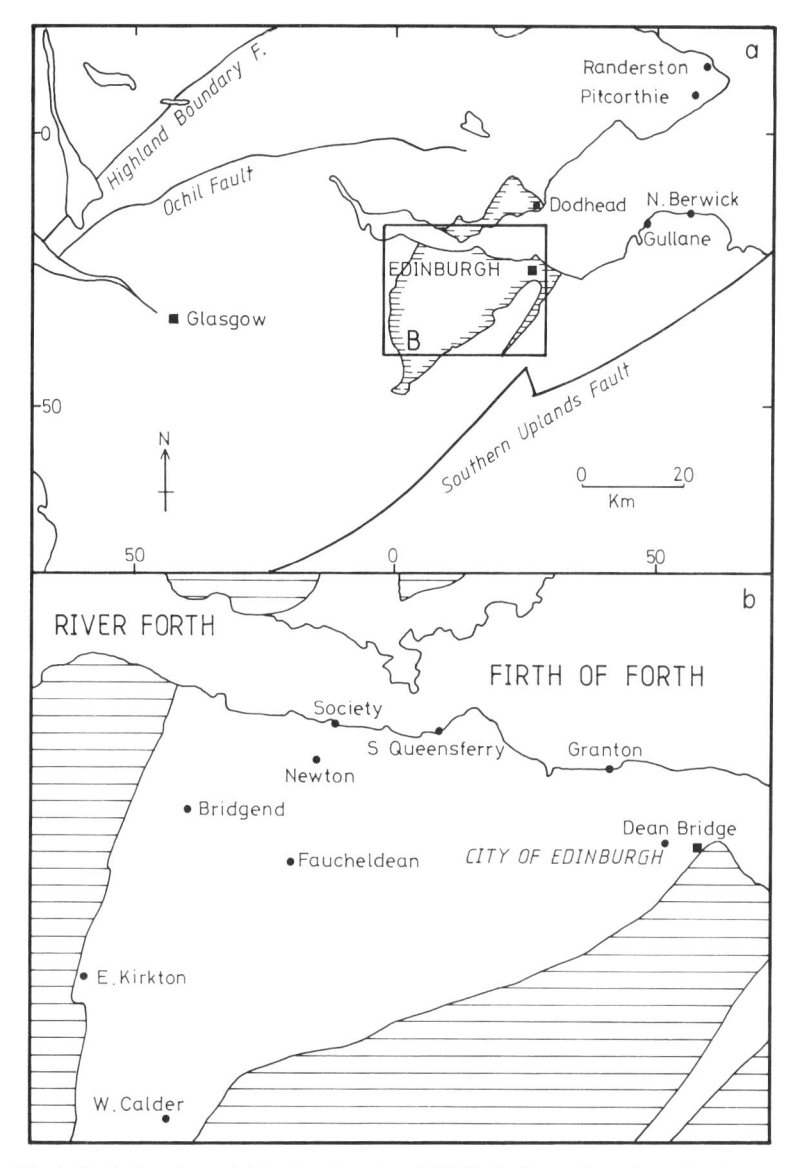

FIG. 2. Localities in Table 1 and text. (a) Regional setting of Oil Shale Group (boundary of main outcrop ruled), (b) Lothian region.

single bed. Several oil shale beds are represented at some localities by comparatively lean shales (Richey 1942; Anderson 1942; Kennedy 1943). Individual beds also vary in thickness from several metres down to several centimetres.

Vertical transitions between lithologies are often indistinct; limestones pass upwards into decreasingly calcareous shales, and oil shales pass upwards into decreasingly kerogenous shales. Some limestones include kerogenous layers (<100 μm) indistinguishable from laminae in the oil shales. Transitional lithologies consist of abundant siliciclastic grains, or sparite lenses, in a groundmass of organic matter. Desiccation cracks are abundant in the shales and limestones, but very rare in the oil shales.

Near the top of the Oil Shale Group a cherty limestone with kerogenous laminae occurs within a series of lavas and tuffs at East Kirkton (Table 1). The limestone is interbedded with black shales

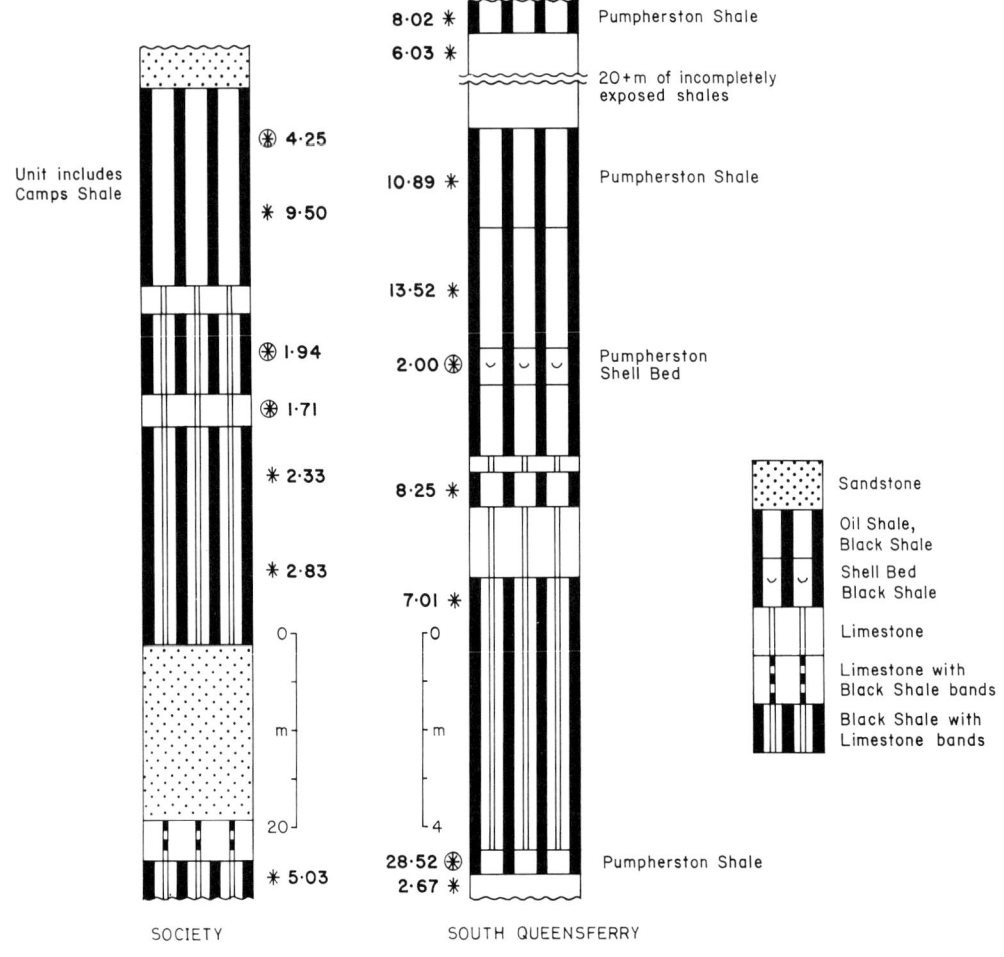

FIG. 3. Oil Shale Group sequences at Society, and a section near South Queensferry including the Pumpherston Oil Shales. Asterisks indicate TOC (%) determinations; encircled asterisks indicate pyrolyses (mg/g; see Table 1).

and tuffs (Muir & Walton 1957; and see log in Howell & Geikie 1861). Kerogenous laminae (<50 μm) are separated by sparite layers (500 μm). Some kerogenous layers consist of several anastomosing laminae, with calcite spherulites up to 2 mm diameter (Fig. 7). The spherulites show nucleii of siliciclastic grains or plant fragments. Sparite layers contain numerous lensoid (50 × 20 μm) pseudomorphs after gypsum. These are the only pseudomorphs after evaporites recorded in the Oil Shale Group.

Laminated dolomitic limestones, well known as fossiliferous 'shrimp beds', occur within black shale sequences at Granton (Tait 1925; Briggs & Clarkson 1983) and Gullane (Traquair 1907; Hesselbo & Trewin 1984). The beds are less than 0.5 m thickness. Organic matter in the beds occurs disseminated in the dolomite and as thin (<10 μm) isolated laminae. There are also horizons of anastomosing organic laminae and carbonate with up to 30 kerogenous laminae per mm thickness, reminiscent of the oil shales. Hesselbo & Trewin (1984) describe and figure laminae of concentrated organic matter associated with dolomite in the Gullane bed. The organic material, which occurs in a similar form at Granton, is in fact a plastic bitumen, which was probably generated within the beds. It is associated with displacive fibrous dolomite (Fig. 6), which may be a consequence of over-pressuring; several other bitumen-enriched beds in the Oil Shale Group sequence contain displacive carbonate. The dolomitic laminites lie below the lowermost oil shales *sensu stricto*, but shales at a similar

TABLE 1. *Rock-Eval pyrolysis data for oil shales, shales and limestones of the Oil Shale Group*

No.	Rock type	Locality	Stratigraphy	TOC (% wt)	S1 (mg/g)	S2 (mg/g)	S2/TOC (mg HC/ g TOC)	S1/(S1+S2)
1	Oil shale	Pitcorthie	Lower OSG	21.10	—	44.06	209	—
2	Oil shale	S. Queensferry	Dalmahoy OS	19.90	—	117.41	590	—
3	Oil shale	Newton	Pumpherston OS	9.88	1.17	41.42	419	0.03
4	Oil shale	S. Queensferry	Pumpherston OS	28.52	1.72	177.71	623	0.01
5	Oil shale	Society	Camps OS	4.25	1.65	14.64	344	0.10
6	Oil shale	W. Calder	Dunnet OS	19.59	1.34	124.36	635	0.01
7	Oil shale	Bridgend	Champfleurie OS	16.30	—	51.96	319	—
8	Oil shale	Faucheldean	Broxburn OS	14.90	0.28	76.43	513	0.00
9	Black shale	Dean Bridge	Wardie Shales	13.78	0.68	16.69	121	0.04
10	Black shale	E. Kirkton	Under Lst.	22.67	2.70	98.36	434	0.03
11	Gypsiferous laminite	E. Kirkton	Under Lst.	1.50	—	—	—	—
12	Black shale	Granton	Granton Sst.	1.67	0.46	2.56	153	0.15
13	Dolomitic laminite	Granton	Granton Sst.	3.21	0.39	1.91	59	0.17
14	Marine black shale	S. Queensferry	Pumpherston SB	4.74	—	3.90	82	—
15	Marine black shale	Randerston	Lower OSG	11.60	—	30.64	264	—
16	Black limestone	Society	Burdiehouse Lst.	1.94	0.60	5.66	292	0.10
17	Grey limestone	Society	Burdiehouse Lst.	1.71	0.27	1.81	106	0.13
18	Black shale adjacent to sill	Dodhead	Burdiehouse Lst.	8.10	0.68	1.83	23	0.27

stratigraphic level are sufficiently kerogenous to have warranted distillation trials (Peach *et al.* 1910).

Depositional environment of source rocks

Oil shales: significance of algal bodies

One of the most widely used models for oil shale formation is an adaptation of the model for torbanite (boghead coal) formation (Cane 1976). Torbanites consist almost entirely of yellow algal bodies which are visualized as colonizing the surface of small shallow water bodies with low drainage input, replenished by rainfall and seepage through lake margin peats, or the centres of larger water bodies (Moore 1968). The Oil Shale Group oil shales also contain yellow algal bodies although in lesser amounts (Conacher

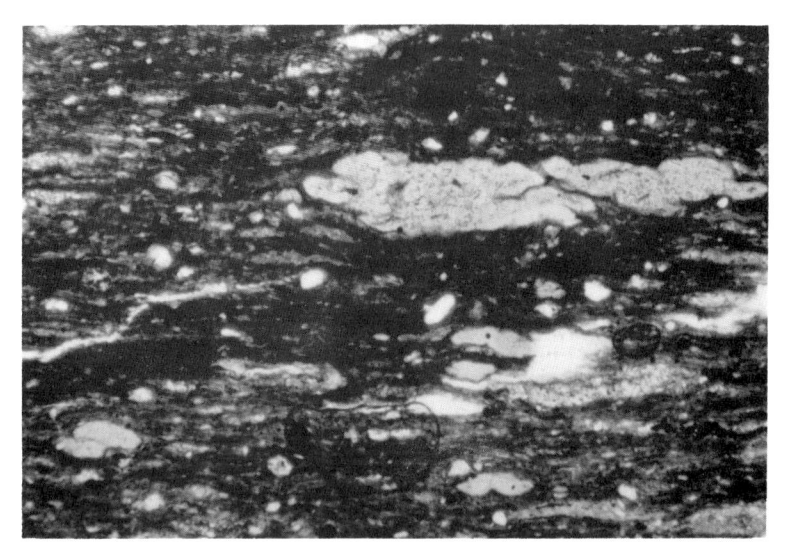

FIG. 4. Discrete *Botryococcus*-type algal bodies (light) in oil shale, Pumpherston Oil Shale, South Queensferry (×47).

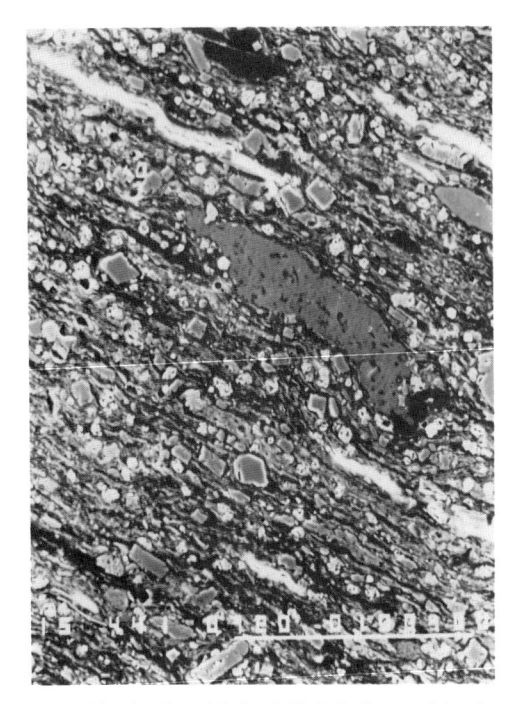

FIG. 5. Fine laminar fabric of oil shale. Large object is coprolite, rhomboid/rectangular grains are authigenic albite. Broxburn Oil Shale, Bridgend. (SEM × 440, courtesy of H. Shaw.)

Carboniferous torbanites and oil shales (Stadnikow & Weizmann 1929; Douglas *et al.* 1969b). Figure 8 demonstrates the relationship between hydrocarbon yield and percentage total alginite (telalginite + lamalginite) in the Oil Shale Group. A similar relationship has been determined for Australian oil shales by Hutton *et al.* (1980).

Laminar organic matter

Although yellow algal bodies are abundant in the sediments, there is also a substantial proportion of laminar organic matter (lamalginite). This organic matter would be described as bituminite (e.g. Stach *et al.* 1982) or dispersed organic matter by some workers. The lamellar form of the lamalginite in sections normal to bedding (Fig. 5) and rounded form parallel to bedding strongly suggest that it too is derived from algal matter (A. C. Hutton, pers. comm.). Stach *et al.* (1982) similarly infer their 'bituminite' to be a decomposition product of algae. The lamalginite can be envisaged as the deposit of flocculent oozes of blue-green planktonic algae. A comparable combination of benthonic and planktonic algal matter is found in Tertiary oil shales in Thailand

1917; MacGregor 1938), and hence oil shale deposition has been considered as similar to that of torbanites but with a greater influx of siliciclastic sediment and plant debris (Carruthers *et al.* 1912).

The yellow algal bodies in the oil shales are of the *Pila*-type (see Stach *et al.* 1982). The similarity of the algal bodies in oil shales and torbanites to the recent alga *Botryococcus braunii* (Zalessky 1914; Blackburn & Temperley 1936) has enabled analogies to be drawn between the ancient sediments and the recent alga-bearing sediments. *Botryococcus braunii* is the principal constituent of a number of highly resilient sapropelic deposits of recent ephemeral lakes, including balkhashite (Kazakhstan) and coorongite (South Australia).

The algal bodies in the oil shales are of particular relevance to the oil potential of the shales. The analogous recent algae and algal sediments have a high hydrocarbon yield (Bergmann 1963; Brown *et al.* 1969), and ancient sediments with a significant content of these algae are good source rocks (Traverse 1955; Gelpi *et al.* 1968). The analogy is supported by organic geochemical studies of balkhashite (Zelinsky 1926) coorongite (Douglas *et al.* 1969a), the isolated alga (Maxwell *et al.* 1968) and the

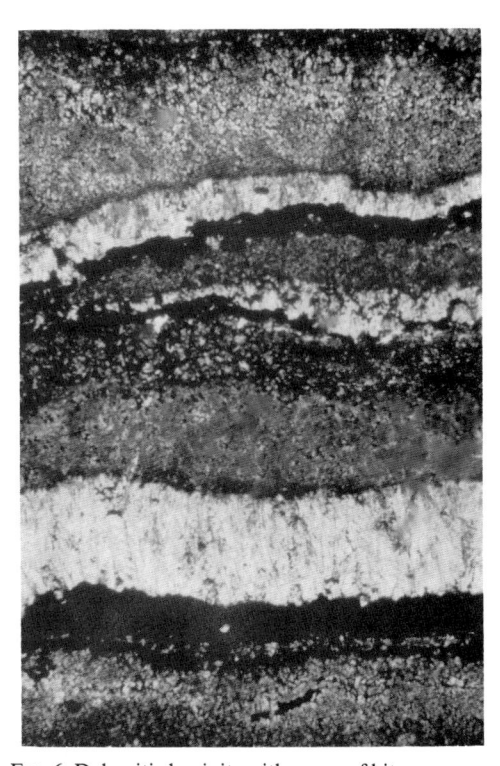

FIG. 6. Dolomitic laminite with seams of bitumen associated with fibrous displacive dolomite and bitumen disseminated in dolosparite. Granton (× 28).

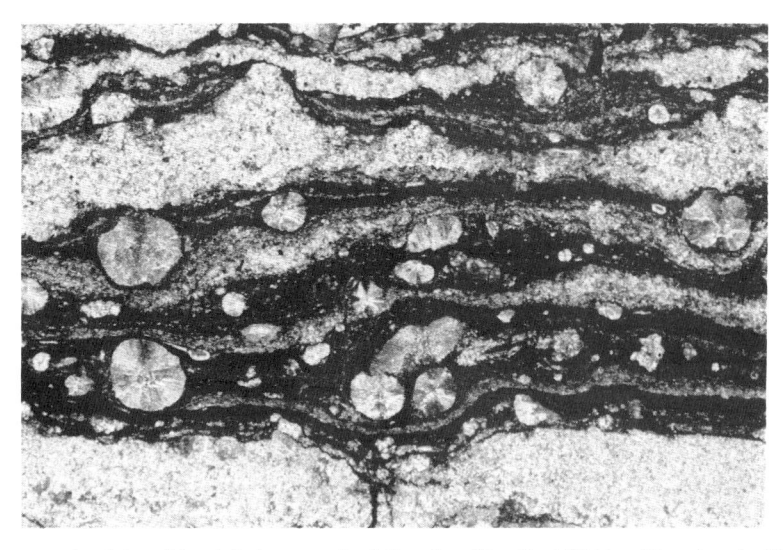

FIG. 7. Kerogenous laminite with calcite lenses and calcite spherulites, East Kirkton Limestone (× 16).

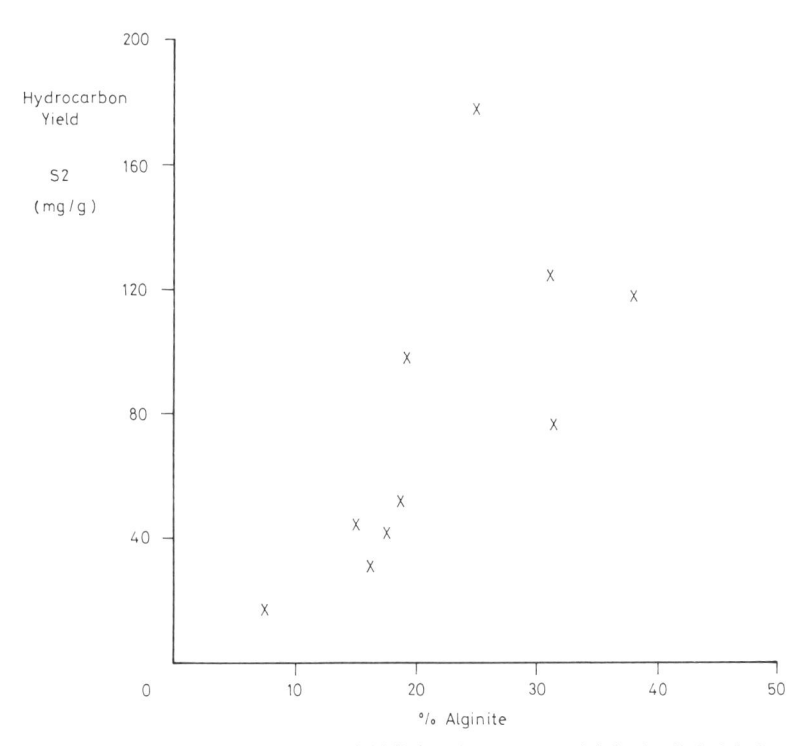

FIG. 8. General correlation between hydrocarbon yield (S2) and percentage alginite in oil shale/other samples (1–4, 6–10, 15). Percentages from 2-D counts of 500 data points.

(Sherwood et al. 1984) and Australia (Hutton et al. 1980). The nature of the organic matter in these oil shales, including those of the Oil Shale Group, is typical of low to moderate salinity lacustrine deposits (see Sherwood et al. 1984).

Vegetable matter: relationship to oil shales

In addition to algal matter, the oil shales and other sediments include a significant proportion of vegetable matter (Douglas et al. (1983) report the presence of vitrinite and inertinite). In some shales, plant debris is the predominant type of organic matter. Several thin coals occur in the Oil Shale Group sequence (Fig. 1). The coals may have occupied a similar palaeogeographic position to the oil shales (Fig. 9) and it is of palaeoclimatic importance that they occur within the same succession. Their palaeogeographic equivalence is exemplified by the Two Foot Coal horizon (Fig. 1), which varies through coal, poor coal, shaly coal, bituminous shale and oil shale (Kennedy 1943). Very thin (~5 cm) coal lenses occur at several localities within limestones, particularly in the Burdiehouse Limestone Formation (Connell 1836; Peach et al. 1910; Anderson 1942; Love 1960).

Similar examples of closely related oil shale and coal deposits are found in Cenozoic intermontane basins in Thailand. Gibling et al. (1985) envisage the Thai oil shales, which have a very similar petrography to those of the Oil Shale Group, as forming when 'peat swamps close to a steep basin margin were flooded by shallow lakes, allowing algae to replace rooted vegetation'.

Lacustrine sedimentation

The oil shales were deposited as algal oozes in shallow lakes which periodically dried out to expose vast expanses of mudflat (see Parnell 1984b for discussion of playa environment in which these mudflat sediments may have been deposited). The well-defined laminae, well-preserved and abundant algal matter, and lack of exposure or significant sediment transport in the oil shales *sensu stricto* suggest their deposition in a permanent lake, possibly stratified such that anaerobic bottom conditions prevailed. Thus elements of both playa-lake and perennial lake models may be applicable. Present-day playa lake deposits are best developed in an arid climate, whereas the Oil Shale Group was deposited in a more tropical climate (see below). Organic productivity reached a maximum during permanent lacustrine phases, but was still high during impermanent phases. The mudflats consisted of carbon-rich mud or carbonate mud or an intermediate between these two end members. Thicknesses of carbonate sediment increase towards the basin margin at some levels (see Parnell 1984b). Whilst carbonate grains deposited through basin-centre anaerobic waters tend to redissolve, grains deposited nearer the aerobic margins of a lake are preserved and accumulate (Degens & Stoffers 1976).

Other lithologies

Gradations from oil shale to other shales or limestones represent further dilution of algal sediment by siliciclastic or carbonate sediment (the interplay of carbonate and carbonaceous sedimentation is discussed by Loftus & Greensmith, this volume). Several dark shales yield a marine fauna, but some (including those from South Queensferry and Randerston) additionally contain *Botryococcus*-type bodies. These algae do *not* occur in marine environments, and it must be concluded that the beds represent marginal environments which allowed restricted access to marine waters. The Pumpherston Shell Bed at

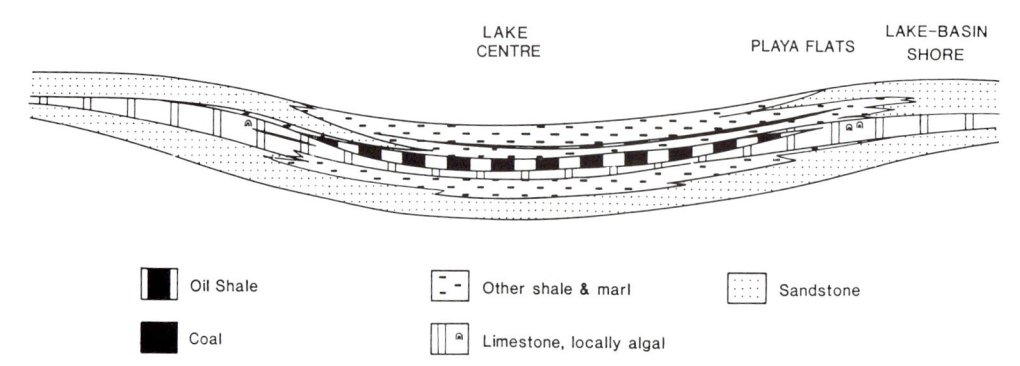

FIG. 9. Schematic relationships between Oil Shale Group lithologies, showing distribution relative to basin centre and margins.

South Queensferry occurs in a series of non-marine shales and oil shales (Fig. 3). As oil shales represent periods of relatively permanent lacustrine conditions, i.e., high lake-stand, it is conceivable that at these times the lake had an outlet to the sea. Atypical fossil assemblages (e.g. Love 1960; Mitchell & Mykura 1962) support the notion that some beds with 'marine' faunas actually represent restricted marginal environments. The dolomitic laminites may similarly have been deposited in a marginal setting. Hesselbo & Trewin (1984) concluded that the bed at Gullane is the deposit of a stratified lake or lagoon, whilst the very similar bed at Granton contains a marine fauna, including conodonts (Briggs & Clarkson 1983). The lithologies are otherwise identical and surely represent environments more similar than the faunas suggest. The mushes of lenticular gypsum crystals in the sediments at East Kirkton are analogous to modern occurrences which grow during early diagenesis below algal mats in sabkha/playa environments.

Palaeoclimate

The Oil Shale Group sediments were deposited in a palaeoclimate transitional between the preceding semi-arid Old Red Sandstone/lowest Dinantian and the later humid tropical climate of the Upper Carboniferous. The Lower Dinantian Cementstone Group (Fig. 1) consists of sabkha sediments with evidence for gypsum and halite deposition. The passage upwards into a sequence including oil shales, the deposits of more permanent water bodies, marks a gradual climatic amelioration. In the Lower Oil Shale Group, limestones are more common than oil shales but this trend is reversed in the Upper Oil Shale Group (Greensmith 1962) as the aridity became less marked. Coals occur at several levels towards the top of the Upper Oil Shale Group. The overlying Lower Limestone Group (latest Dinantian) and Namurian/Westphalian sequence saw an increasing prevalence of coal-bearing sediments under a more humid climatic regime. In the Oil Shale Group, trees are recorded in growth position within oil shale (Conacher 1917) and limestone (Howell & Geikie 1861). The presence of gypsum near the top of the Upper Oil Shale Group suggests that, despite the deposition of coals not far above and below, and the presence of *Lepidodendron* and *Stigmaria* within the gypsiferous beds (Howell & Geikie, 1861) semi-arid conditions could still prevail at those times. However, a direct association between trees and gypsiferous sabkha sediments in present-day North Africa (West et al. 1979) shows that the juxtaposition of vegetation and evaporites in the Oil Shale Group is not anomalous.

Hydrocarbon potential

Existing data

Numerous older accounts report oil yields from the distillation of the oil shales, up to 200 l/tonne (Cameron & McAdam 1978). Little other data directly relating to hydrocarbon potential has been published. Samples of Fells Shale from Westwood (West Calder) yielded 11.2% TOC. (Bitterli 1963) and were the subject of a combined gas chromatography–mass spectrometry investigation by Douglas et al. (1969b) who extracted a range of hydrocarbon molecules. Carbon-13 NMR spectrometry of another sample by Miknis et al. (1982) showed that the oil shale has a high proportion of aliphatic carbon, which indicates a good hydrocarbon generating potential. Cornford (1984) has suggested that these may represent the richest beds, but TOC yields of up to 30% are recorded by the present writer. A mass chromatogram, gas chromatogram and kerogen pyrogram of three different oil shale samples are presented by Douglas et al. (1983).

Pyrolysis data

Data for the hydrocarbon yield of a variety of samples, determined by pyrolysis, are recorded in Table 1 (see stratigraphy in Fig. 1). Measurements were made of hydrocarbons already generated and retained within the rock (S1) and those yielded by kerogen in the rock during pyrolysis (S2). The hydrogen index (S2/TOC) generally increases with source-rock potential. The yield generally increases with percentage alginite (Fig. 8). The stratigraphy of the Lower Oil Shale Group is highly variable and largely omitted from Fig. 1: the samples from the Wardie Shales and Granton Sandstones are from about 400 m and 600 m respectively below the Dalmahoy Shale. Two samples (Nos 1 and 15) are from East Fife (Fig. 2) where correlation with the Lothian region is vague; they are probably from low in the Lower Oil Shale Group. Sample no. 15 is from immediately above Randerston Limestone VII (bed 41 of Kirby 1901). In general, the oil shales *sensu stricto* yield up to 30% TOC, and other black shales up to 15% TOC. Up to 11.4% TOC has been determined in shales containing desiccation cracks. The H/C atomic ratios of oil shales range from 1.5 to 2.0. The quality of a single oil shale horizon can vary markedly (see above): the yields noted here for oil shales may be higher or lower at different localities. These examples however show that the oil shales reach exceptionally rich hydrocarbon potential. Some other black shales, including a 'marine' sample also have rich yields. By comparison the limestones have poor yields but in absolute terms the black limestone has a good potential. Indeed a bitumen distillate was obtained from dark Burdiehouse limestone 150 years ago (Connell 1836). This is not surprising: the limestone contains organic laminae identical to those in shales and oil shales and is essentially a calcite-supported algal deposit. Of three 'oil shale' samples analysed by Douglas et al. (1983), two (10 and 13% TOC) were carbonate-rich (>40%), the other (20% TOC) carbonate-poor (2.5%). The marine samples and limestones contain similar, but fewer, algal bodies to the oil shales.

The S1 and transformation ratio (S1/S1+S2) data suggest that migrated hydrocarbons are generally unimportant in oil shale analyses, but more important in the limestones. The analysis of sample no. 18 confirms thin section observations that shales adjacent to late Carboniferous sills contain a significant quantity of migrated bitumen (see Geikie (1900) for details of intrusions into shales at Dodhead). The present maturity of the oil shales varies with proximity to igneous intrusions and structural highs/lows, but is generally early/mid-mature with respect to oil generation (Cornford 1984).

The main region of oil shale deposition in the Lothians (Fig. 2) from which sample nos 2–8 were collected, occupies about 3000 km^2, but the occurrence of thin, but rich, oil shales in East Fife (sample no. 1) extends the total region of oil shale deposition to at least 10 000 km^2.

No samples from the lowest Dinantian Cementstone Group (Fig. 1) have been analysed in this survey, but the Group contains dark lacustrine shales and dolomitic laminites which may have some hydrocarbon generating potential (e.g. at North Berwick, where black shales and dolomites contain migrated bitumen; and note records of oil/gas in the Spilmersford borehole, East Lothian (Davies 1974)). Extrusive volcanic rocks in the Edinburgh district which Parnell (1984a) concluded had generated hydrocarbons by heat, *pre-date* Oil Shale Group sediments, implying that hydrocarbon source rocks already existed. Furthermore, bitumen-enriched Cementstone Group shales occur adjacent to sills in the North Berwick district (P. Eakin pers. comm.), offering direct evidence that igneous activity generated hydrocarbons from these rocks.

Conclusions

The Oil Shale Group contains rocks with excellent source-rock potential. The richest beds are the oil shales *sensu stricto* with TOC contents of up to 30% and S2 yields of up to 180 mg/g. Other shales in the sequence, with up to 15%, also have significant potential. The organic matter in the oil shales and shales is predominantly algal, a mixture of laminar algal matter and discrete *Botryococcus*-type algal bodies which probably represent the deposits of planktonic and benthonic algal communities respectively. The hydrocarbon yield is greatest in those shales which are richest in alginite. Dark limestones contain organic laminae identical to those in the shales, and themselves have some limited source-rock potential.

The rich oil shales are the deposits of a stratified lake, whilst abundant evidence for exposure in the other shales and limestones suggests a playa environment for their deposition. Widespread evidence for vegetation, including coals, shows that the Oil Shale Group was not deposited in an arid basin like many modern playas, but in a more tropical climate.

ACKNOWLEDGEMENTS: I acknowledge with pleasure discussion with G. Loftus; technical assistance provided by E. Lawson, M. Gill and S. Watters; and H. Shaw for making available Fig. 5.

References

ANDERSON, J. G. S. 1942. The oil-shales of the Lothians: structure: Area II (Pumpherston). *Geological Survey Wartime Pamphlet No. 27.*

BARNARD, P. C. & COOPER, B. S. 1981. Oils and source rocks of the North Sea area. *In:* ILLING, L. V. & HOBSON, G. D. (eds), *Petroleum Geology of the Continental Shelf of North-west Europe*, Institute of Petroleum, London, 169–175.

BERGMANN, W. 1963. Geochemistry of lipids. *International Series of Monographs in the Earth Sciences*, **16**, 503–542.

BITTERLI, P. 1963. Aspects of the genesis of bituminous rock sequences. *Geologie en Mijnbouw*, **42**, 183–201.

BLACKBURN, K. B. & TEMPERLEY, B. N. 1936. Botryococcus and the algal coals. *Transactions of the Royal Society of Edinburgh*, **58**, 841–868.

BRIGGS, D. E. G. & CLARKSON, E. N. K. 1983. The Lower Carboniferous Granton 'Shrimp-Bed', Edinburgh. *Special Papers in Palaeontology*, **30**, 161–177.

BROWN, A. C., KNIGHTS, B. A. & CONWAY, E. 1969. Hydrocarbon content and its relationship to physiological state in the green alga *Botryococcus braunii*. *Phytochemistry*, **8**, 543–547.

CADELL, H. M., GRANT WILSON, J. S., CALDWELL, W. & STEUART, D. R. 1906. The Oil-shales of the Lothians, 2nd edn. *Memoir of the Geological Survey of Scotland*, Her Majesty's Stationery Office, Edinburgh.

CAMERON, I. B. & McADAM, A. D. 1978. Oil-shales of the Lothians, Scotland: present resources and former workings. *Report of the Institute of Geological Sciences, No. 78*.

CANE, R. F. 1976. The origin and formation of oil shale. *Developments in Petroleum Science*, **5**, 27–60.

CARRUTHERS, R. G., CALDWELL, W. & STEUART, D. R. 1912. The Oil-shales of the Lothians, 2nd edn. *Memoir of the Geological Survey of Scotland*, Her Majesty's Stationery Office, Edinburgh.

CONACHER, H. R. J. 1917. A study of oil shales and torbanites. *Transactions of the Geological Society of Glasgow*, **16**, 164–192.

CONNELL, A. 1836. Analysis of coprolites and other organic remains imbedded in the Limestone of Burdiehouse near Edinburgh. *Transactions of the Royal Society of Edinburgh*, **13**, 283–296.

CORNFORD, C. 1984. Source rocks and hydrocarbons of the North Sea. *In:* GLENNIE, K. W. (ed.), *Introduction to the Petroleum Geology of the North Sea*. Blackwell Scientific Publications, Oxford, 171–204.

DAVIES, A. 1974. The Lower Carboniferous (Dinantian) sequence at Spilmersford, East Lothian, Scotland. *Bulletin of the Geological Survey of Great Britain*, **45**, 1–24.

DEGENS, E. T. & STOFFERS, P. 1976. Stratified waters as a key to the past. *Nature*, **263**, 22–27.

DICK, J. R. F. 1981. *Diploselache woodi* gen. et sp. nov., an early Carboniferous shark from the Midland Valley of Scotland. *Transactions of the Royal Society of Edinburgh*, **72**, 99–113.

DOUGLAS, A. G., EGLINTON, G. & MAXWELL, J. R. 1969a. The hydrocarbons of coorongite. *Geochimica et Cosmochimica Acta*, **33**, 569–577.

——, —— & —— 1969b. The organic geochemistry of certain samples from the Scottish Carboniferous Formation. *Geochimica et Cosmochimica Acta*, **33**, 579–590.

——, HALL, P. B. & SOLLI, H. 1983. Comparative organic geochemistry of some European oil shales. *In:* MIKNIS, F. P. & McKAY, J. F. (eds), *Geochemistry and Chemistry of Oil Shales*, American Chemical Society, 59–84.

GEIKIE, A. 1900. The geology of Central and Western Fife and Kinross. *Memoir of the Geological Survey of Scotland*, Her Majesty's Stationery Office, Edinburgh.

GELPI, E., ORO, J., SCHNEIDER, H. J. & BENNET, E. O. 1968. Olefins of high molecular weight in two microscopic algae. *Science*, **161**, 700–702.

GIBLING, M. R., UKAKIMAPHAN, Y. & SRISUK, S. 1985. Oil shale and coal in intermontane basins of Thailand. *Bulletin of the American Association of Petroleum Geologists*, **69**, 760–766.

GREENSMITH, J. T. 1962. Rhythmic deposition in the Carboniferous Oil-Shale Group of Scotland. *Journal of Geology*, **70**, 355–364.

—— 1968. Palaeogeography and rhythmic deposition in the Scottish Oil-Shale Group. *United Nations Symposium on the Development and Utilization of Oil Shale Resources*, Section B, Tallinn, 1–16.

HESSELBO, S. P. & TREWIN, N. H. 1984. Deposition, diagenesis and structures of the Cheese Bay Shrimp Bed, Lower Carboniferous, East Lothian. *Scottish Journal of Geology*, **20**, 281–296.

HOWELL, H. M. & GEIKIE, A. 1861. The geology of the neighbourhood of Edinburgh, 1st edn. *Memoir of the Geological Survey of Scotland*, Her Majesty's Stationery Office, Edinburgh.

HUTTON, A. C., KANTSLER, A. J., COOK, A. C. & McKIRDY, D. M. 1980. Organic matter in oil-shales. *Australian Petroleum Exploration Association Journal*, **20**, 44–67.

KENNEDY, W. Q. 1943. The oil-shales of the Lothians: Structure: Area IV (Philpstoun). *Geological Survey Wartime Pamphlet No. 27*.

KIRBY, J. W. 1901. On Lower Carboniferous strata and fossils at Randerston, near Crail, Fife. *Transactions of the Edinburgh Geological Society*, **8**, 61–75.

LOVE, L. G. 1960. Assemblages of small spores from the Lower Oil-Shale Group of Scotland. *Proceedings of the Royal Society of Edinburgh*, **67**, 99–126.

MACGREGOR, A. R. 1973. *Fife and Angus Geology: An Excursion Guide*. Scottish Academic Press, Edinburgh.

MACGREGOR, M. 1938. Conditions of deposition of the oil-shales and cannel coals of Scotland. *In: Oil Shale and Cannel Coal*. Institute of Petroleum, London, 6–18.

MAXWELL, J. R., DOUGLAS, A. G., EGLINTON, G. & McCORMICK, A. 1968. The botryococcenes—hydrocarbons of novel structure from the alga *Botryococcus brauni* Kützing. *Phytochemistry*, **7**, 2157–2171.

MIKNIS, F. P., NETZEL, D. A., SMITH, J. W., MAST, M. A. & MACIEL, G. E. 1982. ^{13}C NMR measurements of the genetic potentials of oil shales. *Geochimica et Cosmochimica Acta*, **46**, 977–984.

MITCHELL, G. H. & MYKURA, W. 1962. The geology of the neighbourhood of Edinburgh, 3rd edn. *Memoir of the Geological Survey of Scotland*, Her Majesty's Stationery Office, Edinburgh.

MOORE, L. R. 1968. Cannel coals, bogheads and oil-shales. *In:* MURCHISON, D. & WESTOLL, T. S. (eds), *Coal and Coal-bearing Strata*, Oliver & Boyd, London, 19–29.

MUIR, R. O. & WALTON, E. K. 1957. The East Kirkton Limestone. *Transactions of the Geological Society of Glasgow*, **22**, 157–168.

PARNELL, J. 1984a. Hydrocarbon minerals in the Midland Valley of Scotland with particular reference to the Oil-Shale Group. *Proceedings of the Geologists Association*, **95**, 275–285.

—— 1984b. The depositional environment of oil-shales

in the Oil-Shale Group, Midland Valley of Scotland. In: *European Dinantian Environments 1st Meeting Abstracts*, Open University, Milton Keynes, 13–15.

PEACH, B. N., CLOUGH, C. T., HINXMAN, L. W., GRANT WILSON, J. S., CRAMPTON, C. B., MAUFE, H. B. & BAILEY, E. B. 1910. The geology of the neighbourhood of Edinburgh, 2nd edn. *Memoir of the Geological Survey of Scotland*, Her Majesty's Stationery Office, Edinburgh.

RICHEY, J. E. 1942. The oil-shales of the Lothians: Structure: Area I (West Calder). *Geological Survey Wartime Pamphlet No. 27.*

SHERWOOD, N. R. C., COOK, A. C., GIBLING, M. & TANTISUKRIT, C. 1984. Petrology of a suite of sedimentary rocks associated with some coal-bearing basins in northwestern Thailand. *International Journal of Coal Geology*, **4**, 45–71.

STACH, E., MACKOWSKY, M. T., TEICHMULLER, M., TAYLOR, G. H., CHANDRA, D. & TEICHMULLER, R. 1982. *Stach's Textbook of Coal Petrology*, Gebruder Borntraeger, Berlin.

STADNIKOW, G. L. & WEIZMANN, A. 1929. Ein Beitrag zur Kenntnis der Umwandlung der Fettsäuren im Laufe der geologischen Zeitperioden. *Brennstoff Chemie*, **10**, 401–403.

TAIT, D. 1925. Notice of a shrimp-bearing limestone in the Calciferous Sandstone Series at Granton, near Edinburgh. *Transactions of the Edinburgh Geological Society*, **11**, 131–134.

TRAQUAIR, R. H. 1907. Report on fossil fishes collected by the Geological Survey of Scotland from shales exposed on the shore near Gullane, East Lothian. *Transactions of the Royal Society of Edinburgh*, **46**, 103–117.

TRAVERSE, A. 1955. Occurrences of the oil-forming alga *Botryococcus* in lignites and other Tertiary sediments. *Micropalaeontology*, **1**, 343–350.

WEST, I. M., ALI, Y. A. & HILMY, M. E. 1979. Primary gypsum nodules in a modern sabkha on the Mediterranean coast of Egypt. *Geology*, **7**, 354–358.

ZALESSKY, M. D. 1914. On the nature of *Pila*, the yellow bodies of Boghead and on sapropel of the Ala-Kool gulf of Lake Balkhash. *Bulletin Comité Géologique de Petrograd*, **33**, 495–507.

ZELINSKY, N. D. 1926. Künstliche Naptha aus Balchasch-sapropeliten. *Brennstoff Chemie*, **7**, 33–37.

J. PARNELL, Department of Geology, Queen's University, Belfast BT7 1NN, UK.

Lacustrine sequences in an early Mesozoic rift basin: Culpeper Basin, Virginia, USA

P. J. W. Gore

SUMMARY: Lacustrine sequences in the Culpeper Basin, Virginia, USA were deposited in a half-graben dominated by fluvial sheetflood deposition. During periods of increased precipitation and runoff, lakes formed within the basin. Numerous lacustrine sequences, ranging from less than 1 m to more than 35 m thick, document the expansion and contraction of perennial freshwater and saline lakes controlled by Milankovitch-type cyclic climatic change. Many of the lacustrine sequences contain black, laminated shale, deposited in deep, anoxic water in stratified lakes. Most of the lakes were hydrologically closed, as indicated by the presence of evaporite crystal moulds and pseudomorphs in associated mudflat deposits.

Introduction

The rift basins of the Newark Supergroup, eastern North America (Fig. 1), contain a complex mosaic of early Mesozoic lacustrine and fluvial deposits. Lacustrine strata are present in all of the basins, and a wide variety of types of deposits have been described, ranging from playa lakes to extensive, deep, stratified lakes (Hubert *et al.* 1976, 1978; Wheeler & Textoris 1978; Olsen 1980b; Manspeizer & Olsen 1981; Olsen 1984a; Hentz 1985; Demicco & Kordesch 1986). The Culpeper Basin is one of the larger basins in the southern part of the Newark Rift System, and the southernmost basin containing Jurassic rocks. The sedimentary record of the Culpeper Basin spans an interval of about 30 million years from Karnian (Late Triassic) to Pliensbachian (Early Jurassic). Within the Culpeper Basin, several types of lake deposits are present representing deposition in both freshwater and saline lakes, and in both shallow and deep lakes. Lacustrine sequences are commonly interbedded with mudflat and fluvial sheetflood deposits, and numerous episodes of lake expansion and contraction are recorded. In the Early Jurassic Waterfall Formation alone, there were at least 10 episodes of lake expansion and contraction (Hentz 1985). Lake deposits in the Waterfall Formation average about 35 m thick, representing deposition in deep stratified lakes with anoxic bottom waters (Hentz 1985). The Waterfall Formation, however, contains only 16% of the stratigraphic column of the Culpeper Basin. Lacustrine deposits are also present in the underlying Bull Run and Buckland Formations, which together comprise 74% of the stratigraphic column. Lacustrine deposits in these two formations differ significantly from those in the Waterfall Formation in lithology, thickness, and stratigraphic sequences. The purpose of this paper is to present a basin-wide synthesis of lacustrine deposition by describing the palaeoenvironmental setting of the lacustrine deposits, comparing the palaeoenvironmental setting of the Bull Run and Buckland Formations to that of the Waterfall Formation, and discussing reasons for the change in style of lacustrine deposition within the basin. Nine lacustrine facies are described, representing deposition in offshore, nearshore, and mudflat environments in an early Mesozoic rift basin.

Geological setting

The Newark Rift System is a series of early Mesozoic fault-bounded basins in eastern North America which formed during the earliest phases of rifting prior to the opening of the Atlantic Ocean (Fig. 1). The rocks in these basins are dominated by continental sediments deposited in fluvial, alluvial fan, and lacustrine environments. Most of these sediments are red, but grey to black lacustrine deposits are also common. In addition, some basins contain coal, aeolian sandstone, or evaporites. The sedimentary rocks, together with interstratified Early Jurassic basalt flows, comprise the Newark Supergroup (Froelich & Olsen 1984, 1985). Diabase intrusives are present throughout much of the Newark Supergroup, some of which are interpreted as feeder dykes to the Early Jurassic basalt flows (Martello *et al.* 1984; Froelich & Gottfried 1985).

The Culpeper Basin, located in Virginia and Maryland, USA is an elongate half-graben containing approximately 9000 m of interbedded fluvial, alluvial fan, and lacustrine deposits of Late Triassic and Early Jurassic age (Fig. 2). The basin is a maximum of about 20 km wide and is more than 180 km long. Late Triassic rocks crop out in an arcuate pattern in the northern, eastern, and southern parts of the basin, and Early Jurassic rocks are confined to the west-central part of the basin (Fig. 2). Structurally, the basin is bounded on the west by a zone of east-dipping listric normal faults which separate the early Mesozoic strata from Precambrian and lower Paleozoic granitic, metasedimentary, and metavolcanic rocks of the Blue Ridge Province. The eastern border of the basin is in part an unconformable overlap onto Precambrian and lower Paleozoic crystalline rocks of the Appalachian Piedmont Province, and is in part fault-bounded (Leavy 1980; Froelich et al. 1982). Tectonic movement along the western-border fault system tilted the basin floor down toward the west, forming a half-graben. Bedding in the basin dips west toward the fault zone, and dip

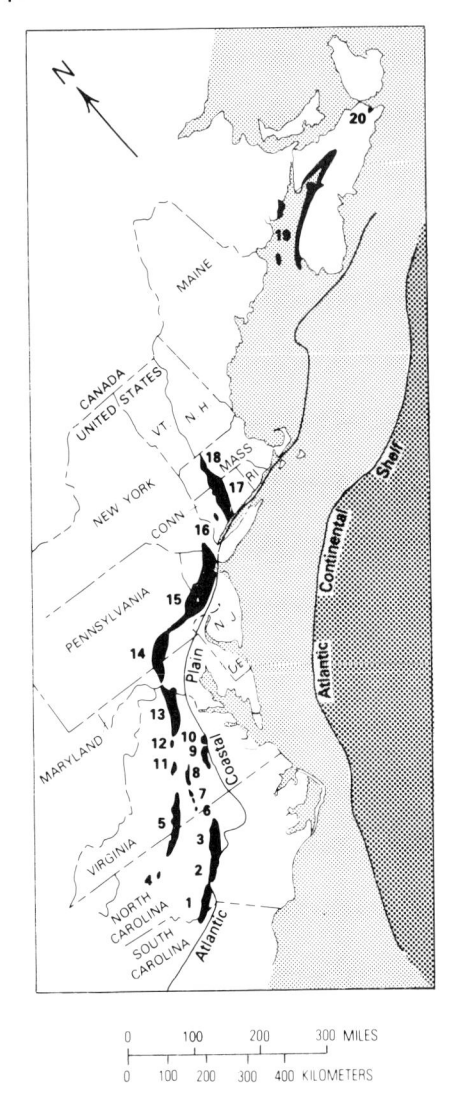

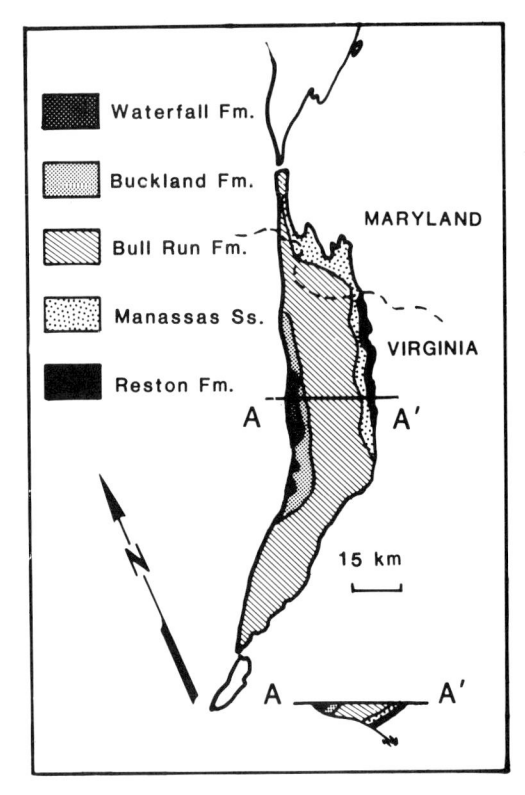

FIG. 1. Distribution of the Newark Supergroup rift basins in eastern North America (from Froelich & Olsen 1984, 1985).

EXPLANATION

1. Wadesboro (N.C. – S.C.)
2. Sanford (N.C.)
3. Durham (N.C.)
4. Davie County (N.C.)
5. Dan River and Danville (N.C. – Va.)
6. Scottsburg (Va.)
7. Basins north of Scottsburg (Va.)
8. Farmville (Va.)
9. Richmond (Va.)
10. Taylorsville (Va.)
11. Scottsville (Va.)
12. Barboursville (Va.)
13. Culpeper (Va. – Md.)
14. Gettysburg (Md. – Pa.)
15. Newark (N.J. – Pa. – N.Y.)
16. Pomperaug (Conn.)
17. Hartford (Conn. – Mass.)
18. Deerfield (Mass.)
19. Fundy or Minas (Nova Scotia – Canada)
20. Chedabucto (Nova Scotia – Canada)

FIG. 2. Generalized geologic map and stratigraphic column for the Culpeper Basin, based on Lindholm (1979). Geologic cross-section is based on Ratcliffe & Burton (1985), Schamel et al. (1986), and Schamel (Earth Sciences & Resources Institute, University of South Carolina, written communication, July 1986). The Reston Formation, Manassas Sandstone, and much of the Bull Run Formation are Late Triassic in age. The upper part of the Bull Run Formation, and the Buckland and Waterfall Formations are Early Jurassic in age. These five formations comprise the Culpeper Group.

steepens westwards (Froelich et al. 1982, p. 57) (Fig. 2). Synsedimentary movement along the fault system led to the formation of alluvial fans along the western basin margin, and the upfaulted highlands to the west supplied coarse clastic sediment to east-flowing fluvial systems. Asymmetric subsidence and westward tilting of the basin floor caused ponding of water against the western edge of the basin, forming large lakes which shallowed eastward. During the latter part of basin filling, a change to west- and north-west-flowing fluvial systems suggests subsidence near the fault zone coupled with uplift in the central part of the basin (Hentz 1985).

Early studies of the Culpeper Basin determined that most of the sediments were fluvial in origin, including alluvial fan and floodplain deposits, and pointed out that some of the sediments were deposited in temporary stagnant bodies of water (Roberts 1928). Between 1928 and the late 1970s there were few geological investigations in the basin. Among those studies are reports on fish-bearing black shales (Baer & Martin 1949; Applegate 1956), and on the sedimentary petrology of a unit of lacustrine, oolitic limestone in a quarry near Culpeper, Virginia (Young & Edmundson 1954; Carozzi 1964). In the 1970s, increasing attention turned toward the stratigraphy and sedimentology of the Culpeper Basin. At this time, the first Early Jurassic spores and pollen were discovered in the Newark Supergroup (Cornet et al. 1973; Cornet 1977). Prior to this, all of the rocks in the Newark Supergroup were believed to be late Triassic in age. The presence of fossiliferous, grey and black lacustrine shales and carbonates was mentioned by Cornet (1977), Lee (1977, 1979), and Lindholm (1979) in their reports on basin stratigraphy. These studies revealed the first information about the distribution and character of lacustrine beds, and stimulated more detailed research on the sedimentary history of the basin by Lindholm et al. (1979), Hentz (1981, 1982, 1985), Froelich et al. (1982), Lindholm et al. (1982), Gore (1983, 1984, 1985a, 1985b, 1985c, 1986, in press), Gore & Lindholm (1983), Sobhan (1985), Gore & Traverse (1986), and Smith & Robison (1986).

Stratigraphic nomenclature in the basin has had a long and complicated history, and is currently under revision by the U.S. Geological Survey (Froelich in prep.). In this report, the nomenclature of Lindholm (1979) is used (Fig. 2). The rocks in the Culpeper Basin belong to the Culpeper Group, which consists of five formations, from oldest to youngest (Lindholm 1979): Reston Formation (a conglomeratic unit), Manassas Sandstone (a unit dominated by red sandstone), Bull Run Formation (a sequence of mudstone and siltstone, with local conglomeratic units), Buckland Formation (a sequence of five basalt-flow units with interbedded sedimentary rocks), and Waterfall Formation (a sequence of interbedded sandstone, shale and conglomerate, with a basalt flow near the top). Lacustrine sediments are present in three of the five formations, the Bull Run, Buckland and Waterfall Formations, which are described below.

Bull Run Formation

The Bull Run Formation (Late Triassic and Early Jurassic) is dominated by greyish-red to reddish-brown plagioclase-rich mudstones and siltstones, with thin beds of claystone and fine-grained, micaceous, pale red sandstone. Most of the rocks are massive or mottled, but some of the coarser units are laminated or cross-laminated. Mud-cracks and bioturbation are common in the claystone and mudstone, and small caliche-like pedogenic calcite nodules and evaporite crystal moulds and pseudomorphs are present locally. Conglomeratic units are present along the faulted western border of the basin and are interpreted as alluvial fan deposits (Lindholm et al. 1979). The Bull Run Formation is the thickest unit in the Culpeper Basin, ranging up to 5100 m (Lindholm 1979). This formation is referred to by several other workers as the Balls Bluff Siltstone (Lee 1977, 1979, 1980; Olsen 1984a).

Buckland Formation

The Buckland Formation (Early Jurassic) consists of a series of basalt flows with interbedded sedimentary units. Five basalt-flow units are mapped, but several of these are composite, consisting of multiple flows separated by relatively thin sedimentary sequences (Lindholm 1979). The sedimentary rocks are dominantly sandstone, mudstone, mudshale, and siltshale (Lindholm 1979). Because of poor exposure, the detailed stratigraphy of the sedimentary units in this formation is not known. The mean thickness of the Buckland Formation is 1670 m (Lindholm 1979), and later workers have subdivided it into several thinner formations (Olsen 1984a; Froelich in prep).

Waterfall Formation

The Waterfall Formation (Early Jurassic), exposed only in the west-central part of the basin, consists of about 1500 m of interbedded sandstone, shale, and conglomerate (Lindholm 1979). Lacustrine sequences in the Waterfall Formation average about 35 m thick (Hentz 1985). These sequences are dominated by nearshore, grey, wavy-bedded sandstone and offshore, black, calcareous shale with turbidites (Hentz 1985). Lacustrine sequences are separated by fluvial deposits ranging from 5 m to approximately 240 m thick (Hentz 1985, p. 105). Mudflat deposits are present locally between the nearshore lacustrine and fluvial beds (Hentz 1985, p. 104). Mudflat deposits are dominated by thin beds of red sandstone, siltstone, or mudshale with mudcracks, burrows, intraclasts, traces of cross-lamination, and small calcite-filled vugs (Hentz 1985). The calcite-filled vugs were probably originally caliche or evaporite minerals, and suggest an arid to semi-arid climate (Hentz 1985, p. 104).

Lacustrine facies

Newark Supergroup lacustrine sediments are typically recognized on the basis of fine grain size, laminations, and presence of non-marine fossils. However, not all lacustrine sediments are fine grained; lacustrine sandstones and conglomerates are present locally. In addition, laminations are not present in all lacustrine deposits; they may have been destroyed by bioturbation. Also, some lacustrine units lack fossils. Lacustrine rocks may be red, brown, green, grey, tan, or black, but grey or black beds generally signal the presence of lacustrine deposits. Red, non-lacustrine sediments can be altered to grey hornfels in contact metamorphic aureoles surrounding diabase intrusives, and may resemble grey lacustrine deposits. Lithology, mineralogy, sedimentary structures, sedimentary sequences, and lateral gradation into red non-lacustrine deposits away from the intrusive indicate that these grey rocks are not lacustrine. Dark grey or black colour in lacustrine shales is due to the presence of organic matter or pyrite (Pettijohn 1975; Potter et al. 1980). In the Newark Supergroup, total organic carbon (TOC) contents of lacustrine beds range from less than 1% to 40% (Ziegler 1983; Olsen 1985b). Lacustrine deposits of the Waterfall Formation have an average TOC of approximately 2% (Smith & Robison 1986). The average TOC of the Bull Run and Buckland Formations is slightly lower (Smith & Robison 1986).

Organic matter in lacustrine sediments is a mixture of allochthonous material (spores, pollen, leaves, needles, seeds, wood, etc.) and autochthonous material (planktonic algae, microbial mats, aquatic plants) (Dean & Fouch 1983). Most lacustrine organic matter is believed to be derived from planktonic algae (Dean & Fouch 1983), or from bottom dwelling blue-green algal and fungal mats (Eugster & Hardie 1975). Organic matter in the Waterfall Formation is typical of non-marine source rocks, and is dominated by unicellular algae and palynomorphs (spores and pollen), with less abundant woody material (Smith & Robison 1986).

Stratification within lake waters leads to a loss of oxygen in the deep water below wave base as the result of respiration and decay, and large quantities of organic matter may be preserved in the sediments (Allen & Collinson 1986). The upper water mass (epilimnion) is saturated or supersaturated with oxygen derived from the atmosphere or from photosynthesis, but in the lower water mass (hypolimnion), oxygen may be depleted or completely eliminated. Lakes may be perennially stratified with permanently stagnant bottom waters (meromictic), overturn infrequently (oligomictic), overturn once yearly (monomictic), or twice yearly (dimictic). Overturn causes the lake waters to be mixed, transporting oxygen to the bottom, and nutrients into surface waters. Organic matter can also accumulate in shallow playa lakes with high organic productivity and low sediment influx (Eugster & Hardie 1975). Stratified lakes and shallow playa lakes are not necessarily mutually exclusive. A shallow saline lake surrounded by extensive playa mudflats can become chemically stratified if there is an increase in freshwater inflow, forming a freshwater cap over saline bottom waters (ectogenic meromictic lake).

Hydrological conditions within a lake affect the nature and arrangement of sedimentary facies. A distinction must be drawn between lakes which have an outlet and are hydrologically open, and lakes which lack an outlet and are hydrologically closed. During its history, a lake may pass through both open and closed stages, depending on local conditions. Perennial inflow to a closed basin will result in a rise in lake level until the water fills the basin and spills out, causing the lake to become hydrologically open (Smoot 1985). Hydrologically open lakes have relatively stable, fixed shorelines because inflow and precipitation are balanced by outflow and evaporation (Eugster & Kelts 1983; Allen & Collinson 1986). Sediments in open lakes are generally fluvial-derived; chemical sediments may form, but are restricted to the lake itself and do not occur in fringing mudflats or spring areas (Eugster & Kelts 1983; Allen & Collinson 1986). Hydrologically closed lakes have a water budget in which inflow and precipitation may be exceeded by evaporation and infiltration (Allen & Collinson 1986). This imbalance results in dramatic water level changes and rapid, frequent shoreline movements resulting in transgressive-regressive sedimentary sequences (Eugster &

Kelts 1983; Allen & Collinson 1986). Sediments in closed basin lakes tend to be a complex mixture of fluvial-derived sediment, reworked intrabasinal sediment, and chemical and biochemical deposits. Chemical sediments in closed basins form not only in the lake itself, but also in adjacent mudflats and spring areas (Eugster & Kelts 1983). Salinity increases as lake levels drop, such that a change from deep, relatively fresh water to shallow saline or hypersaline water may be recorded within only a few centimetres of sediment.

Closed basin deposition has been proposed for at least some lacustrine sediments in all of the Newark Supergroup basins (Smoot 1985). Evidence for closed basin conditions in the Newark includes (Smoot 1985): (1) decrease in grain size toward the basin centre, (2) palaeocurrent patterns indicating flow toward the basin centre, (3) local sources for coarse-grained terrigenous sediments, (4) presence of evaporites, or evaporite crystal moulds and pseudomorphs, and (5) presence of repeated transgressive–regressive lacustrine sequences within the basins. A closed basin model also implies certain relationships between lacustrine and fluvial deposits. For example, the presence of a fluvial sandstone which was deposited in a relatively deep, perennial river near the centre of a basin suggests that the basin was open, or that the present basin is only a small erosional remnant of a much larger basin (Smoot 1985, p. 9). Perennial river deposits should exist only near the margins of closed basins, and should be laterally-equivalent to perennial lake deposits. Ephemeral fluvial deposits may extend far into the basin, but they generally interfinger with lacustrine shale or playa mudstone (Smoot 1985).

Many of the Culpeper Basin lakes were hydrologically closed. This is indicated by (1) grain size distribution within the basin (Lindholm *et al.* 1979), (2) palaeocurrent patterns (Lindholm 1979), (3) sediment provenance (Lindholm *et al.* 1979), (4) presence of evaporite crystal moulds and pseudomorphs in lacustrine and mudflat deposits (Froelich *et al.* 1982; Gore 1983), and (5) presence of transgressive–regressive lacustrine sequences (Lindholm *et al.* 1982; Gore 1983; Hentz 1985). In addition, thick fluvial deposits representing deposition in deep perennial rivers are not present near the centre of the basin. Fluvial sandstones near the centre of the basin (particularly in the Bull Run Formation) are generally thin, and resemble sheetflood deposits (Froelich *et al.*, 1982, p. 61; Sobhan 1985). Both saline and freshwater lacustrine deposits are present, and the sequence of facies records the expansion and contraction (transgression and regression) of perennial lakes. Culpeper Basin lacustrine units are grouped into nine lithofacies, differentiated primarily on the basis of lithology and sedimentary structures, and to a lesser extent on colour, mineralogy, and fossil content. The lithofacies are: (1) black shale, (2) graded sandstone, (3) conglomerate, (4) olive-grey to greyish-red laminated mudshale and clayshale, (5) limestone, (6) grey to tan ripple-cross-laminated to massive sandstone and siltstone, (7) red laminated and cross-laminated sandstone, siltstone, and mudstone, (8) disrupted laminites, and (9) massive siltstone and mudstone.

(1) Black shale (facies BS)

A wide variety of primary (depositional) and secondary (biogenic, compactional, and diagenetic) sedimentary structures are present in the Culpeper Basin black shales. Hentz (1985) recognized three types of laminae in black shales in the Waterfall Formation, unevenly-laminated type 1 (deposited from currents in anoxic water), unevenly-laminated type 2 (deposited by settling from suspension in anoxic water), and finely laminated (deposited below wave base by settling from suspension in periodically oxygenated water). Although there are several varieties of black shale in the basin, two major types are distinguished on the basis of lamination style, authigenic minerals, and fossil assemblages, and are referred to here as type A and type B.

Type A black shales (facies BS-A). Description: These shales have graded and homogeneous laminae, consisting of light-coloured calcite-rich layers alternating with dark-coloured organic-rich layers on a submillimetre scale (Fig. 3). Lenticular laminations and soft-sediment deformation structures, such as convolute laminations and microfaults are present locally, and carbonate-rich sand layers are present in some units. These shales contain pyrite, phosphate nodules (some of which are coprolites), and abundant fossils, including conchostracans (*Cyzicus* sp. and *Cornia* sp.), ostracods (*Darwinula* sp.), plant remains, and well-preserved, phosphatized fish (Fig. 3). Type A black shales are common in the Early Jurassic Buckland and Waterfall Formations, but they are rare in the Bull Run Formation. The three types of black shale laminae described by Hentz (1985) are included with type A black shales.

Interpretation: The presence of submillimetre scale laminations, fine grain size, and non-marine fossils suggests deposition in relatively quiet lacustrine waters. Fine-grained, organic-rich, laminated sediments are accumulating in modern lakes in areas that are protected from bioturba-

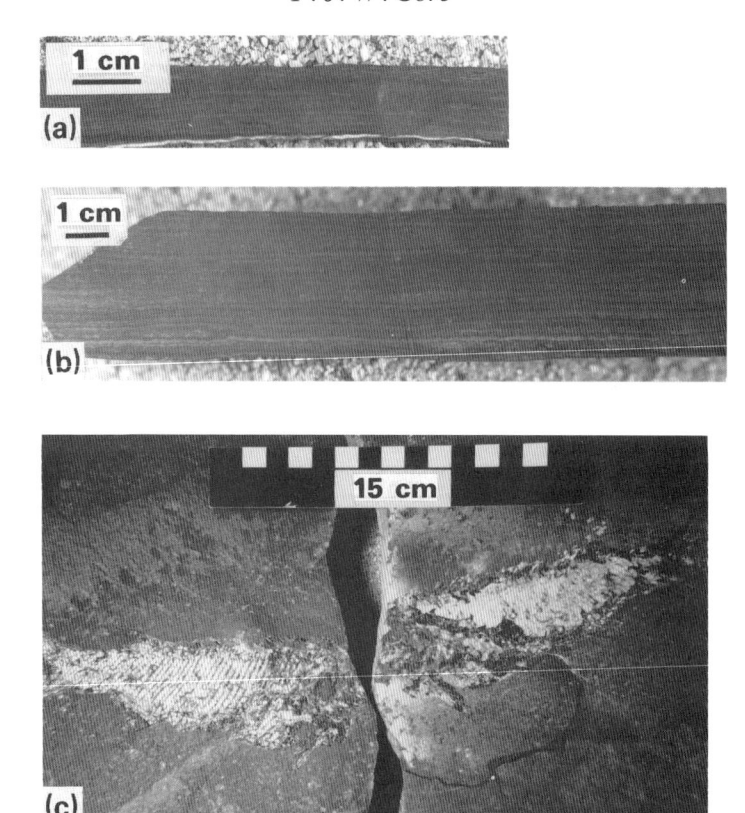

FIG. 3. Photographs of type A black shale facies. (a) Photograph of laminated black shale. Note thin, discontinuous laminations; Buckland Formation, Midland fish-bed (Fig. 13E). (b) Photograph of laminated black shale with thicker, more continuous laminations than in (a), above; Buckland Formation, Midland fish-bed (Fig. 13E). (c) Photograph of fish fossil from the Midland fish-bed; Buckland Formation (Fig. 13E).

tion, high clastic input, strong bottom currents, and oxidation of organic matter. These conditions are found in anoxic bottom waters in modern meromictic to oligomictic lakes, such as Lake Tanganyika, East Africa (Degens et al. 1971; Degens & Stoffers 1976; Stoffers & Hecky 1978; Demaison & Moore 1980), Lake Malawi, East Africa (Degens & Stoffers 1976; Crossley 1984), Lake Kivu, East Africa (Degens et al. 1972, 1973; Degens & Stoffers 1976), Lake Zurich, Switzerland (Kelts & Hsü 1978), Lake Bosumtwi, Ghana (Talbot 1976, 1983; Talbot & Kelts 1985; Talbot et al. 1984), the Dead Sea, Israel–Jordan (Ryder 1980), and Green Lake, New York, USA (Ludlam 1969). Laminated sediments can also be deposited in lakes which are not permanently stratified, such as Lake Turkana, East Africa, which has oxygenated bottom waters (Yuretich 1979; Cohen 1984). Lakes in which the deposition of organic matter (primarily planktonic algae) exceeds the rate of oxygen replenishment for at least part of a year will lack bioturbation (Eugster & Kelts

1983, p. 332). In addition, laminations may be preserved in organic-rich, anoxic sediments beneath an oxygenated water column if the fauna is restricted to small grazers and if burrowers are absent (Eugster & Kelts 1983, p. 332). In the marine environment, however, it seems that laminations and organic carbon are preserved only when the overlying water column is anoxic (<0.1 ml/l dissolved oxygen), such as in the Pacific oxygen minimum zone and in the California borderland basins (Rhoads & Morse 1971; Byers 1974; Berner 1981; Maynard 1982; Waples 1983; Savrda et al. 1984; Thompson et al. 1985, p. 176). Organic content can be used as a rough estimate of oxygen conditions. Black shales with TOC less than approximately 2% were probably deposited in waters with some oxygen (Waples 1983). For example, laminated sediments of Lake Turkana, which were deposited under an oxygenated water column, generally have TOC contents of less than 1% (Yuretich 1976, 1979; Cohen 1984). Sediments with greater amounts of organic matter were probably deposited under anoxic or near anoxic conditions (Waples 1983). Smith & Robison (1986) reported average TOC for Waterfall Formation lacustrine sediments of about 2%. Those samples with TOC greater than 2% were likely deposited in anoxic water, but samples with less than 2% TOC may have been deposited in at least partially oxygenated water.

Pyrite forms under anoxic conditions in association with organic matter decomposition, sulphate reduction by anaerobic bacteria, and the production of hydrogen sulphide (Berner 1970, 1981), but it is not a good indicator of anoxic bottom waters. Pyrite can form in stratified lakes with anoxic bottom waters, but it can also form below the sediment–water interface in organic-rich sediments deposited in oxygenated water, or in anoxic microenvironments in otherwise oxidizing sediments (Goldhaber & Kaplan 1974; Berner 1981; Waples 1983).

Modern marine phosphate nodules form in the transition zone between aerobic and anaerobic waters, or where the water column is oxygenated, but conditions are anoxic and strongly reducing at or below the sediment–water interface (Cook 1976, p. 530). Type A black shales which contain phosphate nodules may have been deposited in dysaerobic waters near the oxic–anoxic transition. Dysaerobic conditions (0.1–1.0 ml/l dissolved oxygen) are capable of supporting the invertebrates commonly present in these black shales. Conchostracans can tolerate waters with dissolved oxygen content as low as 0.3 ml/l at temperatures of 23°C in the laboratory (Moore & Burn 1968).

The exquisite preservation of fish fossils in some beds of this facies (Fig. 3) indicates absence of scavengers, suggesting anoxic bottom conditions. In addition, preservation was probably enhanced by the diagenetic effects of phosphate-rich interstitial waters derived from partial decomposition of organic matter within the sediment (Jenkyns 1986).

On the basis of modern analogues, this facies was probably deposited in anoxic water in a relatively deep, stratified lake. The interpretation of some of the sedimentary features is somewhat ambiguous, however. Laminations generally indicate anoxic sediment, however they are known to form under oxygenated waters. Organic content suggests anaerobic conditions for some samples, but more data are needed to relate TOC to particular lacustrine facies. Pyrite can form under a variety of conditions if the sediment is anoxic, and the presence of phosphate suggests dysaerobic conditions, assuming comparison to marine sediments is valid. Most of these sediments were probably deposited in anoxic waters in a meromictic lake, but it is possible that some of these shales (particularly those beds containing phosphate nodules) may have been deposited under one of the following conditions: (1) in the transition zone (metalimnion) between oxic (dysaerobic) and anoxic (anaerobic) water near the thermocline in a stratified lake, or (2) in dysaerobic water where the oxic–anoxic boundary lay at or below the sediment–water interface, or (3) in anoxic water in a stratified lake into which turbidity currents occasionally introduced oxygen (Demaison & Moore 1980, p. 1187), or (4) under conditions which fluctuated between oxic and anoxic, possibly related to lake overturn, or to an oscillating oxygen–hydrogen-sulphide boundary (Degens & Stoffers 1976).

Type B black shales (facies BS-B). Description: Type B black shales have graded laminae averaging about 1 mm thick, ranging to about 7 mm thick, with planar to scoured bases (Fig. 4). The lower sediment fraction of the graded laminae consists of tan silt, grading up into dark, organic-rich lutite. Some laminae are lenticular and discontinuous (Fig. 4). Most of these shales are calcareous, and some contain remnant evaporite crystal forms which have been dissolved and filled, or replaced by calcite (Fig. 4). X-ray diffraction reveals the presence of small amounts of anhydrite or gypsum, or both (mineralogy by D. A. Textoris, University of North Carolina, 1985). Most of these shales lack fossils but a few bedding planes in some units contain scarce, small, poorly preserved conchostracans (*Cyzicus* sp. and *Paleolimnadia*? sp.), and scarce plant fragments are present locally. Type B shales are restricted to the Bull Run Formation.

Interpretation: The presence of laminations and fine grain size suggest deposition in relatively quiet lacustrine water. The graded laminae are probably the result of introduction of terrigenous silt to the lake by runoff, followed by suspension deposition. Scours and lenticular laminae suggest occasional bottom currents or sporadic stirring of the sediment by waves. Laminations, organic content, and lack of bioturbation all suggest deposition in a stratified lake with anoxic bottom water. The presence of evaporite crystals in some beds indicates that at times these lakes were saline or hypersaline. Elevated salinity promotes organic productivity because of faunal restriction and decreased consumption of algae (Eugster & Kelts 1983). In addition, denser saline waters in a stratified lake become anoxic, and preservation of organic matter is enhanced (Kirkland & Evans 1981; Eugster & Kelts 1983). Both elevated salinity and anoxia were probably responsible for the near absence of invertebrates in this facies.

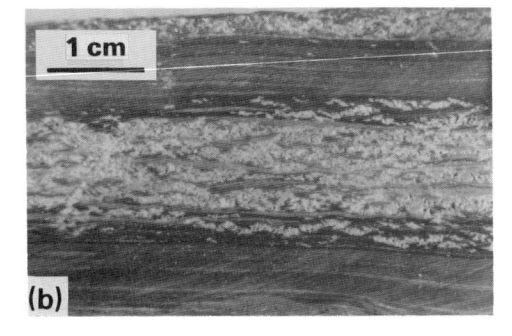

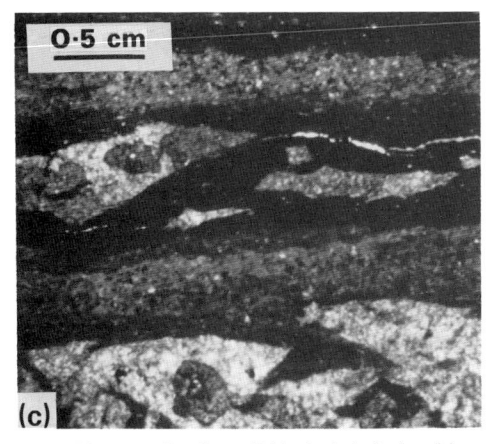

FIG. 4. Photographs of type B black shale facies. (a) Graded and lenticular laminations in black shale; Bull Run Formation, section G-1 (Fig. 13c), Manassas National Battlefield Park, Virginia. (c) Photomicrograph of calcite pseudomorphs of evaporite crystals in (b).

Conchostracans do not tolerate salinities above 6 parts per thousand (Webb 1979), and darwinulid ostracodes inhabit only fresh to brackish water (Benson et al. 1961, p. Q253–254). The presence of scarce invertebrates on a few bedding planes may be related to freshening events or lake overturn.

(2) Graded sandstone (facies GS)

Description: These sandstones are grey, normally-graded to massive, and some are laminated or cross-laminated (Fig. 5). Hentz (1985) described sequences of sedimentary structures in these beds which resemble Bouma sequences (Bouma 1962). The basal parts of the beds are graded to massive sandstone (A), overlain by parallel laminated sandstone (B), ripple cross-laminated or wavy-laminated sandstone and siltstone (C), and topped by shale (D or E). The most common sequences are AE, consisting of graded or massive sandstone layers several centimetres thick, interbedded with facies BS-A (Fig. 5). These beds have scoured and load-marked bases, locally overlain by shale rip-up clasts. Some beds contain a transported biota consisting of plant fragments, conchostracans, and fish scales.

Interpretation: The graded bedding, scour-marked bases, and rip-up clasts in these sandstones suggests deposition from bottom currents. The resemblance between the sedimentary structures in these sandstones and Bouma sequences strongly suggests that these beds are lacustrine turbidites (Hentz 1985). These turbidites were probably deposited from sediment-laden river water entering the lake, resulting in hyperpycnal flow (high density underflow) down a delta front to the deepest part of the lake (Sturm & Matter 1978; Allen & Collinson 1986). Sediment-laden river waters can also introduce oxygen to the anoxic hypolimnion in a stratified lake (Demaison & Moore 1980, p. 1187). Temporary oxygenation can result in short periods of bioturbation by organisms which are transported in by the turbidity current, or which migrate in naturally; bioturbation would cease after the oxygen is used up in respiration and decay (Savrda et al. 1984). In some stratified lakes, sediment-laden river water may be denser than the epilimnion but less dense than the hypolimnion, resulting in an interflow along the thermocline or chemocline (Sturm & Matter 1978). Coarser material would settle throughout the lake, and the fines would remain in suspension until the lake waters overturned (Sturm & Matter 1978). Sandstone beds deposited from interflows would lack scoured bases. Most of the Culpeper Basin graded

FIG. 5. Photograph of graded sandstone facies. Note scoured base (arrow) overlying black shale; Waterfall Formation, stream cut near Millbrook Quarry, Thoroughfare Gap, Virginia.

sandstones have scoured bases and are interpreted as turbidite deposits.

(3) Conglomerate (facies CG)

Description: Both clast- and matrix-supported conglomerates are associated with black shales and graded sandstones. Clast-supported conglomerates are horizontally stratified, normally-graded, or massive, with imbricated clasts, basal erosion surfaces and channel-form shapes up to 120 m wide and 10.5 m deep (Hentz 1985). Some of the conglomerates are overlain by trough-cross-bedded sandstone (Hentz 1985). Matrix-supported conglomerates are normally-graded or massive, with basal erosion surfaces and load structures, and occur as isolated beds in facies BS-A (Hentz 1985). Lacustrine conglomerates are restricted to the Waterfall Formation, although intraclastic limestones are present in the Bull Run Formation (see limestone facies).

Interpretation: Horizontal stratification, clast imbrication, grading, and channel-form shape suggest that clast-supported conglomerates were deposited from currents flowing down channels on a delta front (Hentz 1985). The association of sedimentary features is similar to Walker's (1975) graded-stratified model of conglomerates associated with turbidites. Matrix-supported conglomerates may be due to debris flows or slumps into the deeper part of the lake.

(4) Olive-grey to greyish-red laminated mudshale and clayshale (facies SH)

Description: These shales are olive-grey to greyish-red, with graded and homogeneous laminae (Fig. 6). Mudcracks, scour marks, and small intraformational clasts are present in some beds. Fossils are locally abundant, consisting of conchostracans (*Cyzicus* sp.), ostracods (*Darwinula* sp.), notostracans (*Triops* cf. *cancriformis*), pelecypods, insects, disarticulated fish remains, and plant fragments (including pollen and spores) (Gore & Traverse 1986) (Fig. 6). Organic content is generally low.

Interpretation: The fine grain size and presence of laminations indicate suspension deposition in quiet water away from major sources of coarse terrigenous input. Preservation of laminations may have been aided by the presence of microbial mats (Neumann *et al.* 1970). Scours and intraformational clasts suggest occasional current or wave activity. The presence of mudcracks in some beds indicates occasional subaerial exposure and desiccation, suggesting shallow water deposition (or drying up of a deeper lake culminating in subaerial exposure). The character of the fauna suggests deposition in fresh to brackish, slightly alkaline, oxygenated water. The low organic content also suggests deposition in oxygenated water.

(5) Limestone (facies LS)

Description: Several types of limestone are present, including stromatolitic limestone, mudcracked laminated limestone, limestone with graded laminae, massive bioturbated limestone, intraclastic limestone, pelleted-ostracodal limestone, and oolitic limestone (with some complex reworked ooids) (Fig. 7). Most of the limestone units are medium grey, but the stromatolites are dark grey to black, and are associated with stellate evaporite crystal pseudomorphs.

Interpretation: Calcite in lacustrine deposits originates through four main processes (Kelts & Hsü 1978): (1) inorganic precipitation induced by biological activity (photosynthesis) or physical changes in the environment (temperature, evaporation, or mixing of waters), (2) accumulation of shells, (3) introduction of carbonate from extrabasinal sources, or re-sedimented carbonate derived from fringing playa flats, nearshore, or slope deposits, and (4) diagenetic precipitation. The several types of limestone in the Culpeper Basin each formed under slightly different conditions, but in general, all represent deposition in relatively shallow water, or on lake margin carbonate-flats. Fine-grained carbonate in many of the limestones probably results from inorganic precipitation or bioinduced precipitation by planktonic algae. Laminae and graded laminae were produced by suspension deposition, and the presence of mudcracks in some laminated units indicates occasional subaerial exposure. Sediment-binding by microbial mats may have aided preservation of laminae in shallow water (Neumann *et al.* 1970). Intraclastic limestones contain clasts derived from mudcracked carbonate mudflats. The pelleted-ostracodal limestone is an accumulation of shells and calcareous, organic-rich pellets. The carbonate in the pellets is probably derived from bioinduced precipitation. The presence of oolites indicates current or wave activity in shallow nearshore areas of saline closed basin lakes (Eugster & Kelts 1983, p. 343). The complex nature of some of the ooids in the oolitic limestone indicates reworking, and represents deposition in a shallow lake with fluctuating depth and wave base, probably resulting from climatic change (Young & Edmundson 1954; Carozzi 1964). The association of stromatolites with stellate evaporite crystal pseudomorphs suggests that elevated salinity precluded algal grazers, permitting algal mounds to form in

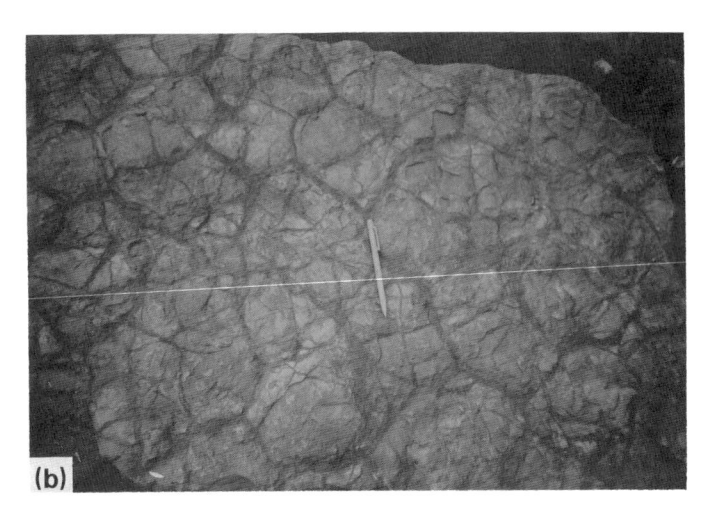

FIG. 6. Photographs of olive-grey to greyish-red laminated mudshale and clayshale facies. (a) Graded and discontinuous lenticular laminations in olive-grey mudshale to clayshale; Bull Run Formation, section G-1 (Fig. 13C), Manassas National Battlefield Park, Virginia. Scale in mm. (b) Mudcracks on bedding plane showing several scales of cracking; Bull Run Formation, Culpeper Crushed Stone Quarry, Culpeper, Virginia. Pencil for scale is 13 cm long. (c) Conchostracans (*Cyzicus* sp.) in olive-grey clayshale; Bull Run Formation, section G-1 (Fig. 13C), Manassas National Battlefield Park, Virginia. Scale in mm.

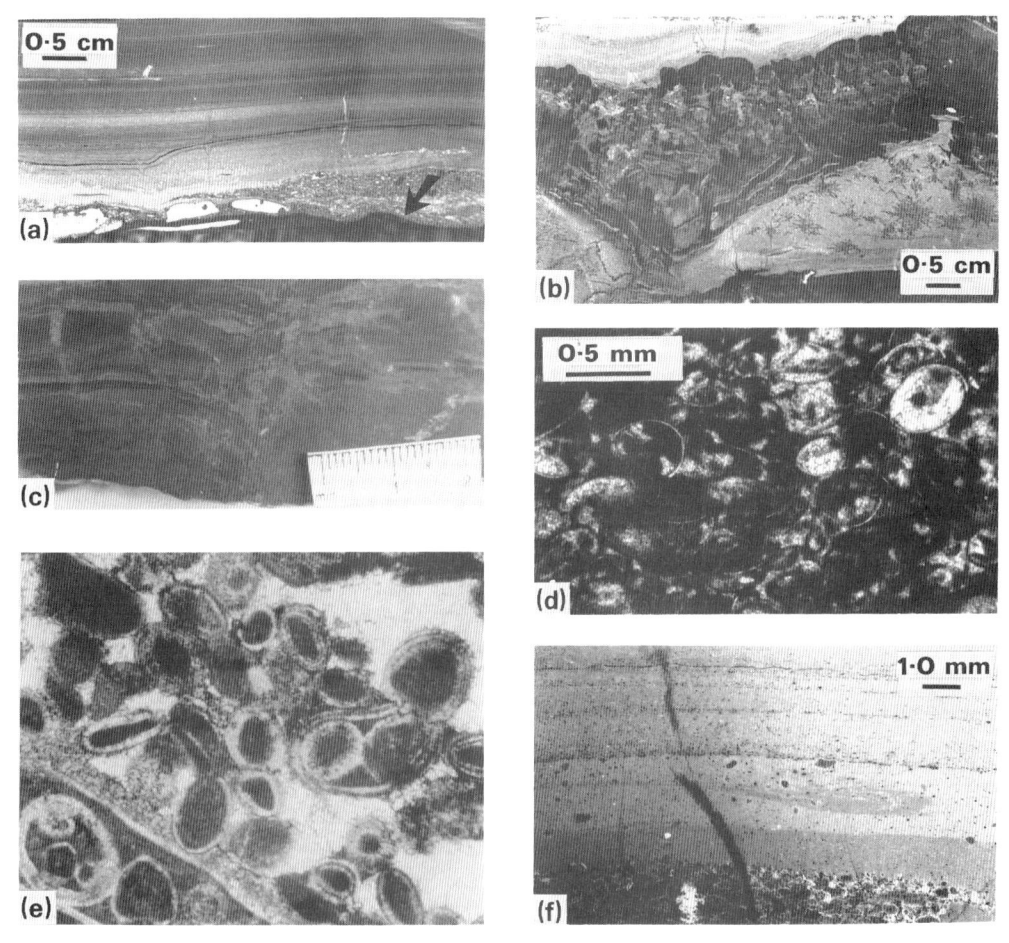

FIG. 7. Photographs of limestone facies. (a) Limestone with graded laminae. Intraclasts and tiny coated grains are present just above the base of this layer. Note scoured base overlying black shale (arrow); Bull Run Formation, section C-1 (Fig. 13D), Cedar Run, near Calverton, Virginia. (b) Stromatolitic limestone (black) overlying discontinuous layer of calcareous silt containing stellate calcite pseudomorphs of evaporite crystals; Bull Run Formation, section G-1 (Fig. 13C), Manassas National Battlefield Park, Virginia. (c) Limestone with graded laminae and mudcracks; Bull Run Formation, section 8/B (Fig. 12A), Manassas National Battlefield Park, Virginia. Scale in mm. (d) Photomicrograph of pelleted ostracodal limestone in early diagenetic concretion; Buckland Formation, Midland fish-bed at Licking Run (Fig. 13E), Midland, Virginia. (e) Photomicrograph of oolitic limestone with oolitic intraclasts, compound nuclei in reworked ooids, and pressure solution stylolitization; Bull Run Formation, Culpeper Crushed Stone Quarry, Culpeper, Virginia (Fig. 18). Photograph courtesy of Tucker Hentz (Texas Bureau of Economic Geology). Scale unknown. (f) Negative print of thin section of limestone with graded laminae shown in (a). Note coated grains with white rims near the bottom of the layer.

shallow water. The dark colour of the stromatolites may be due to algal-derived organic matter.

(6) Grey to tan ripple-cross-laminated to massive sandstone and siltstone (facies SS)

Description: These sandstone and siltstone beds are commonly ripple cross-stratified with wave-rippled tops, or wavy-laminated, and some contain soft-sediment deformation features such as load casts, convolute laminations, or ball-and-pillow structures (Fig. 8). Some beds are nearly massive, but retain wisps and traces of lamination. Mud drapes, mudstone intraclasts, and burrows are present locally. These sandstones and siltstones contain conspicuous muscovite flakes and plant fragments. Stellate crystal moulds or pseudomorphs are present in some beds. This

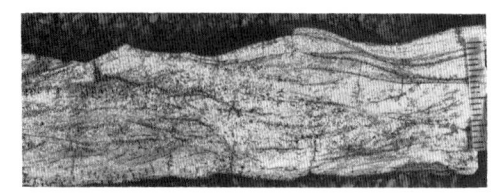

FIG. 8. Photograph of wave-rippled sandstone; Bull Run Formation, section G-7 (Fig. 13A), Manassas National Battlefield Park, Virginia. Scale in mm.

facies is commonly interbedded with facies SH. In the Waterfall Formation, facies SS typically lies between fluvial and sublittoral lacustrine deposits (Hentz 1985).

Interpretation: The wave-generated structures indicate deposition in shallow water near a source of terrigenous sediment input. The local presence of mud drapes indicates fluctuations in wave energy, with mud deposited during times of slack water. Fluctuating wave action and bioturbation suggest deposition in a nearshore, shallow water environment (Hentz 1985). These beds are interpreted as shallow water, lacustrine delta-fill deposits because of the ripple cross-stratification, soft-sediment deformation structures, massive bedding, abundant muscovite, and plant fragments, based on comparisons with a modern analogue from the Atchafalaya Basin, Louisiana, USA (Coleman 1966). The position of these sandstones within the lacustrine sequences also suggests a shallow water, deltaic origin.

(7) Red laminated and cross-laminated sandstone, siltstone, and mudstone (facies RX)

Description: This facies consists of thinly interbedded alternating layers of red sandstone, siltstone, and mudstone to mudshale, with laminations, cross-laminations, climbing-ripple cross-laminations, symmetrical and asymmetrical ripple marks, load casts, small ball-and-pillow structures, mudcracks, burrows, and reptile tracks (Fig. 9). Sandstone beds are generally quite thin, usually less than 10 cm thick. Cross-stratified bed sets average 1–2 cm thick; some of them are lenticular, and some have wave-rippled tops with mud drapes (Fig. 9). Detrital muscovite is common. Rarely, invertebrate fossils (ostracodes or conchostracans) are present in the red mudshale. This facies comprises much of the section in the Culpeper Basin. In the Waterfall Formation, this facies lies between alluvial plain deposits and littoral to sublittoral lacustrine deposits (Hentz 1985). With increased bioturbation and mudcracking, which disrupt primary sedimentary features, this facies is transitional to

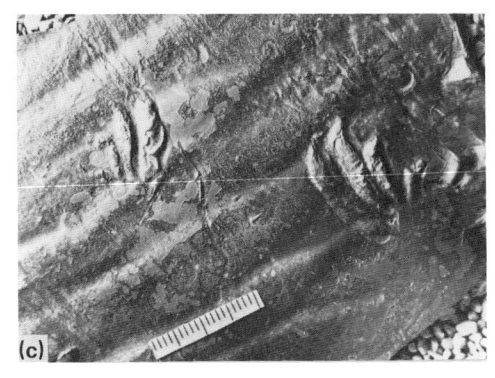

FIG. 9. Photographs of red laminated and cross-laminated sandstone, siltstone, and mudstone facies; Bull Run Formation, section G-3 (Fig. 12F), Manassas National Battlefield Park, Virginia. (a) Wavy laminations and climbing-ripple cross-laminations in coarse siltstone. Scale in mm. (b) Lenticular, ripple-form beds of cross-laminated coarse siltstone with mud drapes. Scale in mm. (c) Cast of reptile tracks (*Rhynchosauroides*) on ripple-marked bedding plane covered by mud drape. Scale in mm. (Identification of tracks by Paul E. Olsen, Lamont-Doherty Geological Observatory, New York, pers. comm. 1983).

disrupted laminites (facies DL) and massive siltstone and mudstone (facies MS).

Interpretation: The thin, red sandstone beds are interpreted as fluvial sheetflood deposits, some of which entered the shallow lacustrine environment. The style of cross-lamination and asymmetrical ripples suggest deposition from currents rather than from waves. Most of these deposits are probably non-lacustrine, but symmetrical ripples in some beds are probably the result of sediment reworking by waves in shallow ponded water (see for example, Hardie *et al.* 1978, p. 17). Mud drapes on ripples indicate suspension deposition during periods of slack water. Scarce invertebrate fossils present in some of the finer-grained beds also suggest lacustrine conditions; either ponded water on mudflats, or shallow water near the edges of lakes into which sheetfloods have entered. Reptile tracks also attest to deposition in shallow water or on mudflats, and mudcracks indicate occasional subaerial exposure.

(8) Disrupted laminites (facies DL)

Description: These rocks are red or grey, and contain graded laminae which are intensely disrupted by mudcracks, burrows, crystal moulds or pseudomorphs, and massive patches (Fig. 10). Conchostracans, ostracods, and plant fragments are present in some beds.

Interpretation: These beds record the early stages of destruction of laminae in sediments of facies RX and SH by mudcracking, bioturbation, evaporite crystal growth, and soil-forming processes on lake margin mudflats. Similar facies are present in the Lower Jurassic East Berlin Formation of the Hartford Basin (Demicco & Kordesch 1986). Given enough time, these processes would alter the sediment to a massive siltstone or mudstone, as in facies MS.

(9) Massive siltstone and mudstone (facies MS)

Description: This facies consists of red and grey massive siltstone and mudstone commonly containing mudcracks, mud intraclasts, and evaporite crystal moulds (stellate and swallow-tailed) (Fig. 11b). Wisps and traces of remnant laminations are present in some beds. The red massive rocks are siltier than the grey rocks, and locally they contain burrows, root marks, calcite-lined vugs, caliche-like pedogenic calcite nodules, clay films lining irregular cracks, and locally abundant detrital muscovite. Some of the grey massive rocks contain scarce ostracods and conchostracans.

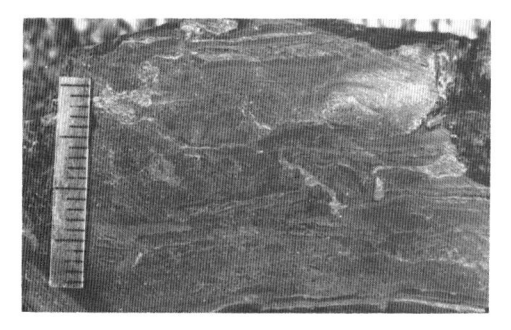

FIG. 10. Photograph of disrupted laminite facies; Bull Run Formation, section G-3 (Fig. 12E). Scale in mm.

Interpretation: Subaerial exposure of nearshore lacustrine or sheetflood deposits on a lake margin mudflat during times of low lake stand led to intense, prolonged mudcracking, bioturbation, crystal growth, and soil-forming processes. The presence of burrows and rootmarks suggests that the massive character of these beds is due in part to bioturbation. Evaporite crystal moulds attest to growth of salts within the sediment, which can also destroy depositional sedimentary structures (Hardie *et al.* 1978). Intense mudcracking formed mud intraclasts. Irregular clay films indicate soil-forming processes, suggesting that this facies is at least in part a palaeosol. Relict traces of lamination suggest that these rocks were originally like facies SH and RX. Locally abundant detrital muscovite, particularly in the red siltstones, further suggests that some of these sediments were derived from the homogenization of facies RX. Scarce invertebrate fossils in some of the grey massive rocks also suggest similarity to facies SH, and mudcracks in these rocks attest to subaerial exposure. The presence of crystal moulds, calcite-lined vugs (possibly pseudomorphs after gypsum or anhydrite nodules), and caliche-like pedogenic calcite nodules suggests an arid to semi-arid climate.

Palaeoenvironmental setting of lacustrine deposits

Culpeper Basin lacustrine rocks were deposited in three major subenvironments, each having its own distinctive lithologic character: offshore (central lake), nearshore (littoral or lake margin), and mudflat. The offshore, deeper water lacustrine rocks consist of black shale (BS), locally with interbedded graded sandstone (GS) and conglomerate (CG). The nearshore, shallow water lacustrine rocks are dominated by olive-grey to greyish-red laminated mudshale and clayshale (SH) with associated limestone (LS) and grey to tan ripple-cross-laminated to massive sandstone and siltstone (SS). The mudflat deposits consist of disrupted laminites (DL), and massive siltstone and mudstone (MS). The red laminated and cross-laminated sandstone, siltstone, and mudstone facies (RX) is interpreted as sheetflood deposits which are locally associated with shallow water lacustrine units and mudflats. These facies are arranged in several types of sequences which document expansion and contraction of perennial lakes.

Lacustrine sequences

Lacustrine facies are arranged in symmetrical and asymmetrical sequences. Symmetrical sequences exhibit the following environmental

FIG. 11. Photographs of massive siltstone and mudstone facies; Bull Run Formation. (a) Massive red siltstone with burrows and white calcite-filled mudcrack; section 8/B (Fig. 12A). Scale in mm. (b) Swallow-tailed evaporite crystal moulds (probably of sulphate minerals) in tan massive claystone to mudstone; section G-3 (Fig. 12F). Scale in mm.

transitions: mudflat → nearshore lacustrine → offshore lacustrine → nearshore lacustrine → mudflat. These sequences document gradual lake expansion and deepening (transgression) followed by gradual shallowing and sediment infilling (regression). The upper, regressive part of these sequences is generally thicker. The asymmetrical sequences are characterized by a diastemic surface abruptly overlain by deeper water deposits, which shallow upward, resembling the regressive parts of the symmetrical sequences.

Three general types of lacustrine sequences are distinguished here on the basis of thickness and facies present: (1) 'very thin' sequences, (2) 'thin' sequences, and (3) 'thick' sequences. These sequences are separated by fluvial red beds and palaeosols ranging from centimetres to tens or hundreds of metres thick.

(1) 'Very thin' lacustrine sequences: These sequences range from 10 cm to 1 m thick, and are dominated by facies SH, LS, and MS, representing deposition in the nearshore, shallow water lacustrine environment and on associated mudflats (Fig. 12). These deposits generally consist of only one or two lacustrine beds in sections dominated by sheetflood deposits and palaeosols, and record relatively shallow water or playa conditions. These deposits are present in both the Bull Run and Buckland Formations, but facies LS is more common in the Bull Run Formation. They have not yet been recognized in the Waterfall Formation.

Many of the 'very thin' lacustrine sequences contain a fossil assemblage consisting of some or all of the following: ostracods, conchostracans, pelecypods, disarticulated fish scales, and plant fragments (including pollen and spores). The presence of the invertebrate fossils and fish indicates relatively fresh water. During the time these beds were deposited, precipitation and freshwater inflow to the basin exceeded evaporation and infiltration. Mudcracked horizons are present at the tops of some of the lacustrine beds, indicating a drop in lake level, associated with decreased precipitation or inflow, or increased evaporation. Stellate and swallow-tailed evaporite crystal moulds and pseudomorphs, probably produced by calcium sulphate crystals, are present locally in the lacustrine beds and in surrounding mudflat or floodplain sediments, indicating that the lake waters occasionally became saline or hypersaline, and that the basin was hydrologically closed (see for example, Hardie et al. 1978; Eugster & Kelts 1983). Closed basin sedimentation is also signalled by the thinness of the lacustrine sequences, indicating rapid shoreline movements. The lacustrine beds are commonly overlain by red mudflat and sheetflood deposits, suggesting a drop in lake level with subaerial exposure of the lake bottom sediments, or progradation of fluvial sheetfloods into the lake. The dominance of shallow water and mudflat deposits in 'very thin' sequences indicates deposition in small, shallow lakes, or at the edges of large lakes; the presence of limestone indicates low terrigenous sediment input.

(2) 'Thin' lacustrine sequences: 'Thin' lacustrine sequences range from 1 to 7 m thick and are dominated by facies BS, SH, SS, and MS, with minor LS, representing deposition in both offshore and nearshore lacustrine environments, and on associated mudflats (Fig. 13). Thin sheetflood deposits of facies RX are interbedded with the shallower water facies in some sections. 'Thin' lacustrine sequences are present in both the Bull Run and Buckland Formations, and first appear approximately 1850–2400 m above the base of the Bull Run. These beds, like the 'very thin' lacustrine sequences, are overlain and underlain by mudflat and fluvial sheetflood deposits of facies MS and RX (Fig. 13).

A typical 'thin' lacustrine sequence consists of 0–1.5 m MS → 0–1 m SH → 0.1–2.3 m BS → 0.6–2.6 m interbedded SH and SS → RX (Fig. 13). 'Thin' lacustrine sequences are generally symmetrical in terms of depositional environment, indicating gradual expansion and contraction of perennial lakes. Black shales comprise the central part of symmetrical sequences (and the lower part of asymmetrical sequences), and may be either type A or B. Sequences with type A shales are generally fossiliferous throughout, indicating relatively fresh water. These sequences are generally overlain by fluvial-dominated sequences (rather than mudflat-dominated sequences or palaeosols), suggesting perennial inflow to the lake which was probably responsible for maintaining low salinities. Some of the fossiliferous, freshwater deposits, particularly where evaporite crystals are absent from associated mudflat deposits, may represent deposition in hydrologically open lakes. The characteristics of associated fluvial deposits need to be examined to determine whether they represent ephemeral or perennial streams (following Smoot 1985).

Fossils are typically scarce in lacustrine sequences which contain type B black shales, and in sequences which contain abundant evaporite crystal moulds or pseudomorphs. The presence of evaporite crystal moulds or pseudomorphs signals a semi-arid to arid climate, and deposition in saline to hypersaline waters. The presence of relict evaporites and the thinness of the lacustrine sequences reflect rapidly-moving shorelines, indicating deposition in a hydrologically-closed basin. In one outcrop in the Bull Run Formation (Fig. 13C), evaporite crystal pseudomorphs are present in facies BS-B (and higher in the sequence in facies SS and in association with stromatolites). Overlying shallow water deposits of facies SH contain an abundant, relatively freshwater fauna (Gore in press; Gore & Traverse 1986). The presence of brines coupled with an overlying, nearshore deposit with a freshwater fauna suggests either a decrease in the salinity of the lake with time, possibly as a result of increased precipitation or inflow, or more likely a stratified lake with a saline or hypersaline anoxic hypolimnion, capped by a freshwater epilimnion (ectogenic meromictic lake).

(3) 'Thick' lacustrine sequences: 'Thick' lacustrine sequences are present only in the Waterfall Formation, and average about 35 m thick according to Hentz (1985, p. 105) (Fig. 14). From section descriptions in Hentz (1981), the thickness of grey (i.e., lacustrine) deposits in the Waterfall Formation ranges from about 7 m to more than 38 m. Most of these sequences are not entirely exposed (top or bottom covered), but of nine lacustrine sequences described, the mean thickness is at least 20 m. These sequences are dominated by facies BS, GS, and SS, with minor CG, representing deep water offshore deposits and nearshore to shoreline deposits (Fig. 14). Some of the thicker sequences studied in more detail by Hentz (1985) consist of 6–8 m SS → 1–4.7 m BS (with associated GS and CG) → 9–20 m BS (with associated GS) → 6–10 m SS → 5–240 m RX, DL, and MS (fluvial and mudflat deposits). These sequences are interpreted to represent the deepening and shallowing of large perennial lakes (Hentz 1985).

The presence of invertebrate and fish fossils in most of these beds indicates relatively freshwater conditions. Small calcite-filled vugs interpreted as caliche or pseudomorphs of evaporite minerals

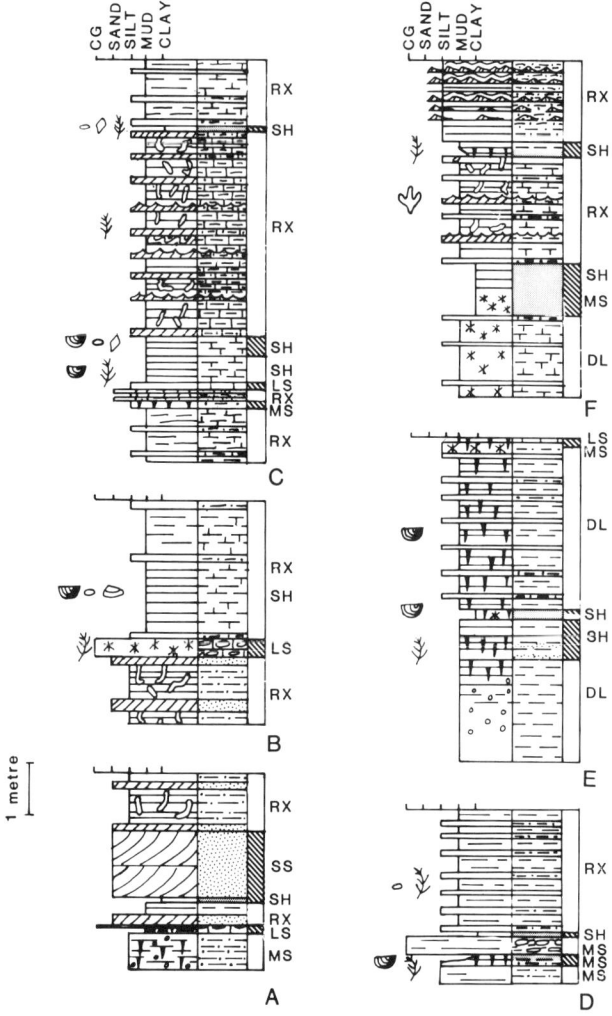

FIG. 12. Stratigraphic sections illustrating 'very thin' lacustrine sequences in the Bull Run Formation. Grey beds and red beds containing fish scales and/or invertebrate fossils are lacustrine in origin. Most of the strata in these sections are non-lacustrine, representing fluvial sheetflood deposits and palaeosols. Sections are presented in stratigraphic order. See Gore (1983) for more detailed section locations. (A) A bed of grey, laminated and stromatolitic, mudcracked limestone with intraclasts (Fig. 7c) deposited in shallow water or on a mudflat, is overlain by a thin red clastic sequence (probably sheetflood deposits near the edge of a shallow lake), overlain by lacustrine grey clayshale. The thick grey sandstone (SS) may be deltaic; section 8/B. Massive siltstone at bottom of the section is illustrated in Fig. 11a. (B) Grey, intraclastic limestone with evaporite crystal moulds or pseudomorphs and plant fragments (nearshore to mudflat) is overlain by red laminated mudshale (shallow water) which locally contains invertebrate fossils; section 8/B. (C) Mudcracked and rooted grey and red siltstone and sandstone (mudflat deposits) are overlain by fossiliferous red to grey laminated mudshale (shallow water lacustrine). Higher in the section, above a series of sheetflood deposits, there is another 'very thin' lacustrine unit consisting of shallow water fossiliferous, grey clayshale; section 8/B. (D) Grey massive siltshale fines upward into laminated mudshale. The top of the bed is mudcracked and the upper few centimetres contain sparse conchostracans (shallow water to mudflat). Higher in the section, some beds of red mudstone interbedded with sheetflood deposits contain sparse ostracodes (shallow water lacustrine); section 8/G. (E) Mudcracked mudflat deposits with associated invertebrate fossils indicate deposition in ponded water. Disrupted laminites at base of section are illustrated in Fig. 10. Evaporite crystal moulds are present in red clayshale and siltstone (mudflat). Conchostracans are present in a thin layer of grey clayshale, indicating deposition in ponded water; section G-3. (F) Grey claystone to mudstone units in this interval were probably deposited in nearshore lacustrine conditions. Grey massive claystone to mudstone with evaporite crystal moulds is illustrated in Fig. 11b. Note reptile tracks (Fig. 9c) in red bed sequence (Fig. 9a–b); section G-3.

GRAIN SIZE

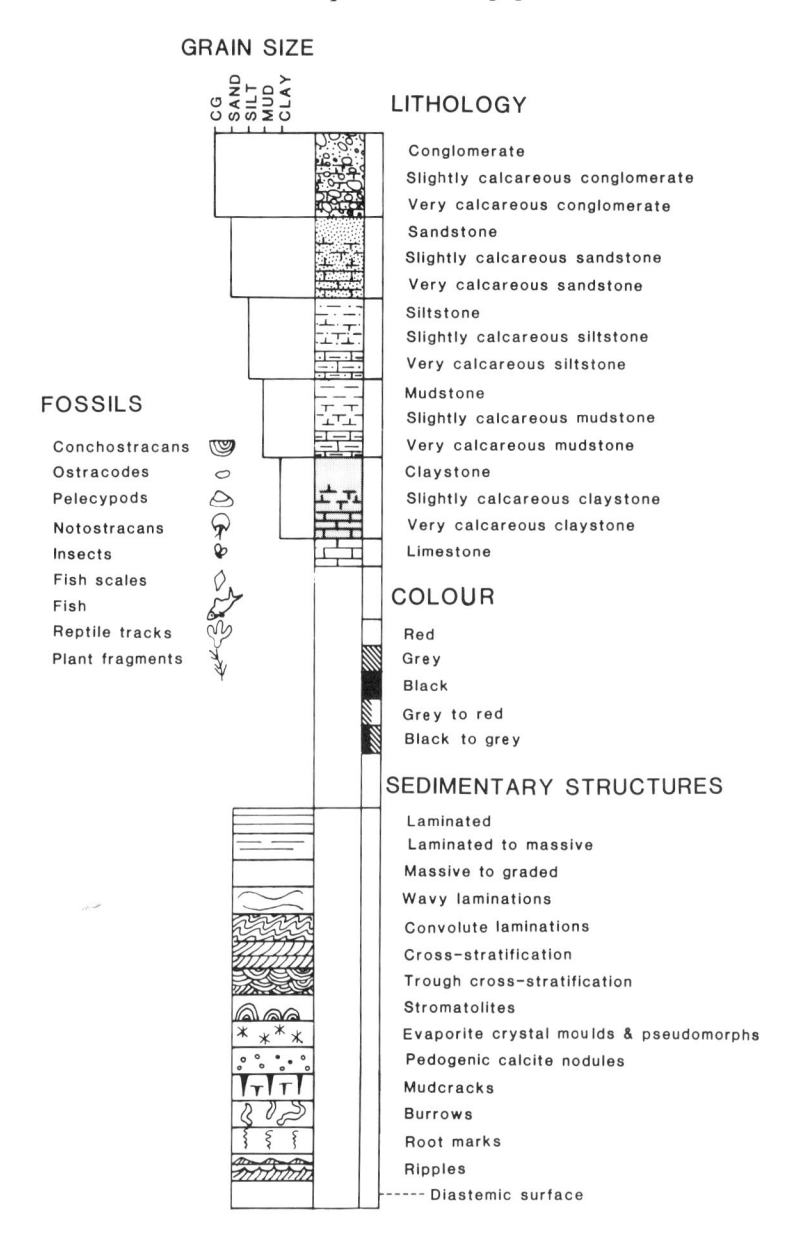

LITHOLOGY

Conglomerate
Slightly calcareous conglomerate
Very calcareous conglomerate
Sandstone
Slightly calcareous sandstone
Very calcareous sandstone
Siltstone
Slightly calcareous siltstone
Very calcareous siltstone
Mudstone
Slightly calcareous mudstone
Very calcareous mudstone
Claystone
Slightly calcareous claystone
Very calcareous claystone
Limestone

FOSSILS

Conchostracans
Ostracodes
Pelecypods
Notostracans
Insects
Fish scales
Fish
Reptile tracks
Plant fragments

COLOUR

Red
Grey
Black
Grey to red
Black to grey

SEDIMENTARY STRUCTURES

Laminated
Laminated to massive
Massive to graded
Wavy laminations
Convolute laminations
Cross-stratification
Trough cross-stratification
Stromatolites
Evaporite crystal moulds & pseudomorphs
Pedogenic calcite nodules
Mudcracks
Burrows
Root marks
Ripples
Diastemic surface

are present in the fluvial/mudflat deposits (Hentz 1985, p. 104). Calcite-filled vugs suggest an arid to semi-arid climate and continued closed-basin deposition. These vugs are present in only a few sequences, however, which suggests that some of the Waterfall lakes may have been hydrologically open. The greater thickness of the Waterfall deposits may indicate greater water depth than the thinner deposits, or greater shoreline stability.

Distribution of lacustrine facies and facies sequences

The lithology and thickness of lacustrine deposits vary across the Culpeper Basin. Lacustrine units are absent from the lower part of the Bull Run Formation. About 1150–1500 m above the base of the Bull Run, 'very thin' lacustrine sequences first appear with sedimentary features indicative

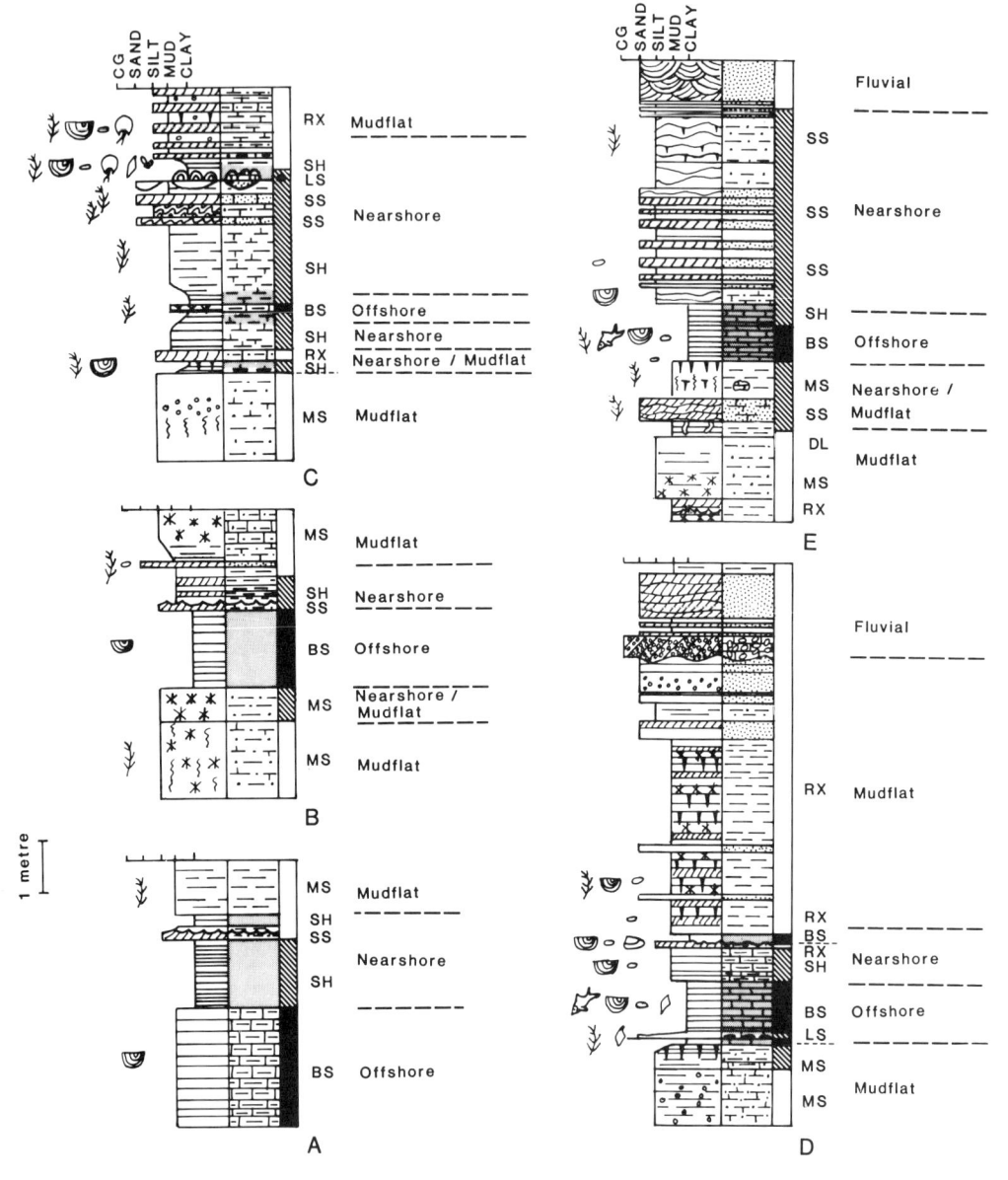

FIG. 13. Stratigraphic sections illustrating 'thin' lacustrine sequences. Sections are presented in stratigraphic order. See Gore (1983) for more detailed section locations. (A) The upper, regressive part of a symmetrical lacustrine sequence is shown, illustrating the transition from offshore black shale (type B with scarce conchostracans), to nearshore grey clayshale with wave-rippled sandstone (Fig. 8), passing upward into relatively massive red palaeosol and mudflat deposits; Bull Run Formation, section G-7. (B) Symmetrical lacustrine sequence with abundant evaporite crystals in the surrounding mudflat deposits, indicating gradual lake expansion and contraction during relatively arid climatic conditions. The black shale is type B, and contains a few scarce conchostracans on isolated bedding planes; Bull Run Formation, section G-8. (C) Near the base of this section, a 'very thin' lacustrine sequence abruptly overlies a mudflat palaeosol (note diastemic surface separating these units). The lacustrine shale is darker at the bottom, contains graded laminae (Fig. 6a), has a mudcracked top, and contains extremely abundant, large conchostracans (Fig. 6c). This 'very thin' sequence represents the abrupt deepening of a lake, followed by shallowing and subaerial exposure. Thin sheetflood deposits separate this lake from a 'thin' symmetrical sequence which dominates this section. The 'thin' sequence represents the gradual deepening and shallowing of a perennial lake with anoxic bottom waters. Evaporite

of deposition in shallow water, nearshore lacustrine and mudflat environments. The dominance of these sequences by shallow water and mudflat deposits indicates shallow lakes, or deposition near the edges of large lakes. Higher in the Bull Run (about 1850–2400 m above the base) 'thin' lacustrine sequences appear which contain sedimentary features indicative of deeper water, offshore deposition in addition to nearshore and mudflat deposits.

Farther west and higher stratigraphically, 'very thin' and 'thin' lacustrine sequences are present in the Buckland Formation. Little is known of these lacustrine sequences because exposures are almost non-existent. A notable exception is the Midland fish-bed, a symmetrical sequence about 6 m thick, which lies sandwiched within a red fluvial and mudflat sequence above the lowermost basalt flow (Fig. 13E). Within the red bed sequence between the Midland fish-bed and the underlying basalt flow, a 35 cm, 'very thin' lacustrine sequence is present. Some workers refer to the sedimentary sequence overlying the lowermost basalt flow which contains the Midland fish-bed, as the Midland Formation (Olsen 1984a). Freshwater fossils are present in both of the Midland lacustrine deposits. Evaporite crystal moulds in underlying mudflat and floodplain deposits indicate a semi-arid to arid climate and closed basin deposition. A unit of trough cross-bedded sandstone, which overlies the Midland fish-bed, was deposited in a fluvial environment. The character of these fluvial deposits needs to be examined in more detail to help determine whether the lake may have deepened to the point that it became hydrologically open (following Smoot 1985).

The Waterfall Formation, located in the west-central part of the basin, contains 'thick' lacustrine sequences, dominated by offshore, deep water deposits. In general, the lacustrine deposits are coarser in the Waterfall Formation than they are in the Bull Run or Buckland Formations. The nearshore lacustrine units in the Waterfall are dominated by sandstone (facies SS), and the offshore units contain sandy turbidites (facies GS) and conglomerate beds (facies CG). In contrast, the Bull Run and Buckland lacustrine sequences are dominated by finer-grained deposits; shallow water, nearshore units in the Bull Run and Buckland are mainly mudshale and clayshale (facies SH), with only a few interbedded sandstone layers (facies SS). Fine-grained mudflat deposits (facies MS and DL) are more abundant in the Bull Run and Buckland than in the Waterfall Formation. Offshore deposits in the Bull Run Formation lack turbidite sandstones, and those in the Buckland Formation have only thin, scattered sandstone interbeds. The difference in the texture of the deposits across the basin is related to proximity to the western source area. Lindholm (1979) documented that most of the palaeocurrents in the Bull Run and Buckland Formations indicate a western source, although Smoot (pers. comm. 1986) noted more scatter in palaeoflow directions. In addition, studies of the location of alluvial fan deposits, clast lithologies, and clast size-distributions indicate dominance of a western source area. Dominance of a western source area is atypical because the west-tilted half-graben basin geometry produces a larger drainage basin to the east than to the west. The dominance of the western source area may be related to palaeoclimatic patterns (see section on climate below). The Waterfall Formation received sediment from both the west and the east. Waterfall Formation conglomerates contain clasts derived from the Blue Ridge Province which lies to the west of the basin, and basalt clasts which were derived from the underlying

---

crystals pseudomorphs are present in the black shale (Fig. 4b–c), in some of the convolute-laminated sandstone and siltstone beds, and in association with the stromatolites (Fig. 7b), indicating saline to hypersaline waters. The presence of a relatively diverse freshwater fauna in the grey and red clayshales and mudshales near the top of the section indicates a change to freshwater conditions, probably associated with ectogenic meromixis (freshwater cap over a saline hypolimnion). Sheetflood deposits are associated with nearshore, shallow water red mudstone near the top of the section; Bull Run Formation, section G-1, Manassas National Battlefield Park, Virginia. (D) Symmetrical(?) lacustrine sequence containing type A black shale. Limestone near the base of sequence is illustrated in Figs 7a and f. Above the main lacustrine sequence, a thin bed of red sheetflood deposits (RX) overlies grey, calcareous shale. The top of this red bed is scoured and is overlain by fossiliferous dark grey to black shale which is capped by a series of red mudflat and sheetflood deposits, grading upward into fluvial deposits; Bull Run Formation, section C-1, Cedar Run, near Calverton, Virginia. (E) Symmetrical lacustrine sequence representing the gradual deepening and shallowing of a lake. The lower part of the section is dominated by red mudflat or floodplain deposits with evaporite crystal moulds, overlain by grey, massive, mudcracked, rooted mudstone with abundant plant fragments. The black shales near the centre of the sequence are known as the Midland fish-bed (Fig. 3). A photomicrograph of a concretion near the base of this bed is shown in Fig. 7d. The sandy character of the upper part of the sequence suggests shallowing due to delta-fill deposits. The sequence is capped by a red to tan fluvial unit; Buckland Formation, Licking Run, Midland, Virginia.

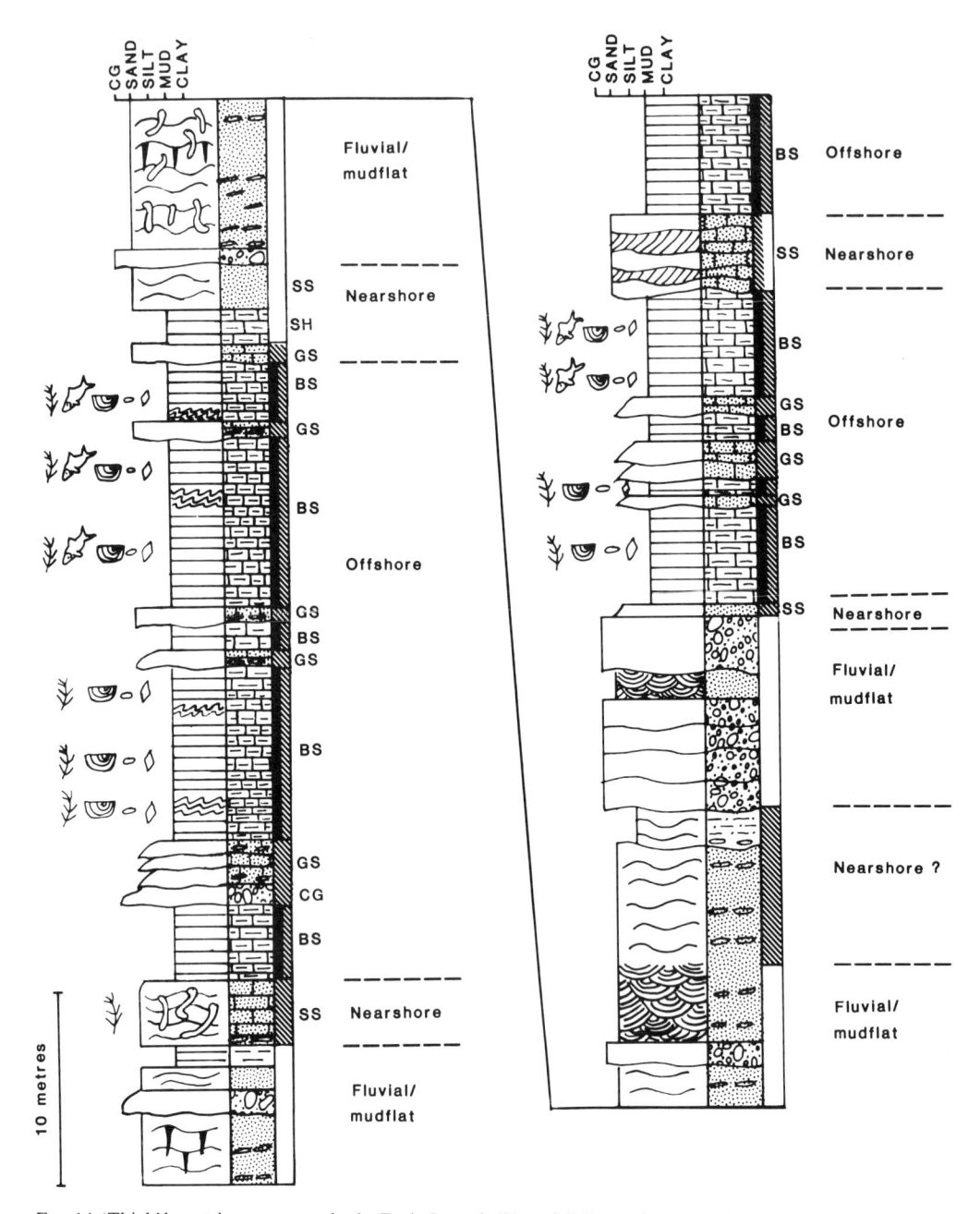

FIG. 14. 'Thick' lacustrine sequences in the Early Jurassic Waterfall Formation near Thoroughfare Gap, Virginia. Data used in compiling this section are from Hentz (1981, 1985), supplemented by my own field observations, sampling, and description. The lacustrine sequences in this section are dominated by grey to black mudshale and siltshale with interbedded sandstone layers interpreted as lacustrine turbidites. Lacustrine sequences are separated by nearshore, mudflat, and fluvial deposits. Note the difference in scale between this figure and Figs 12 and 13. See Hentz (1981) for more detailed section location.

Buckland Formation, presently exposed to the east of the Waterfall outcrop belt (Lindholm 1979, p. 1731; Lindholm et al. 1979; Hentz 1981, p. 112, 128; Hentz 1985). Palaeocurrent measurements in the Waterfall Formation indicate flow primarily to the west and north-west (Hentz 1981, 1985). This suggests uplift in the central part of the basin during the Early Jurassic.

The thickness of lacustrine deposits varies across the basin. In general, lacustrine sequences thicken toward the western-border fault, from less than 1 m in the lower part of the Bull Run Formation in the central part of the basin to more than 35 m in the Waterfall Formation in the western part of the basin (Fig. 15). Bedding in the Culpeper Basin dips westward toward the western-border fault, such that the youngest beds are located in the west-central part of the basin, and older beds are exposed to the east (Fig. 2). Thus, a westward change in the thickness of lacustrine beds in surface exposures is also a change in thickness upsection. The thickest lacustrine beds lie within 2 km of the faulted basin margin. The trend is not gradual; lacustrine sequences in the Bull Run and Buckland Formations are comparable in thickness, but those in the Waterfall Formation are 5–10 times thicker (Fig. 15). The abrupt change in thickness of lacustrine units between the Buckland Formation and the Waterfall Formation is puzzling. The abrupt change may be an artifact of a lack of data from the upper part of the Buckland, or to an abrupt increase in subsidence, or a westward shift in the locus of maximum subsidence during the middle to late Early Jurassic. If lacustrine sequence thickness is related to water depth, then the westward, upsection increase in thickness can be interpreted in several ways. Water depth within the lakes may have increased toward the fault (as is suggested by the westward-tilted basin floor), or climate and tectonic activity may have changed, causing an increase in water depth during the Early Jurassic. Alternatively, the thickness change may be explained by a change from predominantly hydrologically-closed lakes in the Bull Run and Buckland Formations to predominantly hydrologically-open lakes in the Waterfall Formation. Hydrologically-open lakes have stable shorelines and would produce thicker lacustrine sequences. The thickness change is based on surface exposures of separate lacustrine sequences. It is not known whether individual lacustrine sequences thicken systematically in any particular direction because of the lack of subsurface data and poor outcrop continuity; resolution of this problem will require subsurface data.

Lateral trends in lacustrine sequence thickness exist in other Newark Supergroup basins. In the Hartford Basin, located about 600 km to the north (Fig. 1), the thickness of lacustrine black shale beds gradually increases toward the faulted basin margin from less than 1 m to nearly 5 m in three formations (Portland, East Berlin, and Shuttle Meadow Formations) and the thickest lacustrine beds lie within about 1.5 km of the basin margin (LeTourneau 1985). The increase in thickness of these beds toward the fault suggests that the lakes were deepest and most-persistent adjacent to the faulted basin margin, and shallowed away from the fault. In the Newark Basin, which lies about 300 km north of the Culpeper Basin (Fig. 1), lacustrine sequences in the Towaco Formation thicken from a mean of 20 m to a mean of 30 m toward the Ramapo fault which bounds the basin on the northwest (Olsen 1984a). Microlaminated black shales in the Towaco Formation thicken by more than an order of magnitude toward the fault from less than 1 m to 5 m (Olsen 1980b, p. 365–366; Olsen 1984a, p. 379). In contrast, the Lockatong Formation (Newark Basin) is thickest near the basin centre and thins toward the edges of the basin (Olsen 1980a, p. 10; Olsen 1980b, p. 354). Individual sequences thin laterally from 5.7 to 1.5 m, and associated facies changes indicate that water depth was greatest near the basin centre (Olsen 1984a, p. 21). The difference in thickness patterns of lacustrine sequences in the Newark Basin is related to a change in the locus of maximum subsidence. During the Karnian (Late Triassic) when the Lockatong Formation was

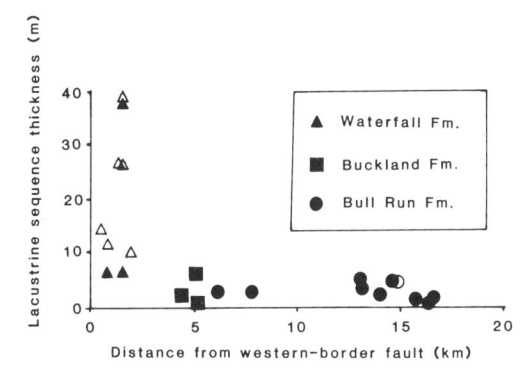

FIG. 15. Graph illustrating the thickness of lacustrine sequences plotted against distance from the western-border fault. Solid black symbols represent sequences which are exposed so that total thickness can be measured. Open symbols represent sequences which are partially covered; total thickness cannot be determined. Thickness of lacustrine sequences increases abruptly within 2 km of the western-border fault.

deposited, subsidence was greatest near the basin centre. During the Hettangian (Early Jurassic) when the Towaco was deposited, subsidence was greatest along the border fault. A similar change in subsidence history may have occurred in the Culpeper Basin, as suggested by the abrupt increase in thickness of lacustrine sequences from the Buckland Formation to the Waterfall Formation. More information is needed on the lateral thickness changes of individual lacustrine sequences in the Bull Run and Buckland Formations to assess whether they display any systematic patterns in thickness.

Water depth

Several methods have been proposed to determine the water depth of Newark Supergroup lacustrine deposits, none of which are entirely satisfactory for the Culpeper Basin. The main methods that have been used to determine water depth include: (1) lake-filling model, (2) basin-floor tilting model, and (3) lateral extent model. These methods and several others are discussed in more detail by Olsen (1984a, p. 269–284).

(1) Lake-filling model: This model is based on the assumption that regressive lacustrine deposits are produced by the gradual, complete filling of a stable (non-subsiding) lake with sediment. If so, then the original, pre-compactional thickness of the regressive part of the section would be approximately equal to the depth of the lake. The amount of compaction is difficult to determine for heterolithic sedimentary sequences. Compaction ratios for some black shale beds range from 3:1 to 5:1 or more, as determined from examination of early diagenetic concretions (Gore 1986). The compaction ratios of the sand units have not been determined, but they are undoubtably less than those of the shale. If a modest compaction factor of 50% (2:1) is assumed (following Burst 1969), then the water depth would have been approximately twice the thickness of the regressive part of the sequence (Conybeare 1967). This method led Hentz (1985) to conclude that the Waterfall Formation lakes were approximately 35 m deep.

The major problem with the lake-filling model is that not all lakes become completely filled with sediment. Lakes dry up in response to increasing aridity, producing an abnormally thin sedimentary record in comparison with water depth. For example, the modern Great Salt Lake (Utah, USA) was about 300 m deep approximately 15 000–17 000 years ago; within about 500–2000 years, lake level dropped to its current depth of about 12 m (Hardie *et al.* 1978; Eugster & Kelts 1983; Spencer *et al.* 1984). Although the lake was once 300 m deep, it is represented by a sedimentary sequence only 0.2 m thick (Spencer *et al.* 1984). On this basis, any of the lacustrine sequences in the Culpeper Basin, regardless of thickness, could have been deposited in an extremely deep lake, just as easily as in a much shallower one. Sediment thickness would have no relationship to original water depth. Thickness of lacustrine sequences is more a function of shoreline fluctuation, resulting from the balance between precipitation plus inflow, versus evaporation, than of water depth. However, few of the Culpeper Basin lacustrine sequences are topped by mudcracked horizons. Many lacustrine sequences coarsen upward and are overlain by fluvial deposits (Figs 12–14), suggesting deltaic filling of the basin. Because of these facies relations, the thickness of the regressive part of some lacustrine sequences (including those in the Waterfall Formation described by Hentz 1985) may be related to original water depth, but no simple relationship exists, and attempts at quantification of depth are misleading.

(2) Basin-floor tilting model: Lake depth may be estimated by assuming that a planar basin floor was tilted toward the border fault, and that the lake was ponded against the upfaulted highlands. Water depth can be calculated trigonometrically by finding the difference in palaeoelevation between lake shore deposits and the deepest part of the lake (probably adjacent to the border fault in most basins, except for the Lockatong Formation in the Newark Basin) for various assumed angles of basin floor tilt (Fig. 16). Manspeizer & Olsen (1981) demonstrated that water depth can vary by several orders of magnitude for basin floor tilt angles ranging from $0.001°$ to $1.0°$ (Fig. 16). Considering a lake about 20 km wide (reasonable for parts of the Culpeper Basin), calculated maximum depths range from less than 1 m to several hundred metres (Manspeizer & Olsen 1981, p. 83) (Fig. 16). This method is considered unreliable because of the wide range of possible water depths.

(3) Lateral extent of subwave base facies: This method to estimate water depth was presented by Manspeizer & Olsen (1981) and Olsen (1984a), and is based on the relationship between lake fetch (distance the wind can blow over the surface of the lake, generating waves), wind speed, and depth to wave base. Assuming that the black shales were deposited below wave base, their geographic extent defines the minimum area of the deepest part of the lake. The size of the lake is determined by the maximum distance over which the black shale facies can be traced in the field. Lake size is used to determine fetch. Wind speeds are assumed to be like those in modern

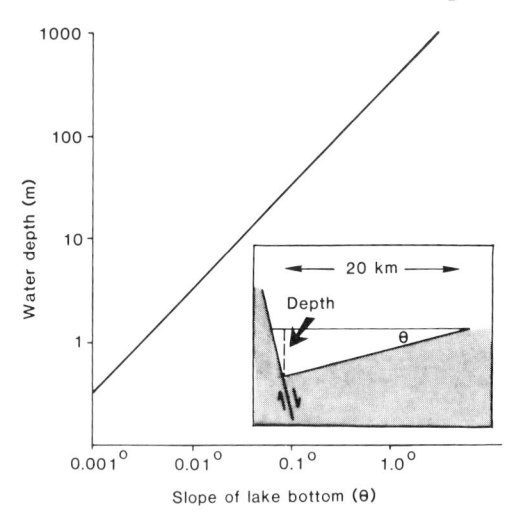

FIG. 16. Graph illustrating the range of possible water depths resulting from varying lake bottom slope angle, using the basin-floor tilting model in a 20 km wide basin. Because of the wide range of possible water depths, this method is unreliable. Modified from Manspeizer & Olsen (1981).

rift valleys, and depth to wave base is calculated (see Manspeizer & Olsen 1981, p. 44; Olsen 1984a). For Lockatong Formation lake beds (Newark Basin), depth to wave base was calculated to be 40–90 m, and Olsen (1984a) stated that water depth was probably in excess of 100 m. This method is not applicable in the Culpeper Basin at this time because the lateral extent of individual lacustrine sequences is not yet known.

Climate and cyclic sedimentation

Interpretation of the palaeoclimate of Newark Supergroup sediments has been based on sedimentologic and palaeontological criteria. Features indicative of both humid and arid to semi-arid conditions exist in the Newark Supergroup. The presence of widespread coal deposits in several of the basins suggests a humid climate. Arid to semi-arid conditions are indicated by features such as caliche nodules (Hubert 1977, 1978), playa lakes with limestone and chert (Wheeler & Textoris 1978), eolian sandstones (Hubert & Mertz 1984), and evaporite crystal moulds and pseudomorphs. Fluctuations in species diversity of pollen and spores have also been used to interpret palaeoclimate (Cornet 1977). High species diversity of spores is interpreted to indicate a more humid climate; low spore diversity and pollen dominance (primarily conifers) is interpreted to indicate a more arid climate (Cornet 1977, p. 69).

Palaeogeographic reconstructions for Late Triassic and Early Jurassic time indicate that the continents were assembled into a single landmass, Pangaea (Robinson 1973; Hay *et al.* 1982; Parrish *et al.* 1982; Ziegler *et al.* 1983). Throughout the early Mesozoic the supercontinent drifted northward (Parrish *et al.* 1982), and the Culpeper Basin moved from about 10°N to 15°N latitude. Palaeoclimatic maps based on atmospheric circulation patterns are successful at predicting the distribution of climatically sensitive sedimentary deposits (such as coals and evaporites), and indicate that during the Late Triassic and Early Jurassic, the Culpeper Basin was located in the intertropic dry belt, and would have been dry year-round (Robinson 1973; Parrish *et al.* 1982; Scotese & Summerhayes 1986). Predicted rainfall for the Culpeper Basin area in the Pliensbachian (Early Jurassic) was 'moderately low' (Parrish *et al.* 1982).

Details of the local climate depend not only on the climatic pattern of Pangaea, but also on topographic effects associated with rifting. Palaeogeographic reconstructions place the Newark Rift System in the belt of easterly trade winds, and assume that the mountains on either side of the rift valley were of unequal height, with the

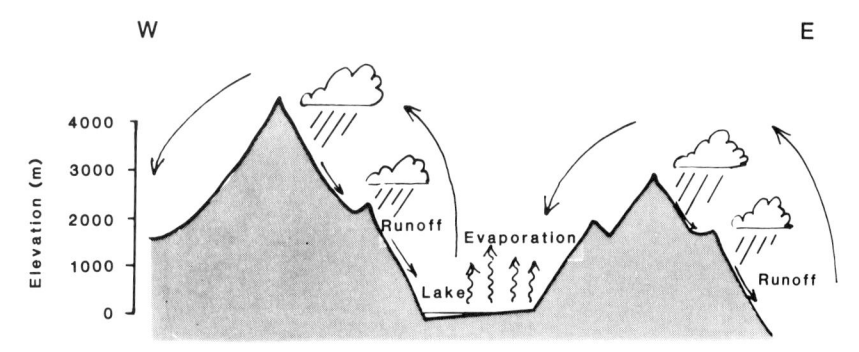

FIG. 17. Sketch illustrating the effects of topography on climate in a rift valley with easterly winds. Modified from Hay *et al.* (1982).

higher range to the west (Hay et al. 1982). This model of local palaeoclimate indicates orographic precipitation on the eastern side of the mountains which lay to the east of the rift valley (Fig. 17). The eastern sides of the rift valley lay in a rain shadow; air descending into the valley from the east heated adiabatically, increasing in evaporation potential and decreasing in relative humidity. As a result, the eastern side of the valley was extremely arid, with no freshwater input from the east (Fig. 17). Evaporation occurred within the rift valley, and as the air rose up the western mountains, it cooled, causing precipitation on the western side of the rift valley. Rainfall on the western slopes caused runoff into the basin, forming lakes. This palaeoclimatic pattern results in lakes on the western side of the rift valley, synchronously with arid conditions on the eastern side of the valley (Hay et al. 1982, p. 28). Wind directions may have changed seasonally, reversing the pattern, as suggested by Manspeizer (1981). The moisture pattern predicted by the model of Hay et al. (1982) agrees in general with the observed distribution of sedimentary deposits within the basin. Lake deposits are thicker in the western part of the basin, reflecting greater depth and shoreline stability. In the eastern part of the basin, lake deposits are thin and associated with extensive mudflats containing abundant evaporite crystal moulds and pseudomorphs. However, these facies changes may be due to a westward-tilted basin floor, to climatic change, or to changes in basin closure through time. The strongest evidence in support of the model of Hay et al. (1982) in the Culpeper Basin is the palaeocurrent pattern, which indicates flow predominantly from the west during the deposition of the Bull Run and Buckland Formations (Lindholm 1979).

Both freshwater and saline lakes existed in the Bull Run Formation. This indicates fluctuations in the balance between precipitation plus inflow, and evaporation plus infiltration during the Late Triassic and Early Jurassic, which may have been controlled by climatic cycles. Lacustrine facies sequences share certain similarities throughout the Culpeper Basin, but whether or not these sequences are repeated on a periodic basis (i.e., whether they are cyclic) is uncertain. The spacing between successive lacustrine sequences generally cannot be determined due to a lack of long, continuous sections.

Thickness patterns of lacustrine sequences in the Lockatong Formation (Newark Basin) have been shown to indicate Milankovitch-type climatic cycles. Lacustrine sequences ranging from 1 to 5 m thick in the Lockatong are interpreted as the products of a 23 000 year cycle, resulting from variations in the precession of the equinoxes (Van Houten 1962, 1964, 1969; Olsen 1984a, 1984b, 1985a). The Lockatong sequences have been termed 'Van Houten cycles' by Olsen (1985b) for Franklyn B. Van Houten, who first reported them. These 'thin' cycles are arranged in clusters to form 25 m and 100 m cycles, interpreted to represent 100 000- and 400 000-year climatic cycles (Olsen 1984a, 1984b, 1985a). The 100 000- and 400 000-year cycles are related to the eccentricity of the Earth's orbit (Olsen 1984b). Cycles of roughly 2 000 000 years are also present (Olsen 1985a). Cyclic lacustrine sediments in the Newark Basin are interpreted to be the result of changes in water depth as a function of precipitation and evaporation rates within the basin. Cornet (1977) also identified 500 000-year climatic cycles in the Late Triassic Richmond, Taylorsville, and Sanford Basins (Fig. 1) on the basis of fluctuations in species diversity of pollen and spores. (The Sanford Basin is also known as the Sanford Sub-basin of the Deep River Basin). Long-term climatic cycles of availability of moisture during the Late Triassic, and presumably during the Early Jurassic, strongly affected plant populations, lacustrine water depth, salinity, duration of lakes, and probably also the hydrocarbon potential of these deposits.

Cycles are also present in other formations in the Newark Basin. The Passaic Formation (also known as the Brunswick Formation) overlies the Lockatong Formation (Olsen 1980a). The Passaic is strikingly similar to the Bull Run Formation in its dominance by red clastics, with associated grey and black lacustrine beds. Both formations occupy similar stratigraphic positions below the Early Jurassic basalt flows within their respective basins, and both have alluvial fan conglomerate along the basin edge, fining basinward into sandstone, siltstone, and mudstone. Clusters of 'thin' lacustrine cycles like those of the Lockatong, but separated by red fluvial units, are present throughout the Passaic (Olsen 1984a). The Early Jurassic Towaco Formation, also in the Newark Basin, contains sedimentary cycles which resemble Lockatong cycles, separated by red fluvial deposits (Olsen 1984a). Towaco cycles are 4–6 times thicker than Lockatong cycles (20–30 m), and the sediments are generally coarser grained (Olsen 1980b, 1984a). Hentz (1985) compared the lacustrine sequences in the Waterfall Formation to the Towaco cycles, noting their similar lithologic sequences, and the presence of lacustrine turbidites and conglomerates interbedded with black shales (see also Olsen 1980b, 1984a).

Cyclic sedimentary sequences are also present in the Hartford Basin. Hubert et al. (1976, 1978) reported symmetrical lacustrine cycles in the

Early Jurassic East Berlin Formation which range from 2 to 7 m thick, and are separated by floodplain red mudstone. These sequences contain fish-bearing black shale, and resemble some of the lacustrine sequences in the Bull Run and Buckland Formations. Demicco & Kordesch (1986) counted 15 lacustrine cycles in the upper 100 m of the East Berlin Formation. Some of these lake beds are very extensive and have been traced over at least 108 km north–south (along strike) and 20 km east–west, covering a minimum area of 2160 km^2 (Hubert *et al.* 1976).

In the Culpeper Basin, a series of 4 m thick asymmetrical lacustrine sequences, similar to Van Houten cycles, is present in the Culpeper Crushed Stone Quarry in the Bull Run Formation (Fig. 18). Olsen (1984a) referred to the sequences in this quarry as cycles, and noted that they occur in clusters which comprise larger scale cycles 20 m thick (Olsen 1984a, p. 48). At the base of each sequence, offshore lacustrine dark grey to black shales (BS-B) abruptly overlie red fluvial or mudflat deposits (RX or MS). These black shales are calcareous, and apparently barren of fossils (Albert J. Froelich, U.S. Geological Survey, pers. comm. August 1985). Grey, wave-rippled, carbonate-rich sandstone, and red, mudcracked shale are present higher in the sequence, and

FIG. 18. Photographs of cyclic lacustrine sequences in the Culpeper Crushed Stone Quarry, Virginia. (a) View of quarry wall. Note thin, continuous layers of black shale. (b) Close-up of a lacustrine sequence. Note black shale overlying a diastemic surface. Black shale passes upward into grey and red sandstone, siltstone, and shale. Bedding thickens upward.

bedding thickens upward (Fig. 18b). Thin beds of oolitic limestone are present in some sequences (Young & Edmundson 1954; Carozzi 1964) (Fig. 7e). At the top of a sequence, the strata are red to brown and mudcracked or massive; dinosaur trackways are present locally (Robert E. Weems, U.S. Geological Survey, pers. comm.). These lacustrine sequences have little or no variation in thickness or lithology across the quarry (Fig. 18a), suggesting deposition in relatively large lakes. The asymmetrical character of the sequences indicates rapid lake deepening or transgression followed by gradual shallowing or regression, and mudcracks indicate subaerial exposure (Fig. 6b). The presence of oolites suggests elevated salinity (Eugster & Kelts 1983), which probably explains the lack of fossils in the black shales.

The periodicity of lacustrine sequences in the Newark Basin corresponds with Milankovitch-type astronomical cycles (Olsen 1984a, 1984b, 1985a). Similar patterns appear to be present in nearly all of the Newark Supergroup basins (Paul E. Olsen, Lamont-Doherty Geological Observatory, pers. comm. November 1985). If the climate during the Late Triassic and Early Jurassic was controlled by astronomical forcing, then the lacustrine sequences in the Culpeper Basin must be arranged in Milankovitch-type cycles, as exposures in the Culpeper Crushed Stone Quarry suggest.

Palaeoenvironmental interpretation

During the Late Triassic and Early Jurassic, the Culpeper Basin was an asymmetrically-subsiding half-graben. Subsidence was greatest along the faulted western basin margin, and the basin floor tilted down toward the west (Fig. 19). Tectonic tilting was probably the dominant control on depositional geometry. A closed basin formed, surrounded by mountains that were higher on the western, faulted side of the basin. The dominant climatic pattern of easterly winds placed the eastern side of the basin in a rain shadow, with a semi-arid to arid climate (orographic desert). Air rising up the higher mountains on the west caused orographic precipitation, forming streams which flowed into the basin (Fig. 17). Although the drainage basin was larger to the east of the half-graben, there was little freshwater inflow or sediment input from the east during much of the basin history because of the effects of topography on the climate.

Alluvial fans developed along the western edge of the basin, and sheetfloods transported sand, silt, and mud to the east across a broad alluvial plain (Fig. 19a). Water ponded locally on the alluvial plain, forming shallow playa lakes. When subsidence occurred along the western margin of the basin, the basin floor tilted and water ponded against the alluvial fans and western highlands, forming extensive lakes which shallowed eastward (Fig. 19b). The slope of the basin floor was probably low because the lacustrine deposits are interspersed with fluvial sheetflood deposits derived from the west. The lakes expanded when inflow exceeded evaporation. Depending on the rate of lake level rise, either asymmetrical or symmetrical lacustrine sequences formed. Asymmetrical sequences formed when lake level rise was rapid, producing a diastemic surface directly overlain by deeper water deposits. Symmetrical sequences formed when lake level rise was gradual, allowing transgressive deposits to accumulate. The lakes were perennially stratified (meromictic to oligomictic) with anoxic bottom waters, like modern Lake Tanganyika or Lake Malawi, and black shales were deposited in the deepest water. Most of the time, the lakes were hydrologically closed. Near the shoreline on the western side of the basin, close to sources of

FIG. 19. Palaeogeographic reconstructions of Culpeper Basin lakes. (a) 'Very thin' lacustrine sequences represent deposition in playa lakes, or near the edges of large lakes. Some of these lakes may be restricted to the central part of the basin, as suggested by an eastward fining of grain size within the Bull Run Formation. Sheetflood deposits are derived from the west. (b) Subsidence in the western part of the basin, coupled with an increase in precipitation leads to the formation of large, deep, stratified lakes. Coarser sediment is concentrated along the western edge of the basin, and turbidity currents carry sand eastward into the deepest part of the lake. From stage (b), the lake proceeds to stage (c) if the climate remains relatively humid, or to stage (d) if the climate becomes more arid. (c) The lake is infilled totally or partially by prograding delta and shoreline deposits from the west. Total infilling leads to an eastward sloping alluvial plain covered by fluvial channel and sheetflood deposits, similar to (a), unless renewed subsidence occurs. (d) If the climate becomes increasingly drier, then the lake level drops, exposing vast mudflats on the east side of the basin, and salinity increases if the basin is closed. Stromatolites grow in the shallow saline waters, and evaporite crystals may be precipitated in the lacustrine sediments and in surrounding mudflat deposits. Palaeosols develop on sediments which remain subaerially exposed for long periods. With an increase in precipitation, or an increase in freshwater inflow, a freshwater cap may cover the dense saline waters, forming an ectogenic meromictic lake, and the lake proceeds to a stage similar to (b). If precipitation remains low or decreases, and the lake dries up almost entirely, then the lake may proceed to a stage similar to (a).

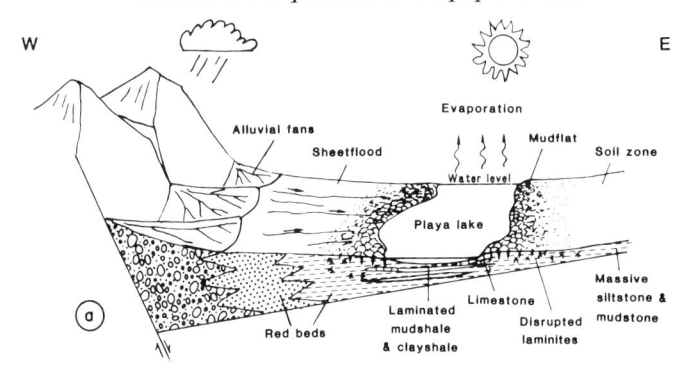

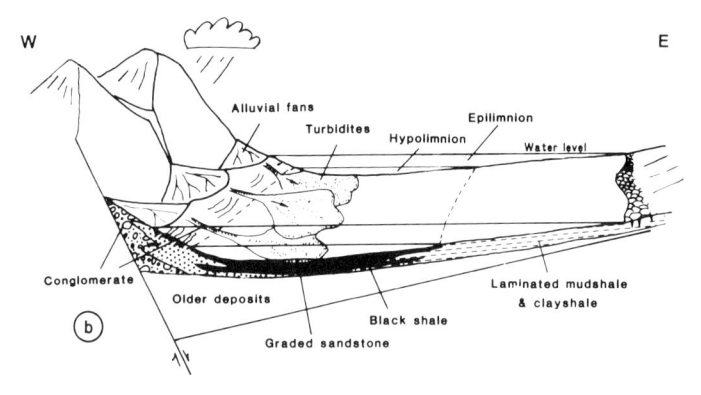

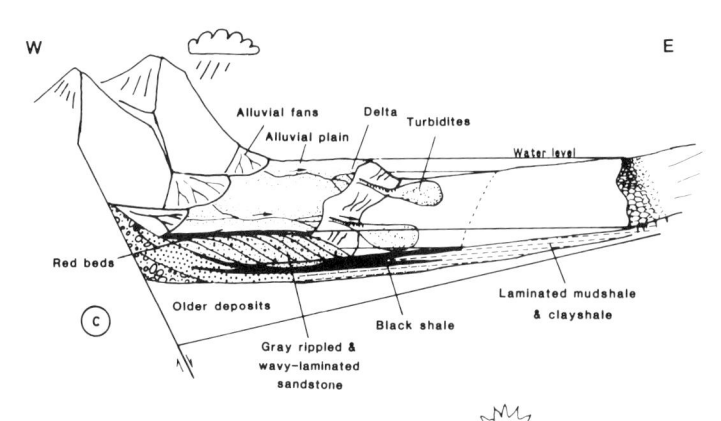

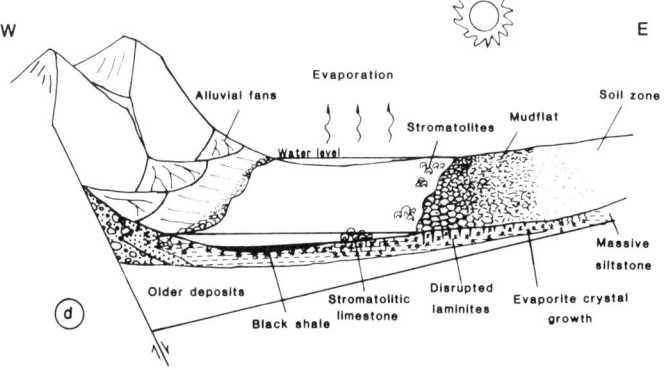

Lacustrine sequences in Culpeper Basin

clastic input, wave-rippled sandstones and siltstones were deposited. Sporadic floods or earthquakes along the western margin of the basin produced turbidity currents, which deposited thin beds of graded sandstone and occasionally conglomerates derived from shallow water (Fig. 19b). Farther from sources of terrigenous input, laminated clayshales, mudshales, and limestones were deposited in shallow water and were probably bound by microbial mats. Fresh water covered extensive shallow areas on the eastern side of the lake, and supported an invertebrate population dominated by ostracods and conchostracans, occasionally associated with pelecypods, notostracans, insects, and fish. As the lakes filled with sediment from fluvial sheetfloods and deltaic input, regressive, coarsening-upward sequences were deposited (Fig. 19c). Or, when the lakes contracted due to increased evaporation rates, salinity increased, and stromatolites grew, and shallow water sediments were subaerially exposed on vast mudflats (Fig. 19d). Mudcracks formed, and evaporite crystals (gypsum and/or anhydrite) precipitated in the mudflat sediments. Locally, extensive bioturbation from roots and infaunal invertebrates homogenized the mudflat deposits into massive sediment, and palaeosols developed, locally containing caliche or caliche-like pedogenic calcite nodules. Renewed freshwater input to saline lakes led to ectogenic meromictic stratification, with a freshwater epilimnion and a saline, anoxic hypolimnion (Fig. 19b).

In the Early Jurassic, subsidence increased along the western margin of the basin, coupled with uplift and erosion in the central part of the basin. At this time, most of the lakes were fresh, and may have been hydrologically open, as suggested by the scarcity or absence of evaporite crystal moulds and pseudomorphs in associated mudflat deposits of the Waterfall Formation.

Conclusions

Culpeper Basin lakes formed in a half-graben dominated by fluvial sheetflood and alluvial fan deposition. As the climate became wetter, vast lakes formed within the basin, probably deepening to the west under the influence of a westward-tilted basin floor. Lacustrine sequences ranging from less than 1 m to more than 35 m thick document the expansion and contraction of perennial lakes that formed in the basin during long-term periods of increased precipitation under the control of Milankovitch-type cyclic climatic change. The waters in the lakes were stratified (meromictic or oligomictic), and black, organic-rich, laminated shales were deposited in the anoxic bottom waters. Some of these lakes were fresh, as indicated by the presence of freshwater fossils, and others were saline or hypersaline, as indicated by the presence of evaporite crystal moulds and pseudomorphs and the absence of freshwater fossils. The saline or hypersaline lakes and some of the freshwater lakes were hydrologically closed, as indicated by the presence of evaporite crystal moulds and pseudomorphs in mudflat deposits. In some sequences, the association of evaporite crystals (gypsum and/or anhydrite) with the offshore black shales, and the presence of a fresh to brackish water fauna in nearshore deposits suggests ectogenic meromixis, in which density stratification originated by flooding a saline lake with fresh water. Thicker lacustrine sequences and scarcity of evaporite crystal moulds and pseudomorphs in the Waterfall Formation suggest greater shoreline stability and open basin deposition.

ACKNOWLEDGEMENTS: This research was supported by the National Geographic Society (grant 3008-84), two research grants from the Emory University Research Committee, a Faculty Development Award from Emory College, two grants-in-aid from Sigma Xi, and a grant from the George Washington University Chapter of Sigma Xi. The cooperation of the U.S. National Park Service for allowing me to collect rock samples in the Manassas National Battlefield Park is appreciated. The Department of Paleobiology, Smithsonian Institution is acknowledged for its assistance during a visit to the excavation of the Midland fish-bed during the summer of 1984. Thanks are extended to those who accompanied me in the field or helped through discussions: Roy C. Lindholm (George Washington University), Daniel A. Textoris & Walter H. Wheeler (University of North Carolina, Chapel Hill), Albert J. Froelich, Joseph P. Smoot, Eleanora I. Robbins & Robert E. Weems (U.S. Geological Survey), Paul E. Olsen (Lamont-Doherty Geological Observatory), Alfred Traverse (Pennsylvania State University), Tucker Hentz (Texas Bureau of Economic Geology), Peter LeTourneau (Connecticut Geological Survey), Thomas H. Shaw (Emory University), Thomas J. Gore III, Khalid Maletan, and many others. Patricia L. Renwick and E. Lynn Zeigler (Emory University) assisted with bibliographic work, and manuscript preparation. Comments of Richard Yuretich (University of Massachusetts), Kerry Kelts (EAWAG, Zurich), and an unknown reviewer are especially appreciated.

References

ALLEN, P. A. & COLLINSON, J. D. 1986. Lakes. *In:* READING, H. G. (ed.), *Sedimentary Environments and Facies*, Blackwell Scientific Publications, Boston, 63–94.

APPLEGATE, S. P. 1956. Distribution of Triassic fish in the Piedmont of Virginia. *Bulletin of the Geological Society of America*, **67**, 1749.

BAER, F. M. & MARTIN, W. H. 1949. Some new finds of fossil ganoids in the Virginia Triassic. *Science*, **110**, 684–686.

BENSON, R. H., BERDAN, J. M., VAN DEN BOLD, W. A. *et al.* 1961. Systematic descriptions. *In:* MOORE, R. C. (ed.), *Treatise on Invertebrate Paleontology, Part Q, Arthropoda 3, Crustacea, Ostracoda*, Geological Society of America and University of Kansas Press, 442 pp.

BERNER, R. A. 1970. Sedimentary pyrite formation. *American Journal of Science*, **268**, 1–23.

—— 1981. A new geochemical classification of sedimentary environments. *Journal of Sedimentary Petrology*, **51**, 359–365.

BOUMA, A. H. 1962. *Sedimentology of some Flysch Deposits: a Graphic Approach to Facies Interpretation*, Elsevier, Amsterdam, 169 pp.

BURST, J. F. 1969. Diagenesis of Gulf coast clayey sediments and its possible relation to petroleum migration. *Bulletin of the American Association of Petroleum Geologists*, **53**, 73–93.

BYERS, C. W. 1974. Shale fissility: Relation to bioturbation. *Sedimentology*, **21**, 479–484.

CAROZZI, A. V. 1964. Complex ooids from Triassic lake deposit, Virginia. *American Journal of Science*, **262**, 231–241.

COHEN, A. S. 1984. Effects of zoobenthic standing crop on laminae preservation in tropical lake sediment, Lake Turkana, East Africa. *Journal of Paleontology*, **58**, 449–510.

COLEMAN, J. M. 1966. Ecological changes in a massive fresh-water clay sequence. *Transactions of the Gulf Coast Association of Geological Societies*, **16**, 159–174.

CONEYBEARE, C. E. B. 1967. Influence of compaction on stratigraphic analysis. *Bulletin of Canadian Petroleum Geology*, **15**, 331–345.

COOK, P. J. 1976. Sedimentary phosphate deposits. *In:* WOLFE, K. H. (ed.), *Handbook of Strata-Bound and Stratiform Ore Deposits*, Elsevier, New York, 7, 505–535.

CORNET, B. 1977. The palynostratigraphy and age of the Newark Supergroup. Ph.D. Thesis, Pennsylvania State University, University Park, 506 pp.

——, TRAVERSE, A. & MCDONALD, N. G. 1973. Fossil spores, pollen and fishes from Connecticut indicate Early Jurassic age for part of the Newark Supergroup. *Science*, **182**, 1243–1247.

CROSSLEY, R. 1984. Controls of sedimentation in the Malawi rift valley, central Africa. *Sedimentary Geology*, **40**, 33–50.

DEAN, W. E. & FOUCH, T. D. 1983. Lacustrine environment. *In:* SCHOLLE, P. A., BEBOUT, D. G. & MOORE, C. H. (eds), *Carbonate Depositional Environments*, Memoir of the American Association of Petroleum Geologists, **33**, 97–130.

DEGENS, E. T. & STOFFERS, P. 1976. Stratified waters as a key to the past. *Nature*, **263**, 22–27.

——, VON HERZEN, R. P. & WONG, H.-K. 1971. Lake Tanganyika: Water chemistry, sediments, geological structure. *Naturwissenschaften*, **58**, 229–241.

—— OKADA, H., HONJO, S. & HATHAWAY, J. C. 1972. Microcrystalline sphalerite in Lake Kivu, East Africa. *Mineralium Deposita*, **7**, 1–12.

—— VON HERZEN, R. P., WONG, H.-K., DEUSER, W. G. & JANNASCH, H. W. 1973. Lake Kivu: Structure, chemistry, and biology of an East African rift lake. *Geologische Rundschau*, **62**, 245–277.

DEMAISON, G. J. & MOORE, G. T. 1980. Anoxic environments and oil source bed genesis. *Bulletin of the American Association of Petroleum Geologists*, **64**, 1179–1209.

DEMICCO, R. V. & KORDESCH, E. G. 1986. Facies sequences of a semi-arid closed basin: The Lower Jurassic East Berlin Formation of the Hartford Basin, New England, USA. *Sedimentology*, **33**, 107–118.

EUGSTER, H. P. & HARDIE, L. A. 1975. Sedimentation in an ancient playa-lake complex: The Wilkins Peak Member of the Green River Formation of Wyoming. *Bulletin of the Geological Society of America*, **86**, 319–334.

—— & KELTS, K. 1983. Lacustrine chemical sediments. *In:* GOUDIE, A. & PYE, K. (eds), *Chemical Sediments and Geomorphology*, Academic Press, London, 321–368.

FROELICH, A. J. & GOTTFRIED, D. 1985. Early Jurassic diabase sheets of the eastern United States—A preliminary overview. *In:* ROBINSON, G. R. JR. & FROELICH, A. J. (eds), *Proceedings of the Second U.S. Geological Survey Workshop on the Early Mesozoic Basins of the Eastern United States,* United States Geological Survey Circular, **946**, 79–86.

—— & OLSEN, P. E. 1984. Newark Supergroup, a revision of the Newark Group in eastern North America. *Bulletin of the United States Geological Survey*, **1537-A**, A55–A58.

—— & —— 1985. Newark Supergroup, a revision of the Newark Group in eastern North America. *In:* ROBINSON, G. R. Jr. & FROELICH, A. J. (eds), *Proceedings of the Second U.S. Geological Survey Workshop on the Early Mesozoic Basins of the Eastern United States*, United States Geological Survey Circular, **946**, 1–3.

—— LEAVY, B. D. & LINDHOLM, R. C. 1982. Geologic traverse across the Culpeper Basin (Triassic–Jurassic) of northern Virginia. *In:* LYTTLE, P. T. (ed.), *Central Appalachian Geology*, NE-SE Geological Society of America Field Trip Guidebook, Field trip 3, 55–81.

GOLDHABER, M. B. & KAPLAN, I. R. 1974. The sulfur cycle. *In:* GOLDBERG, E. D. (ed.), *The Sea*, **5**, John Wiley & Sons, New York, 569–655.

GORE, P. J. W. 1983. Sedimentology and invertebrate paleontology of Triassic and Jurassic lacustrine deposits, Culpeper Basin, northern Virginia. Ph.D. Dissertation, George Washington University, Washington, D.C., 356 pp.

—— 1984. Triassic and Jurassic lacustrine sequences in the Culpeper Basin, northern Virginia. *Geological Society of America, Abstracts with Programs*, **16**, 141.

—— 1985a. Lacustrine sequences of Triassic-Jurassic age in the Culpeper Basin, Virginia and Deep River Basin, N.C. *Geological Society of America, Abstracts with Programs*, **17**, 21.

—— 1985b. Triassic floodplain lake in the Chatham Group (Newark Supergroup) of the Durham Subbasin of the Deep River Basin, N.C.: Comparison to the Culpeper Basin, VA. *Geological Society of America, Abstracts with Programs*, **17**, 94.

—— 1985c. Early Mesozoic lacustrine sedimentation in the Culpeper Basin, Virginia and the Deep River Basin, North Carolina: A comparative study. *Bulletin of the American Association of Petroleum Geologists*, **69**, 1438.

—— 1986. Early diagenetic nodules, compaction, and secondary lamination in Early Jurassic lacustrine black shale, Culpeper Basin, Virginia. *Bulletin of the American Association of Petroleum Geologists*, **70**, 596.

—— in press. Paleoecology and sedimentology of a Late Triassic lake, Culpeper Basin, Virginia. *Palaeogeography, Palaeoclimatology, Palaeoecology*.

—— & LINDHOLM, R. C. 1983. Paleoecology of Triassic and Jurassic lacustrine deposits in the Culpeper Basin of northern Virginia. *Geological Society of America, Abstracts with Programs*, **15**, 122.

—— & TRAVERSE, A. 1986. Triassic notostracans in the Newark Supergroup: Culpeper Basin, northern Virginia, with a contribution on the palynology. *Journal of Paleontology*, **60**, 1086–1096.

HARDIE, L. A. SMOOT, J. P. & EUGSTER, H. P. 1978. Saline lakes and their deposits: A sedimentological approach. *In*: MATTER, A. & TUCKER, M. E. (eds), *Modern and Ancient Lake Sediments*, Special Publication. International Association of Sedimentologists, Blackwell Scientific Publications, London, **2**, 7–41.

HAY, W. W., BEHENSKY, J. F. JR., BARRON, E. J. & SLOAN, J. L. II. 1982. Late Triassic-Liassic paleoclimatology of the proto-central North Atlantic rift system. *Palaeogeography, Palaeoclimatology, Palaeoecology*, **40**, 13–30.

HENTZ, T. F. 1981. Sedimentology and structure of Culpeper Group lake beds (Lower Jurassic) at Thoroughfare Gap, Virginia. M.S. Thesis, University of Kansas, Lawrence, 166 pp.

—— 1982. Sedimentology and structure of Lower Jurassic lake beds in the Culpeper Basin at Thoroughfare Gap, Virginia. *Geological Society of America, Abstracts with Programs*, **14**, 24–25.

—— 1985. Early Jurassic sedimentation of a rift-valley lake: Culpeper Basin, Virginia. *Bulletin of the Geological Society of America*, **96**, 92–107.

HUBERT, J. F. 1977. Paleosol caliche in the New Haven Arkose, Connecticut: record of semi-aridity in Late Triassic-Early Jurassic time. *Geology*, **5**, 302–304.

—— 1978. Paleosol caliche in the New Haven Arkose, Newark Group, Connecticut. *Palaeogeography, Palaeoclimatology, Palaeoecology*, **24**, 151–168.

—— & MERTZ, K. A. JR. 1984. Eolian sandstones in Upper Triassic-Lower Jurassic red beds of the Fundy Basin, Nova Scotia. *Journal of Sedimentary Petrology*, **54**, 798–810.

—— REED, A. A. & CAREY, P. J. 1976. Paleogeography of the East Berlin Formation, Newark Group, Connecticut Valley. *American Journal of Science*, **276**, 1183–1207.

—— REED, A. A., DOWDALL, W. L. & GILCHRIST, J. M. 1978. *Guide to the Redbeds of Central Connecticut*, Field trip guide, Society of Economic Paleontologists and Mineralogists, Eastern Section. Contribution No. 32, Department of Geology & Geography, University of Massachusetts, Amherst, 129 pp.

JENKYNS, H. C. 1986. Pelagic environments. *In*: READING, H. G. (ed.), *Sedimentary Environments and Facies*, Blackwell Scientific Publications, Boston, 343–397.

KELTS, K. & HSÜ, K. J. 1978. Freshwater carbonate sedimentation. *In*: LERMAN, A. (ed.), *Lakes: Chemistry, Geology, Physics*, Springer-Verlag, New York, 295–323.

KIRKLAND, D. W. & EVANS, R. 1981. Source rock potential of evaporitic environment. *Bulletin of the American Association of Petroleum Geologists*, **65**, 181–190.

LEAVY, B. D. 1980. Tectonic and sedimentary structures along the eastern margin of the Culpeper Basin, Virginia. *Geological Society of America, Abstracts with Programs*, **12**, 182.

LEE, K. Y. 1977. Triassic stratigraphy in the northern part of the Culpeper Basin, Virginia and Maryland. *Bulletin of the United States Geological Survey*, **1422-C**, 17 pp.

—— 1979. Triassic–Jurassic geology of the northern part of the Culpeper Basin, Virginia and Maryland. *Open File Report. United States Geological Survey*, **79–1557**, 10 pp.

—— 1980. Triassic–Jurassic geology of the southern part of the Culpeper Basin and the Barboursville Basin. *Open File Report. United States Geological Survey*, **80–468**, 9 pp.

LETOURNEAU, P. M. 1985. The sedimentology and stratigraphy of the Lower Jurassic Portland Formation, central Connecticut. M.S. Thesis, Wesleyan University, Middletown, Connecticut, 247 pp.

LINDHOLM, R. C. 1979. Geologic history and stratigraphy of the Triassic–Jurassic Culpeper Basin, Virginia. *Bulletin of the Geological Society of America*, **90**, 995–997, 1702–1736.

—— HAZLETT, J. M. & FAGIN, S. W. 1979. Petrology of Triassic–Jurassic conglomerates in the Culpeper Basin, Virginia. *Journal of Sedimentary Petrology*, **49**, 1245–1262.

—— GORE, P. J. W. & CROWLEY, J. K. 1982. A lacustrine sequence in the Upper Triassic Bull Run Formation (Culpeper Basin) in northern Virginia.

Geological Society of America, Abstracts with Programs, **14**, 35.

LUDLAM, S. D. 1969. Fayetteville Green Lake, New York. III. The laminated sediments. Limnology and Oceanography, **14**, 848–857.

MANSPEIZER, W. 1981. Early Mesozoic basins of the central Atlantic passive margins. In: BALLY, A. W., WATTS, A. B., GROW, J. A., MANSPEIZER, W., BERNOULLI, D., SCHREIBER, C. & HUNT, J. M. (eds), Geology of Passive Continental Margins: History, Structure and Sedimentologic Record (With Special Emphasis on the Atlantic Margin), American Association of Petroleum Geologists Education Short Course Note Series 19, 4.1–4.60.

—— & OLSEN, P. E. 1981. Rift basins of the passive margin: Tectonics, organic-rich lacustrine sediments, basin analysis. In: HOBBS, G. W. (ed.), Field Guide to the Geology of the Paleozoic, Mesozoic, and Tertiary Rocks of New Jersey and the Central Hudson Valley, Atlantic Margin Energy Symposium, American Association of Petroleum Geologists Eastern Section Meeting, Petroleum Exploration Society of New York, New York, 25–103.

MARTELLO, A. R., GRAY, N. H., PHILPOTTS, A., DOWLING, J. J. & KOZA, D. M. 1984. Mesozoic diabase dikes of southeastern New England. Geological Society of America, Abstracts with Programs, **16**, 48.

MAYNARD, J. B. 1982. Extension of Berner's 'New geochemical classification of sedimentary environments' to ancient sediments. Journal of Sedimentary Petrology, **52**, 1325–1331.

MOORE, W. & BURN, A. 1968. Lethal oxygen thresholds for certain temporary pond invertebrates and their applicability to field situations. Ecology, **49**, 349–351.

NEUMANN, A. C., GEBLEIN, C. D. & SCOFFIN, T. P. 1970. The composition, structure and erodability of subtidal mats, Abaco, Bahamas. Journal of Sedimentary Petrology, **40**, 274–297.

OLSEN, P. E. 1980a. Triassic and Jurassic formations of the Newark Basin. In: MANSPEIZER, W. (ed.), Field Studies of New Jersey Geology and Guide to Field Trips, 52nd Annual Meeting of the New York State Geological Association, Department of Geology, Rutgers University, Newark, New Jersey, 2–39.

—— 1980b. Fossil great lakes of the Newark Supergroup in New Jersey. In: MANSPEIZER, W. (ed.), Field Studies of New Jersey Geology and Guide to Field Trips, 52nd Annual Meeting of the New York State Geological Association, Department of Geology, Rutgers University, Newark, New Jersey, 352–398.

—— 1984a. Comparative paleolimnology of the Newark Supergroup: A study of ecosystem evolution. Ph.D. Dissertation, Yale University, New Haven, Connecticut, 762 pp.

—— 1984b. Periodicity of lake-level cycles in the Late Triassic Lockatong Formation of the Newark Basin (Newark Supergroup, New Jersey & Pennsylvania). In: BERGER, A. L., IMBRIE, J., HAYS, J., KUKLA, G. & SALTZMAN, B. (eds), Milankovitch and Climate, D. Reidel Publishing Company, Boston, 129–146.

—— 1985a. Control of organic-rich lacustrine strata by Milankovitch-type cyclic climate: a new hydrocarbon exploration strategy. Geological Society of America, Abstracts with Programs, **17**, 681.

—— 1985b. Distribution of organic-matter-rich lacustrine rocks in the Early Mesozoic Newark Supergroup. In: ROBINSON, G. R. JR. & FROELICH, A. J. (eds), Proceedings of the Second U.S. Geological Survey Workshop on the Early Mesozoic Basins of the Eastern United States, United States Geological Survey Circular, **946**, 61–64.

PARRISH, J. T., ZEIGLER, A. M. & SCOTESE, C. R. 1982. Rainfall patterns and the distribution of coals and evaporites in the Mesozoic and Cenozoic. Palaeogeography, Palaeoclimatology, Palaeoecology, **40**, 67–101.

PETTIJOHN, F. J. 1975. Sedimentary Rocks. Harper & Row Publishers, New York, 628 pp.

POTTER, P. E., MAYNARD, J. B. & PRYOR, W. A. 1980. Sedimentology of Shale. Springer-Verlag, New York, 310 pp.

RATCLIFFE, N. M. & BURTON, W. C. 1985. Fault reactivation models for origin of the Newark Basin and studies related to eastern U.S. seismicity. In: ROBINSON, G. R. JR. & FROELICH, A. J. (eds), Proceedings of the Second U.S. Geological Survey Workshop on the Early Mesozoic Basins of the Eastern United States, United States Geological Survey Circular, **946**, 36–45.

RHOADS, D. C. & MORSE, J. W. 1971. Evolution and ecologic significance of oxygen-deficient marine basins. Lethaia, **4**, 413–428.

ROBERTS, J. K. 1928. The geology of the Virginia Triassic. Bulletin. Virginia Geological Survey, **29**, 205 pp.

ROBINSON, P. L. 1973. Paleoclimatology and continental drift. In: TARLING, D. H. & RUNCORN, S. K. (eds), Implications of Continental Drift to the Earth Sciences, Academic Press, London, **1**, 449–476.

RYDER, R. T. 1980. Lacustrine Sedimentation and Hydrocarbon Occurrences with Emphasis on Uinta Basin Models. American Association of Petroleum Geologists Fall Education Conference, Houston, Texas, 103 pp.

SAVRDA, C. E., BOTTJER, D. J. & GORSLINE, D. S. 1984. Development of a comprehensive marine biofacies model: evidence from Santa Monica, San Pedro, and Santa Barbara Basins, California continental borderland, Bulletin of the American Association of Petroleum Geologists, **68**, 1179–1192.

SCHAMEL, S., RESSETAR, R., GAWARECKI, S., TAYLOR, G. K., TRAVERSE, A., HOUGHTON, H. F. & LETOURNEAU, P. 1986. Early Mesozoic rift basins of eastern United States. Bulletin of the American Association of Petroleum Geologists, **70**, 644.

SCOTESE, C. R. & SUMMERHAYES, C. P. 1986. Computer model of paleoclimate predicts upwelling in the Mesozoic and Cenozoic. Geobyte, **1**, 28–42.

SMITH, M. A. & ROBISON, C. R. 1986. Lacustrine source rock potential, Culpeper Basin, Virginia. Bulletin of the American Association of Petroleum Geologists, **70**, 650.

SMOOT, J. P. 1985. The closed-basin hypothesis and its use in facies analysis of the Newark Supergroup. *In*: ROBINSON, G. R. JR. & FROELICH, A. J. (eds), *Proceedings of the Second U.S. Geological Survey Workshop on the Early Mesozoic Basins of the Eastern United States*, United States Geological Survey Circular, **946**, 4–10.

SOBHAN, A. N. 1985. Petrology and depositional history of Triassic red beds, Bull Run Formation, eastern Culpeper Basin, Virginia. M.S. Thesis, George Washington University, Washington, D.C., 225 pp.

SPENCER, R. J., BAEDECKER, M. J., EUGSTER, H. P. *et al.* 1984. Great Salt Lake, and precursors, Utah: The last 30,000 years. *Contributions to Mineralogy and Petrology*, **86**, 321–334.

STOFFERS, P. & HECKY, R. E. 1978. Late Pleistocene–Holocene evolution of the Kivu-Tanganyika Basin. *In*: MATTER, A. & TUCKER, M. E. (eds), *Modern and Ancient Lake Sediments*, Special Publication 2, International Association of Sedimentologists, Blackwell Scientific Publications, London, 43–55.

STURM, M. & MATTER, A. 1978. Turbidites and varves in Lake Brienz (Switzerland): Deposition of clastic detritus by density currents. *In*: MATTER, A. & TUCKER, M. E. (eds), *Modern and Ancient Lake Sediments*, Special Publication. International Association of Sedimentologists, Blackwell Scientific Publications, London, **2**, 147–168.

TALBOT, M. R. 1976. Late Quaternary sedimentation in Lake Bosumtwi. *20th Annual Report of the Research Institute of African Geology*, University of Leeds, 69–73.

—— 1983. Late Quaternary sedimentation in a tropical meromictic lake—Lake Bosumtwi, Ghana. *4th International Association of Sedimentologists Regional Meeting*, Split, Yugoslavia, 162.

—— & KELTS, K. 1985. Carbonate mineral genesis in a tropical lake. *6th European Regional Meeting of Sedimentology, International Association of Sedimentologists. LLEIDA'85*, 455–456.

—— LIVINGSTONE, D. A., PALMER, P. G., MALEY, J., MELACK, J. M., DELIBRIAS, G. & GULLIK-SEN, S. 1984. Preliminary results from sediment cores from Lake Bosumtwi, Ghana. *Palaeoecology of Africa and the Surrounding Islands*, **16**, 173–192.

THOMPSON, J. B., MULLINS, H. T., NEWTON, C. R. & VERCOUTERE, T. L. 1985. Alternative biofacies model for dysaerobic communities. *Lethaia*, **18**, 167–179.

VAN HOUTEN, F. B. 1962. Cyclic sedimentation and the origin of analcime-rich Upper Triassic Lockatong Formation, west-central New Jersey and adjacent Pennsylvania. *American Journal of Science*, **260**, 561–576.

—— 1964. Cyclic lacustrine sedimentation, Upper Triassic Lockatong Formation, central New Jersey and adjacent Pennsylvania. *Bulletin. Geological Survey of Kansas*, **169**, 497–531.

—— 1969. Late Triassic Newark Group, north central New Jersey and adjacent Pennsylvania and New York. *In*: SUBITSKY, S. S. (ed.), *Geology of Selected Areas in New Jersey and Pennsylvania and Guidebook*, Rutgers University Press, New Brunswick, New Jersey, 314–347.

WALKER, R. G. 1975. From sedimentary structures to facies models: An example from fluvial environments. *In*: HARMS, J. C., SOUTHARD, J. B., SPEARING, D. R. & WALKER, R. G. (eds), *Depositional Environments as Interpreted from Primary Sedimentary Structures and Stratification Sequences*, Society of Economic Paleontologists and Mineralogists Short Course Notes, **2**, 63–79.

WAPLES, D. W. 1983. Reappraisal of anoxia and organic richness, with emphasis on Cretaceous of North Atlantic. *Bulletin of the American Association of Petroleum Geologists*, **67**, 963–978.

WEBB, J. A. 1979. A reappraisal of the paleoecology of conchostracans (Crustacea: Branchiopoda). *Neues Jahrbuch für Geologie und Paläontologie, Abhandlungen*, **158**, 259–275.

WHEELER, W. H. & TEXTORIS, D. A. 1978. Triassic limestone and chert of playa origin in North Carolina. *Journal of Sedimentary Petrology*, **48**, 765–776.

YOUNG, R. S. & EDMUNDSON, R. S. 1954. Oolitic limestone in the Triassic of Virginia. *Journal of Sedimentary Petrology*, **24**, 275–279.

YURETICH, R. F. 1976. Sedimentology, geochemistry, and geological significance of modern sediments in Lake Rudolf (Lake Turkana) Eastern Rift Valley, Kenya. Ph.D. dissertation, Princeton University, Princeton, New Jersey, 305 pp.

—— 1979. Modern sediments and sedimentary processes in Lake Rudolf (Lake Turkana) Eastern Rift Valley, Kenya. *Sedimentology*, **26**, 313–331.

ZIEGLER, A. M., SCOTESE, C. R. & BARRETT, S. F. 1983. Mesozoic and Cenozoic paleogeographic maps. *In*: BROSCHE, P. & SÜNDERMANN, J. (eds), *Tidal Friction and the Earth's Rotation II*, Springer-Verlag, Berlin, 240–252.

ZIEGLER, D. G. 1983. Hydrocarbon potential of Newark Rift system: Eastern North America. *Northeastern Geology*, **5**, 200–208.

P. J. W. GORE, Department of Geology, Emory University, Atlanta, Georgia 30322, USA.

Organic geochemical characteristics of major types of terrestrial petroleum source rocks in China

Fu Jiamo, Sheng Guoying & Liu Dehan

SUMMARY: Terrestrial source rocks, deposited under non-marine conditions in continental environments, have been divided into five major groups: Group A, sedimentary formations of big lake basins in the interior of the Chinese plate; Group B, lacustrine clastic formations in fault-bounded basins; Group C, salt lake evaporite-clastic formations in fault-bounded basins; Group D, lagoonal-lacustrine volcano-clastic formations of intermontane basins; and Group E, paralic coal-bearing formations of platform areas. The abundance, type and evolutionary characteristics of the organic matter, as well as the biomarker characteristics of various groups of source rocks are discussed briefly. There are several problems relating to evaluation of hydrocarbon prospects in terrestrial sedimentary basins, including the origin of Type I kerogens and their hydrocarbon potentials, immature oils, and coal-bearing formations as potential source rocks.

Introduction

Most oil fields found so far in China are in non-marine sedimentary basins ranging in age from Carboniferous to Palaeogene. Conditions in the basins changed greatly with time.

Because the formation and distribution of hydrocarbons in a sedimentary basin are closely related to its formation, it is especially important in hydrocarbon exploration and prospect evaluation to classify and compare basins. The formation and distribution of source rocks, and the generation of their hydrocarbon products are also controlled by the basin type. Although much has recently been published on crude oils of non-marine origin (Swain 1980; Thomas 1981; Shi Jiyang et al. 1982; Huang Difan & Li Jinchao 1982; Fu Jiamo et al. 1983; Tissot 1984; Wang Tieguan et al. this volume), comparative studies of different source-rock groups and their hydrocarbon potential in terms of basin classification nature have seldom been undertaken. Sedimentary basins can be divided in terms of different factors, of which the most important is tectonism. From the viewpoint of source-rock studies, palaeoclimate, depositional conditions and the geothermal regime must also be taken into account. Tectonic conditions in a basin relate to the global regime and the specific basin-forming processes. For example Mesozoic–Cenozoic basins in eastern China are mainly fault-bounded or 'rift valley' basins, a special type of tensional basin, while those of western China are intermontane basins resulting from compressional forces. Differing palaeoclimatic conditions result in various groups of source rocks, such as evaporites, freshwater sediments and coal-bearing formations, all of which are known to be good sources for crude oil in China. Lithological associations, especially the association of source rocks and reservoir rocks, are controlled by depositional conditions, and markedly affect the primary migration of hydrocarbons. For example quite different petroleum prospects occur in the Jianghan and Dongpu Basins even though their source rocks are similar.

The geothermal regime controls the maturation of source rocks and generation of hydrocarbons. The depth of the 'Moho' is a key factor which has controlled the thermal evolution of Chinese basins.

Source-rock classification

In view of the differing characteristics of oil- and gas-bearing basins in China (Fig. 1), five major groups of terrestrial source rocks are recognized:

(1) Group A: Sedimentary formations of large lake basins in the interior of the Chinese plate. A typical example is the Songlia Basin.
(2) Group B: Lacustrine clastic formations in fault-bounded basins. Almost all important source rocks of Cenozoic oil- and gas-bearing basins in eastern China belong to this group or, sometimes, to Group C.
(3) Group C: Salt lake evaporite-clastic formations in fault-bounded basins. The typical examples of this type are Palaeogene sediments of the Jianghan and Dongpu Basins: both were deposited under hypersaline conditions but differing primary migration led to the latter being more prospective (see above).
(4) Group D: Lagoonal-lacustrine, volcano-clastic formations in intermontane basins. This very special group of source rocks is only known in western China, e.g., the Carboniferous–Permian sediments of the Zhungeer Basin. The source rocks are composed of

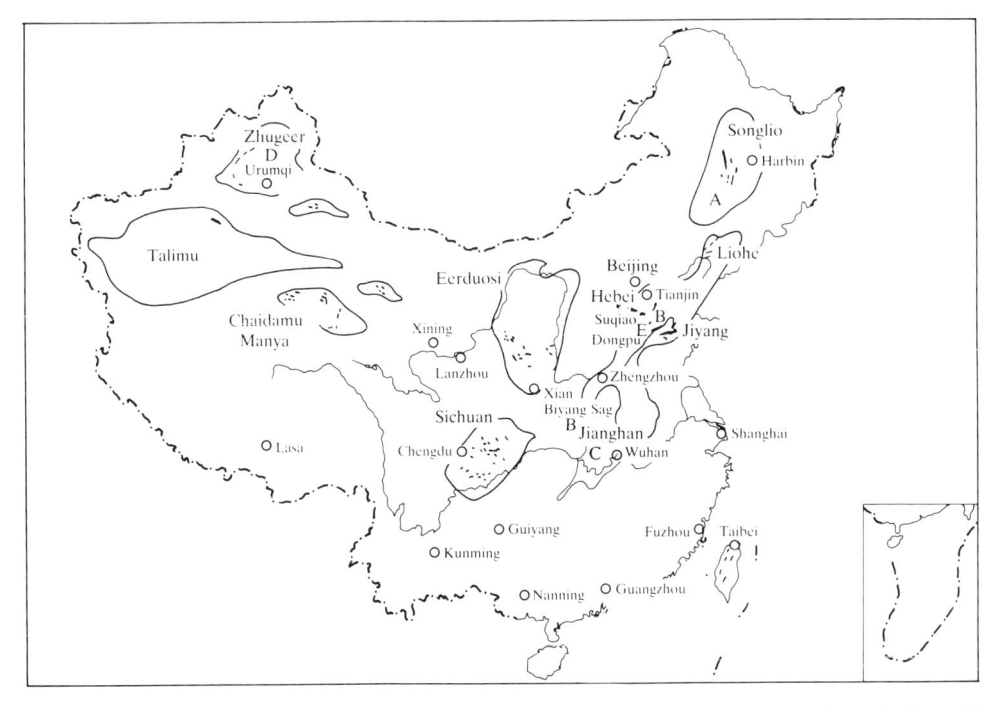

FIG. 1. Map of China, showing distribution of major groups of terrestrial source rocks. A–E: Group A–Group E.

basic tuff, tuffaceous mudstone and mudstone, of which the mineral composition is quite complicated, including montmorillonite, illite, chlorite and mixed layer clays, as well as small amounts of analcime, albite, plagioclase, quartz, dolomite, ankerite, calcite, siderite etc. (unpublished data of Guiyang Geochemistry Institute);

(5) Group E: Paralic coal-bearing formations of the platform. One example of this group is the Carboniferous–Permian coal-bearing formations of the Suqia region, which contain a special kind of coal-liptobiolites considered to be a good source for natural gas and condensate (Liu Dehan et al. 1985).

Abundance, type and evolutionary characteristics of the organic matter of the major source-rock types

This brief discussion of the abundance, type and evolutionary characteristics of organic matter is based on analytical data from different hydrocarbon-bearing basins.

Type and abundance of organic matter

Kerogen is generally divided into Types I, II and III (Tissot & Welte 1978). Subtypes have been adopted recently in China, such as Type I_A, standard sapropelic kerogen; Type I_B, humus-containing sapropelic kerogen; Type II_A, humic-sapropelic kerogen; Type II_B, sapropelic-humic kerogen; Type III_A, sapropel-containing humic kerogen; and Type III_B, standard humic kerogen (Yan Wanli & Li Yongkang 1982).

Group A source rocks are mainly composed of Type I and Type II_A kerogen. The organic matter distribution is controlled by the sedimentary facies in the large lake basin, Type I kerogen being generally developed within the deep lake facies.

Group B source rocks are much more complicated. Most contain Type II kerogen, with some Type III and Type I kerogens (Tables 1 and 2). The main reasons for this complicated situation are the changes in environment related to the fault movements and the distance of the sag from the adjacent area of uplift (Fig. 2). For instance, the source rock of the Ruoyan Sag is farther away from the uplift area than that of the Langgu Sag and so is better developed, being less influenced by clastic input.

Kerogens of the Group C source rocks are normally Type II, and those of Group D source rocks are possibly Type II and Type I. The organic matter of Group E source rocks depends

TABLE 1. *Abundance and type of organic matter major sedimentary formation group in oil- and gas-bearing basin, China*

Location		Epoch	Abundance of types of organic matter					Types of kerogen		
			C_{org} (%)	Bit. A (%)	HC ppm	A/C_{org} (%)	HC/C_{org} (%)	I	II	III
Group A										
Songlio Basin		K	2.40		1467		6.11	I		
		K	0.71		285		4.01		II$_A$	
		K	0.33		28		0.85			III
Group B										
Liohe W.D.		E	2.29	0.35	2002	15.28	8.74	I		
Hanghua D.	Qikuo Sag	E	1.49	0.19		12.75		I		
	Banqiao Sag	E	0.94	0.05		5.32			II	
		E	0.60	0.02		3.33				III
Central Hebei D.	Langgu Sag	E	1.19	0.07	451	6.13	3.79		II	
		E	0.70	0.05	351	7.57	5.01			III
	Ruoyan Sag	E	0.86	0.02	1055	23.84	12.27	I+II		
Jiyang D.	Zhanhuaa	E	2.80	0.28	1500	10.00	5.36	I$_B$		
	Dongying	E	1.30	0.15	250	11.54	1.92		II	
	Dawanzha	E	0.70	0.03	78	4.00	1.11			III
Biyang Sag		E	0.60							
		E	1.33	0.05–0.2	870–1718	8–10		I	II	
		E	1.59							
Group C										
Jianghan Basin	Guanghuasi	E	0.55	0.29	1476	51.82	26.84		II	
		E	0.51	0.33	2230	64.90	43.73		II	
	Zhongsi	E	0.66	0.11	457	15.90	6.92		II	
		E	0.52	0.10	550	19.89	10.58		II	
Jiyang D.		E	1.05	0.28		26.74			II	
Central Hebei		E	0.54						II$_A$	
Chaidamu Manya D.		N	0.05–0.4	0.04–0.4	18–143	8.2–9.82	3.60–8.76			III
Group D										
Zhungeer Basin		P	1.01		1970		19.50	I	II	
Group E										
Hebei‡		C–	*45	1.6–5.0	2900–4000	3–10	0.5–0.8			
Dongpu		P$_2$	†2–15	0.03–0.16	68–180	3.8–20	0.09–0.22		II	III

*Data of this line are from coal.
†Data of this line are from carbonaceous shales.
‡Fushun resinite coal belongs to Type I kerogen.
N:Neogene; E:Palaeogene; K:Cretaceous; P:Permian; P$_2$:Upper Permian; C:Carboniferous.

on their coal maceral composition, liptobiolites are thought to be Type II kerogen, while vitrinite belongs to Type III.

Carbon contents of Group A and Group B source rocks are high, generally 0.6–2.2%, while those of Group D are low, and those of Group C are the lowest, only 0.4–1%. Group E source rocks have the highest organic carbon contents, but these are variable, depending on the sedimentary and diagenetic environment. It is of interest that, although the organic contents of Group C are lowest, their bitumen A/C_{org} ratios (e.g., soluble organic matter/total organic matter) are possibly the highest (Table 1), having certain similarity to those of carbonate source rocks (Fu Jiamo & Jia Rongfen 1984).

Controls on thermal evolution

Key factors which affect the thermal evolution of organic matter are considered to be palaeogeothermal gradient and unconformities. Owing to the different depths to the Moho and other conditions, the average palaeogeothermal gradient of source rocks in eastern China is normally higher than that of western China. Each group of source rock

TABLE 2. *Elemental composition of kerogen of different source rocks*

Group	Location	Epoch		C (%)	H (%)	O (%)	Ratio of atom H/C	O/C	Type kerogen
Group A	Songlio Basin	Cretaceous	K	79–75	9.21	3.61	1.39	0.03	I
			K	72–40	8.06	7.65	1.34	0.08	II
			K	63–80	4.19	17.09	0.79	0.20	III
Group B	Liohe W.D.	Palaeogene	E				1.22–1.31		I
	Central Hebei D. Langgu Sag	Palaeogene	E				1.28	0.12	II
		Palaeogene	E				0.82	0.14	III
	Jiyang D. Zhanhua	Palaeogene	E				1.28–1.41	0.15–0.29	I_B+III
	Dongying	Palaeogene	E	55.5–62.8	5.18–5.76	6.31–11.1	1.07–1.2	0.07–0.13	II
	Danwanzhan	Palaeogene	E	61–61	5.45	22.34	1.06	0.27	III
	Biyang Sag	Palaeogene	E	66–76	6.63	14–68	1.19	0.16	II
		Palaeogene	E	65–73	8.98	13–34	1.63	0.15	I_A
		Palaeogene	E	76–55	8.85	12–91	1.37	0.13	II
Group C	Jianghan Guang*hust	Palaeogene	E				1.16	0.19	II
	Central Hebei	Palaeogene	E				1.50–1.25	0.1–0.2	II_A
	Chaidamu Manya D.	Neogene	N	51–12	3.36	21–54	0.79	0.32	III
		Neogene	N	69–97	3.60	12–56	0.62	0.13	III
		Neogene	N	67–58	4.26	11–75	0.76	0.13	III
Group D	Zhungeer Basin	Permian	P	78–85	8.98	3.66	1.35	0.05	I
			C	78–41	7.17	5.72	1.09	0.05	II
Group E	Hebel Dongpu	Carboniferous to	C–P	78–75	5.10	7.52	0.78	0.10	III
	Hebei Suqiao			74–66	7.27	12–38	1.16	0.12	II
	Zhejiang	Permian		74–44	6.98	7.71	1.12	0.08	

has its own burial history. For example, that of the Group A source rocks of the Songlia Basin is quite simple but that of the Group B and C source rocks of the fault-bounded basins is complicated. Sedimentation rates in the Palaeogene fault-bounded lake basins, including the salt lake basins of eastern China were high, up to 0.2–0.3 mm per year. The average geothermal gradient of six major oil- and gas-bearing basins in the Beijing-Tianjing region is about 3.7°C/100 m. The geothermal gradient in salt lake basins (e.g., the Jianghan Basin) averages 3.1°C/100 m, but

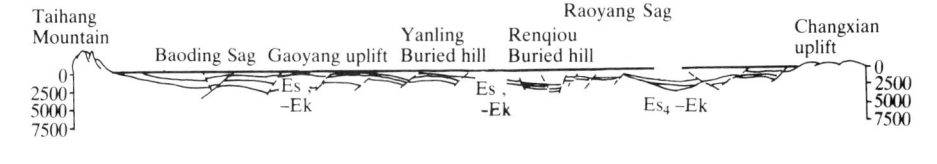

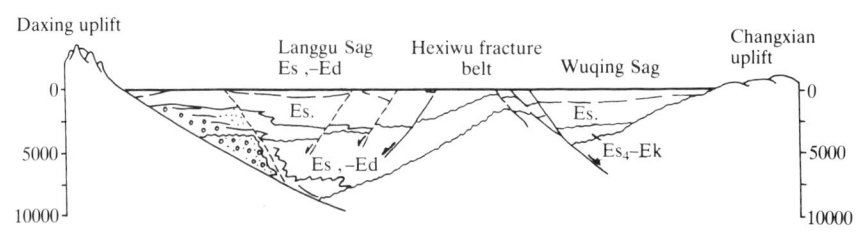

FIG. 2. Pre-Neogene palaeotectonic cross profile of the Raoyang and Langgu Sags, showing the different distances of the two sags from nearby uplifted areas (Group A source rocks).

varies from 2.3°C/100 m (main area of salt beds) up to 3.7°C/100 m (area with almost no salt). Depth to OGT (oil generation threshold), taken as 90–95°C, in fault-bounded basins is generally about 1700–2700 m but the Songlia Basin also has a very high geothermal gradient, with OGT at only 1330 m and a temperature of 66°C.

Main biomarker composition characteristics of major source-rock groups

Group A and Group B

Both Group A and Group B source rocks have similar biomarker compositions. Normal alkanes have wide ranges (up to C_{40}), and bimodal distributions with main peaks at C_{17}, C_{19} and C_{22}, C_{23} or C_{27}. High molecular weight normal alkanes of crude oils sometimes show odd-even predominance, such as OEP values of 1.14–1.29 for oils from the Donying Depression. The main components of the isoprenoid alkanes are pristane and phytane, having very low amounts of isoprenoid alkanes of lower molecular weight than pristane and rare amounts of isoprenoid higher than phytane. The $iC_{18}+iC_{16}+iC_{15}/Pr+Ph$ ratio is usually less than 1, and even as low as 0.2–0.5 for low maturity oils. The Pr/Ph ratios are near or greater than unity for most oils. C_{27}–C_{29} steranes and C_{27}–C_{29} diasteranes are common components of the source rocks and crude oils. 4-methylsteranes and 4-methyldiasteranes are also sometimes found. The distribution of C_{27}–C_{29} steranes varies depending on organic input. Although most crude oils have sterane isomerization ratios reaching their isomerization equilibrium values, low maturity oils often have very low isomerization values, such as 20S/(20S+20R)-5α(H), 14α(H), 17α(H)C_{29} sterane ratios of 9–23% for oils from the Langgu Sag. In addition to hopanoid alkanes (Wang Xieqing & Lang Rengchi 1980), many triterpanes of high plant origin have been detected, including onocerane, oleanane and lupane etc. Oils of the Eerduosi Basin are rich in 18α(H) oleanane, for instance the T_{E3-4} oil of well Ling-55 contains equal amounts of oleanane and $C_{30}αβ$ hopane, which possibly originate from a higher plant precursor: oxygen containing betulinol (Shang Huiyun et al. 1982). Gammacerane, a triterpane originating from aquatic organisms, has also been detected in crude oils and source rocks, such as the Palaeogene oils of the Shengli oil field and the Biyang Depression. Porphyrins detected in oils of the central Hebei and Jiyang Depressions are almost entirely Ni-porphyrins. The highest content of Ni-porphyrins, up to more than 500 ppm, has been analysed in Palaeogene Gaosheng oil from the Liaohe oil field. In addition to porphyrins detected in oils, chlorin and unknown red organic pigments have been found in the source rocks of the Shengli, Renqiu and Changqing fields etc. (Sheng Guoying & Fu Jiamo 1984).

It is not easy to differentiate Group A and Group B source rocks on the basis of their biomarker contents. Group A source rocks generally contain less gammacerane and triterpanes of higher plant origin than Group B, while Group B is very rich in biomarkers, especially hopanoid and steroid hydrocarbons (Fig. 3). Typical examples of oils from Group A source rocks are those of the Songlia Basin which originated from plankton algae of deep lake facies and contain no oleanane or gammacerane. Typical examples for Group B source rocks are those of the Langgu Depression, which contain Type II kerogen (Zhang Zhencai et al. 1983) and those in the Biyang Depression which contain Type I kerogen of bacterial origin. Type I kerogen of the Biyang Depression was shown to contain triter-

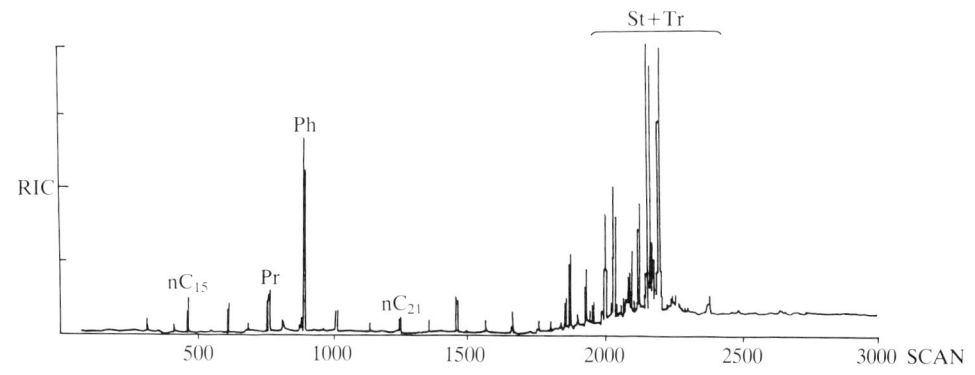

FIG. 3. RIC trace of An-3 dark mudstone of Group A source rocks. Pr: pristane; Ph: phytane; St+Tr: steranes and triterpanes; nC_{15}: C_{15} normal alkane; nC_{21}: C_{21} normal alkane.

panes of high plant origin, such as oleanane, spirotriterpane, taraxstane, serratane etc. (Zhen Xianzhang unpublished data; Kimble *et al.* 1974). According to biomarker quantitation (Rullkötter *et al.* 1984), C_{30} hopane reaches up to 8.04% of the total alkane fraction in the An-3 mudstone (Group B source rocks) (Fig. 3). Botryococcane homologues have been found in the Maoming oil shales (Brassell *et al.* 1986 and this volume).

Group C

Group C source rocks and their oils often show even–odd predominance of n-alkanes (Sheng Guoying *et al.* 1980), very high concentrations of phytane, relatively high gammacerane/$C_{30}\alpha\beta$ hopane ratio (Sinninghe Damsté *et al.* 1986) and abundant metal, mainly nickel, porphyrins. According to a recent study (Fu Jiamo *et al.* 1986), Pr/Ph and gamacerane/$C_{30}\alpha\beta$ hopane ratios for a suite of Jianghan samples have been measured to be 0.05–0.46 and 0.35–2.16 respectively (Fig. 4). In addition to C_{27}–C_{29} steranes, there are also less abundant 4-methyl steranes (C_{28}–C_{30} compounds, possibly from dinoflagellates, Boon *et al.* 1979; Robinson *et al.* 1984) as well as pregnane and homopregane present. In one sample collected from the south of the Jizhong Basin, a whole series of low molecular weight steranes (C_{20}–C_{26}, except the C_{25} compound) have been detected, showing pregnane as the main peak of the alkane fraction (Fig. 5). Occasionally Group C samples have an unusual distribution of hopanoid hydrocarbons, that is, the content of C_{35}, C_{34} homohopanes is equal to those of $C_{30}\alpha\beta$ hopane and higher than those of C_{31}, C_{32} and C_{33} homopanes (Fig. 4).

The biomarker composition of the aromatics is quite unusual (Sinninghe Damsté *et al.* 1986). The major components are sulphur-containing compounds rather than simple hydrocarbons, comprising mainly alkyl-thiolane (Fig. 6), with less abundant sulphur-containing steroids and hopanoids. For example, the aromatic fractions of two oils are composed of five series of alkyl-thiophenes ($C_nH_{2n-4}S$) with base ion m/z 101, 115, 129, 143 and 157 (Fig. 6). Their specific structures and their distribution patterns show a partial similarity with the corresponding alkanes, having n-alkyl and isoprenoid thiophenes and thiolanes, even–odd predominance within ranges of C_{12}–C_{30} and a main peak at C_{20} (Sheng Guoying *et al.* 1986). Jianghan oils, especially oils produced from shallow depths, often contain large amounts of metal porphyrins, mostly Ni-porphyrin (Yang Zhiquiong *et al.* 1983), in which an unusual compound: C_{32}DPEP with one six member ring has been detected recently by Xu Fengfang *et al.* (1985).

Group D

Group D source rocks and related crude oils show normal alkane distributions of algal origin, abundant isoprenoid alkanes and, rarely, high abundances of carotane and its derivatives. A

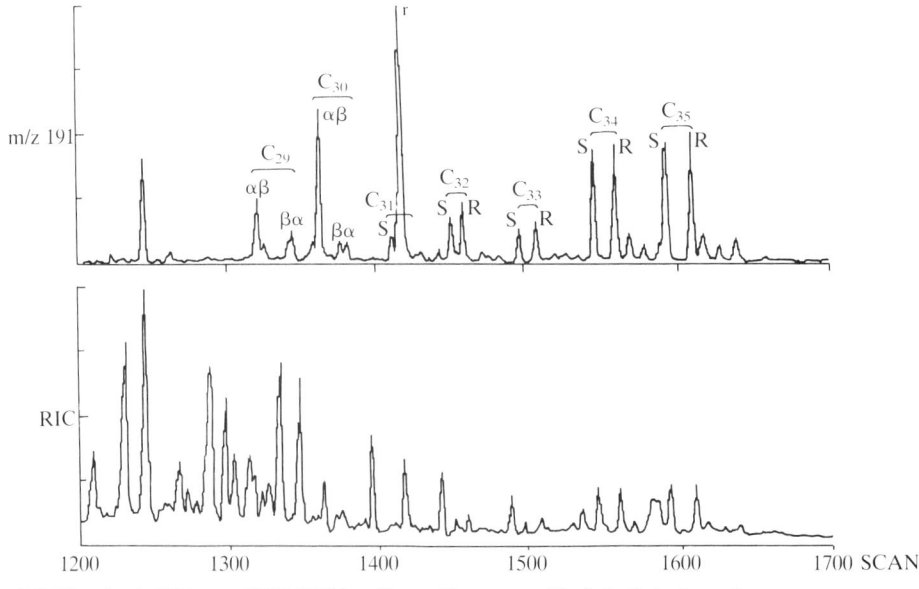

FIG. 4. RIC and m/z 191 trace of E21 (1704 m, Group C source rock) aliphatic hydrocarbons. r = gammacerane, $C_{30} = C_{30}$ hopane etc., $\alpha\beta = 17\alpha(H)$, $21\beta(H)$ hopane, $\beta\alpha = 17\beta(H)$, $21\alpha(H)$ hopane, S = 22S $17\alpha(H)$, $21\beta(H)$ hopane, R = 22R $17\alpha(H)$, $21\beta(H)$ hopane.

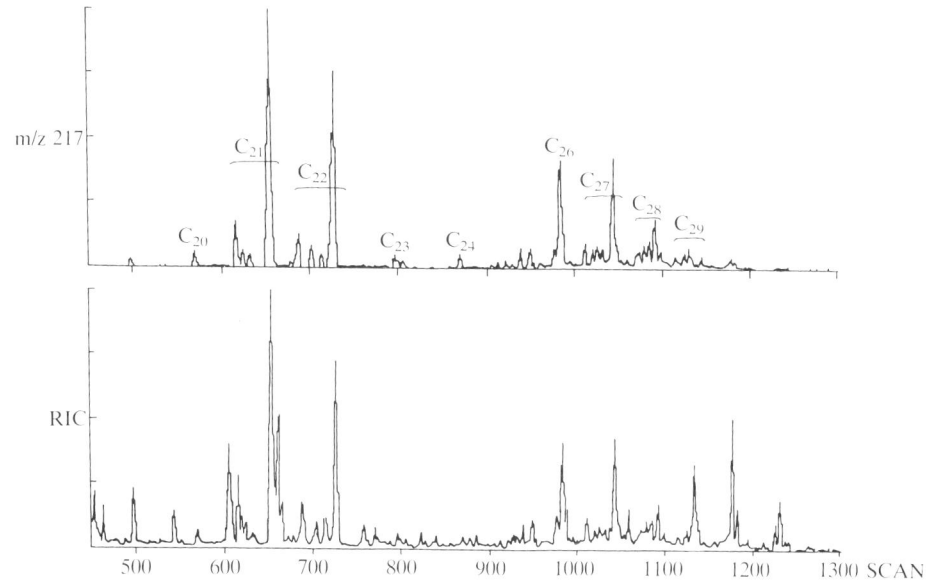

FIG. 5. RIC and m/z 217 trace of CHAO-2 (3015 m, Group C source rock) aliphatic hydrocarbons. C_{20}–C_{29}: C_{20}–C_{29} steranes.

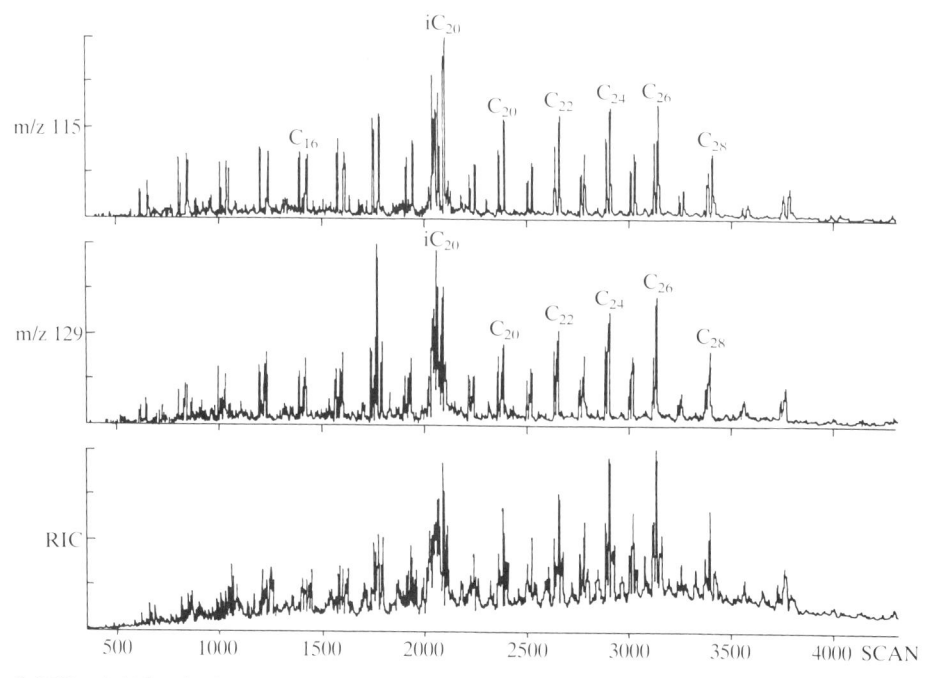

FIG. 6. RIC, m/z 115 and m/z 129 trace of aromatic fraction in J18 oil, Jianghan oil field, showing the distribution of alkylthiolanes.

wide range distribution of n-alkanes in the oils with main peaks at C_{17}, C_{19} and C_{20}, as well as odd–even predominance at C_{15}, C_{17} and C_{19} indicate an algal or bacterial precursor. About 20% of saturated hydrocarbons are composed of isoprenoids, such as 25.8% of the Feng 3 oil from the Zhungeer Basin in which the most abundant compounds are norpristane, pristane and phytane. Also low molecular weight isoprenoids, iC_{13}–iC_{18}/iC_{13}–iC_{20} are very abundant (up to 67–79.5%). Some high molecular weight isoprenoids of iC_{21}–iC_{30}, even lycopane ($C_{40}H_{82}$) have been detected in certain oils. In the saturated alkane chromatograms or TLC traces of the oils generated from this group of source rocks, β-carotane ($C_{40}H_{86}$) and sometimes γ-carotane can be seen clearly (Fig. 7). The mass chromatogram of m/z 125 indicates a whole series of carotane homologues, including C_{13}–C_{40} carotane homologues, in the Feng 3 oil (Jiang Shusheng 1983).

Group E

Group E source rocks have abundant biomarkers of higher plant origin, such as diterpanoid hydrocarbons. According to our recent primary research results on Fushan Palaeogene resinite, the aliphatic fraction is composed mainly of diterpenes and diterpanes in the range of C_{17}–C_{23}. The most important compounds are C_{19}–C_{20} diterpenes and diterpanes, with many isomers, having skeletons of abietic acid and pimaric acid (Fig. 8). Many compounds have been identified, including abietane, fichtelite and iosene etc. In addition, a number of aromatic sesquiterpenoid and diterpanoid compounds have been detected, i.e., 5, 6, 7, 8-tetrahydrocadalene, 13-methylpodocarpa-8, 11, 13-triene, 19-norabieta-8, 11, 17-triene and abieta-8, 11, 17-triene etc. (Simoneit & Mazurek 1982).

Some problems relating to the evaluation of oil and gas prospects with terrestrial source rocks

Origin of Type I kerogens and their hydrocarbon potential

Owing to variation in sedimentary and diagenetic environments, as well as organic input, various groups of source rocks have their own characteristics even though the kerogen types are similar. For example organic matter of Group E source rocks can be divided into Type I, Type II and Type III kerogens as usual, but their origins and characteristics are quite different from that of Group A to Group D source rocks. This is shown by Type I kerogen which has been studied in detail from Group A, Group B and Group E source rocks. Type I kerogen from the Songlia Basin is considered to originate from a planktonic or algal precursor deposited in a deep lake facies, while that from Group E relates to maceral of higher plant origin, such as resinite. Type I kerogen of the Biyang Basin originated from mixed precursors, showing evidence of bacterial

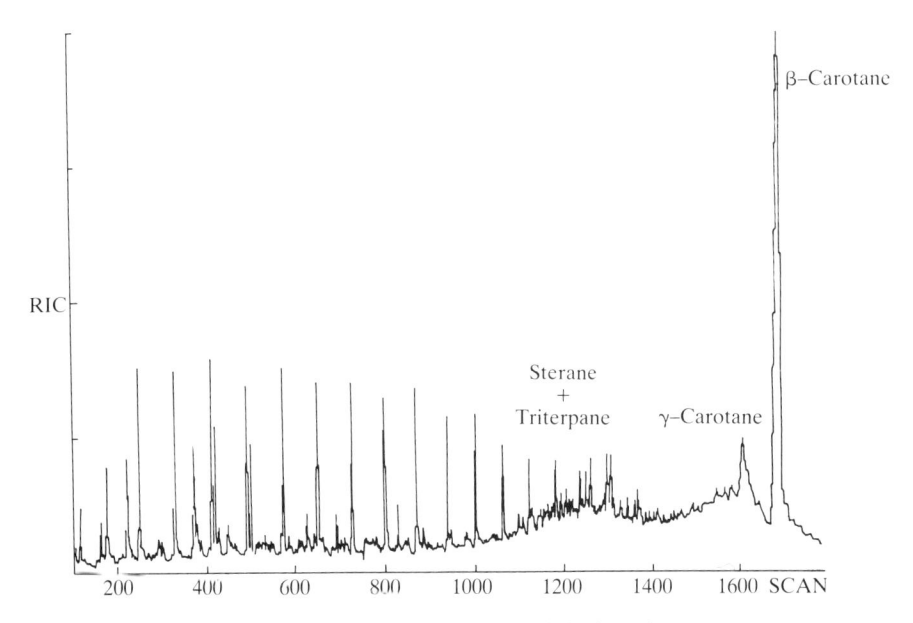

FIG. 7. RIC of Z-114 oil produced by Group D source rocks aliphatic hydrocarbon.

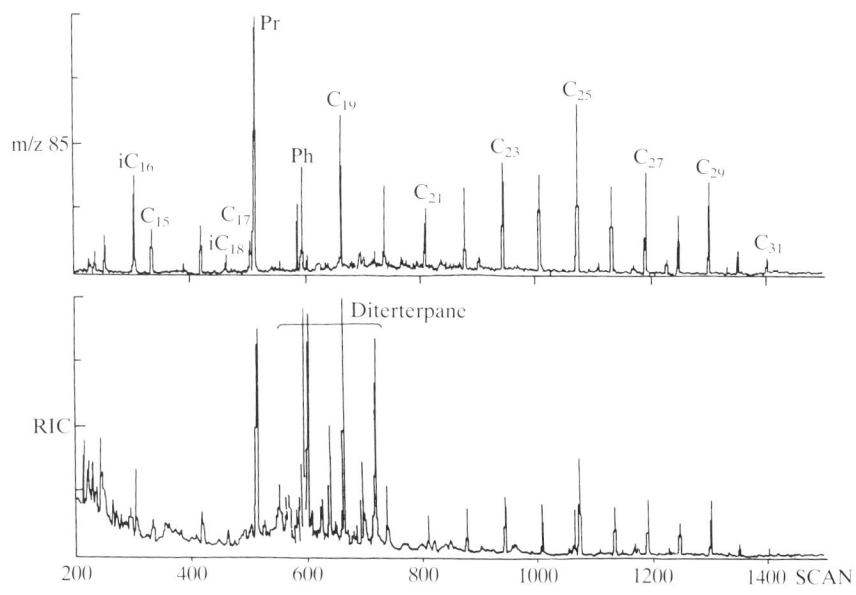

FIG. 8. RIC and m/z 85 trace of Fushun Palaeogene resinite (Group E source rock). Pr: pristane; Ph: phytane; C_{15}:C_{15} normal alkane.

degradation and a certain amount of higher plant input. All of these Type I kerogens have good potential for hydrocarbons, but differ in the relative amounts of gases and liquid hydrocarbons which they can generate, as well as in their compositions.

Immature oils

A number of low maturity or even immature oils have been found in lacustrine and salt lake sedimentary formations of Palaeogene fault-bounded basins in eastern China (Shi Jiyang *et al.* 1982; Fu Jiamo *et al.* 1983). Immature oils usually occur in shallow strata, above the OGT. Because of the low maturity, the content of saturated hydrocarbons in the oils is very low, usually less than 30%. Biological markers can be used to measure the maturity of the oils. The biomarker isomerization values of immature oils are much less than those of commonly occurring oils, and far from equilibrium values. As shown by recent studies conducted by Mackenzie *et al.* (1984) on the maturity of North Sea oils, the sterane ratios of all oils studied have reached their isomerization equilibrium values, i.e. equal or more than 50–55% for 20S/(20S+20R) C_{29} sterane, 65–75% for $\beta\beta/(\alpha\alpha+\beta\beta)$ C_{29} sterane. Sterane ratios of low mature or immature oils from Groups A and B source rocks are much less than those of North Sea oils. The 20S/(20S+20R) C_{29} sterane ratios for oils from the Gudo (Yi-18) and Changqing oil fields (Linshen-2) are as low as 11.5–17%. A commercial immature oil has recently been studied in detail from Jianghan oil-bearing Basin (Fu Jiamo *et al.* 1986). Geological and geochemical data, including molecular ratios and quantitative biomarker data, show that the maturity of the oil can be considered equivalent to about R_0 0.45%. The biomarker character of the oil is quite similar to that of immature source rocks, showing high contents of phytane and other biomarkers, and low ratios of 20S/(20S+20R) $C_{29}\alpha\alpha\alpha$sterances, 22S/(22S+22R) $C_{32}\alpha\beta$hopanes and DPEP/DPEP+ETIO porphyrins.

It would be especially interesting to study the OGT of source rocks of Palaeogene fault-bounded basins in eastern China in more detail, and to understand the commercial prospects and accumulation conditions of immature oils. This may result in a new frontier for petroleum exploration in eastern China and the Chinese continental shelf in future.

Coal-bearing formations as source rocks

Recently coal has been reported as an important potential source for crude oils in Canada and Australia (Snowdon & Powell 1982). An example of similar significance has been discovered and studied in the Suqia region, North China (Liu Dehan *et al.* 1985). Based on fluorescent microscopy studies, analyses of organic compositions and simulation experiments on many core samples and cuttings, it was found that Carboniferous–Permian coal-bearing formations in the Suqia region have intervals rich in spores, cuticles, resins and bark. These intervals have large hydrocarbon potentials, with high S_1+S_2 yields up to about 200 kg/gC, high yields of liquid hydrocarbons, of 73.5 mg/g or more, and of natural gases up to 184 ml/g. Geological and geochemical data suggest that they are important sources of the natural gases and condensates found in the region.

Conclusions

Several preliminary conclusions can be drawn concerning the organic geochemical characteristics of the major types of terrestrial petroleum source rocks in China.

(1) In terms of the type and nature of the oil and gas-bearing basins, five major groups of terrestrial source rocks have been identified: Group A, sedimentary formations of large lake basins in the interior of the Chinese plate; Group B, lacustrine clastic formations in fault-bounded basins; Group C, salt lake, evaporite-clastic formations in fault-bounded basins; Group D, lagoonal-lacustrine, volcano-clastic formations in intermontane basins; and Group E, paralic coal-bearing formations of the platform.

(2) The type and abundance of organic matter are related to the source rock types. They are also related to the sedimentary facies in lacustrine basins, particularly the distance of a depression from the area of uplift and the occurrence of the source rocks within a certain sag. The main factors controlling the thermal evolution of the organic matter are palaeogeothermal gradient and unconformities.

(3) The main biomarker compositions of the source-rock groups have been discussed briefly. Group A source rocks and related oils are characterized by an absence or low abundance of gammacerane and triterpanes of higher plant origin. Group B source rocks and oils have very high concentrations of biomarkers, especially hopanoid and steroid hydrocarbons, with triterpanes of higher plant origin. Group C source rocks have quite high contents of phytane, showing high ratios of Pr/Ph, gammacerane/$C_{30}\alpha\beta$hopane, even odd predominance of n-alkanes and unusual sulphur-containing compounds. Group D source rocks contain abundant isoprenoid alkanes and carotanes, including β-carotane and γ-carotane. Group E source rocks and oils are quite rich in biomarkers of higher plant origin such as diterpanoid hydrocarbons with abietic and pimaric acid skeletons.

(4) Several problems are of special significance in the evaluation of petroleum prospects in the terrestrial sedimentary basins in China. The organic matter of different source-rock groups may contain similar kerogen, e.g. Type I kerogen, but have different characteristics and hydrocarbon potentials. A number of low mature or even immature oils have been found in lacustrine and salt lake sedimentary formations of Palaeogene fault-bounded basins in eastern China. Studies of the commercial prospects and accumulation conditions of these immature oils may result in a new frontier for petroleum exploration in eastern China and the China continental shelf. A new natural gas and condensate field has been explored recently and a preliminary study on related liptobiolites has been conducted to confirm the hydrocarbon potential of coal-bearing formations.

ACKNOWLEDGEMENTS: The authors acknowledge members of Organic Geochemistry Unit, University of Bristol U.K. and Organic Geochemistry Division, Institute of Geochemistry, Academia Sinica, especially Professor G. Eglinton, Dr S. C. Brassell, Mrs A. P. Gowar, Dr Zhou Zhongyi, Dr Wang Benshan and Mrs Ye Jisun for assistance with the analytical work, supplying unpublished data, helpful discussion and in the preparation of the manuscript. UNESCO and UNDP are thanked for partial financial support (CPR/84/005). We are also grateful to Drs A. J. Fleet, N. J. L. Bailey and P. J. D. Park for reviewing and assistance in preparation of this paper.

References

BOON, J. J., RIJPSTRA, W. I. C., DE LANGE, F., DE LEEUW, J. W., YOSHIOKA, M. & SHIMIZU, Y. 1979. Black Sea sterol—A molecular fossil for dinoflagellate blooms. *Nature*, 277, 125–127.

BRASSELL, S. C., EGLINTON, G. & FU JIAMO 1986. Biological marker compounds as indicators of the depositional history of the Maoming oil shale. *In*: LEYTHAEUSER, D. & RULLKÖTTER, J. (eds), *Advances in Organic Geochemistry, 1985*, Pergamon, Oxford, 927–941.

FU JIAMO & JIA RONGFEN 1984. Existing forms of dispersed organic matter in carbonate rocks, their evolution characters and relationship to evaluation of carbonate source rocks. *Geochimica Sinica*, 1–9.

——, WANG BENSHAN, SHI JIYANG, JIA RONGFEN & SHENG GUOYING 1983. Evolution of organic matter and origin of sedimentary ore deposits — (1) Origin and evaluation of oil and gas. *Acta Sedimentologica Sinica*, **1**, 40–59.

——, SHENG GUOYING & JIANG JIGANG 1985. Immature oil generated from a saline deposit-bearing basin. *Oil and Gas Geology, Jiangling*, **6**, 150–158.

——, ——, PEN PINGAN, BRASSELL, S. C., EGLINTON, G. & JIANG JIGANG 1986. Peculiarities of salt lake sediments as potential source rocks in China. *In*: LEYTHAEUSER, D. & RULLKÖTTER, J. (eds), *Advances in Organic Geochemistry, 1985*, Pergamon, Oxford, 119–126.

HUANG DIFAN & LI JINCHAO 1982. *Formation of Oil-pools in Continental Facies, China*, Ganshu People Press, Lanzhou.

JIANG SHUSHENG 1983. Carotane and its geochemical characteristics in crude oil, Kalamayi oil field. *Oil and Gas Geology, Jiangling*, **4**, 151–159.

KIMBLE, B. J., MAXWELL, J. R., PHILP, R. P. & EGLINTON, G. 1974. Identification of steranes and triterpanes in geolipid extracts by high-resolution gas chromatography and mass spectrometry. *Chemical Geology*, **14**, 173–198.

LIU DEHAN, FU JIAMO, QIN JIANZHONG & SHI XIHUI 1985. The discovery of liptobiolites in Suqia region—on the determination of coal-generating gases and coal-generating oil and their prospects. *Geochimica Guiyang*, **4**, 313–322.

MACKENZIE, A. S., MAXWELL, J. R., COLEMAN, M. L. & DEEGAN, C. E. 1984. Biological marker and isotope studies of North Sea crude oil and sediments. *Proceedings of the 11th World Petroleum Congress*, John Wiley & Sons, New York, 45–56.

ROBINSON, N., EGLINTON, G., BRASSELL, S. C. & CRANWELL, P. A. 1984. Dinoflagellate origin for sedimentary 4α-methylsteroids and 5α(H)-sternols. *Nature*, **308**, 439–442.

RULLKÖTTER J., MACKENZIE, A. S., WELTE, D. M., LEYTHAEUSER, D. & RADKE, M. 1984. Quantitative gas chromatography–mass spectrometry analysis of geological samples. *Organic Geochemistry*, **6**, 817–827.

SHANG HUIYUN, TONG YUYING, JIANG NAIHUANG & ZHU BAOQUAN 1982. Geochemical characteristics of steranes and triterpanes in Erduosi basin. *Petroleum Exploration and Development*, **2**, 19–25.

SHENG GUOYING & FU JIAMO 1984. A novel kind of fossil pigment found in sediments, China. *Scientia Sinica (Series B)*, 73–79.

——, FAN SHANFA, LIU DEHAN, SU NENGXIAN & ZHOU HONGMING 1980. The geochemistry of n-alkanes with an even-odd predominance in Tertiary Shahejie formation of Northern China. *In*: DOUGLAS, A. G. & MAXWELL, J. R. (eds), *Advances in Organic Geochemistry, 1979*, Pergamon, Oxford, 115–123.

——, FU JIAMO, BRASSELL, S. C., EGLINTON, G. & JIANG JIGANG 1986. Long chain alkyl-thiophene compounds found in sulphur-high crude oil from a hypersaline basin. *Geochimica Sinica*, **2**, 138–146.

SHI JIYANG, MACKENZIE, A. S., ALEXANDER, R., EGLINTON, G., GOWAR, A. P., WOLFF, G. A. & MAXWELL, J. R. 1982. A biological marker investigation of petroleums and shales from the Shengli oilfield, the People's Republic of China. *Chemical Geology*, **35**, 1–31.

SIMONEIT, B. R. T. & MAZUREK, M. A. 1982. Organic matter of the troposphere—II. Natural background of biogenic lipid matter in aerosols over the rural western United States. *Atmospheric Environment*, **16**, 2139–2159.

SINNINGHE DAMASTÉ, J. S., TEN HAVEN, H. L., DE LEEUW, J. W. & SCHENCK, P. A. 1986. Organic geochemical studies of a Messinian evaporitic basin, northern Apennines (Italy)—II. Isoprenoid and n-alkyl thiophenes and thiolanes. *In*: LEYTHAUSER, D. & RULLKÖTTER, J. (eds), *Advances in Organic Geochemistry, 1985*, Pergamon, Oxford, 791–805.

SNOWDEN, L. R. & POWELL, T. G. 1982. Immature oil and condensate-modification of hydrocarbon generation model for terrestrial organic matter. *Bulletin of the American Association of Petroleum Geologists*, **66**, 775–778.

SWAIN, F. M. 1980. Petroleum in continental facies. *In*: MASON, J. F. (ed.), *Petroleum Geology in China*, Pennwell Books, Tulsa, 1–25.

THOMAS, B. M. 1981. Land plant source rocks for oil and their significance in Australian basins. *Australian Petroleum Exploration Association Journal*, **22**, 164–178.

TISSOT, B. P. 1984. Recent advances in petroleum geochemistry applied to hydrocarbon exploration. *Bulletin of the American Association of Petroleum Geologists*, **68**, 545–563.

—— & WELTE, D. H. 1978. *Petroleum Formation and Occurrence*, Springer-Verlag, New York.

WANG XIEQING & LANG RENGCHI 1980. Biological marker hydrocarbons in source rock and crude oil of some Chinese basins. *Acta Petrolei Sinica*, **1**, 52–62.

XU FENFANG, SHENG GUOYING, FU JIAMO, EVERSHED, R. P. & JIANG JIGANG 1985. A new kind of deoxophylloerythroetioporphyrin found in crude oil from the gypsum-salt environment. *Geochimica Sinica*, **4**, 358–362.

YAN WANLI & LI YONGKANG 1982. Type and evolution model of continental kerogen in the Songliao basin. *Scientia Sinica (Series B)*, **25**, 304–317.

YANG ZHIQUIONG, TONG RUYING & FAN ZHAOAN 1983. Biological marker compounds of source rocks and crude oils from Jianghan salt basin and their geological significance. *Oil and Gas Geology, Jiangling*, **4**, 269–282.

ZHANG ZHENICAI, WANG ZHONGRAN, CHEN XIANGHONG & ZHEN XIANGHAN 1983. Steranes and terpanes from crude oil and source rocks in Jizhong depression. *Oil and Gas Geology, Jiangling*, **4**, 89–99.

FU JIAMO, SHENG GUOYING & LIU DEHAN, Institute of Geochemistry, Academia Sinica, Guiyang, Guizhou Province, The People's Republic of China.

Characteristics of Mesozoic and Cenozoic non-marine source rocks in north-west China

Luo Binjie, Yang Xinghua, Lin Hejie & Zheng Guodong

SUMMARY: The formation and distribution of Mesozoic and Cenozoic lacustrine source rocks in north-west China is determined by climatic factors and the tectonics of terrestrial sedimentary basins. The most suitable condition for the formation of non-marine source rocks was a humid to semi-humid climate and a deep to intermediate lacustrine basin during the expansion stage of the lakes. Dominantly fresh to brackish water bodies with sluggish hydrodynamic bottom environments were conducive to the accumulation and preservation of organic matter. The non-marine source rocks show low to high organic matter contents, mainly from a mixture of terrestrial and aquatic sources. Thus, they are characterized by high wax, low sulphur, high nickel and low vanadium. They also have a series of corresponding organic-geochemical signatures. All of these are important indicators in studies of terrestrial source rocks, crude oil and oil-source correlations. A chart with the $\delta^{13}C$ value of the crude oil against the log plot of pristane/phytane values can be used to discriminate the environment of deposition of the source rock of the oil.

It is well known that the Mesozoic–Cenozoic petroleum in north-west China mostly formed in continental strata (e.g. Yang Wanli et al. 1985). Palaeolakes are important sites for generating oil and gas, as are palaeolake-bog complexes. The distribution and characteristics of non-marine source rocks depend on the climate, generation and development of basins, sedimentary processes and hydrodynamic conditions. These factors influence the types of organic source material and hydrocarbon formed. The main aspects of non-marine source beds have been given in Luo Binjie et al. (1981) and Fan Pu et al. (1980). This paper provides a short summary of present views without repeating data.

Distribution of terrestrial source rocks in north-west China

Mesozoic–Cenozoic continental basins

The Zhong-Chao Palaeoland, and Chaidamu and Talimu Platforms were situated between the Asia–Europe and Gondwana Plates. These continental pieces grew and then coalesced after the Caledonian and Hercynian movements (Lanzhou Institute of Geology, 1982). Differential basin subsidence was controlled by NW, NE, NS and EW faults related to the Sino-Indian movement. As a result, a series of rifts, tilted basins, platform-sags and intermontane basins formed with various tectonic styles, lithologies and degrees of consolidation of the strata. During steady subsidence of the basins, the lakes formed and terrestrial source rocks were deposited. The age of the strata depends on the time of the formation of the basins.

The oldest continental basin in north-west China is the Permian Zhungeer Basin. Palaeolakes and palaeobogs developed in the Permian, Triassic, Jurassic and Tertiary periods. Four main sequences of source rocks were correspondingly deposited. Lakes developed in the Shanxi–Gansu–Ningxia Basin in the Triassic period. Lake-bogs developed, and corresponding source rocks formed in the Jurassic and Cretaceous periods. The Tulufan, Chaidamu, Jiuquan, Chaoshui and Minhuo Basins contained lakes from the Jurassic period and the main source rocks of the Jiuquan Basin formed in the Cretaceous. Lakes and bogs of the Chaidamu Basin developed in the Tertiary and Quaternary periods with corresponding source rocks. (Fig. 1).

Climate control

The palaeoclimate in north-west China underwent a cyclic evolution from the Permian to Quaternary. The distribution of organic matter was controlled by these palaeoclimatic cycles. We can conveniently divide the period into five general palaeoclimatic intervals. Within the first, from the Carboniferous–Permian humid phase to the Early–Middle Triassic semi-humid phase, the source rocks of the Permian series developed. Within the second interval, from the Late Triassic semi-humid phase to the Early–Middle Jurassic humid phase and then to the Late Jurassic semi-arid phase, source rocks formed during the Late Triassic and Middle Jurassic. Within the third interval, from the Early Cretaceous semi-humid phase to the Palaeocene and Eocene arid phase,

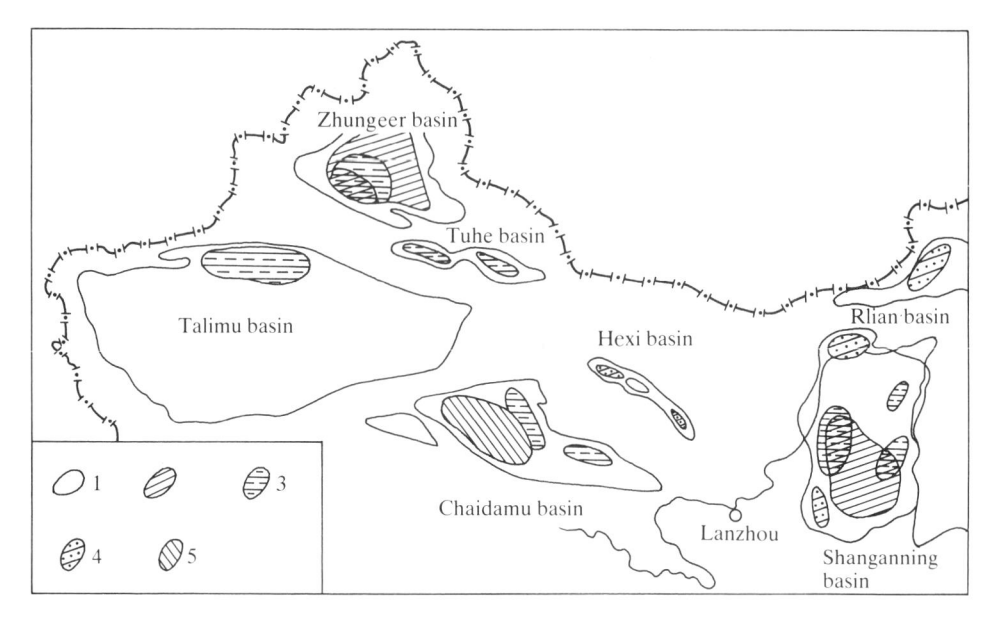

FIG. 1. Distribution of non-marine basins in Northwest China. 1. Oil–gas basin; 2. T_3–depression; 3. J_3–depression; 4. K–depression; 5. E–depression.

Early Cretaceous source rocks were deposited. Within the fourth interval, from the Oligocene semi-humid phase to the Miocene arid phase, source rocks formed mainly in the Oligocene epoch. The fifth climatic interval was short; spanning the Pliocene semi-humid phase to the Holocene arid phase, during which time natural gas was generated in the Pleistocene of Chaidamu Basin. It should be pointed out that the first and second intervals were only of regional character.

Sedimentary environment of source rocks

Salinity of the lakes

The source rocks of north-west China were deposited under a variety of conditions. In the Zhungeer Basin, they formed in a freshwater environment during Permian, Middle–Late Triassic and Middle–Late Jurassic times. They were deposited under fresh–brackish conditions during the Jurassic in the Talimu Basin and during the Late Cretaceous in the Jiuxi Basin. However, during the formation of Tertiary source rocks in the Chaidamu Basin the lake environment was saline. The non-marine source rocks of north-west China are therefore mostly fresh–brackish in origin though salinities tend to increase from Mesozoic to Cenozoic times. The chloride ion content of the source rocks is, in general, less than 0.02% for those of Permian, Early Triassic and Late Jurassic age. Middle–Early Jurassic to Early Cretaceous source rocks average 0.02–0.03%, but are locally 0.03–0.10%. The Tertiary source rocks formed in the brackish and salt-water environment and have chloride ion contents of 0.06–0.55%. The growth of aquatic biota and the properties of the deposited sedimentary organic matter were influenced by the salinity of the lakes (Scientific Research Institute, 1982).

Lake depth

In general, the evolution of the palaeolakes in the north-west of China shows a progressive sequence of generation, development, expansion, contraction and extinction. In this process the lake body changed from small to large and then back to small; the lake depth from shallow to deep, then to shallow again; the sediments from coarse to fine, then to coarse again; the colour of the sediments from pale to dark grey, then to pale grey again. The source rocks developed mostly in the middle period of the sedimentary cycles in the deep or intermediate stages of lake expansion, while oil shale and coal layers formed in the lake-bogs which were present in the early and late stages of sedimentary cycles. For example, the Early Cretaceous of the Jiuxi Basin contains deep-intermediate lacustrine deposits formed as the size of the lake increased and the basin steadily subsided. The lithology is mostly black and grey-green mudstone, shale and marl (accounting for 80% of the strata) with plant fragments and abundant estherites and ostracodes. It is the main source-rock section in the basin, with a thickness of 800–1000 m.

New lakes followed the extinction of earlier lakes as a result of later tectonic movements and palaeoclimatic variation. Lakes with multiple sedimentary cycles occurred because of frequent tectonic movements and gave rise to multiple generations of source rock–reservoir–cap rock sequences in the Mesozoic–Cenozoic. For example, in the Zhungeer Basin, the source rocks of the Upper Permian are deep and intermediate black and grey-green lacustrine mudstones with oil shale. The Upper Triassic Baijiantan Group is grey-green conglomeratic mudstone with pelecypoda and plant fossils which may have the source potential. The Middle Jurassic Shangonghe Group is a lake-bog facies which consists of grey and yellow-green silt rocks and silt-bearing mudstone and makes up the source rocks developed in the southern depression of the basin. The Oligocene Wulungu Formation is a relatively good reservoir containing grey-green lacustrine sandstone and mudstone with shelly limestone.

The Shanxi–Gansu–Ningxia Basin which covers an area of 230 000 km^2 is the second largest sedimentary basin in China. The Upper Triassic strata (T_3Y_2–T_3Y_3) of the basin represent the expanded stage of a lake. They consist of an enormous volume of thin bedded, dark mudstones, with abundant organic matter, which in places are oil shales. They are deep and intermediate lacustrine deposits and the main source rocks of the basin. The other principal source rocks of the basin are Middle–Lower Jurassic and Upper Cretaceous in age; there are also lake-bog source rocks.

Hydrodynamic stability

Studies of the hydrodynamics and sediments in the modern Qinghai Lake (Fig. 2), which is 27 m deep and brackish (Luo Binjie 1980) show that the organic carbon and nitrogen contents of the sands and silts of the high energy, shallow marginal area are less than 1% and 0.1% respectively. In the mud and silt-sludge of the deep-water, low energy, central area, they are more than 2% and 0.2% respectively. The high-energy belt extends into the mouths of the main rivers which feed the lake. In the central, most restricted area of circulation, the organic carbon content is more than 2.8%, especially in the southern part of the lake which is a stable hydrodynamic area of the lake. Here H$_2$S appears

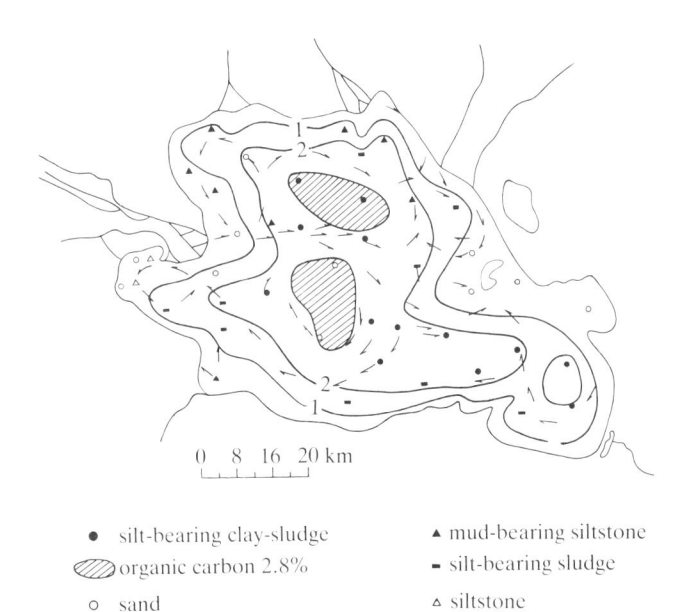

FIG. 2. The modern, 5000 km^2, 27 m deep, brackish Qinghai Lake as a model of circulation and organic matter accumulation (data from Lanzhou Institute of Geology 1974). Contours of organic carbon contents of sediment in percent. Data for lithology from bottom samples.

in the bottom water, the Eh value of the sediments is lower than -200 mV, and the asphaltene content is higher than that in other areas. The organic carbon content more than doubles in the north of the stagnant-water area in which there is the same depth of circulating water. It can be concluded that lake currents control the distribution of sediments and the accumulation and preservation of organic matter (Fig. 2).

In general, a lake has only one or a few main supply rivers, which control these hydrodynamic conditions. The source rocks of the Cretaceous Qingshankou Formation in the Songliao Basin were influenced by the palaeo-Renjiang River. The basin contains two areas where the organic carbon content is 1.5–4.0%. Of these two areas of source rock the western one, which has better potential for the generation of oil, is interpreted to have experienced more stable hydrodynamic conditions.

The main source rock unit T_3Y_2 (Tertiary 3) of the Shanxi–Gansu–Ningxia Basin was divided into five areas which have more than 5% mudstone in the strata by several rivers. Three of these areas are distributed in the west and south of the basin, and contain more than 60–70% mudstone. they are the main oil-forming areas and reflect stable anoxic lake conditions.

The western parts of the northerly trending basins and the southern part of easterly trending basins of China generally experienced long-term hydrodynamic stability.

Multiple depressions and oil-forming centres

Within basins, source rocks are found in the depressions. Although some basins were deepest in the centre, others had central upwarps with depressions on either side. Block faulting, with alternate upheaval and depression influenced sedimentation. The deep depressions with deep water and a thick stagnant water-layer were those conducive to the accumulation and preservation of organic matter.

The long-term and uniform development of deep depressions was favourable to the development of very thick source-rock sequences. For example, there are four north-westerly trending depressions in the Lower Cretaceous strata in the Jiuxi Basin, among which the Qingxi depression with the greatest subsidence and the most fine-grained sediment is the main source unit. The source rocks are as thick as 1000 m. In the Permian period of the Zhungeer Basin, four depressions, Malasihu, Zhongyang, Wucaiwan and Jimusar, developed.

Migration of the oil-forming centre in the basins

The depocentres of many basins frequently shifted resulting in movement of the centres of organic accumulation. For instance, in the Wulumuqi pediment depression of the southern margin of the Zhungeer Basin, the depression during the Triassic was in the area east of Wulumuqi, during the Jurassic and Cretaceous periods at Malas, and then in the Tertiary at Anjihai and Dushanzi, west of Malas. The depressions of the Chaidamu Basin were at the north-western and southern margins of the basin in the Jurassic and Cretaceous but the Oligocene depression was in the south-west part of the basin, while, in the Pliocene epoch, the depression gradually shifted to the north-east, and then to the eastern part of the basin in the Quaternary (Fig. 3). In T_3Y_2 of the Shanxi–Gansu–Ningxia Basin, the depression was located in the south-west part of the basin, and in T_3Y_3, T_3Y_4, T_3Y_5, it gradually shifted towards the north-east. Although many separate depressions formed in the Jurassic period, the centre of the depressions shifted to the west of the basin. The shift of the sedimentary centre controlled the position of the oil-prone area centre.

Coexistence of oil-prone source rocks and coal

In some cases the distribution of oil-prone source rocks in time and space in the continental basins is associated with coal and in others with salt deposits. In the freshwater lake-bogs of humid climatic periods, the organic-rich rocks are zoned from the margin to the centre of the basin. The zones are, in turn, coal, oil shale and oil-prone source rocks. Moreover, the coexistence of coal and oil is common. At the top of the Upper Triassic series in the Shanxi–Gansu–Ningxia Basin, the margin of the basin forms swamp facies with abundant coal. Towards the middle part of the basin, kaolinite clays, kaolinton, pyrite and oil shales occur. Their contents of gallium, germanium and uranium are relatively high. The oil-prone sediments exhibit a gradual transition into lacustrine facies with more dark mudstone and organic matter due to well-developed reducing conditions. In the lower-middle Yanan Formation of the Lower–Middle Jurassic series, the Lingwu-Yianchi area in the west of the basin contains swamp facies having coal beds with an accumulated thickness of 30 m. Coal beds gradually pinch out and the lacustrine oil-prone source rocks appear towards the east of Huachi and

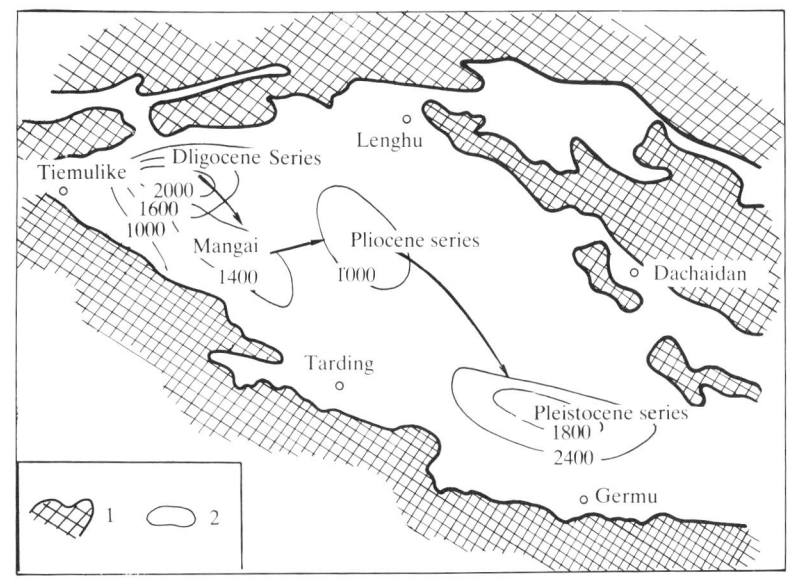

FIG. 3. Examples of the Eocene to Quaternary shift in the lacustrine depocentres of Chaidamu Basin, and thus source-rock area, as a consequence of large-scale shear tectonics within the northern Tibetan Plateau. 1. Basement; 2. Isopachs of source rocks (m).

Qing Yang. The middle and upper Yanan Formation forms a set of strongly rhythmical coal strata and dark mudstone with a thickness of 120 m; the area which is favourable for source rocks is 46 000 km^2. The palaeogeographic landscape contained river networks and groups of lake-bogs. The distribution of coal and oil-prone source rocks thus forms a network with vertical or lateral variations.

In the arid and semi-arid climatic periods, saline lakes contain rock-salt deposits formed in the marginal area, but organic-rich mudstone in the centre; the oil-prone sediments formed in the relatively freshened region, while the salt-rock formed in the saline region. In addition, different source-rock associations can have formed where environmental conditions varied with time. Different source-rock associations have been observed, e.g.

coal series	gypsum, salt rock
oil shale	carbonate
clastic rock	clastic rock
carbonate	oil shale
gypsum, salt rock	coal series

Dark mudstone and shale are considered to be the main lacustrine source rocks, associated with sandstone, or interbedded sandstone and mudstone. Dark siltstone and marlstone are regarded as possible source rocks.

Geochemical characteristics of continental source rocks in north-west China

Abundance of organic matter

According to studies of petroleum generation in the continental basins in China, gas can be generated when organic carbon contents are 0.3–0.4%, but such conditions are unfavourable for oil formation (Huang Difan & Wang Jie 1980). In general, source rocks should have organic carbon contents higher than 0.4–0.8% for oil to be generated, while the contents in good source rocks are 1–6% (locally 10%). Lacustrine source rocks with suitable types of organic matter have organic carbon contents of 1–4%. The contents of 10–27% in the carbonaceous rocks of bog facies are favourable for the formation of gas, and only a certain amount of oil can be formed. The organic matter content in the source rocks of north-west China is usually high (1–10%). The organic matter forming oil and gas is mostly dispersed organic matter. The lower limit of chloroform asphalt A and the content of total hydrocarbons in the source rocks are of the order of 10^{-5}% and 100 ppm, respectively, while the contents of chloroform asphalt A and total hydrocarbon in good source rocks are of the order of 10^{-4}% and more than 500 ppm.

Types of organic matter

The fossils in the Mesozoic and Cenozoic terrestrial source rocks of north-west China are predominantly of aquatic lacustrine organisms, such as ostracods, algae, chara, fish, estherites, pelecypoda, gastropoda, etc. with ostracods especially abundant. Spores, pollen and various kinds of plant are common. The principal feature is that aquatic and terrestrial plants are mixed, in variable amounts depending on the period. Similar mixing of modern organisms from different environments also occurs in the modern Qinghai Lake, the largest lake in China.

In the non-marine source rocks of big freshwater lakes, the kerogen is Type I and Type II$_A$. In brackish lakes, it is Type II$_A$ and Type II$_B$, while it is Type II–Type III in saline lakes (Tissot & Welte 1978). It is mainly Type III in deposits from freshwater swamps (Yang Wanli et al. 1985). The kerogen in the source rocks in north-west China predominantly belongs to Type II, but Type III is common, and Type I is found locally. The maturity threshold for petroleum generation depends on the type of organic matter undergoing maturation. In north-west China the degree of maturity varies from basin to basin and between different regions in the same basin (Institute of Geochemistry 1982).

Other environmental indicators

Authigenic pyrites formed within non-marine kerogen can be used as mineralogical indicators for kerogen classification. Spheroids with cubic and octahedral monocrystals are associated with sapropelite kerogen (Type I), spheroids with pentagonaldodecahedra are associated with huminite (Type III). Mixtures of both forms of mono-crystals indicate mixed kerogen (Type II).

Element composition and organic compounds from different environments

Non-marine sourced crude oil in northern China can be distinguished from marine-sourced oil by high wax, low sulphur, high nickel and low vanadium contents. The original organic matter contained a large proportion of cutinite, spores, resinite, phyllem and other organic components with high contents of wax which are produced by both aquatic and terrestrial plants. The low sulphur contents of non-marine crude oils reflects deposition of their source rocks in fresh–brackish water environments (Huang Difan & Li Jinchao 1984).

In non-marine sourced oils, saturated hydrocarbon contents can be as high as 60–90%, those of aromatic hydrocarbons 10–20% and of asphaltenes 4–20%. These values can vary with environment and reflect the organic composition of source rocks formed in different environments. During the transition lakes to lake-bogs to swamps, the organic matter of source rocks and the compositions of the crude oils generated from the source rocks change progressively. The saturated hydrocarbon compounds of the organic matter in the non-marine source rocks decrease gradually, while aromatic hydrocarbons, asphaltenes and compounds containing oxygen increase gradually. The condensation of aromatic hydrocarbons also increases, but alkyl-replacement is reduced. There can be similar changes of the organic matter composition of source rocks when lacustrine conditions change from freshwater to brackish to saline.

The structure of aromatic hydrocarbons in organic matter varies with depositional environment. Based on triangular graphs of aromatic structure there is a linear increase in aromatic content progressing from swamp, through freshwater lacustrine, brackish-water lacustrine and saline lacustrine facies.

Due to the flourishing of aquatic organisms, such as algae and plankton, in fresh and brackish lakes, the content of low-molecular alkanes is rather high in their source rocks and crude oils. In particular those less than C_{22} are highly abundant. Inputs of terrestrial organisms promote high-molecular alkanes (C_{22+}). Owing to the flourishing of both aquatic and terrestrial organisms in the lake-bog facies, two peaks located either side of C_{22} appear on the n-alkane distribution. Odd-numbered carbon compounds predominate over even-numbered ones in freshwater sediments, while the opposite is the case in saline lacustrine sediments.

The distribution of iso-alkanes in organic matter and crude oil shows an obvious difference in source rocks formed in different environments. On a triangular graph of Pr/nC$_{17}$, Pr/Ph and Ph/nC$_{18}$ (Fig. 4), freshwater swamp, fresh-brackish lacustrine, brackish lacustrine and saline lacustrine oil-prone source rocks show a clear separation. Crude oils show corresponding distributions but the range is narrower, and the distinction between each group is more apparent. In the swamp facies, the content of pristane is more than that of phytane and the ratio of Pr/Ph is more than 2.5–3; in freshwater facies, the pristane content is equal to phytane, and the ratio of Pr/Ph is 0.8–1; in the saline lacustrine facies, pristane is less than phytane, and the ratio of Pr/Ph is less than 0.7.

Biological markers, such as steranes, triterpanes, hopane, norhopane and γ- and β-carotane, have been found in the non-marine source rocks

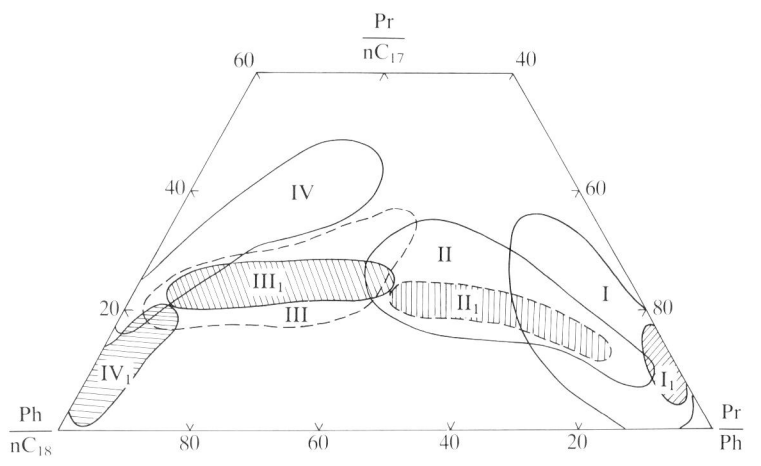

Characteristics of source rocks in NW China 297

FIG. 4. Triangular plot of isoprenoids for different types of Chinese lacustrine source rocks and crude oils: I. Freshwater swamp source rock; I$_1$. Freshwater swamp crude oil; II. Fresh to brackish source rocks; II$_1$. Fresh to brackish crude oil; III. Brackish-water source rocks; III$_1$. Brackish crude oil; IV. Saline lake source rocks; IV$_1$. Saline lake crude oil.

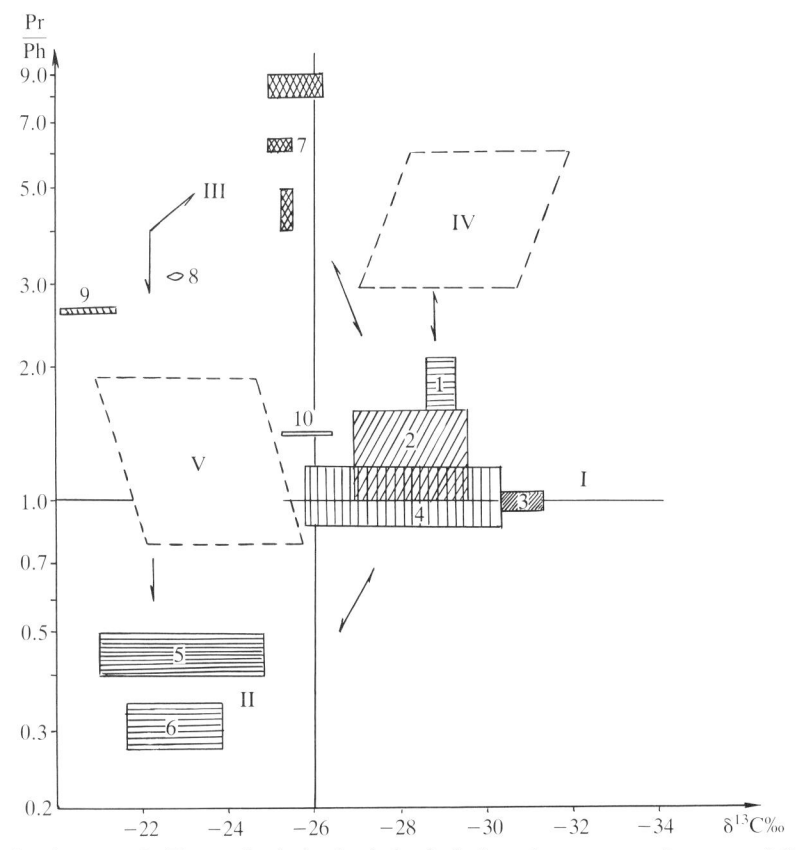

FIG. 5. Carbon isotope vs Pr/Ph as a discrimination index for hydrocarbon source environments: I. Fresh to brackish lake environment. I. 1. Shanganning, 2. Songliao, 3. Jiouquan, 4. Zhungeer. II. Saline lake environment; 5. Chaidamu, 6. Jianghan. III. River–lake-bog environment; 7. Eromanga–Cooper Basin (Australia), 8. Shanganning (Permian), 9. Dongqioulitake. IV. Peat-swamp environment. V. Marine environment, 10. Kekeya.

and crude oil of China (Fan Pu & King 1984). 18α(H)-oleanane is also found, commonly associated with terrestrial angiosperms.

Where environments changed from lake facies into swamp facies, biomarkers such as C_{27} and C_{28} sterane and paraffinic hydrocarbons gradually decrease, these being derived from plankton, bacteria and algae. Biomarkers such as C_{29} sterane, oleanane and Y-lupane, however, gradually increase, these being derived from terrestrial organisms and higher plants. In the crude oil from freshwater or brackish-water sources, the ratio of C_{27} cholestane to total steranes is nearly 0.1, while in the saline-water crude oil, it is 0.2–0.3.

Trisnorhopane (Tm) and benzo-hopane reflect swamp environments. Due to the fact that the influence of environment on the percentage of aromatic hydrocarbon proton (PAP) and aromatic degree is stronger than that of thermal evolution, PAP and aromatic degree can be considered to be a parameter for source material.

The mean $\delta^{13}C$ value of terrestrial plants is $-25.5‰$; it is -27 to $-32‰$ for the freshwater plankton, and -9 to $-23‰$ for marine organisms. Generally speaking, the $\delta^{13}C$ value in the different types of kerogen are as follows: Type I $< -27‰$, Type II -27 to $-26‰$ and Type III $> -26‰$ (Huang Difan & Li Jinchao 1984).

The $\delta^{13}C$ of the saline-water source rocks is -25 to $-29‰$, and that of crude oil derived from these source rocks is -23 to $-26‰$. The $\delta^{13}C$ of the freshwater and brackish-water source rocks and their crude oils is -27 to $-32‰$ and -26 to $-33‰$, respectively.

We propose the following method for identifying the environments of deposition of the source rocks of non-marine crude oils. As on Fig. 5, we plot the $\delta^{13}C$ value for the crude oil as the abscissa and the logarithmic value of pristane/phytane as the ordinate. The different regions in Fig. 5 represent the different environments in which the source rocks of the various types of crude oils were deposited. In Fig. 5, I indicates fresh–brackish-water lacustrine environments (e.g. crude oils from Shanganning, Songliao, Jiouquan and Zhungeer), II indicates the saline-water lacustrine environments (e.g. Chaidamu, Jianghau), III the river–lake-bog environments (e.g. the crude oils from the Eromanga–Cooper, Australia: Powell 1984; the Jurassic of Shanganning and Dongqioulitake), IV the peat-swamp environments, and V the marine environments.

As the types of crude oil in China are many and complex, the study of the environment of deposition of the source rocks of the different types of crude oil is profitable for the development of the theories of hydrocarbon formation, especially for non-marine sources of oil and gas.

ACKNOWLEDGEMENT: We thank Dr. K. Kelts of the Geological Institute, ETH-2, CH 8092 Zurich, for assistance with preparation of the manuscript.

References

FAN PU & KING, J. D. 1984. Characteristics of biomarker compounds (cycloakanes) in oil formed under various depositional environments. *Acta Sedimentologica Sinica*, **2**, 1–17.

HUANG DIFAN & LI JINCHAO. 1984. Kerogen types and study on effectiveness, limitation and interrelation of their identification. *Acta Sedimentologica Sinica*, **2**, 18–33.

—— & WANG JIE. 1980. Genesis of oil and gas of continental origin in the Mesocenozoic basins. *Acta Petroleum Sinica*, **1**, 31–42.

INSTITUTE OF GEOCHEMISTRY, ACADEMIA SINICA. 1982. *Organic Geochemistry*. Science Press, Beijing, 95–109.

LANZHOU INSTITUTE OF GEOLOGY, ACADEMIA SINICA. 1982. *Terrestrial Oil and Gas Formation, Evolution and Accumulation in China*. The Gansu People's Publishing House, Lanzhou, 25–64.

LUO BINJIE. 1980. Hydrodynamics of the lake water. In: *Comprehensive Investigation of Qinhai Lake*. Science Press, Beijing, 67–88.

——, YANG XINGHUA & WANG YOUXIAO. 1981. Evolution of organic matter and formative stage of petroleum in the continental sediment. *Proceedings of Lanzhou Institute of Geology, Academia Sinica*. Science Press, Beijing, 1–62.

LUO BINJIE, HUANG RUCHANG, SHEN PING, HUI RONGYAD, SHAO HONGSHUN, WANG YOUXIAO & RONG GUANGHUA. 1980. Formation and migration of continental oil and gas in China. *Scientia Sinica*, **23**, 1286–1295, 1417–1427.

POWELL, T. G. 1984. Advances in concept of hydrocarbon generation by terrestrial organic matter. *Petroleum Geology Conference, Beijing*.

SCIENTIFIC RESEARCH INSTITUTE OF PETROLEUM EXPLORATION AND DEVELOPMENT. 1982. *Terrestrial Oil and Gas Formation in China*. Petroleum Industry Press, Beijing, 26–44.

TISSOT, B. P. & WELTE, D. H. 1978. *Petroleum Formation and Occurrence*. Springer-Verlag, New York, 111–131.

YANG WANLI, LI YONG KANG & GAO RUIGI. 1985. Formation and evolution of non-marine petroleum in Songliao Basin, China. *Bulletin of the American Association of Petroleum Geologists*, **69**, 1112–1122.

LUO BINJIE, YANG XINGHUA, LIN HEJIE & ZHENG GUODONG, Lanzhou Institute of Geology, Academia Sinica, Lanzhou, China.

Biological markers in lacustrine Chinese oil shales

S. C. Brassell, Sheng Guoying, Fu Jiamo & G. Eglinton

SUMMARY: A major objective in molecular organic geochemistry is the assessment and characterization of sedimentary depositional environments from the occurrence of specific biological marker compounds related to particular source organisms. The aliphatic hydrocarbon distributions of petroleums and/or sediments from Chinese freshwater and hypersaline lacustrine environments are described and contrasted in the search for diagnostic features among their biological marker distributions. The Eocene Maoming oil shale, which crops out in Guangdong Province, shows major differences in hydrocarbon composition through its stratigraphic succession. These variations reflect changes in lithology and in the origins of the sedimentary organic matter, as seen from petrographic examination. The lower lignite and vitrinite layers contain significant contributions of biological markers from bacterial sources, plus components derived from terrigenous higher plants, such as C_{29} steroids. In contrast, the overlying shales show a dominance of dinoflagellate-derived 4-methylsteroids and culminate in an horizon with botryococcane homologues as their dominant alkanes. Thus, the geochemical profile suggests that a swampy environment, in which peats accumulated, deepened to support populations of dinoflagellate and, finally, blooms of *Botryococcus* green algae. The aliphatic hydrocarbons of a series of sediments and petroleums from the Eocene hypersaline sequence of the Jianghan Basin in southern Hubei Province shows various features that may be typical of such facies. These include a dominance of even-numbered n-alkanes, high amounts of phytane and gammacerane, and an abundance of C_{35}-hopanes. Similar characteristics are observed in shales from the Eocene Green River Formation and in Rozel Point crude oil, a petroleum thought to originate from sediments deposited under hypersaline conditions. The same features also appear in a lower Tertiary evaporitic mudstone from south of the Renqui oil field in Hebei Province, together with 4-methylsteranes, β-carotane and 18α(H)-oleanane, a marker for terrigenous higher plants. These data illustrate the potential of biological markers in the differentiation of different lacustrine environments and suggest that such characteristics may provide signatures for environmentally similar, but geographically distant, deposits.

Introduction

The biological marker characteristics of petroleums, like their isotopic signatures (e.g. Sofer 1984), are largely inherited from their source rocks. Hence, a major objective in molecular organic geochemical studies is the recognition of diagnostic features among the biological marker distributions of petroleums, which can help describe and distinguish the original depositional environments of their source rocks, for example, whether marine or lacustrine. This goal can be addressed empirically through the examination of the characteristics of well-defined sedimentary environments. Thus, for example, the presence of C_{30} steranes (24-propylcholestanes) is held to be indicative of contributions from marine organic matter (Moldowan *et al.* 1985). Alternatively, the assemblage and abundance of compounds derived from organisms of restricted habitat can provide circumstantial evidence of aspects of the sedimentary depositional environment. For example, the presence of botryococcane is indicative of inputs from *Botryococcus* green algae (Moldowan & Seifert 1980), which only live in fresh or brackish waters.

Biological marker information can therefore contribute to the task of recognizing the significance of, and differentiating between, the various depositional environments that can be responsible for the formation of petroleum source rocks. In particular, the growing recognition of the importance of lacustrine sediments (Powell 1986), prompts a search for specific features that may serve to characterize such environments. The geographic and temporal diversity, variety and size of the lacustrine deposits in the People's Republic of China provide many excellent examples for the study of such sedimentary environments and their associated petroleums. In this paper the characteristics of specific biological marker components among the aliphatic hydrocarbon distributions of two Chinese lacustrine environments are examined and compared with each other and other environments. The first example is the freshwater Maoming oil shale sequence (Brassell *et al.* 1986) from southern China (Fig. 1). The second principal suite of

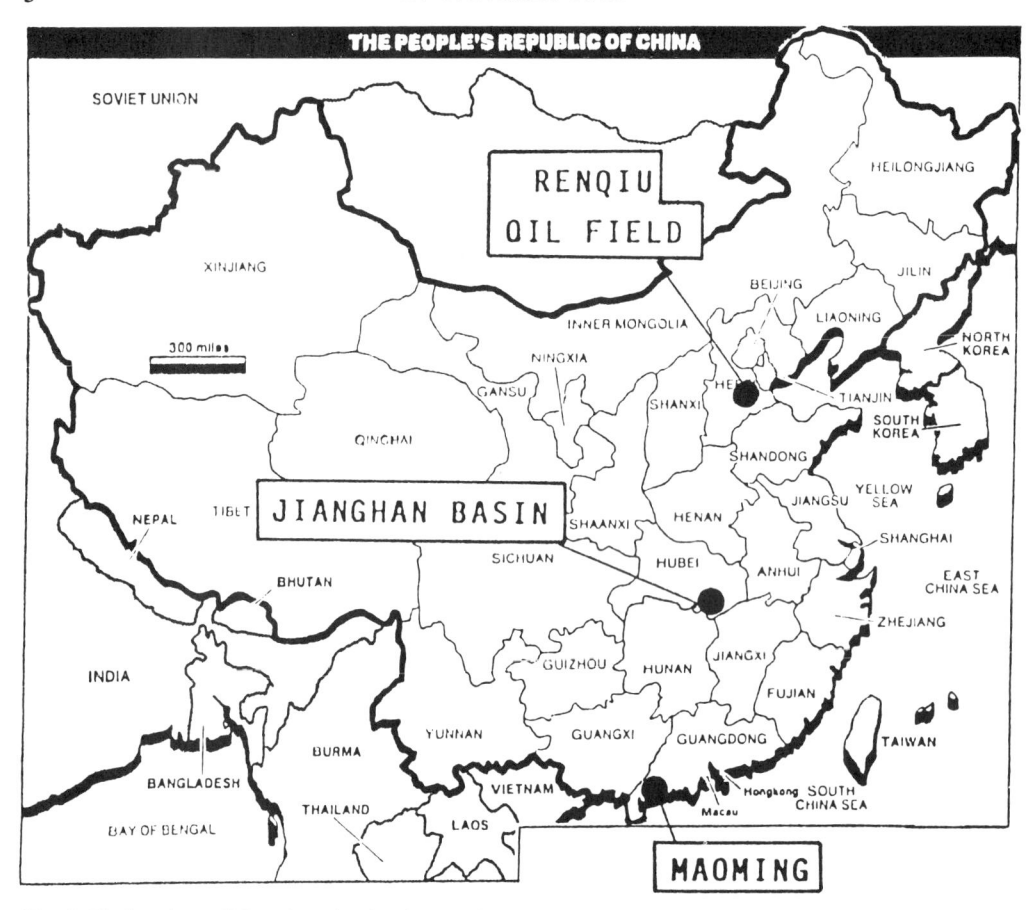

FIG. 1. The locations of Maoming, the Jianghan Basin and the Renqiu oil field in Guangdong, Hubei and Hebei Provinces, respectively, in the People's Republic of China.

sediments and petroleums are those from the hypersaline Jianghan Basin (Fig. 1; Xie Taijun et al. 1984; Fu Jiamo et al. 1986).

Samples and experimental procedures

Samples

The basal lignite of the Eocene Maoming oil shale sequence lies unconformably on Cretaceous red sandstones. Six sediment samples from the stratigraphic section, including the basal lignite (M-4), a vitrinite lens (M-5) from the overlying claystone and four sections of oil shale from increasingly higher levels (M-1, M-2, M-3 and M-M, respectively) have been studied. This paper concentrates on the interpretation and comparison of the biological marker composition of the lignite (M-4) and the two uppermost shales (M-3 and M-M). The visual descriptions of all six samples and a more comprehensive survey and evaluation of their biological marker characteristics are detailed elsewhere (Brassell et al. 1986).

For the Jianghan Basin, the distributions of aliphatic hydrocarbons of several sediments and petroleums have been previously documented (Fu Jiamo et al. 1986). Here, these data are complemented by additional information from other sediments, comprising mudstones (coded WX63, GX715 & WX113a) and a glauber salt (mirabilite; $Na_2SO_4 \cdot 10H_2O$) sample (WX113b), and petroleums (B1, W1349) reservoired at shallow depths (ca. 600 m and 1200 m, respectively). Such accumulated petroleums only occur in the Qianjiang depression of the Jianghan Basin where saline lake sediments are buried in concert with reservoir sandstones (Fig. 2; Fu Jiamo et al. 1986).

Other samples considered here include the Rozel Point crude oil (e.g., ten Haven et al. this volume), the Green River shale (e.g., Murphy et

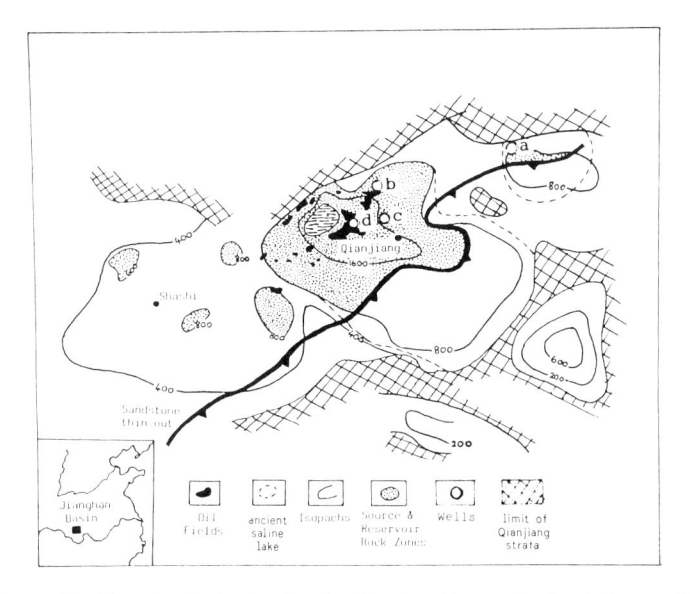

FIG. 2. A map of part of the Jianghan Basin showing the Qianjiang Depression (central area with oil fields) and well sites of: (a) B1 oil, (b) WX sediments, (c) W1349 oil and (d) G33 oil.

al. 1967), plus a Tertiary hypersaline sediment (Z2) from ca. 40 km south of the Renqiu oil field in Hebei Province (Fig. 1).

Experimental procedures

All sediment and petroleum samples were extracted with $CH_2CL_2/MeOH$ by Soxhlet or ultrasonication, and the extracts separated by chromatographic methods to yield aliphatic hydrocarbon fractions which were analysed by gas chromatographic (GC) and gas chromatographic–mass spectrometric (GC-MS) techniques as previously described for these, or comparable, sample suites (Brassell et al. 1986; Fu Jiamo et al. 1986).

Maoming oil shale

The distributions of aliphatic hydrocarbons (Fig. 3) for the oil shales (M-M and M-3) and the basal lignite (M-4) are significantly different, as are their petrographic characteristics (Brassell et al. 1986). The uppermost shale (M-M) contains abundant colonial algae and exhibits a dominance of C_{31} and C_{33} botryococcane homologues among its aliphatic hydrocarbons (Fig. 3a; Brassell et al. 1986). These components are probably related to those with similar mass spectral features observed in Sumatran crude oils (Seifert & Moldowan 1981). The prominence of these compounds in the M-M shale can be taken to indicate that their presumed source organisms, *Botryococcus* green algae, were major contributors of its organic matter. 4-methylsteranes are the major components of the M-3 oil shale as seen in its RIC trace (Fig. 3b). They are sufficiently abundant to be readily recognized in the m/z 217 mass fragmentogram (Fig. 4b), although this typically only depicts sterane distributions. Indeed, this sample contains markedly higher proportions of 4-methylsteranes, notably the C_{30} members, than steranes. The reverse is true for the M-M oil shale (Fig. 4a). Significantly, petrographic inspection of M-3 shows an abundance of hydrocarbon wisps attributable to dinoflagellates (Brassell et al. 1986). Hence, since 4-methylsteranes are held to derive from 4-methylsterol precursors contributed by dinoflagellates (Boon et al. 1979; Robinson et al. 1984; Wolff et al. 1986), both visual and biological marker information suggest a major contribution from dinoflagellate algae to M-3.

The n-alkane distributions of the oil shales are dominated by higher, odd-numbered members (C_{27} and C_{29}), presumably derived from higher plant sources. Both samples also possess slightly greater amounts of pristane than phytane (Fig. 3a and b). The M-3 oil shale contains marginally higher proportions of these, and other, acyclic isoprenoid alkanes, including lycopane (peak c in Fig. 3b). The distributions of hopanoids in M-M and M-3 are generally similar (Fig. 5a and b, respectively), except for the higher amounts of

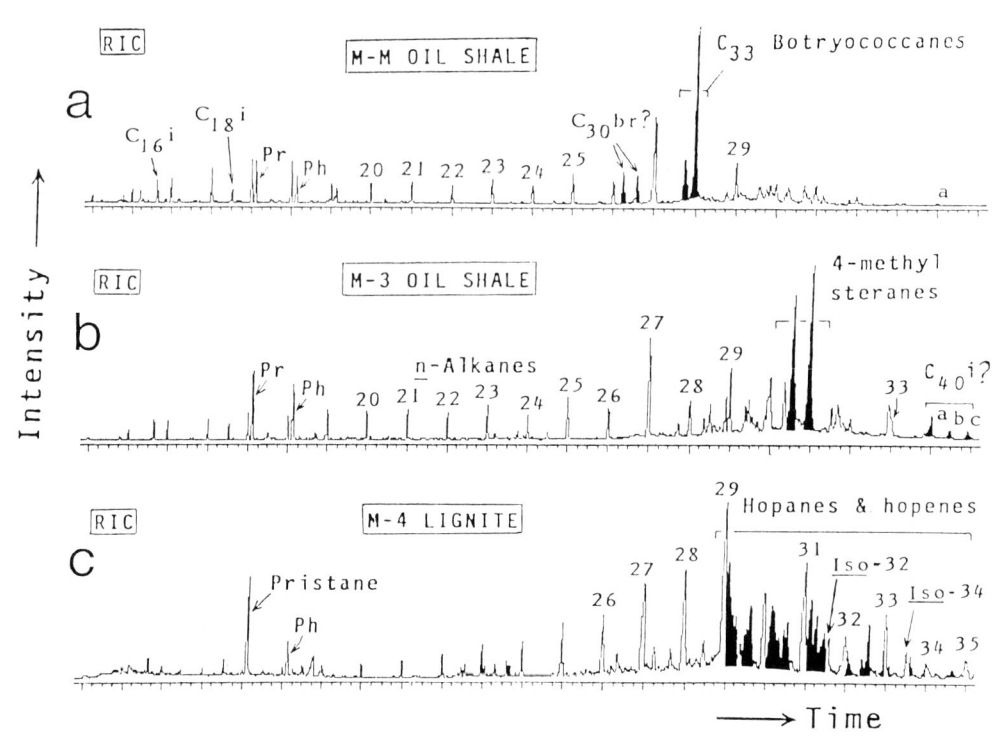

FIG. 3. Annotated reconstituted ion chromatogram (RIC) from GC-MS analysis of the aliphatic hydrocarbons of: (a) M-M oil shale, (b) M-3 oil shale and (c) M-4 lignite. n-alkanes are numbered according to their chain length. $C_{16}i$, $C_{18}i$, Pr, Ph and $C_{40}i$ denote C_{16}, C_{18}, C_{19}, C_{20} and C_{40} isoprenoids (Brassell et al. 1986), respectively. C_{30}br indicates the peaks corresponding to branched isoprenoids (Brassell et al. 1986). iso-32 and iso-34 refer to 2-methylhentriacontane and 2-methyltritriacontane, respectively.

22R C_{31} 17α(H),21β(H)-hopane and C_{30} 17β(H), 21β(H)-hopane in the latter.

Higher n-alkanes (e.g., C_{27}) with a low CPI are prominent components of the aliphatic hydrocarbon distributions of the basal lignite (M-4). It contains significant amounts of iso alkanes (notably C_{32} and C_{34}) and its hopanoids are in markedly higher abundance than in the other samples. In addition, the proportion of individual hopanoids in the lignite differs from that of the other samples, especially in terms of its higher amounts of hop-17(21)-enes and neohop-13(18)-enes (Fig. 5c). Such features presumably stem from the differences in the contributions from bacteria to the shales and lignite. The lignite also has a dominance of pristane among its acyclic isoprenoids and C_{29} diasterenes among its major steroidal components, whereas 4-methylsteranes were barely detectable among its constituents. All of these characteristics, namely high proportions of pristane, hopanes and hopenes (van Dorsselaer et al. 1977), and 24-ethylsteroids (C_{29}), are recognized features of lignites and coals (Hoffmann et al. 1984).

The combined evidence from petrographic and biological marker data suggests that the Maoming oil shale sequence was laid down in a swampy environment giving rise to peats. These waters may then have deepened to support populations of dinoflagellate algae, and, finally, blooms of botryococcoid green algae.

The combination of the occurrence of thermally unstable components (e.g., hopenes, diasterenes, 5β-steranes, 4β-methylsteranes and 17β(H), 21β(H)-hopanes), together with the low concentrations or absence of biological markers produced during catagenesis (e.g., 20S steranes and 22S extended hopanes) indicate the low maturity of the Maoming oil shale sequence (Brassell et al. 1986). Such information from the biological marker distributions is consistent with the low vitrinite reflectance values of 0.34 and 0.38 for the M-2 oil shale and M-4 lignite, respectively (Brassell et al. 1986).

Botryococcane and 4-methylsteranes co-occur in various proportions in non-marine crude oils from the western Otway Basin (McKirdy et al. 1986). Hence, the association of botryococcane

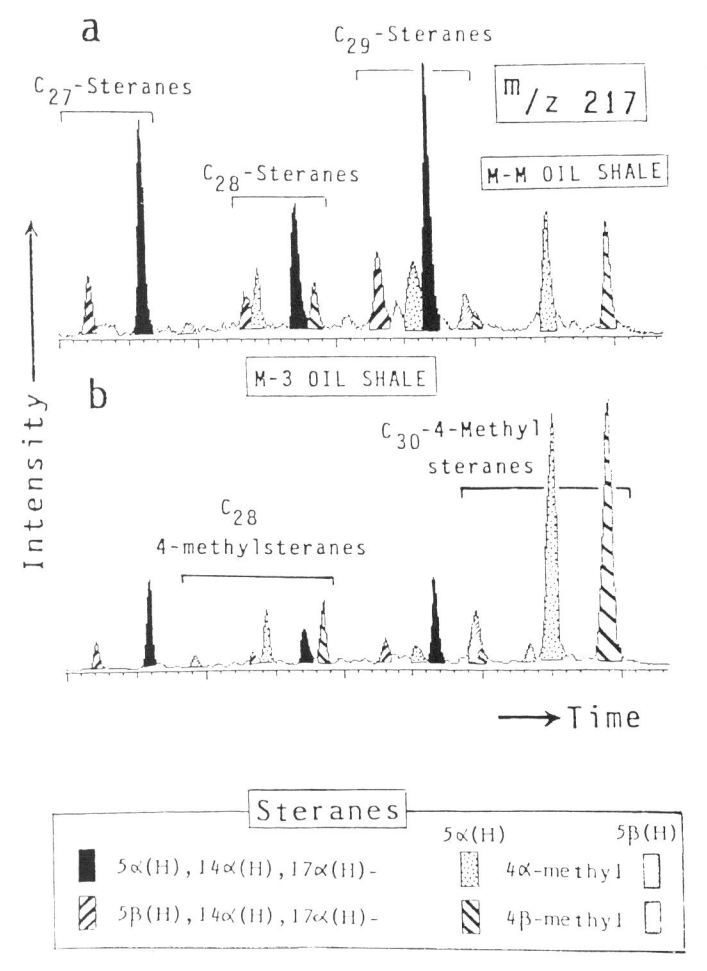

FIG. 4. Annotated partial m/z 217 chromatograms depicting the distributions of steranes and 4-methylsteranes among the aliphatic hydrocarbons of: (a) M-M oil shale and (b) M-3 oil shale.

homologues and 4-methylsteranes in the Maoming oil shale seems comparable, especially as any petroleum generated from these strata would most probably be pooled from the various levels, thereby mixing their hydrocarbon characteristics. A dominance of 4-methylsteranes, such as in the M-3 horizon of the Maoming oil shale, is observed in other lacustrine sediments, including the Messel shale (Kimble et al. 1974) and Oligocene Chinese shales from Hebei Province (e.g., Wolff et al. 1986). Although the presumed precursors of 4-methylsteranes, namely 4-methylsterols, are prominent components of both recent marine (de Leeuw et al. 1983) and lacustrine (Robinson et al. 1984) environments, it appears that 4-methylsteranes only dominate steranes in non-marine environments. Lycopane is present in M-3 and is also a prominent constituent of the Messel shale (Kimble et al. 1974). However, it has little diagnostic value for lacustrine environments since it can occur as a significant component of marine sediments, such as those from the Cariaco Trench (Brassell et al. 1981). Overall, the major diagnostic biological markers in freshwater lake sediments, like the Maoming oil shales, appear to be homologues of botryococcane and 4-methylsteranes.

Jianghan Basin

The aliphatic hydrocarbon distributions of the petroleums and mudstones from the Jianghan Basin show a number of common features (cf. Fu Jiamo et al. 1986), although those of the glauber salt differ in several respects. First, the n-alkane distributions often show an even over odd predominance (EOP) (Fig. 6a–c); a feature

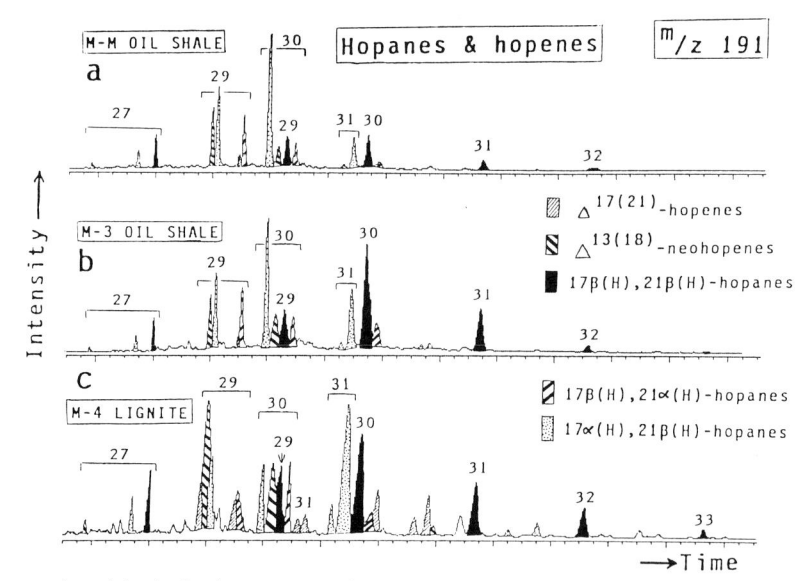

FIG. 5. Annotated partial m/z 191 chromatograms illustrating the distributions of hopanes and hopenes among the aliphatic hydrocarbons of: (a) M-M oil shale, (b) M-3 oil shale and (c) M-4 lignite. Peak numbers refer to the carbon number of the individual hopanoids.

observed previously for other samples from this basin (Xie Taijun et al. 1984; Fu Jiamo et al. 1986). This EOP in the n-alkanes may reflect their derivation from functionalized precursors (e.g., n-alkanoic acids, n-alkanols), rather than an origin from direct biological contributions of n-alkanes. However, it is uncertain whether the EOP in n-alkanes is principally a function of specific biological inputs to these environments or as a result of their peculiar diagenetic pathways.

Second, phytane (Ph) is always markedly more abundant than pristane (Pr; Fig. 6), and is usually the dominant branched/cyclic aliphatic hydrocarbon. Low Pr/Ph ratios can be interpreted as an indication of anoxic depositional conditions (Didyk et al. 1978), but in this instance it seems most probable that the high amounts of phytane reflect contributions from archaebacteria (e.g., halophiles; Kates 1978), rather than chlorophyll degradation products. Related contributions of C_{25} (both 2,6,10,15,19- and 2,6,10,14,18-pentamethyleicosanes) and C_{30} (both 2,6,10,14,18,22-hexamethyltetracosane and squalane) acyclic isoprenoid alkanes are also observed, especially in the glauber salt (Fig. 6), and they further indicate significant lipid inputs from archaebacteria (e.g., Holzer et al. 1979; Brassell et al. 1981). The prominence of such components in the glauber salt (Fig. 6d) illustrates the apparent tolerance of the archaebacteria to high salinities, such as those necessary for the precipitation of evaporite minerals, including glauber salt.

Third, gammacerane is a prominent component in the m/z 191 mass chromatograms (Fig. 7), typically dominating the $17\alpha(H),21\beta(H)$-hopanes. Abundant gammacerane, as observed here in the Jianghan Basin (Xie Taijun et al. 1984; Fu Jiamo et al. 1986), was first reported in the Green River shale (Hills et al. 1966) and is a feature common to other hypersaline sediments (e.g., ten Haven et al. 1985) and oils, including Rozel Point (ten Haven et al. this volume). Its absence in the glauber salt prompts the suggestion that its source organisms, unknown at present, may be unable to tolerate highly saline regimes, as implied by the deposition of glauber salt. Gammacerane, however, is not restricted to saline lake sediments; it also occurs as a minor component in many marine environments (e.g., Moldowan et al. 1985).

Fourth, C_{35} and/or C_{34} hopane homologues are often dominant over the C_{31} to C_{33} (Fig. 7; ten Haven et al. 1985, this volume; Fu Jiamo et al. 1986), in contrast to their usual observed decrease in relative abundance with increasing carbon number. Such a feature suggests either different distributions of their precursor polyhydroxybacteriohopanes, or diagenetic reactions that favour the retention and survival of the C_{35} and/or C_{34} hopane skeleton.

Finally, β-carotane occurs in several of the samples from this location, including the glauber salt and one of the oils (G33). This feature is even more marked in sediments and oils from Xinjiang province (Fu Jiamo et al. unpubl. data), and in

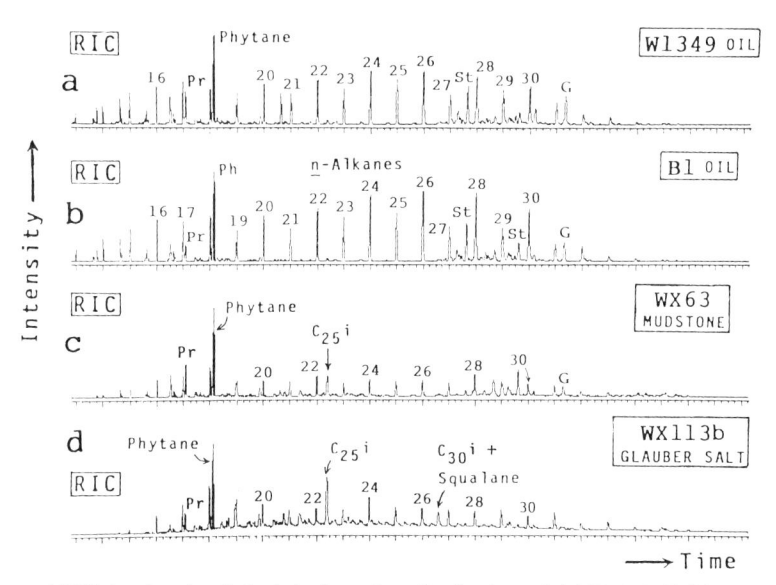

FIG. 6. Annotated RIC showing the aliphatic hydrocarbon distributions of: (a) W1349 oil, (b) B1 oil, (c) WX63 mudstone and (d) WX113b glauber salt. Selected n-alkanes are numbered according to their chain length. Pr, $C_{25}i$ and $C_{30}i$ refer to pristane and C_{25} and C_{30} acyclic isoprenoids, respectively. St and G denote steranes and gammacerane, respectively.

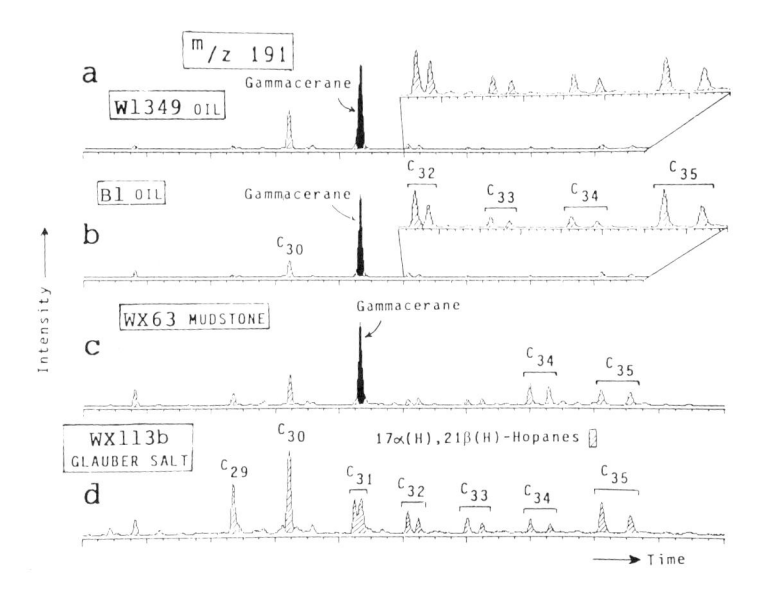

FIG. 7. Annotated partial m/z 191 chromatograms showing the triterpenoid distributions of: (a) W1349 oil, (b) B1 oil, (c) WX63 mudstone and (d) WX113b glauber salt. The carbon numbers of the suite of C_{29} to C_{35} $17\alpha(H),21\beta(H)$-hopanes are given.

the Green River shale, wherein β-carotane was first identified (Murphy et al. 1967). It is, however, an intermittent characteristic, rather than a uniform one, probably reflecting an autochthonous input from a specific class of bacteria.

These features, except for the presence of β-carotane, all appear in Rozel Point crude oil (ten Haven et al. this volume), in a marl from the northern Apennines in Italy (ten Haven et al. 1985) and in seep oils from Sicily (ten Haven et al. this volume). Hence, the biological marker characteristics of the Jianghan Basin samples seem to be generally representative of hypersaline depositional environments.

The maturity of the sediment samples is relatively low (Fu Jiamo *et al.* 1986), as indicated by their biological marker distributions, such as the EOP for n-alkanes. Similarly, the oils are all early-generated petroleums.

Renqiu oil field

A hypersaline sediment from south of the Renqiu oil field in north-east China (Fig. 1) exhibits all (Fig. 8) of the characteristic features described above for the Jianghan Basin. It has a high phytane content, an EOP in its n-alkanes (Fig. 8a), possesses a significant amount of gammacerane and high relative proportions of C_{35} hopanes (Fig. 8b), and also contains β-carotane (Fig. 8a). In addition, it contains minor proportions of 4-methylsteranes (significantly less than steranes; Fig. 8c) and 18α(H)-oleanane (Fig. 8b), a characteristic higher plant triterpenoid found in Nigerian crude oils (Whitehead 1974; Ekweozor *et al.* 1979).

Hence, this sample shows the features characteristic of a hypersaline depositional environment

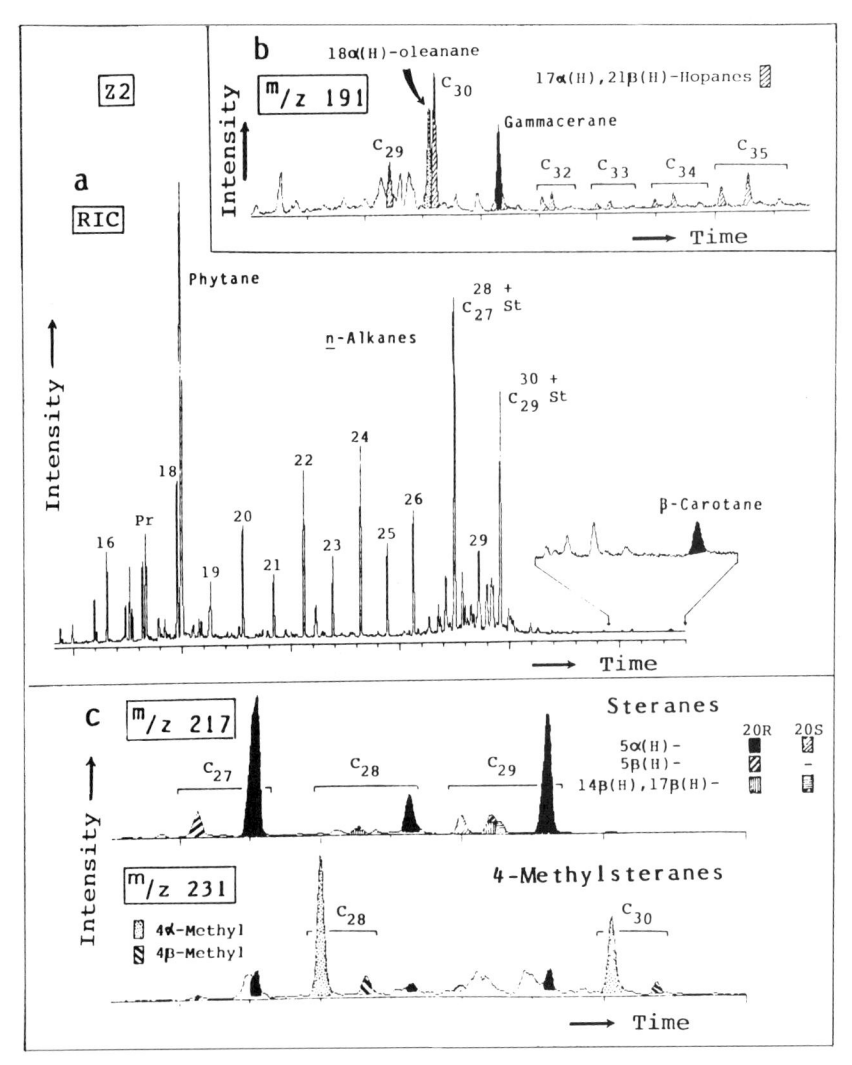

FIG. 8. Annotated illustrations of the distributions of aliphatic hydrocarbons in Z2. (a) RIC with n-alkanes numbered, phytane and β-carotane denoted, and with the coelution of C_{28} and C_{30} n-alkanes with C_{27} and C_{29} steranes, respectively, also shown. (b) Partial m/z 191 chromatogram showing triterpanes, including the series of C_{29}–C_{35} 17α(H),21β(H)-hopanes. (c) Partial chromatograms of m/z 217 and m/z 231 depicting steranes and 4-methylsteranes, respectively.

coupled with evidence of terrigenous contributions of organic matter. Perhaps it was deposited close to the shoreline of a saline lake and therefore received inputs from the surrounding vegetation.

The biological marker composition of Z2, notably its low relative concentrations of 20S and 14β(H),17β(H)-steranes, indicate its immaturity, comparable to samples from the Jianghan Basin.

Conclusions

The major diagnostic features in the biological marker distributions of the freshwater Maoming shale are the presence of botryococcane homologues and the dominance of 4-methylsteranes over steranes in the oil shales. The lignite is characterized by high amounts of pristane, n-alkanes and hopanes and hopenes. Such characteristics contrast directly with the major features of the Jianghan Basin samples, namely an EOP for n-alkanes, a dominance of phytane, high amounts of gammacerane and high proportions of C_{35} and/or C_{34} hopanes. Carotanes may also be present. On this basis it becomes possible to differentiate such environments. However, the Z2 sample shows a combination of these features, reflecting its inferred deposition in a hypersaline lake receiving terrigenous inputs.

A further consideration is that all of the sediments, both freshwater and hypersaline, are Tertiary. It therefore remains to be seen whether the diagnostic features in their biological marker distributions hold for pre-Tertiary sediments and petroleums, given that evolutionary changes may influence the biological populations of such environments and, hence, their geochemical characteristics. The one exception is the observed association between 4-methylsteranes and botryococcane, since the lacustrine oils from the Otway Basin which contain these compounds are Cretaceous.

An interesting observation regarding the occurrence of botryococcane homologues, 4-methylsteranes, and phytane samples, respectively, in their tendency to dominate the biological marker distributions. Similar behaviour is shown by other branched/cyclic compounds, such as 28,30-bisnorhopane which dominates the cyclic fraction of many immature sediments and petroleums (e.g., Monterey shales, Seifert et al. 1978). This tendency for certain terpenoids to dominate biological marker distributions has analogies (Brassell et al. 1983) to the existence of monospecific fossil assemblages in certain sediments. Hence, there is evidence from both conventional and molecular fossils that the establishment of particular depositional conditions may be conducive to the blooming of individual opportunistic species.

ACKNOWLEDGEMENTS: SCB and GE thank the NERC for support of the C-GC-MS facility at Bristol (GR3/2951 & GR3/3758). SG and FJ acknowledge the support of the United Nations Development Project (CPR/84/005). This contribution stems from a collaborative research programme between the Institute of Geochemistry in Guiyang and the Organic Geochemistry Unit in Bristol, a joint project supported by the Academia Sinica and the Royal Society.

References

BOON, J. J., RIJPSTRA, W. I. C., DE LANGE, F., DE LEEUW, J. W., YOSHIOKA, M. & SHIMIZU, Y. 1979. Black Sea sterol—a molecular fossil for dinoflagellate blooms. *Nature*, **277**, 125–127.

BRASSELL, S. C., WARDROPER, A. M. K., THOMSON, I. D., MAXWELL, J. R. & EGLINTON, G. 1981. Specific acyclic isoprenoids as biological markers of methanogenic bacteria in marine sediments. *Nature*, **290**, 693–696.

——, EGLINTON, G. & MAXWELL, J. R. 1983. The geochemistry of terpenoids and steroids. *Biochemical Society Transactions*, **11**, 575–586.

——, —— & FU JIAMO 1986. Biological marker compounds as indicators of the depositional history of the Maoming oil shale. *In*: LEYTHAEUSER, D. & RULLKÖTTER, J. (eds), *Advances in Organic Geochemistry 1985*, Pergamon Press, Oxford, 927–941.

DE LEEUW, J. W., RIJPSTRA, W. I. C., SCHENCK, P. A. & VOLKMAN, J. K. 1983. Free, esterified and residual bound sterols in Black Sea Unit 1 sediments. *Geochimica et Cosmochimica Acta*, **47**, 455–465.

DIDYK, B. M., SIMONEIT, B. R. T., BRASSELL, S. C. & EGLINTON, G. 1978. Organic geochemical indicators of palaeoenvironmental conditions of sedimentation. *Nature*, **272**, 216–222.

EKWEOZOR, C. M., OKOGUN, J. I., EKONG, D. E. U. & MAXWELL, J. R. 1979. Preliminary organic geochemical studies of samples from the Niger Delta, Nigeria: I. Analyses of crude oils for triterpanes. *Chemical Geology*, **27**, 11–28.

FU JIAMO, SHENG GUOYING, PENG PINGAN, BRASSELL, S. C., EGLINTON, G. & JIANG JIGANG 1986. Peculiarities of salt lake sediments as potential source rocks in China. *In*: LEYTHAEUSER, D. & RULLKÖTTER, J. (eds), *Advances in Organic Geochemistry 1985*, Pergamon Press, Oxford, 119–126.

HILLS, I. R., WHITEHEAD, E. V., ANDERS, D. E., CUMMINS, J. J. & ROBINSON, W. E. 1966. An optically active triterpane, gammacerane, in Green

River, Colorado, oil shale bitumen. *Journal of the Chemical Society, Chemical Communications*, 752–754.

HOFFMANN, C. F., MACKENZIE, A. S., LEWIS, C. A., MAXWELL, J. R., OUDIN, J. L., DURAND, B. & VANDENBROUCKE, M. 1984. A biological marker study of coals, shales and oils from the Mahakam Delta, Kalimantan, Indonesia. *Chemical Geology*, **42**, 1–23.

HOLZER, G., ORO, J. & TORNABENE, T. G. 1979. Gas chromatographic—mass spectrometric analysis of neutral lipids from methanogenic and thermoacidophilic bacteria. *Journal of Chromatography*, **186**, 795–809.

KATES, M. 1978. The phytanyl ether-linked polar lipids and isoprenoid neutral lipids of extremely halophilic bacteria. *Progress in the Chemistry of Fats and Other Lipids*, **15**, 301–342.

KIMBLE, B. J., MAXWELL, J. R., PHILP, R. P., EGLINTON, G., ALBRECHT, P. ENSMINGER, A., ARPINO, P. & OURISSON, G. 1974. Tri- and tetraterpenoid hydrocarbons in the Messel oil shale. *Geochimica et Cosmochimica Acta*, **38**, 1165–1181.

MCKIRDY, D. M., COX, R. E., VOLKMAN, J. K. & HOWELL, V. J. 1986. Botryococcane in a new class of Australian non-marine crude oils. *Nature*, **320**, 57–59.

MOLDOWAN, J. M. & SEIFERT, W. K. 1980. First discovery of botryococcane in petroleum. *Journal of the Chemical Society, Chemical Communications*, 912–914.

——, —— & GALLEGOS, E. J. 1985. Relationship between petroleum composition and depositional environment of petroleum source rocks. *American Association of Petroleum Geologists Bulletin*, **69**, 1255–1268.

MURPHY, M. T. J., MCCORMICK, A. & EGLINTON, G. 1967. Perhydro-β-carotene in Green River shale. *Science*, **157**, 1040–1042.

POWELL, T. G. 1986. Petroleum geochemistry and depositional setting of lacustrine source rocks. *Marine and Petroleum Geology*, **3**, 200–219.

ROBINSON, N., EGLINTON, G., BRASSELL, S. C. & CRANWELL, P. A. 1984. Dinoflagellate origin for sedimentary 4α-methylsteroids and 5α(H)-stanols. *Nature*, **308**, 439–441.

SEIFERT, W. K. & MOLDOWAN, J. M. 1981. Paleoreconstruction by biological markers. *Geochimica et Cosmochimica Acta*, **45**, 783–794.

——, ——, SMITH, G. W. & WHITEHEAD, E. V. 1978. First proof of structure of a C_{28}-pentacyclic triterpane in petroleum. *Nature*, **271**, 436–437.

SOFER, Z. 1984. Stable carbon isotope compositions of crude oils: application to source depositional environments and petroleum alteration. *Bulletin of the American Association of Petroleum Geologists*, **68**, 31–49.

TEN HAVEN, H. L., DE LEEUW, J. W. & SCHENCK, P. A. 1985. Organic geochemical studies of a Messinian evaporitic basin, northern Apennines (Italy) I: hydrocarbon biological markers for a hypersaline environment. *Geochimica et Cosmochimica Acta*, **49**, 2181–2191.

VAN DORSSELAER, A., ALBRECHT, P. & CONNAN, J. 1977. Changes in the composition of polycyclic alkanes by thermal maturation (Yallourn lignite, Australia). *In*: CAMPOS, R. & GONI, J. (eds), *Advances in Organic Geochemistry 1975*, Enadimsa, Madrid, 53–59.

WHITEHEAD, E. V. 1974. The structure of petroleum pentacyclanes. *In*: TISSOT, B. & BIENNER, F. (eds), *Advances in Organic Geochemistry 1973*, Editions Technip, Paris, 225–243.

WOLFF, G., LAMB, N. A. & MAXWELL, J. R. 1986. The origin and fate of 4-methylsteroid hydrocarbons. I. Diagenesis of 4-methylsteranes. *Geochimica et Cosmochimica Acta*, **50**, 335–342.

XIE TAIJUN, WU LIZHEN & JIANG JIGANG 1986. Formation of oil and gas fields in Jianghan saline lacustrine basin. *In*: *Proceedings of Beijing Petroleum Symposium, September 1984*.

S. C. BRASSELL* & G. EGLINTON, Organic Geochemistry Unit, University of Bristol, School of Chemistry, Cantock's Close, Bristol BS8 1TS, UK.
SHENG GUOYING & FU JIAMO, Institute of Geochemistry, Academia Sinica, Guiyang, Guizhou Province, The People's Republic of China.

*Present address: Department of Geology, Stanford University, Stanford, California 94305-2115, U.S.A.

Geochemical characteristics of crude oils and source beds in different continental facies of four oil-bearing basins, China

Wang Tieguan, Fan Pu & F. M. Swain

SUMMARY: In order to correlate Chinese continental crude oils with their sources and, thus, the different depositional facies of the sources, the geochemical characteristics of crude oils and of extracts from related source rocks in four different types of continental facies were studied. The four types of facies were those of (1) inland and paralic freshwater to brackish lake basins, (2) saline lake basins, (3) inland swampy lake and swamp basins, and (4) terrestrial and marine intertonguing facies basins. Oils from these facies show both similarities and differences. Some specific biomarkers, such as the stereoisomers of onocerane, serratane and long chain acyclic polymethylalkanes which could be useful indicators of a palaeoenvironment, are also discussed. All the components were identified using mass spectrometry and a few retention index data: therefore, the identification of some compounds must be tentative.

Introduction

Crude oil fields in continental sediments have been discovered in many countries. These have formed under a variety of geological conditions. Most Chinese oil fields, however, formed in freshwater lake basins ranging in age from Permian to Neogene.

In recent years, many authors have studied the formation of continental freshwater basins (Swain 1980; Huang Ruchong *et al.* 1981; Huang Difan *et al.* 1982), but the characteristics of crude oils in different basins and from source rocks deposited in various types of aqueous environment have seldom been compared in detail.

The source rocks of Chinese continental oil formed under a variety of lacustrine conditions.

(1) Freshwater and fresh–brackish water lake basins are widely distributed in China. Freshwater lake basins have an abundant supply of both terrestrial organic material and freshwater during wet climatic periods. Under conditions of rapid deposition, oil source beds, such as the Upper Triassic ones of the Shanganning Basin develop in deeper parts of the lake. During the semi-arid climatic periods, the lake water may become brackish. Overall, this type of lake basin has source rocks consisting mainly of dark-coloured mudstones. Paralic freshwater basins are somewhat different from inland ones. Temporary marine transgressions result in mixing of freshwater and seawater. This has a profound influence on the characteristics of the accumulated organic and inorganic matter. The Shahejie Formation of Eocene–Oligocene age in Huonghua Depression, Bohai Bay Basin, provides an example of paralic lake deposits. Glauconite was discovered in the northern part of the depression from section 3 to section 1 of Shahejie Formation (He Jinyu 1977). Glauconite is scarce in the underlying section 3, but it is relatively enriched in the lower section 1, reaching up to 1.6% of the heavy minerals. Within the section, glauconite coexists with freshwater ostracods, which have smooth surfaces and thin shells, and a brackish water form (*Chinocythere*) which has complex ornament. A few foraminifera (?), one star-shaped crinod fragment and a *Pecten* fragment (?) were also reported. Thus, it appears that the Shahejie Formation is a fresh–brackish water deposit. On the other hand, the Oligocene lower Gancaigou Group, Chaidamu Basin is an example of an inland fresh–brackish water basin, free of seawater influence. In this case, it was the arid or semi-arid conditions that controlled the salinity of lake water.

(2) Inland saline lake basins contain only a small number of oil fields in China. The most typical is the Jianghan Basin (Qiangjiang Formation, Oligocene). Another example is Chaidamu Basin (Miocene to Pliocene sequence). In the deposits of salt lakes, petroleum source rocks are interbedded with salt and gypsum-bearing rocks owing to the cyclic changes of dry and semi-arid climate. Dark-coloured mudstones and gypsiferous mudstones appear to be the most likely source intervals and probably develop when the saline water of the lake became freshened by runoff. Although this kind of lake contains relatively few organisms and the organic content of any resulting source rock is only 0.5–0.8% lower than that in other kinds of lacustrine source rocks, the organic material is well preserved because of the high salinity and strong reducing environment. Good depositional conditions for source rocks are therefore a possibility

in inland saline lake basins (Huang Difan et al. 1982).

(3) Inland swampy lake basins with conditions suitable for coal formation commonly occurred during Jurassic and Tertiary times in China. Jurassic rocks in Chaidamu Basin are developed as freshwater lake deposits. Coal-forming conditions existed here principally in the early and the late stages of the lake existence. The source rocks occur mainly in the middle part of the succession and consist of dark grey to black mudstones and siltstones. Peat developed along the margins of the lake basin, while oil-prone sediments formed particularly in the deeper parts of the basins, resulting in intertonguing of the oil-prone and coaly layers (Swain 1980).

(4) Interdigitated terrestrial and marine deposits can form during intervals of transgression and regression. Marine plankton and terrigenous flora can therefore be found mixed in the same sediments. Such sediments are preserved in the Shanganning Basin, and are represented by the Benxi and Taiyuan Formations of Middle to Upper Carboniferous age. The Benxi Formation consists of typical intertonguing marine and terrestrial paralic swamp deposits, while the Taiyuan Formation formed in a terrestrial swampy deposit with a marine transgressive unit. Source intervals formed in both formations and the oils found in the Permian reservoir of the Shanganning Basin are believed to come from these source rocks (Huang Ruchong et al. 1981).

In this study, a total of nine typical oil and seven source rock samples were selected for comparison from the Huanghua Depression in the Bohai Bay Basin, Jianghan Basin, Chaidamu Basin and Shanganning Basin (Figs 1 and 2, Table 1). The purpose of the study was to determine the similarities and differences of continental crude oils formed in the four depositional palaeoenvironments discussed above.

Experimental

Sample preparation

The oil samples studied were separated on silica gel columns into an n-hexane eluate, a benzene eluate and a pyridine and methanol eluate. A

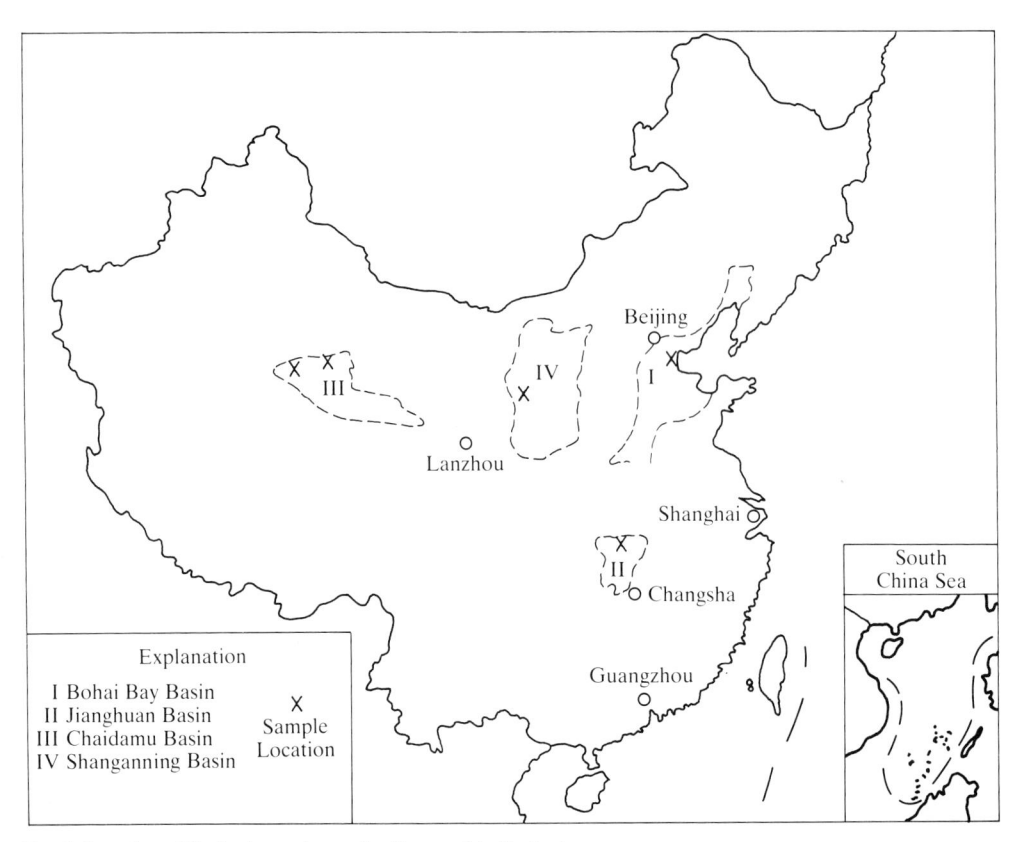

FIG. 1. Location of the basins and samples discussed in the text.

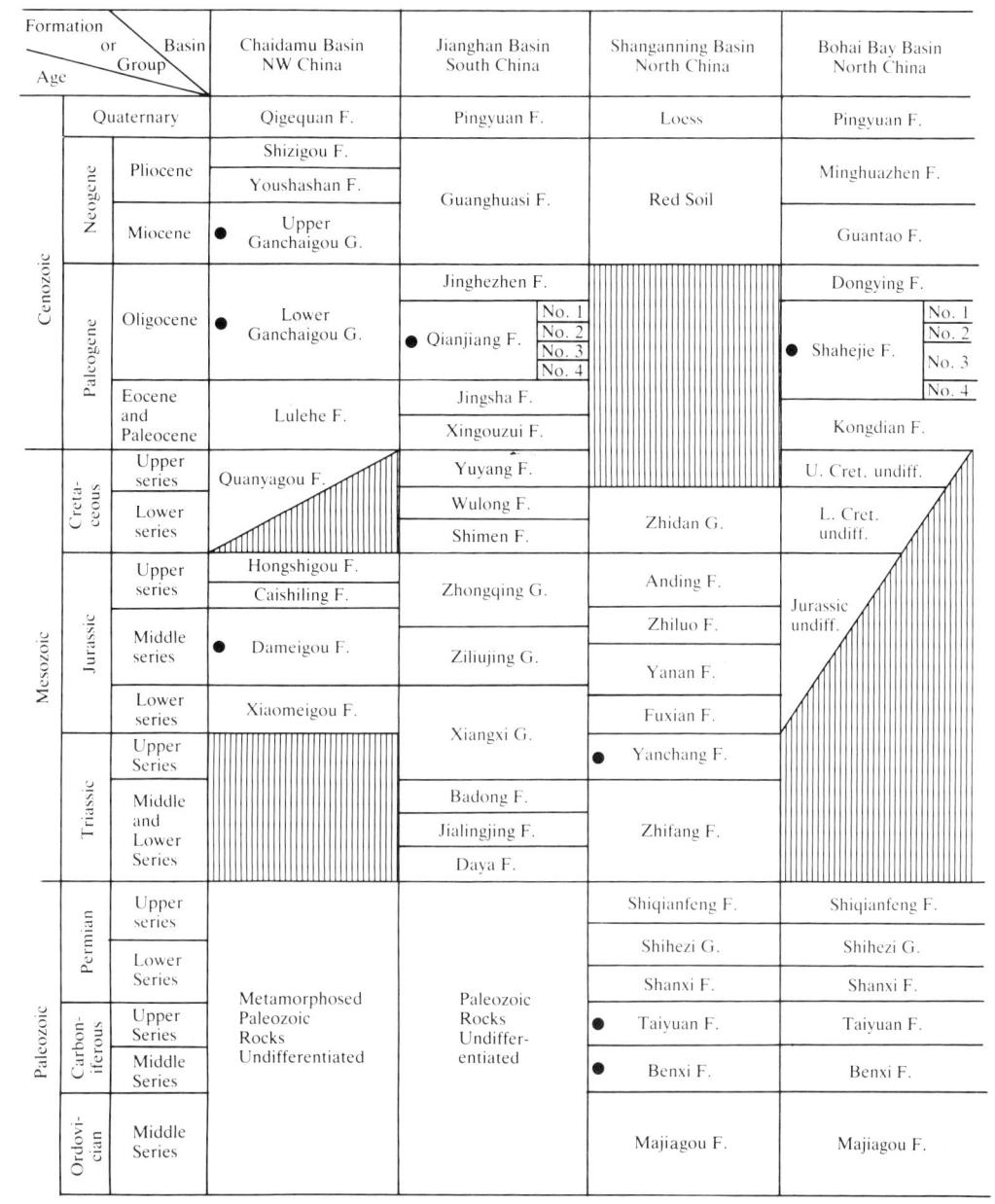

Geochemical characteristics of crude oils 311

Age	Formation or Group	Chaidamu Basin NW China	Jianghan Basin South China	Shanganning Basin North China	Bohai Bay Basin North China
Cenozoic — Quaternary		Qigequan F.	Pingyuan F.	Loess	Pingyuan F.
Cenozoic — Neogene — Pliocene		Shizigou F. / Youshashan F.	Guanghuasi F.	Red Soil	Minghuazhen F.
Cenozoic — Neogene — Miocene		● Upper Ganchaigou G.			Guantao F.
Cenozoic — Paleogene — Oligocene		● Lower Ganchaigou G.	Jinghezhen F. / ● Qianjiang F. (No. 1, No. 2, No. 3, No. 4)		Dongying F. / ● Shahejie F. (No. 1, No. 2, No. 3, No. 4)
Cenozoic — Paleogene — Eocene and Paleocene		Lulehe F.	Jingsha F. / Xingouzui F.		Kongdian F.
Mesozoic — Cretaceous — Upper series		Quanyagou F.	Yuyang F.	Zhidan G.	U. Cret. undiff.
Mesozoic — Cretaceous — Lower series			Wulong F. / Shimen F.		L. Cret. undiff.
Mesozoic — Jurassic — Upper series		Hongshigou F. / Caishiling F.	Zhongqing G.	Anding F.	Jurassic undiff.
Mesozoic — Jurassic — Middle series		● Dameigou F.	Ziliujing G.	Zhiluo F. / Yanan F.	
Mesozoic — Jurassic — Lower series		Xiaomeigou F.	Xiangxi G.	Fuxian F.	
Mesozoic — Triassic — Upper Series				● Yanchang F.	
Mesozoic — Triassic — Middle and Lower Series			Badong F. / Jialingjing F. / Daya F.	Zhifang F.	
Paleozoic — Permian — Upper series				Shiqianfeng F.	Shiqianfeng F.
Paleozoic — Permian — Lower Series				Shihezi G. / Shanxi F.	Shihezi G. / Shanxi F.
Paleozoic — Carboniferous — Upper Series		Metamorphosed Paleozoic Rocks Undifferentiated	Paleozoic Rocks Undifferentiated	● Taiyuan F.	Taiyuan F.
Paleozoic — Carboniferous — Middle Series				● Benxi F.	Benxi F.
Paleozoic — Ordovician — Middle Series				Majiagou F.	Majiagou F.

FIG. 2. Stratigraphic system in the four Chinese basins studied.

100-g sample of each source rock was extracted in Soxhlets with a benzene and methanol mixture (80:20). The extracts after removal of sulphur were separated on silica gel columns as was done with the oil samples.

Gas chromatography

The saturated hydrocarbon (n-hexane eluate) analysis were carried out on a Hewlett Packard 5750G equipped with a 72 × ⅛ inch stainless steel column (3% Apiezo L on 80/100 mesh supeleoport solid phase) programmed from 80 to 280°C (6°C/min) with a flow rate of 50 ml/min He.

Computerized gas chromatography–mass spectrometry (C-GC-MS)

The saturated hydrocarbon samples were analysed by the Department of Biochemistry, Uni-

TABLE 1. *Description of samples*

No.	Sample	Age†	Formation or group	Stratigraphic unit	Depositional environment
I *Huanghua Depression, Bohai Bay Basin*					
H-3	Crude oil	Oligocene	No. 2 Shahejie Formation	Es2	Paralic fresh–brackish water lake
H-2	Crude oil	Oligocene	No. 3 Shahejie Formation	Es3	
H-1	Mudstone	Oligocene	No. 3 Shahejie Formation	Es3	
II *Jianghan Basin*					
J-1	Mudstone	Oligocene	No. 4 Qianjiang Formation	Eq4	Inland brackish–brine lake
J-2	Crude oil	Oligocene	No. 4 Qianjiang Formation	Eq4	
J-3	Crude oil	Oligocene	Qianjiang Formation	Eq	
III *Chaidamu Basin*					
C-1	Crude oil	Miocene	Upper Ganchaigou Group	N1	Inland brackish lake
C-2	Crude oil	Oligocene	Lower Ganchaigou Group	E3	Inland fresh–brackish water lake
C-3	Crude oil	Middle Jurassic	Dameigou Formation	J2	Inland swampy lake
IV *Shanganning Basin*					
S-4	Crude oil	Late Triassic	Yanchang Formation	T3	Inland freshwater lake
S-10	Crude oil	Permian	—	P	Continental facies
S-5	Calcareous mudstone	Late Carboniferous	Taiyuan Formation	C3t	Inland swampy lake and marine
S-6	Siltstone	Late Carboniferous	Taiyuan Formation	C3t	Inland swampy lake and marine
S-7	Mudstone	Carboniferous	—	C	Terrestrial and marine intertonguing paralic swamp
S-8	Mudstone	Middle Carboniferous	Benxi Formation	C2b	
S-9	Mudstone	Middle Carboniferous	Benxi Formation	C2b	

† Age of reservoir in the case of oil.

versity of Minnesota using a LKB 9000 C-GC-MS with a two-foot packed 3% OV-1 column. The GC was programmed from 70–300°C at 6°C/min.

Results

Organic content

Differences between various aquatic environments have a profound influence on the accumulation and preservation of organic matter and development of source rocks (Swain 1970). A relatively greater amount of organic matter is found in the source rocks of both inland swampy lake facies (Chaidamu Basin, Jurassic) and terrestrial and marine intertonguing facies (Shanganning Basin, Carboniferous) than in other types. This is because conditions in these two depositional environments favoured the formation of coal. In addition to plant material, the source material in the Carboniferous may have included marine aquatic organisms. For instance, a few layers of biogenic shelly limestone, which are rich in fusulinid foraminifera, have organic carbon contents of up to 2.89% and bitumen contents which reach 2770 ppm. In general, though, these layers are less than 10 m thick. In contrast, about half of the black mudstone in the Benxi Formation, Middle Carboniferous, which reaches up to more than 330 m in thickness, belongs to marine facies. In the mudstone, organic carbon contents can be as high as 3.11% (Huang Ruchong *et al.* 1981).

In freshwater or fresh–brackish water facies, the organic content of source rocks are less than that in the two types mentioned above. In this type of basin in China, the source rocks are rich in various kinds of charophytes, algae, ostracods, bivalves and gastropods. These organisms may favour the formation of oil and gas.

In comparison with the Upper Triassic source rocks in Shanganning Basin, the Oligocene source rocks in Huanghua Depression have somewhat higher organic contents. Thus it appears that the conditions in paralic lakes are favourable for the accumulation of organic matter from both terrestrial and marine sources. On the other hand, inland saline lake source rocks have few fossils and low organic contents. Overall, it seems that saline lake conditions are not suitable for the accumulation of organic matter. Furthermore, diagenetic processes in inland saline lakes, involving the consumption of organic matter by sulphate-reducing bacteria contribute to this situation.

Bitumen group composition

The bitumen group compositions of the extracts from source rocks also reflect the difference in sedimentary palaeoenvironments (Table 2, Fig. 3). The bitumen group compositions of clayey source rocks from different lake facies are much more closely related to those of crude oil than are those of terrestrial and marine intertonguing swampy facies. The former have more saturates and less aromatics than the latter though both may have similar asphaltene and resin contents. This distinction can be attributed to differences in source materials. The large amount of lignin in higher plants may be the major source of aromatics in swamp facies source rocks. The mixture of terrestrial plants and aquatic organisms is the most likely parent material of lake facies oils, including lipids, wax, etc. which can form a considerable quantity of saturates.

There are no significant differences in the bitumen group composition of various oil types studied (Fig. 3), but the maturity of oils differ significantly from sample to sample. Among them, the S-10 oil from the Permian, Shanganning Basin, and C-3 oil from Jurassic, Chaidamu Basin, are from the oldest reservoir, representing the most mature oils, having more than 80% saturates. All the Triassic and Tertiary oils seem to have the same degree of maturity.

Normal alkanes

The distributions of n-alkanes differ little among the various samples. However, there are some differences in the number of major carbon peaks, $(C_{21}+C_{22})/(C_{28}+C_{29})$ ratios and CPI values (Table 2, Figs 4 and 5). The S-10 and C-3 oils are of relatively high maturity as indicated by their major peaks and $(C_{21}+C_{22})/(C_{28}+C_{29})$ ratios (4.3 and 3.9 respectively). This is consistent with their CPI values (1.0 and 0.95 respectively), high saturates contents, and older ages. The other oils are less mature (see Fig. 4), possibly reflecting some characteristics of their depositional palaeoenvironments. The examples are: the C-1 oil from inland brackish–brine lake basin (Chaidamu Basin), and the H-2 oil from paralic fresh–brackish lake basin (Huanghua Depression), which have the same $(C_{21}+C_{22})/(C_{28}+C_{29})$ ratio (2.6), but different major peaks (C_{22} and C_{17}). One possible explanation is that if the $(C_{21}+C_{22})/(C_{28}+C_{29})$ ratio (2.6) means the two oils have an equal maturity, the difference of major peak numbers may indicate that paralic lake basins like Huanghua Depression received more aquatic organic material, including some marine material, than the inland lake basins like the Chaidamu Basin. According to the $(C_{21}+C_{22})/(C_{28}+C_{29})$ ratio, the sample J-2 is the least mature oil among the samples examined (Fig. 4d) and has the highest major peak number (C_{24}). This implies that, in inland brine lakes, more organic material was supplied in the form of terrestrial plants and lesser amounts came from aquatic organisms than in inland and paralic freshwater lakes because the brine lakes are not suitable for aquatic productivity.

TABLE 2. *Column and gas chromatography analysis data*

No.	Sample	Age	Bitumen group composition %			Normal alkane			Branched alkanes pristane/phytane
			Saturates	Aromatics	Resins and asphaltenes	Carbon number distribution	Major peak	CPI	
H-3	Crude oil	Es2	62.92	14.06	23.02	C_{15}–C_{30}	C_{15}	0.98	2.31
H-2	Crude oil	Es3	66.22	28.86	4.92	C_{11}–C_{33}	C_{17}	0.98	1.57
H-1	Mudstone	Es3	54.60	9.59	35.81	C_{11}–C_{32}	C_{26}	0.97	n.d.
J-1	Mudstone	Eq4	38.46	8.77	52.82	C_{12}–C_{32}	C_{22}	0.91	n.d.
J-2	Crude oil	Eq4	70.26	11.59	18.15	C_{13}–C_{32}	C_{24}	0.78	0.08
J-3	Crude oil	Eq	69.93	11.82	12.25	C_{14}–C_{32}	C_{22}	0.60	0.18
C-1	Crude oil	N1	40.55	13.07	46.57	C_{12}–C_{30}	C_{22}	0.74	0.27
C-2	Crude oil	E3	68.35	14.38	17.27	C_{13}–C_{30}	C_{22}	0.72	0.30
C-3	Crude oil	J2	83.40	11.82	4.78	C_{13}–C_{31}	C_{17}	1.0	2.90
S-4	Crude oil	T3	65.21	22.62	12.17	C_{11}–C_{31}	C_{15}	1.0	1.25
S-10	Crude oil	P	85.16	7.42	7.42	C_{12}–C_{31}	C_{15}	0.95	4.63
S-5	Calcareous mudstone	C3t	13.85	37.26	48.89	C_{15}–C_{32}	C_{18}, C_{19}	1.02	1.91
S-6	Siltstone	C3t	11.06	24.67	64.27	C_{13}–C_{31}	C_{19}	0.95	1.78
S-7	Mudstone	C	13.59	34.22	52.19	C_{13}–C_{28}	C_{19}	1.02	2.80
S-8	Mudstone	C2b	20.79	36.98	42.33	C_{14}–C_{32}	C_{19}	1.01	2.00
S-9	Mudstone	C2b	18.40	39.06	42.54	C_{15}–C_{26}	C_{19}	1.03	1.03

n.d. = no data.

314   *Wang Tieguan* et al.

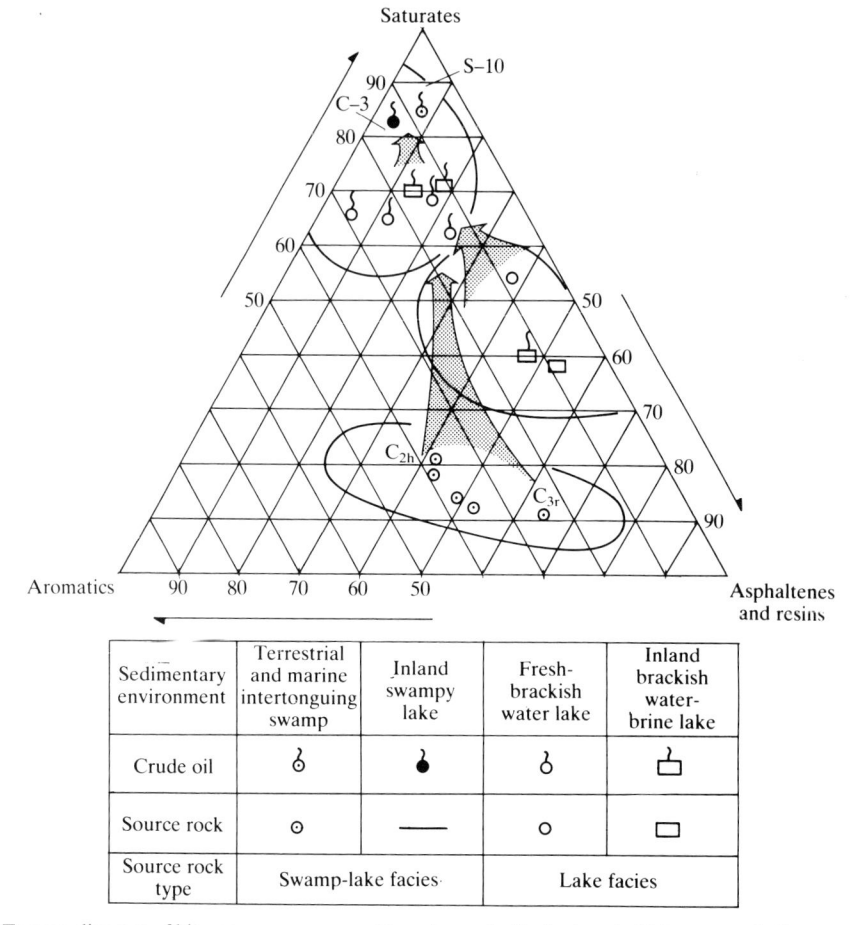

FIG. 3. Ternary diagram of bitumen group composition. Arrow indicates trend of bitumen evolution.

Four oil samples, including J-2, J-3 (Oligocene, Jianghan Basin) and C-1, C-2 (Miocene and Oligocene respectively, Chaidamu Basin), show a notable even–odd predominance (EOP) of n-alkane within the C_{20}–C_{28} range (Table 2, Fig. 4). Welte & Waples (1973) suggested that under very reducing conditions, reduction of n-fatty acids, alcohols from waxes and phytanic acid or phytol is prevalent over decarboxylation, resulting in predominance of even carbon-numbered n-alkane molecules (CPI < 1) and a predominance of phytane over pristane. Dembicki *et al.* (1976) indicated that bacterial activity in a hypersaline environment could produce an organic-rich deposit containing a predominance of even carbon-numbered C_{20}–C_{30} n-alkanes. Shen Guoying *et al.* (1980) reported an example of remarkable preponderance of even over odd alkanes in the source rocks and crude oils from Shahejie Formation, Shengli Oil Field, Bohai Bay Basin. They also pointed out that those samples occur in which evaporites, gypsum and carbonates are closely associated. We have obtained similar results for our Oligocene samples from Jianghan Basin, suggesting similar palaeoenvironmental conditions to those described by Shen Guoying *et al.* (1980).

GC traces of the Oligocene–Miocene samples from the Chaidamu Basin which were deposited in a fresh–brackish water environment also show even–odd n-alkane predominance. Thus, it is possible that even-carbon predominant oils also form from source rocks deposited in brackish water environments of inland basins under dry climatic conditions (Wang Youxiao *et al.* 1983).

Branched alkanes

Based on the data obtained by GC-MS analysis, the different type of isoprenoids, including head-to-tail linked 'regular' isoprenoids (2, 6, 10, ... and 3, 7, 11, ... polymethylalkane series) and

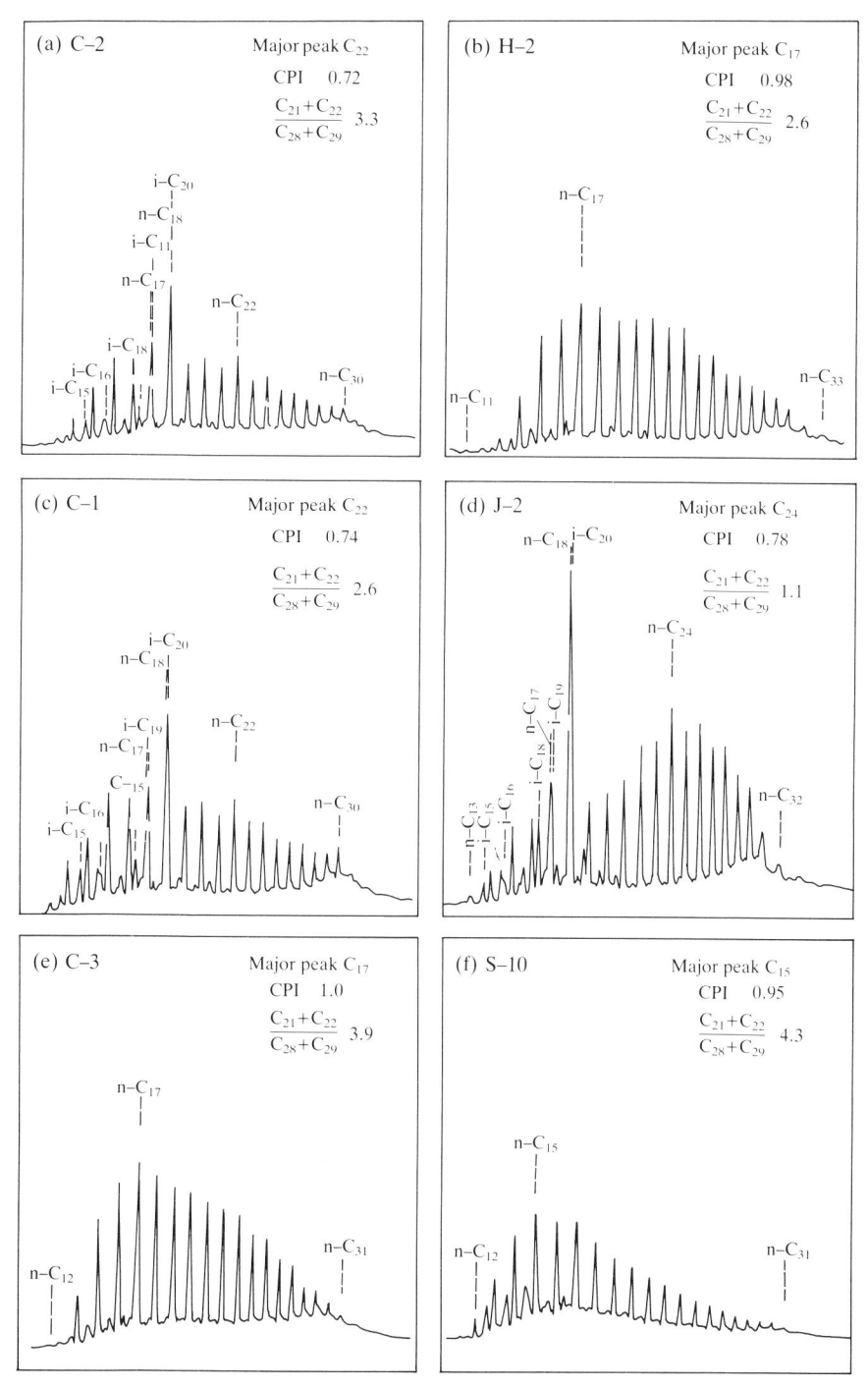

FIG. 4. Saturates hydrocarbon gas chromatograms of typical oils.

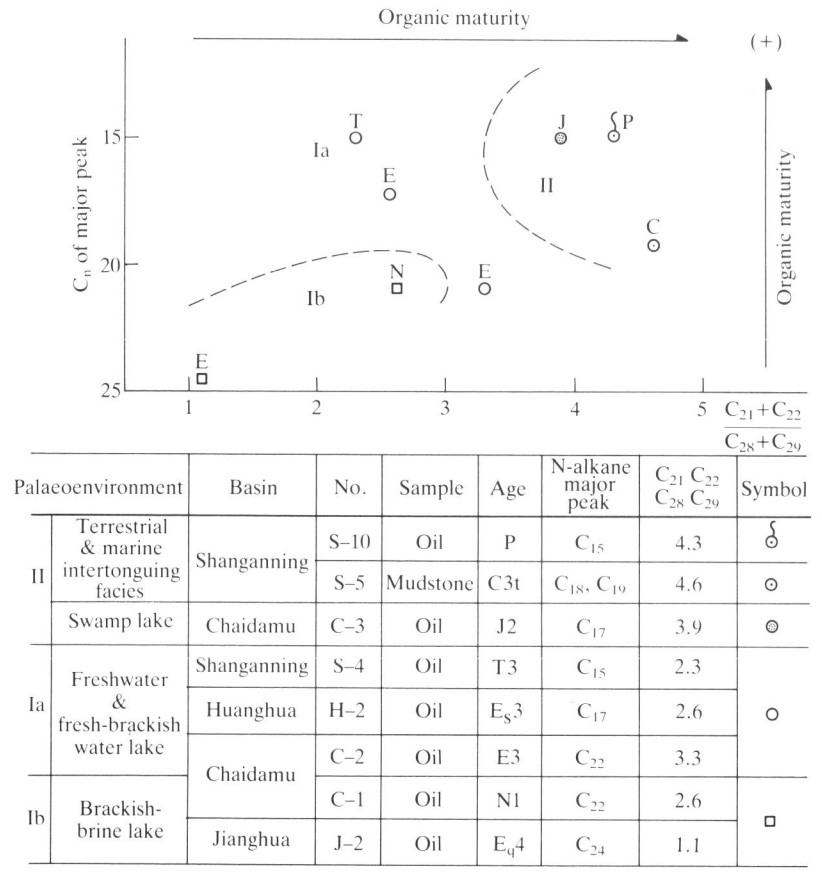

Palaeoenvironment		Basin	No.	Sample	Age	N-alkane major peak	$\frac{C_{21} C_{22}}{C_{28} C_{29}}$	Symbol
II	Terrestrial & marine intertonguing facies	Shanganning	S–10	Oil	P	C_{15}	4.3	⧖
			S–5	Mudstone	C3t	C_{18}, C_{19}	4.6	⊙
	Swamp lake	Chaidamu	C–3	Oil	J2	C_{17}	3.9	⊚
Ia	Freshwater & fresh-brackish water lake	Shanganning	S–4	Oil	T3	C_{15}	2.3	○
		Huanghua	H–2	Oil	E_s3	C_{17}	2.6	
		Chaidamu	C–2	Oil	E3	C_{22}	3.3	
Ib	Brackish- brine lake		C–1	Oil	N1	C_{22}	2.6	□
		Jianghua	J–2	Oil	E_q4	C_{24}	1.1	

FIG. 5. The relationship of n-alkane distribution and organic maturity in different palaeoenvironments designated Ia, Ib, and II.

head-to-head and tail-to-tail linked 'irregular' isoprenoids, were identified (Table 3, Figs 6 and 7).

Among the 'regular' isoprenoids, pristane and phytane (Fig. 6a, b) are the most important members. The pristane-to-phytane (Pr/Ph) ratio is an effective organic geochemical indicator of palaeoenvironment. Didyk *et al.* (1978) suggested that anoxic conditions favour preservation of C_{20} isoprenoid skeletons, thus giving low Pr/Ph values, while oxic conditions cause greater degradation so that the C_{20} skeleton is less likely to survive intact in the sediment, leading to a high Pr/Ph ratio. In the case of the four Chinese basins studied in this paper, the Pr/Ph ratio of crude oils displays a close relationship with their sedimentary palaeoenvironments (Fig. 8). Paralic swamp and inland swampy lake have higher Pr/Ph ratios, freshwater and fresh–brackish water basins have intermediate Pr/Ph ratios, while brackish–brine lakes have considerably lower ratios.

These are due to the redox conditions and the relative salinities. Moreover, higher salinites in the aqueous environments favour reducing conditions and some lipids of halophilic bacteria are believed to be a source for phytane in saline environment (Kaplan & Baedecker 1970). The decrease of the crude oil Pr/Ph ratio with the increase of the relative salinity in the water body is illustrated in Figure 8.

The crude oils of the paralic environments, which would have had higher salinities than inland freshwater environments, have higher Pr/Ph ratios. This could be due to the influence of seawater. Transgression would bring some marine organisms into the paralic environment as previously mentioned. Although both pristane and phytane are mainly thought to originate by geochemical diagenesis of phytol (Welte & Waples 1973; Didyk *et al.* 1978), pristane can also be generated by certain marine organisms (Blumer *et al.* 1971). In addition, the S-10 oil

TABLE 3. *GC-MS analysis data*

Sample no.	Branched alkane — Regular isoprenoid head-to-tail linkage type — 2, 6, 10,... polymethyl-	Regular isoprenoid head-to-tail linkage type — 3, 7, 11,... polymethyl-	Irregular isoprenoid — Head-to-head linkage type	Irregular isoprenoid — Tail-to-tail linkage type	Cyclic alkane — Alkyl-cyclohexane series	Cyclic alkane — Dicyclohexyl-alkane series	Cyclic alkane — Specific pentacyclic triterpane & sterane
H-3	C_{14}, C_{15}, C_{16}, C_{18}, C_{19}*1, (C_{20})*2	—	(C_{17})*1			C_{15}	Onocerane-II, hopane, 17(H), 21(H)-homohopane
H-2	(C_{14})*1, C_{16}, C_{18}, (C_{19}, C_{20})*2, C_{21}, C_{23}, C_{24}, C_{25}	C_{15}, C_{14}	C_{15}, C_{18}, C_{21}, C_{23}	(C_{22})*1, C_{23}, C_{25}, C_{26}, C_{27}	C_{12}, C_{16}, C_{20}	C_{15}	Onocerane-I, onocerane-II, onocerane-III, serratane-II
J-1	C_{15}, C_{16}, C_{18}, C_{19}, C_{20}, C_{21}, (C_{23})*1, C_{24}, C_{28}	(C_{19})*1	C_{19}, C_{20}, C_{21}, C_{27}, C_{28}	(C_{22})*1	C_{16}, C_{21}		Hopane
J-2	C_{14}, C_{15}, C_{16}, C_{18}, C_{19}, C_{20}, C_{21}, C_{23}, C_{24}, (C_{24})*1, (C_{25})*1, C_{25}	C_{15}	C_{19}, C_{20}	(C_{22})*1, C_{24}	C_{16}, C_{21}		n.d.
J-3	(C_{19}, C_{20})*2	(C_{25})*1	C_{23}		n.d.	n.d.	Onocerane-II
C-1	(C_{13})*1, C_{14}, C_{19}, (C_{20})*2	C_{14}, (C_{15})*1	C_{18}, (C_{19})*1		C_{16}	C_{15}	Onocerane-II, serratane-II
C-2	C_{18}, (C_{19})*2, C_{20}		C_{15}, C_{16}, C_{26}		n.d.	n.d.	Onocerane-III, serratane-I
C-3	(C_{19}, C_{20})*2	C_{15}	(C_{18})*1, C_{19}		C_{19}		Onocerane-III, serratane-II
S-10	C_{14}, C_{16}, C_{19}	C_{14}, C_{18}			C_{15}		Onocerane-III, serratane-II
S-5	C_{19}, (C_{20})*2, C_{26}		C_{20}, C_{22}, C_{23}	C_{27}	C_{21}		Onocerane-II, serratane-II, hopane series
S-8	C_{19}, (C_{20})*2, C_{21}		C_{19}		C_{17}, C_{18}, C_{19}, C_{23}		n.d.
S-9	C_{19}, (C_{20})*2		C_{18}, C_{22}		n.d.	n.d.	Onocerane-III, serratane-I

*1 = demethylated isoprenoids.
*2 = (C_n) was only detected by GC, but unexamined by MS.
n.d. = no data.

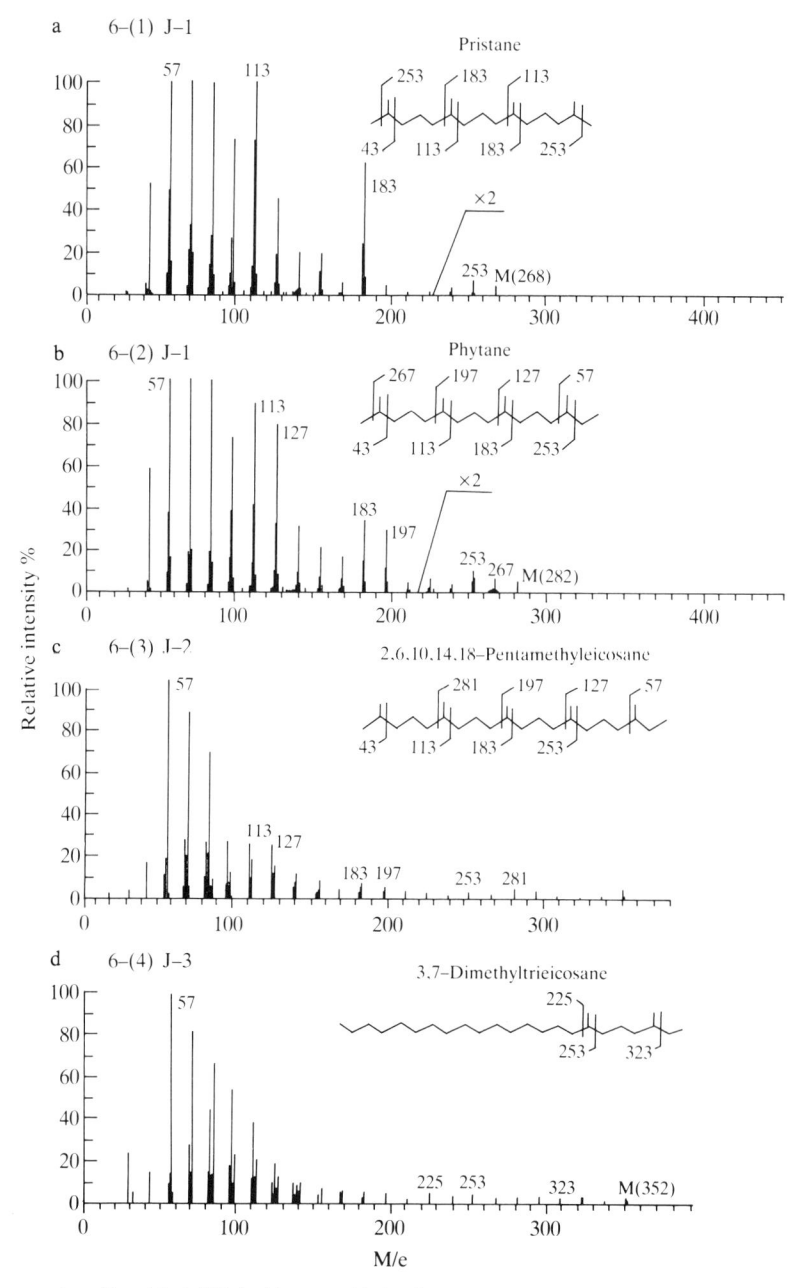

FIG. 6. Mass spectra of head-to-tail linked isoprenoids (a–d).

sample from a paralic swamp palaeoenvironment has the highest Pr/Ph ratio, up to 4.63. The latter may also be caused by weakly reducing sedimentary conditions in swamp environments. Supporting evidence is provided by the correspondence of the CPI of n-alkane and Pr/Ph curves. Although the CPI and Pr/Ph data are based on the different organic compounds, the CPI values in swamp environments discussed are about 1 which indicates the same redox conditions as the Pr/Ph ratio does.

Besides pristane and phytane, the C_{25} isoprenoid hydrocarbon (2,6,10,14,18-pentamethyleicosane, Fig. 6c) among the 2, 6, 10, ... series of

Geochemical characteristics of crude oils

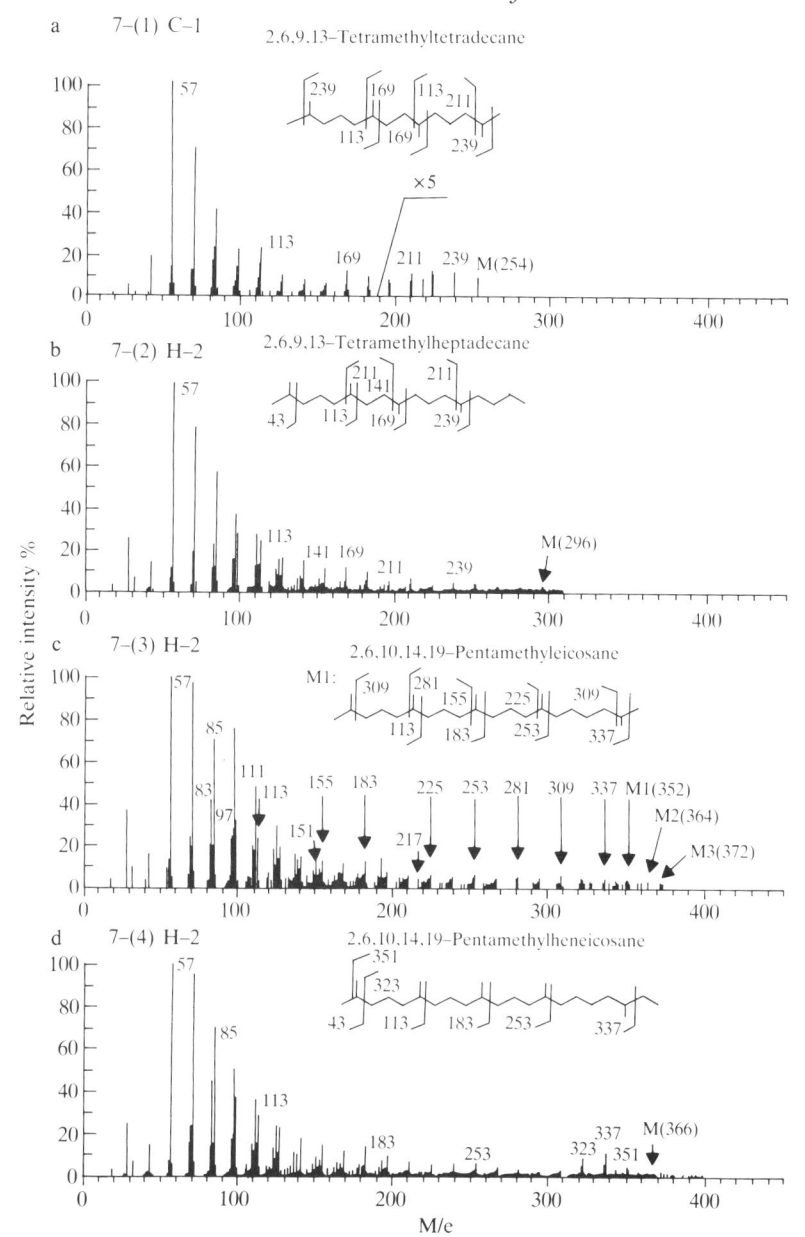

FIG. 7. Mass spectra of head-to-head and tail-to-tail linked isoprenoids (a–d).

regular head-to-tail isoprenoids has previously been reported as a constituent of various oils, shales and other sediments (Brassell *et al.* 1981). Waples *et al.* (1974) suggested that this C_{25} isoprenoid may represent a biological marker, possibly typical of lagoonal-type, saline environment. In the present study, the 2,6,10,14,18-pentamethyleicosane occurs in oils both from the paralic fresh–brackish water (H-2) and inland brackish–brine (J-2) environments. This is related to the halophilic bacteria. Its homologous compounds, including C_{21}, C_{23}, C_{24}, C_{26} and C_{28}, isoprenoid hydrocarbons, have also been found in the samples examined.

As a minor compound, the 3, 7, 11, . . . series of regular isoprenoid hydrocarbons exist in most of the oils and source rocks studied (Table 3) and range from C_{14} to C_{25}. The compound 3,7-dimethyltrieicosane is shown in Figure 6d as an example of this series. Albaigés (1980) proposed

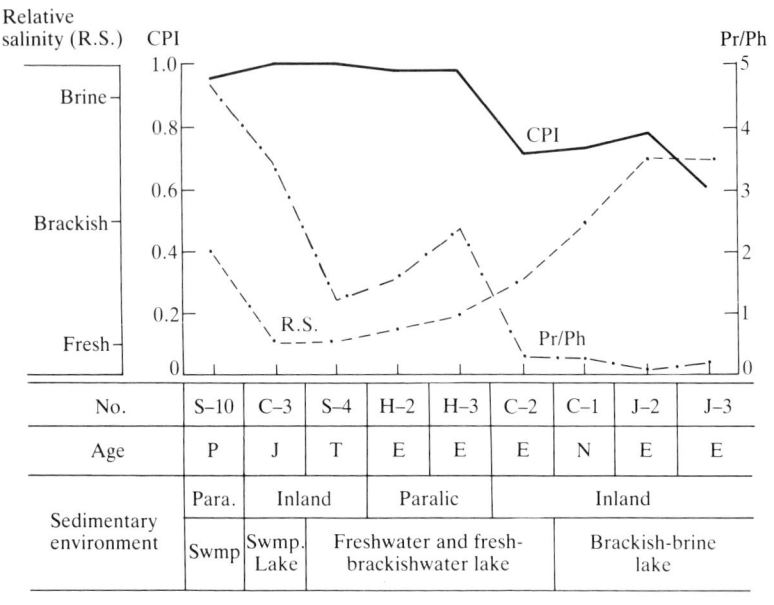

FIG. 8. Comparison of pristane-to-phytane ratio, CPI of crude oil, and relative salinity in aqueous palaeoenvironments.

that thermocatalytic degradation could account for the formation of the regular 3, 7, 10, ... series.

Irregular head-to-head isoprenoid hydrocarbons, which range from C_{15} to C_{28}, are only tentatively identified. For example, 2,6,9,13-tetramethyltetradecane and 2,6,9,13-tetramethylheptadecane are shown in Fig. 7a, b. Moldowan & Seifert (1979) indicated that the head-to-head structural feature has only one known biological natural product analogue, $\omega,\omega°$-biphytanediol, which is found in cell wall membranes of thermoacidophilic bacteria of the *Calderiella* series. Thus the discovery of head-to-head linked isoprenoids in petroleum provides compelling evidence for a ubiquitous and substantial contribution of bacterial cell wall lipids to crude oils. In our case, the head-to-head linked isoprenoids are found in almost every sample examined. They may indicate a microbial source input and have no specific sedimentary facies significance.

The irregular tail-to-tail polymethylalkanes have also been tentatively identified in four samples from paralic and inland lakes. But the C_{25} 2,6,10,14,19-pentamethyleicosane and C_{26} 2,6,10,14,19-pentamethylheneicosane occur only in a paralic brackish water lake. Both of them may be derived from the degradation of lycopane (Fig. 7c, d). Lycopane is thought to be found in significant quantities in recent and ancient marine sediments and also could have been directly derived from bacteria (Brassell *et al.* 1981). It seems that the 2,6,10,14,19-pentamethyleicosane may be related to halophilic or methanogenic archaebacteria. One of its isomers, 2,6,10,15,19-pentamethyleicosane has been designated as a biological marker for methanogenic activity in marine sediments of different ages (Brassell *et al.* 1981; Rowland *et al.* 1982). Preliminary results of our study indicate that the distribution and evolution of C_{25} 2,6,10,14,19-pentamethyleicosane are indeed worth additional investigation.

Cyclic alkanes

The cyclic alkanes, encompassing lower molecular weight alkylcyclohexanes, dicyclohexylalkanes and higher molecular weight steranes and pentacyclic triterpanes, have been recognized in the crude oils and source rocks examined. The alkyl cyclohexanes consist of C_{12}, C_{15} to C_{21}, and C_{23}, but the dicyclohexylalkanes only of C_{15}. The C_{16} alkylcyclohexane and C_{15} dicyclohexyl alkane are shown in Fig. 9. Such structures are reported to be uncommon in the biosphere and thus it is difficult to postulate biological precursors for these compounds, although intramolecular cyclization of unsaturated fatty acids could account for their presence in geological conditions (Maxwell *et al.* 1971). Consequently the formation of these compounds cannot be linked to conditions in the sedimentary palaeoenvironment.

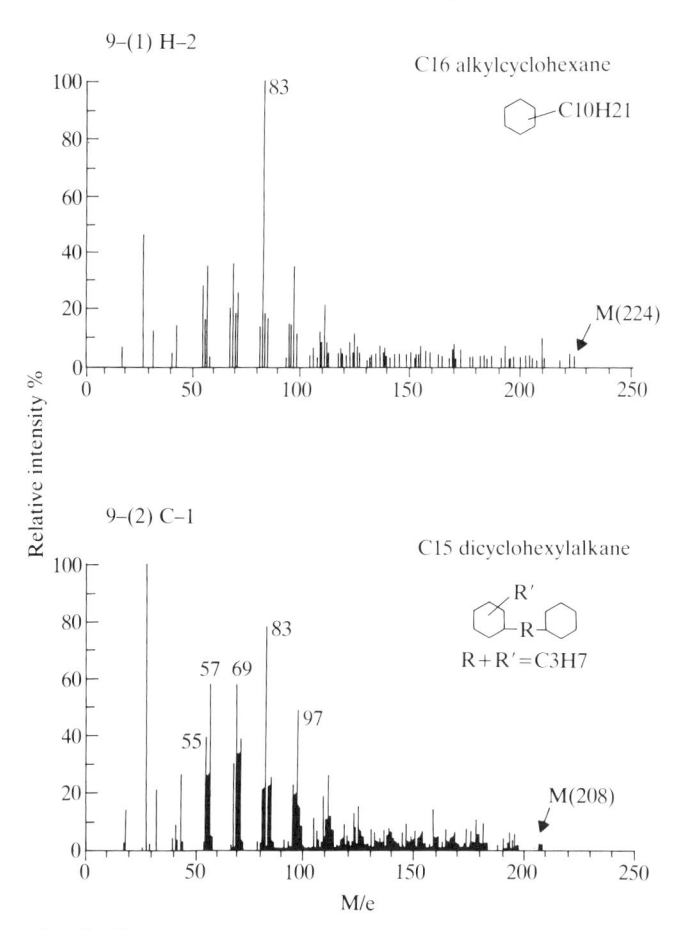

FIG. 9. Mass spectra of cyclic alkanes.

Steranes have been found in most of the samples studied. Table 4 shows a relatively complete list of steranes found in samples from Jianghan Basin. A more complete sterane series was reported from the same basin by Yang Zhiquiong et al. (1983).

A hopane series is present in S-5, a black, fossiliferous calcareous mudstone of the Taiyuan Formation, Late Carboniferous, Shanganning Basin (Table 5).

Although sometimes the individual pentacyclic triterpane may not always be separated from other components under the chromatographic conditions in the GC-MS system in the present study, it is still possible to identify onoceranes and serratanes in the oils and source rocks from the different palaeoenvironments, although, up to now, only onoceranes have been reported from one formation, the Oligocene Shahejie Formation of the Jizhong Depression, which is about 100 km west of Huanghua Depression in the Bohai Bay Basin (Zhang Zhencai et al. 1983). The various stereoisomers of serratane and onocerane seem to have different distributions. The results of our study suggest that serratane-I may be limited to freshwater palaeoenvironment, and onocerane-III to relatively low salinity environment. Serratane-II (Fig. 10) and onocerane-II (Fig. 11) appear to be present in environments ranging from freshwater to brackish–brine environments. In brine environment, the two biomarkers probably are not developed (Fig. 12).

Discussion and Conclusions

The four basins discussed in this report contain predominantly continental sediments and terrestrial and non-marine aquatic organic matter. However, sediments in two of the basins also contain paralic sediments and subordinate amounts of marine organisms.

TABLE 4. *Steranes and pentacyclic triterpanes isolated from crude oils and their source rocks in Jianghan Basin*

No.*	Assignment	Molecular formula	Molecular weight	Sample no.
1	Coprostane 20R	$C_{27}H_{48}$	372	J-2
2	24-Methylcoprostane 20R	$C_{28}H_{50}$	386	J-1, J-2, J-3
3	Ergostane 20R	$C_{28}H_{50}$	386	J-1
4	Unknown C_{29} sterane	$C_{29}H_{52}$	400	J-1
5	Unknown C_{29} sterane	$C_{29}H_{52}$	400	J-1
6	24-Ethylcholestane 20S	$C_{29}H_{52}$	400	J-1
7	24-Ethylisocholestestane 20R	$C_{29}H_{52}$	400	J-1
8	24-Ethylcoprostane 20R	$C_{29}H_{52}$	400	J-2, J-3
9	Onocerane-II	$C_{30}H_{54}$	414	J-3
10	Stigomastane	$C_{29}H_{52}$	400	J-1
11	Hopane	$C_{30}H_{52}$	412	J-1

*The relative retention time is that referred to by Seifert & Moldowan (1979).

TABLE 5. *Hopane series isolated from S-5 source rock, Taiyuan Formation, Shanganning Basin*

No.	Assignment	Molecular formula	Molecular weight
1	18α(H)-22,29,30-trisnorhopane II	$C_{27}H_{46}$	370
2	17α(H),21β(H)-30-homohopane	$C_{27}H_{46}$	370
3	17α(H),21β(H)-hopane	$C_{30}H_{52}$	412
4	17α(H),21β(H)-30-homohopane 22S	$C_{31}H_{54}$	426
5	17α(H),21β)-30-homohopane 22R	$C_{31}H_{54}$	426

Reducing conditions in lacustrine depositional environments are a necessary prerequisite for the formation of continental petroleum source rocks. Although these conditions prevailed in all the basins studied, the oils formed in different basins display both some similarities and some differences in their organic geochemical characteristics.

Firstly, the four oil-bearing basins contain enough organic source material to form continental crude oils even though their depositional palaeoenvironments were quite different, ranging from inland and paralic swamp and freshwater lake to inland salt lakes. The results of our study indicate that inland and paralic fresh–brackish water basins are best suited for petroleum source

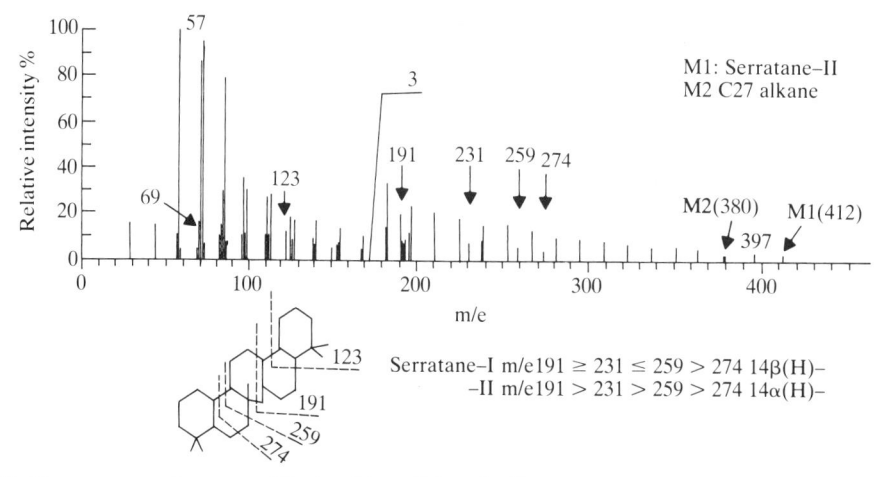

FIG. 10. Mass spectrum of serratane-II isolated from C-3 crude oil.

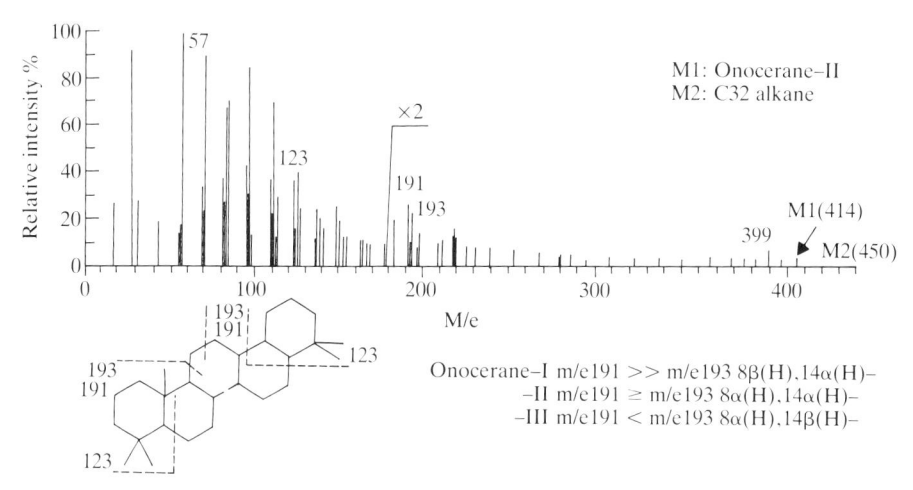

FIG. 11. Mass spectrum of onocerane-II isolated from J-3 crude oil.

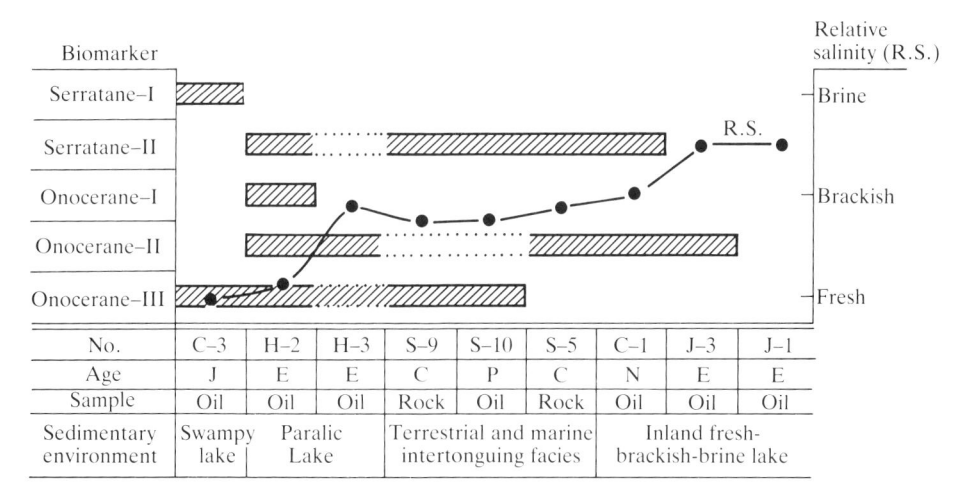

Onocerane–I m/e 191 $\gg$ m/e 193 8β(H),14α(H)–
         –II m/e 191 $\geq$ m/e 193 8α(H),14α(H)–
         –III m/e 191 $<$ m/e 193 8α(H),14β(H)–

FIG. 12. Serratane and onocerane distribution and sedimentary palaeoenvironments.

rock formation. In fact, almost all large oil fields discovered so far in China are located in these kinds of basins. Continental oils can also occur in salt lake and swampy lake basins, but conditions in these basins are less favourable for petroleum occurrence. In salt lake basins, the organic carbon content is smaller than in swampy lake basins although the quality of bitumen is better and conditions are more suitable for preservation of organic matter. Only middle-sized oil fields have been discovered in salt lake basins, and a few oil-pools occur in the swamp and swampy lake basins in China.

Secondly, the compositions of the bitumen groups of the oils formed in the different depositional palaeoenvironments are similar. Differences only relate to maturity. The carbon number distributions of the n-alkanes are very similar and range from C_{11} to C_{33}. This range in n-alkane distribution implies the existence of two kinds of sources of organic matter in the four Chinese oil-forming basins: terrestrial higher plant material giving rise to the alkanes with higher carbon numbers, and aquatic lower organisms providing alkanes with the lower carbon numbers. Based on the $(C_{21}+C_{22})/(C_{28}+C_{29})$ ratio and the carbon number of n-alkane major peaks on the gas chromatograms, one can clearly distinguish both different oils and different source contributions in inland and paralic environments. Moreover, both pristane-to-phytane ratio and CPI values are effective parameters for palaeoreconstruction, particularly of redox conditions and salinities.

Thirdly, steranes and pentacyclic triterpanes can be used as reliable palaeoenvironmental indicators. In China, some complete sterane and hopane series have been identified in the oils from Shanganning Basin (Wang Xieqing et al. 1980; Shang Huiyun et al. 1982) and Jianghan Basin (Yang Zhiqiong et al. 1983). In addition, biomarker compounds have been found to be indicators of continental organic inputs; for example, oleanane in Shanganning Basin (Wang Xieqing et al. 1980) and perylene in Jianghan Basin (Yang Zhiqiong et al. 1983). Due to the limited data, we were unable to generate additional information on these compounds. However, it appears that some stereoisomers of onocerane and serratane may prove to be important as specific biomarkers for some palaeoenvironments.

Last but not least, in recent years long chain ($>C_{20}$) polymethylisoprenyl hydrocarbons have become of considerable interest to scientists, because they have been shown to be the major compound in the lipids of archaebacteria. In our study, we identified 21 polymethylisoprenyl hydrocarbons (Table 3), including regular head-to-tail linkage (38.1%), irregular head-to-head linkage (28.6%), and tail-to-tail linkage (33.3%), in the oils and source rocks from the four Chinese basins. Up to now only a few analogues, such as perhydrocarotene, have been reported from Chinese crude oil (Jiang Zhusheng 1983). The polymethylisoprenyl alkanes, which exist as the principal lipid fraction in so-called archaebacteria (i.e., thermoacidophilic, halophilic, and methanogenic bacteria), belong to the C_{25} pentaisoprenyl, C_{30} squalenyl and C_{40} lycopenyl series. Archaebacteria seem to be an important contributor of lipids to both sediments and crude oils. The specific acyclic isoprenoid alkanes could also be palaeoenvironmental indicators. Recently, it has been proved that the C_{25} alkane 2,6,10,15,19-pentamethyleicosane is only known to exist in methanogenic bacteria and in marine sediments of Recent to Cretaceous ages (Brassell et al. 1981; Rowland et al. 1982). Its regular isomer 2,6,10,14,18-pentamethyleicosane is found in thermoacidophilic bacteria (Brassell et al. 1981). In our examples, the C_{25} 2,6,10,14,18-pentamethyleicosanes are found in the oils from the non-marine saline palaeoenvironment and the irregular C_{25} 2,6,10,14,19-pentamethyleicosane and C_{26} 2,6,10,14,19-pentamethylheneicosane are tentatively recognized in the oils from paralic palaeoenvironment. It is possible that future studies will show them to be useful indicators of continental and marine palaeoenvironments.

ACKNOWLEDGEMENTS: J. F. Wehmiller, University of Delaware, assisted with gas chromatographic analyses; Thomas Krick, University of Minnesota, performed the C-GC-MS analyses. Burnaby Munson, University of Delaware, provided advice on interpretation of mass spectra. The Department of Geology, University of Delaware, provided laboratory facilities and research funds in support of the study.

References

ALBAIGÉS, J. 1980. Identification and geochemical significance of long chain acyclic isoprenoid hydrocarbons in crude oils. In: DOUGLAS, A. G. & MAXWELL, J. R. (eds), Advances in Organic Geochemistry 1979, Pergamon Press, Oxford, 19–28.

BLUMER, M., GUILLARD, R. R. & CHASE, T. 1971. Hydrocarbons of marine phytoplankton. Marine Biology, 8, 183–189.

BRASSELL, S. C., WARDROPER, A. M. K., THOMSON, I. D., MAXWELL, J. R. & EGLINTON, G. 1981. Specific acyclic isoprenoids as biological markers of methanogenic bacteria in marine sediments. Nature, 290, 693–696.

DEMBICKI, H., MEISCHEIN JR, W. G. & HATTIN, D. E. 1976. Possible ecological and environmental significance of the predominance of even-carbon number C_{20}–C_{30} n-alkanes. Geochimica et Cosmochimica Acta, 40, 203–208.

DIDYK, B. M., SIMONEIT, B. R. T., BRASSELL, S. C. & EGLINTON, G. 1978. Organic geochemical indicators of palaeoenvironmental conditions of sedimentation. Nature, 272, 216–222.

HE JINGYU 1977. The glauconite of Lower Tertiary in the northern part of Huanghua Depression. Internal publication, PRC.

HUANG DIFAN, LI JINCHAO et al. (Scientific Research Institute of Petroleum Exploration and Development, Beijing) 1982. Formation of Oil-pools in Continental Facies, China. Petroleum Industry Press, Beijing.

HUANG RUCHONG, LUO BINJIE, FAN PU et al. (Lanzhou Institute of Geology, Academia Sinica) 1981. Formation, Evolution and Migration of Continental Oil and Gas in China. Ganshu People Press, Lanzhou.

JIANG ZHUSHENG 1983. Carotane and its geochemical characteristics in crude oil, Kalamyi oil field. Oil and Gas Geology, Jiangling, 4, 151–159.

KAPLAN, I. R. & BAEDECKER, M. J. 1970. Biological productivity in the Dead Sea, Part II evidence for phosphatidyl glycerophosphate lipid in sediment, Israel Journal of Chemistry, 8, 529–533.

MAXWELL, J. R., PILLINGER, C. T. & EGLINTON, G. 1971. Organic geochemistry, Quarterly Review, 25, 571–628.

MOLDOWAN, J. M. & SEIFERT, W. K. 1979. Head-to-head linked isoprenoid hydrocarbons in petroleum. *Science*, **224**, 169–171.

ROWLAND, S. J., LAMB, N. A., WILKINSON, C. F. & MAXWELL, J. R. 1982. Confirmation of 2,6,10,15,19-pentamethyleicosane in methanogenic bacteria and sediments. *Tetrahedron Letters*, **23**, 101–104.

SEIFERT, W. K. & MOLDOWAN, J. M. 1979. The effect of biodegradation on steranes and terpanes in crude oils. *Geochimica et Cosmochimica Acta*, **43**, 111–126.

SHANG HUIYUN, TONG YUYING, JIANG NAIHUANG & ZHU BAOQUAN 1982. Geochemical characteristics of steranes and triterpanes. *In*: *Erduosi Basin, Petroleum Exploration and Development*, **2**, 19–25.

SHEN GUOYING, FAN SHANFA, LIN DEHAN, SU NENGXIAN & ZHOU HONGMING 1980. The geochemistry of n-alkanes with an even–odd predominance in the Tertiary Shahejie Formation of northern China. *In*: DOUGLAS, A. G. & MAXWELL, J. R. (eds), *Advances in Organic Chemistry 1979*. Pergamon Press, Oxford, 115–122.

SWAIN, F. M. 1970. *Non-marine Organic Geochemistry*, Cambridge University Press, 445 pp.

—— 1980. Petroleum in continental facies. *In*: MASON, J. F. (ed.), *Petroleum Geology in China*. PennWell Books, Tulsa, 1–25.

WANG XIEQING, LI KE & LANG RENCHI 1980. Biological marker hydrocarbons in source rock and crude oil of some Chinese basins. *Acta Petrolei Sinica, Beijing*, 52–62.

WANG YOUXIAO, CHENG XUCHUI, WU YIHUA & CHUEN LANLAN 1983. Geochemical characteristics of crude oils from Qaidam Basin. *Oil and Gas Geology, Jiangling*, **4**, 121–127.

WAPLES, D. W., PAT HAUG & WELTE, D. H. 1974. Occurrence of a regular C_{25} isoprenoid hydrocarbon in Tertiary sediments representing a lagoonal-type, saline environment. *Geochimica et Cosmochimica Acta*, **38**, 381–387.

WELTE, D. H. & WAPLES, D. W. 1973. Uber die Bevorzugung geradzahlier n-Alkane in Sedimentgesteinen. *Naturwissenschaften*, **60**, 516–517.

YANG ZHIQUIONG, TONG RUYING & FAN ZHAOAN 1983. Biological marker compounds of source rocks and crude oils from Jianghan salt basin and their geological significance. *Oil and Gas Geology, Jiangling*, **4**, 269–282.

ZHANG ZHENCAI, WANG ZHONGRAN, CHEN XIANGHONG & ZENG XIANJHANG 1983. Gonanes and terpanes from crude oil and source rocks in Tizhong Depression. *Oil and Gas Geology, Jiangling*, **4**, 89–99.

WANG TIEGUAN, Department of Geology, Jianghan Petroleum Institute, Jiangling, Hubei, People's Republic of China.
FAN PU, Lanzhou Institute of Geology, Academia Sinica, Lanzhou, Ganzu, People's Republic of China.
F. M. SWAIN, Department of Geology, University of Delaware, Newark, DE 19716, USA.

Biological marker, isotopic and geological studies of lacustrine crude oils in the western Otway Basin, South Australia

D. M. McKirdy, R. E. Cox & J. G. G. Morton

Oil seeps, real and illusory, have played an important role in the history of petroleum exploration in the Otway Basin. An alleged oil seep at Alfred Flat near Salt Creek, SA led to the drilling of Australia's first oil exploration well in 1892. The dark green to black rubbery material mistaken for weathered oil was in fact coorongite. This enigmatic substance originated from blooms of the freshwater green alga *Botryococcus braunii* which stranded around the edges of small ephemeral lakes east of the Coorong Lagoon. These lakes formed between Pleistocene beach ridges after periods of exceptional rainfall.

Coorongite was first discovered in the type area in 1852, and last reported there in the mid 1950s. By 1915, some explorers had turned their attention to the bitumen and fresh waxy crude oil periodically washed ashore on beaches near Kingston and at localities further south. The existence of adjacent submarine oil seepage was correctly inferred, and in 1916 wildcat wells were drilled close to the coast at Kingston and Robe. Subsequent drilling onshore and offshore has failed to locate commercial oil. Nevertheless, continued widespread strandings of coastal bitumen and the existence of abundant reservoir bitumen throughout the Pretty Hill Sandstone in Esso Crayfish-1, constitute unequivocal proof that oil has been generated in the subsurface of the western Otway Basin.

A suite of low gravity (5–27° API) coastal bitumens was recently collected from 18 separate strandings sites between Kingston, SA and Portland, Victoria. These bitumens range from paraffinic to aromatic-intermediate in bulk composition, and may be divided into four families on the basis of their sulphur contents (S = 0.3–3.3%) and carbon and sulphur isotopic signatures. Three of the bitumen families are weathered high wax crudes which have undergone slight to moderate biodegradation. The fourth family comprises non-waxy asphaltite. The carbon isotopic composition of the C_{12+} saturated and aromatic hydrocarbons, in conjunction with low pristane/phytane ratios (Pr/Ph ⩽ 2), and C_{27}–C_{29} sterane distributions in which C_{27} (cholestane) is a major compound, reflect the primary algal source of all four bitumen families. Sterane and triterpane-based biomarker parameters show that the bitumens are mature crude oils, several of which also display evidence of fractionation effects attributable to migration. The presence in the coastal bitumens of botryococcane, the saturated equivalent of an unusual C_{34} acyclic triterpenoid hydrocarbon synthesized only by *B. braunii*, signifies that their parent oils were generated from *lacustrine* source rocks (McKirdy *et al.* 1986). Preservation of the remains of *Botryococcus* in source beds deposited under anoxic conditions in a deep stratified lake with an increasingly saline hypolimnion would account for both the waxy character of the three main bitumen types and their botryococcane and sulphur contents. Source rocks rich in lacustrine organic matter (Type I kerogen) of algal and bacterial origin so far have not been penetrated by the drill but are suggested here to occur in presumed Early Cretaceous rift-valley sediments (Otway Basin) of the offshore western Otway Basin. The existence of at least four chemically and isotopically distinct oil families implies considerable vertical and/or lateral changes in the organic facies of this lacustrine source bed sequence. Major oil accumulations have been discovered in ostensibly similar Early Cretaceous lacustrine source-reservoir sequences within rift basins along the South Atlantic margin in Brazil (Reconcavo Basin), Gabon (Gabon Basin) and Angola (Cabinda Basin).

Reference

McKirdy, D. M., Cox, R. E., Volkman, J. K. & Howell, V. J. 1986. Botryococcane in a new class of Australian non-marine crude oils. *Nature*, **320**, 57–59.

D. M. McKirdy* & R. E. Cox, Australian Mineral Development Laboratories, Frewville, South Australia.

J. G. G. Morton, South Australian Department of Mines and Energy, Parkside, South Australia.

*Present address: Department of Geology and Geophysics, University of Adelaide, Adelaide, South Australia.

The lacustrine Condor oil shale sequence

A. C. Hutton

SUMMARY: Organic matter in sedimentary rocks is derived from numerous types of organisms that lived in many different environments. Australian Tertiary oil shales, for example the Condor deposit, are termed lamosites and were formed in a lacustrine environment with the bulk of the organic matter derived from freshwater planktonic algae such as *Pediastrum*. The Condor oil shale deposit, located near Proserpine, Queensland, contains four oil shale units and a brown coal unit. The oil shales contain abundant lamalginite (or lamellar alginite), derived from planktonic freshwater algae, with a small quantity of vitrinite derived from terrestrial plants and other minor constituents. The lower oil shale unit contains a dark brown oil shale whereas the upper three oil shale units contain predominantly brown oil shale. The coal unit contains liptinite-rich cannel coals. Reflectance data for vitrinite indicate that the lower oil shale unit and the underlying coal unit are within the oil generation window.

Introduction

The Tertiary Condor oil shale deposit is located 15 km south of Proserpine (Fig. 1) in the northern part of the Hillsborough Basin which is a sediment-filled north-west–south-east trending graben, most of which lies under Repulse Bay and the Hillsborough Channel of Central Eastern Queensland, Australia. The graben was formed by Late Cretaceous or Early Tertiary block faulting (Paine 1972) and is bounded on the eastern and western edges by the volcano-sedimentary Midgeton Block and on the northern edge by the volcano-sedimentary Airlie Block. Seismic and magnetic surveys have shown that the basin is an asymmetrically synclinal basin

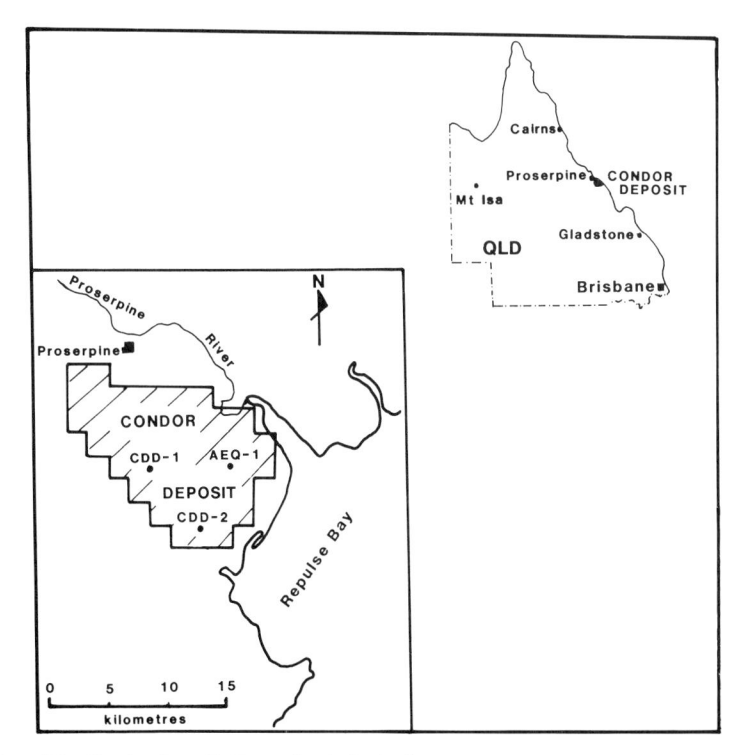

FIG. 1. Location of the Condor deposit, Queensland, Australia.

From: FLEET, A. J., KELTS, K. & TALBOT, M. R. (eds), 1988, *Lacustrine Petroleum Source Rocks*, Geological Society Special Publication No. 40, pp. 329–340.

(Ampol Exploration (Queensland) Pty Ltd, 1965) with dips of 10–14° along the western edge of the basin and 12–19° along the eastern edge.

The Hillsborough Basin was tested for petroleum potential with two drill holes in 1956–1957 and a further hole in 1965. Subsequently the Geological Survey of Queensland drilled two stratigraphic holes in 1971. During the late 1970s Southern Pacific Petroleum NL and Central Pacific Minerals NL surveyed the area as a prospective oil shale deposit and drilling commenced in 1979.

Green et al. (1984) defined three stratigraphic units which were subdivided into seven subunits (Fig. 2). The deposit extends 16 km along strike and 4 km downdip from the southwestern margin. The sequence, which has a maximum thickness of 1000 m near the centre of the deposit, has been extensively drilled (156 fully cored drill holes totalling 16 000 m). Dominant lithologies are oil shale (termed lamosite) brown coal, sandstone and claystone.

In situ resources total 9.65×10^9 barrels of shale oil using a cutoff grade of 50 l per tonne at 0% moisture (LTOM).

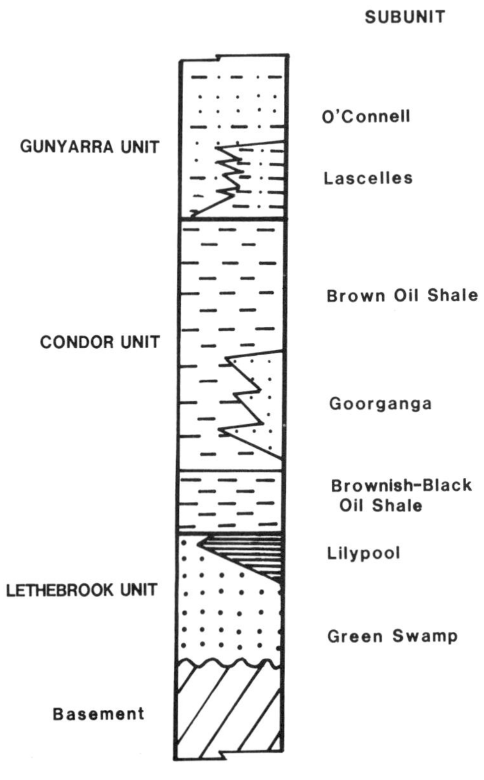

FIG. 2. Stratigraphic section, Condor deposit (after Green et al. 1984).

Experimental

Three hundred and fifty core samples from 12 Condor drill holes were examined for included organic matter with particular reference to the liptinite macerals. Reflectance from vitrinite was measured using monochromatic light of 546 nm wavelength and in oil immersion ($N_c^{23} = 1.5180$ $\gamma_e = 44$) at a temperature of $23 \pm 1°C$. The photometer was calibrated against glass and synthetic spinel standards of 0.42–1.82% reflectance. The illuminated field was 0.03 mm^2 and the back-projected image of the measuring stop 0.002 mm^2.

Fluorescence examination was carried out in oil immersion using a Leitz Ortholux microscope which incorporated a TK400 dichroic mirror fitted in an Opak vertical illuminator. Fluorescence colours quoted relate to a filter system comprising a BG3 excitation filter with BG38 and K490 suppression filters in conjunction with an objective of × 50 nominal magnification.

Maceral analyses were carried out in fluorescence mode but with repeated checking using white light whenever nonfluorescing matter was encountered. A minimum of 500 points per sample was obtained using a grid with 0.05 mm between steps and 0.2 mm between traverses.

Terminology

Sedimentary rocks contain organic matter which is usually derived from a variety of precursors which include terrestrial vascular and non-vascular plants, freshwater algae (planktonic and benthonic) as well as marine algae, dinoflagellates and acritarchs. Oil shales are organic-rich rocks which produce liquid hydrocarbons when subjected to pyrolysis. Five per cent organic matter is commonly given as the lower limit for oil shales (Tissot & Welte 1978). Oil shales contain insoluble organic matter, commonly called kerogen, and usually, a small quantity of soluble organic matter called bitumen. In many oil shales the kerogen is derived from a variety of precursors and thus, in these oil shales, it is a mixture of different types of organic matter that, in some cases, are not closely related chemically.

Macerals

Maceral terminology, as defined by the International Committee for Coal Petrology (ICCP, 1963, 1971, 1975) and Stach et al. (1975, 1982) is useful when describing the insoluble organic components of oil shales with the exception of organic matter derived from algae. Two distinctive types of algal-derived organic matter are recognized. Lamalginite is derived from thin-walled unicellular or colonial algae that has a characteristic lamellar shape in sections perpendicular to bedding. Telalginite is derived from thick-walled unicellular or colonial algae that is rounded to disc-shaped in sections parallel to bedding and exhibits internal structure. Lamalginite has weaker fluorescence intensity than telalginite where the two occur in the same sample. A list of macerals found in oil shales is given below (Table 1).

Rock types

Australian Tertiary oil shale deposits, such as the Condor deposit, are well-bedded sequences that contain seven rock types: lamosite, carbonaceous lamosite, coal, carbonaceous shale, claystone, sandstone and limestone.

(1) Lamosite (first defined by Hutton *et al.* 1980) is a brown to dark greyish-brown laminated or massive oil shale composed of at least 5 volume per cent (vol.%) lamalginite with accessory (generally <2%) telalginite, vitrinite, inertinite, sporinite and rare bitumen.

(2) Carbonaceous lamosite is a dark brown to black laminated or massive lamosite that contains not less than 5 vol.% vitrinite and/or inertodetrinite; two types occur: that composed of interlayered lamosite and coal and that which is predominantly lamosite but which contains abundant ubiquitous vitrinite and/or inertinite.

(3) Coal is a dark greyish-brown to black rock that contains not less than 50 vol.% of vitrinite and/or inertinite with accessory to abundant liptinite derived from terrestrial plants such as resinite, cutinite, sporinite, liptodetrinite but not alginite; coal with >5 vol.% liptinite is termed a cannel coal.

(4) Carbonaceous shale is a dark grey to black shale or claystone that contains <50 vol.% vitrinite/inertinite/liptinite.

(5) Claystone is a laminated or massive rock containing <5 vol.% organic matter; where organic matter occurs vitrinite is the dominant organic matter; claystone in the lamosite-bearing units contains minor lamalginite whereas that occurring in coaly units contains liptinite derived from terrestrial plants.

(6) Limestone is a massive carbonate-rich rock containing less than 5 vol.% organic matter; siderite is a common constituent of many limestones of the Condor deposit.

(7) Sandstone is a massive rock of similar organic composition to claystone but is coarser grained. It mostly occurs in the lower units of the sequence. Rocks with silt-sized grains are included in this category because they contain the same types and abundance of organic matter.

Organic petrography of the Condor deposit

The Condor oil shale deposit has been divided into three units which have been further subdivided to give seven informally-named, conformable units (Green *et al.* 1984; Fig. 2). The lower unit, the Lethebrook unit, is dominated by coal, sandstone, claystone and carbonaceous shale. The younger Condor and Gunyarra units have lamosite, claystone, limestone and minor coaly layers.

Lethebrook unit

The Lethebrook unit is divided into the Green Swamp subunit which unconformably overlies basement, and the Lilypool subunit.

TABLE 1. *Simplified table of commonly-occurring macerals in oil shale and brown coals*

Maceral	Submaceral	Normal fluorescence colours	Precursors
Vitrinite Group			
Ulminite		Nil to rarely dull brown	Cell walls of higher plants
Corpohuminite		Nil	Cell excretions and humic gels
Vitrodetrinite		Nil	Unoxidized attrital plant matter
Inertinite Group		Nil	Charred wood, oxidized cell walls, fungal remains
Liptinite Group			
Sporinite		Yellow-green to orange-brown	Spores, pollen
Cutinite		Yellow to orange-brown	Leaf and stem cuticle
Resinite		Yellow to brown	Resins, fats, waxes and oils
Bituminite		Yellow to brown	Humic and bituminous matter
Alginite	Telalginite	Green to orange	Thick-walled colonial and unicellular algae
	Lamalginite	Green to orange	Thin-walled colonial and unicellular algae, dinoflagellates and acritarchs
Liptodetrinite			Detrital liptinite

(1) *Green Swamp subunit.* This basal subunit comprises interbedded volcanolithic sandstone and claystone with minor coal and carbonaceous shale. Organic matter in the sandstone and claystone is rare and comprises sparse to rare primary vitrinite with very rare resinite, liptodetrinite and secondary coal grains up to 0.3 mm diameter. These coal grains contain vitrinite with included inertinite, sporinite, cutinite and resinite. It is most abundant in the sandstone and this mode of occurrence together with grain size, are consistent with an allochthonous origin, possibly from the Early Permian Calen Coal Measures which occur to the west and the east of the Hillsborough Basin. Primary vitrinite is probably hypautochthonous, that is, it has been derived from vegetation growing close to that part of the basin in which the sediments were deposited.

(2) *Lilypool subunit.* The Lilypool subunit is transitional to the older Green Swamp subunit and contains dull black cannel coal, coal, carbonaceous shale, sandstone and claystone.

Coals of the Lilypool subunit and the underlying Green Swamp subunit are composed of vitrite and clarite layers. Vitrite layers are composed of vitrinite (ulminite and corpohuminite) with < 5% liptinite (dominantly resinite and cutinite with rare liptodetrinite). Clarite layers contain vitrinite (mostly attrinite with ulminite) and up to 20% liptinite (resinite, liptodetrinite and minor cutinite and sporinite).

Condor unit

The Condor unit is subdivided into the Brownish-black Oil Shale, Goorganga and Brown Oil Shale subunits. A 200 m section, comprising the lower part of the Brown Oil Shale subunit and the upper part of the Brownish-black Oil Shale subunit, constitutes the main economic oil shale. Average shale oil yield for this section is 65 LTOM (Green & Bateman 1981).

(1) *Brownish-black Oil Shale subunit.* This subunit contains yellowish-brown to brownish-black lamosite and claystone. Minor lithologies include carbonaceous lamosite and coal. Boundaries with the Lilypool and Brown Oil Shale subunits are gradational. Details of the organic matter are given in Table 2.

The groundmass in samples from the Brownish-black Oil Shale subunit is a darker brown than that in samples from the overlying subunit and this probably contributes significantly to the hand-specimen colour. A weak fluorescence emanates from the groundmass and with prolonged irradiation for 1 hour or more intensity increases and a fine network of organic matter is visible. Much of this organic matter is similar in shape and size to lamalginite and an algal precursor for some, if not all, is probable.

(2) *Brown Oil Shale subunit.* The Brown Oil Shale subunit ranges in thickness from 300 to 400 m and is characterized by massive dusky, yellow-brown to brown lamosite with 20-100 m thick layers of claystone.

Hand-specimen colour is darkest near the base of the subunit where highest lamalginite (the dominant organic matter, Table 2) contents, and correspondingly, highest shale oil yields occur. Modified Fischer assays, over 2-m intervals, average 55 LTOM for the subunit as a whole and 63 LTOM for the lower half of the subunit.

The groundmass in most lamosite lacks the darker brown colour that is characteristic of the groundmass in lamosite from the Brownish-black Oil Shale unit although the groundmass becomes darker towards the base of the subunit. Prolonged irradiation of samples with a pale groundmass does not significantly increase the fluorescence from the groundmass. It is assumed that less organic matter occurs in the groundmass of paler lamosite from the Brown Oil Shale subunit than in the groundmass of lamosite from the Brownish-black Oil Shale subunit.

(3) *Goorganga subunit.* A sandstone wedge, termed the Goorganga subunit occurs within the Brown Oil Shale unit. It comprises up to 22 m of sandy oil shale and fine- to medium-grained sandstone overlying up to 163 m of graded, fine- to very coarse-grained sandstone with minor shaley layers. This subunit represents a localized facies variant and may represent the pathway of much of the clastic detritus which entered the basin during deposition of the Brown Oil Shale subunit.

Gunyarra unit

The Gunyarra unit conformably overlies the Condor Unit and consists of the Lascelles and O'Connell subunits. Both subunits are characterized by interbedded lamosite, claystone, sandstone and limestone.

(1) *Lascelles subunit.* This unit ranges in thickness from 80 to 170 m. Sandstone layers are generally < 150 mm thick and contain much less than 1% organic matter. Claystone and limestone contain the same organic assemblages as in corresponding lithologies of other subunits. Lamosite is of two types. In the first type, lamalginite is ubiquitous within a clay-sized groundmass (as

TABLE 2. *Organic matter in lamosite from the Brown Oil Shale and Brownish-black Oil Shale subunits*

	Brownish-black Oil Shale subunit	Brown Oil Shale subunit
Dominant Lamalginite		
Features	lamellae 0.05–0.1 mm long; up to 0.005 mm thick; (O; m) derived in part from *Pediastrum*	lamellae 0.05–0.2 mm long; up to 0.005 mm thick; (O; m) derived in part from *Pediastrum*
Abundance	generally 5–20%	5–40%, generally 10–30%
Other Lamalginite		
Features	0.04–0.01 mm long; <0.005 mm thick; Y; m to i) derived in part from *Septodinium*) minor component only	0.04–0.1 mm long; <0.005 mm thick; (Y; m to i); derived in part from *Septodinium*; constitutes up to 5%, rarely 10% of samples
	also lamellae <0.01 mm long; (O; m; S); may be derived in part from algal spores;	also lamellae <0.01 mm long; (O; m; S to A); may be derived in part from algal spores;
	0.05–0.2 mm long; up to 0.01 mm thick; (G to Y; i; S); single to several lamellae	0.05–0.3 mm long; up to 0.04 mm thick; (G to Y; iS) single to many lamellae
***Botryococcus* Telalginite**		
	0.05–0.1 mm diameter, (Y; m to i; S to A); partly replaced by pyrite; structure generally poorly-preserved	0.005–0.1 mm diameter; (Y; m to i; S to A); partly replaced by pyrite; structure generally poorly preserved; minor replacement by carbonate
Other Organic Matter		
	sporinite (Y; m; R); liptodetrinite derived from lamalginite (Y to O; w to m; S to A); cutinite (O to B; w), humodetrinite (S to A); ulminite (S); stems, twigs and leaves; inertodetrinite (R); bitumen; (YO to O; w; R)	sporinite (Y; m; R); liptodetrinite derived from lamalginite (Y to O; w to m; S to A); cutinite (O; m; R); humodetrinite (S); ulminite (S); twigs and leaves; inertodetrinite (S); sclerotinite (R); bitumen (YO to O; w; S); most abundant in the lower half of the unit
	groundmass with abundant fine-grained organic matter some of which fluoresces; also brown non-fluorescing alginite-like organic matter	groundmass generally with little fine-grained organic matter

Fluorescence: O—orange, YO—yellowish-orange, Y—yellow, G—green, w—weak intensity, m—moderate intensity, i—intense intensity. *Abundance*: A—abundant (1 to 5% of bulk rock), S—sparse (<1% of bulk rock), R—rare (observed but not counted in point count of 500 points).

in lamosite from other subunits); the lamosite commonly grades into claystone.

A second type of lamosite is microlaminated and is composed of interlaminated lamosite, claystone, coal and limestone laminae. Microlaminae range in thickness from 0.01 to 0.2 mm. Boundaries between lamosite and coal or claystone microlaminae are generally gradational whereas boundaries between limestone and other microlaminae are generally sharp. Lamalginite content of lamosite laminae ranges from 5 to 80%. No apparent relationship between alginite content and microlaminae thickness was observed.

Bitumen occurs in all oil shale subunits but is much more abundant in the Lascelles subunit than underlying subunits.

(2) *O'Connell subunit*. This subunit contains the same lithologies as the underlying Lascelles subunit and is made up of fining upwards, cyclic sequences of sandstone–lamosite–claystone– limestone. Lamosite constitutes 30–60% of each cycle. However the overall lamosite:other rock type ratio is less than that in the Lascelles subunit.

Organic assemblages of each rock type are the same as those for corresponding rock types in the Lascelles subunit.

Reflectance data

Drill hole AEQ Proserpine 1 (AEQ-1), which was drilled as a petroleum wildcat hole, has mostly lamosite above 460 m, and below this depth, sandstone and claystone. The latter two rock types contain rare to sparse primary vitrinite and secondary coal grains (in which the vitrinite reflectance ranges from 0.80 to 1.08%).

The reflectance from primary vitrinite (mostly vitrodetrinite) in claystone and sandstone from below the lamosite ranges from 0.35–0.58% (Fig. 3) with a reflectance gradient of 0.15% per 1000 m.

Vitrinite reflectance data for two drill holes,

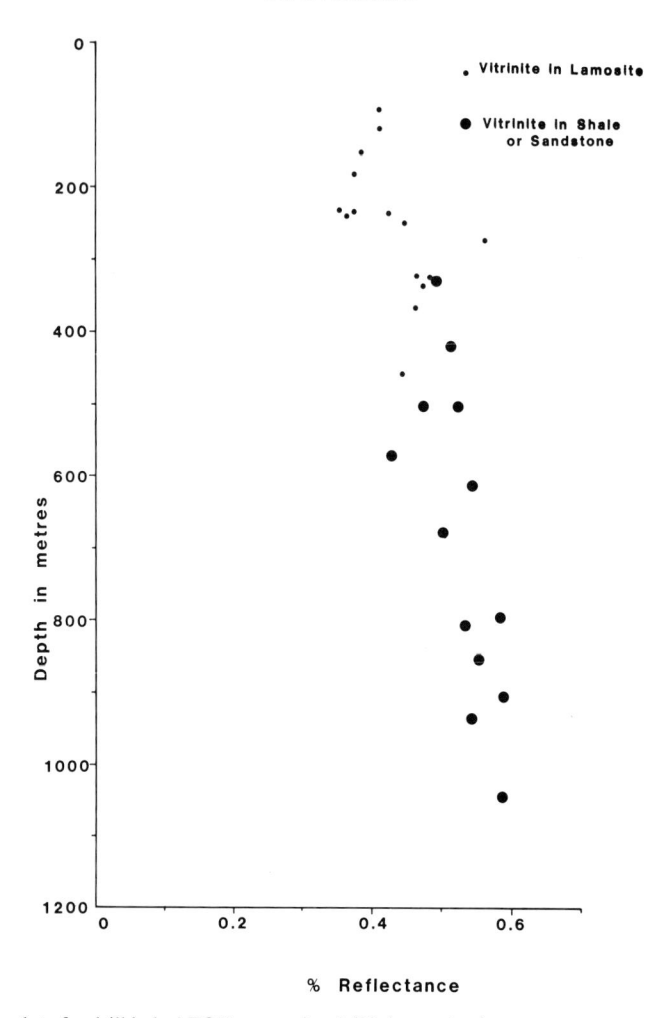

FIG. 3. Reflectance data for drill hole AEQ Prosperpine-1 (28 data points).

which did not intersect basement, are presented in Figs 4 and 5. Above the base of the Brown Oil Shale subunit, values are generally between 0.3 and 0.42%. However, within the Brownish-black Oil Shale and Lilypool subunits, reflectance values rapidly increase to 0.6–0.7%. Maximum vitrinite reflectance (taken on corpohuminite) for the Lilypool subunit was 0.78%. Reflectance data for ten other drill holes (Hutton 1982) show similar trends.

In lamosite samples from the Brown Oil Shale and Brownish-black Oil Shale subunits, reflectance values are related, in part, to the size of the vitrinite (larger vitrodetrinite has highest reflectance values), alginite content (increased lamalginite is associated with decreased reflectance) and depth (Hutton 1982). Increased reflectance with depth would occur if deeper parts of the section had been buried at the respective depths for a considerable time or the samples from greater depth had been exposed to a higher geothermal gradient.

The limited data available indicate that vitrinite reflectance values are lower, at corresponding depths, on the eastern side of the deposit.

Discussion

Origin of alginite

Samples from the Condor deposit contain five types of lamalginite:

(1) well-preserved to poorly-preserved lamalginite derived from *Pediastrum*;
(2) discrete lamellar lamalginite of similar size

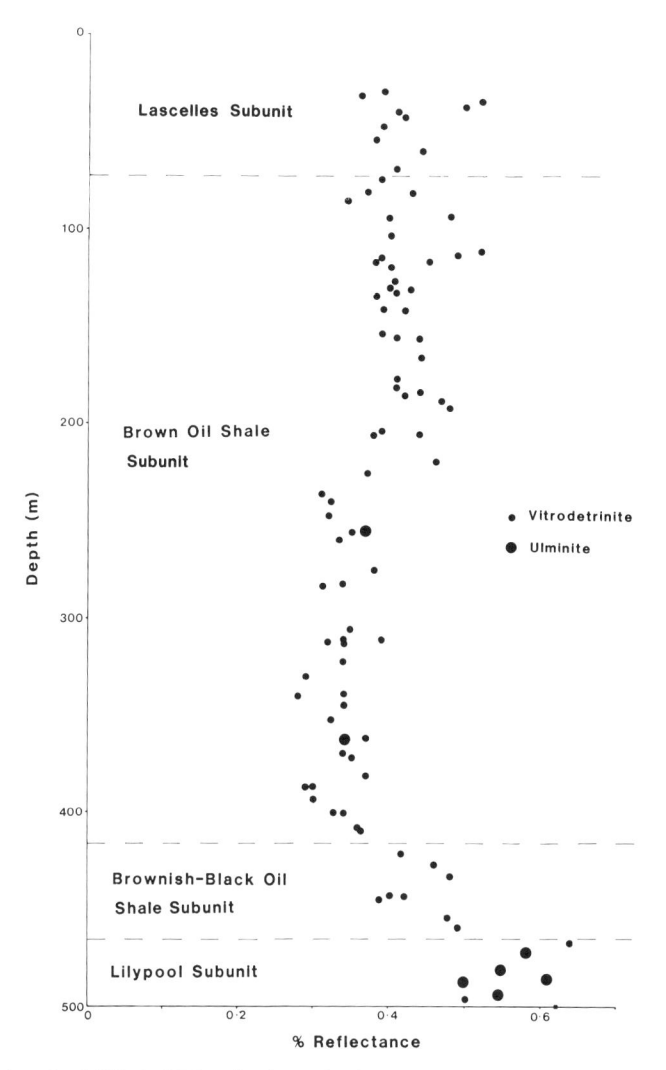

FIG. 4. Reflectance data for drill hole CDD-1 (92 data points).

and similar fluorescence properties to *Pediastrum* but derived from a different parent alga;
(3) lamalginite derived from smooth-walled dinoflagellates including *Septodinium*;
(4) lamalginite derived from abundant small (<0.01 mm diameter), well-preserved algae; and
(5) liptodetrinite, fragments of larger lamalginite.

Palynological studies by Foster (1980) showed that the acritarchs *Leiofusa* and *Micrystridium* are also found in Condor oil shale.

Telalginite is a minor component in most lamosite and claystone samples. It is derived from the colonial green alga *Botryococcus*.

Pediastrum as an indicator of the environment of deposition

Pediastrum, a common component of freshwater plankton, is a colonial stellate, but disc-shaped alga with constituent cells arranged in a single layer. Constituent cells of the coenobium or colony are multinucleate. In laboratory cultures, cell division takes place during the night and swarming occurs during the day or a little later in cooler weather (Harper 1918). Such characteristics are conducive to a bloom-forming alga. Fifteen extant species of *Pediastrum* have been recognized in the United States of America (Smith 1950).

Fossil *Pediastrum* is known from: Cretaceous

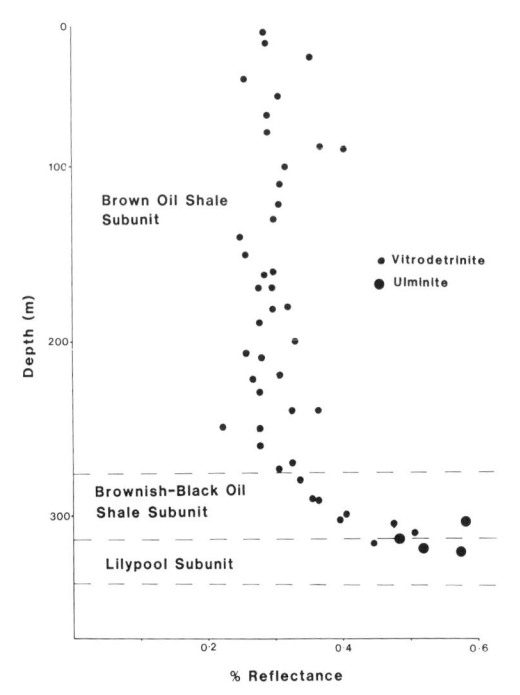

FIG. 5. Reflectance data for drill hole CDD-2 (47 data points).

sediments of Pakistan and California, USA (Evitt 1963); Tertiary deposits of India (Mathur 1963, 1964; Salujha et al. 1969; Singh et al. 1973; Singh & Khanna 1978); Tertiary sediments from Australia (Cookson 1953); Tertiary sediments from Sumatra (Wilson & Hoffmeister 1958); and Pleistocene deposits of Japan (Matsuoka & Hase 1977).

The Indian and Australian occurrences were stated to be from freshwater lacustrine deposits and in some cases *Botryococcus* was present (Cookson 1953; Mathur 1964; Singh & Khanna 1978).

Cretaceous occurrences were in marine strata and Evitt (1963) offered two hypotheses as to their occurrence: 'either some Cretaceous species of *Pediastrum* were marine or more likely, the resistant cell walls enabled the remains of coenobia to be transported from bodies of fresh water in which the organisms lived, into bodies of marine waters in whose sediments they were buried or preserved'. Singh et al. (1973) and Singh & Khanna (1978) discussed both alternatives and suggested that occurrences in the two different environments could be explained if the marine and freshwater bodies were interconnected. Matsuoka & Hase (1977) found an admixture of marine and freshwater diatoms as well as estuarine molluscs in the same sediments as *Pediastrum*. They presumed a river mouth environment with a strong seawater influence. Singh et al. (1973) interpreted admixed marine and freshwater assemblages as indicating 'near-shore conditions of deposition, perhaps estuarine or inland seas'. Mathur (1964) surmised a subtropical to tropical climate from the microfloral assemblages which included hystrichosphaerids and pollen grains as well as *Pediastrum*.

If the occurrence of *Pediastrum* in the Condor deposit is analogous to Tertiary deposits from India and Japan the occurrence indicates a freshwater, near-shore environment with possible marine influence and tropical to subtropical climate.

Assuming that conditions needed for the growth of *Pediastrum* are the same as those when it grew in Lake George (New South Wales, Australia) during the Cainozoic, the presence of *Pediastrum*, which has a salinity tolerance of 1.7 g/l, is an indicator of water depths of at least 7 m for 'reasonably long periods' (Singh et al. 1981). Thus high grade lamosite intervals with abundant *Pediastrum* (which alternate with lower grade intervals throughout the Brown Oil Shale subunit, Fig. 6) probably represent high water levels, maximum growth of algae and highest rates of accumulation. The paucity of *Pediastrum* in carbonate-dominated layers possibly indicates shallower, more saline waters (with salt levels above that of the tolerance of *Pediastrum*) during the deposition of those layers.

Because *Pediastrum* lamalginite is more persistent throughout the deposits, *Pediastrum* may have been more tolerant to small changes in conditions than the minor algal components. Changes in the abundance of minor alginite components may reflect minor fluctuations in the depositional environment better than *Pediastrum*.

Petroleum generation potential

Vitrinite reflectance is regarded as one of the more reliable indicators of coal rank and maturation levels of petroleum source rocks. Many recent studies (e.g., Kantsler et al. 1978; Smith & Cook 1984; Smyth et al. 1984) have used vitrinite reflectance as an indicator of the maturity of a petroleum source rock. In the cited studies the type and abundance of organic matter are used in conjunction with maturation levels to predict likely oil generation zones.

Tissot & Welte (1978, 1984) and Thomas (1982) concluded that significant oil generation from land-plant dominated organic matter occurs where vitrinite reflectance is in the range of 0.7–0.9%. However, Snowdon & Powell (1982) sug-

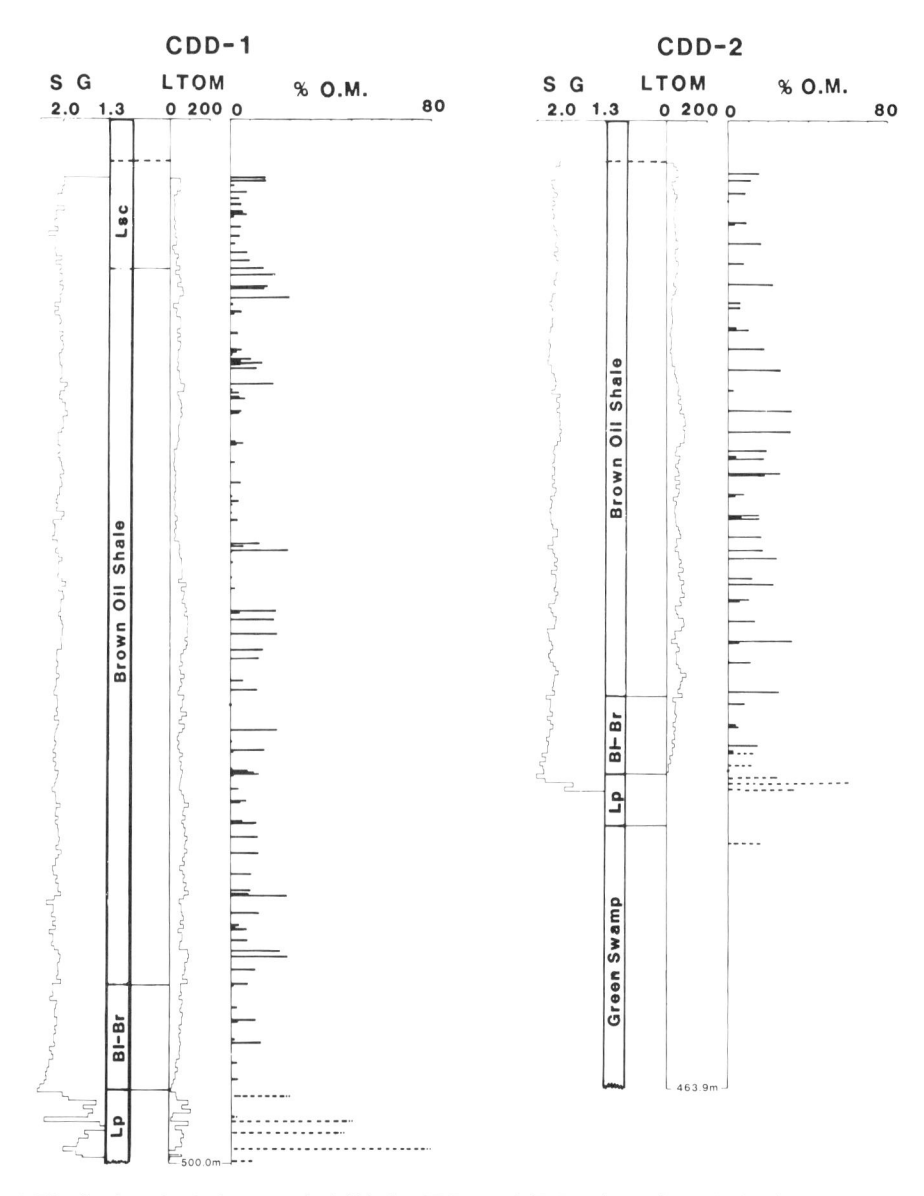

FIG. 6. Distribution of organic matter in drill holes CDD-1 and CDD-2. (Lp: Lilypool subunit, Bl-Br: brownish-black oil shale subunit, Lsc: Lascelles subunit. SG: specific gravity of the oil shale, LTOM: Fischer assay in l/t at 0% moisture, O.M.: organic matter. ——— = alginite, ----- = vitrinite, ····· = terrestrial plant liptinite.)

gested that naphthenic oil generation occurs in resinite-rich rocks at a maturation level corresponding to 0.4% vitrinite reflectance. Similarly, Smith & Cook (1984) suggested that in the Gippsland Basin (Australia), hydrocarbons were derived from vitrinite and liptinite at shallow depths and low maturation levels (vitrinite reflectances of 0.4–0.8%) with vitrinite capable of producing hydrocarbons at lower levels of maturity than liptinite.

Care should be exercised when interpreting maturation levels from vitrinite reflectance as factors other than temperature and time may affect vitrinite reflectance. For example, Hutton & Cook (1980) showed that in a torbanite–coal seam, vitrinite reflectance was higher in coal than in torbanite with reflectance inversely related to the percentage of telalginite.

The Condor oil shale deposit is a thick sequence with two types of petroleum source rocks: the liptinite-rich lamosites of the Brownish-black Oil Shale subunit and younger subunits and the vitrinite/liptinite-rich cannel coals of the Lilypool subunit. Vitrinite in lamosite from the Condor deposit is dominantly vitrodetrinite for which reflectance values are related partly to the size of the vitrodetrinite and partly to depth (Hutton 1982). Thus reflectance of vitrinite in lamosite is probably not a useful indicator of maturity. However, maturation studies can be carried out using vitrinite reflectance for claystone, sandstone and coal samples from below the lamosite units.

On the western side of the Condor deposit, the lower limit of the oil generation window (0.4% vitrinite reflectance) is within the Brownish-black Oil Shale subunit (for example at a depth of approximately 300 m in CDD-2 and 450 m in CDD-1). In addition, for the nine drill holes sampled, the Lilypool subunit has mean maximum reflectances ($R_v$max) of greater than 0.5%. Thus the western part of the deposit has both liptinite-rich and vitrinite-rich source rocks that are at present well within the oil generation window.

Reflectance data for drill hole AEQ Proserpine-1 (Fig. 3), drilled on the eastern side of the basin, suggests that the lower limit of the oil generation window occurs at a similar depth to that on the western side. However the uniform reflectance profile for AEQ Proserpine-1 (which is located on the eastern side of the deposit) of 0.15% per 1000 m is quite different to the sharply increasing profiles of CDD-1 and CDD-2 (which were located on the western side of the deposit). Mean maximum reflectance values at 1000 m depth in AEQ Proserpine-1 are 0.6%—values that occur at 500 m in CDD-1 and 300 m in CDD-2. Thus the western part of the Hillsborough Basin may have higher geothermal gradients and therefore higher rates of heat flow than the eastern part of the basin. This would indicate that the western part of the Condor deposit could have greater oil generation potential than the eastern part of the deposit.

Bitumen is a minor constituent of the lamosites, especially those in the Lascelles and O'Connell subunits. It is interstitial to framework grains and also fills microfractures, solution cavities and spaces within and between bones.

Bitumen is generally derived by the secondary alteration of other organic matter. Four possible modes of formation of the bitumen in the Condor deposit are as follows:

(1) It may represent heavy residual, viscous fractions of an oil that has migrated through the sequence.
(2) It may have been derived from degraded animal tissue.
(3) It may have been formed by bacterial or other biogenic degradation of alginite during or after diagenesis.
(4) It may have been formed by thermal alteration of alginite after lithification of the sediments.

Bitumen was formed during thermal alteration of lamalginite adjacent to a sill in the Stuart oil shale deposit. This occurs where the vitrinite reflectance values are approximately 0.4–0.5% (Hutton & Henstridge 1985). If reflectance values of >0.4–0.5% are an indicator of bitumen formation, the bitumen in the Condor oil shale may have been formed by a similar mechanism. Within the Brownish-black Oil Shale subunit reflectance values are 0.4–0.5%; below this subunit, reflectance values are up to 0.6%. Lamalginite (in the Brownish-black Oil Shale subunit) and resinite and cutinite derived from terrestrial plants (Lilypool subunit) may be sources of the bitumen.

The loss of volatile components from lamalginite during oil generation would also account for the weaker fluorescence intensities and brownish-orange fluorescence colour of alginite in samples from greater than 600 m in another of the sampled drill holes.

It is most likely that much of the bitumen was derived from alginite or resinite by thermal alteration that occurred as a result of the deeper burial of the lower units. More mobile petroleum fractions may have migrated from the onshore sediments through the south eastern offshore extension of Hillsborough basin.

Conclusions

The Condor oil shale deposit is a thick sequence of Tertiary lacustrine sedimentary rocks located on the coastal plain near Proserpine, Queensland (Australia). It overlies a thick section comprising sandstone and claystone topped by a resinite- and cutinite-rich cannel coal with a vitrinite reflectance of 0.6%.

Four oil shale subunits contain interbedded lamosite and claystone. Condor lamosite contains predominantly orange fluorescing lamalginite, derived from *Pediastrum*, a discrete, smooth-walled lamalginite derived from *Septodinium*, a freshwater dinoflagellate, and minor telalginite derived from the colonial green alga *Botryococcus*.

Reflectance data show that the lower oil shale

subunit and the cannel coal are within the oil generation window. The occurrence of bitumen throughout the lamosite, the subdued fluorescence intensity from some of the lamalginite, and the presence of other weakly fluorescing organic matter in the lower oil shale subunit may indicate that petroleum has been generated from these sections of the sequence.

References

AMPOL EXPLORATION (QUEENSLAND) PTY LTD. 1965. Proserpine No. 1 well completion report (CR 1783). Geological Survey of Queensland Record (unpubl.).

COOKSON, I. C. 1953. Records of the occurrence of *Botryococcus braunii*, *Pediastrum* and the *Hystrichasphaerideae* in Cainozoic deposits of Australia. *Memoirs of the Natural Museum, Melbourne*, **18**, 107-123.

EVITT, W. R. 1963. Occurrence of freshwater alga *Pediastrum* in Cretaceous marine sediments. *American Journal of Science*, **261**, 890-893.

FOSTER, C. B. 1980. Report on Tertiary spores and pollen from core samples from the Hillsborough and Duaringa Basins. Geological Survey of Queensland Record, 1980/2 (unpubl.).

GREEN, P. W. & BATEMAN, R. J. 1981. The geology of the Condor oil shale deposit—onshore Hillsborough Basin. *Australian Petroleum Exploration Association Journal*, **21**, 24-32.

GREEN, P. A., MCIVER, R. G. & O'DEA, T. R. 1984. Revised geology of the Condor Oil Shale Deposit. *Proceedings of the Second Australian Workshop on Oil Shale*, 33-37.

HARPER, R. A. 1918. Organization, reproduction and inheritance in *Pediastrum*. *Proceedings of the American Philosophical Society*, **57**, 375-438.

HUTTON, A. C. 1982. Organic petrology of oil shales. Ph.D. thesis, University of Wollongong.

—— & COOK, A. C. 1980. Influence of alginite on the reflectance of vitrinite from Joadja, N.S.W., and some other coals and oil shales containing alginite. *Fuel*, **59**, 711-714.

—— & HENSTRIDGE, D. A. 1985. Pyrolysis of Tertiary Oil Shales by a dolerite intrusion, Stuart Deposit, Queensland, Australia. *Fuel*, **64**, 546-552.

——, KANTSLER, A. J., COOK, A. C. & MCKIRDY, D. M. 1980. Organic matter in oil shales. *Australian Petroleum Exploration Association Journal*, **20**, 44-67.

INTERNATIONAL COMMITTEE FOR COAL PETROLOGY. 1963. *International Handbook of Coal Petrology*, 2nd edn, Centre National de la Recherche Scientifique, Paris.

—— 1971. *International Handbook of Coal Petrology*, 2nd edn, Centre National de la Recherche Scientifique, Paris.

—— 1975. *International Handbook of Coal Petrology*, 2nd Supplement, 2nd edn, Centre National de la Recherche Scientifique, Paris.

KANTSLER, A. J., SMITH, G. C. & COOK, A. C. 1978. Lateral and vertical rank variation: Implications for hydrocarbon exploration. *Australian Petroleum Exploration Association Journal*, **18**, 143-156.

MATHUR, K. 1963. Occurrence of *Pediastrum* in Subathu Formation (Eocene) of Himachal Pradesh, India. *Science and Culture*, **29**, 250.

—— 1964. On the occurrence of *Botryococcus* in Subathu Beds of Himachal Pradesh, India, *Science and Culture*, **30**, 607-608.

MATSUOKA, K. & HASE, K. 1977. Fossil *Pediastrum* from the Pleistocene Hamamatsu Formation around Lake Hamana, Central Japan. *Transactions and Proceedings of the Palaeontological Society of Japan*, **104**, 432-441.

PAINE, A. G. L. 1972. Proserpine, Queensland. Explanatory notes 1:250 000 Geological Series. *Bureau of Mineral Resources. Geology and Geophysics*, 1-25.

SALUJHA, S. K., SRIVASTAVA, N. C. & RAWAT, M. S. 1969. Microfloral assemblage from Subathu sediments of Simla Hills. *Journal of the Palaeontological Society India*, **12**, 25-40.

SINGH, G., OPDYKE, N. D. & BOWLER, J. M. 1981. Late Cainozoic stratigraphy, palaeomagnetic chronology and vegetational history from Lake George, N.S.W. *Journal of the Australian Geological Society*, **28**, 435-452.

SINGH, H. P. & KHANNA, A. K. 1978. Some fossil species of *Pediastrum* in the Subathu Formation of Himachal Pradesh. *Palaeobotanist*, **25**, 466-474.

——, —— & SAH, S. C. D. 1973. Problems and prospects of Tertiary palynology in Northern India. *Bulletin. India Geologists Association*, **6**, 71-77.

SMITH, G. M. 1950. *The Fresh-water Algae of the United States*. McGraw-Hill Book Co., New York.

SMITH, G. C. & COOK, A. C. 1984. Petroleum occurrence in the Gippsland Basin and its relationship to rank and organic matter type. *Australian Petroleum Exploration Association Journal*, **24**, 196-216.

SMYTH, M., COOK, A. C. & PHILP, R. P. 1984. Birkhead revisited: petrological and geochemical studies of the Birkhead Formation, Eromanga Basin. *Australian Petroleum Exploration Association Journal*, **24**, 230-242.

STACH, E., MACKOWSKY, M.-TH., TEICHMULLER, M., TAYLOR, G. H., CHANDRA, G. & TEICHMULLER, R. 1975. *Stach's Textbook of Coal Petrology*, 2nd edn, Gebruder Borntraeger, Berlin, 428 pp.

——, ——, ——, ——, —— & ——. 1982. *Stach's Textbook of Coal Petrology*, 3rd edn, Gebruder, Berlin, 535 pp.

SNOWDON, L. R. & POWELL, T. G. 1982. Immature oil and condensate, modification of hydrocarbon generation model for terrestrial organic matter. *Bulletin of the American Association of Petroleum Geologists*, **66**, 775-788.

THOMAS, B. M. 1982. Land plant source rocks for oil and their significance in Australian Basins. *Australian Petroleum Exploration Association Journal*, **22**, 164-178.

TISSOT, B. & WELTE, D. H. 1978. *Petroleum Formation and Occurrence*, Springer-Verlag, Berlin, 538 pp.

—— & —— 1984. *Petroleum Formation and Occurrence*. Springer-Verlag, Berlin, 2nd edn, 538 pp.

WILSON, L. R. & HOFFMEISTER, W. S. 1953. Four new species of fossil *Pediastrum*. *American Journal of Science*, **251**, 753-760.

A. C. HUTTON, Department of Geology, University of Wollongong, PO Box 1144, Wollongong, NSW 2500, Australia.

Cenozoic lacustrine basins of South-east Asia, their tectonic setting, depositional environment and hydrocarbon potential

M. R. Gibling

SUMMARY: Cenozoic strike-slip tectonism, which caused local crustal extension in South-east Asia, generated many short-lived but deeply subsiding basins. Under tropical conditions and assisted by intensive karstic weathering, the basins formed ideal sites for lakes during early basinal history in Oligocene to Miocene times. In several basins in northern Thailand, a shallow-lacustrine assemblage of mudstone and coal contains thin, alginite-rich oil shales of moderate grade which formed under oxic or anoxic conditions during the initial flooding of the lake, which was accompanied by high organic productivity. In the Mae Sot Basin, a carbonate-rich lacustrine assemblage contains thick, alginite-rich oil shales of high grade which formed under anoxic conditions in deeper, probably stratified lakes where the organic material was well preserved. The kerogen in both assemblages is Type I, hydrogen-rich. Under conditions of rapid sedimentation and abnormally high heat flow associated with active tectonism, the shallow-lacustrine oil shales have generated hydrocarbons which are exploited in several basins.

Introduction

The recent hydrocarbon developments in onshore Mesozoic to Cenozoic basins of China have stimulated great interest because giant reservoirs have been ascribed to lacustrine source rocks (Hu Chao Yuan & Qiao Hanseng 1983; Li Desheng 1984; Yang Wanli et al. 1985; Chen Quanmao & Dickinson 1986), previously considered of limited importance. The adjoining region of South-east Asia contains many Cenozoic lacustrine basins which, despite several hydrocarbon discoveries, remain relatively unexplored. The present paper documents lacustrine assemblages in basins in Thailand, and briefly discusses similar basins over an area of 2 million km^2 in peninsular Malaysia, Indochina and Burma.

Tectonic setting

South-east Asia, located at the junction of the Eurasian, Indian and Pacific plates, has undergone regional uplift since the mid-Mesozoic. Its geological development since that time is related primarily to the northward movement of India and its collision with the Eurasian continent in the late Mesozoic and early Cenozoic to form the Himalayan Mountain chains (Johnson et al. 1976). The subsequent transcurrent displacement of large tectonic segments (Fig. 1) produced strike-slip faulting throughout the region during the Cenozoic (Tapponnier & Molnar 1977; Tapponnier et al. 1982; Le Dain et al. 1984). Sea-floor spreading in the Andaman Sea (Rodolfo 1969) and the South China Sea (Ben-Avraham & Uyeda 1973) probably was also related to these regional events.

Strike-slip faults are presently active in South-east Asia (Le Dain et al. 1984) and some major faults are shown in Figs 1 and 2. A strike-slip regime has been operative in the region since the Palaeogene or early Neogene, and many sedimentary basins are closely related to the fault systems. Pull-apart basins have formed along fault zones in the Gulf of Thailand (Fig. 1). The Phitsanulok Basin lies at the junction of the Mae Ping and Uttaradit Fault Zones (Fig. 2), where crustal extension was accompanied by downwarping and

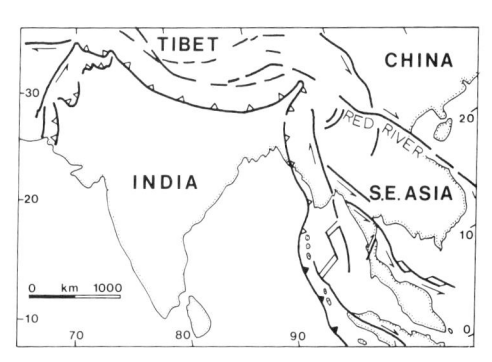

FIG. 1. Map of South-east Asia to show position of major tectonic features. Heavy lines show major faults or plate boundaries. Solid barbs indicate subduction, open barbs indicate intracontinental thrusts. Thin paired lines show spreading centres (Andaman Sea) and pull-apart basins (Gulf of Thailand). Arrows on faults in western Malaysia and Gulf of Thailand do not correspond to present-day motions. Modified from Tapponnier et al. (1982).

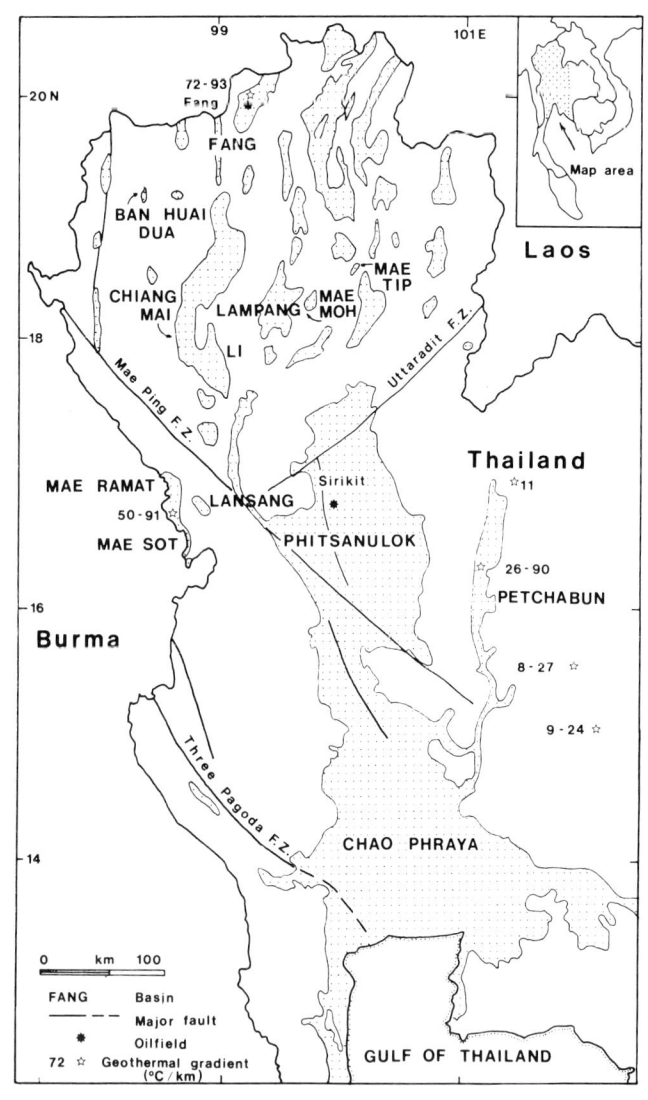

FIG. 2. Major basins of north-central Thailand. Basin areas are taken from Gibling & Ratanasthien (1980); fault lines from Knox & Wakefield (1983) and Chantaramee (1978); geothermal gradients from reliable data in Thienprasert *et al.* (1978).

listric faulting (Knox & Wakefield 1983). The Lansang Basin lies in a complex strike-slip zone (Chantaramee 1978). Other basins may have developed in a similar manner, although the lack of exposure makes this difficult to prove.

Heat flow was probably high throughout the Cenozoic in Thailand. Rb-Sr dating of granitic rocks suggests that major intrusive events took place in the Lower and Middle Cretaceous, and Cenozoic K-Ar dates have been attributed to reheating through hydrothermal circulation (Beckinsale *et al.* 1979). Basaltic volcanics have erupted since the Middle Miocene in many areas (Barr & Macdonald 1981; Knox & Wakefield 1983). Geothermal gradients in western Thailand are locally high, up to 93°C/km (Fig. 2) and abundant hot springs are present (Barr *et al.* 1978; Thienprasert *et al.* 1978). Geothermal gradients average about 40–60°C/km and up to 73°C/km in some intervals under the Gulf of Thailand (Paul & Lian 1975; Trevena & Clark 1986), but they average only 10–40°C/km under northeastern Thailand apart from high values in the Petchabun Basin (Fig. 2) (Thienprasert *et al.*

1978). Heat flow thus appears to decrease eastward from the tectonically active region bordering the Andaman Sea into the more stable interior of South-east Asia.

The sedimentary basins are elongate north–south following the tectonic strike of the basement rocks, and range in present-day area from the Chao Phraya Basin (60 200 km^2 onshore) to small, high-altitude basins (less than 1 km^2). They are bordered by mountains up to 2500 m in elevation. In general, because of the relative youth of the basins, their present topography can be used as a guide to their topography at the time of deposition. The basins contain as much as 10 km of strata in the Gulf of Thailand and 8 km onshore, although even the smallest basins can contain hundreds of metres of strata. The strata are dated from Oligocene to Holocene (Woollands & Haw 1976; Hamilton 1979; Gibling & Ratanasthien 1980; Knox & Wakefield 1983).

Lacustrine sediments were deposited in many onshore basins during the Oligocene to mid-Miocene, after which thick sequences of coarse alluvial sediments were deposited. The unconformable contact observed in many basins between the two depositional units suggests that tectonic forces were the prime cause. Changes in tectonic stress patterns related to the progressive Indo–Eurasian collision, such as those documented by Le Dain *et al.* (1984), could have resulted in the evolution of depositional style within the basins. Many authors have attempted to relate stratigraphic sequences to evolving structural style in strike-slip basins, but the only conclusion that can presently be drawn for the immature South-east Asian basins is that lacustrine conditions were common in the early stages of basinal history.

Lacustrine assemblages

Two lacustrine assemblages are recognized in the Thai basins. Fluvial assemblages with minor lacustrine strata, such as occur in the Ban Huai Dua Basin (Gibling *et al.* 1982) and in the upper strata of many other basins, are not described here.

Facies assemblage A: mudstone–coal–oil shale

Description

This terrigenous-organic assemblage is found in many basins and consists of the three main rock types, listed in order of abundance. The mudstones, commonly in units as much as 30 m thick, are pale to dark grey and brown and vary from laminated to non-bedded. Megafossils include gastropods, bivalves, shelled arthropods, trace fossils, fish, turtles, charophytes and vascular plants. Calcareous nodules associated with roots are locally abundant enough to form thick zones. Laminae composed of vascular plant fragments are common.

Coal occurs in seams over 30 m thick in at least four basins, and in seams several metres thick in many others (Fig. 3). Preliminary data, summarized by Gibling & Ratanasthien (1980), show that the coal is of lignite to high volatile bituminous rank with a heating value from 10.5 to 35.6 MJ/kg. The rank appears to be higher in the hills between the Burmese border and Lampang than further east. Coal at Mae Moh, which consists of finely detritic humic components, with small amounts of woody tissue, resinite, sclerotinite and palynomorphs, was ascribed to herbaceous plants by Gardner (1967), and is unusual in containing well-preserved gastropod shells and carbonate grains. Mudstone partings are common, and the coals are autochthonous, resting on rooted mudstones. At Mae Tip, a coal seam grades laterally towards the basin margin into reddish carbonaceous mudstone (Gibling *et al.* 1985b), and hence terminates by increase in terrigenous content rather than by splitting.

Oil shales are known from at least nine basins, where they form units several metres thick. They are grey-brown, poorly laminated, and are composed of clay and quartz-rich silt with fish and plant fragments. The oil shales are interbedded abruptly to gradationally with coal (Fig. 4) and mudstone, and locally can be traced for more than 1 km. Most beds contain little benthonic fauna, but at Ban Pa Kha, Li, a thin oil shale that separates coal from an overlying thick claystone unit consists of well-defined beds each with a distinctive benthonic assemblage that includes gastropods, bivalves, and burrows. *Planolites* burrows are present in oil shale at Ban Huai Dua (Gibling *et al.* 1982). The organic matter consists of lamalginite (filamentous algal material including *Pediastrum*) with minor telalginite (including the colonial algal *Botryococcus*), liptodetrinite, sporinite, resinite and bitumen; huminite or inertinite form less than 10% of the organic component, apart from one canneloid oil shale from Ban Pa Kha which contains abundant huminite (Sherwood *et al.* 1984). The oil shales yield up to 153 l/tonne, with average yields of about 50 l/tonne (Gibling *et al.* 1985b), and are equivalent to lean, low and moderate grades of oil shale in the Green River Formation (Cole & Picard, 1975).

Minor rock types in some basins include non-

FIG. 3. Mudstone and coal of facies assemblage A in the Mae Moh basin, Thailand.

FIG. 4. Oil shale unit within coal seam, facies assemblage A, in the Mae Tip Basin, Thailand (Hammer (30 cm) for scale).

laminated marlstone and chert in thin beds, and quartz-rich siltstone and sandstone.

Environment of deposition

The assemblage is present in basins hundreds of km² in area to basins as small as Ban Po, Li, where a 35-m coal seam was mined in a basin 1 km². As water levels fluctuated, swamps with rooted vegetation (coal) alternated with shallow lakes too deep for vegetation to take root (mudstone and oil shale). The small size and narrow outlets of the fault-controlled basins allowed water levels to fluctuate rapidly.

The development of many basins was undoubtedly assisted by karstic weathering of the widespread bedrock carbonates, the present-day topography of which is extremely irregular. These rocks have been subjected to prolonged tropical weathering as Thailand lay close to its present position throughout the Cenozoic (Jurdy & van der Voo 1975). Sudden drainage changes have been known to produce lakes within closed depressions in karstic regions of Thailand. The persistence of carbonate shells and grains in the coals, and the unusual co-occurrence of coal and nodular calcrete, may reflect high levels of dissolved carbonate in the groundwaters.

Fluvio-lacustrine coarse clastics are absent from lacustrine sections in many basins, probably because fluvial inflow was limited to small creeks. The Phitsanulok Basin is exceptional in that the major zone of lacustrine rocks contains fluvio-lacustrine sandstones and lacks coals (Knox & Wakefield 1983).

Distribution of bituminous material

The oil shales of Facies Assemblage A are commonly interbedded with coal or intercalated between coal and mudstone, showing that the organic material accumulated during the shallowest stage of lake development. In contrast, studies of some modern African lakes suggest that maximum organic accumulation takes place in the deep anoxic zones of the lakes (Demaison & Moore 1980; Talbot this volume). The shallow water, organic-rich beds in the Thai basins may be explained by abnormally high organic productivity during the initial flooding of the swamps. During the early flooding of manmade Lake Kariba, the waters became nutrient-rich due to the submergence of poorly leached soils and the death and decay of the drowned biota; vast algal blooms, explosive growth of water ferns and huge increases in the fish population resulted (Harding 1966; Mitchell 1969). Alternatively the terrigenous input in the early transgressive stages of the Thai lakes may have been relatively low compared to the organic input. The presence of benthonic organisms and burrows in some beds shows that the bottom waters were oxygenated during deposition, and the preservation of the organic material probably reflects rapid sedimentation with the deposition of an overlying mudstone seal, rather than anoxic conditions.

Facies assemblage B: marlstone–oil shale–sandstone

Description

This mixed carbonate-terrigenous-organic assemblage has so far been documented only in the Mae Sot basin on the Thai-Burmese border, where it was studied in detail at Ban Huai Kalok (Gibling *et al.* 1985a). The three main rock types are listed in order of abundance, although their proportion varies with position in the basin. The basinal fill, at least 800 m thick, contains repeated alternations of marlstone with sandstone and oil shales, in sequences a few metres to tens of metres thick.

The marlstone is pale yellow-grey and ranges from poorly laminated to non-bedded in individual units up to several metres thick. The units contain polygonal desiccation cracks 10 cm deep and visible on the basal surfaces of interbedded sandstones, during the deposition of which the cracks were scoured and infilled with sand and fine gravel. The scarcity of lamination and the random orientation of mica flakes reflect pervasive bioturbation, as trace fossils are commonly seen on the basal surfaces of the sandstones, and a benthonic shelled fauna of gastropods, and rare ostracods is present. The marlstone is composed of calcite and terrigenous minerals with small amounts of dolomite, but contains little organic material.

The oil shale forms units up to 10 m thick that can be traced for several kilometres with little change in thickness or internal arrangement (Fig. 5). The beds are rhythmically laminated, with couplets of carbonate and organic/terrigenous laminae associated with loop bedding. Although the oil shales are predominantly carbonate rocks rich in dolomite, the higher grades tend to be enriched in silicate minerals including analcime. The distinctive planktonic and nektonic biota is composed of fish, plant fragments, insects, coprolites and snakes, the latter found in highly sinuous form on bedding surfaces at two localities. The insect material, which includes thick-bodied hymenopterids, wings and faceted eyes, is preserved in a few beds of lower-grade oil shale and marlstone, along with unidentified soft-bodied organisms, probably larval forms.

The organic material consists of lamalginite and minor telalginite (up to 20% of *Botryococcus* in some laminae) with liptodetrinite, bitumens and humic components (Sherwood *et al.* 1984). Bitumen is present within some porous fish bones. The filamentous algae are believed to be planktonic, but the presence of benthonic mats cannot be discounted.

Ultimate analysis of one oil shale sample showed a relatively high H/C ratio and low O/C ratio characteristic of Type I kerogen (Sherwood *et al.* 1984). From Rock Eval analysis, the hydrogen index (ratio of amount of hydrocarbons produced to amount to total organic carbon) of most samples is higher than values for the Green

FIG. 5. High-grade oil shale unit forming a continuous scarp. Facies assemblage B, Mae Sot Basin, Thailand (hammer (30 cm) for scale).

River Formation (Cole & Picard 1975). The maximum yield from Fischer assay analysis was 341 l/tonne, with an average yield exceeding 150 l/tonne over the richest metre-thick zones.

Thin sandstone and siltstone units are intercalated with the marlstone and rarely with the oil shale. They show groove and flute casts and small-scale cross-beds indicative of current action, and adjacent to the basin margin the units thicken, increase in proportion and show lensoid channel fills up to 60 cm thick (Fig. 6).

Beds of fibrous calcite a few centimetres thick within the marlstone (Gregory 1923) have probably replaced evaporites, but although bands and lenses of gypsum are present in the basin (Brown et al. 1951) drilling has failed to indicate the presence of bedded evaporites. The presently known occurrences are probably of diagenetic origin. Thus the Mae Sot oil shales resemble those of the Green River Formation in their lamination and organic material, but the strata lack the thick evaporites which are a prominent feature of some members and regions of the Green River Formation (Bradley & Eugster 1969).

Environment of deposition

The Mae Sot basin is 65 km long by 35 km wide, and oil shale sequences are known to crop out for at least 30 km along the basin axis. The oil shale sequences with their well-defined lamination, abundant algal material, nektonic biota, lack of benthonic fauna, and preservation of delicate fossils suggest deposition in perennial lakes, with anaerobic conditions below the sediment surface or within the water column. Desiccation cracks were not observed, and the ubiquitous loop bedding is interpreted as a syneresis or compactional feature (Cole & Picard 1975). The marlstone-sandstone sequences with their poorly stratified nature, benthonic fauna and desiccation cracks, suggest shallow aerated lakes and exposed marginal flats. The sandstone and siltstone beds, which are interbedded with desiccation-cracked marlstone and in a few cases with the highest grade oil shales, suggest that runoff from the surrounding hills periodically inundated the exposed, marginal flats and continued as density currents into the deeper parts of the lakes. It is presumed that the marlstone-sandstone and oil shale sequences were laterally equivalent in the marginal and central parts of the basin, respectively; however, no lateral transition has yet been documented and rapid changes in depositional conditions could have resulted in the basin-wide extent of each lithological association in turn.

The steep margins and narrow outlet of the large basin would have allowed the formation of lakes sufficiently deep and long-lived for stratification to develop, while the limited seasonal changes in water temperature under tropical conditions would have encouraged stable strati-

FIG. 6. Sandstone (above and below 30 cm hammer) interbedded with massive, gastropod-bearing marlstone near the basin margin. Note the erosional base of the sandstone at upper right. Facies assemblage B, Mae Sot Basin, Thailand.

fication. Although the lakes could have been many metres deep at their maximum extent, the abrupt contact of the laminated oil shale units with desiccation-cracked marlstones suggests that the oil shales began to form under relatively shallow conditions. Estimates from couplet thickness suggest that perennial-lake conditions persisted for about 25 000–50 000 years (Gibling et al. 1985a), and climatic changes with long-term fluctuations in rainfall are the most probable cause of the fluctuating lake levels. The scarcity of woody material in the marlstones and oil shales suggests that the lakes were not surrounded directly by dense growths of trees and shrubs, although a herbaceous cover might have left little direct trace. Coals and rooted clays, not seen in the area studied, are present in apparently coeval strata near the northern basin margin at Mae Ramat.

The freshwater biota and the scarcity of evaporites within the basin suggest fresh to brackish waters. Oxygen-isotope data indicate that the lake was fresh during oil shale deposition but became more brackish during marlstone deposition (K. Kelts, pers. comm. 1985). The presence of analcime, which is common in modern saline and alkaline lakes (Hay 1978), probably reflects high alkalinity of the Mae Sot lakes. The abundance of dolomite in the oil shales and its scarcity in the shallower water marlstones suggests diagenetic dolomite formation under alkaline, anoxic conditions within the lakes (Baker & Kastner 1981; Kelts & McKenzie 1982), rather than formation from saline groundwaters within the marginal flats followed by transport into the perennial lake. The relative enrichment of silicate and depletion of carbonate minerals in the higher grade oil shales suggests relatively greater carbonate productivity near the lake margins (Dean 1981) with relatively greater terrigenous input to the central parts of the lakes, or possibly dissolution of the carbonate under anaerobic conditions offshore (Degens & Stoffers 1976).

The carbonate-rich strata of facies assemblage B in the Mae Sot basin are of similar age to the terrigenous strata of facies assemblage A in other basins. The local occurrence of carbonate sediments can be explained by the transport of dissolved carbonate to the Mae Sot lakes in great amounts, from Triassic and older bedrock carbonates around the basin which show intensive karsting, coupled with a low terrigenous input.

Distribution of bituminous material

The oil shales were deposited in shallow to moderate-depth lakes under anoxic bottom conditions, and their organic material shows a marked periodicity that represents pulses of high organic productivity coupled with optimal conditions for preservation.

Laminae less than 1 mm thick reflect seasonal phytoplankton blooms or possibly benthonic mats; the latter interpretation would imply an aerobic water column with anaerobic conditions confined to the subsurface sediment. Some unusually thick organic laminae which overlie sandstones and carbonate microturbidites reflect organic accumulation associated with 'random' density-flow events. The reasons for this are unclear, but as some laminae contain abundant woody fragments, it is probable that the flows transported large amounts of terrestrial organic matter along with material reworked from the lake floor. Additionally, the flows could have promoted phytoplankton blooms by recirculating nutrients trapped within stratified bottom waters or by bringing additional nutrients into the lakes. In either case, a significant connection exists between terrestrial runoff and organic accumulation.

Beds of high-grade oil shale, a few centimetres to 1 m thick, within the oil shale sequences reflect periods of maximum organic accumulation within the stratified lake, probably at its highest levels, and contain the lowest proportion of mineral matter and the thinnest couplets. The sequences of laminated oil shale up to ten metres thick reflect periods when the lakes were stratified or when subsurface sediments were anoxic.

Other basins in South-east Asia

Cenozoic lacustrine deposits similar to those of Thailand are known in Burma (Chhibber 1934; Krishnan 1968; Goossens 1978), Malaysia (Stauffer 1973) and Laos, Vietnam and Kampuchea (Fontaine & Workman 1978). Most deposits contain mudstone and coal and resemble facies assemblage A. Laminated oil shale of facies assemblage B is reported only from the Mae Sot basin and contiguous strata in Burma (Gregory 1923). Other lacustrine deposits occur in China and probably throughout the Himalayan region, as well as on the Pacific margin of Asia where the basins are more directly related to subduction (Koesoemadinata 1978). The Gulf of Thailand, the Irrawaddy Valley of Burma and the Bay of Bengal are major structural troughs that were occupied by large southward-flowing rivers during the Cenozoic, so that their deposits include deltaic and marine strata (Paul & Lian 1975). Several basins with oil shales that abut the Andaman Sea near Krabi in southern Thailand also contain marine beds (Brown et al. 1951).

Hydrocarbons

The geological data collectively suggest that maturation and entrapment of hydrocarbons could have occurred widely within the South-east Asian basins, which subsided deeply during active Cenozoic tectonism. Lakes formed periodically within the steep-sided and confined basins, and the high productivity of Type I, algal-rich kerogen allowed source rocks to be deposited and preserved during very shallow stages (assemblage A) and deep anoxic stages (assemblage B) of lake evolution.

Twenty-four samples of oil shale from the upper strata of seven basins show huminite reflectances from 0.22 to 0.53%, indicating brown coal to sub-bituminous rank (Sherwood et al. 1984), comparable to the rank of the associated coals. These maturation indicators (Hood et al. 1975) suggest that, especially at deeper levels, the strata are locally approaching the oil window, despite their young geological age, a condition attributed to the high Cenozoic geothermal gradients. The general westward increase in rank of the coals suggests that heat flow was consistently higher in the west throughout the Cenozoic. The fluvial and fluvio-lacustrine sandstones interbedded with mudstones in many basins form suitable reservoir rocks.

Oil and gas are currently produced from two onshore basins in Thailand. In the Phitsanulok Basin, almost 14 000 BOPD was produced in 1984 in the Sirikit Field (Fig. 2) (Soeparjadi et al. 1985). The reservoirs are fluvio-lacustrine sandstones at 1500–2000 m depth. The associated organic-rich clays are the source rocks, and observations of oil and vitrinite reflectance data suggest that generation took place at depths greater than 4000 m (Knox & Wakefield 1983). In the Fang Basin (Fig. 2) oil seepages have long been collected for medicinal use, and the basin has produced small amounts of oil since 1963 (Piyasin 1979). The hydrocarbons are produced from sandstone and coal interbedded with bituminous shale in the lower part of the basinal fill, which is estimated at 2800 m thick. Thin, oil-bearing sandstones were also drilled in the Chiang Mai basin at 880–1070 m depth (Piyasin 1979). The source rocks in the Phitsanulok and Fang Basins are bituminous mudstones, closely associated with coal at Fang, and belong to facies assemblage A.

Offshore in the Pattani Basin of the Gulf of Thailand, Tertiary sandstones from 1370 to 2745 m depth contain hydrocarbons, and production in 1985 was 15 000 BOPD of condensate, as well as large amounts of gas (Trevena & Clark

1986). Hydrocarbons (23 000 BOPD in 1985) are produced from the central Tertiary basin of Burma and its offshore extension (Paul & Lian 1975; Soeparjadi et al. 1985). All these basins contain marine strata which may have acted as source rocks. Minor amounts of hydrocarbons are produced in the Red River Basin of Vietnam (Fig. 1) but few data are available. Trevena & Clark (1986) noted a severe deterioration of reservoir quality with depth in the Pattani Basin, which they attributed to diagenetic effects promoted by the high geothermal gradients.

Conclusions

Lakes were widely distributed in South-east Asia during the Oligocene to Miocene. Their presence was largely a result of strike-slip faulting, related to the collision of India and Eurasia, which generated confined basins where standing water accumulated under the prevailing humid tropical climate. Karstic weathering assisted in creating basinal topography locally.

Most basins contained relatively shallow and short-lived lakes which alternated with peat swamps, but algae flourished especially during the initial stages of lake expansion. Only one basin contained a lake deep and persistent enough to develop perennially stagnant bottom waters. The periodic fluctuation of the lake levels was probably a consequence of regional climatic changes on the order of a few tens of thousands of years, superimposed on local basinal factors. Fluctuations in lake depth and area exerted a fundamental control on the accumulation of organic material: algal material accumulated preferentially when land surfaces were periodically submerged, and when the lakes were deep enough for anoxic bottom conditions to develop. The coexistence of carbonate lakes and terrigenous lakes in adjacent basins indicates that the local clastic supply and groundwater chemistry strongly influenced the nature of the lake sediments.

Hydrocarbons which occur in several basins were generated from the aliphatic-rich oil shales and bituminous mudstones under conditions of rapid burial and high heat flow associated with the active strike-slip tectonism.

ACKNOWLEDGEMENTS: I thank Kerry Kelts, Michael Talbot and Larry Wakefield for discussion; Paul Buchheim and John Marshall for their helpful review of the manuscript; and Norma Keeping for typing. Funds were provided by Natural Science and Engineering Research Council of Canada Grant A8437.

References

BAKER, P. A. & KASTNER, M. 1981. Constraints on the formation of sedimentary dolomite. *Science*, **213**, 214–216.

BARR, S. M. & MACDONALD, A. S. 1981. Geochemistry and geochronology of late Cenozoic basalts of southeast Asia. *Bulletin of the Geological Society of America*, **92**, 1069–1142.

——, RATANASTHIEN, B., BREEN, D., RAMINGWONG, T. & SERTSRIVANIT, S. 1978. Hot springs and geothermal gradients in northern Thailand. *Open-File Report, Department of Geological Sciences, Chiang Mai University, Thailand*, 35 pp.

BECKINSALE, R. D., SUENSILPONG, S., NAKAPADUNGRAT, S. & WALSH, J. N. 1979. Geochronology and geochemistry of granite magmatism in Thailand in relation to a plate tectonic model. *Journal of the Geological Society of London*, **136**, 529–540.

BEN-AVRAHAM, Z. & UYEDA, S. 1973. The evolution of the China Basin and the Mesozoic paleogeography of Borneo. *Earth and Planetary Science Letters*, **18**, 365–376.

BRADLEY, W. H. & EUGSTER, H. P. 1969. Geochemistry and paleolimnology of the trona deposits and associated authigenic minerals of the Green River Formation of Wyoming. *Professional Paper. United States Geological Survey*, **496-B**, 1–71.

BROWN, G. F., BURAVAS, S., CHARALJAVANAPHET, J., JALICHANDRA, N., JOHNSTON, W. D., SRESTHAPUTRA, V. & TAYLOR, G. C. 1951. Geologic reconnaissance of the mineral deposits of Thailand. *Bulletin of the United States Geological Survey*, **984**, 183 pp.

CHANTARAMEE, S. 1978. Tectonic synthesis of the Lansang area and discussion of regional tectonic evolution. In: NUTALAYA, P. (ed.), *Proceedings of the Third Regional Conference on the Geology and Mineral Resources of Southeast Asia, Bangkok, Thailand*, 177–186.

CHEN QUANMAO & DICKINSON, W. R. 1986. Contrasting nature of petroliferous Mesozoic–Cenozoic basins in eastern and western China. *Bulletin of the American Association of Petroleum Geologists*, **70**, 263–275.

CHHIBBER, H. L. 1934. *The Geology of Burma*, Macmillan, London, 538 pp.

COLE, R. D. & PICARD, M. D. 1975. Primary and secondary sedimentary structures in oil shale and other fine grained rocks, Green River Formation (Eocene), Utah and Colorado. *Utah Geology*, **2**, 49–67.

DEAN, W. E. 1981. Carbonate minerals and organic matter in sediments of modern north temperate hard-water lakes. In: ETHRIDGE, F. G. & FLORES, R. M. (eds), *Recent and Ancient Non-Marine*

Depositional Environments: Models for Exploration, Society of Economic Paleontologists and Mineralogists Special Publication, **31**, 213–231.

DEGENS, E. T. & STOFFERS, P. 1976. Stratified waters as a key to the past. *Nature*, **263**, 22–27.

DEMAISON, G. J. & MOORE, G. T. 1980. Anoxic environments and oil source bed genesis. *Bulletin of the American Association of Petroleum Geologists*, **64**, 1179–1209.

FONTAINE, H. & WORKMAN, D. R. 1978. Review of the geology and mineral resources of Kampuchea, Laos and Vietnam. *In:* NUTALAYA, P. (ed.), *Proceedings of the Third Regional Conference on the Geology and Mineral Resources of Southeast Asia, Bangkok, Thailand*, 541–603.

GARDNER, L. S. 1967. The Mae Mo lignite deposit in northwestern Thailand. *Department of Mineral Resources, Royal Thai Government, Report of Investigation*, **12**, 72 pp.

GIBLING, M. R. & RATANASTHIEN, B. 1980. Cenozoic basins of Thailand and their coal deposits: a preliminary report. *Bulletin of the Geological Society of Malaysia*, **13**, 27–42.

——, SUCHARITPORNCHAIKUL, O. & SUPERTIPANISH, S. 1982. The Cenozoic coal-bearing basin at Ban Huai Dua, Mae Hongson Province, Thailand. *In:* FERNANDEZ, J. C. (ed.), *Proceedings of the Fourth Regional Conference on the Geology and Mineral Resources of Southeast Asia, Manila, Philippines*, 809–815.

——, TANTISUKRIT, C., UTTAMO, W., THANASUTHIPITAK, T. & HARALUCK, M. 1985a. Oil shale sedimentology and geochemistry in Cenozoic Mae Sot Basin, Thailand. *Bulletin of the American Association of Petroleum Geologists*, **69**, 767–780.

——, UKAKIMAPHAN, Y. & SRISUK, S. 1985b. Oil shale and coal in intermontane basins of Thailand. *Bulletin of the American Association of Petroleum Geologists*, **69**, 760–766.

GOOSSENS, P. J. 1978. The metallogenic provinces of Burma: their definitions, geologic relationships and extension into China, India and Thailand. *In:* NUTALAYA, P. (ed.), *Proceedings of the Third Regional Conference on the Geology and Mineral Resources of Southeast Asia, Bangkok, Thailand*, 431–492.

GREGORY, J. W. 1923. The geological relations of the oil shales of southern Burma. *Geological Magazine*, **60**, 152–159.

HAMILTON, W. 1979. Tectonics of the Indonesian Region. *Professional Paper. United States Geological Survey*, **1078**, 345 pp.

HARDING, D. 1966. Lake Kariba, the hydrology and development of fisheries. *In:* LOWE-MCCONNELL, R. H. (ed.), *Man-made Lakes*, Symposia of the Institute of Biology, Academic Press, London, **15**, 7–20.

HAY, R. J. 1978. Geologic occurrence of zeolites. *In:* SAND, L. B. & MUMPTON, P. A. (eds), *Natural Zeolites*, Oxford, Pergamon Press, 135–143.

HOOD, A., GUTJAHR, C. C. M. & HEACOCK, R. L. 1975. Organic metamorphism and the generation of petroleum. *Bulletin of the American Association of Petroleum Geologists*, **59**, 986–996.

HU CHAO YUAN & QIAO HANSENG, 1983. Characteristics of oil and gas distribution in the North China Basin and the adjacent seas. *Proceedings of the Eleventh World Petroleum Congress, London*, **2**, 111–119.

JURDY, D. M. & VAN DER VOO, R. 1975. True polar wander since the Early Cretaceous. *Science*, **187**, 1193–1196.

JOHNSON, B. D., POWELL, C. MCA. & VEEVERS, J. J. 1976. Spreading history of the eastern Indian Ocean and Greater India's northward flight from Antarctica and Australia. *Bulletin of the Geological Society of America*, **87**, 1560–1566.

KELTS, K. & MCKENZIE, J. A. 1982. Modern deep sea and lacustrine examples of dolomite formed during early diagenesis in anoxic conditions. *Eleventh International Congress, International Association of Sedimentologists, Hamilton, Canada, Abstracts*, 125 pp.

KNOX, G. J. & WAKEFIELD, L. L. 1983. An introduction to the geology of the Phitsanulok Basin. *Conference on Geology and Mineral Resources of Thailand, Bangkok, November 1983*, 1–8.

KOESOEMADINATA, R. P. 1978. Sedimentary framework of Tertiary coal basins of Indonesia. *In:* NUTALAYA, P. (ed.), *Proceedings of the Third Regional Conference on the Geology and Mineral Resources of Southeast Asia, Bangkok, Thailand*, 621–639.

KRISHNAN, M. S. 1968. *Geology of India and Burma*. Higginbothams, Madras, India, 5th edn, 536 pp.

LE DAIN, A. Y., TAPPONNIER, P. & MOLNAR, P. 1984. Active faulting and tectonics of Burma and surrounding regions. *Journal of Geophysical Research*, **89**, 453–472.

LI DESHENG, 1984. Geologic evolution of petroliferous basins on continental shelf of China. *Bulletin of the American Association of Petroleum Geologists*, **68**, 993–1003.

MITCHELL, D. S. 1969. The ecology of vascular hydrophytes on Lake Kariba. *Hydrobiologia*, **34**, 448–464.

PAUL, D. D. & LIAN, H. M. 1975. Offshore Tertiary basins of Southeast Asia, Bay of Bengal to South China Sea. *Proceedings of the Ninth World Petroleum Congress, Tokyo*, **3**, 107–121.

PIYASIN, S. 1979. Petroleum exploration in the north. *Geological Society of Thailand Special Paper, Year 2522*, 18–24 (in Thai).

RODOLFO, K. S. 1969. Bathymetry and marine geology of the Andaman Basin, and tectonic implications for Southeast Asia. *Bulletin of the Geological Society of America*, **80**, 1203–1230.

SHERWOOD, N. R., COOK, A. C., GIBLING, M. & TANTISUKRIT, C. 1984. Petrology of a suite of sedimentary rocks associated with some coal-bearing basins in northwestern Thailand. *International Journal of Coal Geology*, **4**, 45–71.

SOEPARJADI, R. A., VALACHI, L. Z. & SOSROMIHARDJO, S. 1985. Oil and gas developments in Far East in 1984. *Bulletin of the American Association of Petroleum Geologists*, **69**, 1780–1855.

STAUFFER, P. H. 1973. Cenozoic deposits of West Malaysia and Singapore. *In:* GOBBETT, D. J. & HUTCHINSON, C. S. (eds), *Geology of the Malay*

Peninsula, John Wiley-Interscience, New York, 143-176.
TAPPONNIER, P., MATTAUER, M., PROUST, F. & CASSAIGNEAR, C. 1981. Mesozoic ophiolites, sutures and large scale tectonic movements in Afghanistan. *Earth and Planetary Science Letters*, **52**, 355-371.
—— & MOLNAR, P. 1977. Active faulting and Cenozoic tectonics of China. *Journal of Geophysical Research*, **82**, 2905-2930.
——, PELTZER, G., LE DAIN, A. Y., ARMIJO, R. & COBBOLD, P. 1982. Propagating extrusion tectonics in Asia: new insights from simple experiments with plasticine. *Geology*, **10**, 611-616.
THIENPRASERT, A., GALOUNG, W., MATSUBAYASHI, O., UYEDA, S. & WATANABE, T. 1978. Geothermal gradients and heat flow in northern Thailand. *Committee for Coordination of Joint Prospecting for Mineral Resources in Asian Offshore Areas (CCOP) Technical Bulletin*, **12**, 17-31.
TREVENA, A. S. & CLARK, R. A. 1986. Diagenesis of sandstone reservoirs of Pattani Basin, Gulf of Thailand. *Bulletin of the American Association of Petroleum Geologists*, **70**, 299-308.
WOOLLANDS, M. A. & HAW, D. 1976. Tertiary stratigraphy and sedimentation in the Gulf of Thailand. *Offshore Southeast Asia Conference, SEAPEX Programme*, **7**, 22 pp.
YANG WANLI, 1985. Daqing oil field, People's Republic of China: a giant field with oil of nonmarine origin. *Bulletin of the American Association of Petroleum Geologists*, **69**, 1101-1111.
——, LI YONGKANG & GAO RUIQI. 1985. Formation and evolution of nonmarine petroleum in Songliao Basin, China. *Bulletin of the American Association of Petroleum Geologists*, **69**, 1112-1122.

M. R. GIBLING, Department of Geology, Dalhousie University, Halifax, Nova Scotia, Canada, B3H 3J5.

Anoxic–oxic cyclical lacustrine sedimentation in the Miocene Rubielos de Mora Basin, Spain

P. Anadón, L. Cabrera & R. Julià

SUMMARY: The Rubielos de Mora Basin is a Miocene half-graben (3 × 10 km) with a sedimentary fill of over 600 m of alluvial and lacustrine sequences. Three main units make up this basin infill: lower (alluvial) and middle and upper lacustrine units. The upper lacustrine unit is mainly formed in the western part of the basin, by cyclical sequences. These sequences consist of organic-poor, non-laminated mudstones and marls cyclically alternating with thinly laminated facies such as sandy mudstones, bioclastic laminae, oil shales, rhythmites (carbonate-clay, varve-like couplets) and marls. Low-Mg calcite and non-stoichiometric (Mg-poor) dolomite are the dominant carbonate minerals in the non-laminated facies, whereas variable amounts of low-Mg calcite, high-Mg calcite, aragonite and non-stoichiometric dolomite have been recorded in most of the laminated facies. Low-Mg calcite is the main carbonate mineral in sandy mudstones and bioclastic laminae which occur at the lower part of laminated intervals in the cycles. Analysis of oil shales reveals a major macrophyte contribution of organic matter, displaying an early diagenetic evolutionary stage. Cyclical sequences record alternating oxic–anoxic bottom conditions in marginal zones of a meromictic lake, due to cyclical changes in lake water volume. This feature is recorded in the cyclical sequences of mineralogical changes and facies transitions. Strong tectonic subsidence (not balanced by sedimentation and taking place in an active rift setting) and palaeoclimate were the most striking features which could favour the establishment of a meromictic lake during this stage of the basin history.

Introduction

The occurrence of oil shale deposits in the lacustrine Mora de Rubielos Basin was already recorded by the beginning of this century (Fernández-Navarro 1914; Gavala 1921). From then on, although in a very discontinous way, the basin-fill deposits have been studied from diverse points of view, in order to establish the age as well as the potential economic interest of the bituminous facies (Crusafont *et al.* 1966; De Bruijn & Moltzer 1974; IGME–CGS unpubl., IGME 1986).

A broad two-fold subdivision of the oil shale deposits in the basin-fill sequence can be established on the basis of their sequential arrangement:

(1) Thick, monotonous laminated oil shale deposits with frequent rhythmic intervals,
(2) thinner oil shale and other laminated deposits which cyclically alternate with non-laminated, organic-poor, terrigenous and carbonate deposits.

This paper deals with these cyclical sequences observed in the western sectors of the Mora de Rubielos Basin. The aims are the description of the main sedimentological features of the oil shales and associated, organic-poor deposits in order to outline the evolving palaeoenvironmental conditions which resulted in their cyclical sedimentation.

Geological setting

The Rubielos de Mora Basin is located in the south-east part of the Iberian Chain (north-east Spain). The NW–SE oriented Iberian Chain is composed of a Hercynian basement unconformably overlain by a thick sedimentary cover of Mesozoic and Palaeogene rocks. An array of NW–SE orientated folds and thrusts, mainly developed during the late Palaeogene, stretches along this chain. During the Neogene a widespread rifting affected the north-east Iberian Peninsula giving rise to several half-grabens which were superimposed on the pre-existing structural features (Vegas *et al.* 1980).

The Miocene Rubielos de Mora Basin is a half-graben which belongs to this rift system. The basin is bounded by ENE–WSW to NE–SW striking faults and from its present exposures it can be inferred that it was at least 10 × 3 km (Fig. 1).

The Lower Cretaceous outcrops which surround the Rubielos de Mora Basin are mainly formed by a lower thick red-bed sequence (sandstones and mudstones) overlain by thick marine limestones with minor interbedded dolostones, carbonate mudstones and sandstones.

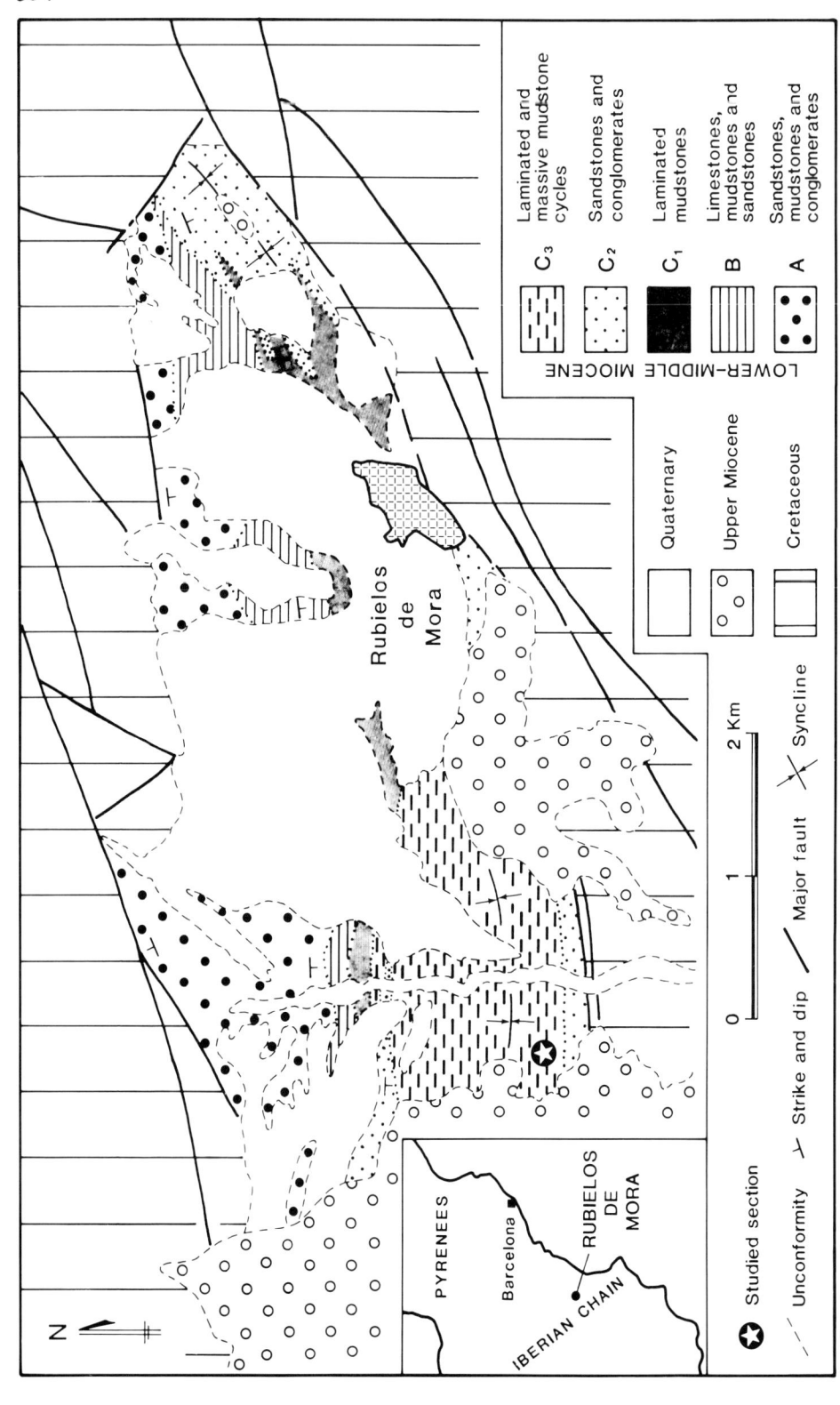

FIG. 1. Generalized geologic map of Rubielos de Mora Basin and location of the studied section.

This limestone unit underlies an upper arkosic sandstone unit. The Miocene basin-fill sequence is made up of over 600 m of terrigenous and minor carbonate deposits of alluvial and lacustrine origin. The whole sequence forms an overall synclinal structure. Two fossil mammal localities in the middle and upper parts of the basin-fill sequence indicate an early–middle Miocene age (Crusafont et al. 1966; De Bruijn & Moltzer 1974). Upper Miocene to Pliocene alluvial fan deposits (conglomerates, red sandstones and mudstones) make up the upper sequences which unconformably overlie the basin-fill deposits as well as the surrounding Cretaceous exposures, in the western part of the basin.

Stratigraphy of the Rubielos de Mora Basin

The Miocene basin-fill sequences in the Rubielos de Mora half-graben are to a large extent concealed by alluvial and colluvial Quaternary deposits, thereby defining two main outcrop zones (Figs 1 and 2). Exposures are not good enough to allow precise correlation between the sequences observed in each zone. Three main depositional units have been defined: Lower (A), Middle (B) and Upper (C). These major units can be traced through the basin (Figs 1 and 2) and record three different, successive evolutionary stages of the basin infilling (Anadón 1983). The two lower units crop out in the northern part of the basin. The upper unit, which crops out in the southern basin sectors, records the latest evolutionary stage of the basin infill. It shows distinctive facies changes which enables one to establish a three-fold mapping division (Fig. 1). The main depositional features of each unit can be summarized:

(A) Lower unit: Formed by up to 300 m of sandstones with minor interbedded mudstones and conglomerates. This unit unconformably overlies Lower Cretaceous limestones and sandstones, but in some places the contact is faulted.

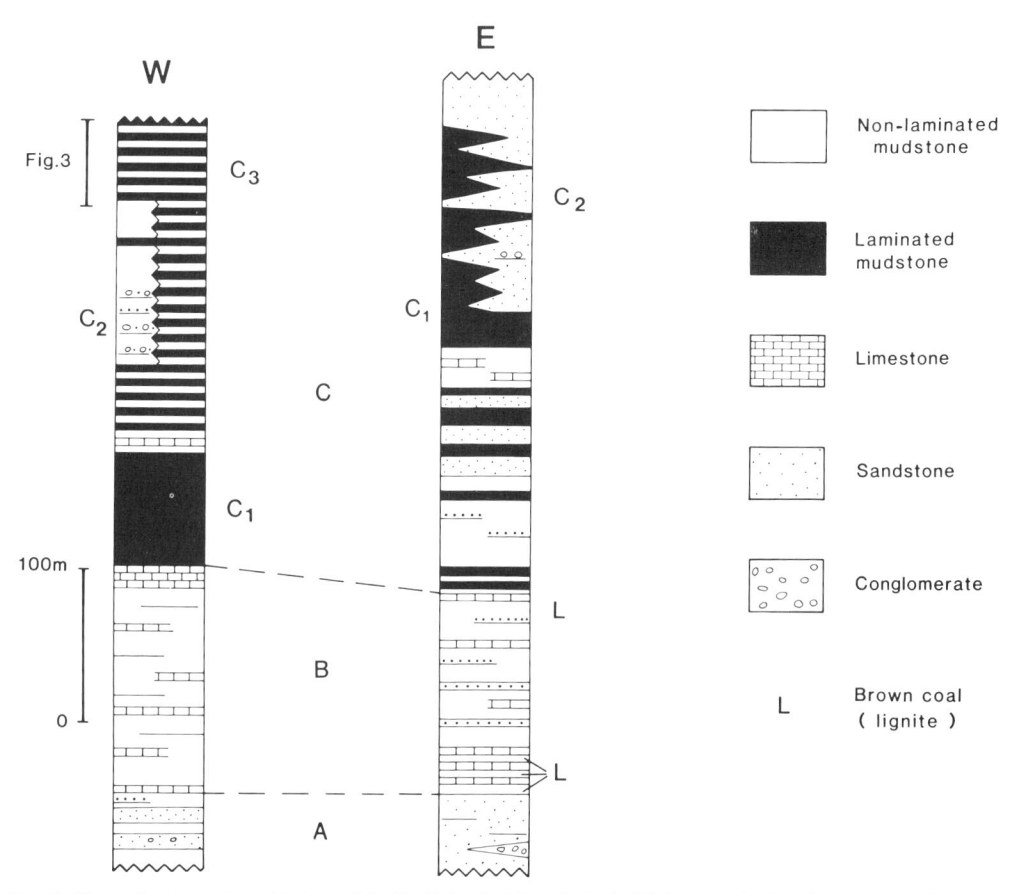

FIG. 2. Generalized stratigraphic logs of the Rubielos de Mora Basin infill showing the location of the studied section.

This unit records an early evolutionary stage of the basin, largely dominated by alluvial sedimentation.

(B) Middle unit: This unit is up to 70 m thick and mainly consists of lacustrine limestones with interbedded mudstones, sandstones and brown coals (lignites). Limestones show a high content of bioclastic remains (gastropods, bivalves, ostracods and charophytes).

(C) Upper unit: This unit is up to 300 m thick and it has been divided into three minor units (Figs 1 and 2).

(C1) This minor unit is up to 250 m thick and consists mainly of very thinly laminated mudstones (oil shales) and interbedded rhythmite beds showing a very scarce or absent bioturbation. The rhythmites consist of alternating laminae of carbonate (mainly aragonite) and clays. The laminated mudstones have a high organic matter content and delicately preserved fossil remains (leaves, insects, amphibia, etc.) are often recorded. The deposits of C1 unit are very uniform and cyclical sequences have not been observed.

(C2) Sequences up to 150 m thick made up this minor unit which mainly consists of sandstones, mudstones and conglomerates. These terrigenous deposits interfinger with the C1 and C3 minor units.

(C3) This minor unit is up to 200 m thick and consists of laminated marls and organic-rich shales cyclically alternating with non-laminated mudstones and marls. Rhythmite beds (carbonate–clay couplets) also occur, although less frequently than in the C1 unit. The oil shale and rhythmite deposits show the same general features as those observed in C1 unit. Organic matter content in the laminated mudstones is variable but they are often true oil shales. Outcrops of these units are restricted to the western part of the basin (Fig. 1).

Despite these minor units showing close stratigraphic relationships, they record rather diverse palaeoenvironmental conditions, varying from alluvial (unit C2) to lacustrine (C1 and C3). In broad outline when compared with unit B, the features of C1 and C3 deposits (oil shales and thinly laminated rhythmites) reflect the occurrence of relatively 'deeper' lacustrine conditions.

Cyclical sequences in the upper unit

Well-developed, small-scale cyclical sequences a few metres thick characterize the upper unit (C3) in the western part of the basin. A 50-m thick section composed of these cyclical sequences (Cerro Porpol section) has been studied in detail (Fig. 3). This succession overlies a terrigenous sequence made up of mudstones, sandstones and conglomerates, which corresponds to a tongue of unit C2. Exposures are not good enough to trace the complete lateral extent of the cyclical sequences, but in some cases they can be traced for about 1 km.

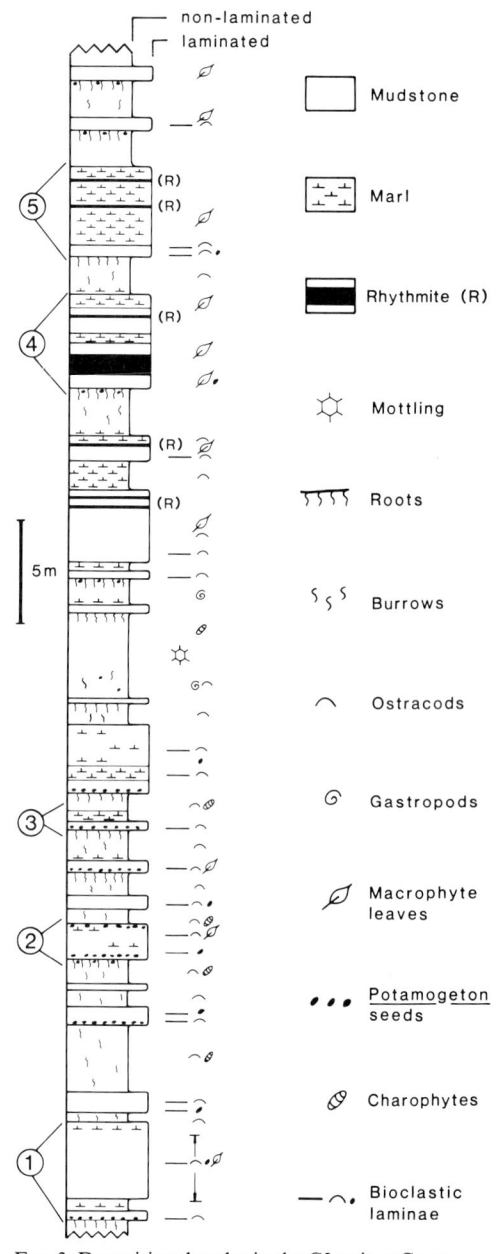

FIG. 3. Depositional cycles in the C3 unit at Cerro Porpol. See Figs 1 and 2 for location.

Two main alternating facies assemblages have been recognized in the cyclical sequences. One of them consists of massive, non-laminated facies (a) and the other is mainly made up by laminated facies (b). These facies are sometimes intergradational and their recognition is based on mineralogical, textural, sedimentological and sequential criteria.

Facies description

Non-laminated facies assemblage (a)

Two main rock types constitute this facies assemblage: white marls (a1) and grey bioturbated mudstones (a2). They mainly differ from each other in their carbonate content, colour and relative position in the cyclical sequences.

White marls (a1). This facies makes up whitish beds ranging from 0.1 to 0.7 m thick. The main carbonate minerals are low magnesium calcite (=LMC) and non-stoichiometric dolomite in varying proportion. Low magnesium calcite is usually dominant but dolomite content can be up to 60%. Scattered very fine quartz grains can be observed in the matrix. Marl beds usually occur between an underlying laminated facies (b) and overlying grey bioturbated mudstones (a2). They are rarely found as interbedded in laminated facies. The upper contacts of the marl beds are usually gradational into the overlying facies (a2).

Bioturbation is frequent and burrows are often filled by green clays. The biogenic content mainly consists of exotic carbonaceous macrophyte debris and ostracod shells, often reduced to casts by dissolution.

Grey bioturbated mudstones (a2). This facies mainly consists of massive, lumpy mudstones which make up beds from 0.5 to 4.5 m thick and show a typical light greenish grey colour. Non-stoichiometric dolomite and LMC are the main carbonate minerals. The former is clearly dominant. Pyrite is frequent either as spots scattered in mudstone or linked to bioturbation traces. Quartz, feldspar and clay minerals (illite and kaolinite) make up the bulk of the finer terrigenous fraction in this rock type. Coarser quartz grains up to 3 mm in size often appear floating in the mudstone.

Bioturbation is always widespread and burrows can display a high bioclastic content. Moreover at the top of the massive carbonate mudstone beds string-like, vertically arranged and sometimes bifurcated ferruginous traces, up to 1 mm in diameter are common (Fig. 4). These bioturbation traces can occur at intervals of 0.03–0.3 m and may be attributed to subaqueous macrophyte plant roots. At the top of these bioturbated beds often occur scattered fossil seeds of *Potamogeton*. Thus root traces may be attributed to this subaqueous plant. Other carbonaceous remains of macrophytes are often recorded in the carbonate

FIG. 4. Sharp, erosive surface at the top of a non-laminated mudstone (a2) bed with subaqueous plant roots. Erosive surface is crowded by *Potamogeton* seeds included in a thin sandy mudstone bed (b1). Scale bar = 1 cm.

mudstones. Moreover, charophyte oogonia, minor gastropod bioclasts and ostracods are the usual biogenic remains, sometimes preserved just as casts. Bioclasts can be scattered in the fine grained matrix or, as mentioned above, concentrated in burrows.

Laminated facies assemblage (b)

This facies assemblage includes a variety of finely laminated rock types. The presence of unstable carbonate minerals (aragonite and high-magnesium calcite (HMC)) distinguishes them also from the homogeneous marls and mudstones. The sequences made up by these facies show a sharp and sometimes slightly erosive base and they range from 0.3 to 4.5 m in thickness.

The following rock types have been included in the laminated facies assemblage (b):

Sandy mudstones (b1). Thin, faintly laminated sandy mudstone beds with a relatively high content of quartz grains commonly occur at the base of laminated sequences. The base of these beds is sharp, sometimes even slightly erosive, and overlies grey bioturbated mudstones (Fig. 4). The carbonate content of this facies consists of LMC, although contents of HMC of up to 35% have been recorded. Traction laminae, containing ostracod shells, charophyte oogonia, vegetal debris and large amounts of *Potamogeton* seeds are very common.

Bioclastic layers (b2). Laminae crowded mainly by ostracod shells and other carbonate bioclasts (charophyte remains) frequently occur. In some cases high concentrations of *Potamogeton* seeds and other plant remains are selectively accumulated too. Clastic laminae range from less than 1 to 4 mm in thickness and they occur either grouped and alternating with dark clay laminae or isolated, interbedded in thick beds of organic-rich shales (b3). The carbonate content in this facies is mainly biogenic. Low-magnesium calcite is the most frequent mineral but minor amounts of HMC have also been recorded.

Shales (b3). The bulk of the laminated sequences consists of organic-rich, brown to dark grey shales. These occur either as very thin laminae alternating with clastic accumulations of ostracods and *Potamogeton* seeds (facies b2) or as individual beds up to 3 m thick. Organic matter contents can be quite high and they can be considered oil shales. Bioturbation traces are very scarce and horizontal burrows with iron sulphide infilling have been observed locally. Carbonaceous vegetal debris, including very well preserved fossil leaves (which are often concentrated in thin beds or laminae), are abundant. Occasionally thin shelled ostracods, rare gastropod shells and *Potamogeton* seeds are scattered in the mudstone. Articulated skeletons of *Chelotriton paradoxus* (amphibia, Salamandridae) have also been recorded (B. Sanchiz pers. comm.).

Rhythmites (b4). Rhythmites mainly consist of couplets of very thin white and dark laminae, respectively made up of carbonates and clays. The beds of this facies are from a few mm to 0.85 m thick (Fig. 5). Carbonate lamina thickness ranges from 35 μm to 0.1 mm, while clay laminae are from 0.01 to 1 mm thick. Sometimes rhythmite facies grades into laminated marls (b5) with interbedded clay laminae (see logs 1 and 5 in Fig. 6). Carbonate mineralogy in the white laminae is not uniform. Low-magnesium calcite is in some cases the only mineral present. More often it occurs together with aragonite and/or HMC. Aragonite can be in some cases the dominant mineral phase, while HMC is always subordinate. Dolomite, when present, only accounts for a very minor part of the carbonate content.

Marls (b5). Finely stratified to laminated white marls often occur in the laminated parts of the studied sequences. These marls make up laminae or beds ranging from 1 mm to several metres in thickness. Scattered ostracod shells and bioclasts

FIG. 5. Rhythmite facies (b4) with a interbedded thin laminated marl bed. Rhythmite intervals consist of carbonate–clay couplets. In this case calcite (LMC) is the dominant carbonate mineral, but minor amounts of aragonite prisms have been recorded in SEM. Scale bar = 1 cm.

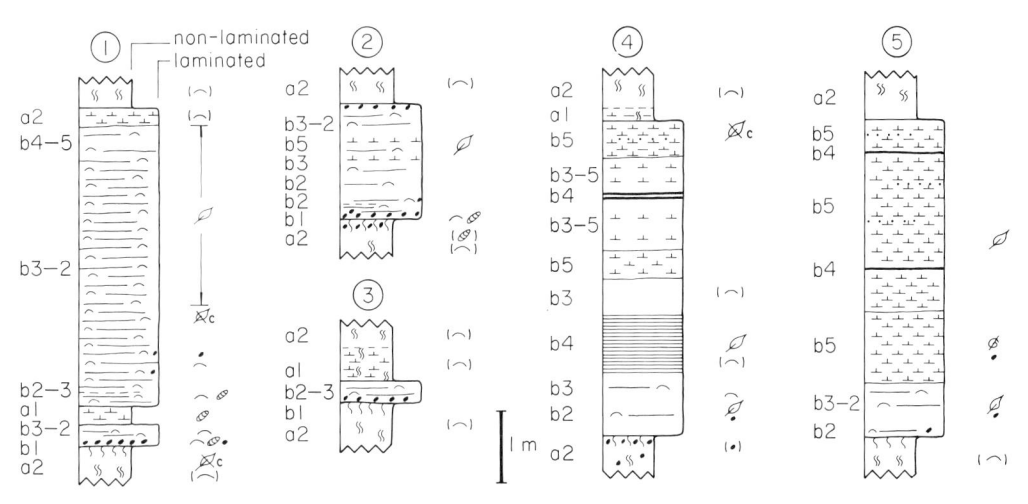

FIG. 6. Detailed logs of several cycles in the Cerro Porpol section. See Fig. 3 for legend and location in the sequence.

and minor poorly preserved vegetal debris are the most frequent palaeontological remains in these deposits.

Thinner marl laminae and beds (up to 5 cm thick) are always interbedded in the brown to dark grey shales which form most of the laminated sequences. Some of these thin marl beds show a chalky appearance and are devoid of lamination. Carbonate content mainly consists of dolomite (usually dominant) and LMC. Minor amounts of aragonite also occur.

Thicker laminated marl beds are interbedded in the dark shales or represent the dominant facies at the upper part of the laminated intervals. In the latter case, the upper part of the laminated marl beds grades upwards into the base of the non-laminated facies assemblage sequences. Green clay and very thin detrital quartz laminae are often interbedded in the marl beds. Carbonate mainly consists of LMC and dolomite which can be dominant. Aragonite and/or HMC also occur frequently.

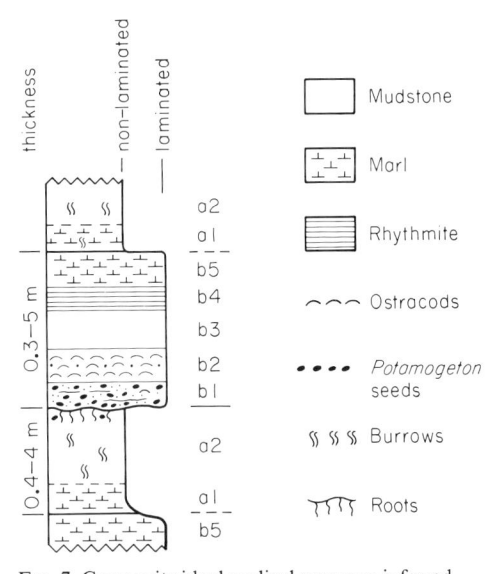

FIG. 7. Composite ideal cyclical sequence inferred from the observed cycles in the Cerro Porpol and other outcrops in the western part of the basin.

Sequential facies analysis

The diverse facies here described succeed one another cyclically (Fig. 6). From the detailed analysis of the studied section and overall observations made on other cyclical sequences, a composite ideal cycle can be deduced, including all the facies here described (Fig. 7).

The real cycles recorded in the studied section differ from the composite or ideal cycle, on the basis of absence and or recurrence of some facies (Figs 3 and 6). The grey bioturbated mudstone facies (a2) is always present and its contact to the laminated facies is always sharp and marked by detrital and skeletal grain accumulations. Sandy mudstones (b1) bioclastic accumulations (b2) and shales (b3) are always the lowermost laminated facies in each cycle. Rhythmites (b4) and laminated marls (b5) never occur in this sequential position. The sequential distribution pattern of these carbonate facies seems also related to the relative development of the cycles. Thus the thinner laminated episodes are in general devoid of carbonate deposits. On the contrary rhythmite deposits (b4) and laminated marls (b5) are especially well developed in the thicker cycles at the upper part of the section (Fig. 3). Non-laminated marls (a1) can be considered as transitional deposits between the laminated marls (b5) and the non-laminated grey mudstones (a2).

Organic matter and carbonates in the cycles

Carbonate mineralogy and organic matter analysis provided useful data in order to establish a first approach to the palaeolimnological conditions which resulted in the sedimentation of the lacustrine sequences as well as in their early diagenetic evolution.

Organic matter

Organic matter content in the studied facies is very variable. Organic matter pyrolysis tests made on samples of non-laminated facies show a content of less than 5 l/t. The average organic matter content in the oil shales ranges from 20 to 40 l/t, with a maximum content of 70 l/t (IGME–CGS unpubl. data). Reyes & Feixas (1984) attributed kerogen included in the oil shales to the Type I of Tissot & Welte (1978).

Several features reveal the source of the organic matter accumulated in the Rubielos de Mora lacustrine oil shales. Reyes & Feixas (1984) record *Botryococcus*, undetermined algal remains, pollen, spores and amorphous materials in the oil shales. Moreover, macrophyte leaves and other carbonaceous vegetal debris are abundant. Several tree genera (*Ulmus*, *Salix*, *Quercus* and *Glyptostrobus*) have been recorded on the basis of their fossil leaves (Fernández-Navarro 1914). On the other hand the widespread occurrence through the studied section of debris and seeds of littoral subaqueous macrophytes (*Potamogeton*) is outstanding. It may be especially significant that at the base of the laminated sequences occur dense accumulations of *Potamogeton* seeds (facies b1 and b2).

Preliminary studies of organic geochemistry carried out on the extractable organic matter from two samples of oil shales (facies b4) have shown (X. de las Heras pers. comm.) a high content of n-alkanes in the C_{25}–C_{33} range, maximizing at C_{31}. This data enables one to postulate a significant contribution from terrestrial plant wax (Eglinton & Farrington 1967). The carbon preference index values (Bray & Evans 1961):

$$CPI = \frac{\sum_{n=7}^{17} C_{2n+1}}{\sum_{n=7}^{17} C_{2n}}$$

are about 5. This high odd over even n-alkanes predominance points to a good organic matter preservation and poorly advanced diagenetic evolution (McKirdy et al. 1983, Didyk et al. 1978). Presence of sterenes in the extractable organic matter corroborates the CPI interpretation (Gagosian & Farrington 1978). The pristane/phytane ratio has been considered as indicative of the oxidizing or reducing conditions of the depositional environment. This ratio in the studied samples is about 0.5, showing that transformation took place in an anoxic environment (Powell & McKirdy 1973; Cardoso & Chicarelli 1983) and recording a still early diagenetic evolutionary stage (Costa 1983).

Carbonates

Carbonate mineralogy

Sediment mineralogy was determined by means of X-ray diffraction (XRD) supplemented by SEM and EDAX investigations. Clay mineralogy was studied in the <2 µm size by XRD. A semi-quantitative estimation of the individual carbonate minerals was made by comparing peak heights with prepared standards. Clay minerals (illite, kaolinite) quartz and feldspars are the main components recorded among the terrigenous clastics. Minor amounts of early diagenetic pyrite have also been identified (Fig. 8).

Moreover, diverse carbonate contents have been recorded through the studied section in the non-laminated, homogeneous facies as well as in the laminated ones. Paper-thin carbonate laminations in rhythmites, laminated marls and homogeneous marl layers are the most conspicuous carbonate deposits. Calcite, dolomite, aragonite and magnesium calcite are the carbonate minerals recorded in these deposits (Table 1). These minerals are especially interesting in order to establish a first approach to the study of the palaeolimnological conditions which resulted in the sedimentation of the lacustrine sequences.

Calcite (LMC) is present in all the described lacustrine facies. Carbonate content in some of the carbonate rhythmite laminae and thicker laminae, as well as in most of the bioclastic laminae, exclusively consists of this mineral. In the mainly terrigenous facies LMC is largely biogenic and is restricted to bioclastic laminae and small-scale lenses. In the rhythmite laminations and in the thicker carbonate laminae, LMC makes up very fine grained (less than 1 µm), sub-

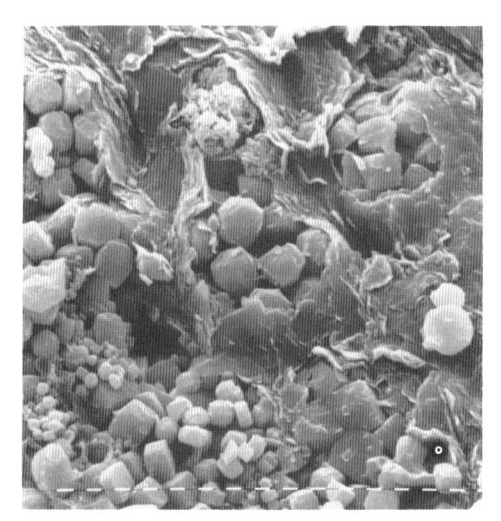

FIG. 8. Early diagenetic framboidal pyrite in oil shale facies (b3). Scale bars = 10 μm.

euhedral to dominantly anhedral crystals. These crystals may form small-scale aggregates. Large polyhedral or blocky calcite crystals have not been observed.

Magnesium calcite (HMC up to 14% mol Mg) and aragonite are not present either in the non-laminated terrigenous and carbonate facies or in the bioclastic laminae. However they are often present in carbonate laminations.

Aragonite is more frequent than HMC, whose occurrence is rather scarce. Carbonates in some of the carbonate rhythmite laminae and marl laminae can consist of nearly 70% aragonite. Aragonite occurs associated with dolomite and/or HMC. In those samples where XRD shows a clear predominance of aragonite, characteristic subeuhedral to euhedral aragonite prisms up to 3 μm long, have been observed in SEM (Fig. 9). In a few cases diagenetic change from aragonite prisms to very fine anhedral to subeuhedral LMC crystals have been observed.

Non-stoichiometric dolomite (Mg-poor dolomite: 42–46% Mg) is usually present in significant amounts in the non-laminated facies, where it is exclusively associated to LMC. Carbonate laminae interbedded in the oil shale layers display a wider range of dolomite content (4–93%). In these laminae dolomite always occurs together with calcite. Aragonite and/or HMC may be present too. Dolomite occurrence is rather scarce in bioclastic laminae and the thinner carbonate rhythmite laminae. No euhedral or subeuhedral dolomite crystals have been detected by SEM in samples where XRD showed that dolomite was present in significant amounts. Anhedral crystals are generally dominant.

Origin

Although a certain clastic input of allochthonous carbonates was likely to be fed into the lake it is here assumed that most of these carbonates are not terrigenous, but resulted from biogenic and/

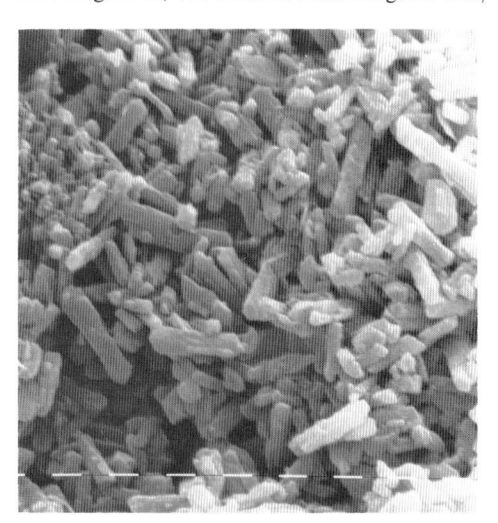

FIG. 9. Aragonite prisms in a carbonate laminae from rhythmite facies (b4). Scale bars = 1 μm.

TABLE 1. *Summary of carbonate content and mineralogy ranges*

	Non-laminated facies		Laminated facies				
	a1	a2	b1	b2	b3	b4	b5
% Carbonate	64–71	40–48	25–60	20–65	10	42–90	45–75
LMC	39–88	65–100	(65)*–100	100	†	9–100	10–100
HMC	—	—	–(35)*	—	†	0–27	0–40
Dolomite	16–61	0–35	–(2.6)*	—	†	0–7	0–93
Aragonite	—	—	—	—	†	0–70	0–68

* Values within brackets correspond to percentages recorded only in one sample.
† Presence of carbonate in μm thick interbedded laminae seen by SEM.

or inorganic precipitation in the water column as well as from early diagenetic reactions in the lake bottom sediment. This assumption is based on the following points.

(1) The carbonate mineralogical content of the nearer source area rocks (calcite and minor dolomite) does not relate to the changes in the carbonate mineral assemblages through the sequences.
(2) Unstable mineral phases as HMC and aragonite could not be derived from the Mesozoic source rocks. Their preservation in the studied lacustrine sequences must result from special diagenetic conditions.
(3) The frequently varved or more crudely laminated arrangement of the carbonate dominated deposits, cannot be interpreted as the result of terrigenous event contributions.
(4) There is an absence of carbonate terrigenous grains in SEM, despite the occurrence of very fine detrital quartz grains among the sheet-like clay grains.
(5) There is an absence of noticeable amounts of dolomite (at least more than 1%), not detected by XRD analysis, in those lacustrine facies where clear terrigenous inputs have been recorded.

Aragonite and/or HMC have been recorded in several recent to late Neogene lacustrine sequences (Neev & Emery 1967; Begin et al. 1974; Kelts & Hsü 1978; Spencer et al. 1984) but these minerals (specially HMC), are rather unstable and rarely preserved in older sedimentary sequences despite its likely early occurrence (Desborough 1978). The frequent occurrence of aragonite and minor amounts of HMC in the early Miocene lacustrine sequences described here is a remarkable phenomenon. This may be attributed to the sheltering of the early carbonate mineral assemblages by the 'dry' conditions resulting from the presence of oil shale deposits and the clay deposits dominance (Lippmann 1973). The very low thermal gradient which has affected the lacustrine sequences, as deduced from the organic matter data, has probably contributed too. These conditions have been emphasized to favour preservation of unstable carbonate minerals (Lippmann 1973; Koch & Rothe 1985).

At present it has not been possible to precisely evaluate the amounts of carbonate linked either to early diagenetic or to endogenetic precipitation (Desborough 1978; Talbot & Kelts 1985). In order to determine more accurately the real importance of each kind of process the relative influence exerted on the carbonate assemblages by diverse late and early diagenetic processes must be established (i.e., diagenetic changes induced by recent meteoric waters, possible contribution of bicarbonate resulting from bacterial sulphate reduction and methanogenesis). Anyway, according to the above mentioned facts, aragonite, some HMC and some LMC can be interpreted as endogenic deposits precipitated in the upper zone of the lacustrine water column. Dolomite, LMC and some HMC may be early diagenetic and formed in the bottom sediment of the lake. Finally some calcite may have formed in late post-depositional diagenetic stages.

Carbonate mineral assemblages similar to those described here have been recorded in several ancient and modern lacustrine deposits (Müller et al. 1972; Degens et al. 1973; Kelts & Hsü 1978; Spencer et al. 1984; Talbot & Kelts 1985). Despite the rather large quantitative variations observed, our carbonate mineral assemblages fit quite well those resulting from sedimentary and early diagenetic processes developed in alkaline lakes. Mineral composition of the carbonate deposits suggests the occurrence of transition from LMC to HMC and aragonite mineral assemblages (b1, b2 to b5, b6). These changes usually reflect evaporative concentration of lake waters as well as increasing Mg concentration (Müller et al. 1972; Degens et al. 1973; Hsü & Kelts 1978; Spencer et al. 1984) although early diagenetic processes in organic-rich bottom sediments affecting carbonate mineralogy must also be taken into account (Desborough 1978; Talbot & Kelts 1985).

Model of cyclical sedimentation

The idealized lacustrine cycle defined above (Fig. 7) enables us to establish a model of the evolutionary trends of the lacustrine sedimentation in the basin during its latest lacustrine stage. This model is suggested on the basis of a complete, ideal cycle which would be formed by all the recorded facies (Fig. 10). Of course more complex and varied trends can be deduced from the real cycles observed in the studied section. These cycles are very variable and so truncations and recurrences can alter the proposed ideal cyclical lacustrine evolution (Fig. 6).

Intense burrowing, low organic matter content, absence of unstable carbonate minerals (HMC, aragonite) and a widespread occurrence of dolomite are the characteristic features of the non-laminated facies (a). In broad outline, these facies developed under oxygenated bottom conditions as is shown by the intense burrowing. Disturbance of sediment by bioturbation favoured the organic matter decay under oxidizing conditions. These early diagenetic conditions precluded the devel-

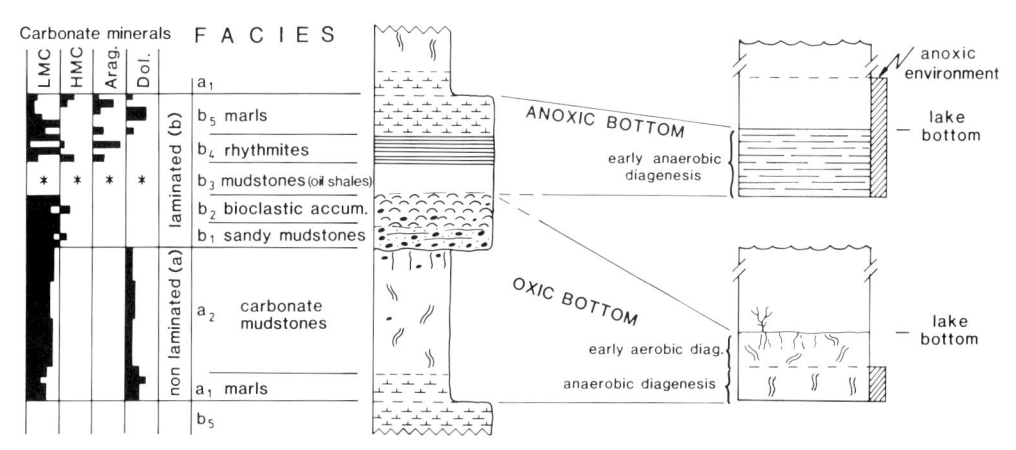

FIG. 10. Main sedimentological features and interpretation of the ideal composite cyclical sequence resulting from alternating oxic–anoxic bottom conditions. * Indicates occasional occurrence of carbonate minerals in interbedded laminae within oil shales.

opment of the 'dry conditions' needed for preservation of early unstable carbonate minerals possibly formed (Lippmann 1973; Koch & Rothe 1985). The dolomite content of these facies may indicate the early occurrence of these unstable carbonate precursors. The striking occurrence of root bioturbation and the abundance of *Potamogeton* seeds at the top of the non-laminated grey mudstone beds (a2) record occasional shallowing upwards trends involving a change from oxic, aphotic bottom conditions to oxic, photic conditions under which littoral subaqueous macrophytes could proliferate.

The beds consisting of non-laminated facies (a) are overlain by laminated facies deposits (b). The contact between these facies is sharp, flat, sometimes slightly erosive. This fact is emphasized by the occurrence in the lower parts of the laminated episodes of larger amounts of reworked components (mainly quartz grains, skeletal debris and *Potamogeton* seeds) than in the underlying grey bioturbated mudstones (a2). Laminated facies closely related with the contact (b1–b2) were deposited by tractive currents and resulted from a mechanical accumulation under high energy conditions. On the other hand an early exclusive occurrence of LMC and very low amounts of dolomite (perhaps reworked) characterize these facies too. These features suggest that b1 and b2 facies developed as a consequence of a lacustrine 'transgression' which in turn could result from an increase in the volume of the lake waters. Increasing volume probably led to an increase of depth and to a decrease in the solute concentration in the lake water, the latter resulting in the nearly exclusive occurrence of LMC in the facies developed during early transgressive stages.

Similar facies relationships, revealing increasing energy conditions linked to transgressive lacustrine stages, have been previously described. Singh *et al.* (1981) record, as related to lacustrine transgressive events, the sandy and shelly deposits overlying disconformities in Lake George Quaternary deposits (south-east Australia). Flat pebble conglomerates and lime sandstones underlying oil shale facies were also interpreted in a similar way in the cycles of the Wilkins Peak Member of the Green River Formation (Eugster & Hardie 1975).

The detrital episodes recorded by b1 and b2 laminated facies are followed by oil shale (b3), rhythmite (b4) and laminated marl (b5) facies. Thin laminations, sometimes arranged as rhythmites, very scarce or absent bioturbation and preservation of delicate fossil remains (macrophyte leaves, insects, articulated amphibia skeletons, etc.) show that these facies were deposited under anoxic bottom conditions. High contents of organic matter (preserved in an early diagenetic stage under the mentioned anaerobic bottom conditions) and preservation of unstable carbonate minerals (HMC and aragonite) are also distinctive features of these facies. The noticeable thickness of some of the beds made up by these laminated facies (up to 4.5 m) suggests that the lake water mass was quite permanently stratified (meromictic lake). However short episodes of intensive burrowing, very sparsely found in the laminated beds, could record inputs of oxygenated water from the mixolimnion, linked to turbiditic flows. These flows might be related with mass flow deposits similar to those recorded in older lacustrine facies from eastern basin areas (subaqueous debris flows and olisthostromes; Anadón 1983).

The occurrence of oil shale facies (b3) overlying the detrital laminated episodes made up of facies b1 and b2 sustain the development of transgressive stages in the lacustrine basin. The rising of water level would favour stratification of the bottom water mass and/or rising of the oxycline. As a consequence, the sedimentation of organic-rich mudstones develops in the anoxic lake bottom.

On the other hand the transition from oil shales to rhythmite (b4) and laminated marl facies (b5) record a progressive concentration of solute in the lake waters. The carbonate mineralogy (HMC, aragonite and dolomite) of the laminated facies which made up the upper part of the cycles can be interpreted as a result of evaporative concentration in an alkaline lake. In some cases oil shales with interbedded carbonate laminae (rhythmites) pass upwards to laminated marl deposits (b5). High-magnesium calcite and aragonite, which probably record a more concentrated water stage, form laminated deposits. The frequent varved-like pattern of the carbonate–clay couplets could be caused by the alternation of carbonate precipitation and very fine terrigenous contributions. Carbonate precipitation could be generated in the water column either by algal-blooms or by mixing of waters of diverse salinity. In any case, it is difficult to be precise about the possible seasonality and/or cyclical frequence of these processes.

Evaporative concentration was probably linked to a water level drop, with reduction of water volume, as is indicated by the overlying non-laminated bioturbated white marls (a1) as well as by widespread quartz sand laminae interbedded in the laminated marls (b5). These sandy laminae probably record the basin-ward spreading of the terrigenous contributions which resulted from the lowering of the base level in the lacustrine basin. The occurrence of oil shale deposits overlain by facies which record lacustrine contraction has been recorded by several authors. In extreme cases evaporites are the deposits which result from contraction from larger, deeper and fresher lakes to more saline conditions (Eugster 1985, p. 621) and dessication features can appear at the top of the sequences.

In the Rubielos de Mora lacustrine cycles the bioturbated marls (a1) reflect the pronounced change from anoxic bottom conditions to oxic conditions during the culminating phase of the lake level drop. At this point the initial conditions of the ideal cycle would have been reached again.

Low organic matter content and absence of unstable carbonate minerals enable one to record the rather different sedimentary conditions in relation to the underlying laminated facies. In any case the sedimentary record observed in the studied sequences reveals neither any kind of evaporitic saline deposits nor dessication features. So a permanent lake remained in the basin even during the culminating stages of the lacustrine shrinkage.

Discussion

In broad outline the lacustrine sequences studied in the Rubielos de Mora Basin display an evolutionary trend similar to others recorded in several modern and ancient lakes (Eugster & Hardie 1975, 1978; Surdam & Stanley 1979; Ryder et al. 1976; Last & Slezak in press). Cyclically arranged lacustrine sequences from the Wilkins Peak Member (Eocene Green River Formation; Eugster & Hardie 1975) show a certain resemblance when compared with the studied sequences. In both cases organic-rich oil shales occur sequentially related with deposits which record retraction phases during lacustrine evolution. However several conspicuous differences can be established between both case studies. Sedimentation of Wilkins Peak Member oil shales was frequently terminated by rapid desiccation resulting in evaporite deposition or in sedimentation of carbonate mudstones displaying well-developed subaerial exposure features (Eugster & Hardie 1975, 1978; Eugster 1985). On the contrary, in the Rubielos de Mora Basin laminated oil shales are overlain by deposits with a gradually larger carbonate content. These deposits record a process of evaporative concentration linked to a water level drop. However, evaporite deposits, emersion traces or any other sedimentological features which could indicate extremely severe lacustrine restriction, have not been recorded in the studied cyclical sequences.

The cyclically alternating oxic and anoxic bottom conditions in some of the areas of the Rubielos de Mora Basin have alternative interpretations.

(1) Sequences record alternation of holomictic (seasonally stratified) and meromictic (permanently stratified) stages during the history of the lake.
(2) Sequences are the result of alternate rises and falls of the oxycline which would have affected the sublittoral, marginal zone of a meromictic lake.

Some ancient examples referable to the first alternative have been discussed by Boyer (1981), who postulated the occurrence of ectogenic meromixis to explain the alternation of holomictic and meromictic conditions in several shallow lacustrine sequences. However, the stratigraphic relationships between the diverse sequences observed in the basin, suggest that the second alternative is more convincing, at least for the lower parts of the unit C (Fig. 2). In fact, these lower cyclical sequences in the western zones of the basin are laterally equivalent to thick, uniform laminated oil shale sequences. This lateral relationship would reflect the existence of deep, permanently stratified waters, fringed by shallower sublittoral areas into which anoxic bottom conditions occasionally spread (shifting anaerobiosis).

From this point of view the environmental evolution which led to the sedimentation of the lacustrine sequences of the unit C would show a better resemblance with the oil shale bearing sequences of the Piceance Creek and Uinta Basins and the Laney Member of the Green River Basin (Eugster 1985; Ryder et al. 1976). In these sequences evaporitic episodes were less frequent and less intense than in the Wilkins Peak Member, and lacustrine evolution was steady enough to result in sedimentation of thick oil shales deposits (Eugster 1985). In any case, the regressive stages observed in the Rubielos de Mora sequences, although resulting in saline conditions which enhanced precipitation of aragonite and HMC, did not lead to strong evaporite sedimentation. On the other hand, taking into account the size and general geometric features of the Rubielos de Mora Basin (a relatively small, assymetrical half-graben) it is improbable that extensive playa areas developed around the inner lacustrine zone during lacustrine regression. This fact is a quite distinguishing feature between the general settings of the sequences developed at the Wilkins Peak Member and the Rubielos de Mora Basin.

The sedimentary evolution as well as the diagenetic conditions in the Rubielos de Mora Basin were strongly influenced by tectonic evolution. Alluvial and lacustrine deposits of units A and B record the beginning of differential subsidence along the faults which bounded the basin. An increase in tectonic activity along the southern, steep basin margin, during sedimentation of unit C, is recorded by frequent occurrence of slumps, olistholiths and chaotic breccia facies affecting the thick, non-cyclical oil shale deposits deposited in the eastern parts of the basin. This increasing tectonic activity may have resulted in a larger subsidence rate unbalanced by sedimentation. As a result a deepening of the basin, could favour the stratification of the water column.

During the early and middle Miocene the general tectonic and climatologic setting of the lacustrine Rubielos de Mora Basin was suitable for the development of permanently stratified (meromictic) lakes. Rifting processes (Vegas et al. 1980) resulted in formation of rapidly subsiding half-grabens where relatively deep lacustrine basins could develop. Moreover stratification of lacustrine waters could be enhanced by the climatological conditions. Thus, warm subtropical to tropical climatic regimes were well established by the early to middle Miocene in the north-western Mediterranean area (Bessedik 1985).

The later tectonic evolution of the Rubielos de Mora Basin did not favour the generation of mature oil. Low pressure and temperature conditions prevented oil maturation. However, in structural settings where adequate burial conditions could develop, small basins similar to the Rubielos de Mora half-graben could give rise to significant amounts of mature oil.

ACKNOWLEDGEMENTS: We wish to thank X. de las Heras for organic geochemistry analyses. We are grateful to Drs. P. A. Allen, K. Kelts and A. Matter who provided thoughtful suggestions and criticisms on an earlier version of the manuscript. This work has received financial support from the CSIC–CAICVT Research no. ID 851.

References

ANADÓN, P. 1983. Características generales de diversas cuencas lacustres terciarias con pizarras bituminosas del NE de la Península Ibérica. *Comunicaciones del X Congreso Nacional de Sedimentología, Menorca*, **1**, 9–12.

BEGIN, Z. B., EHRLICH, A. & NATHAN, Y. 1974. Lake Lisan, the Pleistocene precursor of the Dead Sea. *Bulletin of the Geological Survey of Israel*, **63**, 30 pp.

BESSEDIK, M. 1985. Reconstitution des environnements miocénes des regions Nord-ouest Méditerraneennes à partir de la Palynologie. Thèse Université des Sciences et Techniques du Languedoc, Montpellier.

BOYER, B. W. 1981. Tertiary lacustrine sediments from Sentinel Butte, North Dakota and the sedimentary record of ectogenic meromixis. *Journal of Sedimentary Petrology*, **51**, 429–440.

BRAY, E. E. & EVANS, E. D. 1961. Distribution of n-paraffins as a clue to recognition of source beds. *Geochimica et Cosmochimica Acta*, **22**, 2–15.

CARDOSO, J. N. & CHICARELLI, M. I. 1983. The organic geochemistry of the Paraíba Valley and Maraú oil-shales. *In:* BJOROY, M. (ed.) *Advances in Organic Geochemistry 1981.* John Wiley, Chichester, 823–833.

COSTA, C. 1983. Geochemistry of Brazilian oil shales. *In:* MIKNIS, F. P. & MCKAY, J. F. (eds) *Chemistry of Oil Shales.* Advanced Chemistry Series Symposium, **230**, 13–15.

CRUSAFONT, M., GAUTIER, F. & GINSBURG, L. 1966. Mise en évidence du Vindobonien inférieur continental dans l'Est de la province de Teruel (Espagne). *Compte Rendu Sommaire de la Société Géologique de France*, **1966**, 30–32.

DE BRUIJN, H. & MOLTZER, J. G. 1974. The rodents from Rubielos de Mora; the first evidence of the existence of different biotopes in the Early Miocene of Eastern Spain. *Koninklijke Nederlandse Akademie van Wetenschappen, Proceedings, Series B*, **77**, 129–145.

DEGENS, E. T., VON HERZEN, R. P. WONG, H. K. & JANNASCH, H. W. 1973. Lake Kivu; structure, chemistry and biology of an East African rift lake. *Geologische Rundschau*, **62**, 245–277.

DESBOROUGH, G. A. 1978. A biogenic-chemical stratified lake model for the origin of oil shale of the Green River Formation: an alternative to the playa-lake model. *Bulletin of the Geological Society of America*, **89**, 961–971.

DIDYK, B. M., SIMONEIT, B. R. T., BRASSELL, S. C. & EGLINTON, G. 1978. Organic geochemical indicators of palaeoenvironmental conditions of sedimentation. *Nature*, **272**, 216–222.

EGLINTON, G. & HAMILTON, R. J. 1967. Leaf epicuticular waxes. *Science*, **156**, 1322.

EUGSTER, H. P. 1985. Oil shales, evaporites and ore deposits. *Geochimica et Cosmochimica Acta*, **49**, 619–635.

—— & HARDIE, L. A. 1975. Sedimentation in an ancient playa-lake complex: the Wilkins Peak member of the Green River Formation of Wyoming. *Bulletin of the Geological Society of America*, **86**, 319–334.

—— & —— 1978. Saline lakes. *In:* LERMAN, A. (ed.) *Lakes: Chemistry, Geology, Physics.* Springer-Verlag, New York, 237–293.

FERNÁNDEZ-NAVARRO, L. 1914. La cuenca petrolifera de Rubielos de Mora. *Revista de la Academia de Ciencias*, **13**, 273–255.

GAGOSIAN, R. B. & FARRINGTON, J. W. 1978. Sterenes in surface sediments from the southwest African shelf and slope. *Geochimica et Cosmochimica Acta*, **42**, 1091–1101.

GAVALA, J. 1921. Nota acerca de los yacimientos de lignitos y pizarras bituminosas de Rubielos de Mora (Teruel). *Boletín del Instituto Geológico y Minero de España*, **42**, 263–302.

HSÜ, K. J. & KELTS, K. 1978. Late Neogene chemical sedimentation in the Black Sea. *In:* MATTER, A. & TUCKER, M. E. (eds), Modern and Ancient Lake Sediments. *Special Publication of the International Association of Sedimentologists*, **2**, 129–145.

I.G.M.E. 1986. *Mapa Geologico de España. E. 1:50,000. (2 serie). Hoja n°591. Mora de Rubielos.* Instituto Geológico y Minero de España, 55 pp.

KELTS, K. & HSÜ, K. J. 1978. Freshwater carbonate sedimentation. *In:* LERMAN, A. (ed.), *Lakes: Chemistry, Geology, Physics.* Springer-Verlag, New York, 295–323.

KOCH, R. & ROTHE, P. 1985. Recent meteoric diagenesis of Miocene Mg-calcite (Hydrobia Beds, Mainz Basin, Germany). *Facies*, **13**, 271–286.

LAST, W. M. & SLEZAK, L. A. in press. The Salt Lakes of Western Canada: A Paleolimnological Overview. *Hydrobiologica*.

LIPPMANN, F. 1973. *Sedimentary Carbonate Minerals.* Springer-Verlag, New York, 228 pp.

MCKIRDY, D. M., ALDRIDGE, A. K. & YPINA, P. J. M. 1983. A geochemical comparison of some crude oils from pre-Ordovician carbonate rocks. *In:* BJOROY, M. (ed.), *Advances in Organic Geochemistry 1981*, John Wiley, Chichester, 99–107.

MÜLLER, G., IRION, G. & FÖRSTNER, U. 1972. Formation and diagenesis of inorganic Ca-Mg carbonates in the lacustrine environment. *Naturwissenschaften*, **59**, 158–164.

NEEV, D. & EMERY, K. O. 1967. Dead Sea: depositional processes and environments of evaporites. *Bulletin of the Geological Survey of Israel*, **41**, 147 pp.

POWELL, T. G. & MCKIRDY, D. M. 1973. Relationships between ratio of pristane to phytane, crude oil composition and geological environment in Australia. *Nature*, **243**, 37–39.

REYES, J. L. & FEIXAS, J. C. 1984. Las pizarras bituminosas: definición, composición y clasificación. *I Congreso Español de Geologia*, **2**, 817–827.

RYDER, R. T., FOUCH, T. D. & ELISON, J. H. 1976. Early Tertiary sedimentation in the western Uinta basin, Utah. *Bulletin of the Geological Society of America*, **87**, 496–512.

SINGH, G., OPDYKE, N. D. & BOWLER, J. M. 1981. Late Cainozoic stratigraphy, palaeoclimatic chronology and vegetational history from Lake George, N.S.W. *Journal of the Geological Society of Australia*, **28**, 435–452.

SPENCER, R. J., BAEDECKER, M. J., EUGSTER, H. P., et al. 1984. Great Salt Lake, and precursors, Utah: the last 30,000 years. *Contributions to Mineralogy and Petrology*, **86**, 321–334.

SURDAM, R. C. & STANLEY, K. O. 1979. Lacustrine sedimentation during the culminating phase of Eocene Lake Gosiute, Wyoming (Green River Formation). *Bulletin of the Geological Society of America*, **90**, 93–110.

TALBOT, M. R. & KELTS, K. 1985. Carbonate mineral genesis in a tropical lake. *6th European Regional Meeting of the International Association of Sedimentologists, Lleida* (Abstr.), 455–456.

TISSOT, B. P. & WELTE, D. H. 1978. *Petroleum Formation and Occurrence.* Springer-Verlag, New York, 538 pp.

VEGAS, R., FONTBOTÉ, J. M. & BANDA, E. 1980. Widespread Neogene rifting superimposed on Alpine regions of the Iberian Peninsula. Proceedings of the European Geophysical Society Symposium; Evolution and Tectonics of the Western Mediterranean and Surrounding Areas. *Publicaciones del Instituto Geografico Nacional, Madrid*, **201**, 109–128.

P. ANADÓN & R. JULIA, Institut 'Jaume Almera' C.S.I.C., C/.Martí i Franquès s.n., 08028 Barcelona, Spain.

L. CABRERA, Departament de Geologia Dinàmica Geofisica i Paleontologia, Facultat de Geologia, Universitat de Barcelona, 08028 Barcelona, Spain.

Sand turbidites and organic-rich diatomaceous muds from Lake Malawi, Central Africa

R. Crossley & B. Owen

SUMMARY: Seven cores taken from 350 m water depth are described, near the foot of a rift valley boundary fault, in the anoxic hypolimnion of Lake Malawi. The sediments average 29% sand turbidites, 18% largely undisturbed 'pelagic' mud and 53% 'pelagic' mud disturbed by slumping and bottom currents. Diatom-rich laminae can be correlated between cores and comparison of the lithostratigraphy with limnological records permits age estimation. Sand turbidite bed thickness/maximum grain size ratios are also examined. Blooms of phytoplankton production appear to have made a major contribution to the organic-carbon content in the undisturbed 'pelagic' muds.

Introduction

Biologically productive lakes with extensive areas of anoxic bottom waters, such as Lakes Malawi and Tanganyika, seem to offer good sites for deposition of sediments with source-rock potential (Desmaison & Moore 1980; Crossley 1984).

A mean organic carbon content of 4.7% (SD = 1.01) is reported by Heckey & Degens (1973) for a 2.4 m core obtained from 550 m depth in Lake Malawi (Fig. 1a; site x). They also describe a 2.9 m core, from site 12 in a water depth of 1190 m in the North Basin of Lake Tanganyika, with a mean organic carbon content of 5.8% (SD = 1.66). The lithologies of both cores reportedly comprise finely laminated units interbedded with turbidites.

The organic carbon percentages suggest that these sediments might offer potential as economic hydrocarbon sources. Interestingly, an African Lakes Corporation steamer captain reportedly referred to 'the presence of jets of petroleum in the vicinity of Lake Tanganyika . . .' (Anon 1896) and a seep of crude oil has recently been found on that lake (Rosendahl 1984).

We obtained a box core and six gravity cores from 350 m water depth off Nkhata Bay in Lake Malawi, at sites approximately 10 km from the core locality of Heckey & Degens (1973) (Fig. 1). The coring sites are close to the Nkhata Bay standard limnological station and are within the anoxic hypolimnion.

The aims are: (1) to describe the sediments in order to assist identification of equivalents in the geologic record; (2) to examine whether diatom sedimentation can be correlated with recorded limnological/biological data, and whether diatom-rich laminae can be used to correlate between cores; and (3) to interpret the deposits in terms of sedimentary processes.

Methods

The gravity corer consisted of 2 m long, 50 mm diameter, PVC pipes and a weight stand made from Landrover brake-drums. The pipe ran through the weight stand to prevent excess penetration and no valve or other core-retention system was used lest it cause elevated water pressure within the pipe that might 'blow away'

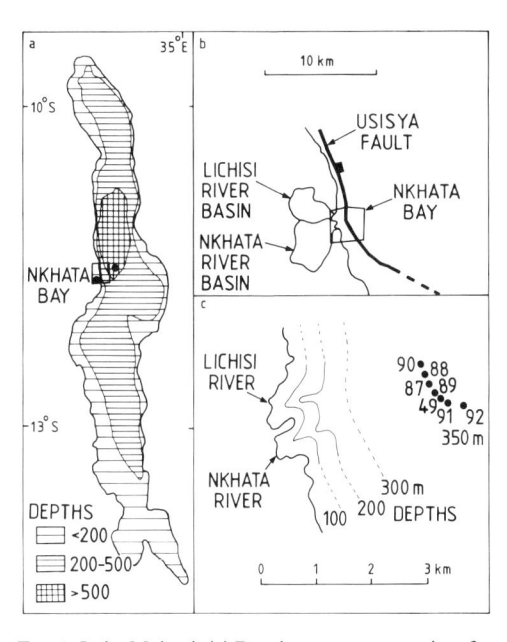

FIG. 1. Lake Malawi. (a) Depth contours, core site of Heckey & Degens (1973) indicated by filled circle at site x. (b) Main drainage basins entering the lake at Nkhata Bay and location of Usisya fault (Crossley, 1984). (c) Depth contours off Nkhata Bay and location of core sites.

the sediment surface. The top few millimetres of the gravity cores were disturbed, but not lost, during lake to ship transfer. The box core was obtained using a 150 mm square Eckman dredge. The top few millimetres of the box core were lost.

The cores were kept vertical during transport to the laboratory. They were allowed to air-dry and the box core was additionally dried in an oven at 60°C. Differential shrinkage caused the cores to split along lithological boundaries and also to crack approximately perpendicular to lamination. The lithology was described by R.C. using a binocular microscope, by scanning the entire surface of successive peeled-off laminae, or by examining several cross-sections of successive layers where 'peeling' proved impractical. All colours are described for dry sediment using Munsell colour charts.

Larger clastic grains, arthropod remains and diatom frustules tended to protrude from the shrunken surfaces, thereby facilitating identification. Size measurements were made using an eyepiece graticule. Identification and qualitative estimates were made of relative abundances of the diatoms from selected subsamples. (The researcher was not provided with the stratigraphic context of the subsamples since this might have influenced his abundance estimates.)

Results

The gravity cores can be subdivided into mud and sand units (Fig. 2). The box core, and the top 70 mm of each gravity core, was examined in detail and three principal types of lithologic units were differentiated: sands, laminated muds and unlaminated muds (Figs 3 and 4). The main features of these sedimentary units are described below.

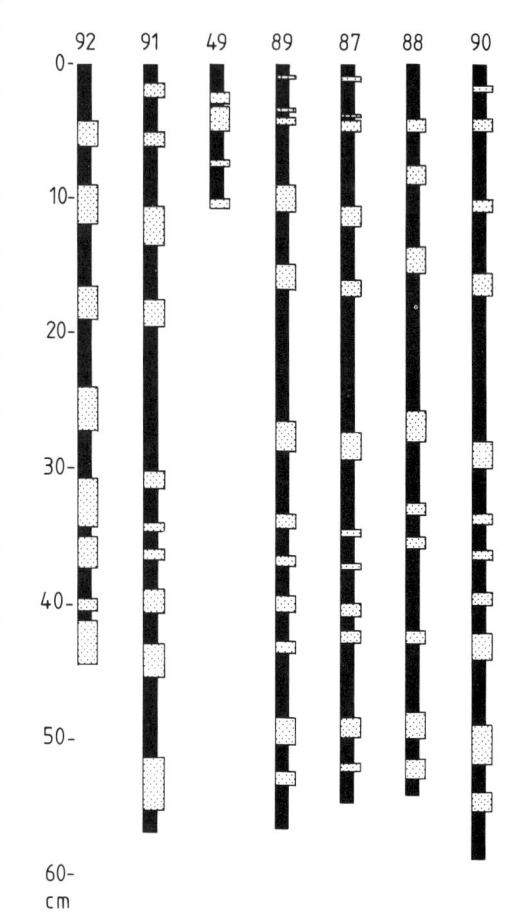

FIG. 2. Core logs showing mud (black) and sand (stippled).

Sand beds

Sand beds fine upwards, in some cases into fine sand and in others into mud. Two of the beds show erosional bases which locally cut at least 2 mm into underlying laminae. The lower part of each sand unit is unlaminated and contained mica flakes are often imbricated at steep angles. The mid-parts of some sand beds show continuous or discontinuous parallel lamination and cross lamination occurs in the upper part of two beds.

The sands are composed primarily of mica (mainly biotite), quartz and feldspar. Quartz and feldspar grains range up to 0.6 mm diameter whilst mica flakes are up to 0.8 mm across. Grains of pyroxene, amphibole and garnet are present. The quartz and feldspar grains are usually angular, though a few sub-rounded grains were noted. Mica flakes are generally angular, but heavily abraded flakes are also present.

Carbonized and non-carbonized plant debris, chitinous arthropod exoskeletons and fish bones are present though usually in fragmented states. Complete remains of arthropods are preserved, however, in the muddy tops of two units. Partially decomposed plant fragments are abundant in some sand units. Planktonic diatom frustules occur throughout. Frustules of the littoral diatoms *Cymbella turgida*, *Epithemia zebra*, *Rhopalodia gracilis*, *Amphora ovalis*, *Hantzchia amphioxis* and *Synedra ulna* are rare. Littoral diatoms were not found in the other types of deposits.

Laminated mud units

Laminae thicknesses reach 1 mm. Three types of laminae are distinguished: (1) white, pure diatom deposits to pale-grey diatom-rich detrital muds;

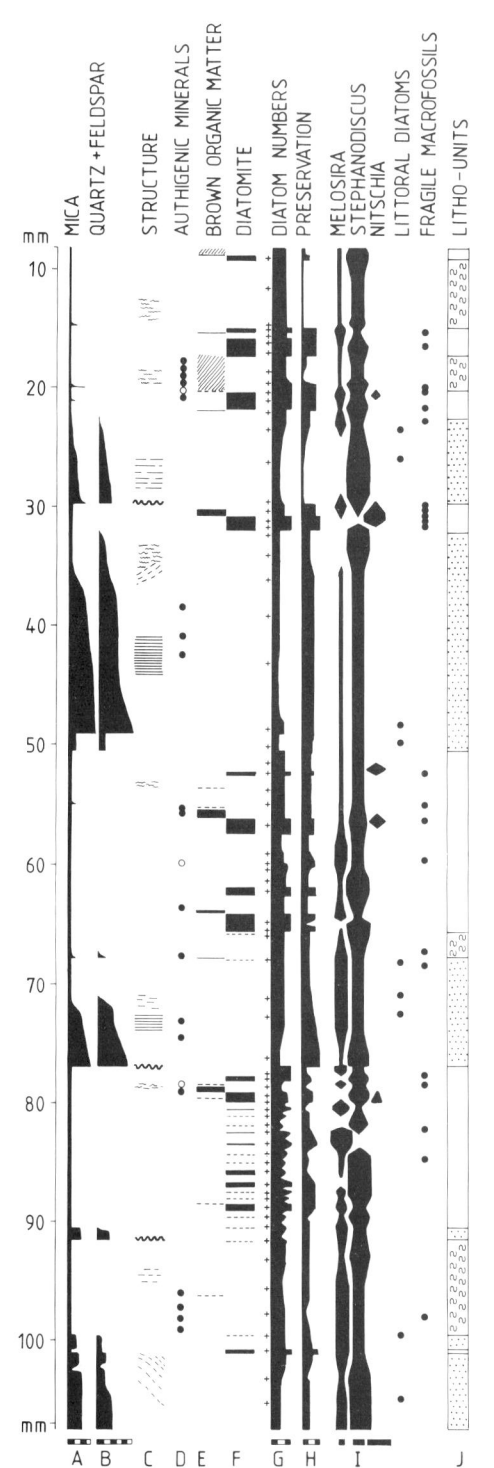

(2) pale olive to grey diatom-rich detrital mud; and (3) rare, dark brown organic matter.

The pure diatom laminae generally contain rare flakes of mica up to 0.15 mm diameter. The dark brown laminae suffered shrinkage in excess of 70% upon drying, which produced conchoidal fracturing. When placed in 20% hydrogen peroxide, the dark brown components of the organic matter laminae oxidize rapidly, leaving an orange, birefringent, flakey residue which oxidizes slowly. Mica flakes up to 0.15 mm diameter and diatom frustules occur only rarely in these laminae.

The diatomaceous muds are the richest in non-diatom fossils; fragile carbonized plant remains are common, insect and copepod exoskeletons retaining fine hair-like appendages are present, pupal cases of *Chaoborus* or *Chironomus*, insect wings, fish bones and pollen grains occur rarely. Faecal pellets composed of remains of diatoms, crustacea and fish are also present.

Other mud units

Two types can be distinguished. Type A is pinkish to yellowish brown stained, and shows parallel orientation of micas and other platey or elongate clasts. Diatom preservation is poor. Mica grain sizes are coarsest (0.5 mm) at the base of some beds. Fragile organic remains are absent.

Type B is pale grey and either homogeneous or shows indistinct wavy lamination defined by irregular patches of diatom enrichment. Deformed diatomite laminae occur near the bases of some beds. Diatom preservation varies from good to poor, and fragile organic remains are rare.

FIG. 3. Detailed log of box core 49. Key: A—size of mica clasts, scale in 0.2 mm units; B—size of quartz and feldspar clasts, scale in 0.1 mm units; C—sedimentary structures, thick wavy lines = erosive base, parallel lines = continuous and discontinuous parallel lamination, wavy lines = discontinuous wavy lamination; D—filled circles = vivianite, open circles = carbonate; E—black = dark brown organic matter laminae; F—black = diatomite and diatom-rich laminae; G—diatom subsamples and abundance, with units representing sparse, common and abundant; H—diatom preservation, with units representing all broken, some whole, many whole, most whole; I—abundance of main diatom genera, with units representing sparse, common, abundant; J—interpretation, blank = undisturbed 'pelagic' sediment, stipple = sand turbidites, squiggles = disturbed 'pelagic' sediment.

Authigenic minerals

Blue-green vivianite generally occurs as irregular patches, sometimes coating fragments of fish bone. It is particularly common on or in the dark brown organic matter laminae and in Type A mud units. A 2.5 mm diameter 'rosette' of vivianite occurs in one Type A unit and a 0.5 mm diameter 'rosette' occurs in a sand unit. Carbonate, occurring as 0.5 mm diameter clusters of white polyhedra about 0.03 mm across, rests directly on diatomite laminae at 20 and 79 mm depths in the box core. Carbonate also occurs at 60 mm depth in the box core, as a single lamina about 0.2 mm thick.

Correlation with limnological records

The limited data available suggest that the diatoms of Lake Malawi bloom during April to September (Talling 1969; Heckey & Kling in press). It is therefore possible that the undisturbed diatomite laminae in the cores represent these blooms. Scant records suggest that the diatoms *Stephanodiscus, Melosira* and *Surirella* are common members of the phytoplankton community of Nkhata Bay (Illes 1960; Eccles 1964). Degnbol & Mapila (1982) also record *Melosira* off Nkhata Bay in 1978 and 1979, but Heckey & Kling (in press) report that in 1980 *Nitschia* and *Stephanodiscus* were prominent, and *Melosira* was rare.

Our coring was undertaken at the end of December, 1985, so undisturbed diatomite lamina number 5 (Fig. 4) might be expected to represent the 1980 diatom population maxima. The prominence of *Nitschia* in this lamina is compatible with the 1980 records, though *Stephanodiscus* seems to be under represented in the deposit. It may be that the lamina is largely derived from a major *Nitschia* bloom that occurred between the monthly sampling intervals used by Heckey & Kling (in press).

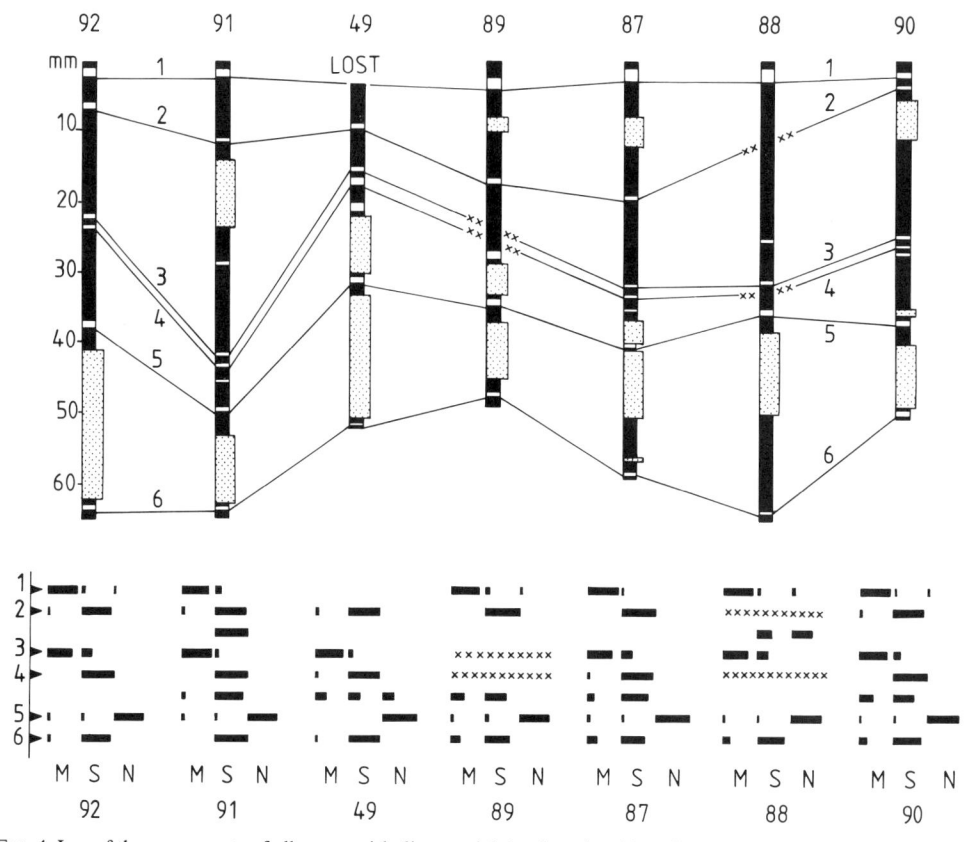

FIG. 4. Log of the upper parts of all cores, with diatom-rich laminae in white, diatomaceous mud in black, and sand stippled. The relative percentages of *Melosira* (M), *Stephanodiscus* (S) and *Nitschia* (N), in each diatom-rich lamina are also shown. On the basis of these percentages, and positions in the cores, six diatom laminae can be correlated between cores. Absence of any of these six laminae in a core is indicated by crosses.

Correlation between cores

Correlations were made between the upper parts of the cores using diatomite laminae (Fig. 4). Six laminae were used in the correlation, but they were not all present in all cores. The absence of laminae in cores 88 and 89 is attributed to erosion or slumping since the muds at the positions where the laminae might be expected are of Type A or Type B character (see interpretations later).

Additional diatomite laminae are also present in some cores. These laminae are thought to be the result of resuspension/redeposition by bottom currents because: the diatoms they contain are often poorly preserved; the proportions of genera in the laminae may be evenly mixed (rather than dominated by a single genus as in the other six laminae); the proportions of genera in the laminae tend to vary substantially from core to core.

Interpretation of the lithologic units

Sand beds

These are interpreted as turbidites in the light of their fining-upwards textures and sedimentary structures. The presence in some units of heavily abraded mica flakes, partially decomposed plant debris and littoral diatoms, suggest that these turbidity currents originated in shallow water.

Laminated muds

The laminated muds represent largely undisturbed sediments deposited from suspension (though evidence for some redeposition of diatoms within the laminated mud units was noted earlier). Correlation with the limnological records suggests that the diatomite laminae represent annual blooms. Rapid sedimentation from a single major diatom bloom, rather than the integration of several months of diatom 'rain', is suggested by the monogeneric dominance of the laminae, and by the paucity of remains of arthropods, fish, carbonized plant debris and mica flakes. The presence of carbonate polyhedra on two diatomite laminae may indicate that the photosynthesizing diatoms occasionally caused supersaturation of carbonate in the epilimnion.

Dark brown organic laminae contain few diatom frustules and mica flakes, suggesting rapid deposition. Determining the nature of the organic matter was not achieved, although chlorophyta, cyanophyta or bacteria all seem possible. The four thickest dark laminae all succeed important diatomite laminae perhaps suggesting unusual bursts of biological activity.

The diatomaceous muds contain most of the arthropod remains and faecal pellets. Limnological records indicate that total pelagic zooplankton numbers did not, in 1980, show any major seasonal variations (Degnbol & Mapila 1982). If this is typical, their association with the diatomaceous muds would suggest that these units represent much of the annual pelagic 'rain'. The rivers entering the lake in the Nkhata Bay area carry significant quantities of sediment only during and immediately after major rainstorms. Their sediment inputs are therefore limited to the November to April wet season. It seems probable that much of the terrigenous mud fraction is deposited during this period. However, strong easterly winds in August and September drive waves that remobilize substantial amounts of sediment in bays protected from the prevailing southerly and northerly winds and this might provide a second mud source.

Other mud units

Type A mud units are interpreted as the results of reworking by bottom currents. Muds of this type occur above and below diatomite number 5 in all seven cores, which suggests that bottom currents may at times be widespread and may follow the bathymetric contours. There is some evidence that a deep south-bound return current may develop off Nkhata Bay during periods of strong southerly winds: a Fisheries Department long-line, anchored in 150 m of water off Nkhata Bay, broke free and drifted several kilometres southwards on one such occasion (J. Mwenifumbo, pers. comm.) despite strong southerly winds.

Type B mud units might be the result of slumping of pelagic deposits, possibly from the submerged section of the Nkhata Bay fault scarp.

Discussion

In terms of position relative to sediment source, and in terms of basin floor gradient, the core sites can be considered proximal. The bed thickness/grain size relationships found (Fig. 5) straddle the field of thin proximal and distal turbidites reported by Sadler (1982). In contrast however, Sadler's proximal examples showed a graded Bouma A-division with a rippled top and no internal lamination, whereas the sands off Nkhata Bay often additionally show internal parallel lamination.

We had expected that diatom frustules would suffer substantial damage during turbidity current flow, but the diatoms in some sand units show moderate to good preservation. This suggests that grain–grain collisions within the turbidity currents are insufficiently powerful or numerous to break a significant proportion of diatom frustules.

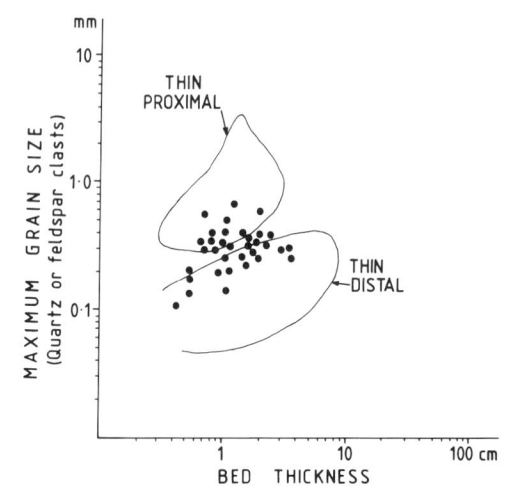

FIG. 5. Maximum grain size (quartz and feldspar clasts) and bed thickness relations for turbidites off Nkhata Bay. Fields of 'proximal' and 'distal' thin turbidites reported by Sadler (1982) are also shown.

The average annual sedimentation rate over the last 5 years at these sites was 8 mm (dry thickness). The composition of the sediment at the sites averages 29% sand turbidites, 18% largely undisturbed 'pelagic' mud and 53% disturbed 'pelagic' mud (on dry thickness basis). Recognition of disturbed mud units is difficult. Failure to appreciate the large quantity of reworked microfossils in these sediments might cause erroneous palaeoenvironmental/palaeoclimatic interpretations.

The average contents of the different types of deposit forming in the largely undisturbed 'pelagic' mud are: 30% diatomite, 5% dark brown organic laminae and 65% diatomaceous mud. Many forms of organic carbon are deposited in these sediments: diatoms; dark brown organic matter; arthropod exoskeletons, wings and pupal cases; carbonized plant remains; and faecal pellets. We have presented evidence that the diatomites and organic matter laminae result from events of short duration, such as major plankton blooms. Since the resulting deposits form 35% of the total 'pelagic' sediment thickness, these events are very important—particularly in depositing organic carbon.

Consequently, part of the explanation for the high organic carbon content of these sediments may lie in 'bursts' of phytoplankton production that overwhelm the feeding capacity of the zooplankton. As a result, an unusually large quantity of carbon settles into the hypolimnion. The relative inefficiency of freshwater fermentation processes (Desmaison & Moore 1980) then permits a large proportion of this carbon to be preserved in the sediment.

ACKNOWLEDGEMENTS: We are grateful to the University of Malawi, Amoco Production Co., and Bob Raynolds, for funding and equipment. The Malawi Fisheries Department kindly allowed us to use their boat and Tony Seymour provided crucial assistance at Nkhata Bay. Comments by G. Lister materially improved the paper and Robertson Research Ltd. helped with manuscript preparation.

References

ANON 1896. The proposed railway route Chiromo to Blantyre. *The Central African Planter*, 132 pp.

CROSSLEY, R. 1984. Controls of sedimentation in the Malawi Rift Valley, Central Africa. *Sedimentary Geology*, **40**, 33–50.

DEGNBOL, P. & MAPILA, S. 1982. Limnological observations on the pelagic zone of Lake Malawi from 1970 to 1981. In: *Biological Studies on the Pelagic Ecosystem of Lake Malawi*, FAO, Rome, ref: FI:DP/MLW/75/019, 5–47.

DESMAISON, G. T. & MOORE, G. T. 1980. Anoxic environments and oil source genesis. *Bulletin of the American Association of Petroleum Geologists*, **64**, 1179–1209.

ECCLES, D. H. 1964. Research results, Lake Nyassa. In: *Annual Report of the Joint Fisheries Research Organisation, Northern Rhodesia*, **10**, 51–58.

HECKEY, R. E. & DEGENS, E. T. 1973. Late Pleistocene–Holocene chemical stratigraphy and palaeolimnology of the Rift Valley Lakes of Central Africa. *Technical Report of Wood's Hole Oceanographic Institute*, **73-28**.

—— & KLING, H. J. (in press). Phytoplankton of the great lakes in the western rift valley of Africa. *Archiv fur Hydrobiologie, Bichäfte*.

ILLES, T. D. 1960. Activities of the Organisation in Nyasaland. In: *Annual Report of the Joint Fisheries Research Organisation, Northern Rhodesia*, **9**, 7–41.

ROSENDAHL, B. R. 1984. Exploring the rift lakes of Africa. *Topic*, **148**, 9–13.

SADLER, P. M. 1982. Bed-thickness and grain size of turbidites. *Sedimentology*, **29**, 37–51.

TALLING, J. F. 1969. The incidence of vertical mixing, and some biological and chemical consequences, in tropical African lakes. *Verhandlungen der Internationale Vereinigung fur theoretische und angewandte Limnologie*, **17**, 998–1012.

R. CROSSLEY* & B. OWEN, Department of Geography & Earth Sciences, University of Malawi, P.O. Box 280, Zomba, Malawi, Central Africa.
*Present address: Robertson Research International Limited, Llandudno, Gwynedd LL30 1SA, UK.

Index

Pages on which figures appear are printed in *italic*, and those with tables in **bold**

Achanarras fish bed, Orcadian Basin   176
aeolian activity, ephemeral lakes   48
aeolian dust, wind change records   32
aeolian processes, and dry lakes   48
African lakes
  modern phytoplankton production in   27
  recent   viii
  tropical, stratification state   33
Albert, Lake
  clay mineral composition   141
  mineral assemblages   *144*
Aleksinac Formation   89
algae   13, 48, 49, 107, 108, 283, 296, 301, 312, 330, 349
  benthic   233
  blue-green   13, 37, 88, 195, 240, 250
  calcified   226
  chlorophytic   226
  cyanophitic   226
  detrital   226
  dinoflagellate   301
  and dolomicrite genesis   208
  epipelic   174
  green   16, 21, 37, 117, 299, 301, 327, 335, 338
  periphytonic   *231*
  planktonic   214, 250
  salt-tolerant   16
  as sterol source   113–14
  suitable preservation conditions   233
  unicellular   111, 250
algal blooms   49, 62, 69, 153, 174, 206, 302
  and carbonate precipitation   364
  mass mortality   221
algal bodies, yellow   239, 240
  relevance to shale oil potential   240
algal genera, Burdiehouse Limestone Formation   *232*
algal influence, Burdiehouse Limestone Formation   226, 229
algal mat lake model (shallow), microlaminated lake sediments   174
algal mats   30, 51, 174, 250
  low preservation potential   19
  modern   175
  see also bacterial mats; cyanobacterial mats; microbial mats
algal matter, laminar   244
algal oozes, accumulation of   233
alginite   87, 221, 236, 240, 243
  biodegradation of   338
  origin of   334–5
alluvial fans   265, 272, 274, 355
Amadeus, Lake (Australia)   53
anaerobic bottom conditions   242
anaerobic decay and magadiite precipitation   214–15
anaerobic oxidation   66
analcime   280, 345, 347
  syngenetic   179
anhydrite   135, 253
ankerite   82, 280

anoxia   14, 253
  deep meromictic oligo- to hypersaline lakes   20
  seasonal   15
  short term   19
anoxic bottom conditions   347, 349, 364
anoxic bottom waters   15, 252, 253, 272, 274, 369
anoxic conditions   316
  Thai Basin lakes   347
anoxic depositional conditions   304
anoxic event, late Miocene   123
anoxic lake (deep) model, lacustrine source rock accumulation   29–30
anoxic processes   19
anoxic/oxic bottom conditions, interpretations   364
aquatic productivity   13–14
aragonite   356, 358, 360, 361, 362, 363, 364
aragonite precipitation   365
Aranguadi, Lake (Ethiopia Rift)   15
archaebacteria   110–11, 320
  methanogenic   187
  tolerant of high salinity   304
arctic lakes   14
Artemia   18
arthropods   370, 373, 374
  shelled 343
asphaltanes   71
asphaltenes   71, 184, 296, 313
asphaltite, non-waxy   327
attrinite   332
Australian lakes
  evolution of   53–7
  large, petroleum source rock or metal ore deposition   45–57
  sites for hydrocarbon and/or metal deposits   51, 53
authigenic minerals, Lake Malawi diatomaceous muds   372

Bacillus acidocaldarius   112
bacteria   174
  autotrophic   62
  biological markers for   109–12, 117
  chemolithic   16
  fermentative   67
  green sulphur   112
  haloalkaliphilic   111
  halophilic   111, 124, 319
  heterotrophic   16
  and lakes   16–17
  methanogenic   110, 111, 117, 124, 324
    nutritional requirements   65
  phototrophic   16, 17
    purple   22
  possible source of pentacyclic triterpenoids   111
  purple
    photosynthetic   62–3, 67–8
    sulphur   61, 62–3
  source of lipids   109
  as sterol source   113–14
  sulphate reducing   312

375

Index

bacterial degradation 286–7
bacterial mats 117
  chemotrophic 22
  see also algal mats; cyanobacterial mats; microbial mats
bacterial plates 17, 61, 62–3
bacterial productivity 61
bacterial reworking 185, 194
balkhashite 240
ball-and-pillow structures 258
barite 135
basin centre brines, Orcadian Basin 181
basin subsidence, differential 291
basin type, controlling formation and distribution of source rocks and hydrocarbon products 279
basins
  closed 5, 48–9
    brine composition fluctuations 6
  continental, Mesozoic–Cenozoic, northwest China 291
  hypersaline, Australia 126
  intermontane 291
  lacustrine, SE Asia 341–49
  open 5
    nutrient potential 14
    primary productivity 14
  restricted, lateral changes in clay mineral composition 140
  rift, Culpeper, Newark Supergroup 247–74
  sedimentary 279
    clay-mineral distribution 140–3
    SE Asia 341–49
Batchawana, Lake (Canada), pyrite concentration 133–4
Beatrice oil 195–7
  Devonian contribution 196–7
  Jurassic source problems 196
Beggiatoa 17, 132
Beggiatoaceae 22
Belfry Lake beds, clay mineral variability 145–7, *148*
Belfry Member, Fort Union Formation
  facies 145–6
  stratigraphy 146
benthonic mats 345, 348
Benxi Formation 310, 312
Big Soda Lake
  microbiological and biogeochemical processes in 59–74
    organic geochemistry of the pelagic sediments 68–73
    a modern analogue of lacustrine oil source rocks 73–4
    seasonal variation in limnological properties *61*
Bighorn Basin (Montana and Wyoming) 145–7
biogenic calcite 360–1
biogenic degradation, of alginite 338
biogenic reworking 226
bioherm preservation 51
bioherms 7
  potential hydrocarbon sources 51
bioinduced precipitation 255
biological marker compounds 187–91
biological markers viii, 103
  for bacteria 109–12, 117

Chinese lacustrine oil shales 299–307
diagnostic of freshwater lake sediments 302–3
higher plants 117
hypersaline environments 115–16
for lagoonal-type saline environment 318, 319
measuring maturity of oils 287
methanogenic bacteria 117
for methanogenic activity in marine sediments 320
northwest China oils 296, 298
for phytoplankton 106–9
upwelling environments 116–17
used in recognition of palaeohypersaline environments 123–8
for vascular plants 104–6, 117
see also biomarkers
biomarker composition characteristics, Chinese basin source rocks 283–6
biomarker features, carbonate-specific 195
biomarker parameters, relating oils and possible source rocks 103
biomarkers 96
  polycyclic 96
  showing environmental changes 298
  western Otway Basin 327
biomass 14, 15, 16
  microbial 62–3
  photosynthetic bacteria 61
  phytoplankton 27, 62
  plankton 15
bioturbation 18, 226, 249, 254, 258, 259, 274, 345, 357, 362
  root 363
bioturbators, lacustrine 18
bisnorhopane 111, 307
bituminous material, Thai basins, distribution of 345, 347–8
bitumen 330, 333, 339, 343, 345
  and Botrycoccane (Australia) 107–8
  coastal 327
  formed during thermal alteration of lamalginite 338
  in lamosites 338
  migrated 209, 244
  plastic 237
  possible modes of formation 338
bitumen characterization, East African lakes and Lake Uinta 85, 88
bitumen content, Chinese Basins 312
bitumen group composition, Chinese Basins 313, *314*
bitumen groups 323–4
bitumen impregnation 197
bitumen ratio 184
bituminite 240
bivalves 312, 343
Biyang Depression, China 283
  oils 283–4
Black Sea 20
Blelham Tarn (Lake District), pyrite concentrations 135
bog lakes, dystrophic 15
Bogoria, Lake 31, 36, 37, 39, 92
  recent analogue for occurrence of magadiite and trona crystals 214
Bohai Bay Basin (China) 309

Index

Bonneville, Lake 175
borates 213
botryococcane 107–9, 302–3, 327
botryococcane homologues 284, 301, 302–3, 307
Botryococcus 13, 21, 30, 37, 49, 108, 299, 301, 335, 336, 338, 343, 345, 360
Botryococcus braunii 240, 327
Botryococcus-rich deposits 51
Bosumtwi, Lake 36, 37, 38
bottom fauna 18
Bouma sequences 254, 373
breccias, mud-flake 211, 214
brecciation 46
brines
    lake
        evolution of 6–7
        rapid fluctuations of concentration and depth 10
        parent to trona 213
    seepage reflux 207
    sodium carbonate 213
Broxburn Shale 221
Buckland Formation 247, 249
    lacustrine sequences 260, 262, 265, 268
Bull Run Formation 247, 249
    black shales 253
    freshwater and saline lakes 270
    lacustrine sequences 260, 261, 263, 265, 267, 268
        asymmetrical (Van Houten cycles) 271–2
        'very thin' *262–3*
Burdiehouse Limestone Formation 222–9
    algal influence 226, 229
    bitumen distillate 244
    diagenesis 226
    fauna and flora 225–6
    key to deposition of Dinantian Oil Shales 219–33
    lithological distribution 226
    lithology 224–5, *227*
    and Oil Shale Group, similar accumulation environments 230, 233
burial diagenesis 143–5
burrowing 362
burrows 258, 259

Cadagno, Lake, bacterial 20
Cadell, Lake 219, *220*, 233
    during Burdiehouse Limestone Formation times *228*
    palaeolimnology 229–31
calcimorphs 224–5
calcite 82, 154, 155–7, 173, 177, 253, 280, 345, 360
    dissolution of 157
    drusy *231*
    fibrous 206, 346
    high magnesium (HMC) 358, 361, 362, 363, 364
        precipitation of 180
    in lacustrine deposits 255
    low magnesium (LMC) 357, 358, 360–1, 362, 363, 364
calcite grains 224–5
calcite nodules 249
    pedogenic 259, 274
calcite spear authogenesis 226
calcite spherules 238
calcium carbonates, early precipitation of 6

calcium dissolution 154
Calderiella 320
caliche nodules 269, 274
Camps Shale 222, 233
cannel coal 331, 332, 339
    Lilypool subunit 338
carbon content
    Chinese basin source rocks 281
    organic
        in carbon-rich beds 208–9
        of lake basins 11–12
carbon cycle
    and alkaline lakes 15–16
    anaerobic 20
    bacterial effects 16–17
    lacustrine 3, 11–22
carbon isotopes in lake organic matter 22
carbon–sulphur (C/S) ratios *180*
    distinguishing freshwater from marine sedimentary rocks 135
    uses of 178–9
carbonate 372
    anoxic 20
    authigenic 16
    ferroan 178
    from calcium dissolution 154
    micritic 173
    stromatolitic 7
carbonate chemistry, palaeosalinity determinations, Orcadian Basin 177
carbonate facies, lacustrine 13
carbonate laminites 205–7, 215
    hydrocarbon traces 208
carbonate minerals, early diagenetic 236
carbonate polyhedra 373
carbonate precipitates 48
carbonate precipitation 180, 364
carbonate productivity 347
carbonate-clay couplets 356
carbonate-rich beds, Middle Old Red Sandstone, Orcadian Basin 205–8
carbonates 314
    cyclical sequences, Rubielos de Mora Basin 360–2
    origin of 361–2
    in rhythmites 358
Carboniferous source beds 10–11
carnotite mineralization 51
carotanes 184, 185, 194, 284, 286, 288, 296, 304–5, 306, 307
cement stratigraphy, limestone members, Burdiehouse Limestone Formation **230**
Cementstone Group (Dinantian), sabkha sediments 243
Cerro Porpol section, Rubielos de Mora Basin 356–60
Chad, Lake 13, 36
Chaidamu Basin (China) 291
    coal-forming conditions 310
    depressions 294
    inland saline lake basin, Miocene–Pliocene sequence 309
    organic matter 312
    source-rock environment of deposition 292
    spore/pollen indicators 165–6

Chaidamu Platform 291
chalcedony 210, 236
Chao Phraya Basin (Thailand) 343
Chaoshui Basin (China) 291
chara 296
charophyte chalks 7
charophyte oogonia 358
charophytes 312, 343
Chelotriton paradoxus 358
chemical sediments 250
 in closed basins 251
chemical weathering
 and nutrients 38
 and Orcadian Basin solute inflow 180
chemoautotrophy 61
 Big Soda Lake 63
chemocline 17
 Big Soda Lake 60, 63
 interflow along 254
chemosynthesis 17
chert 209–15, 344
 in carbonate laminites 206
 magadi-type 213, 215
 model 178, 209–11
 nodular, as marker horizon 209
 Orcadian, precipitated by ground waters 213
 pseudomorphs within 211–13
 stratigraphic and geographical distribution, Orcadian Basin 209
chert laminae, intermittent nature of 214
chert lenses 226
chert nodule development, early/primary diagenetic 177, 178
chert nodules 178, *212*
 in carbonate-poor beds 209
 magadi-type 214
China 139
 biological markers in lacustrine oil shales 301–7
 continental petroliferous basins *160*
 lacustrine source facies for hydrocarbon production 81
 Northwest, Mesozoic and Cenozoic source rocks
  coexistence of oil-prone source rocks and oil-prone coal 294–5
  distribution of terrestrial source rocks 291–2
  geochemical characteristics of continental source rocks 295–8
  sedimentary environment of source rocks 292–4
 organic geochemical characteristics, terrestrial petroleum source rocks (major types) 279–88
Chinocythere 309
chlorin 283
chlorite 143, 146, 280
chlorite-smectite 143
Chlorobacteriaeceae 17
chlorophytes 13, 373
cholestane 288
Chromatiaceae 22
clarite layers 332
Clarks Fork Basin 145–7
clastic dilution, mesotrophic temperate lakes 18
clastic laminae 358
clastic sediment yields reduced, tropical Africa 38

clay-mineral distributions, sedimentary basins 140–3
 lacustrine environments 141–3
 marine environments 140–1
claystone 331, 332, 338
climate
 and cyclic sedimentation, Culpeper Basin 269–72
 humid viii
  tropical Africa 38, 39
 influence on organic matter preservation 41
 local, Culpeper Basin 269–70
 and *Pediastrum* 336
climate control, terrestrial source rocks, northwest China 291–2
climatic changes 37–8
 and fluctuating lake levels 347
climatic cycles, and petroleum source rocks 309
climatic fluctuations, effect on clay mineral suites 143
climatic patterns, Culpeper Basin 272
climatic variation, and stacked cyclic sequences 173
CO_2
 bacterial generation of 215
 from dissolution of calcite 157
 lake water 157
 sources, Lake of Laach 154–5
 volcanic 154, 155
CO_2 springs, Lake of Laach 153, 154
coal 242, 331, 332
 Thai basins 343
coal deposits 269
coal seams 221
coal-bearing formations as source rocks 287–88
collophanite 206
compaction 143
 difficulties in determination 268
conchostracans 251, 253, 254, 255, 259, 260, 274
Condor oil shale sequence, Australia 329–39
 organic petrography of 331–3
 petroleum generation potential 336–8
Condor unit, Condor oil shale sequence 332
conglomerate beds 265
conglomerates, Culpeper Basin, clast- or matrix-supported 255
connate waters and trona precipitation 214
coorongite 49, 240, 327
coprolites 236, 251, 345
Corangamite, Lake (Australia) 53
corpohumite 332, 334
Cretaceous source beds 11
cross-laminations 258, 370
 climbing-ripple 258
 ripple 254
crude oil seeps 369
crude oils
 from carbonate-evaporite sources 116
 hopanes in 115–16
 measurement of biodegradation of 117
 non-marine, northwest China 296
 paralic environments 316, 318
 progressive change in composition, northwest China 296
 sterane distribution 114
cryptalgal clasts 226
Culpeper Basin, lacustrine units 251–9

Culpeper Group   249
cutinite   331, 332, 338
cyanobacteria   51, 107, 115–16, 194
cyanobacterial mats   10, 13, 48
   in ephemeral lakes   49, 51
   trapping metal ions   51
   see also algal mals; bacterial mats; microbial mats
cyanophytes   13, 373
cycles, fluviolacustrine and lacustrine, Orcadian Basin   215
cyclic sedimentation   221–2
   and climate, Culpeper Basin   269–72
   and distribution of organic carbon   205–9
   model of   362–4
cyclic sequences, stacked   173
cyclical sequences, Rubielos de Mora Basin   356–65
cyclization, intramolecular, of fatty acids   320

Dalmahoy Shale   243
dawsonite   82
de-watering, of clay   145
Dead Sea   16, 20, 111
delta progradation   235
denitrification   67
   active   68
depositional environments
   biological markers used to differentiate   299
   carbon cycle in lakes   11–22
   hypersaline
      Jianghan Basin biological markers characteristic   305–6
      Rengui oilfield   306–7
   lacustrine   106, 141–3, *149*
      Culpeper Basin   259–72
   marine   104, 140–1, *149*
   oil shales, Scotland   221
   source rocks
      Dinantian Oil Shale Group   243
      models   7–11
      of non-marine crude oils, proposed identification method   *297*, 298
      Thai basins   344–5, 346–7
   use of sterols or steranes to distinguish between   113
depressions, multiple, and oil-forming centres, northwest China   294
desiccation cracks   205, 207, 210, 211, 214, 235, 236, 243
   polygonal   345
Desulphovibrio desulphuricans   16
Devonian lacustrine source beds   10
diagenesis
   and biomarkers   104
   Burdiehouse Limestone Formation   226
   burial 143–5
   early   177, 243
      related to oil and gas formation   59
      and trona precipitation   214
   formation of pyrite   132, 135
   geochemical, of phytol   316
   in high salinity-alkalinity conditions   148
   incorporation of bacterially formed sulphur   116
   progressive   141
diagenetic concretions, early   268
diagenetic effects, phosphate-rich interstitial waters   253
diagenetic modification, carbonate laminites   206
diagenetic precipitation, calcite   255
diagenetic pressure-solution effects   173
diagenetic processes
   and clay mineral stratigraphy   142–3
   early   362
   inland saline lake   312
diastemic surface   260
diasteranes   191, *192*, 283
diasterenes   302
diatom bloom, Lake Malawi, correlation with cores   372–3
diatom laminae   371
diatom rain   373
diatomaceous muds   371, 374
diatomite laminae   372–3
diatoms   13, 37, 62, 72, 116, 155, 157, 336, 374
   littoral   370
   pennate   68–9
dicarboxylic acids   104
dimethyltricosane   319
dimethysulphide   64
dinoflagellates   13, 113, 115, 117, 301, 330
dissolved inorganic carbon (DIC)   13, 16
dissolved organic carbon (DOC)   13, 16
diterpanes   72, 286
   tri- and tetracyclic   106
diterpenes   286
diterpenoid acids   104
docosane, marker for a hypersaline environment   124
dolomicrite beds   206, 207–8, 215
dolomicrite domes   207, 208
dolomite   82, 173, 179, 280, 345, 358, 359, 360, 362, 363
   diagenetic formation   347
   diagenetic origin for   177
   fibrous displacive   237
   non-stochiometric   357, 361
   in rhythmic laminites   10
dolomite formation   180
dolomite laminites   238, 243
dolomite precipitation   208
   primary   177
dolomite-calcite pairs   177
dolomitic laminates   244
Dongpu Basin (China), salt lake evaporite-clastic formations, fault-bounded basin   279
Donying Depression (China)   283
dry lakes   45
   Australian, chemical characteristics (water and sediments)   48–9
Dunaliella   16
Durness Group   180
dysaerobic conditions   253

East African Rift Lakes   92, 139, 175
   changes in clay mineral suites   141

East Berlin Formation 259
    symmetrical lacustrine cycles 270–1
    trona 214
ectogenic meromixis 250, 261, 365
Ectothiorhodospira vacuolata 61
Edward, Lake 36, 37, 82
Eerduosi Basin (China)
    oils 283
    spore/pollen indicators 161–2
efflorescent crusts 214
efflorescent salts 48
element composition and organic compounds from different environments, northwest China 296–8
elemental analyses, East African lakes and Lake Uinta 85–6
elemental ratios as palaeosalinity indicators 131
Elko Formation 88
Emiliania huxleyi 109
Ennerdale Water, pyrite concentrations 134–5
Enteromorpha prolifera 110, 126
environmental analysis, using algae, Burdiehouse limestone 226, 229
environmental change, Tropical Africa, influence on lacustrine systems 31–2
environmental indicators, northwest China 296
ephemeral lake model, lacustrine source rock accumulation 30
ephemeral lakes 45
    Australian, chemical characteristics (water and sediments) 48
    sapropelic deposits 240
epilimnion 13, 17, 60, 215, 250
    anoxic 29
    freshwater 261
    oxic 214
    recycling in 17–18
    supersaturation of carbonate in 373
epiphytes, cyanobacterial 67
erosion, and ephemeral lakes 48
estherites 293, 296
euphotic zone 193
eutrophic lakes, Windermere and Blelham Tarn 135
evaporative concentration 362, 364
evaporative pumping 48
evaporite crystal moulds 259, 261, 265, 269
evaporite crystals 253, 274
evaporite dissolution 180–1
evaporite facies, Orcadian Basin 181
evaporites 8, 22, 251, 314, 364
evaporitic facies 115
event lake basins 5, 7
Eyre, Lake (Australia) 53

faecal pellets 17, 18, 371, 373, 374
Fang Basin (Thailand) 348
Fells Shale 243
fermentative reactions 64, 67
fernenes 111
FeS, amorphous 131–2
fichtelite 286
fish fossils 211, 274, 296, 343, 345
    exquisite preservation of 253

'Fish laminites', Orcadian Basin 173
Flagstone Group 196
    absence of evaporite minerals 180
flamingoes 13
flooding, and sedimentation of organic matter 17
fluorescence 331, 339
fluvial deposits, ephemeral 251
foraminifera, fusulinid 312
Fort Union Formation 145–7
fossils
    fish and invertebrate 261
    scarce 261
fractionation
    of bitumens due to migration 327
    of recent sediments (analytical procedure) 91–2
freshwater systems, little known about pyrite formation 132–5
fringing swamps 12
Frome, Lake (Australia) 51, 53
Fucoid Beds 180
fulvic acids, pyrolysis of 93, 95, 97

gallium 294
gammacerane 115, 126, 283, 306, 307
    in hypersaline sediments 304
    resistant to biodegradation 115
gammacerane index 115
gammacerane/hopane ratio 284, 288
Gancaigou Group, Chaidamu Basin, arid conditions 309
gastropods 7, 225, 296, 312, 343, 345
genetic potential 37
George, Lake (eastern Australia) 13, 51, 363
geostrophic currents and upwelling 77
geothermal gradients
    Chinese salt lake basins 282–3
    SE Asian basins 348
    Thailand 342–3
geothermal regime 279
germanium 294
Ghyben-Herzberg interface 48, *49*, 51
Gippsland Basin 337
glauber salts 300, 304
glauberite 179
glauconite 309
Glyptostrobus 360
Gondwanaland (Permian), and lake deposits 11
Gooranga subunit, Condor unit, Condor oil shale sequence 332
Gosiute, Lake 39, *40*, 82
grainstones, algalclastic 224
Granton Sandstone 243
Great Lakes, the 14
Great Salt Lake 5, 16, 19, 268
Green River Basin 39
    accumulation of rich oil shales 41
Green River Formation 7, 8, 10, 29, 30, 39, 139, *144*
    clay mineralogy 142
    deposition of 82
    playa lake model 174
    trona deposits 213–14
Green River model, lacustrine source rock accumulation 29

Index

Green River shale 115, 139, 300
  carotanes 304–5
  gammacerane 304
Green Swamp subunit, Lethebrook unit, Condor oil shales 332
Gregory Rift Lakes (E Africa) 15
griegite 132
groundwater concentration 48
groundwater seepage 51, 53
Gunyarra unit, Condor oil shale sequence 332–3
gypsum 48, 135, 179, 213, 243, 253, 314, 346
  absence of 180
gypsum precipitation 208
gypsum sedimentation 51, 53

halite 48, 180
halite pseudomorphs 179
halite saturation 5
halite sedimentation 51, 53
haloalkaliphiles 115
halobacterium 16
Ham-Scarfe Subgroup 209
Hartford Basin (Newark Supergroup basins)
  lateral trends in lacustrine sequence thickness 267
  sedimentary sequences 270–1
heat flow, high, Thai basins 342–3
Hebei depression, central (China) 283
Helmsdale Boulder Beds 196
Helmsdale Fault 196
high potential environments 10
Hillsborough Basin 330
  geothermal gradient and heat flow 338
Holocene (early-to-mid), accumulation of rich sediments, tropical Africa 38
homopregnanes 126, 284
hopane homologues 304
hopane/sterane ratio 96, 107, 196
  fluctuations controlled 192
  source and maturity information 189, 191
hopanes 96, 111, 115, 192–3, 284, 296, 302, 307, 321
  bacterially derived 187, *189*
  pentacyclic 187
hopanoids 301–2
  hypersaline environments 126
hopenes 115, 302, 307
hot springs 342
Huanghua Depression, Bohai Bay Basin 309, 310, 313
  organic content 312
humic acids
  pyrolysates enriched in biomarkers 96–7
  pyrolysis of 93, 95
huminite 343
huminite reflectance 348
hydrocarbon maturation, SE Asian basins 348
hydrocarbon potential, Oil Shale Group 243–4
hydrocarbon provinces, lacustrine, potential *4*
hydrocarbon yield
  carbonate laminites, and TOC 208–9
  East African lakes and Lake Uinta 88
hydrological balance 3, 5
hydrodynamic stability, modern and palaeolakes, northwest China 293–4

hydrothermal circulation 342
hydrous pyrolysis, organic matter, Big Soda Lake 71
hyperpycnal flow 254
hypersaline environments 111, 126
  bacterial activity 314
  biological markers for 115–16, 123
hypolimnion 17, 46, 215
  aerobic 60
  anaerobic 60, 61
  anoxic 54, 174, 261, 369
    and alkaline 214
  oxygen loss 250
  preservation in 235
  saline 214

illite 141, 146, 280, 357, 360
inertinite 242, 331, 332, 343
inorganic precipitation, calcite 255
inorganic processes in anoxic sediments 20
insects 345
interstitial water, in clay mineral transformation 147
intraclastic limestone 255
intraformational clasts 255
intrastratal deformation features 173
ionic discharge, buffering Orcadian lake system salinity level 181
ionic ratios 6
iron sulphide, formation of 131
iron-sulphur compound 131
isoprenoid alkanes
  acyclic 111, 192, 301, 304
  as palaeoenvironmental indicators 324
isoprenoid compounds 108
isoprenoid hydrocarbons 318–20
isoprenoid/n-alkane ratios 185–6
isoprenoids 314–20
  acyclic 109, 110, 112

Jianghan Basin (China) 303–6, 310, 314, 324
  hypersaline 299–300
  inland saline lake basin 309
  salt lake evaporite-clastic formations, fault-bounded basin 279
  steranes and pentacyclic triterpanes, crude oils and source rocks **322**
Jiuquan Basin (China) 291
  spore/pollen indicators 164–5
Jiuxi Basin (China)
  depressions in 294
  lacustrine deposits in a subsiding basin 292–3
  source-rock environment of deposition 292
Jiyang depression (China) 283
Jizhong Basin (China) 284
Jurassic source beds 11

kaolinite 141, 143, 146, 357, 360
kaolinite clays 294
Kerio River 141
kerogen composition
  lacustrine oil source rocks 30
  a product of preservation state and source input 86

kerogenites, laminated 22
kerogenous laminae 237
kerogenous laminites *241*
kerogens 71, 83, 91, 95, 330
  algal 107
  Chinese basin source rocks 280-1, **282**
  Green River 86
  precursor composition **34-5**, 37
  precursors, African lakes 37
  Type I 10, 21, 30, 83, 101, 184, 280, 281, 296, 348
    indicative of sedimentation in anoxic waters 21-2
    lacustrine origin, composition of 31
    origin of, and hydrocarbon potential 286-7
    origin unclear 22
    precursors 286-7
    sapropelic 296
  Type II 39, 41, 83, 184, 280, 281, 296
    characteristics found in E African basins 21
    precursors in African lakes 36
  Type IIa 30, 280
  Type III 83, 280, 281, 296
    lignitic 15
ketones 109
Kivu, Lake 15, 20, 36, 37
  permanently stratified 82

Laach, Lake of (W Germany), palaeo-environment information from deep water siderite 153-7
lacustrine assemblages
  Thai basins 343-8
    marlstone-oil shale-sandstone 345-7
    mudstone-coal-oil shale 343-5
  lacustrine clastic formations, fault-bounded Chinese Basins 279
lacustrine environments 104
  clay-mineral distributions, sedimentary basins 141-3
  petroleum generation and migration *149*
lacustrine facies
  and facies sequences, distribution of, Culpeper Basin 263-8
  Middle Old Red Sandstone 205, *206*
  Newark Supergroup 250-9
lacustrine kerogen problems 21-2
lacustrine petroleum accumulations, characteristics of 139-40
lacustrine sediments vii
  Buckland, Bull Run and Waterfall Formations 249-50
lacustrine sequences
  Culpeper Basin 259-63, *264-5, 266*
  lateral trends in thickness of, Culpeper Basin 267, 268
  Newark Supergroup basins 267-9
  restriction of petroleum generation and migration 147-9
  Rubielos de Mora Basin, evolutionary trend 364
  symmetrical and asymmetrical, Culpeper Basin 272
  thickness of 268
    Lockatong Formation, showing Milankovitch-type climatic cycles 270

lacustrine source rocks vii, viii, 295
  case histories ix-x
  composition of 30-1
  Devonian Orcadian Basin 203
  Dinantian Oil Shale Group 235-44
  models of accumulation 29-30
  for Otway Basin coastal bitumens 327
  spores and pollens as indicators of 159-67
lacustrine systems, clastic and carbonate 81-9
Lady's Walk Shale 197
lake basins 5
  ancient 3
  freshwater, China 309
  inland 313
    and paralic fresh-brackish as petroleum source rocks 322-3
    saline 309-10
    swampy 310
  organic carbon content of 11-12
  origins and examples 7
  paralic 5, **7**
    freshwater, China 309
  tectonic 5, **7**, 341-3
lake biomass 14
Lake Cadell Basin *227*
lake, defined 3
lake depth, palaeolakes, northwest China 292-3
lake floor relief, and water depth estimates 175
lake level changes 31, *32*
lake sediment mineralogy, influence of local source areas 141
lake types, basic 45-6
  related to presence of organic matter and metals and their preservation *52*
  relationship with climatic conditions and salinity fluctuations *46*
lake water CO_2 157
lake waters
  chemistry of 53-4, 213
  geochemistry of 6-7
lakes
  as above-ground component of hydrological basis 48
  acid, carbon-limited 15
  alkaline 15-16
    and carbonate mineral assemblages 362
  anoxic, favourable environment for source rock deposition 18
  biologically productive 369
  brine 313
  buffering capacity of 6-7
  carbon cycle in 11-22
  and climate 77
  closed 272, 274
  deep, distribution of organisms 46, *47*
  dimictic 250
  dry 45
  ectogenic meromictic 250, 261
  environmental sensitivity of 7, *10*
  ephemeral 45
  evolution of under Australian conditions 53-7
  existence of vii
  factors affecting condition of 13
  freshwater 313

groundswater-seepage areas 53
homictic-dimictic 153
hypersaline 10, 16, 17
    Na-Cl 16, 48
large
    anoxic, as source environment 10
    economic aspects of 45
    temperate 14–15
meromictic 17, 59–74, 174, 233, 250, 363, 365
modern
    chemical classification 7, 8
    distribution of 4
    examples of aqueous productivity **14**
    large, distribution of 3
    oligosaline and mesosaline 13
    study of 3
monomictic 250
multiple sedimentary cycles 293
oligomictic 250
oligosaline 17, 18
oligotrophic 14, 134–5
open and closed 5, 250–2
    Lake Cadell 220, 230, 233
organic matter and metals (Australia) 49–51
paralic 312
permanent 45–6, 48
rapid fluctuations in size and shape 141
rates of change 3
rift 142–3, *145*
saline 46, 48, *50*
salinity of, northwest China 292
stratified 254, 261
in tectonically active areas 57
thermally stratified 206
lamalginite 236, 330–1, 332, 334–5, 338, 343, 345
    derived from algal matter 240
laminae
    anastomosing 237
    bioclastic 361
    in black shales 251, 252
    carbonate 358
    clastic 358
    clay 358
    graded 259
        and lenticular 253
    kerogenous 237
    Lake Malawi mud units 370–1
    marl 359, 361
    mudshales and clayshales, Culpeper Basin 255
    organic 374
    rhythmite 361
    and suspension deposition 255
    traction 358
laminates
    disrupted 258, 259
    dolomitic 244
lamination 363
    indicating anoxic sediment 253
    parallel 370
laminites 226
    carbonate 205–7, 208, 215
    dolomitic 238, 243
    kerogenous *241*

Orcadian Basin
    deep stratified lake origin supported by organic geochemistry 194–5
    sulphur enrichment 179
lamosite 331, 332, 333, 334, 336, 338
    carbonaceous 331, 332
    liptinite-rich 338
    microlaminate 333
Laney shale basin 13
Laney Shale Member, Green River Formation 174, 179, *180*, 365
Langgu sag (China) 283
    oils 283
    palaeotectonic cross profile *282*
    source rocks 280
Lansang Basin (Thailand) 342
Lascelles subunit, Gunyarra unit, Condor oil shale sequence 332–3
Latheron Subgroup 209
Leiofusa 335
Leman, Lake 92, 95
Lepidodendron 243
Lethebrook unit, Condor oil shales 331–2
Liaohe oil field (China), Gaosheng oil 283
lignite, basal, Maoming oil shale 300
    hopanoids in 302
Lilypool subunit, Lethebrook unit, Condor oil shales 332, 338
limestone 331
    cherty 237
    Culpeper Basin 255, 257
    dark 244
    lacustrine 356
    laminated, dolomitic 238
    Oil Shale Group 243
    oolitic 255
    ostracodal 222, 255
    stromatalitic 222
limestone couplets 224
limestone unit, Rubielos de Mora Basin 353, 355
lipid composition, phytoplankton 106
lipid signature 104
lipids 21, 22, 103, 111, 117, 304
    of archaebacteria 324
    from bacteria 109
    halophilic bacteria, as source for phytane 316
    polar 115
    study of 117
liptinite 331, 337
liptobiolites 281
liptodetrinite 331, 332, 335, 343, 345
lithological and bulk geochemical parameters, Middle Devonian lacustrine laminites, Orcadian Basin 182–4
littoral water plants 12
Lockatong Formation, Newark Supergroup 179, 267–8
loop bedding 345, 346
Lower Limestone Group 243
Lower Rousay Group 209
Lower Stromness Group, organic-rich laminites 209
Lugano, Lake, temperate lake model 14–15
lupane 283, 298
lycopane 286, 301

lycopane degradation 320

maceral analyses, Condor shale sequence 330
macerals 330–1
  commonly occurring in oil shale and brown coal **331**
mackinawite 132
macrophyte plant roots 357
Mae Sot Basin (Thailand) 345, 346
Magadi, Lake 213, 214
magadiite 209–10, 213
magadiite precipitation 213
  and anaerobic decay 214–15
  reasons for 214
magnesium concentrations 177
Malawi, Lake 20, 36
  sand turbidites and organic-rich diatomaceous muds 369–74
Manassas Sandstone 249
Manyara, Lake, phosphate concentration 15
Maoming oil shale 284, 299, 301–3
  biological market distributions, diagnostic 307
  low maturity 302
marine systems, and pyrite formation 132
marine vs. non-marine environments 7, **9**
marls
  bioturbated 364
  Cerro Porpol section 356–60
  laminated 356, 360
marlstone 345
  laminated 344
maturity discrepancy, Italian samples 126
maturity measurements, Beatrice crude and deep Orcadian lacustrine samples 196
Melosira 372
meromictic conditions 41
meromictic lakes 17, 174, 233, 250, 363
  Big Soda Lake, microbial and biogeochemical processes 59–74
Messel shale 303
Messinian Formation 115, 116
metal accumulation, Australian lakes 53
metal remobilization 51
metalliferous deposits 51
methane, Big Soda Lake 64–6
methane oxidation 73
  Big Soda Lake 66
methanogenesis 17, 18, 20, 22, 68, 362
  Big Soda Lake 64–6
  microbial 154
  stimulation of (experimental) 65
methanogenic activity, use of inhibitors (experimental) 65–6
methylsteranes 301, 302, 302–3, 306, 307
Mey Subgroup 209
mica 370, 371
mica flakes 370, 371
micrite, lacustrine basinal 7
microbial biomass 62–3
microbial decomposition reactions 20
microbial degradation 18
microbial heterotrophy, Big Soda Lake 63–8
  retarded by chemical factors 64

microbial mats 14, 255, 274
  deep water 17
  hypolimnial 17
  ideal precursor source-bed qualities 19
  see also algal mats; bacterial mats; cyanobacterial mats
microfossils, reworked 374
microlaminations, organic 173–5
micronodular textures 173
microspar 206
microturbidites, carbonate 348
Micrystridium 335
Midgeton Block 329
Midland fish-bed 265
Midland Formation 265
migration, primary 279
Milankovitch-type climatic forcing model 173
Milankovitch-type cycles 274
  Lockatong Formation 270
  Newark Basin 272
mineral abundances, changes with depth 141
mineralogical alteration 143
Minhuo Basin (China) 291
mirabilite 300
mixed-layer clays 146, 280
mixolimnion
  aerobic 63
  anaerobic 65
  anoxic 63, 64, 66
  bacterial density in 63
  Big Soda Lake 60, **61**
    production and vertical fluxes of organic matter, summer and winter 60–2
  sulphate reduction 67
mixotrophy 132, *133*
Mobutu, Lake 36, 37, 38
  permanent mixing 32
Moho 279, 281
molecular geochemistry, Middle Devonian lacustrine laminites, Orcadian Basin 184–91
molluscs 7
monimolimnion 63, 65
  permanently anoxic 59
  sulphate reduction 66–7
monimolimnion sediments
  Big Soda Lake 68–73
  and methane 64
monolimnion 17
monomixis *see* stratification
monosulphides 131, 132
Monterey Formation 111
Monterey shale 123
montmorillonite 280
mud drapes 257, 258
  on ripples 258–9
mud intraclasts 259
mud units, laminated, Lake Malawi 370–1
mudcracks 249, 255, 258, 259, 261, 272, 274
mudrocks, black 208
mudshales and clayshales, Culpeper Basin 255, *256*
mudstone
  bituminous 224
  grey bioturbated, Cerro Porpol section 356–60
  micritic *231*

Index

muscovite 258
  detrital 259
  flakes 257

nahcolite 82
Napperby, Lake (Australia) 53
Newark Basin (Newark Supergroup basins), trends in lacustrine sequence thickness 267–8
Newark Rift System 247
Newark Supergroup
  closed basin deposition for some lacustrine sediments 251
  lacustrine facies 250–9
  rift basins 247
Niandt Limestone (Caithness) 209
nickel porphyrins 284
Niger Delta Region, clay mineral composition 140
nitrification, at oxic/anoxic interface of the mixolimnion 68
nitrogen contents, Qinghai Lake 293–4
nitrogen cycle, Big Soda Lake 67–8
nitrogen fixation 67
  littoral zone sites 67–8
Nitzschia 372
Nitzschia palea 62
norhopane 296
norpristane 286
North Sea, clay mineralogy 140–1
notostracans 255, 274
nutrient recycling 39
nutrients 13, 14, 60
  recirculated 348

O'Connell subunit, Gunyarra unit, Condor oil shale sequence 333
Officer Basin 192
  pseudomorphed sodium carbonate minerals 214
oil seeps
  Fang Basin 348
  Otway Basin 327
oil shale deposition 29
oil shale facies, occurrence of 364
Oil Shale Group 219, *220*
  alluvial fan-delta complex *223*
  ideal cycle *222*
  lacks evaporites 221
  lacustrine petroleum source rocks 235–44
  relationships between lithologies *242*
Oil Shale Group sediments 235–6
oil shales 139, 294, 330, 356, 365
  accumulation of 41
  Australian, rock types 331
  Chinese lacustrine, biological markers in 299–7
  deposited as algal oozes 242
  Green River Basin 41
  high oil potential 235
  laminated, Type I 22
  and related sediments, petrography of 236–8
  rhythmically laminated, Thai basins 345, 346–7
  Thai basins 343
    formed under shallow conditions 347
  Thailand, Tertiary 240, 242

oil shales, Scotland
  composition and lithology 219, 221
  environments of deposition 221
  as flocculent phytoplanktonic oozes 221
  kerogenous components 221
oil-forming centres, northwest China 294, *295*
oil-prone sediments 295
Old Red Sandstone 196, 205
oleanane 283, 284, 298, 306, 324
oligosaline lakes
  biogenetic methanogenesis 18
  modern 17
oligotrophic lakes 14
  Ennerdale Water 134–5
olistoliths 365
oncolites 7
onocanane 283
onocerane 321, *323*
ooids 46
oolites 255, 272
oolitic limestone 255
opaline sediment distribution 77
Orcadian Basin
  hydrocarbon source rocks, thermal maturity and burial history 203
  lacustrine cherts, significant for environment of source rock deposition 205–15
  Middle Devonian palaeolimnology and organic geochemistry 173–97
  depositional models 173–7
  palaeosalinity determinations 177–81
organic carbon 69, *70*, 155, 157
  East African lakes and Lake Uinta 83, *84*
  and pyrite formation 135
  Qinghai Lake 293–4
organic carbon content 11–12, 295
  carbonate-rich beds 208–9
  Chinese basin source rocks 281
  Lake Tanganyika 369
organic enrichment
  East African lakes and Lake Uinta 88
  limitation of 179
organic fractions, recent sediments, comparison of 93–5
organic matter 103
  accumulation in shallow playa lakes 250
  adsorption of metal ions 51
  anaerobic decay of 215
  changes in accumulation through time 35–7
  characteristics of, East African lakes and Lake Uinta 88
  Chinese basins 312
    abundance, type and evolutionary characteristics of major source-rock types 280–1
  conditions favourable to preservation of 41
  cyclical sequences, Rubielos de Mora Basin 360
  decay under oxidizing conditions 362
  as estimate of oxygen conditions 252
  from the aquatic system 13
  fungal/bacterial 235
  hydrogen-depleted 87
  internal cycling of 14
  lacustrine, carbon distribution 95
  lacustrine sediments 250

organic matter (cont.)
  lamalginite 332
  laminar 240-2
  lamosite, Brown Oil Shale and Brownish Black Oil Shale subunits, Condor unit, Condor oil shale sequence **333**
  main sources in sediments 104
  and metals in lakes 49-51
  Middle Devonian laminites, Orcadian Basin 181-95
  nature of, Big Soda Lake 69, 71-3
  northwest China
    abundance of, continental source rocks 295
    preservation in deep depressions 294
    types of 296
  recent sediments, geochemical characterization of 91-101
  sedimentation of 17-18
  terrestrial 96
    carbon distribution 95-6
    wind-blown 48
  Thai basins 345
  Waterfall Formation 250
organic matter accumulation and palaeoclimates 33-9
organic matter composition 36-7
organic matter content 358
organic matter preservation conditions, deep lakes 49
organic microlaminations 173-5
organic preservation vii, viii
orthophosphorous, inorganic 13-14
ostracods 7, 225, *231*, 236, 251, 255, 259, 260, 274, 293, 296, 309, 312, 345, 358
Otway Basin (western), South Australia 327
oxalate, anaerobic decomposition of 68
oxic, aphotic and photic bottom conditions 363
oxic conditions 316
oxic environments (deep), benthic life 18-19
oxic/anoxic bottom conditions, interpretations 364
oxidation, anaerobic 66
oxidation episode, Permo-Carboniferous 203
oxidizing capacity, in lakes 19
oxycline 364
  Big Soda Lake 60
oxygen depletion 153
oxygen disappearance depth, Big Soda Lake 61
oxygen index, East African lakes and Lake Uitna 83, *85*
oxygen loss 250
oxygenated bottom conditions 362
oxygenation, temporary 254

palaeoclimate models, effects of local variability 78
palaeoclimates viii, 279
  Australia 51, 53
  Newark Supergroup sediments 269-72
  northwest China, cyclic evolution 291-2
  Oil Shale Group 243
  prediction of 77-8
palaeoclimatic cycles 270, 272, 274, 291-2
palaeoclimatic maps

prediction of climatically sensitive sedimentary deposits 269
uses of 77
palaeoclimatic patterns and source, Culpeper Basin 265, 267
palaeocurrents, Culpeper Basin 265, 267
  pattern of 270
palaeoenvironmental conditions
  assessed by biomarkers, importance of 103-4
  indicators of 117
palaeoenvironmental indicators viii-ix
palaeoenvironmental interpretation, Culpeper Basin 272-4
palaeoenvironments
  paralic swamp 316, 318
  pristane/phytane (Pr/Ph) ratios as geochemical indicators 316, 318
palaeogeographic controls on lacustrine environments of deposition and production/preservation of organic matter *11*
palaeogeographic reconstructions 10
  Culpeper Basin lakes *272-3*
palaeogeothermal gradient 281, 288
palaeolake-bog complexes 291
palaeolakes 291
palaeopressure maps 77
palaeosalinity determinations, Orcadian Basin 177-81
palaeosalinity indicators **136**
palaeosols 259, 260, 274
palynomorphs 250
Paraparchites 225
particulate inorganic carbon (PIC) 13
particulate organic carbon (POC) 13, 17
particulate organic matter, vertical fluxes of 62
Passaic Formation (Brunswick Formation) 270
Pattani Basin (Gulf of Thailand) 348, 349
Pediastrum 13, 21, 37, 334, 338, 343
  as an indicator of the environment of deposition 335-6
pedogenesis 226, 259
pedogenic reworking 226
pelagic mud 374
pelagic rain 7
pelagic sediments
  Big Soda Lake, organic geochemistry of 68-73
  potential mechanism for layer formation 62
pelecypods 255, 260, 274, 296
pentacyclic compounds 111
pentamethyleicosanes 324
perhydrocarotene 324
perialpine lakes, freshwater, sulphate concentrations 19-20
Peridium 13
permanent (deep) lake model (Bradley), environments of deposition 8
permanent lakes 45
  Australian, chemical characteristics (water and sediments) 48
  implications of changes over time 45-6
Permian source beds 11
perylene 324
petroleum exploration, use of biological markers 123
petroleum migration 143, 145, *146*

Chinese continental petroliferous basins 160–7
  in marine sequences 147
petroleum source rocks, related to deep lakes 51
Phitsanulok Basin (Thailand) 341, *342*
  lacks coal 345
  oil production 348
phosphate 206
phosphate nodules 251, 253
phosphorous 14
photosynthesis 17
  anoxygenic 61, 62
photosynthetic carbon assimilation, rates of 27
phytane 96, 111, 128, 283, 284, 286, 288, 296, 301, 304, 306, 314, 316
phytol, geochemical diagenesis of 316
phytoplankton 117, 194
  biological markers for 106–9
phytoplankton biomass 27, 62
phytoplankton blooms 348
phytoplankton productivity 62
  fresh and permanent lakes 49
  modern, Africa lakes 27
Piceance Creek Basin 365
plankton 296
Planolites 343
plant debris 242, 370, 371
plant fragments 255, 257, 258, 259, 260, 293, 345, 370
plant material, rotting 68
playa characteristics, Triassic source beds 11
playa lake deposits 242
playa lake origin, Green River Formation 29, 82, 174
playa lakes 51
  nutrient recycling important 39
  Western Australia 51
playa-lake complex model, environments of deposition 8, 10
playa-lake model
  microlaminated lake sediments 174
  trona precipitation 214
playas, modern, fluctuations in brine chemistry 214
polygonal cracking 48, 51
polymethylalkanes 320
pore waters
  chemistry of 215
  concentrated, expulsion of 143
  marine 147
  meteoric 226
porosity and petroleum migration 143, 145, *146*
porphyrins
  metal 284
  Ni-porphyrins 283
Potamogeton 360
Potamogeton seeds 357, 358, 360, 363
precipitation, orographic *269*, 270
precipitation–evaporation balance, African lakes 31
pregnane 126, 284
preservation potential 17, 49
  lake basins 5
  reduced 37–8
Pretty Hill Sandstone 327
primary migration 279
pristane 96, 283, 286, 296, 301, 302, 304, 314, 316
pristane alkanes, n-alkanes 307

pristane/phytane (Pr/Ph) ratios 71, 112, 115, 185, 283, 288, 296, 323–4, 360
  carbonate source rocks 193
  geochemical indicator of palaeoenvironment 316, 318
  low, indicating anoxic depositional conditions 304
  and redox potential 111
product character and generative capacity, East African lakes and Lake Uinta 88
product-precursor relationships, biological markers 107
productivity
  high
    East African Rift lakes 82
    upwelling areas 116
  organic 242
proglacial lakes 14
pseudomorphs, within cherts 211, 213
  identity of 213
Pumpherston Shell Bed 236, 242
pycnocline 17
pyrite 19, 236, 251, 294, 357
  diagenetic formation of 132, 135
  early diagenetic 360, *361*
  formation of 131–5
    at sediment/water interface 135
    under varying conditions 253
  framboidal 178, 221, *361*
  sedimentary 178
pyrites, authigenic 298
pyrolized compounds, analytical procedures 93
pyrolysates, from stable residues 95, **96**
  distribution of alkanes in 95–6
pyrolysis
  and generative capacity of oil source rocks 81
  Rock-Eval
    data for Oil Shale Group **241**, 245–6
    East African lakes and Lake Uinta 83–7
pyrolysis-gas chromatography, East African lakes and Lake Uinta 86–7

Qianjiang depression, Jianghan Basin 302, *303*
Qiketai oil field 169
Qinghai Lake (China) 298
  hydronamics and sediments 295–6
quartz 282, 359, 360, 362, 372
Quercus 362

rainfall maps, Triassic 77
rainfall predictions, from palaeopressure maps 77–8
rainwater composition 6
Ramapo Fault 267
red beds 209, 260, 353
Red River Basin (Vietnam) 349
redox boundary 19
reducing conditions 69, 131, 314, 360
reducing environments 309
reflectance data, Condor Oil Shale sequences 333–4
remineralization 117
Rengiu oil field (China) 301, 306–7
reptile tracks 258, 259

resedimentation processes 39
resinite 331, 332, 338, 343
 Palaeogene Fushan (China) 286
Reston Formation 249
retene 105
Rhizodus hibberti 225
rhythmic laminations 8
 Lake Van 15–16
rhythmic sedimentation 235–6
rhythmite beds, Rubielos de Mora Basin 356
rhythmites 356, 359, 360, 361, 364
rift lakes, Connecticut River valley, clay-mineral relationships 142–3, *145*
rifting, topographic effects from affecting local climate, Culpeper Basin 269–70
ring cleavage 105
rip-up clasts 254
ripple marks 258
Robbery Head Limestone 209
Rozel Point oil 123, 128, 300, 305
Rubielos de Mora Basin (Spain) 353–65
 facies description 357–9
 sequential facies analysis 359–62
 stratigraphy 355–7
Ruoyan sag (China)
 palaeotectonic crossprofile *282*
 source rocks 280
Ruppia beds 68

saline/salt lakes *49*
 Australian, chemical characteristics (water and sediments) 48–9, *50*
 petroleum source rocks 309
 solar heated 14
 species diversity low 46, 48
salinity viii
 elevated 253, 255
 and lake level 251
 past, indicators of 135–6
Salix 360
salt crusts 49, 51
Salt Lake Group 123, 128
salt-casts 210
sand beds, Lake Malawi 370
sandstone 331, 332
 Culpeper Basin 254–5
 fluvial 251
 siltstone and mudstone, Culpeper Basin 258–9
 wave-rippled 274
sandstones and siltstones, Culpeper Basin 257–8
Sandwick Fish Bed cycle, pseudomorph-bearing chert laminae 211
Sandwick Fish Bed (Niandt Limestone) 209
Sanford Basin 270
Sanshui Basin (China), spore/pollen indicators 166–7
Scottish Calciferous Sandstone Series 235
seasonal mixing, and microbial processes 60
sediment facies sequences, reversed 5, 6
sediment reworking 258, 373
sedimentary basins
 clay-mineral distribution in 140–3
sedimentation rates
 Chinese basins 282

 and preservation of organic matter 54
seepage zone, dry lakes 48
seepage-spring zone 48
Septodinium 335, 338
serratane 284, 321, *322*, *323*
Shahejie Formation 309, 314, 321
Shanganning Basin (China) 309, 310, 321, 324
 organic matter 312
Shangonghe Group 293
Shanxi-Gansu-Ningxi Basin (China) 291
 depressions 294
 main source rock unit 294
 source rocks 293
  oil-prone coexisting with oil-prone coal 294–5
sheetfloods, Culpeper Basin 272, 274
Shengli oil field, China 283
shrimp beds 237
siderite 280
 from ancient, varved sediments 155–7
 models for formation 155
 modern lake environment 154–7
siderite precipitation, chemical model *154*
siltstones (massive) and mudstones, Culpeper Basin 259
slumping 373
smectite 141, 143, 146
smectite-illite conversion 143, 145, 147
soda lakes 15, 59–74
sodium ions 213
soft-sediment deformation 210, 251, 257
solar input 13
Solar Lake 17
solute concentration 364
 decrease in 363
solute inflow, Orcadian Basin 180
solutes, change in seasonal distribution 60
Songlia Basin (China) 282
 oils 283
 sedimentary formations 279
 source rocks influenced by palaeo-Renjiang River 294
 Type I kerogen 286
source differences, effects of 187
source rock associations, northwest China 295
source rocks
 Chinese basins 30–1
 Chinese continental oil, formed under varying lacustrine conditions 309–10
 of continental lacustrine petroliferous basin 161
 inland saline lake 312
 northwest China
  continental, geochemical characteristics 295–8
  sedimentary environment 292–4
 oil-prone coexisting with oil-prone coal 294–5
 shale, primary migration from 143
 terrestrial, problems relating to the evaluation of oil and gas prospects 286–8
 Thai basins 348
 within Oil Shale Group 235
source-rock classification, Chinese oil- and gas-bearing basins 279–80
sparite layers 238
spirorbids 225

Spirorbis pusillus 225
spirotriterpane 284
Spirulina 13
spore colour, variation in and maturity 203
spore colouration data **183**, 184
spores and pollens 296
    fluctuations in species diversity and palaeoclimates 269
    indicating geological age of original rock formation 160
    indicating source rocks 161
    Jurassic, Newark Supergroup 249
    of reservoir rocks identifiable 160–1
    as 'trace element' for petroleum migration pathway 160
sporinite 331, 332, 343
squalane 110, 124, 126
squalenes 110
stable residues, recent sediments, pyrolysis of 93, **94**, 95–7
stellate crystal moulds 257, 259
Stephanodiscus 372
sterane ratios 287
A-nor steranes 115
steranes 72, 96, 99, 112–15, 191, 283, 284, 296, 298, 301, 320, 321, 324
    distributions
        Australian sediments, environmental assessment 192
        in crude oils 114
        Italian samples 126
        source and maturity information 187, 189, *192, 193*
    as source rock markers 114–15
sterenes 112, 360
steroid aromatization 189
sterols 72, 106, 107
    algal origin 187
    diatom-derived 116–17
    sources 113–14
    and steranes 112–15
Stigmaria 243
stratabound ore deposits 51
stratification 49, 253
    bottom water 364
    ectogenic meromictic 274
    enhanced by climatological conditions 365
    influence of on sediment organic content 36
    influence of wind shear in African lakes 32
    meromictic, Lake Cadell 230
    Orcadian lake 214
    and oxygen loss 250
    permanent 82, 235, 365
    regular periods needed for source rock accumulation 37
    seasonal 19
    and source rock accumulation 30
    stable viii, 39, 41
        Thai basin lakes 346–7
    thermal 60–1, 206
    and trona precipitation 214
stratified lake model, organic microlaminations 174–5
    and wave base approximations 176

stratigraphic variability, lake sediments 141
stromatalite domes and mats 175
stromatolites 174, 195, 211, 226, 255, 274
stromatolitic carbonates 7
stromatolitic deposits 207
stromatolitic limestone 255
stylolitization 173
subsidence
    asymmetric 249
    basin 291
    SE Asian basins 348
Suigetsu, Lake (Japan), sulphate reduction 17
sulphate 179
    converted to sulphide 131
    high concentrations in marine conditions 132
sulphate reducers 64, 117
sulphate reduction 16, 17
    bacterial 48, 178, 362
    Big Soda Lake 64, 66–7
    and microbial decomposition 19
sulphides
    acid volatile 135
    concentrations high, Big Soda Lake 59
    from bacterial activity 51
sulphur
    elemental 132
    in freshwater sedimentary rocks 135–6
    organic 136
    pyrite/acid, as salinity indicator 136
sulphur compounds, organic 116, 128
sulphur concentrations **20**, 193
sulphur content, non-marine sourced oils 296
sulphur enrichment 179
sulphur reduction, rates of **20**
Suqia region (China), paralic coal-bearing formations 280
surface productivity, not a reliable guide to organic preservation 39
Surirella 372
suspension deposition 255, 258–9
swelling clay component 178
syneresis (synaeresis) 211, 346
syneresis (synaeresis) cracks 178, 179
synsedimentary movement, Newark Rift System 249

Taiyuan Formation 310, 321
    hopane series **322**
Talimu Basin (China)
    source-rock environment of deposition 292
    spore/pollen indicators 162–4
Talimu Platform 291
Tanganyika, Lake 10, 20, 31, 36, 37, 92
    deep anoxic lake model 29–30
    permanently stratified 82
    plankton biomass 15
taraxstane 284
tectonic evolution, affecting Rubielos de Mora Basin 365
tectonic tilting, Culpeper Basin 272
teepee structures 48, 51
telalginite 236, 331, 335, 343, 345
    and vitrinite reflectance 337–9

terpanes 72
terrestrial organisms 296
terrestrial plant wax 360
terrestrial plants 298, 330, 338
terrestrial runoff and organic accumulation 348
terrigenous material 345
Tertiary source beds 11
tetracosane 124
thermal alteration, of alginite and lamalginite 338
thermal stratification 60–1, 206
thermal stress 187, 189
thermocatalytic degradation 320
thermocline 17, 29
   Big Soda Lake 60, 62
   interflow along 254
   permanent, high TOC values 36
Thiobacillus 17
thiophene 128
thrombolites 226
time windows 10–11
tocopherols 111
Todilto (Limestone) Formation 89
torbanite 239, 240
   vitrinite reflectance 337–8
Torrens, Lake (Australia) 53
Towaco Formation 267, 268
   sedimentary cycles 270
trace fossils 343
   lack of 19
traction laminae 358
tractive currents 363
transgressions
   lacustrine 214, 363, 364
   leading to starved basin conditions 38
   and regressions
      Culpeper Basin 260–3
      cyclic sedimentation 205
      interdigitated terrestrial and marine deposits 310
trisnorhophane 298
triterpanes 72, 96, *98*, 187, *190*, 191, 283, 296
   higher plant origin 288
   pentacyclic 105–6, 320, 324
triterpenoids, pentacyclic 104–5
trona 8, 180, 213
   solubility of 214, 215
trona precipitation 213–14, 215
   reasons for 214
   sub-lacustrine 214
turbiditic flows 363
Tulufan Basin (China) 291
   spores and pollens 167
tundra lakes, shallow 14
turbidite sedimentation, oligosaline lakes 17
turbidites 15, 39, 176, 373
   lacustrine 254
   sandy 265, 374
turbidity currents 274
Turkana, Lake 18, 36
   laminated sediments 252
   major sediment sources 141, *142*, *143*
Turkey, Lake (Canada), pyrite concentration 133–4
Turkwel River 141

Uinta Basin 365
   oil fields 140
Uinta, Lake 82, 83, 88
   generative capacity and potential for organic enrichment 87
ulminite 332
Ulmus 360
Upper Rousay Group 209
Upper Stromness Group 209
upwarps, in depressions 294
upwelling environments, biological markers 116–17
upwelling, prediction of 77
uranium 294
Urmia, Lake (Iran) 18, 19

vadose fabrics 226
Van Houten cycles, Lockatong Formation 270
Van, Lake, soda lake 15
Vanda, Lake (Antarctica) 14
vascular plant remains 226
vascular plants 343
   biological markers for 104–6, 117
   sterol source 113
Vechten, Lake, sulphate concentration 16
vegetable matter, related to oil shales 242
vegetal debris 358
vegetation cover
   heavy 41
   restricted during Devonian 184
vegetational change, Tropical Africa 31, 38
vertebrate remains 225
vertical mixing 60
Victoria, Lake 37, 38
   lacustrine environment 12
   modern analogue of palaeolake Gosiute 39, *40*, 41
   shallowness of 39
vitrinite 242, 300, 331, 332, 333, 337
vitrinite reflectance 302, 330, 333, 336–8, 348
vitrite layers 332
vitrodetrinite 333, 334
vivianite 372
volcanic springs, CO source 157
vugs
   calcite-filled 249, 261, 263
   calcite-lined 259

Wadi Natrun (Egypt), hypersaline lake 16
Waldsea lake 17
Walls Sandstone 203
Wardie Shales 243
water depth, Culpeper Basin
   basin floor tilting model 268, *269*
   lake-filling model 268
   lateral extent of sub-wave base facies method 268–9
water depth estimates, in ancient lakes 175–7
water loss, porosity increase and petroleum migration 143, 145, *146*
water table 48
water/sediment composition, changed with time 53, *54*

Index

Waterfall Formation 247, 249
  lacustrine conglomerates 255
  lacustrine sequences 265, 267, 268
  sandstone, siltstone and mudstone 258–9
  sandstones and siltstones 258
  thick lacustrine sequence 261, 263
  TOC of lacustrine sediments 253
wave action, fluctuating 258
wave fetch, wave base and wind speed, and water depth estimates 175–7
waxy bitumens 108
waxy crude oil 88, 89, 108, 327
  algal origin for 109
  Beatrice oil 196
waxy kerogens 21, 22, 30, 31
waxy *n*-alkanes, Big Soda Lake sediments 71
weathering
  chemical 38, 180
  karstic 344–5, 349
  tropical 345
Wilkins Peak Member, Green River Formation 29, 363, 364, 365
wind, influencing tropical lakes 31–2
wind shear 32

Windermere, Lake, pyrite concentrations 135
Wulumuqi Pediment Depression, Zhungeer Basin 294
Wulungu Formation 293

Xialiaohe Depression (China), spore/pollen indicators 166

Yanan Formation 294–5

zeolites 142
Zhungeer Basin (China)
  depressions in 294
  lagoon-lacustrine, volcano-clastic formations, intermontane basins 279–80
  palaeolakes and palaeobogs 291
  saturated hydrocarbons 286
  source rocks
    formed in freshwater environment 292
    Upper Permian 293
  spore/pollen indicators 164
Zong-Chao Palaeoland 291
Zurich, Lake 8, 13–14